ADVANCES IN HAZARDOUS INDUSTRIAL WASTE TREATMENT

ADVANCES IN HAZARDOUS INDUSTRIAL WASTE TREATMENT

EDITED BY

LAWRENCE K. WANG

NAZIH K. SHAMMAS

YUNG-TSE HUNG

CRC Press
Taylor & Francis Group
Boca Raton   London   New York

CRC Press is an imprint of the
Taylor & Francis Group, an **informa** business

CRC Press
Taylor & Francis Group
6000 Broken Sound Parkway NW, Suite 300
Boca Raton, FL 33487-2742

© 2009 by Taylor & Francis Group, LLC
CRC Press is an imprint of Taylor & Francis Group, an Informa business

Library of Congress Cataloging-in-Publication Data

Advances in hazardous industrial waste treatment / edited by Lawrence K. Wang, Nazih K. Shammas,
Yung-Tse Hung.
   p. cm.
Includes bibliographical references and index.
ISBN-13: 978-1-4200-7230-3
ISBN-10: 1-4200-7230-7
   1. Factory and trade waste. 2. Hazardous waste. 3. Hazardous waste site remediation. I. Wang,
Lawrence K. II. Shammas, Nazih K. III. Hung, Yung-Tse. IV. Title.

TD897.A38 2008
628.4'2--dc22   2008008265

Visit the Taylor & Francis Web site at
http://www.taylorandfrancis.com

and the CRC Press Web site at
http://www.crcpress.com

Contents

Preface

Environmental managers, engineers, and scientists who have had experience with industrial and hazardous waste management problems have noted the need for a handbook series that is comprehensive in its scope, directly applicable to daily waste management problems of specific industries, and widely acceptable by practicing environmental professionals and educators.

Many standard industrial waste treatment and hazardous waste management texts adequately cover a few major industries, for conventional in-plant pollution control strategies, but no one book, or series of books, focuses on new developments in innovative and alternative cleaner production technologies, waste minimization methodologies, environmental processes, design criteria, effluent standards, performance standards, pretreatment standards, managerial decision methodology, and regional and global environmental conservation.

The entire *Industrial and Hazardous Wastes Treatment* book series emphasizes in-depth presentation of environmental pollution sources, waste characteristics, control technologies, management strategies, facility innovations, process alternatives, costs, case histories, effluent standards, and future trends for each industrial or commercial operation, such as the metal plating and finishing industry or the photographic processing industry, and in-depth presentation of methodologies, technologies, alternatives, regional effects, and global effects of each important industrial pollution control practice that may be applied to all industries, such as industrial ecology, pollution prevention, in-plant hazardous waste management, site remediation, groundwater decontamination, and stormwater management.

In a deliberate effort to complement other industrial waste treatment and hazardous waste management texts published by Taylor & Francis and CRC Press, this book, *Advances in Hazardous Industrial Waste Treatment*, covers many new industries and new waste management topics, such as characteristics of industrial hazardous wastes, soil remediation, treatment of metal finishing industry wastes, leachate treatment using bioremediation, remediation of sites contaminated by hazardous wastes, enzymatic removal of aqueous pentachlorophenol, remediation of sites contaminated by underground storage tank releases, biological treatment of wastes containing urea and formaldehyde, hazardous waste deep-well injection, waste management in the pulp and paper industry, and treatment of nickel-chromium plating waste, are presented in detail. Special efforts were made to invite experts to contribute chapters in their own areas of expertise. Since the field of industrial hazardous waste treatment is very broad, no one can claim to be an expert in all industries, and so collective contributions are better than a single author's presentation for a handbook of this nature.

This book is to be used as a college textbook as well as a reference book for the environmental professional. It features the major metal manufacturing, forming, coating and finishing industries and hazardous pollutants that have significant effects on the environment. Professors, students, and researchers in environmental, civil, chemical, sanitary, mechanical, and public health engineering and science will find valuable educational materials here. The extensive bibliographies for each metal-related industrial waste treatment or practice should be invaluable to environmental managers or researchers who need to trace, follow, duplicate, or improve on a specific industrial hazardous waste treatment practice.

A successful modern industrial hazardous waste treatment program for a particular industry will include not only traditional water pollution control but also air pollution control, noise control, soil conservation, site remediation, radiation protection, groundwater protection, hazardous waste management, solid waste disposal, and combined industrial–municipal waste treatment and management. In fact, it should be a total environmental control program. Another intention of this handbook is to provide technical and economical information on the development of the most feasible total environmental control program that can benefit both industry and local municipalities. Frequently, the most economically feasible methodology is a combined industrial–municipal waste treatment.

Lawrence K. Wang, New York
Nazih K. Shammas, Massachusetts
Yung-Tse Hung, Ohio

About the Editors

Lawrence K. Wang has over 25 years of experience in facility design, plant construction, operation, and management. He has expertise in water supply, air pollution control, solid waste disposal, water resources, waste treatment, hazardous waste management and site remediation. He is a retired dean/director of both the Lenox Institute of Water Technology and Krofta Engineering Corporation, Lenox, Massachusetts, and a retired VP of Zorex Corporation, Newtonville, New York. Dr. Wang is the author of over 700 papers and 17 books, and is credited with 24 U.S. patents and 5 foreign patents. He received his BSCE degree from National Cheng-Kung University, Taiwan, ROC, his MS degrees from both the University of Missouri at Rolla and the University of Rhode Island at Kingston, and his PhD degree from Rutgers University, New Brunswick, New Jersey.

Nazih K. Shammas is an environmental expert, professor and consultant for over forty years. He is an ex-dean and director of the Lenox Institute of Water Technology, and advisor to Krofta Engineering Corporation, Lenox, Massachusetts. Dr. Shammas is the author of over 250 publications and eight books in the field of environmental engineering. He has experience in environmental planning, curriculum development, teaching and scholarly research, and expertise in water quality control, wastewater reclamation and reuse, physicochemical and biological treatment processes and water and wastewater systems. He received his BE degree from the American University of Beirut, Lebanon, his MS from the University of North Carolina at Chapel Hill, and his PhD from the University of Michigan at Ann Arbor.

Yung-Tse Hung has been a professor of civil engineering at Cleveland State University since 1981. He is a Fellow of the American Society of Civil Engineers. He has taught at 16 universities in eight countries. His primary research interests and publications have been involved with biological wastewater treatment, industrial water pollution control, industrial waste treatment, and municipal wastewater treatment. He is now credited with over 450 publications and presentations on water and wastewater treatment. Dr. Hung received his BSCE and MSCE degrees from National Cheng-Kung University, Taiwan, and his PhD degree from the University of Texas at Austin. He is the editor of *International Journal of Environment and Waste Management*, *International Journal of Environmental Engineering*, and *International Journal of Environmental Engineering Science*.

Contributors

Donald B. Aulenbach
Lenox Institute of Water Technology
Lenox, Massachusetts
and
Rensselaer Polytechnic Institute
Troy, New York

Khim Hoong Chu
Department of Chemical and
    Process Engineering
University of Canterbury
Christchurch, New Zealand

Nicholas L. Clesceri
Rensselaer Polytechnic Institute
Troy, New York

Anuska Mosquera Corral
Department of Chemical Engineering
School of Engineering
University of Santiago de Compostela
Santiago de Compostela, Spain

José Luis Campos Gómez
Department of Chemical Engineering
School of Engineering
University of Santiago de Compostela
Santiago de Compostela, Spain

Yung-Tse Hung
Department of Civil and Environmental
    Engineering
Cleveland State University
Cleveland, Ohio

Azni Idris
Department of Chemical and Environmental
    Engineering
Universiti Putra Malaysia, Serdang
Selangor, Malaysia

Eui Yong Kim
Department of Chemical Engineering
University of Seoul
Seoul, Korea

Robert LaFleur
Rensselaer Polytechnic Institute
Troy, New York

Ramón Méndez Pampin
Department of Chemical Engineering
School of Engineering
University of Santiago de Compostela
Santiago de Compostela, Spain

Nymphodora Papassiopi
School of Mining Engineering and Metallurgy
National Technical University of Athens
Athens, Greece

Ioannis Paspaliaris
School of Mining Engineering and Metallurgy
National Technical University of Athens
Athens, Greece

Katayon Saed
Building and Environmental Division
School of Engineering
Ngee Ann Polytechnic
Singapore

William A. Selke
Lenox Institute of Water Technology and
Krofta Engineering Corporation
Lenox, Massachusetts

Nazih K. Shammas
Lenox Institute of Water Technology and
Krofta Engineering Corporation
Lenox, Massachusetts

Lawrence K. Wang
Lenox Institute of Water Technology and
Krofta Engineering Corporation
Lenox, Massachusetts
and
Zorex Corporation
Newtonville, New York

Ping Wang
Center of Environmental Sciences
University of Maryland
Annapolis, Maryland

Anthimos Xenidis
School of Mining Engineering and
   Metallurgy
National Technical University of Athens
Athens, Greece

1 Characteristics of Hazardous Industrial Waste

Nazih K. Shammas

CONTENTS

1.1 INTRODUCTION

The improper management of hazardous waste poses a serious threat to both the health of people and the environment. When the United States Environmental Protection Agency (U.S. EPA) began developing the hazardous waste management regulations in the late 1970s, the Agency estimated that only 10% of all hazardous waste was managed in an environmentally sound manner.

Proper identification of a hazardous waste can be a difficult and confusing task, as the Resource Conservation and Recovery Act (RCRA) regulations establish a complex definition of the term "hazardous waste." To help make sense of what is and is not a hazardous waste, this chapter presents the steps involved in the process of identifying, or "characterizing," a hazardous waste.

This chapter will introduce the entire hazardous waste identification process, but will focus particularly on the final steps and the characteristics and properties of hazardous wastes. After reading this chapter, one will be able to understand the hazardous waste identification process and the definition of hazardous waste, and be familiar with the following concepts:

1. Hazardous waste listings
2. Hazardous waste characteristics
3. The "mixture" and "derived-from" rules
4. The "contained-in" policy
5. The Hazardous Waste Identification Rules (HWIR)

1.2 HAZARDOUS WASTE IDENTIFICATION PROCESS

A hazardous waste is a waste with a chemical composition or other properties that make it capable of causing illness, death, or some other harm to humans and other life forms when mismanaged or released into the environment.[1] Developing a regulatory program that ensures the safe handling of such dangerous wastes, however, demands a far more precise definition of the term. U.S. EPA therefore created a series of hazardous waste identification regulations, which outline the process to determine whether any particular material is a hazardous waste for the purposes of RCRA.

Proper hazardous waste identification is essential to the success of the hazardous waste management program. The RCRA regulations require that any person who produces or generates a waste must determine if that waste is hazardous. For this purpose, the RCRA includes the following steps in the hazardous waste identification process[2]:

1. Is the waste a "solid waste"?
2. Is the waste specifically excluded from the RCRA regulations?
3. Is the waste a "listed" hazardous waste?
4. Does the waste exhibit a characteristic of hazardous waste?

Hazardous waste identification begins with an obvious point: in order for any material to be a hazardous waste, it must first be a waste. However, deciding whether an item is or is not a waste is not always easy. For example, a material (like an aluminum can) that one person discards could seem valuable to another person who recycles that material. U.S. EPA therefore developed a set of regulations to assist in determining whether a material is a waste. RCRA uses the term "solid waste" in place of the common term "waste." Under RCRA, the term "solid waste" means any waste, whether it is a solid, semisolid, or liquid. The first section of the RCRA hazardous waste identification regulations focuses on the definition of solid waste. For this chapter, you need only understand in general terms the role that the definition of solid waste plays in the RCRA hazardous waste identification process.

Only a small fraction of all RCRA solid wastes actually qualify as hazardous wastes. According to U.S. EPA estimates, of the 12 billion tons (metric) of industrial, agricultural, commercial, and household wastes generated annually, 254 million tons (2%) are hazardous, as defined by RCRA regulations.[3] At first glance, one would imagine that distinguishing between hazardous and non-hazardous wastes is a simple matter of chemical and toxicological analysis. Other factors must be considered, however, before evaluating the actual hazard posed by a waste's chemical composition. Regulation of certain wastes may be impractical, unfair, or otherwise undesirable, regardless of the hazards they pose. For instance, household waste can contain dangerous chemicals, such as solvents and pesticides, but making households subject to the strict RCRA waste management regulations would create a number of practical problems. Congress and U.S. EPA have exempted or excluded certain wastes, including household wastes, from the hazardous waste definition and regulations. Determining whether or not a waste is excluded or exempted from hazardous waste regulation is the second step in the RCRA hazardous waste identification process. Only after determining that a solid waste is not somehow excluded from hazardous waste regulation should the analysis proceed to evaluate the actual chemical hazard of a waste.

The final steps in the hazardous waste identification process determine whether a waste poses a sufficient chemical or physical hazard to merit regulation. These steps in the hazardous waste identification process involve evaluating the waste in light of the regulatory definition of hazardous waste. The remainder of this chapter explains the definition, characteristics, and properties of hazardous wastes.

1.3   EXCLUSIONS FROM SOLID AND HAZARDOUS WASTES

The statutory definition points out that whether a material is a solid waste is not based on the physical form of the material (i.e., whether or not it is a solid as opposed to a liquid or gas), but rather that the material is a waste. The regulations further define solid waste as any material that is discarded by being either *abandoned*, *inherently waste-like*, a certain *military munition*, or *recycled* (Figure 1.1). These terms are defined as follows:

1. *Abandoned.* This simply means "thrown away." A material is abandoned if it is disposed of, burned, or incinerated.
2. *Inherently waste-like.* Some materials pose such a threat to human health and the environment that they are always considered solid wastes; these materials are considered to be inherently waste-like. Examples of inherently waste-like materials include certain dioxin-containing wastes.

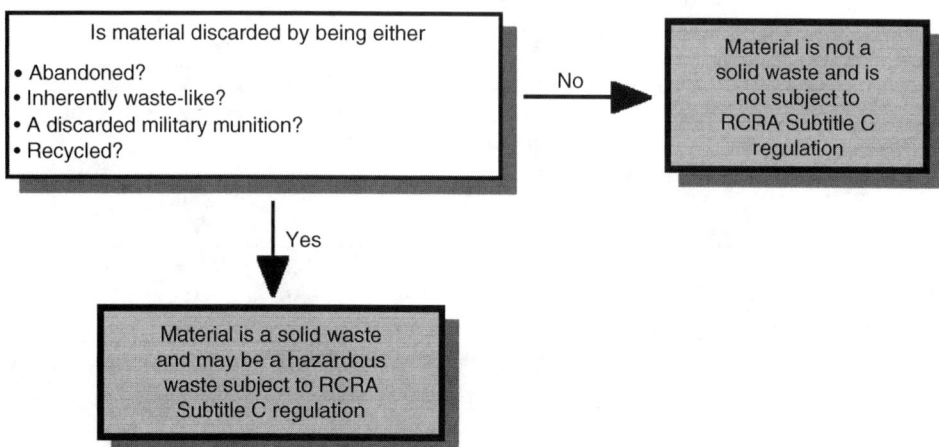

FIGURE 1.1   Determination of whether a waste is a solid waste. *Source:* U.S. EPA, Resource Conservation and Recovery Act—Orientation Manual, Report EPA 530-R-02-016, U.S. EPA, Washington, DC, January 2003.

3. *Military munitions.* Military munitions are all ammunition products and components produced for or used by the U.S. Department of Defense (DOD) or U.S. Armed Services for national defense and security. Unused or defective munitions are solid wastes when abandoned (i.e., disposed of, burned, incinerated) or treated prior to disposal; rendered nonrecyclable or nonuseable through deterioration; or declared a waste by an authorized military official. Used (i.e., fired or detonated) munitions may also be solid wastes if collected for storage, recycling, treatment, or disposal.
4. *Recycled.* A material is recycled if it is used or reused (e.g., as an ingredient in a process), reclaimed, or used in certain ways (used in a manner constituting disposal, burned for energy recovery, or accumulated speculatively).

1.3.1 RECYCLED MATERIALS

Materials that are recycled are a special subset of the solid waste universe. When recycled, some materials are not solid wastes, and therefore not hazardous wastes, but others are solid and hazardous waste, but are subject to less-stringent regulatory controls. The level of regulation that applies to recycled materials depends on the material and the type of recycling (Figure 1.2). Because some types of recycling pose threats to human health and the environment, RCRA does not exempt all recycled materials from the definition of solid waste. As a result, the manner in which a material is recycled will determine whether or not the material is a solid waste, and therefore whether it is

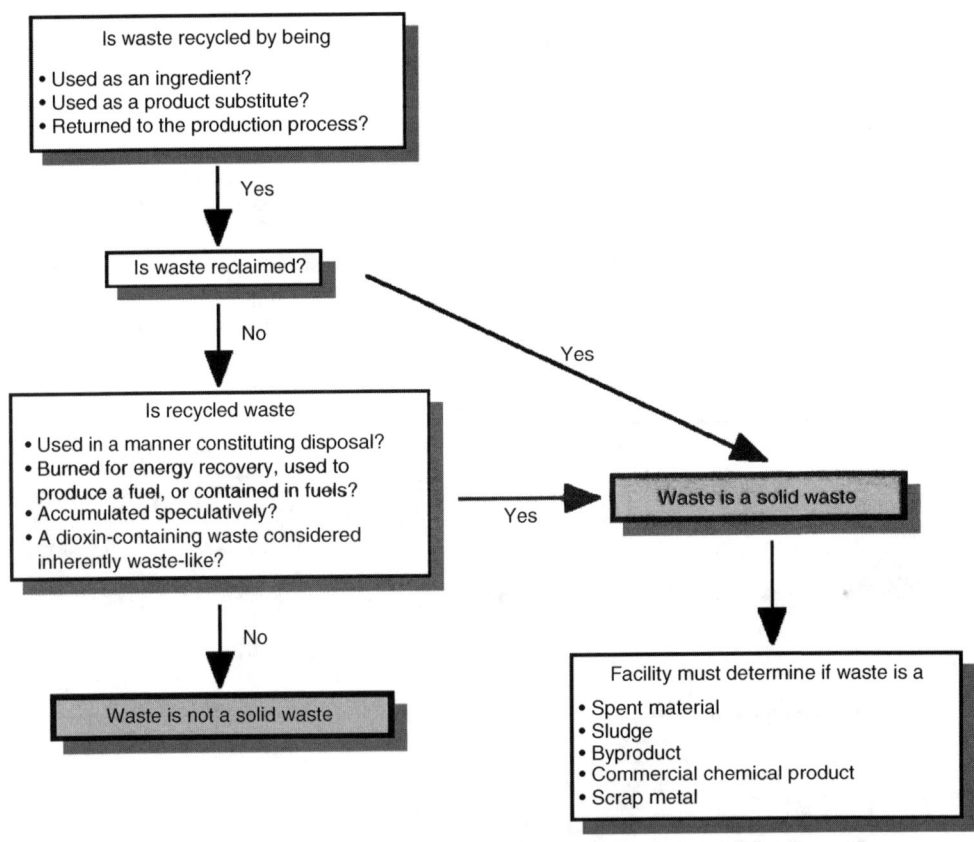

FIGURE 1.2 Determination of whether recycled wastes are hazardous wastes. *Source:* U.S. EPA, Resource Conservation and Recovery Act—Orientation Manual, Report EPA 530-R-02-016, U.S. EPA, Washington, DC, January 2003.

regulated as a hazardous waste. In order to encourage waste recycling, RCRA exempts three types of wastes from the definition of solid waste[2]:

1. *Wastes used as an ingredient.* If a material is directly used as an ingredient in a production process without first being reclaimed, then that material is not a solid waste.
2. *Wastes used as a product substitute.* If a material is directly used as an effective substitute for a commercial product (without first being reclaimed), it is exempt from the definition of solid waste.
3. *Wastes returned to the production process.* When a material is returned directly to the production process (without first being reclaimed) for use as a feedstock or raw material, it is not a solid waste.

Materials, however, are solid wastes and are not exempt if they are recycled in certain other ways. For example, materials recycled in the following ways are defined as solid wastes:

1. *Waste used in a manner constituting disposal.* Use constituting disposal is the direct place- ment of wastes or products containing wastes (e.g., asphalt with petroleum-refining wastes as an ingredient) on the land.
2. *Waste burned for energy recovery, used to produce a fuel, or contained in fuels.* Burning hazardous waste for fuel (e.g., burning for energy recovery) and using wastes to produce fuels are regulated activities. However, commercial products intended to be burned as fuels are not considered solid wastes. For example, off-specification jet fuel (e.g., a fuel with minor chemical impurities) is not a solid waste when it is burned for energy recovery, because it is itself a fuel.
3. *Waste accumulated speculatively.* In order to encourage recycling of wastes as well as to ensure that materials are recycled and not simply stored to avoid regulation, U.S. EPA estab- lished a provision to encourage facilities to recycle sufficient amounts in a timely manner. This provision designates as solid wastes those materials that are accumulated speculatively. A material is accumulated speculatively (e.g., stored in lieu of expeditious recycling) if it has no viable market or if the person accumulating the material cannot demonstrate that at least 75% of the material is recycled in a calendar year, commencing on January 1.
4. *Dioxin-containing wastes considered inherently waste-like.* Dioxin-containing wastes are considered inherently waste-like because they pose significant threats to human health and the environment if released or mismanaged. As a result, RCRA does not exempt such wastes from the definition of solid waste even if they are recycled through direct use or reuse without prior reclamation. This is to ensure that such wastes are subject to the most protective regulatory controls.

1.3.2 SECONDARY MATERIALS

Not all materials can be used directly or reused without reclamation. A material is reclaimed if it is processed to recover a usable product (e.g., smelting a waste to recover valuable metal constituents), or if it is regenerated through processing to remove contaminants in a way that restores them to their useable condition (e.g., distilling dirty spent solvents to produce clean solvents). If secondary materials are reclaimed before use, their regulatory status depends on the type of material. For this solid waste determination process, U.S. EPA groups all materials into five categories: spent materials, sludges, byproducts, commercial chemical products (CCPs), and scrap metal.

1.3.2.1 Spent Materials

Spent materials are materials that have been used and can no longer serve the purpose for which they were produced without processing. For example, a solvent used to degrease metal parts will eventually become contaminated such that it cannot be used as a solvent until it is regenerated. If a spent material must be reclaimed, it is a solid waste and is subject to hazardous waste regulation. Spent materials are also regulated as solid wastes when used in a manner constituting disposal, when burned for energy recovery, when used to produce a fuel or contained in fuels, or when accumulated speculatively (see Table 1.1).

1.3.2.2 Sludges

Sludges are any solid, semisolid, or liquid wastes generated from a wastewater treatment plant, water supply treatment plant, or air pollution control device (e.g., filters, baghouse dust). Sludges from specific industrial processes or sources (known as listed sludges) are solid wastes when reclaimed, when used in a manner constituting disposal, when burned for energy recovery, used to produce a fuel, or contained in fuels, or when accumulated speculatively. On the other hand, characteristic sludges (those that exhibit certain physical or chemical properties) are not solid wastes when reclaimed, unless they are used in a manner constituting disposal, are burned for energy recovery, used to produce a fuel, or contained in fuels, or are accumulated speculatively (Table 1.1).

1.3.2.3 Byproducts

Byproducts are materials that are not one of the intended products of a production process. An example is the sediment remaining at the bottom of a distillation column. *Byproduct* is a catch-all term and includes most wastes that are not spent materials or sludges. Listed byproducts are solid wastes when reclaimed; used in a manner constituting disposal; burned for energy recovery, used to produce a fuel, or contained in fuels; or accumulated speculatively. On the other hand, characteristic byproducts are not solid wastes when reclaimed, unless they are used in a manner constituting disposal; burned for energy recovery, used to produce a fuel, or contained in fuels; or accumulated speculatively (Table 1.1).

1.3.2.4 Commercial Chemical Products

These are unused or off-specification chemicals (e.g., chemicals that have exceeded their shelf-life), spill or container residues, and other unused manufactured products that are not typically considered

---

TABLE 1.1
Regulatory Status of Secondary Materials

	These Materials are Solid Wastes When			
	Reclaimed	Used in a Manner Constituting Disposal	Burned for Energy Recovery, Used to Produce a Fuel, or Contained in Fuels	Accumulated Speculatively
Spent materials	X	X	X	X
Listed sludges	X	X	X	X
Characteristic sludges		X	X	X
Listed byproducts	X	X	X	X
Characteristic byproducts		X	X	X
Commercial chemical products		X[a]	X[a]	
Scrap metal	X	X	X	X

[a] If such management is consistent with the product's normal use, then commercial chemical products used in a manner constituting disposal or burned for energy recovery, used to produce a fuel, or contained in fuels are not solid wastes.

Source: U.S. EPA, Resource Conservation and Recovery Act—Orientation Manual, Report EPA 530-R-02-016, U.S. EPA, Washington, DC, January 2003.

chemicals. Commercial chemical products are not solid wastes when reclaimed, unless they are used in a manner constituting disposal; or burned for energy recovery, used to produce a fuel, or contained in fuels (Table 1.1).

1.3.2.5 Scrap Metal

Scrap metal comprises worn or extra bits and pieces of metal parts, such as scrap piping and wire, or worn metal items, such as scrap automobile parts and radiators. If scrap metal is reclaimed, it is a solid waste and is subject to hazardous waste regulation. Scrap metal is also regulated as a solid waste when used in a manner constituting disposal; burned for energy recovery, used to produce a fuel, or contained in fuels; or accumulated speculatively. This does not apply to processed scrap metal, which is excluded from hazardous waste generation entirely.

1.3.3 SHAM RECYCLING

For all recycling activities, the above rules are based on the premise that legitimate reclamation or reuse is taking place. U.S. EPA rewards facilities recycling some wastes by exempting them from regulation, or by subjecting them to less stringent regulation. Some facilities, however, may claim that they are recycling a material in order to avoid being subject to RCRA regulation, when in fact the activity is not legitimate recycling. U.S. EPA has established guidelines for what constitutes legitimate recycling and has described activities it considers to be illegitimate or sham recycling. Considerations in making this determination include whether the secondary material is effective for the claimed use, if the secondary material is used in excess of the amount necessary, and whether or not the facility has maintained records of the recycling transactions. Sham recycling may include situations when a secondary material falls into the following categories:

1. It is ineffective or only marginally effective for the claimed use (e.g., using certain heavy metal sludges in concrete when such sludges do not contribute any significant element to the concrete's properties).
2. It is used in excess of the amount necessary (e.g., using materials containing chlorine as an ingredient in a process requiring chlorine, but in excess of the required chlorine levels).
3. It is handled in a manner inconsistent with its use as a raw material or commercial product substitute (e.g., storing materials in a leaking surface impoundment as compared to a tank in good condition that is intended for storing raw materials).

1.3.4 EXEMPTIONS FROM HAZARDOUS WASTES

Not all RCRA solid wastes qualify as hazardous wastes. Other factors must be considered before deciding whether a solid waste should be regulated as a hazardous waste. Regulation of certain wastes may be impractical or otherwise undesirable, regardless of the hazards that the waste might pose. For instance, household waste can contain dangerous chemicals, such as solvents and pesticides, but subjecting households to the strict RCRA waste management regulations would create a number of practical problems. As a result, Congress and U.S. EPA exempted or excluded certain wastes, such as household wastes, from the hazardous waste definition and regulations. Determining whether or not a waste is excluded or exempted from hazardous waste regulation is the second step in the RCRA hazardous waste identification process. There are five categories of exclusions[2]:

1. Exclusions from the definition of solid waste
2. Exclusions from the definition of hazardous waste
3. Exclusions for waste generated in raw material, product storage, or manufacturing units
4. Exclusions for laboratory samples and waste treatability studies
5. Exclusions for dredged material regulated under the Marine Protection Research and Sanctuaries Act or the Clean Water Act

If the waste fits one of these categories, it is not regulated as an RCRA hazardous waste, and the hazardous waste requirements do not apply.

1.3.4.1 Solid Waste Exclusions

A material cannot be a hazardous waste if it does not meet the definition of a solid waste. Thus, wastes that are excluded from the definition of solid waste are not subject to the RCRA Subtitle C hazardous waste regulation. There are 20 exclusions from the definition of solid waste:

1. *Domestic sewage and mixtures of domestic sewage.* Domestic sewage, or sanitary waste, comes from households, office buildings, factories, and any other place where people live and work. These wastes are carried by sewer to a municipal wastewater treatment plant (called a publicly owned treatment works [POTW]). The treatment of these wastes is regulated under the Clean Water Act (CWA). Mixtures of sanitary wastes and other wastes (including hazardous industrial wastes) that pass through a sewer system to a POTW are also excluded from Subtitle C regulation once they enter the sewer. In certain circumstances, this exclusion may be applied to domestic sewage and mixtures of domestic sewage that pass through a federally owned treatment works (FOTW).
2. *Industrial wastewater discharges (point source discharges).* Another exclusion from RCRA designed to avoid overlap with CWA regulations applies to point source discharges. Point source discharges are discharges of pollutants (e.g., from a pipe, sewer, or pond) directly into a lake, river, stream, or other water body. CWA regulates such discharges under the National Pollutant Discharge Elimination System (NPDES) permitting program. Under this exclusion from the definition of solid waste, wastewaters that are subject to CWA regulations are exempt from Subtitle C regulation at the point of discharge. Any hazardous waste generation, treatment, or storage prior to the discharge is subject to RCRA regulation. Many industrial facilities that treat wastewater on site utilize this point source discharge exclusion.
3. *Irrigation return flows.* When farmers irrigate agricultural land, water not absorbed into the ground can flow into reservoirs for reuse. This return flow often picks up pesticide or fertilizer constituents, potentially rendering it hazardous. Because this water may be reused on the fields, it is excluded from the definition of solid waste.
4. *Radioactive waste.* Radioactive waste is regulated by either the Nuclear Regulatory Commission or the U.S. Department of Energy (DOE) under the Atomic Energy Act (AEA). To avoid duplicative regulation under RCRA and AEA, RCRA excludes certain radioactive materials from the definition of solid waste. However, RCRA excludes only the radioactive components of the waste. If a radioactive waste is mixed with a hazardous waste, the resultant mixture is regulated by both AEA and RCRA as a mixed waste. Similarly, if a facility generates a hazardous waste that is also radioactive, the material is a mixed waste and is subject to regulation under both RCRA and AEA.
5. In situ *mining waste. In situ* mining of certain minerals may involve the application of solvent solutions directly to a mineral deposit in the ground. The solvent passes through the ground, collecting the mineral as it moves. The mineral and solvent mixtures are then collected in underground wells where the solution is removed. Such solvent-contaminated earth, or any nonrecovered solvent, is excluded from the definition of solid waste when left in place.
6. *Pulping liquors.* Pulping liquor, also called black liquor, is a corrosive material used to dissolve wood chips for the manufacturing of paper and other materials. To promote waste minimization and recycling, U.S. EPA excluded pulping liquors from the definition of solid waste if they are reclaimed in a recovery furnace and then reused in the pulping process. If the liquors are recycled in another way, or are accumulated speculatively, they are not excluded.

7. *Spent sulfuric acid.* Spent sulfuric acid may be recycled to produce virgin sulfuric acid. To promote waste reduction and recycling, such recycled spent sulfuric acid is excluded from the definition of solid waste, unless the facility accumulates the material speculatively.

8. *Closed-loop recycling.* To further promote waste reduction and recycling, spent materials that are reclaimed and returned to the original process in an enclosed system of pipes and tanks are excluded from the definition of solid waste, provided that the following conditions are met:

   a) Only tank storage is involved, and the entire process, through reclamation, is closed to the air (i.e., enclosed).

   b) Reclamation does not involve controlled flame combustion, such as that which occurs in boilers, industrial furnaces, or incinerators.

   c) Waste materials are never accumulated in tanks for more than 12 months without being reclaimed.

   d) Reclaimed materials are not used to produce a fuel, or used to produce products that are used in a manner constituting disposal.

   An example of such a closed-loop system might include a closed solvent recovery system in which the dirty solvents are piped from the degreasing unit to a solvent still where the solvent is cleaned, and then piped back to the degreasing unit.

9. *Spent wood preservatives.* Many wood-preserving plants recycle their wastewaters and spent wood-preserving solutions. These materials are collected on drip pads and sumps, and are in many cases returned directly to the beginning of the wood-preserving process where they are reused in the same manner. Although the process resembles a closed-loop recycling process, the closed-loop recycling exclusion does not apply because drip pads are open to the air. Consistent with their objective to encourage recycling hazardous waste, U.S. EPA developed two specific exclusions for spent wood-preserving solutions and wastewaters containing spent preservatives, provided that the materials have been reclaimed and are reused for their original purpose. In addition, wood-preserving solutions and wastewaters are excluded from the definition of solid waste prior to reclamation. To use this exclusion, a facility is required to reuse the materials for their intended purpose and manage them in a way that prevents releases to the environment.

10. *Coke byproduct wastes.* Coke, used in the production of iron, is made by heating coal in high-temperature ovens. Throughout the production process many byproducts are created. The refinement of these coke byproducts generates several listed and characteristic wastestreams. However, to promote recycling of these wastes, U.S. EPA provided an exclusion from the definition of solid waste for certain coke byproduct wastes that are recycled into new products.

11. *Splash condenser dross residue.* The treatment of steel production pollution control sludge generates a zinc-laden residue, called dross. This material, generated from a splash condenser in a high-temperature metal recovery process, is known as a splash condenser dross residue. Because this material contains 50 to 60% zinc, it is often reclaimed, reused, or processed as a valuable recyclable material. Facilities commonly handle this material as a valuable commodity by managing it in a way that is protective of human health and the environment, so U.S. EPA excluded this residue from the definition of solid waste.

12. *Hazardous oil-bearing secondary materials and recovered oil from petroleum-refining operations.* Petroleum-refining facilities sometimes recover oil from oily wastewaters and reuse this oil in the refining process. In order to encourage waste minimization and recycling, U.S. EPA excluded such recovered oil from the definition of solid waste when it is returned to the refinery. Oil-bearing hazardous wastes that are recycled back into the petroleum-refining process are also excluded. In 2002, U.S. EPA proposed to conditionally exclude oil-bearing secondary materials that are processed in a gasification system to

produce synthesis gas fuel and other nonfuel chemical byproducts. Condensates from the Kraft process steam strippers, the most commonly used pulping process today, utilizes various chemicals to break down wood into pulp. This process generates overhead gases that are condensed and often recycled as fuel. To encourage the recycling of these condensates, U.S. EPA excluded them from the definition of solid waste provided the condensate is combusted at the mill that generated it.

13. *Comparable fuels.* In order to promote the recycling of materials with high fuel values, certain materials that are burned as fuels are excluded from the definition of solid waste, provided that they meet certain specifications (i.e., are of a certain degree of purity). This is to ensure that the material does not exceed certain levels of toxic constituents and physical properties that might impede burning. Materials that meet this specification are considered comparable to pure or virgin fuels.

14. *Processed scrap metal.* Scrap metal includes, but is not limited to, pipes, containers, equipment, wire, and other metal items that are no longer of use. To facilitate recycling, scrap metal that has been processed to make it easier to handle or transport and is sent for metals recovery is excluded from the definition of solid waste. Unprocessed scrap metal is still eligible for an exemption from hazardous waste regulation when recycled.

15. *Shredded circuit boards.* Circuit boards are metal boards that hold computer chips, thermostats, batteries, and other electronic components. Circuit boards can be found in computers, televisions, radios, and other electronic equipment. When this equipment is thrown away, these boards can be removed and recycled. Whole circuit boards meet the definition of scrap metal, and are therefore exempt from hazardous waste regulation when recycled. On the other hand, some recycling processes involve shredding the board. Such shredded boards do not meet the exclusion for recycled scrap metal. In order to facilitate the recycling of such materials, U.S. EPA excluded recycled shredded circuit boards from the definition of solid waste, provided that they are stored in containers sufficient to prevent release to the environment, and are free of potentially dangerous components, such as mercury switches, mercury relays, nickel–cadmium batteries, and lithium batteries.

16. *Mineral processing spent materials.* Mineral processing generates spent materials that may exhibit hazardous waste characteristics. Common industry practice is to recycle these mineral processing wastes back into the processing operations to recover mineral value. U.S. EPA created a conditional exclusion from the definition of solid waste for these spent materials when recycled in the mineral processing industry, provided the materials are stored in certain types of units and are not accumulated speculatively.

17. *Petrochemical recovered oil.* Organic chemical manufacturing facilities sometimes recover oil from their organic chemical industry operations. U.S. EPA excluded petrochemical recovered oil from the definition of solid waste when the facility inserts the material into the petroleum-refining process of an associated or adjacent petroleum refinery. Only petrochemical recovered oil that is hazardous because it exhibits the characteristic of ignitability or exhibits the toxicity characteristic for benzene (or both) is eligible for the exclusion.

18. *Spent caustic solutions from petroleum refining.* Petrochemical refineries use caustics to remove acidic compounds such as mercaptans from liquid petroleum streams to reduce produced odor and corrosivity as well as to meet product sulfur specifications. Spent liquid treating caustics from petroleum refineries are excluded from the definition of solid waste if they are used as a feedstock in the manufacture of napthenic and cresylic acid products. U.S. EPA believes that spent caustic, when used in this manner, is a valuable commercial feedstock in the production of these particular products, and is therefore eligible for exclusion.

19. *Glass frit and fluoride-rich baghouse dust generated by vitrification.* In July 2000, U.S. EPA proposed that glass frit and fluoride-rich baghouse dust generated by vitrification be classified as products and excluded from the definition of solid waste. Glass frit is useable

as a commercial chemical product, and fluoride-rich baghouse dust can be recycled back into the aluminum reduction pots as electrolyte or sold as a product for other industrial uses such as steel making.

20. *Zinc fertilizers made from recycled hazardous secondary materials.* U.S. EPA promulgated a conditional exclusion from the definition of solid waste for hazardous secondary materials that are recycled to make zinc fertilizers or zinc fertilizer ingredients. Zinc, an important micronutrient for plants and animals, can be removed from zinc-rich manufacturing residue and used to produce zinc micronutrient fertilizer. A second conditional exclusion applies to the zinc fertilizer products made from these secondary materials.

1.3.4.2 Hazardous Waste Exclusions

U.S. EPA also exempts certain solid wastes from the definition of hazardous waste. If a material meets an exemption from the definition of hazardous waste, it cannot be a hazardous waste, even if the material technically meets a listing or exhibits a characteristic. There are 16 exemptions from the definition of hazardous waste:

1. *Household hazardous waste.* Households often generate solid wastes that could technically be hazardous wastes (e.g. solvents, paints, pesticides, fertilizer, poisons). However, it would be impossible to regulate every house in the U.S. that occasionally threw away a can of paint thinner or a bottle of rat poison. Therefore, U.S. EPA developed the household waste exemption. Under this exemption, wastes generated by normal household activities (e.g., routine house and yard maintenance) are exempt from the definition of hazardous waste. U.S. EPA has expanded the exemption to include household-like areas, such as bunkhouses, ranger stations, crew quarters, campgrounds, picnic grounds, and day-use recreation areas. Although household hazardous waste is exempt from Subtitle C, it is regulated under Subtitle D as a solid waste.

2. *Agricultural waste.* To prevent overregulation of farms and promote waste recycling, solid wastes generated by crop or animal farming are excluded from the definition of hazardous waste provided that the wastes are returned to the ground as fertilizers or soil conditioners. Examples of such wastes are crop residues and manures.

3. *Mining overburden.* After an area of a surface mine has been depleted, it is common practice to return to the mine the earth and rocks (overburden) that were removed to gain access to ore deposits. When the material is returned to the mine site, it is not a hazardous waste under RCRA.

4. *Bevill and Bentsen wastes.* In the Solid Waste Disposal Act Amendments of 1980, Congress amended RCRA by exempting oil, gas, and geothermal exploration, development, and production wastes (Bentsen wastes); fossil fuel combustion wastes; mining and mineral processing wastes; and cement kiln dust wastes (Bevill wastes) from the definition of hazardous waste pending further study by U.S. EPA. These wastes were temporarily exempted because they were produced in very large volumes, were thought to pose less of a hazard than other wastes, and were generally not amenable to the management practices required under RCRA. Items 5 to 8 (following) describe these exemptions in detail.

5. *Fossil fuel combustion waste.* In order to accommodate effective study, fossil fuel combustion wastes were divided into two categories, large-volume coal-fired utility wastes and the remaining wastes. After studying these wastes, in 1993 U.S. EPA decided to permanently exempt large-volume coal-fired utility wastes, including fly ash, bottom ash, boiler slag, and flue gas emission control waste from the definition of hazardous waste. Further study by U.S. EPA, in 2000, indicated that all remaining fossil fuel combustion wastes need not be regulated under RCRA Subtitle C. However, U.S. EPA determined that national

nonhazardous waste regulations under RCRA Subtitle D are appropriate for coal combustion wastes disposed in surface impoundments and landfills and used as mine-fill. These regulations have now been proposed and subsequently finalized by U.S. EPA.

6. *Oil, gas, and geothermal wastes.* Certain wastes from the exploration and production of oil, gas, and geothermal energy are excluded from the definition of hazardous waste. These wastes include those that have been brought to the surface during oil and gas exploration and production operations, and other wastes that have come into contact with the oil and gas production stream (e.g., during removal of waters injected into the drill well to cool the drill bit).

7. *Mining and mineral processing wastes.* Certain wastes from the mining, refining, and processing of ores and minerals are excluded from the definition of hazardous waste.

8. *Cement kiln dust.* Cement kiln dust is a fine-grained solid byproduct generated during the cement manufacturing process and captured in a facility's air pollution control system. After study, U.S. EPA decided to develop specific regulatory provisions for cement kiln dust. Until U.S. EPA promulgates these new regulatory controls, however, cement kiln dust will generally remain exempt from the definition of hazardous waste.

9. *Trivalent chromium wastes.* The element chromium exists in two forms, hexavalent and trivalent. U.S. EPA determined that, although hexavalent chromium poses enough of a threat to merit regulation as a characteristic hazardous waste, trivalent chromium does not. Therefore, to prevent unnecessary regulation, U.S. EPA excluded from the definition of hazardous waste trivalent chromium-bearing hazardous wastes from certain leather-tanning, shoe-manufacturing, and leather-manufacturing industries.

10. *Arsenically treated wood.* Discarded arsenically treated wood or wood products that are hazardous only because they exhibit certain toxic characteristics (e.g., contain harmful concentrations of metal or pesticide constituents), are excluded from the definition of hazardous waste. Once such treated wood is used, it may be disposed of by the user (commercial or residential) without being subject to hazardous waste regulation. This exclusion is based on the fact that the use of such wood products on the land is similar to the common disposal method, which is landfilling. This exclusion applies only to end-users and not to manufacturers.

11. *Petroleum-contaminated media and debris from underground storage tanks (USTs).* USTs are used to store petroleum (e.g., oil) and hazardous substances (e.g., ammonia). When these tanks leak, the UST program under RCRA Subtitle I provides requirements for cleaning up such spills. To facilitate the corrective action process under the UST regulations, contaminated media (soils and groundwater) and debris (tanks and equipment) at sites undergoing UST cleanup that are hazardous only because they exhibit certain toxic characteristics (e.g., contain a harmful concentrations of leachable organic constituents) are excluded from the definition of hazardous waste.

12. *Spent chlorofluorocarbon refrigerants.* Chlorofluorocarbons (CFCs) released to the atmosphere damage the stratospheric ozone layer. To promote recycling and discourage the practice of venting used CFCs to the atmosphere as a means of avoiding Subtitle C regulation, U.S. EPA excluded recycled CFCs from the definition of hazardous waste because the refrigerants are generally reclaimed for reuse.

13. *Used oil filters.* In order to promote the recycling and recovery of metals and other products from used oil filters, U.S. EPA exempted used oil filters that have been properly drained to remove the used oil.

14. *Used oil distillation bottoms.* When used oil is recycled, residues (called distillation bottoms) form at the bottom of the recycling unit. To promote the recycling of used oil and the beneficial reuse of waste materials, U.S. EPA excluded these residues from the definition of hazardous waste when the bottoms are used as ingredients in asphalt paving and roofing materials.

15. *Landfill leachate or gas condensate derived from listed waste.* Landfill leachate and landfill gas condensate derived from previously disposed wastes that now meet the listing description of one or more of the petroleum refinery listed wastes would be regulated as a listed hazardous waste. However, U.S. EPA temporarily excluded such landfill leachate and gas condensate from the definition of hazardous waste provided their discharge is regulated under the CWA. The exclusion will remain effective while U.S. EPA studies how the landfill leachate and landfill gas condensate are currently managed, and the effect of future CWA effluent limitation guidelines for landfill wastewaters.

16. *Project XL pilot project exclusions.* U.S. EPA has provided two facilities with site-specific hazardous waste exclusions pursuant to the Project XL pilot program. The waste generated from the copper metallization process at the IBM Vermont XL project is excluded from the listing. Byproducts resulting from the production of automobile air bag gas generants at the Autoliv ASP Inc. XL project in Utah are also exempt from regulation as hazardous waste. In addition to these finalized exclusions, in July 2001 U.S. EPA proposed a site-specific exclusion for mixed wastes generated at the Ortho-McNeil Pharmaceutical Inc. facility in Spring House, PA, under the Project XL program.

1.3.4.3   Raw Material, Product Storage, and Process Unit Waste Exclusions

Hazardous wastes generated in raw material, product storage, or process (e.g., manufacturing) units are exempt from Subtitle C hazardous waste regulation while the waste remains in such units. These units include tanks, pipelines, vehicles, and vessels used either in the manufacturing process or for storing raw materials or products, but specifically do not include surface impoundments. Once the waste is removed from the unit, or when a unit temporarily or permanently ceases operation for 90 days, the waste is considered generated and is subject to regulation.

1.3.4.4   Sample and Treatability Study Exclusions

Hazardous waste samples are small, discrete amounts of hazardous waste that are essential to ensure accurate characterization and proper hazardous waste treatment. In order to facilitate the analysis of these materials, RCRA exempts characterization samples and treatability study samples from Subtitle C hazardous waste regulation:

1. *Waste characterization samples.* Samples sent to a laboratory to determine whether or not a waste is hazardous are exempt from regulation. Such samples (typically less than one gallon of waste) are excluded from Subtitle C regulation, provided that these samples are collected and shipped for the sole purpose of determining hazardous waste characteristics or composition. Storage, transportation, and testing of the sample are excluded from RCRA regulation even when the laboratory testing is complete, provided the sample is returned to the generator, and other specific provisions are met. When shipping the sample to or from the laboratory, the sample collector must comply with certain labeling requirements, as well as any applicable U.S. Postal Service (USPS) or U.S. Department of Transportation (DOT) shipping requirements.

2. *Treatability study samples.* To determine if a particular treatment method will be effective on a given waste or what types of wastes remain after the treatment is complete, facilities send samples of waste to a laboratory for testing. U.S. EPA conditionally exempts those who generate or collect samples for the sole purpose of conducting treatability studies from the hazardous waste regulations, provided that certain requirements, including packaging, labeling, and record-keeping provisions, are met. In addition, under specific conditions, laboratories conducting such treatability studies may also be exempt from Subtitle C regulation.

1.3.4.5 Dredge Materials Exclusions

Dredge materials subject to the permitting requirements of Section 404 of the Federal Water Pollution Control Act of Section 103 of the Marine Protection, Research, and Sanctuaries Act of 1972 are not considered hazardous wastes.

1.4 DEFINITION OF HAZARDOUS WASTE

According to Congress, the original statutory definition of the term *hazardous waste* is as follows:

> A solid waste, or combination of solid waste, which because of its quantity, concentration, or physical, chemical, or infectious characteristics may (a) cause, or significantly contribute to, an increase in mortality or an increase in serious irreversible, or incapacitating reversible, illness; or (b) pose a substantial present or potential hazard to human health or the environment when improperly treated, stored, transported, or disposed of, or otherwise managed.

This broad statutory definition provides a general indication of which wastes Congress intended to regulate as hazardous, but it obviously does not provide the clear distinctions necessary for industrial waste handlers to determine whether their wastes pose a sufficient threat to warrant regulation or not. Congress instructed U.S. EPA to develop more specific criteria for defining hazardous waste. There are therefore two definitions of hazardous waste under the RCRA program: a statutory definition and a regulatory definition. The statutory definition cited above is seldom used today. It served primarily as a general guideline for U.S. EPA to follow in developing the regulatory definition of hazardous waste. The regulatory definition is an essential element of the current RCRA program. It precisely identifies which wastes are subject to RCRA waste management regulations.

Congress asked U.S. EPA to fulfill the task of developing a regulatory definition of hazardous waste by using two different mechanisms: by listing certain specific wastes as hazardous and by identifying characteristics that, when present in a waste, make it hazardous. Following its statutory mandate, U.S. EPA developed a regulatory definition of hazardous waste that incorporates both listings and characteristics. In regulatory terms, a RCRA hazardous waste is a waste that appears on one of the four hazardous wastes lists (F list, K list, P list, or U list), or exhibits at least one of four characteristics—ignitability, corrosivity, reactivity, or toxicity[4].

1.4.1 HAZARDOUS WASTE LISTINGS

A hazardous waste listing is a narrative description of a specific type of waste that U.S. EPA considers dangerous enough to warrant regulation. Hazardous waste listings describe wastes from various industrial processes, wastes from specific sectors of industry, or wastes in the form of specific chemical formulations. Before developing a hazardous waste listing, U.S. EPA thoroughly studies a particular wastestream and the threat it can pose to human health and the environment. If the waste poses enough of a threat, U.S. EPA includes a precise description of that waste on one of the hazardous waste lists in the regulations. Thereafter, any waste fitting that narrative listing description is considered hazardous, regardless of its chemical composition or any other potential variable. For example, one of the current hazardous waste listings is: "API separator sludge from the petroleum refining industry." An API separator is a device commonly used by the petroleum-refining industry to separate contaminants from refinery wastewaters. After studying the petroleum-refining industry and typical sludges from API separators, U.S. EPA decided these sludges were dangerous enough to warrant regulation as hazardous waste under all circumstances. The listing therefore designates all petroleum-refinery API separator sludges as hazardous. Chemical composition or other factors about a specific sample of API separator sludge are not relevant to its status as hazardous waste under the RCRA program.

Using listings to define hazardous wastes presents certain advantages and disadvantages. One advantage is that listings make the hazardous waste identification process easy for industrial waste handlers. Only knowledge of a waste's origin is needed to determine if it is listed; laboratory analysis is unnecessary. By comparing any waste to narrative listing descriptions, one can easily determine whether or not the waste is hazardous. U.S. EPA's use of listings also presents certain disadvantages. For example, listing a waste as hazardous demands extensive study of that waste by U.S. EPA. U.S. EPA lacks the resources to investigate the countless types of chemical wastes produced in the U.S., and therefore the hazardous waste listings simply cannot address all dangerous wastes. Another disadvantage of the hazardous waste listings is their lack of flexibility. Listings designate a waste as hazardous if it falls within a particular category or class. The actual composition of the waste is not a consideration as long as the waste matches the appropriate listing description. For instance, some API separator sludges from petroleum refining might contain relatively few hazardous constituents and pose a negligible risk to human health and the environment. Such sludges are still regulated as hazardous, however, because the listing for this wastestream does not consider the potential variations in waste composition. Thus, the hazardous waste listings can unnecessarily regulate some wastes that do not pose a significant health threat. It is also possible for industries to substantially change their processes so that wastes would no longer meet a listing description in spite of the presence of hazardous constituents. The hazardous waste characteristics provide an important complement to listings by addressing most of the shortcomings of the listing methodology of hazardous waste identification.

1.4.2 HAZARDOUS WASTE CHARACTERISTICS

A hazardous waste characteristic is a property that, when present in a waste, indicates that the waste poses a sufficient threat to merit regulation as hazardous. When defining hazardous waste characteristics, U.S. EPA does not study particular wastestreams from specific industries. Instead, U.S. EPA asks the question, "What properties or qualities can a waste have which cause that waste to be dangerous?" For example, U.S. EPA found that ignitability, or the tendency for a waste to easily catch fire and burn, is a dangerous property. Thus, ignitability is one of the hazardous waste characteristics, and a waste displaying that property is regulated as hazardous, regardless of whether the waste is listed. When defining hazardous waste characteristics, U.S. EPA identifies, where practicable, analytical tests capable of detecting or demonstrating the presence of the characteristic. For instance, U.S. EPA regulations reference a laboratory flash-point test to be used when deciding if a liquid waste is ignitable. Whether or not a waste displays a hazardous characteristic generally depends on how it fares in one of the characteristics tests. Therefore, the chemical makeup or other factors about the composition of a particular waste typically determine whether or not it tests as hazardous for a characteristic.

Using characteristics to define hazardous wastes presents certain advantages over designating hazardous wastes by listings. One advantage is that hazardous characteristics and the tests used to evaluate their presence have broad applicability. Once U.S. EPA has defined a characteristic and selected a test for use in identifying it, waste handlers can evaluate any wastestream to see if it is classified as a hazardous waste. Furthermore, use of characteristics can be a more equitable way of designating wastes as hazardous. Instead of categorizing an entire group of wastes as hazardous, characteristics allow a waste handler to evaluate each waste sample on its own merits and classify it according to the actual danger it poses. Aware of these advantages, U.S. EPA originally planned to use characteristics as the primary means of identifying hazardous waste. U.S. EPA hoped to define and select test methods for identifying all hazardous characteristics, including organic toxicity, mutagenicity (the tendency to cause mutations), teratogenicity (the tendency to cause defects in offspring), bioaccumulation potential, and phytotoxicity (toxicity to plants). U.S. EPA encountered problems, however, when trying to develop regulatory definitions of these properties. One primary problem was that no straightforward testing protocols were available for use in determining if a

TABLE 1.2
Typical Hazardous Wastes Generated by Selected Industries

Waste Generators	Waste Type
Chemical manufacturers	Strong acids and bases
	Reactive wastes
	Ignitable wastes
	Discarded commercial chemical products
Vehicle maintenance shops	Paint wastes
	Ignitable wastes
	Spent solvents
	Acids and bases
Printing industry	Photography waste with heavy metals
	Heavy metal solutions
	Waste inks
	Spent solvents
Paper industry	Ignitable wastes
	Corrosive wastes
	Ink wastes, including solvents and metals
Construction industry	Ignitable wastes
	Paint wastes
	Spent solvents
	Strong acids and bases
Cleaning agents and cosmetic manufacturing	Heavy metal dusts and sludges
	Ignitable wastes
	Solvents
	Strong acids and bases
Furniture and wood manufacturing and refinishing	Ignitable wastes
	Spent solvents
	Paint wastes
Metal manufacturing	Paint wastes containing heavy metals
	Strong acids and bases
	Cyanide wastes
	Sludges containing heavy metals

Source: U.S. EPA, Hazardous Waste, available at http://www.epa.gov/epaoswer/osw/
hazwaste.htm#hazwaste, 2008.

waste possessed any of these characteristics. For example, deciding if a particular wastestream poses an unacceptable cancer risk demands extensive laboratory experimentation. Requiring such analysis on a routine basis from industrial waste handlers would be impractical. Therefore, U.S. EPA developed a hazardous waste definition that relies on both listings and characteristics to define hazardous wastes. Table 1.2 shows some typical hazardous wastes generated by selected industries.[3]

1.4.3   LISTED HAZARDOUS WASTES

U.S. EPA has studied and listed as hazardous hundreds of specific industrial wastestreams. These wastes are described or listed on four different lists that are found in the regulations (RCRA Part 261, Subpart D). These four lists are as follows[2]:

1. *The F list.* The F list designates as hazardous particular wastes from certain common industrial or manufacturing processes. Because the processes producing these wastes

can occur in different sectors of industry, the F list wastes are known as wastes from nonspecific sources.

2. *The K list.* The K list designates as hazardous particular wastestreams from certain specific industries. K-list wastes are known as wastes from specific sources.
3. *The P list and the U list.* These two lists are similar in that both list as hazardous pure or commercial grade formulations of certain specific unused chemicals.

These four lists each designate anywhere from 30 to a few hundred wastestreams as hazardous. Each waste on the lists is assigned a waste code consisting of the letter associated with the list followed by three numbers. For example, the wastes on the F list are assigned the waste codes F001, F002, and so on. These waste codes are an important part of the RCRA regulatory system. Assigning the correct waste code to a waste has important implications for the management standards that apply to the waste.

1.4.4 Listing Criteria

Before listing any waste as hazardous, U.S. EPA developed a set of criteria to use as a guide when determining whether or not a waste should be listed. These listing criteria provide a consistent frame of reference when U.S. EPA considers listing a wastestream. Remember that U.S. EPA only uses these criteria when evaluating whether to list a waste; the listing criteria are not used by waste handlers, who refer to the actual hazardous waste lists for hazardous waste identification purposes. There are four different criteria upon which U.S. EPA may base its determination to list a waste as hazardous. Note that these four criteria do not directly correspond to the four different lists of hazardous waste. The four criteria U.S. EPA may use to list a waste as follows[1]:

1. The waste typically contains harmful chemicals, and other factors indicate that it could pose a threat to human health and the environment in the absence of special regulation. Such wastes are known as toxic listed wastes.
2. The waste contains such dangerous chemicals that it could pose a threat to human health and the environment even when properly managed. Such wastes are known as acutely hazardous wastes.
3. The waste typically exhibits one of the four characteristics of hazardous waste described in the hazardous waste identification regulations (ignitability, corrosivity, reactivity, or toxicity).
4. When U.S. EPA has a cause to believe for some other reason, the waste typically fits within the statutory definition of hazardous waste developed by Congress.

U.S. EPA may list a waste as hazardous for any and all of the above reasons. The majority of listed wastes fall into the toxic waste category. To decide if a waste should be a toxic listed waste, U.S. EPA first determines whether it typically contains harmful chemical constituents. An appendix to RCRA contains a list of chemical compounds or elements that scientific studies have shown to have toxic, carcinogenic, mutagenic, or teratogenic effects on humans or other life forms. If a waste contains chemical constituents found on the appendix list, U.S. EPA then evaluates 11 other factors to determine if the wastestream is likely to pose a threat in the absence of special restrictions on its handling. These additional considerations include a risk assessment and study of past cases of damage caused by the waste.

Acutely hazardous wastes are the second most common type of listed waste. U.S. EPA designates a waste as acutely hazardous if it contains the appendix constituents that scientific studies have shown to be fatal to humans or animals in low doses. In a few cases, acutely hazardous wastes contain no appendix constituents, but are extremely dangerous for another reason. An example is a listed waste that designates unused discarded formulations of nitroglycerine as acutely hazardous. Although nitroglycerine is not an appendix hazardous constituent, wastes containing unused nitroglycerine are so unstable that they pose an acute hazard. The criteria for designating a waste as acutely hazardous require only that U.S. EPA considers the typical chemical makeup of the wastestream. U.S. EPA is not required to study other factors, such as relative risk and evidence of harm, when listing a waste as acutely hazardous.

To indicate its reason for listing a waste, U.S. EPA assigns a hazard code to each waste listed on the F, K, P, and U lists. These hazard codes are listed below. The last four hazard codes apply to wastes that have been listed because they typically exhibit one of the four regulatory characteristics of hazardous waste. There will be more about the four characteristics of hazardous waste later in this chapter. The hazard codes indicating the basis for listing a waste are as follows[2]:

1. Toxic waste (T)
2. Acute hazardous waste (H)
3. Ignitable waste (I)
4. Corrosive waste (C)
5. Reactive waste (R)
6. Toxicity characteristic waste (E)

The hazard codes assigned to listed wastes affect the regulations that apply to handling the waste. For instance, acute hazardous wastes accompanied by the hazard code (H) are subject to stricter management standards than most other wastes.

1.4.5   THE F LIST: WASTES FROM NONSPECIFIC SOURCES

The F list designates as hazardous particular wastestreams from certain common industrial or manufacturing processes. F-list wastes usually consist of chemicals that have been used for their intended purpose in an industrial process. That is why F-list wastes are known as "manufacturing process wastes." The F list wastes can be divided into seven groups, depending on the type of manufacturing or industrial operation that creates them. The seven categories of F-listed wastes are as follows[1]:

1. Spent solvent wastes (F001 to F005)
2. Wastes from electroplating and other metal-finishing operations (F006 to F012, F019)
3. Dioxin-bearing wastes (F020 to F023 and F026 to F028)
4. Wastes from the production of certain chlorinated aliphatic hydrocarbons (F024, F025)
5. Wastes from wood preserving (F032, F034, and F035)
6. Petroleum refinery wastewater treatment sludges (F037 and F038)
7. Multisource leachate (F039)

1.4.5.1   Spent Solvent Wastes

Spent solvent wastes apply to wastestreams from the use of certain common organic solvents. Solvents are chemicals with many uses, although they are most often used in degreasing or cleaning. The solvents covered by the F listings are commonly used in industries ranging from mechanical repair to dry cleaning to electronics manufacturing. U.S. EPA decided that only certain solvents used in certain ways produce wastestreams that warrant a hazardous waste listing. Therefore, a number of key factors must be evaluated in order to determine whether the spent solvent wastes apply to a particular waste solvent. First, one or more of the 31 specific organic solvents designated in the spent solvent wastes listing description must have been used in the operation that created the waste. Second, the listed solvent must have been used in a particular manner; it must have been used for its "solvent properties," as U.S. EPA defines that expression. Finally, U.S. EPA decided that only a wastestream created through the use of concentrated solvents should be listed. Thus, the concentration of the solvent formulation or product before its use in the process that created the waste is also a factor in determining the applicability of the spent solvent wastes listing.

The spent solvent listings provide a good illustration of a principle common to all listed hazard-ous wastes. To determine whether a waste qualifies as listed, knowledge of the process that created the waste is essential, and information about the waste's chemical composition is often irrelevant. For example, the F005 listing description can allow two different wastes with identical chemical contents to be regulated differently because of subtle differences in the processes that created the wastes. A waste made up of toluene and paint is F005 if the toluene has been used to clean the paint from brushes or some other surface. A waste with the same chemical composition is not F005 if the toluene has been used as an ingredient (such as a thinner) in the paint. U.S. EPA considers use as a cleaner to be "use as a solvent"; use as an ingredient does not qualify as solvent use. As can be seen, knowledge of the process that created a waste is the key in evaluating whether a waste can be a hazardous spent solvent or other listed hazardous waste.

1.4.5.2   Wastes from Electroplating and Other Metal-Finishing Operations

The listed hazardous wastes from electroplating and other metal-finishing operations are wastes commonly produced during electroplating and other metal-finishing operations. Diverse industries use electroplating and other methods to change the surface of metal objects in order to enhance the appearance of the objects, make them more resistant to corrosion, or impart some other desirable property to them. Industries involved in plating and metal finishing range from jewelry manufacture to automobile production. A variety of techniques can be used to amend a metal's surface. For example, electroplating uses electricity to deposit a layer of a decorative or protective metal on the surface of another metal object. Chemical conversion coating also amends the surface of a metal, but does so by chemically converting (without use of electricity) a layer of the original base metal into a protective coating. Because each of these processes produces different types of wastes, U.S. EPA only designated wastes from certain metal-finishing operations as hazardous. The first step in determining whether one of the wastes from electroplating and other metal-finishing operations listings applies to a waste is identifying the type of metal-finishing process involved in creating the waste from the following list:

1. Electroplating operations
2. Metal heat-treating operations
3. Chemical conversion coating of aluminum

1.4.5.3   Dioxin-Bearing Wastes

The listings for dioxin-bearing wastes describe a number of wastestreams that U.S. EPA believes are likely to contain dioxins, which are considered to be among the most dangerous known chemi-cal compounds. The dioxin listings apply primarily to manufacturing process wastes from the production of specific pesticides or specific chemicals used in the production of pesticides. One listing (F027) deserves special notice, because it does not apply to used manufacturing wastes. It applies only to certain unused pesticide formulations. This is in fact the only listing on the F list or the K list that describes an unused chemical rather than an industrial wastestream consisting of chemicals that have served their intended purpose. With the exception of one other listing (F028), all of the dioxin-bearing wastes are considered acute hazardous wastes and are designated with the hazard code (H). These wastes are therefore subject to stricter management standards than other hazardous wastes.

1.4.5.4   Wastes from the Production of Certain Chlorinated Aliphatic Hydrocarbons

Wastes from the production of certain chlorinated aliphatic hydrocarbons listings designate as hazardous certain wastestreams produced in the manufacture of chlorinated aliphatic hydro-carbons. These listings stand out on the F list (the list of wastes from nonspecific sources), because they focus on wastes from a very narrow industrial sector. Many other wastestreams from the manufacture of organic chemicals are listed on the K list, the list of wastes from specific sources, including two chlorinated aliphatic wastes.

1.4.5.5 Wood-Preserving Wastes

The wood-preserving wastes listings apply to certain wastes from wood-preserving operations. Many types of wood used for construction or other nonfuel applications are chemically treated to slow the deterioration caused by decay and insects. Such chemical treatment is commonly used in telephone poles, railroad ties, and other wood products prepared to withstand the rigors of outdoor use. Wood preservation typically involves pressure treating the lumber with pentachlorophenol, creosote, or preservatives containing arsenic or chromium. (It should be noted that, from January 1, 2004, many wood treaters have no longer been using arsenic- or chromium-based inorganic preservatives.) The wood-preserving process creates a number of common wastestreams containing these chemicals. For example, once wood has been treated with a preservative excess preservative drips from the lumber. The wood-preserving wastes listings designate this preservative drippage as listed hazardous waste. These listings also apply to a variety of other residues from wood preserving. Whether these listings apply to a particular wood-preserving waste depends entirely on the type of preservative used at the facility (waste generated from wood-preserving processes using pentachlorophenol is F032, waste from the use of creosote is F034, and waste from treating wood with arsenic or chromium is F035). The K list also includes the waste code K001, which applies to bottom sediment sludge from treating wastewaters associated with processes using pentachlorophenol or creosote.

1.4.5.6 Petroleum Refinery Wastewater Treatment Sludges

The petroleum refinery wastewater treatment sludges listings apply to specific wastestreams from petroleum refineries. The petroleum-refining process typically creates large quantities of contaminated wastewater. Before this wastewater can be discharged to a river or sewer, it must be treated to remove oil, solid material, and chemical pollutants. Gravity provides a simple way of separating these pollutants from refinery wastewaters. Over time, solids and heavier pollutants precipitate from wastewaters to form sludge. Other less dense pollutants accumulate on the surface of wastewaters, forming a material known as float. These gravitational separation processes can be encouraged using chemical or mechanical means. Some of the listings apply to sludge and float created by gravitational treatment of petroleum refinery wastewaters; other listings apply to sludge and float created during the chemical or physical treatment of refinery wastewaters. The K list also includes waste for certain petroleum wastestreams generated by the petroleum-refining industry.

1.4.5.7 Multisource Leachate

The multisource leachate listing applies to the liquid material that accumulates at the bottom of a hazardous waste landfill. Understanding the natural phenomenon known as leaching is essential to understanding a number of key RCRA regulations. Leaching occurs when liquids such as rainwater filter through soil or buried materials, such as wastes placed in a landfill. When this liquid comes into contact with buried wastes, it leaches or draws chemicals out of those wastes. This liquid (called leachate) can then carry the leached chemical contaminants further into the ground, eventually depositing them elsewhere in the subsurface or in groundwater. The leachate that percolates through landfills, particularly hazardous waste landfills, usually contains high concentrations of chemicals, and is often collected to minimize the potential that it may enter the subsurface environment and contaminate soil or groundwater.

1.4.6 THE K LIST: WASTES FROM SPECIFIC SOURCES

The K list of hazardous wastes designates particular wastes from specific sectors of industry and manufacturing as hazardous. The K list wastes are therefore known as wastes from specific sources. Like F list wastes, K list wastes are manufacturing process wastes. They contain chemicals that have been used for their intended purpose. To determine whether a waste qualifies as K-listed, two primary questions must be answered. First, is the facility that created the waste within one of the industrial or manufacturing categories on the K list? Second, does the waste match one of the specific K list waste descriptions? There are 13 industries that can generate K-list wastes[1]:

1. Wood preservation
2. Inorganic pigment manufacturing
3. Organic chemicals manufacturing
4. Inorganic chemicals manufacturing
5. Pesticides manufacturing
6. Explosives manufacturing
7. Petroleum refining
8. Iron and steel production
9. Primary aluminum production
10. Secondary lead processing
11. Veterinary pharmaceuticals manufacturing
12. Ink formulation
13. Coking (processing of coal to produce coke, a material used in iron and steel production)

It should be noted that not all wastes from these 13 industries are hazardous, only those specifically described in the detailed K-list descriptions.

In general, the K listings target much more specific wastestreams than the F listings. For example, U.S. EPA has added a number of listings to the petroleum-refining category of the K list. U.S. EPA estimates that 100 facilities nationwide produce wastestreams covered by these new K listings. In contrast, F-listed spent solvent wastes are commonly generated in thousands of different plants and facilities. It should also be noticed that industries generating K-listed wastes, such as the wood-preserving and petroleum-refining industries, can also generate F-listed wastes. Typically, K listings describe more specific wastestreams than F listings applicable to the same industry. For example, two K listings designate as hazardous two very specific types of petroleum refinery wastewater treatment residues: wastewater treatment sludges created in API separators and wastewater treatment float created using dissolved air flotation (DAF) pollution control devices. There are two F listings that complement these two K listings by designating as hazardous all other types of petroleum refinery wastewater treatment sludges and floats. These petroleum refinery listings illustrate that the K listings are typically more specific than the F listings. They also illustrate that the two lists are in many ways very similar.

1.4.7 THE P AND U LISTS: DISCARDED COMMERCIAL CHEMICAL PRODUCTS

The P and U lists designate as hazardous pure or commercial-grade formulations of certain unused chemicals. The P and U listings are quite different from the F and K listings. For a waste to qualify as P- or U-listed, a waste must meet the following three criteria:

1. The waste must contain one of the chemicals listed on the P or U list.
2. The chemical in the waste must be unused.
3. The chemical in the waste must be in the form of a "commercial chemical product."

It has already been explained that hazardous waste listings are narrative descriptions of specific wastestreams and that a waste's actual chemical composition is generally irrelevant to whether a listing applies to it. At first glance, the P and U listings seem inconsistent with these principles. Each P and U listing consists only of the chemical name of a compound known to be toxic or otherwise dangerous; no description is included. U.S. EPA adopted this format because the same narrative description applies to all P and U list wastes. Instead of appearing next to each one of the hundreds of P and U list waste codes, this description is found in the regulatory text that introduces the two lists.

The generic P and U list waste description involves two key factors. First, a P or U listing applies only if one of the listed chemicals is discarded unused. In other words, the P and U lists do not apply to manufacturing process wastes, as do the F and K lists. The P and U listings apply to unused chemicals that become wastes. Unused chemicals become wastes for a number of reasons. For example, some unused chemicals are spilled by accident. Others are intentionally discarded because they are off-specification and cannot serve the purpose for which they were originally produced.

The second key factor governing the applicability of the P or U listings is that the listed chemical must be discarded in the form of a "commercial chemical product." U.S. EPA uses the phrase commercial chemical product to describe a chemical that is in pure form, that is in commercial-grade form, or that is the sole active ingredient in a chemical formulation. The pure form of a chemical is a formulation consisting of 100% of that chemical. The commercial-grade form of a chemical is a formulation in which the chemical is almost 100% pure, but contains minor impurities. A chemical is the sole active ingredient in a formulation if that chemical is the only ingredient serving the function of the formulation. For instance, a pesticide made for killing insects may contain a poison such as heptachlor as well as various solvent ingredients that act as carriers or lend other desirable properties to the poison. Although all of these chemicals may be capable of killing insects, only the heptachlor serves the primary purpose of the insecticide product. The other chemicals involved are present for other reasons, not because they are poisonous. Therefore, heptachlor is the sole active ingredient in such a formulation, even though it may be present in low concentrations.

As can be seen, the P and U listings apply only to a very narrow category of wastes. For example, an unused pesticide consisting of pure heptachlor is listed waste P059 when discarded. An unused pesticide consisting of pure toxaphene is listed waste P123 when discarded. An unused pesticide made up of 50% heptachlor and 50% toxaphene as active ingredients, while being just as deadly as the first two formulations, is not a listed waste when discarded. That is because neither compound is discarded in the form of a commercial chemical product. The reason U.S. EPA chose such specific criteria for designating P- or U-listed chemicals as hazardous is that when U.S. EPA was first developing the definition of hazardous waste, it was not able to identify with confidence all the different factors that can cause a waste containing a known toxic chemical to be dangerous. It was obvious, however, that those wastes consisting of pure, unadulterated forms of certain chemicals were worthy of regulation. U.S. EPA used the P and U lists to designate hazardous wastes consisting of pure or highly concentrated forms of known toxic chemicals. As shall be seen in the following section of this chapter, wastes that remain unregulated by listings may still fall under protective hazardous waste regulation due to the four characteristics of hazardous waste.

1.5 CHARACTERISTIC HAZARDOUS WASTES

A hazardous waste characteristic is a property that indicates that a waste poses a sufficient threat to deserve regulation as hazardous. U.S. EPA tried to identify characteristics that, when present in a waste, can cause death or illness in humans or ecological damage. U.S. EPA also decided that the presence of any characteristic of hazardous waste should be detectable by using a standardized test method or by applying general knowledge of the waste's properties. U.S. EPA believed that unless generators were provided with widely available and uncomplicated test methods for determining whether their wastes exhibited hazardous characteristics, this system of identifying hazardous wastes would be unfair and impractical. Given these criteria, U.S. EPA only finalized four hazardous waste characteristics. These characteristics are a necessary supplement to the hazardous waste listings. They provide a screening mechanism that waste handlers must apply to all wastes from all industries. In this sense, the characteristics provide a more complete and inclusive means of identifying hazardous wastes than do the hazardous waste listings. The four characteristics of hazardous waste are as follows[4]:

1. Ignitability
2. Corrosivity
3. Reactivity
4. Toxicity

The regulations explaining these characteristics and the test methods to be used in detecting their presence are found in RCRA (Part 261, Subpart C). Note that although waste handlers can use the test methods referenced in Subpart C to determine whether a waste displays characteristics, they are not required to do so. In other words, any handler of industrial waste may apply knowledge of the waste's properties to determine if it exhibits a characteristic, instead of sending the waste for expensive laboratory testing. As with listed wastes, characteristic wastes are assigned waste codes. Ignitable, corrosive, and reactive wastes carry the waste codes D001, D002, and D003, respectively. Wastes displaying the characteristic of toxicity can carry any of the waste codes D004 through D043.

1.5.1 IGNITABILITY

Ignitable wastes are wastes that can readily catch fire and sustain combustion. Many paints, cleaners, and other industrial wastes pose such a fire hazard. Most ignitable wastes are liquid in physical form. U.S. EPA selected a flash point test as the method for determining whether a liquid waste is combustible enough to deserve regulation as hazardous. The flash point test determines the lowest temperature at which a chemical ignites when exposed to flame. Many wastes in solid or nonliquid physical form (e.g., wood, paper) can also readily catch fire and sustain combustion, but U.S. EPA did not intend to regulate most of these nonliquid materials as ignitable wastes. A nonliquid waste is only hazardous due to ignitability if it can spontaneously catch fire under normal handling conditions and can burn so vigorously that it creates a hazard. Certain compressed gases and chemicals called oxidizers can also be ignitable. Ignitable wastes are among the most common hazardous wastes.

1.5.2 CORROSIVITY

Corrosive wastes are acidic or alkaline (basic) wastes which can readily corrode or dissolve flesh, metal, or other materials. They are also among the most common hazardous wastestreams. Waste sulfuric acid from automotive batteries is an example of a corrosive waste. U.S. EPA uses two criteria to identify corrosive hazardous wastes. The first is a pH test. Aqueous wastes with a pH greater than or equal to 12.5, or less than or equal to 2 are corrosive under U.S. EPA's rules. A waste may also be corrosive if it has the ability to corrode steel in a specific U.S. EPA-approved test protocol.

1.5.3 REACTIVITY

A reactive waste is one that readily explodes or undergoes violent reactions. Common examples are discarded munitions or explosives. In many cases, there is no reliable test method to evaluate a waste's potential to explode or react violently under common handling conditions. Therefore, U.S. EPA uses narrative criteria to define most reactive wastes and allows waste handlers to use their best judgment in determining if a waste is sufficiently reactive to be regulated. This is possible because reactive hazardous wastes are relatively uncommon and the dangers they pose are well known to the few waste handlers who deal with them. A waste is reactive if it meets any of the following criteria:

1. It can explode or violently react when exposed to water or under normal handling conditions.
2. It can create toxic fumes or gases when exposed to water or under normal handling conditions.
3. It meets the criteria for classification as an explosive under DOT rules.
4. It generates toxic levels of sulfide or cyanide gas when exposed to a pH range of 2 to 12.5.

1.5.4 Toxicity Characteristics

The leaching of toxic compounds or elements into groundwater drinking supplies from wastes disposed of in landfills is one of the most common ways the general population can be exposed to the chemicals found in industrial wastes. U.S. EPA developed a characteristic designed to identify wastes likely to leach dangerous concentrations of certain known toxic chemicals into groundwater. In order to predict whether any particular waste is likely to leach chemicals into groundwater in the absence of special restrictions on its handling, U.S. EPA first designed a laboratory procedure that replicates the leaching process and other effects that occur when wastes are buried in a typical municipal landfill. This laboratory procedure is known as the "toxicity characteristic leaching procedure" (TCLP). Using the TCLP on a waste sample creates a liquid leachate that is similar to the liquid U.S. EPA would expect to find in the ground near a landfill containing the same waste. Once the leachate is created in the laboratory, a waste handler must determine whether it contains any of 39 different toxic chemicals above specified regulatory levels. If the leachate sample contains a sufficient concentration of one of the specified chemicals, the waste exhibits the toxicity characteristic (TC). U.S. EPA used groundwater modeling studies and toxicity data for a number of common toxic compounds and elements to set these threshold concentration levels. Much of the toxicity data were originally developed under the Safe Drinking Water Act (SDWA).

However, there is one exception to using the TCLP to identify a waste as hazardous. The DC Circuit Court, in *Association of Battery Recyclers vs. U.S. EPA*, vacated the use of the TCLP to determine whether manufactured gas plant (MGP) wastes exhibit the characteristic of toxicity. As previously stated, the TCLP replicates the leaching process in municipal landfills. The court found that U.S. EPA did not produce sufficient evidence that co-disposal of MGP wastes from remediation sites with municipal solid waste (MSW) has happened or is likely to happen. On March 13, 2002, in response to the court vacatur, U.S. EPA codified language exempting MGP waste from the toxicity characteristic regulation.

To recap, determining whether a waste exhibits the toxicity characteristic involves two principal steps[1]:

1. Creating a leachate sample using the TCLP
2. Evaluating the concentration of 39 chemicals in that sample against the regulatory levels listed in Table 1.3.

If a waste exhibits the TC, it carries the waste code associated with the compound or element that exceeded the regulatory level. Table 1.3 presents the toxicity characteristic waste codes, regulated constituents, and regulatory levels.

TABLE 1.3
Toxicity Characteristic Constituents and Regulatory Levels

Waste Code	Contaminants	Concentration (mg/L)
D004	Arsenic	5.0
D005	Barium	100.0
D018	Benzene	0.5
D006	Cadmium	1.0
D019	Carbon tetrachloride	0.5
D020	Chlordane	0.03
D021	Chlorobenzene	100.0
D022	Chloroform	6.0
D007	Chromium	5.0
D023	o-Cresol[a]	200.0
D024	m-Cresol[a]	200.0
D025	p-Cresol[a]	200.0
D026	Total cresols[a]	200.0
D016	2,4-D	10.0
D027	1,4-Dichlorobenzene	7.5
D028	1,2-Dichloroethane	0.5
D029	1,1-Dichloroethylene	0.7
D030	2,4-Dinitrotoluene	0.13
D012	Endrin	0.02
D031	Heptachlor (and its epoxide)	0.008
D032	Hexachlorobenzene	0.13
D033	Hexachlorobutadiene	0.5
D034	Hexachloroethane	3.0
D008	Lead	5.0
D013	Lindane	0.4
D009	Mercury	0.2
D014	Methoxychlor	10.0
D035	Methyl ethyl ketone	200.0
D036	Nitrobenzene	2.0
D037	Pentachlorophenol	100.0
D038	Pyridine	5.0
D010	Selenium	1.0
D011	Silver	5.0
D039	Tetrachloroethylene	0.7
D015	Toxaphene	0.5
D040	Trichloroethylene	0.5
D041	2,4,5-Trichlorophenol	400.0
D042	2,4,6-Trichlorophenol	2.0
D017	2,4,5-TP (Silvex)	1.0
D043	Vinyl chloride	0.2

[a] If o-, m-, and p-cresols cannot be individually measured, the regulatory level for total cresols is used.

Source: U.S. EPA, Introduction to Hazardous Waste Identification (40 CFR, Part 261), Report U.S. EPA 530-K-05-012, U.S. EPA, Washington, DC, September 2005.

1.6 WASTES LISTED SOLELY FOR EXHIBITING THE CHARACTERISTIC OF IGNITABILITY, CORROSIVITY, OR REACTIVITY

Hazardous wastes listed solely for exhibiting the characteristic of ignitability, corrosivity, or reactivity are not regulated in the same way that other listed hazardous wastes are regulated under RCRA. When wastes are generated that meet a listing description for one of the 29 wastes listed only for exhibiting the characteristic of ignitability, corrosivity, or reactivity, the waste is not hazardous if it does not exhibit a characteristic.[5,6] This concept is consistent with the mixture and derived-from rules, which will be discussed in the following section. For example, F003 is listed for the characteristic of ignitability. If a waste is generated and meets the listing description for F003 but does not exhibit the characteristic of ignitability, it is not regulated as a hazardous waste. However, such wastes are still subject to the land disposal restrictions unless they do not exhibit a characteristic at the point of generation.

1.7 THE MIXTURE AND DERIVED-FROM RULES

So far, this chapter has introduced the fundamentals of the hazardous waste identification process and an overview of the hazardous waste listings and characteristics. One should now be able to explain in general terms which solid wastes are hazardous wastes. What remains to be explained is when these hazardous wastes cease being regulated as hazardous wastes. The regulations governing this issue are commonly known as the mixture and derived-from rules.

1.7.1 BACKGROUND

When U.S. EPA first developed the RCRA regulations and the definition of hazardous waste in the late 1970s, the Agency focused on establishing the listings and characteristics, criteria allowing industry to identify which wastes deserved regulation as hazardous wastes. Commenters on U.S. EPA's original proposed regulations brought up other key questions about the hazardous waste identification process. For example, these commenters asked, "Once a waste is identified as hazardous, what happens if that waste changes in some way? If the hazardous waste is changed, either by mixing it with other wastes or by treating it to modify its chemical composition, should it still be regulated as hazardous?" U.S. EPA developed a fairly simple and strict answer and presented it in the mixture and derived-from rules.

1.7.2 LISTED HAZARDOUS WASTES

The mixture and derived-from rules operate differently for listed wastes and characteristic wastes. The mixture rule for listed wastes states that a mixture made up of any amount of a nonhazardous solid waste and any amount of a listed hazardous waste is considered a listed hazardous waste. In other words, if a small vial of listed waste is mixed with a large quantity of nonhazardous waste, the resulting mixture bears the same waste code and regulatory status as the original listed component of the mixture. This principle applies regardless of the actual health threat posed by the waste mixture or the mixture's chemical composition. The derived-from rule governs the regulatory status of materials that are created by treating or changing a hazardous waste in some way. For example, ash created by burning a hazardous waste is considered "derived-from" that hazardous waste. The derived-from rule for listed wastes states that any material derived from a listed hazardous waste is also a listed hazardous waste. Thus, ash produced by burning a listed hazardous waste bears that same waste code and regulatory status as the original listed waste, regardless of the ash's actual properties.

The net effect of the mixture and derived-from rules for listed wastes can be summarized as follows: once a waste matches a listing description, it is forever a listed hazardous waste, regardless of how it is mixed, treated, or otherwise changed. Furthermore, any material that comes in contact with the listed waste will also be considered listed, regardless of its chemical composition.

Although the regulations do provide a few exceptions to the mixture and derived-from rules, most listed hazardous wastes are subject to the strict principles outlined above. To understand the logic behind the mixture and derived-from rules, one must consider the fact that if U.S. EPA relied solely on the narrative listing descriptions to govern when a waste ceased being hazardous, industry might easily circumvent RCRA's protective regulation. For example, a waste handler could simply mix different wastes and claim that they no longer exactly matched the applicable hazardous waste listing descriptions. These wastes would no longer be regulated by RCRA, even though the chemicals they contained would continue to pose the same threats to human health and the environment. U.S. EPA was not able to determine what sort of treatment or concentrations of chemical constituents indicated that a waste no longer deserved regulation. U.S. EPA therefore adopted the simple, conservative approach of the mixture and derived-from rules, while admitting that these rules might make some waste mixtures and treatment residues subject to unnecessary regulation. Adopting the mixture and derived-from rules also presented certain advantages. For instance, the mixture rule gives waste handlers a clear incentive to keep their listed hazardous wastes segregated from other nonhazardous or less dangerous wastestreams. The greater the volumes of hazardous waste the more expensive it is to store, treat and dispose.

1.7.3 CHARACTERISTIC WASTES

The mixture and derived-from rules apply differently to listed and characteristic wastes. A mixture involving characteristic wastes is hazardous only if the mixture itself exhibits a characteristic. Similarly, treatment residues and materials derived from characteristic wastes are hazardous only if they themselves exhibit a characteristic. Unlike listed hazardous wastes, characteristic wastes are hazardous because they possess one of four unique and measurable properties. U.S. EPA decided that once a characteristic waste no longer exhibits one of these four dangerous properties, it no longer deserves regulation as hazardous. Thus, a characteristic waste can be made nonhazardous by treating it to remove its hazardous property; however, U.S. EPA places certain restrictions on the manner in which a waste can be treated. One can learn more about these restrictions in the U.S. EPA *Land Disposal Restrictions Module.*[7] Handlers who render characteristic wastes nonhazardous must consider these restrictions when treating wastes to remove their hazardous properties.

1.7.4 WASTE LISTED SOLELY FOR EXHIBITING THE CHARACTERISTIC OF IGNITABILITY, CORROSIVITY, OR REACTIVITY

All wastes listed solely for exhibiting the characteristic of ignitability, corrosivity, or reactivity (including mixtures, derived-from, and as-generated wastes) are not regulated as hazardous wastes once they no longer exhibit a characteristic.[5,8] U.S. EPA can list a waste as hazardous if that waste typically exhibits one or more of the four hazardous waste characteristics. If a hazardous waste listed only for the characteristics of ignitability, corrosivity, or reactivity is mixed with a solid waste, the original listing does not carry through to the resulting mixture if that mixture does not exhibit any hazardous waste characteristics. For example, U.S. EPA listed the spent solvents as hazardous because these wastes typically display the ignitability characteristic. If this waste is treated by mixing it with another waste, and the resulting mixture does not exhibit a characteristic, the listing no longer applies.

If a waste derived from the treatment, storage, or disposal of a hazardous waste listed for the characteristics of ignitability, corrosivity, or reactivity no longer exhibits one of those characteristics, it is not a hazardous waste. For example, if sludge is generated from the treatment of a listed waste and that sludge does not exhibit the characteristic of ignitability, corrosivity, or reactivity, the listing will not apply to the sludge.

1.7.5 Mixture Rule Exemptions

There are a few situations in which U.S. EPA does not require strict application of the mixture and derived-from rules. U.S. EPA determined that certain mixtures involving listed wastes and certain residues from the treatment of listed wastes typically do not pose enough of a health or environmental threat to deserve regulation as listed wastes. The principal regulatory exclusions from the mixture and derived-from rules are summarized below.

There are eight exemptions from the mixture rule. The first exemption from the mixture rule applies to mixtures of characteristic wastes and specific mining wastes excluded under RCRA. This narrow exemption allows certain mixtures to qualify as nonhazardous wastes, even if the mixtures exhibit one or more hazardous waste characteristics. The mining waste exclusion is explained in more detail in a U.S. EPA module.[9]

The remaining exemptions from the mixture rule apply to certain listed hazardous wastes that are discharged to wastewater treatment facilities. Many industrial facilities produce large quantities of nonhazardous wastewaters as their primary wastestreams. These wastewaters are typically discharged to a water body or local sewer system after being treated to remove pollutants, as required by the CWA. At many of these large facilities, on-site cleaning, chemical spills, or laboratory operations also create relatively small secondary wastestreams that are hazardous due to listings or characteristics. For example, a textile plant producing large quantities of nonhazardous wastewater can generate a secondary wastestream of listed spent solvents from cleaning equipment. Routing such secondary hazardous wastestreams to the facility's wastewater treatment system is a practical way of treating and getting rid of these wastes. This management option triggers the mixture rule, however, as even a very small amount of a listed wastestream combined with very large volumes of nonhazardous wastewater causes the entire mixture to be listed. U.S. EPA provided exemptions from the mixture rule for a number of these situations where relatively small quantities of listed hazardous wastes are routed to large-volume wastewater treatment systems. To qualify for this exemption from the mixture rule, the amount of listed waste introduced into a wastewater treatment system must be very small relative to the total amount of wastewater treated in the system, and the wastewater system must be regulated under the CWA.

1.7.6 Derived-from Rule Exemptions

There are five regulatory exemptions from the derived-from rule. The first of these derived-from rule exemptions applies to materials that are reclaimed from hazardous wastes and used beneficially. Many listed and characteristic hazardous wastes can be recycled to make new products or be processed to recover useable materials with economic value. Such products derived from recycled hazardous wastes are no longer solid wastes. Using the hazardous waste identification process discussed at the beginning of this chapter, if the materials are not solid wastes, then whether they are derived from listed wastes or whether they exhibit hazardous characteristics is irrelevant. A U.S. EPA module[10] explains which residues derived from hazardous wastes cease to be wastes and qualify for this exemption.

The other four exemptions from the derived-from rule apply to residues from the treatment of specific wastes using specific treatment processes. For example, one listing describes spent pickle liquor from the iron and steel industry. Pickle liquor is an acid solution used to finish the surface of steel. When pickle liquor is spent and becomes a waste, it usually contains acids and toxic heavy metals. This waste can be treated by mixing it with lime to form sludge. This treatment, called stabilization,[11,12] neutralizes the acids in the pickle liquor and makes the metals less dangerous by chemically binding them within the sludge. U.S. EPA studied this process and determined that the waste treated in this manner no longer poses enough of a threat to warrant hazardous waste regulation. Therefore, lime-stabilized waste pickle liquor sludge derived from the listed waste is not a listed hazardous waste. The other exemptions from the derived-from rule for listed wastes are also quite specific and include waste derived-from the burning of exempt recyclable fuels, biological treatment sludge derived-from treatment of listed waste, catalyst inert support media separated from a listed waste, and residues from high-temperature metal recovery of listed wastes, provided certain conditions are met.

1.7.7 DELISTING

The RCRA regulations provide another form of relief from the mixture and derived-from rule principles for listed hazardous wastes. Through a site-specific process known as "delisting," a waste handler can submit to U.S. EPA a petition demonstrating that while a particular wastestream generated at their facility may meet a hazardous waste listing description, it does not pose sufficient hazard to deserve RCRA regulation. If U.S. EPA grants such a petition, the particular wastestream at that facility will not be regulated as a listed hazardous waste. Because the delisting process is difficult, time-consuming, and expensive, it is not considered a readily available exception to the mixture and derived-from from rules.

The hazardous waste listings, the hazardous waste characteristics, and the mixture and derived-from rules are all essential parts of the definition of hazardous waste, but these key elements are all described in different sections of the RCRA regulations. Only one regulatory section unites all four elements to establish the formal definition of hazardous waste. This section is entitled "Definition of Hazardous Waste," which states that all solid wastes exhibiting one of the four hazardous characteristics are hazardous wastes. This section also states that all solid wastes listed on one of the four hazardous waste lists are hazardous wastes. Finally, this section explains in detail the mixture and derived-from rules and the regulatory exemptions from these rules. Thus, although the section is entitled Definition of Hazardous Waste, it serves primarily as a guide to the mixture and derived-from rules.

1.8 THE CONTAINED-IN POLICY

The contained-in policy is a special, more flexible version of the mixture and derived-from rules that applies to environmental media and debris contaminated with hazardous waste. *Environmental media* is the term U.S. EPA uses to describe soil, sediments, and groundwater. *Debris* is a term U.S. EPA uses to describe a broad category of larger manufactured and naturally occurring objects that are commonly discarded. Examples of debris include the following:

1. Dismantled construction materials such as used bricks, wood beams, and chunks of concrete
2. Decommissioned industrial equipment such as pipes, pumps, and dismantled tanks
3. Other discarded manufactured objects such as personal protective equipment (e.g., gloves, coveralls, eyewear)
4. Large, naturally occurring objects such as tree trunks and boulders

Environmental media and debris are contaminated with hazardous waste in a number of ways. Environmental media are usually contaminated through accidental spills of hazardous waste or spills of product chemicals that, when spilled, become hazardous wastes. Debris can also be contaminated through spills. Most debris in the form of industrial equipment and personal protective gear becomes contaminated with waste or product chemicals during normal industrial operations. Contaminated media and debris are primary examples of "remediation wastes." In other words, they are not wastestreams created during normal industrial or manufacturing operations. They are typically created during cleanups of contaminated sites and during the decommissioning of factories. Handlers of contaminated media and debris usually cannot control or predict the composition of these materials, which have become contaminated though accidents or past negligence. In contrast, handlers of "as-generated wastes," the term often used to describe chemical wastestreams created during normal industrial or manufacturing operations, can usually predict or control the creation of these wastes through the industrial process. Examples of as-generated wastes include concentrated spent chemicals, industrial wastewaters, and pollution control residues such as sludges.

The hazardous waste identification principles, including the mixture and derived-from rules, apply to as-generated industrial wastes. U.S. EPA decided that a more flexible version of these principles should apply to the primary remediation wastes: environmental media and debris. In particular, U.S. EPA determined that strict application of the mixture and derived-from rules was inappropriate for media and debris, especially when listed wastes were involved. Applying the mixture and derived-from rules to media and debris would present certain disadvantages, as the following examples illustrate. First, under the traditional mixture and derived-from rules, environmental media and debris contaminated with any amount of listed hazardous waste would be forever regulated as hazardous. Such a strict regulatory interpretation would require excavated or dismantled materials to be handled as listed hazardous wastes and could discourage environmental cleanup efforts. Second, most spills of chemicals into soil or groundwater produce very large quantities of these media containing relatively low concentrations of chemicals. Strict application of the mixture and derived-from principles to media would therefore cause many tons of soil to be regulated as listed hazardous waste, despite containing low concentrations of chemicals and posing little actual health threat. Finally, one of the main benefits of the mixture and derived-from rules is not relevant to media and debris. The mixture and derived-from principles encourage handlers of as-generated wastes to keep their listed wastes segregated from less hazardous wastestreams to avoid creating more listed wastes. Handlers of contaminated media and debris generally have no control over the process by which these materials come into contact with hazardous waste.

For all of the above reasons, U.S. EPA chose to apply a special, more flexible, version of the mixture and derived-from rules to environmental media and debris. Contaminated soil, groundwater, and debris can still present health threats if they are not properly handled or disposed. Therefore, U.S. EPA requires that any medium and debris contaminated with a listed waste or exhibiting a hazardous characteristic be regulated like any other hazardous waste. Media and debris contaminated with listed hazardous wastes can, however, lose their listed status and become nonhazardous. This occurs after a demonstration that the particular medium or debris in question no longer poses a sufficient health threat to deserve RCRA regulation. The requirements for making this demonstration are explained below. Once the demonstration is made, the medium or debris in question is no longer considered to contain a listed hazardous waste and is no longer regulated. In addition, contaminated media that contain a waste listed solely for the characteristics of ignitability, corrosivity, or reactivity, would no longer be managed as a hazardous waste when no longer exhibiting a characteristic.[5,13] This concept that media and debris can contain or cease to contain a listed hazardous waste accounts for the name of the policy.

The contained-in policy for environmental media is not actually codified in the RCRA regulations. In legal terms, it is merely a special interpretation of the applicability of the mixture and derived-from rules to soil and groundwater that has been upheld in federal court. These principles for the management of contaminated media are therefore known as a policy instead of a rule. The terms of the contained-in policy are relatively general. In order for environmental media contaminated with a listed waste to no longer be considered hazardous, the handler of that media must demonstrate to U.S. EPA's satisfaction that it no longer poses a sufficient health threat to deserve RCRA regulation. Although handlers of listed media must obtain U.S. EPA's concurrence before disposing of such media as nonhazardous, the current contained-in policy provides no guidelines on how this demonstration to U.S. EPA should be made. The contained-in policy is a far easier option for eliminating unwarranted hazardous waste regulation for low-risk listed wastes than the process of delisting a hazardous waste mentioned previously. The delisting process demands extensive sampling and analysis, submission of a formal petition, and a complete rulemaking by U.S. EPA. A determination that an environmental medium no longer contains a listed hazardous waste can be granted on a site-specific basis by U.S. EPA officials without any regulatory procedure.

Debris contaminated with hazardous waste has traditionally been governed by the same nonregulatory contained-in policy explained above. In 1992, U.S. EPA codified certain aspects of the contained-in policy for debris in the definition of hazardous waste regulations.[14,15] In particular, U.S. EPA included a regulatory passage that explains the process by which handlers of debris contaminated with listed hazardous waste can demonstrate that the debris is nonhazardous. This passage also references certain treatment technologies for decontaminating listed debris so that it no longer contains a listed waste. Thus, the term contained-in policy is now something of a misnomer for contaminated debris, since a contained-in rule for debris now exists.

1.9   REGULATORY DEVELOPMENTS

The hazardous waste identification process is subject to critical review, and adjusted accordingly to reflect technology changes and new information. The hazardous waste listings are particularly dynamic as U.S. EPA conducts further research to incorporate new listings. The following is a brief discussion of several developments to hazardous waste identification.

1.9.1   THE HAZARDOUS WASTE IDENTIFICATION RULES

U.S. EPA proposed to significantly impact the RCRA hazardous waste identification process through a rulemaking effort called the Hazardous Waste Identification Rules (HWIR). The first rule, HWIR-media, was finalized on November 30, 1998, and addressed contaminated media.[16] The second rule, HWIR-waste, was finalized on May 16, 2001, and modified the mixture and derived-from rules, as well as the contained-in policy for listed wastes.[5] Both the HWIR-media rule and the HWIR-waste rule attempt to increase flexibility in the hazardous waste identification system by providing a regulatory mechanism for certain hazardous wastes with low concentrations of hazardous constituents to exit the RCRA Subtitle C universe.

The final HWIR-media rule addresses four main issues:

1. The Agency promulgated a streamlined permitting process for remediation sites that will simplify and expedite the process of obtaining a permit.
2. U.S. EPA created a new unit, called a "staging pile," that allows more flexibility when storing remediation wastes during cleanups.
3. U.S. EPA promulgated exclusion for dredged materials permitted under the CWA, or the Marine Protection, Research, and Sanctuaries Act.
4. The rule finalized provisions that enable states to more easily receive authorization when their RCRA programs are updated in order to incorporate revisions to the federal RCRA regulations.

On July 18, 2000, the Agency released HWIR-waste exemption levels for 36 chemicals that were developed using a risk model known as the Multimedia, Multi-pathway and Multi-receptor Risk Assessment (3MRA) Model.[17] The May 16, 2001, HWIR-waste rule revised and retained the hazardous waste mixture and derived-from rules as previously discussed in this module. In addition, the rule finalized provisions that conditionally exempt mixed waste (waste that is both radioactive and hazardous), if the mixed waste meets certain conditions in the rule.[5]

1.9.2   FINAL HAZARDOUS WASTE LISTING DETERMINATIONS

U.S. EPA first signed a proposed consent decree with the Environmental Defense Fund (EDF), following a suit concerning U.S. EPA's obligations to take certain actions pursuant to RCRA. A consent decree is a legally binding agreement, approved by the Court, which details the agreements of the parties in settling a suit. The proposed consent decree, commonly known as the "mega-deadline," settles some of the outstanding issues from the case by creating a schedule for U.S. EPA to take action on its RCRA obligations. The consent decree, which has been updated periodically, requires U.S. EPA to evaluate specified wastestreams and determine whether or not to add them to the hazardous waste listings.

On November 8, 2000, U.S. EPA listed as hazardous two wastes generated by the chlorinated aliphatics industry.[18] The two wastes are wastewater treatment sludges from the production of ethylene dichloride or vinyl chloride monomer (EDC/VCM), and wastewater treatment sludges from the production of vinyl chloride monomer using mercuric chloride catalyst in an acetylene-based process.

On November 20, 2001, U.S. EPA published a final rule listing three wastes generated from inorganic chemical manufacturing processes as hazardous wastes.[19] The three wastes are baghouse filters from the production of antimony oxide, slag from the production of antimony oxide that is speculatively accumulated or disposed, and residues from manufacturing and manufacturing-site storage of ferric chloride from acids formed during the production of titanium dioxide using the chloride-ilmenite process.

U.S. EPA proposed a concentration-based hazardous waste listing for certain waste solids and liquids generated from the production of paint on February 13, 2001.[20] Following a review of the public comments and supplemental analyses based on those public comments, U.S. EPA determined that the paint wastes identified in the proposal do not present a substantial hazard to human health or the environment. Therefore, U.S. EPA did not list these paint production wastes as hazardous. See the April 4, 2002, final determination regarding these hazardous waste listings for additional information.[21]

On February 24, 2005, U.S. EPA published a final rule listing nonwastewaters from the production of certain dyes, pigments, and food, drug, and cosmetic colorants[22] as hazardous, using a mass loading-based approach. Under the mass loading approach, these wastes are hazardous if they contain any of the constituents of concern at annual mass loading levels that meet or exceed the regulatory levels. The listing focuses on seven hazardous constituents: aniline, o-anisidine, 4-chloroaniline, p-cresidine, 1,2-phenylenediamine, 1,3-phenylenediamine, and 2,4-dimethylaniline. Waste that contains less than the specified threshold levels of constituents of concern are not hazardous. This listing is U.S. EPA's final obligation under the consent decree.

1.9.3 PROPOSED REVISION TO WASTEWATER TREATMENT EXEMPTION FOR HAZARDOUS WASTE MIXTURES

On April 8, 2003, U.S. EPA proposed to add benzene and 2-ethoxyethanol to the list of solvents whose mixtures with wastewater are exempted from the definition of hazardous waste.[23] U.S. EPA is proposing to provide flexibility in the way compliance with the rule is determined by adding the option of directly measuring solvent chemical levels at the headworks of the wastewater treatment system. In addition, U.S. EPA is proposing to include scrubber waters derived from the combustion of spent solvents to the headworks exemption. Finally, U.S. EPA is finalizing the "Headworks Rule," as follows[24]:

- adds benzene and 2-ethoxyethanol to the list of solvents whose mixtures with wastewaters are exempted from the definition of hazardous waste,
- exempts scrubber waters derived from the combustion of any of the exempted solvents,
- adds an option to allow generators to directly measure solvent chemical levels at the headworks of the wastewater treatment system to determine whether the wastewater mixture is exempt from the definition of hazardous waste, and
- extends the eligibility for the de minimis exemption to other listed hazardous wastes (beyond discarded commercial chemical products) and to non-manufacturing facilities.

REFERENCES

1. U.S. EPA, Introduction to Hazardous Waste Identification (40 CFR, Part 261), Report U.S. EPA 530-K-05-012, U.S. EPA, Washington, DC, September 2005.
2. U.S. EPA, Resource Conservation and Recovery Act—Orientation Manual, Report EPA 530-R-02-016, U.S. EPA, Washington, DC, January 2003.
3. U.S. EPA, RCRA: Reducing Risk from Waste, Report EPA530-K-97-004, U.S. EPA, Washington, DC, September 1997.
4. U.S. EPA, Hazardous Waste, available at http://www.epa.gov/epaoswer/osw/hazwaste.htm#hazwaste, 2008.
5. U.S. EPA, Hazardous Waste Identification Rule (HWIR-waste), 66 FR 27266, U.S. EPA, Washington, DC, May 16, 2001.
6. U.S. EPA, Rule 66 FR 27283, U.S. EPA, Washington, DC, May 16, 2001.
7. U.S. EPA, Land Disposal Restrictions Module, U.S. EPA, Washington, DC, March 2006.
8. U.S. EPA, Rule 66 FR 27268, U.S. EPA, Washington, DC, May 16, 2001.
9. U.S. EPA, Solid and Hazardous Waste Exclusions Module, U.S. EPA, Washington, DC, 2006.
10. U.S. EPA, Definition of Solid Waste and Hazardous Waste Recycling Module, U.S. EPA, Washington, DC, April 2006.
11. Wang, L.K., Shammas, N.K. and Hung, Y.T., Eds., *Biosolids Treatment Processes*, Humana Press, Totowa, NJ, 2007.
12. Wang, L.K., Shammas, N.K. and Hung, Y.T., Eds., *Biosolids Engineering and Management*, Humana Press, Totowa, NJ, 2008.
13. U.S. EPA, Rule 66 FR 27286, U.S. EPA, Washington, DC, May 16, 2001.
14. U.S. EPA, Rule 57 FR 37194, U.S. EPA, Washington, DC, August 18, 1992.
15. U.S. EPA, Rule 57 FR 34225, U.S. EPA, Washington, DC, August 18, 1992.
16. U.S. EPA, Hazardous Waste Identification Rule (HWIR-media), 63 FR 65874, U.S. EPA, Washington, DC, November 30, 1998.
17. U.S. EPA, Hazardous Waste Identification Rule (HWIR-waste), 65 FR 44491, U.S. EPA, Washington, DC, July 18, 2000.
18. U.S. EPA, Rule 65 FR 67068, U.S. EPA, Washington, DC, November 8, 2000.
19. U.S. EPA, Rule 66 FR 58257, U.S. EPA, Washington, DC, November 20, 2001.
20. U.S. EPA, Rule 66 FR 10060, U.S. EPA, Washington, DC, February 13, 2001.
21. U.S. EPA, Rule 67 FR 1626, U.S. EPA, Washington, DC, April 4, 2002.
22. U.S. EPA, Rule 70 FR 9138, U.S. EPA, Washington, DC, February 24, 2005.
23. U.S. EPA, Rule 70 68 FR 17234, U.S. EPA, Washington, DC, April 8, 2003.
24. U.S. EPA, Revision of Wastewater Treatment Exemptions for Hazardous Waste Mixtures ("Headworks Exemptions"), http://www.epa.gov/epaoswer/hazwaste/id/headworks/index.htm, November 14, 2007.

2 Soil Remediation

Ioannis Paspaliaris, Nymphodora Papassiopi,
Anthimos Xenidis, and Yung-Tse Hung

CONTENTS

2.1 INTRODUCTION

2.1.1 CONTAMINATION OF SOILS

Soil can be defined as the top layer of the Earth's crust, consisting of mineral particles, organic matter, water, air, and living organisms. As the interface between the Earth's atmosphere and the lithosphere, soil performs a number of diverse functions essential for life preservation and human activities; it is the substrate necessary for the growth of plants and animals and the basis for all agricultural production, and it serves as a protection and filtering layer necessary for clean ground-water supplies. The rate of soil formation and regeneration is very slow, so soil is practically a nonrenewable resource. In view of the high rates of soil degradation, it has become essential that soil resources be protected against the factors that degrade its quality and limit its availability. Human activities can greatly affect the geochemical cycles of soil constituents, resulting in the contamination of soil with heavy metals and other toxic compounds. Soil contamination is mainly the result of improper environmental management in chemical industries, mining and mineral processing operations, industrial waste disposal sites, municipal landfills, and other facilities, both during operation and after closure. Additionally, widespread soil contamination may occur as a result of emissions from transport and industry, which re-deposit onto the soil surface, as well as from overuse of agricultural chemicals. The result of this diffuse soil contamination is the accumulation of the various contaminants in the soil surface layer and their dissolution and transportation into deeper soil layers and groundwater under the effect of the infiltrating water. In some cases, uncontrolled urban expansion has led to changes in land use, and former mining or industrial sites have been gradually transformed into residential, recreational, or even agricultural areas; in these cases, contaminated land may pose a high risk to human health and agricultural production.

Soil contamination was not perceived as a problem until the 1970s, when incidents in the U.S. and Europe (Love Canal, NY; Times Beach, MO; Lekkerkerk, the Netherlands) awakened public awareness about the serious threats posed to human health and the environment by abandoned or improperly managed hazardous wastes. In response to the growing public concern, the U.S., the Netherlands, and a number of other European countries started a systematic effort beginning in 1980 to identify potentially contaminated sites, assess the level of contamination, establish priorities for remediation based on risk assessment studies and gradually implement the required remedial actions.

In the U.S., three federal programs are currently in progress for identifying and cleaning up contaminated sites[1]:

1. In 1980, Congress passed the Comprehensive Environmental Response, Compensation, and Liability Act (CERCLA). Commonly known as Superfund, the program under this law is focused on the remediation of *abandoned or uncontrolled hazardous waste sites*. Since 1980, Superfund has assessed nearly 44,400 sites. To date, 33,100 sites have been removed from the Superfund inventory to aid their economic redevelopment, and 11,300 sites remain active with the site assessment program or are included in the National Priorities List (NPL) for the implementation of remedial actions. By September 2000, 1509 sites were included in the NPL with ongoing or completed cleanup activities.
2. The second program is directed at corrective actions at *currently operating industrial facilities*. This program is authorized by the Resource Conservation and Recovery Act (RCRA) of 1980 and its subsequent amendments. At the time of writing, there are no statistical data about the progress of this program. Approximately 2000 sites were included in the RCRA Corrective Action Baseline by the end of September 2007. Amongst these sites, remedy constructions were completed for 560 sites and remedy decisions were made for 726 sites.
3. The third cleanup program, also authorized by the RCRA, addresses contamination resulting from leaks and spills (mainly petroleum products) from *underground storage tanks* (USTs). This law has compelled cleanup activities at many UST sites. By February 1999, over 385,000 releases had been reported, 327,000 cleanup projects initiated, and 211,000 projects completed.

Many policies and practices have been adopted by European countries for the management of contaminated sites. Information about the various national polices, the technical approaches for risk assessment, and the progress of rehabilitation activities in Europe has been compiled in the framework of two European networks—CARACS (Concerted Action for Risk Assessment for Contaminated Sites) and CLARINET (Contaminated Land Rehabilitation Network for Environmental Technologies)—which were funded by the European Commission. A detailed description of European national policies can be found in relevant publications[2,3] and in the CLARINET website (http://www.clarinet.at).

Table 2.1 summarizes the available data related to the registration, assessment, and remediation of contaminated sites in the U.S. and several European countries. The number of sites presented in

TABLE 2.1

Available Data for the Registration, Assessment, and Remediation of Contaminated Sites in the U.S. and Europe

	Number of Sites			
Country	Suspected Contamination	Confirmed Contamination	Cleanup Initiated or Completed	Data Till
U.S., Superfund	44,400	11,300	1,509	2000
U.S., UST		385,000	211,000	1999
U.K.	100,000			1995
The Netherlands	110,000			1998
Belgium	10,500		86	1998
France	250,000	896	125	1997
Spain	18,000	4,900	77	1995
Italy	9,000	1,570		1997
Germany	300,000			1997
Austria	2,476	145	97	1999
Switzerland	50,000	3,000	200	1998
Denmark	14,500	4,048	800	1997
Norway	3,350	2,100	99	1999
Finland	25,000	1,200		1995
Sweden	20,000	12,000		1999
Hungary	10,000	200		1998
Czech Republic	12,000	1,000	210	1998

Source: From NATO/CCMS, Evaluation of Demonstrated and Emerging Technologies for the Treatment and Clean Up of Contaminated Land and Groundwater, NATO CCMS Pilot Study, Phase III, 1999 Annual Report, EPA 542/R-99/007, no. 235, 1999; Ferguson, C. and Kasamas, H., Eds., *Risk Assessment for Contaminated Sites in Europe, Vol. 2. Policy Frameworks*, LGM Press, Nottingham, UK, 1999. With permission.

the table changes yearly, because the entire process is in a state of continuous progression. It has been suggested that the real extent of the problem has become clear only recently. For example, in 1980 about 350 sites in the Netherlands were thought to be contaminated. This number increased to 1600 in 1986 and 110,000 in 1999. The estimated costs[1] for rehabilitation of these sites was 0.5 billion Euros in 1980, 3 billion Euros in 1986, and between 15 and 25 billion Euros in 1999.

2.1.2 REMEDIATION TECHNOLOGIES

Until recently, a common practice for the remediation of contaminated sites was to excavate the contaminated soil, replace it with clean soil, and then dispose of the contaminated material at municipal waste landfills. This practice, however, was gradually discouraged by the environmental authorities, which issued very strict regulations for landfilling and increased the corresponding disposal costs. In many industrial countries, the cost for disposal in a municipal waste landfill ranges from 80 to 150 USD/ton. If contaminated soil is characterized as hazardous waste, landfilling in state-of-the-art hazardous waste landfills may cost[4] between 500 and 800 USD/ton. The high disposal costs and the limited availability of clean soil has led to the development of alternative remediation methods, which permit the reuse of treated soil following the removal or immobilization of contaminants.

Soil remediation technologies can be classified according to the type of treatment processes taking place[5–7]:

1. *Biological processes*. These are based on the use of living organisms (e.g., microorganisms or plants).

2. *Chemical processes.* These destroy, fix, or remove toxic compounds by using one or more types of chemical reactions.
3. *Physical processes.* These separate contaminants from the soil matrix by exploiting physical differences between the soil and the contaminants (e.g., volatility) or between contaminated and uncontaminated soil particles.
4. *Solidification and stabilization processes.* These immobilize the contaminants through physical or chemical processes. Solidification involves the entrapment of contaminants into a consolidated mass and stabilization is the conversion of contaminants to a chemical form that is less available.
5. *Thermal processes.* These exploit physical and chemical processes at elevated temperatures.

Another classification of remediation technologies describes where the action is taking place. *Ex situ* methods are those applied to excavated soil and *in situ* processes are those applied to the soil in its original location. *On-site* techniques are those that take place on the contaminated site; they can be either *ex situ* or *in situ*. *Off-site* processes treat the excavated soil in fixed industrial facilities, away from the contaminated site.

The following categories of technologies are predominately *ex situ*:

1. Soil washing and related chemical treatment techniques
2. Solidification–stabilization
3. Thermal processes
4. Vitrification
5. Bioremediation using landfarming or biopile techniques

The most common *in situ* technologies are as follows:

1. Soil vapor extraction (SVE)
2. Air sparging
3. *In situ* bioremediation techniques combined with SVE and air sparging
4. Soil flushing
5. Electroremediation
6. Phytoremediation

Currently, most remediation projects are carried out using *ex situ* technologies, both in the U.S. and in Europe. However, there is an increasing trend toward the application of *in situ* technologies because of their considerable advantages over *ex situ* techniques, such as less disturbance of the site, lower treatment costs, and so on.

Published data for the cost of remediation technologies are highly variable. One reason for this variability is that remediation costs depend on several case-specific parameters, such as type of contaminants, geotechnical and geochemical characteristics of the soil matrix, and the hydrogeology of the site for *in situ* techniques. Differences in the reported cost data for the same technology between two countries may also reflect a different degree of commercialization for the specific technology. Indicative cost ranges for characteristic remediation technologies are presented in Table 2.2, based on the U.S. and European Union (EU) experiences.

This chapter presents a detailed description of five technologies: soil vapor extraction, bioremediation, phytoremediation, soil washing, and soil flushing. Information about other categories of proven or emerging technologies is available on several websites. An overview of the technologies currently applied in the U.S., with detailed cost and performance data from characteristic case studies, can be found at the FRTR (Federal Remediation Technologies Roundtable) website (http://www.frtr.gov). Detailed information on several soil remediation technologies can also be found on the United States Environmental Protection Agency's (U.S. EPA) Cleanup Information site (http://www.clu-in.org).

TABLE 2.2
Indicative Costs of Remediation Technologies

Remediation Technology	Range of Costs in the U.S.[a] (USD/t)	Range of Costs in the EU[b] (Euro/t)
Bioremediation	50–150	20–40
Soil washing	80–120	20–200
Stabilization–solidification	240–340	80–150
Thermal treatment	120–300	30–100
Incineration	200–1500	170–350
Soil vapor extraction	20–220	20–60
Phytoremediation	10–35	

[a]*Source*: Schnoor, J.L., Phytoremediation. Technology Evaluation Report TE-98-01, Ground-Water Remediation Technologies Analysis Center, Pittsburgh, PA, 1997. With permission.

[b]*Source*: Vic, E.A. and Bardos, P., Remediation of Contaminated Land. Technology Implementation in Europe, Federal Environmental Agency, Austria. CLARINET Report, available at www.clarinet.at, 2002. With permission.

2.2 SOIL VAPOR EXTRACTION

2.2.1 GENERAL DESCRIPTION

Soil vapor extraction (SVE) is a relatively new yet widely applied technology for the remediation of soils contaminated with volatile organic compounds (VOC) in the unsaturated zone above the water table (vadose zone). The process consists of generating an airstream through the contaminated soil subsurface in order to enhance the volatilization of organic contaminants and thus remove them from the soil matrix.[9–13]

Figure 2.1 presents the main components of a typical *in situ* SVE system.[9,10] Vertical extraction wells are installed inside the contaminated zone at appropriate distances from one another. The SVE wells are typically constructed of PVC pipe, with a screened interval, which is placed within the contaminated zone. The wells are connected to blowers or vacuum pumps, which induce a continuous airflow through the pores of the unsaturated zone. The soil surface is sometimes covered with an impermeable seal, made from high density polyethylene (HDPE) or bentonite clay for example, to prevent the vertical influx of air from the surface, which might cause short-circuiting problems, and promote horizontal gas flow through the contaminated area. The airstream, which contains the contaminant vapors, passes initially through an air–water separation unit to remove the entrained moisture and is then directed to the gas treatment unit, where the contaminants are thermally destroyed or removed by adsorption.

There are three main prerequisites for the successful application of SVE technology:

1. The contamination should be trapped in the vadose zone.
2. The contaminants should have high volatility.
3. The contaminated zone should have high permeability.

A general simple rule is that SVE can be applied successfuly for contaminants with vapor pressure greater than 0.5 mmHg and for soils with air permeability coefficients ranging between 1×10^{-2} and 1×10^{-5} cm/sec.[11]

Many modifications and additional treatment options have been proposed to enhance the performance and extend the applicability of SVE systems, examples of which include the following:

1. Pumping of the groundwater to lower the water table and enlarge the vadose zone, with simultaneous treatment of contaminated groundwater.[10]
2. The combination of SVE with *air sparging* technology. Air sparging involves the injection of air into the saturated zone of contaminated groundwater. The air bubbles enhance the

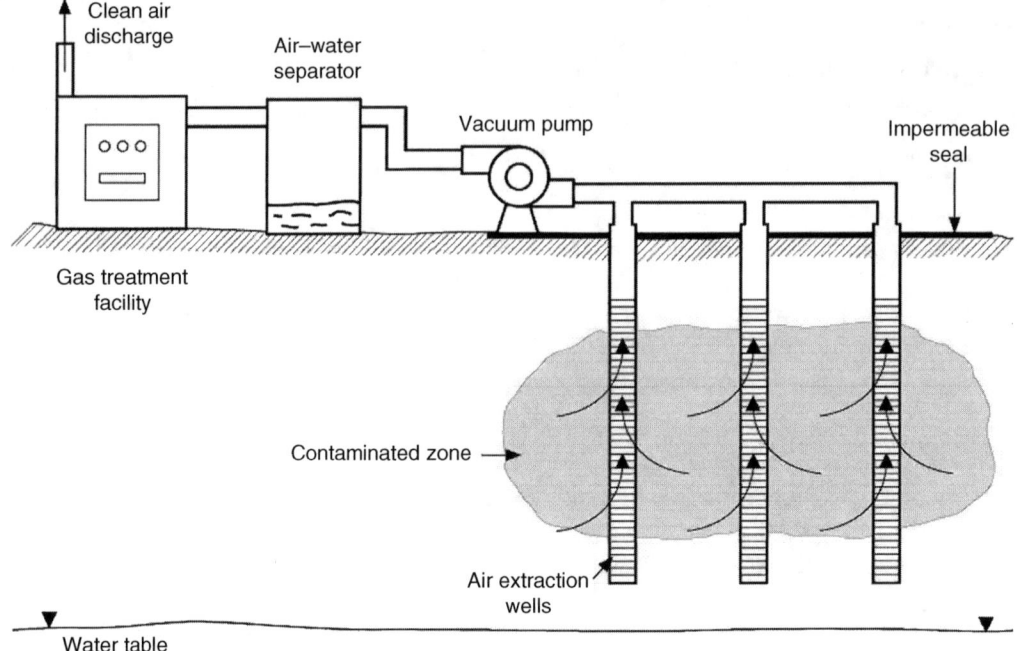

FIGURE 2.1   Schematic representation of an SVE system.

volatilization of dissolved contaminants, especially those with low solubility in water, and then migrate upward to the vadose zone to be captured by the SVE system.[14–17]

3. Combination of SVE with the *bioventing* technology.[17–19] Bioventing uses a system configuration similar to SVE but with a different objective. In bioventing, the induced airflow aims to provide sufficient oxygen for the aerobic biodegradation of contaminants. It is thus possible to remove contaminants with relatively low volatility and high biodegradability.

4. Thermal enhancement of volatilization.[19–21] Volatility of contaminants increases greatly with temperature, so several techniques have been developed to raise soil temperature, including the injection of hot air or steam, electrical resistance heating, and radio frequency heating.

Soil vapor extraction has become a very popular technology since the mid-1990s, because it has several important advantages:

1. It is an *in situ* technology and can even be applied below existing buildings, roads, and so on, thus causing minor disturbance to ongoing site operations.

2. The whole installation may be achieved using low-cost and easily available equipment, and the operation of the system is quite simple.

3. Although it is focused on the treatment of volatile contaminants trapped in the vadose zone, SVE can be integrated easily with other technologies targeting the remediation of groundwater or less volatile compounds, and this flexibility enables the application of the technology to a broad range of sites.

2.2.2   DESIGN CONSIDERATIONS

The most important parameters for the preliminary design of an SVE system are the VOC concentration in the extracted air, the air flow rate, and the radius of influence of each extraction well. These parameters determine the number of wells that must be installed to remediate the whole contaminated area, the time required to obtain the cleanup goals, the size and characteristics of the gas treatment facility and auxiliary equipment, and finally the cost of the whole remediation project.

The design of SVE systems can be based on relatively simple mathematical models that describe the two basic phenomena governing the performance of SVE technology: the phase distribution of the organic contaminants and the characteristics of the airflow in the vadose zone.[11–13,22,23] A simplified modeling approach, providing valuable tools for preliminary design calculations, will be presented in the following sections.

2.2.2.1 Phase Distribution of Organic Contaminants in the Vadose Zone

Organic contaminants can be present in the vadose zone in four distinct phases (Figure 2.2):

1. As an immiscible organic liquid retained by capillary forces in the pore space between the soil particles. This free organic phase is often referred to with the abbreviation NAPL (nonaqueous phase liquid).
2. As dissolved compounds in soil pore water.
3. As an adsorbed film on the surface of soil particles.
4. As vapor in soil air present in the pore space.

The distribution of a contaminant among the four phases depends on (1) the physical and chemical properties of the compound and (2) the characteristics of the soil, and can be described by relatively simple equations (see Table 2.3).

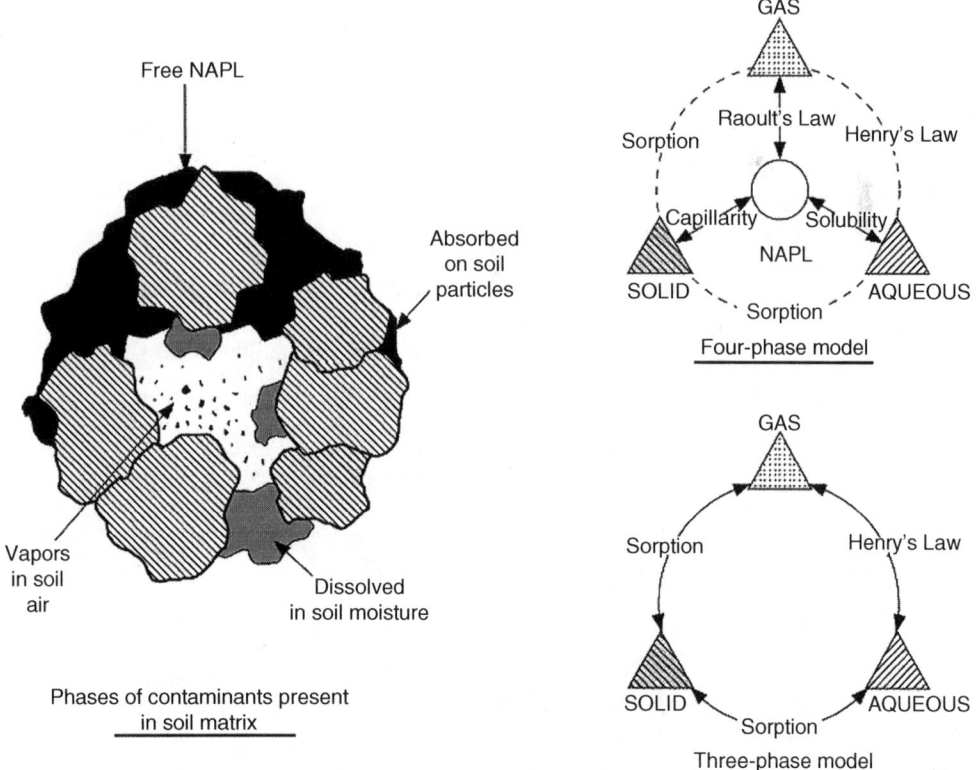

FIGURE 2.2 Phase distribution of organic contaminants in the vadose zone. The solid arrows in the three- and four-phase models represent the equilibria taken into consideration in the equations of Table 2.3.

TABLE 2.3

Basic Equations and Required Data for Calculating the Phase Distribution of Contaminants under Equilibrium Conditions

In the presence of NAPL in the soil matrix

$$C_t = \rho_b C_s + \theta_w C_w + \theta_a C_a + \theta_{or} C_{or} \qquad (1)$$

$$C_a = P^o \times X \times \gamma_i \times MW/(RT) \qquad (2)$$

$$C_{or} = m \times \rho_{or} \times 10^6 \qquad (3)$$

$$C_w = C_w^o \times X \times \gamma_i \qquad (4)$$

$$C_s = K_{oc} \times f_{oc} \times C_w \qquad (5)$$

$$\theta_t = \theta_w + \theta_a + \theta_{or} \qquad (6)$$

Without NAPL in the soil matrix

$$C_t = \rho_b C_s + \theta_w C_w + \theta_a C_a \qquad (7)$$

$$C_a = K_H \times C_w \qquad (8)$$

$$C_s = K_{oc} \times f_{oc} \times C_w \qquad (5)$$

$$\theta_t = \theta_w + \theta_a \qquad (9)$$

Phase distribution values for calculation

C_s = adsorbed concentration of contaminant in the soil particle (mg/kg)

C_w = dissolved concentration in pore water (mg/L)

C_a = vapor concentration in pore air (mg/L)

C_{or} = concentration of contaminant in NAPL (mg/L)

θ_a = pore volume occupied by the gas phase (L/L)

θ_{or} = pore volume occupied by NAPL (L/L)

Required data

Contaminant properties[a]

MW = molecular weight (g/mol)

P^o = vapor pressure of the compound (mmHg)

C_w^o = water solubility (mg/L)

K_H = Henry's constant (dimensionless)

K_{oc} = organic carbon partitioning coefficient (L/kg)

Soil characteristics[b]

ρ_b = soil bulk density (kg/L)

θ_t = total porosity of soil (L/L)

θ_w = pore volume occupied by water (L/L)

f_{oc} = fraction of organic carbon in soil

Contamination data[b]

C_t = total quantity of contaminant per unit soil volume (mg/L)

ρ_{or} = specific density of the NAPL mixture (kg/L)

m = mass fraction of contaminant in the NAPL mixture

X = moles of contaminant in the NAPL mixture

γ_i = activity coefficient of contaminant in NAPL

[a]Data for these properties and for a long list of organic compounds can be found in several environmental engineering textbooks and handbooks.[9,10,24-27]

[b]Soil characteristics and data related to the concentration levels and the composition of organic contaminants should be collected during the investigation of the specific contaminated site.

When a single organic compound is present in the soil matrix as NAPL, its concentration in soil air (C_a) can be directly calculated from the vapor pressure of this compound (P^o) and the Ideal Gas Law:

$$C_a = P^o \times MW/(RT) \qquad (2.1)$$

where MW is the molecular weight of the compound, R is the ideal gas constant, and T is the absolute temperature. For a mixture of compounds, such as gasoline, the partial pressure (P_i) of each constituent i in the soil air depends on the composition of the mixture according to Raoult's Law:

$$P_i = P_i^o \times X_i \times \gamma_i \qquad (2.2)$$

where P_i^o is the vapor pressure of the pure constituent, X_i is the mole fraction of the constituent, and γ_i is the activity coefficient, representing the deviation from the properties of an ideal mixture.

Temperature has a strong influence on the vapor pressure of the contaminants. This effect can be described by the Clausius–Clapeyron equation:

$$\ln\left(\frac{P}{P^o}\right) = \frac{\lambda}{R}\left(\frac{1}{T^o} - \frac{1}{T}\right), \qquad (2.3)$$

where P is the vapor pressure at T, P^o the vapor pressure at T^o, and λ is the molar heat of vaporization.

In the presence of NAPL, the concentration of contaminants in the soil moisture (C_w) can be calculated simply from the solubility of the compounds (equation 3 in Table 2.3). Adsorption of contaminants to the soil particles is a much more complex phenomenon, which depends both on contaminant properties and on soil characteristics. The simplest model for describing adsorption is based on the observation that organic compounds are preferentially bound to the organic matter of soil, and the following linear equation is proposed for calculating the adsorbed concentration (C_s):

$$C_s = K_{oc} \times f_{oc} \times C_w, \qquad (2.4)$$

where K_{oc} is the organic carbon partitioning coefficient of the contaminant and f_{oc} is the fraction of organic carbon in the soil.

When the SVE technology is applied in a contaminated site, the NAPL is gradually removed. Towards the end of the remediation and when NAPL is no longer present, a three-phase model should be considered to calculate the phase distribution of contaminants (see Table 2.3). In this case, the vapor concentration in pore air (C_a) is calculating using the Henry's Law equation (equation 2.5), which describes the equilibrium established between gas and aqueous phases:

$$C_a = K_H \times C_w, \qquad (2.5)$$

where K_H is the Henry's Law constant of the contaminant. Note, however, that during this phase the process is often governed by nonequilibrium rate-limiting conditions.

2.2.2.2 Basic Airflow Equations

The movement of air in the subsurface during the application of SVE is caused by the pressure gradient that is applied in the extraction wells. The lower pressure inside the well, generated by a vacuum blower or pump, causes the soil air to move toward the well. Three basic equations are required to describe this airflow: the mass balance of soil air, the flow equation due to the pressure gradient, and the Ideal Gas Law.

The mass balance of soil air may be described by the classic continuity equation for compressible fluids:

$$\theta_a \frac{\partial \rho_a}{\partial t} = -\left(\frac{\partial(\rho_a u_x)}{\partial x} + \frac{\partial(\rho_a u_y)}{\partial y} + \frac{\partial(\rho_a u_z)}{\partial z}\right) = -\nabla(\rho_a u), \qquad (2.6)$$

where θ_a is the pore volume occupied by the gas phase, ρ_a is the density of air, which is not constant due to air compressibility, and u_x is the air velocity in the x-direction.

For a radial flow from a circumference of radius r toward the well, equation 2.6 may be simplified as follows:

$$\theta_a r \frac{\partial \rho_a}{\partial t} = - \frac{\partial(\rho_a u_r r)}{\partial r}. \tag{2.7}$$

The air velocity due to the pressure gradient can be described by Darcy's Law:

$$u_r = - \frac{K}{\mu_a} \times \frac{dP}{dr}, \tag{2.8}$$

where K is the intrinsic permeability of soil, which is independent of the fluid properties, μ_a is the viscosity of air, and dP/dr is the pressure gradient in the radial r direction.

Finally, the Ideal Gas Law can be used to describe the relationship between air density and pressure:

$$\rho_a = \frac{P \times MW}{RT}, \tag{2.9}$$

where MW is the molecular weight of air, R is the Ideal Gas Law constant, and T is the absolute temperature.

Combining equations 2.7–2.9, a differential equation, with pressure as the single variable, can be derived:

$$\theta_a \frac{\partial P}{\partial t} = \frac{K}{\mu_a} \times \frac{\partial \left[Pr (\partial P / \partial r) \right]}{\partial r}. \tag{2.10}$$

Under steady-state conditions, equation 2.10 has a simple analytical solution, which allows the calculation of the pressure P_r at several radial distances from the well:

$$P_r = \left[P_w^2 + \left(P_I^2 - P_w^2 \right) \times \frac{\ln\left(r/R_w\right)}{\ln\left(R_I/R_w\right)} \right]^{1/2}, \tag{2.11}$$

where P_w is the pressure at the extraction well, R_w is the radius of the well, R_I is the radius of influence of the well, and P_I is the pressure at distance R_I.

2.2.2.3 Radius of Influence and Number of Wells

Equation 2.11 introduces the notion of radius of influence, which is one of the important design parameters of SVE systems. Theoretically, the maximum radius of influence of a well is the distance at which the pressure becomes equal to the ambient atmospheric pressure, i.e., $P_I = P_{atm}$. In practice, R_I is determined as the distance at which a sufficient level of vacuum still exists to induce airflow, e.g., 1% of the vacuum in the extraction well.[9,12] The extraction wells are usually constructed using pipes with a standard radius, e.g., $R_w = 5.1$ cm (2 in.) or 10.2 cm (4 in.), and the vacuum applied in the wells typically ranges from 0.05 to 0.15 atm, i.e., $P_w = 0.95$–0.85 atm.[9,12] If the vacuum required in the radius of influence is 1% of the vacuum in the extraction well, the

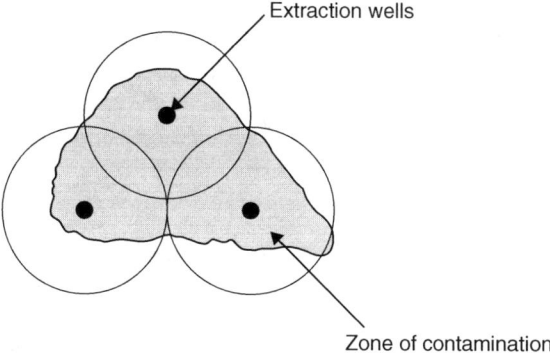

Extraction wells

Zone of contamination

FIGURE 2.3 Determination of the required number of wells from the radius of influence.

corresponding P_I values will range from 0.9985 to 0.9995 atm. The radius of influence R_I is usually determined with preliminary field tests. Vacuum is applied in a test extraction well and the pressure P_r is measured in a monitoring well, installed at a distance r from the well. In practice, pressure drawdown is monitored at two or three points at varying radial distances from the well. Using the field-test data and equation 2.11, it is possible to determine the radius of influence of the well R_I at various operating vacuum values P_w.

Once the radius of influence has been determined, the number of wells N_{wells} required to remediate the entire contaminated area can be calculated from equation 2.12:

$$N_{wells} = 1.2 \times \frac{A_{contam}}{\pi R_I^2}, \tag{2.12}$$

where A_{contam} is the surface area corresponding to the contaminated zone. The factor 1.2 is arbitrarily chosen to account for the overlapping of the areas of influence between the wells and the fact that peripheral wells may reach outside the contaminated zone (Figure 2.3).[12]

2.2.2.4 Air Flow Rates

The flow rate of extracted air can be determined by considering the air velocity, as determined by Darcy's Law (equation 2.8), and the radial distribution of pressure (equation 2.11). The solution for air velocity as a function of the radial distance is given in equation 2.13:

$$u_r = -\frac{K}{2 \times \mu_a} \frac{\left[\dfrac{P_w}{r \times \ln\left(R_w/R_I\right)}\right] \times \left[1 - \left(\dfrac{P_I}{P_w}\right)^2\right]}{\left[1 + \left(1 - \left(\dfrac{P_I}{P_w}\right)^2\right)\dfrac{\ln\left(r/R_w\right)}{\ln\left(R_w/R_I\right)}\right]^{1/2}}. \tag{2.13}$$

Using equation 2.13, one can easily calculate the volumetric flow rate Q_w of the air extracted from the well:

$$Q_w = 2\pi R_w u_w H = \pi H \frac{K}{\mu_a} \frac{P_w \left[1 - \left(P_I/P_w\right)^2\right]}{\ln\left(R_w/R_I\right)}, \tag{2.14}$$

where u_w is the velocity at the wellbore and H is the thickness of the vadose zone though which air is removed. The volumetric flow rate Q_w corresponds to the pressure P_w of the well. To convert this flow rate to equivalent standard conditions, the following relationship can be applied:

$$Q_w^* = Q_w \frac{P_w}{P_{atm}}. \tag{2.15}$$

It is obvious from equation 2.14 that the most important parameter determining the volumetric air flow rate Q_w is the intrinsic permeability K of soil. At this point it is important to stress the difference between water permeability (or hydraulic conductivity) k_w, air permeability k_a, and intrinsic permeability K. In most cases, when permeability data are provided for a type of soil or geological formation, these data are based on hydraulic conductivity measurements and describe how easily the water can flow through this formation. However, the flow characteristic of a fluid depends greatly on its properties, e.g., density ρ and viscosity μ. Equation 2.16 describes the relationship between permeability coefficient k and fluid properties ρ and μ:

$$k = K \times \frac{\rho \times g}{\mu}, \tag{2.16}$$

where K is the geometric or intrinsic permeability of the soil, which depends only on the geometric characteristics of the soil (e.g., particle size distribution), and g is the gravity acceleration constant ($g = 9.81$ m/sec^2). Note that water and air permeability coefficients have units of velocity (cm/sec), but K has units of surface (cm^2).

When the hydraulic conductivity k_w of a soil is known, one can easily estimate the corresponding values of intrinsic and air permeabilities, taking into consideration the properties of water and air under usual environmental conditions: e.g., $\rho_w = 1.0$ g/cm^3, $\mu_w = 1 \times 10^{-2}$ g/(cm·s), $\rho_a = 1.2 \times 10^{-3}$ g/cm^3, and $\mu_a = 1.83 \times 10^{-4}$ g/(cm·s) ($T = 20°C$, $P = 1$ atm). For instance, a soil with hydraulic conductivity $k_w = 1 \times 10^{-3}$ cm/sec has an approximate intrinsic permeability of $K = 1 \times 10^{-8}$ cm^2, and its permeability to airflow under normal conditions will be $k_a = 6.6 \times 10^{-5}$ cm/sec.

The airflow equations presented above are based on the assumption that the soil is a spatially homogeneous porous medium with constant intrinsic permeability. However, in most sites, the vadose zone is heterogeneous. For this reason, design calculations are rarely based on previous hydraulic conductivity measurements. One of the objectives of preliminary field testing is to collect data for the reliable estimation of permeability in the contaminated zone. The field tests include measurements of air flow rates at the extraction well, which are combined with the vacuum monitoring data at several distances to obtain a more accurate estimation of air permeability at the particular site.

2.2.2.5 Removal Rate of Contaminants and Required Cleanup Time

The contaminants removal rate R_{rem} can be calculated by multiplying the flow rate of air extracted from all the wells by the concentration of contaminant in the soil air C_a:

$$R_{rem} = N_{wells} \times Q_w^* \times C_a. \tag{2.17}$$

The required cleanup time T_{clean} is directly related to the removal rate:

$$T_{clean} = M_{spill}/R_{rem}, \tag{2.18}$$

where M_{spill} is the estimated total amount of spill.

Equations 2.17 and 2.18 are very simple, but the accuracy of the predictions depends greatly on the realistic estimation of C_a, which varies with time during the operation of the SVE system. For the start of the SVE project and considering that the free organic phase, NAPL, is present in the subsurface, a first approximation is to calculate C_a from the vapor pressure data of the contaminants (equation 2 in Table 2.3 or equation 2.1). The actual concentration, however, will be lower than this value for two main reasons: (1) the extracted airstream does not pass only through the contaminated zone and (2) limitations on mass transfer exist. An effectiveness factor η should be considered to take into account the effect of these phenomena on removal rates. The value of this factor can be determined by comparing the calculated concentration with data obtained from the preliminary pilot tests at the site:

$$\eta = \frac{C_{a,\text{field}}}{C_{a,\text{equil}}}, \qquad (2.19)$$

where $C_{a,\text{field}}$ is the concentration in extracted air measured during the field tests and $C_{a,\text{equil}}$ is the value calculated from the vapor pressure data.

Practical experience from the application of SVE at sites contaminated with a single type of contaminant (e.g., trichloroethylene, TCE) indicates that the removal of contaminants follows a trend in two distinct phases. During the initial phase, which covers the period from the project startup to the exhaustion of NAPL in the subsurface, the removal rate is almost linear. The second phase is characterized by a constant decrease in removal rates.

This trend can be explained with the following mechanism. In the presence of NAPL, the extracted vapor concentration depends mainly on the vapor pressure of the contaminant. After the disappearance of free NAPL, the extracted vapor concentration becomes dependent on the partitioning of contaminants among the three other phases (see Table 2.3). As the air passes through the pores, the dissolved contaminants volatilize from the soil moisture to the gas phase, causing the desorption of contaminants from the surface of soil particles into the aqueous phase. As a result, the concentration in all three phases decreases, with a consequent decrease in removal rates.

For the initial linear phase of remediation, the pilot test data and equations 2.17 to 2.19 can provide relatively good predictions for the required cleanup time. For the second phase, it is necessary to use more sophisticated models combining airflow, equilibrium, and mass transfer equations 2.13 to 2.16, in order to obtain sufficiently accurate predictions. To obtain a first rough estimation, the methodology proposed by Kuo[12] can be applied. Kuo's approach is based on the observation that the VOC concentrations of extracted air decrease exponentially with time during the second stage of remediation. To simulate the exponential decrease in removal rates, the following procedure is suggested:

1. The mass of contaminant that must be removed during the second stage is divided into two or three equal parts, corresponding to successive cleanup time intervals.
2. Initial $C_{a,i}$ and final $C_{a,f}$ vapor concentrations are calculated for each interval using the phase distribution equations in the absence of NAPL (see Table 2.3).
3. A mean vapor concentration $C_{a,m}$ representing each time interval is determined from the geometric average of the two concentrations:

$$C_{a,m} = \sqrt{C_{a,i} \times C_{a,f}}. \qquad (2.20)$$

4. The successive cleanup time intervals are calculated using the mean concentration values, and they are summed to determine the total required time.

This procedure is illustrated in the practical example presented in Section 2.2.2.6.

Note that the initial linear phase is observed only in sites containing a single contaminant. For sites contaminated with mixtures of contaminants there is a decreasing rate of removal from the beginning of the project due to the different volatility of the components. The more volatile constituents are extracted with a higher rate from all the phases, and as a consequence the total VOC content of extracted air decreases constantly with time. This effect should be considered during the design phase.

2.2.2.6   A Practical Example

A tank containing $20\,m^3$ toluene ruptures, contaminating an area of $1250\,m^2$ in the vadose zone with an average depth of $4\,m$. The soil in the subsurface has the following characteristics: bulk density $\rho_b = 1.7$ g/cm^3, total porosity $\theta_t = 0.4$, moisture corresponding to porosity $\theta_w = 0.2$, and organic carbon content $f_{oc} = 0.01$.

Owing to the high volatility of toluene (vapor pressure $P^\circ = 22$ mmHg), the decision was to use SVE technology. Preliminary field tests were conducted in the area using an extraction well with $R_w = 5.1$ cm (2 in.) and total perforated length inside the contaminated zone $H = 4\,m$. The tests were carried out applying a vacuum of 0.1 atm (i.e., $P_w = 0.9$ atm) in the well, and the pressure was measured at a distance of 6 m and found to be 0.99 atm after reaching steady-state conditions. The flow rate of extracted air, as measured in the exhaust of the vacuum pump, was $Q_w^* = 0.2\,m^3$/min, and the air contained 78 mg/L toluene. The temperature of the subsurface was 25°C.

To determine some important design parameters for this SVE project, the following procedure could be used:

Step 1: Obtain the physicochemical data of the compound of concern. Important sources for this type of data are references 9, 10, and 24–27. From tables included in reference 9, the following properties of toluene were obtained: MW = 92.14 g/mol, $P^\circ = 22$ mmHg = 0.0289 atm, $K_H = 0.276$ (dimensionless), $C_w^\circ = 490$ mg/L, log $K_{oc} = 2.06$, and $\rho_{or} = 0.866$ g/cm^3.

Step 2: Calculate the initial distribution of toluene in the subsurface. The initial distribution of toluene can be calculated using equations 1 to 5 from Table 2.3 and taking into consideration that the organic phase is a pure compound, i.e., $X = 1$, $m = 1$, and $\gamma = 1$. The total quantity of contaminant per unit soil C_t can be estimated from the known amount of spill M_{spill} and the volume of the contaminated zone:

$$C_t = \frac{M_{spill}}{A_{contam} \times H} = \frac{20{,}000\,L \times 0.866\,kg/L}{(1250 \times 4)\,m^3} = \frac{17{,}320\,kg}{5000\,m^3} = 3.464\,kg/m^3 = 3464\,mg/L.$$

The concentrations and the mass distribution of toluene in the four phases, as calculated from this set of equations, are presented in Table 2.4. As seen in the table, the major part of the toluene, i.e., 68.9%, remains in the vadose zone as free NAPL, 27.6% is adsorbed on the surfaces of solid particles, and only 3.5% is distributed between the aqueous and gas phases. Free NAPL occupies only a small part of the available pore volume, and it is not expected to disturb the movement of air through the contaminated zone.

Step 3: Calculate from the field test data the radius of influence, the required number of wells, and the required capacity of the gas treatment facility. The radius of influence R_I can be calculated from equation 2.11 using the pressure monitoring data at $r = 6\,m$ while considering that the minimum required vacuum at R_I should be 0.001 atm, i.e., $P_I = 0.999$ atm. With these values R_I is found to be 9.91 m. The number of wells is calculated from equation 2.12, $N_{wells} = 4.86$. This means that five wells must be installed to remediate the entire contaminated area. Once the number of wells has been determined, the required capacity of the gas treatment facility can be defined from the flow rate data obtained during the field tests. In this case, the gas treatment unit should be able to treat $N_{wells} \times Q_w^* = 5 \times 0.2 = 1.0$ m^3/min of toluene-laden air.

TABLE 2.4
Concentrations, Mass Distribution of Toluene, and Volume Occupied by the Four Phases in the Vadose Zone

	Total	Free NAPL	Aqueous	Gas	Solids
Toluene concentrations	3464 mg/L	866×10^3 mg/L	490 mg/L	109 mg/L	563 mg/kg
Toluene mass distribution	17.32 t	11.94 t	0.49 t	0.11 t	4.78 t
Volume of four phases	5000 m³	14 m³	1000 m³	986 m³	3000 m³

The flow rate data can also be used to estimate the permeability of the subsurface. The required additional parameter is the value of air viscosity, i.e., $\mu_a = 1.83 \times 10^{-4}$ g/(cm·s). The intrinsic permeability of soil is calculated from equations 2.16 and 2.15 and is found to be $K = 1.34 \times 10^{-8}$ cm². Care should be taken to perform the appropriate unit conversions when using equation 2.15.

Step 4: *Estimate the effectiveness factor η for the removal and the cleanup time required to obtain a residual toluene concentration of 150 mg/L.* The phase distribution calculations carried out in Step 2 indicate that the equilibrium concentration of toluene in the gas phase is $C_{a,equil} = 109$ mg/L (see Table 2.4). The concentration measured in the extracted air during the field tests is lower, at $C_{a,field} = 78$ mg/L, indicating that the removal effectiveness is limited either as a result of mass transfer phenomena or the existence of uncontaminated zones in the airflow pattern. The corresponding effectiveness factor is $\eta = 78/109 = 0.716$.

The amount of toluene that must be removed from the soil M_{rem} can be calculated by considering the initial total amount of spill M_{spill} and the residual acceptable quantity corresponding to the cleanup objectives M_{final}:

$$M_{rem} = M_{spill} - M_{final} = 17.32\ t - (150\ g/m^3) \times 5000\ m^3 \times (10^{-6}\ t/g) = (17.32 - 0.75)\ t = 16.57\ t.$$

The removal of toluene is assumed to take place in two stages. The first stage corresponds to the removal of free NAPL, which, according to the phase distribution calculations (Step 2; Table 2.4) represents a mass of $M_{rem1} = 11.94$ t. The second stage corresponds to the removal of toluene, which is distributed among the other three phases, and represents a mass of $M_{rem2} = 16.57 - 11.94 = 4.63$ t.

As this site is contaminated with a single compound, the removal of free NAPL is expected to follow a linear trend with constant removal rate. The required time can be calculated from equations 2.17 and 2.18:

$$T_1 = \frac{M_{rem1}}{\eta \times N_{well} \times Q_w^* \times C_{a,equil}} = \frac{(11.94\,t) \times (10^6\,g/t)}{0.716 \times 5 \times (0.2\,m^3/min) \times (109\,g/m^3) \times (1440\,d/min)} = 106.5\ d$$

The second stage of treatment is assumed to follow an exponential decrease in removal rates. Applying the approach of Kuo, this stage is divided into two time intervals, T_{2-1} and T_{2-2}, representing the successive removal of equivalent amounts of toluene, $M_{rem2-1} = M_{rem2-2} = 2.315$ t. The initial theoretical concentration in the gas phase for the time interval T_{2-1} is equal to the vapor pressure of toluene, $C_{a,i} = 109$ mg/L. The final vapor concentration for this interval $C_{a,f}$ can be calculated from the total residual concentration $C_{t,f}$ and the phase distribution equations 5 and 7 to 9 in Table 2.3:

$$C_{t,f} = \frac{(M_{spill} - M_{rem1} - M_{rem2-1})}{5000\,m^3} = \frac{(17.32 - 11.94 - 2.315)t}{5000\,m^3} = 0.613 \times 10^{-3}\,t/m^3 = 613\,mg/L$$

$$C_{a,f} = \frac{C_{t,f}}{\rho_b K_{oc} f_{oc}/K_H + \theta_w/K_H + \theta_t - \theta_w} = \frac{613\,mg/L}{8.0} = 76.7\,mg/L$$

The mean vapor concentration $C_{a,m}$ for the time interval $T_{2\text{-}1}$ is calculated from the geometric average of $C_{a,i}$ and $C_{a,f}$ (equation 2.20), i.e., $C_{a,m} = 91.4$ mg/L, and the required treatment time from equations 2.17 and 2.18:

$$T_{2\text{-}1} = \frac{M_{rem2-1}}{\eta \times N_{well} \times Q_w^* \times C_{a,equil}} = \frac{(2.315t) \times (10^6\,g/t)}{0.714 \times 5 \times (0.2\ m^3/min) \times (91.4\ g/m^3) \times (1440\ d/min)} = 24.6\ d$$

The same procedure is applied for the last time interval $T_{2\text{-}2}$, and the following values are calculated:

$$C_{a,i} = 76.7 \text{ mg/L}, \ C_{a,f} = 18.8 \text{ mg/L}, \ C_{a,m} = 37.9 \text{ mg/L}, \ T_{2\text{-}2} = 59.4 \text{ d}$$

The total cleanup time, as estimated with this approach, will be

$$T_{clean} = T_1 + T_{2\text{-}1} + T_{2\text{-}2} = 106.5 + 24.6 + 59.4 = 190.5 \text{ d}$$

As seen from these calculations, the removal of free NAPL, representing almost 70% of the total toluene spill, takes approximately 106 days. The operation of the SVE system should continue for an additional 84 days in order to achieve the cleanup objectives and remove the final 30% of the toluene spill.

2.3   BIOREMEDIATION

2.3.1   INTRODUCTION

The bioremediation techniques exploit the biological activity of microorganisms to degrade or detoxify environmentally hazardous compounds. Traditionally, biological treatment has been applied for the remediation of sites contaminated with organic contaminants. Most organic compounds can be degraded through the action of appropriate microbial communities towards more simple and less harmful inorganic or organic molecules. The degree of degradation determines whether mineralization or biotransformation has occurred. Mineralization is the complete degradation of organic compounds into inorganic final products, such as carbon dioxide and water, whereas biotransformation is the partial degradation of the compound to more simple organic molecules.

Bioremediation is not restricted only to biodegradable organic contaminants. New techniques are currently under development for the bioremediation of metal-contaminated sites. Microbial activity can alter the oxidation state of some elements, reducing or increasing their mobility, and this transformation can be used for remediation purposes.

Bioremediation systems in operation today rely on microorganisms indigenous to contaminated sites. The two main approaches, based on the actions of native microbial communities, are biostimulation and intrinsic bioremediation. In biostimulation, the activity of native microbes is encouraged, creating (*in situ* or *ex situ*) the optimum environmental conditions and supplying nutrients and other chemicals essential for their metabolism. The vast majority of bioremediation projects are based on this biostimulation approach. Intrinsic bioremediation is a remedial option that can be applied when there is strong evidence that biodegradation will occur naturally over time without any external stimulation; i.e., a capable microbial community exists at the site, the required nutrients are available, and the environmental conditions are favorable. An additional prerequisite is that the naturally occurring rate of biodegradation is faster than the rate of contaminant migration towards sensitive environmental receivers, e.g., a well used for abstraction of drinking water. In that case, and if sufficient supportive data are provided, the regulatory authorities may issue a permit to pursue the intrinsic bioremediation option for a particular site. This remediation strategy is not a "no action" alternative. It requires the design and implementation of a systematic monitoring procedure to follow closely the progress of this natural process and prevent any undesirable side effects, such as the generation of toxic bioproducts due to unexpected changes in redox conditions.

In some cases the indigenous microorganisms are not able to degrade or detoxify the specific contaminants to acceptable levels. The use of nonnative microbes or even genetically engineered microorganisms especially suited to degrading the contaminants of concern is another bioremediation option known as bioaugmentation, that is currently under development. An important research effort has been devoted since the mid-1990s to discover microbial species capable of destroying or detoxifying specific hazardous pollutants, and to isolate them in pure cultures in order to exploit their efficiency in bioremediation projects. Such pure specific degrading microbial populations have been successfully used for the treatment of contaminated soils under laboratory conditions, but to date there are no known cases of full-scale projects applying the bioaugmentation principle.

Regardless of whether the microbes are native or artificially introduced into the soil, it is important to understand the mechanisms by which they degrade or detoxify hazardous pollutants through their metabolic activity. Understanding these mechanisms is essential for the proper design of bioremediation systems that provide the optimum conditions and the required nutritional supplements for the specific microbial process.

2.3.2 Principles of Bioremediation

2.3.2.1 Basic Microbial Metabolism

The microbial degradation of organic contaminants occurs because the organisms can use the pollutants for their own growth and maintenance. A microbial cell operates two critical types of metabolic processes, referred to as anabolic (cell-building) and catabolic (energy-releasing) processes. Anabolic processes involve the production of new cells and require a source of carbon, which is the most important constituent of cellular mass. Catabolic processes are energy-producing chemical reactions and require a source of energy.

Organic contaminants are used by microorganisms both as a source of carbon and as a source of energy. The microbes gain energy from the contaminants through their oxidation, which involves the breaking of chemical bonds and transfer of electrons away from the contaminant. To complete the chemical reaction, another compound is needed to receive the electrons. The contaminant, which is oxidized, is called the electron donor and the chemical, which is reduced, is called the electron acceptor. The microorganisms use the energy produced from these electron transfers to build new cells or simply to maintain the existing cells. The electron donor and the electron acceptor are essential for cell growth and maintenance and are commonly called the primary substrates.

Depending on the type of electron acceptor, the metabolic modes are broadly classified into three main categories: aerobic respiration, anaerobic respiration, and fermentation. Aerobic respiration is the term used to describe the metabolism in which molecular oxygen (O_2) serves as the electron acceptor. Many microorganisms follow the mode of aerobic respiration, and most bioremediation projects exploit this particular type of metabolism. There is, however, a wide variety of microorganisms that are able to survive and grow under anaerobic conditions using several inorganic or organic compounds other than oxygen as electron acceptors. This form of metabolism is called anaerobic respiration. The most commonly used electron acceptors under anaerobic conditions are nitrates (NO_3^-) and sulfates (SO_4^{2-}), which are soluble constituents in the aqueous phase, and the oxidized forms of iron (Fe[III]) and manganese (Mn[IV]), which are common constituents of soil particles, mainly in the form of oxides. A type of metabolism that can play an important role under strictly

TABLE 2.5

Typical Benzene Biodegradation Reactions under Various Electron Acceptor and Redox Conditions

Indicative Redox Conditions, E_h	Electron Acceptors	Biodegradation Reactions	Refs
> +200 mV	O_2	$C_6H_6 + 7.5\,O_2 \rightarrow 6\,CO_2 + 3\,H_2O$	(28)
< +200 mV	NO_3^-	$C_6H_6 + 6\,NO_3^- + 6\,H^+ \rightarrow 6\,CO_2 + 3\,N_2 + 6\,H_2O$	(29)
< 0 mV	Fe(III)	$C_6H_6 + 30\,Fe^{3+} + 12\,H_2O \rightarrow 6\,CO_2 + 30\,Fe^{2+} + 30\,H^+$	(30)
< –100 mV	SO_4^{2-}	$C_6H_6 + 3.75\,SO_4^{2-} + 7.5\,H^+ \rightarrow 6\,CO_2 + 3.75\,H_2S + 3\,H_2O$	(31)
< –200 mV	C_6H_6	$C_6H_6 + 12\,H_2O \rightarrow 2.25\,CO_2 + 3.75\,CH_4$	(32)

anaerobic conditions is fermentation. During fermentation there is no need for an external electron acceptor, because the organic contaminant serves as both electron donor and electron acceptor.

The typical biodegradation reactions under various electron acceptor conditions are presented in Table 2.5 for the simple case of benzene. Which type of electron acceptor will be used is closely related to the prevailing redox conditions. Under aerobic conditions, with redox potential greater than 200 to 220 mV, biodegradation is mainly performed by aerobic microorganisms. When oxygen is depleted but the redox potential remains relatively high, biodegradation can proceed through the metabolic activity of nitrate-reducing bacteria. The Fe(III) oxides of soil can be used as electron acceptors over a wide range of redox values, depending upon their crystallinity. Finally, sulfate-reducing and methanogenic bacteria are active only under strongly reducing conditions.

2.3.2.2   Co-Metabolism

In some cases, microorganisms can transform a contaminant, but they are not able to use this compound as a source of energy or carbon. This biotransformation is often called co-metabolism. In co-metabolism, the transformation of the compound is an incidental reaction catalyzed by enzymes, which are involved in the normal microbial metabolism.[33] A well-known example of co-metabolism is the degradation of (TCE) by methanotrophic bacteria, a group of bacteria that use methane as their source of carbon and energy. When metabolizing methane, methanotrophs produce the enzyme methane monooxygenase, which catalyzes the oxidation of TCE and other chlorinated aliphatics under aerobic conditions.[34] In addition to methane, toluene and phenol have been used as primary substrates to stimulate the aerobic co-metabolism of chlorinated solvents.

Tetrachoroethylene (perchloroethylene, PCE) is the only chlorinated ethene that resists aerobic biodegradation. This compound can be dechlorinated to less- or nonchlorinated ethenes only under anaerobic conditions. This process, known as reductive dehalogenation, was initially thought to be a co-metabolic activity. Recently, however, it was shown that some bacteria species can use PCE as terminal electron acceptor in their basic metabolism; i.e., they couple their growth with the reductive dechlorination of PCE.[35] Reductive dehalogenation is a promising method for the remediation of PCE-contaminated sites, provided that the process is well controlled to prevent the buildup of even more toxic intermediates, such as the vinyl chloride, a proven carcinogen.

2.3.2.3   Microbial Transformation of Toxic Elements

It has long been known that certain microbes can alter the oxidation state of some toxic metals, mainly by reducing them to a lower oxidation state, and this chemical transformation can be used for the bioremediation of contaminated soils. Three main mechanisms are involved in the bioreduction of toxic elements: dissimilatory (respiratory) reduction, direct enzymatic reduction (not supporting growth), and indirect chemical reduction induced by metabolic byproducts. Dissimilatory reduction has been demonstrated for uranium, selenium, and arsenic.[36–38] Recently, many bacteria species able to couple their growth with the reduction of Ur(VI), Se(VI), and As(V) have been isolated and characterized.[37,38] Direct enzymatic reduction is a kind of co-metabolism, i.e., a reaction that is catalyzed by microbial enzymes but cannot support biomass growth. Enzymatic reduction is one of the mechanisms involved in the bioreduction of Cr(VI).[39] The third mechanism is the indirect chemical reduction by metabolic byproducts. The most characteristic case is the reduction of Cr(VI) by H_2S, which is the main byproduct of the basic metabolism of sulfate-reducing bacteria.

Biological activity can be used in two ways for the bioremediation of metal-contaminated soils: to immobilize the contaminants *in situ* or to remove them permanently from the soil matrix, depending on the properties of the reduced elements. Chromium and uranium are typical candidates for *in situ* immobilization processes. The bioreduction of Cr(VI) and Ur(VI) transforms highly soluble ions such as CrO_4^{2-} and UO_2^{2+} to insoluble solid compounds, such as $Cr(OH)_3$ and UO_2. The selenate anions SeO_4^{2-} are also reduced to insoluble elemental selenium Se^0. Bioprecipitation of heavy metals, such as Pb, Cd, and Zn, in the form of sulfides, is another *in situ* immobilization option that exploits the metabolic activity of sulfate-reducing bacteria without altering the valence state of metals. The removal of contaminants from the soil matrix is the most appropriate remediation strategy when bioreduction results in species that are more soluble compared to the initial oxidized element. This is the case for As(V) and Pu(IV), which are transformed to the more soluble As(III) and Pu(III) forms. This treatment option presupposes an installation for the efficient recovery and treatment of the aqueous phase containing the solubilized contaminants.

2.3.3 ENGINEERING FACTORS

Biological treatment consists of promoting and maintaining the metabolic activity of a microbial population, which then is able to degrade or detoxify the target contaminants. In order to properly design a bioremediation system, it is important to control a number of factors that are crucial for maintaining microbial activity at efficient levels. Some important engineering factors affecting the performance of bioremediation systems include the availability of electron acceptors and nutrients, and environmental conditions such as moisture content, temperature, pH, and redox conditions.

2.3.3.1 Electron Acceptor

The great majority of bioremediation projects involve the aerobic degradation of organic contaminants, and the limiting factor is often the availability of oxygen. The mass of oxygen required by aerobic systems can be calculated based on stoichiometric considerations or laboratory measurements. Both anabolic and catabolic reactions require an electron acceptor. The stoichiometry of catabolic reactions can be easily determined by considering the end products. For instance, the catabolic complete mineralization of toluene, C_7H_8, is described by the following reaction:

Catabolic reaction

$$C_7H_8 + 9\,O_2 \longrightarrow 7\,CO_2 + 4\,H_2O \qquad (2.21)$$

For anabolic reactions, which result in the production of new cells, it is important to know the approximate chemical composition of the biomass. The bacterial protoplasm comprises 75 to 80% water. The solid material is composed of several complex organic molecules, such as proteins, carbohydrates, and DNA. The mean composition of these molecules can be approximated by a relatively simple empirical formula, $C_{60}H_{87}O_{23}N_{12}P$, or in an even more simple form as $C_5H_7O_2N_{10}$. Numerous other elements such as sulfur, sodium, potassium, calcium, magnesium,

TABLE 2.6

Calculation of the Requirements in Oxygen and Macronutrients (N, P) for the Aerobic Biodegradation of Toluene

		Oxygen Required	CO_2 Produced	Biomass Produced	Nitrogen Required	Phosphorus Required
Catabolic reaction (50%) C_7H_8 +		$9\,O_2$ → $7\,CO_2$				
Anabolic reaction (50%) C_7H_8 +		$4\,O_2$ → $2\,CO_2$ + $C_5H_7O_2N$		(~1/12 $C_{60}H_{87}O_{23}N_{12}P$)		
Moles per mole of C_7H_8	1	6.5	4.5	0.5	0.5	0.042
Grams per mole of C_7H_8	92	208	198	56.5	7	1.3
Grams per gram of C_7H_8	1	2.26	2.15	0.61	0.076	0.014

Note: Calculations assume 50% degradation for energy production and 50% for biomass production.

chlorine, iron, and various trace metals are also contained in the biomass, but the sum of these elements represents approximately 5% of the total dry biomass. Using the simple formula for the composition of cellular mass, the assimilation of toluene to build new cells can be described by the following reaction:

Anabolic reaction

$$C_7H_8 + 4\,O_2 + NH_4^+ \longrightarrow C_5H_7O_2N + 2\,CO_2 + 2\,H_2O + H^+ \tag{2.22}$$

Based on established experience with aerobic degradation of organic contaminants, environmental engineers customarily assume that half of the organic compound is converted to cellular mass and half oxidized for energy. With this assumption the amount of oxygen required to biodegrade 1 mol of toluene corresponds to 6.5 mol, i.e., 4.5 mol for energy production and 2 mol for biomass production. An example of these calculations is presented in Table 2.6.

2.3.3.2 Nutrients

As is evident from the empirical formulas describing the typical composition of the cellular mass, nitrogen and phosphorous are important components, often referred to as macronutrients. All other elements are characterized as micronutrients. Most soil and aquifer systems contain a sufficient amount of micronutrients, but very often nitrogen and phosphorus are in shortage and must be added, usually in the form of soluble ammonium and orthophosphate salts. The formula $C_{60}H_{87}O_{23}N_{12}P$ provides a basis for calculating the theoretical amount of nitrogen and phosphorous required to produce new cellular mass. The C:N:P molar ratio for anabolic reactions is 60:12:1, but the actual demand in N and P for the biodegradation of organic carbon is lower, as only part of the organic carbon is used for the production of new biomass. In the previous example for the biodegradation of toluene (see Table 2.6), only 2.5 of the 7 carbon atoms of one toluene molecule were assumed to be assimilated into the new biomass and the total C:N:P ratio for both anabolic and catabolic reactions is 168:12:1. A general rule-of-thumb[12] usually applied by environmental engineers to estimate N and P requirements is the molar ratio C:N:P = 120:10:1.

For bioremediation, an initial feasibility study is always recommended, and the determination of nutrient requirements should be part of this study. The actual requirements are very much dependent upon the type of contaminants, which are often a mixture of compounds of variable biodegradability, and on the availability of nutrients in the specific contaminated soil, and should be determined with appropriate laboratory tests. However, there are guidelines that provide a useful basis for initial economic evaluations and for calculating ranges to be tested during the laboratory tests.

2.3.3.3 Moisture

Moisture is necessary for biodegradation for two reasons:

1. For cellular growth, because water constitutes 75 to 80% of cellular tissue
2. As a medium for the movement of microorganisms to the organic contaminants and vice versa[10]

A moisture ranging between 25 and 85% of complete saturation is considered to be adequate for soil bioremediation.[12] In many cases, the soil moisture in the vadose zone is below or at the lower end of this range, so the addition of water is often needed to maintain good operating conditions.

2.3.3.4 Temperature

Temperature has a major influence on metabolic activity, and microorganisms are classified into three main categories based upon the optimum temperature for their growth. Psychrophiles are microorganisms that can grow at temperatures below 20°C, mesophiles are characterized as organisms with an optimum growth temperature between 25 and 40°C, and thermophiles are those preferring temperatures above 45°C. Very often, when the temperature increases a few degrees above the optimum value, growth declines precipitously; long exposure to the higher temperatures may even result in cell death. Lower temperatures, on the other hand, are not usually lethal—the cells remain dormant, but their activity can restart if the optimum temperature conditions are reestablished. Most soil bioremediation projects are based on the activity of mesophilic bacteria, which are the most common and abundant microorganisms in the subsurface.

2.3.3.5 pH

Another important factor affecting microbial activity is pH. Microorganisms that can grow under acidic (pH < 4) or alkaline (pH > 10) conditions are termed acidophiles or alkalophiles, respectively. Most bacteria, however, are neutrophiles. Neutrophiles can tolerate pH levels between 5 and 9, but their optimum growth is observed in a relatively narrow range around neutrality, i.e., between 6.5 and 7.5.

Microbial activity, which is often stimulated during bioremediation projects, can alter the external pH. For instance, the anaerobic degradation of chlorinated compounds produces organic acids and HCl and the pH may drop to acidic values if the soil has a low buffering capacity. In this case, control of the external pH will be required in order to maintain biodegradation activity at satisfactory levels.

2.3.3.6 Redox Potential

The most critical issue to be investigated during the initial biofeasibility study is the determination of which metabolic mode—aerobic or anaerobic—is more appropriate for the specific contaminants. As shown in Table 2.5, the redox potential is closely related to the metabolic mode, and careful control of this parameter is required to maintain the optimum metabolic mode during bioremediation. A general rule is that the redox potential should be above 50 mV to maintain the activity of aerobic and facultative anaerobic microorganisms and below that value for strictly anaerobic microorganisms.[12]

2.3.4 *In Situ* Methods for the Biological Treatment of Organic Contaminants

Bioremediation methods may be applied either *in situ* or *ex situ*. In this section, the most important *in situ* treatment methods will be examined.

2.3.4.1   Bioremediation in the Vadose Zone

When contamination exists substantially above the water table, i.e., in the vadose zone, a very efficient *in situ* technique that may be used is bioventing, which is similar in many ways to SVE technology. In many SVE applications, it has been observed that air circulation through the porosity of the vadose zone stimulates the biodegradation of organic contaminants. Based on these observations, bioventing technology was developed using a system configuration similar to SVE but optimizing the design in order to promote the aerobic biodegradation of contaminants. In practice, SVE and bioventing are usually combined in an integrated treatment scheme, where the highly volatile compounds are removed by volatilization and the biodegradable constituents are biologically destroyed.

The main requirements for the design of a bioventing system are the following[9]:

1. An O_2 flow must be maintained through the contaminated zone at a level sufficient for the aerobic biodegradation of contaminants. Note that during bioventing the main aim is the maximum utilization of O_2 by the microbial cultures. For this reason, air flow rate is usually an order of magnitude lower than that applied in simple SVE systems. A simple empirical rule is that the mean residence time of air in the contaminated soil pore volume should be between 1 and 2 days.
2. The moisture of the soil should be maintained at an optimum value for microbial activity. As previously mentioned, a minimum level of soil moisture is necessary for successful biodegradation. The continuous circulation of air during bioventing results in the evaporation of soil moisture. For this reason, the design of these systems must include an appropriate installation for adding water to the contaminated zone. Care must be taken to avoid the addition of excess water. If soil moisture is significantly increased, e.g., above the limit of 85%, air circulation is no longer effective due to the decrease in free soil porosity.
3. Macro- and micronutrients should be provided as needed. Soils usually contain sufficient levels of micronutrients, but very often there is a lack of nitrogen and phosphorus. The addition of N and P is particularly important during the initial stages of treatment, in order to stimulate the growth of indigenous bacteria. After the initial development of a critical microbial mass, N and P are constantly recycled due to the lysis of dead microbial cells.[9]

A schematic of a bioventing installation, including a system for the addition of water and nutrients, is depicted in Figure 2.4. When the contaminated zone is near the surface and the soil is sufficiently permeable, the addition of water, together with dissolved nutrients, can be carried out using a simple surface irrigation system. When the contamination is located at lower horizons, an underground infiltration system or a network of wells may be more appropriate.

2.3.4.2   Bioremediation in the Water-Saturated Zone

When the contaminated zone is located below the water table, the availability of oxygen becomes a critical problem due to the low solubility of oxygen in water. In adding the required oxygen, two kinds of systems are usually applied:

1. Water circulation systems, where groundwater is pumped, oxygenated in surface installations and reinjected in the contaminated aquifer
2. Air sparging systems, involving the injection of air directly into the groundwater

Water circulation systems
A typical water circulation system is presented in Figure 2.5. Groundwater is pumped to the surface from a well, which is located downgradient of the contaminated zone, and directed to an installation

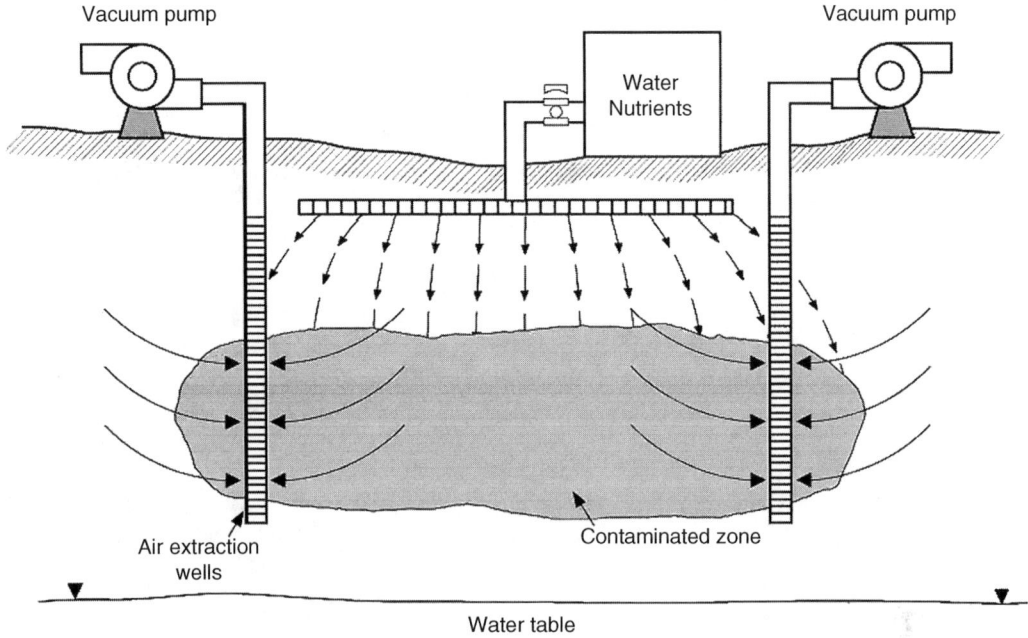

FIGURE 2.4   Schematic diagram of a bioventing system.

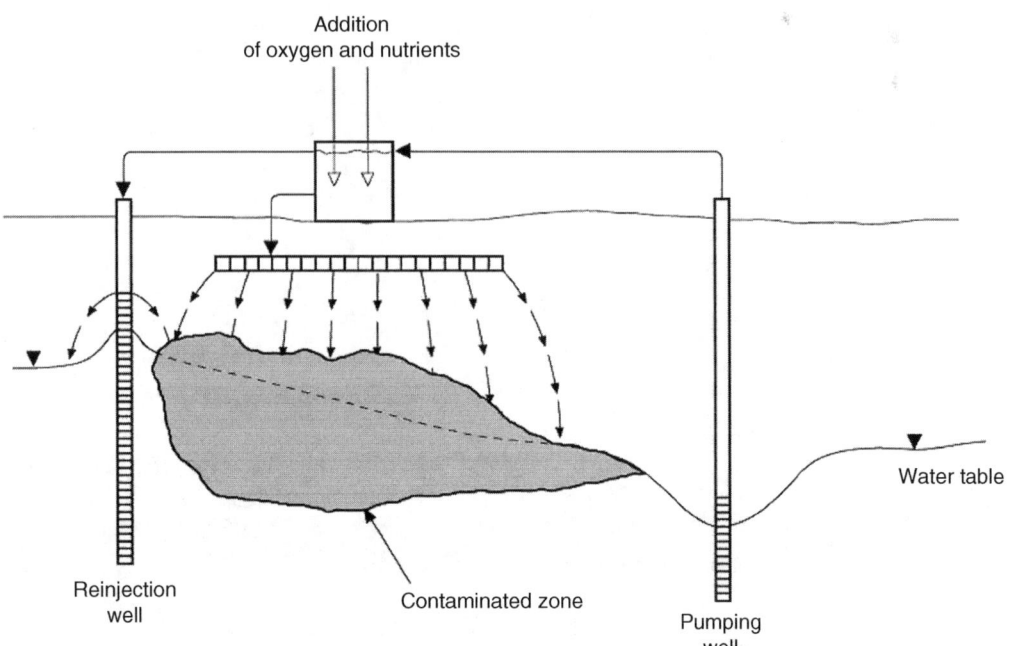

FIGURE 2.5   Representation of a water circulation bioremediation system.

where it is amended with nutrients and oxygen. Enriched groundwater is then reinjected into the aquifer using wells, trenches, or injection galleries, depending on the situation.

The addition of oxygen can be carried out by sparging the groundwater with air or pure oxygen or by adding hydrogen peroxide (H_2O_2). When the oxygenation is carried out by simple aeration, the concentration of dissolved oxygen may not exceed 8 to 12 mg/L, but with pure oxygen sparging, concentrations of up to 40 mg/L of dissolved oxygen can be attained. The H_2O_2 produces O_2 according to the following dissociation reaction:

$$2\ H_2O_2 \rightarrow 2\ H_2O + O_2 \tag{2.23}$$

As a result of this reaction, dissolved H_2O_2 serves as a continuous source of oxygen. Thus the amount of available oxygen is greater than the limit of 40 mg/L, corresponding to the solubility of pure O_2, and depends on the concentration of dissolved H_2O_2. It has been established, however, that H_2O_2 has a toxic effect on microorganisms at concentrations greater than 1000 mg/L. For this reason, concentrations between 100 and 500 mg/L are generally used.

Simple aeration of groundwater in surface installations cannot meet the demand for oxygen supply in most bioremediation projects. This is illustrated in the following example (adapted from reference 10).

Example: Alternative modes of supplying oxygen in a water circulation system
Leakage from a toluene tank resulted in the contamination of an aquifer with approximately 5000 kg of toluene. *In situ* measurements and laboratory tests confirmed the existence of indigenous bacteria, able to biodegrade toluene in the presence of oxygen. It was decided to carry out bioremediation of the site using a water circulation system. Estimate the time required for the remediation of this site if the groundwater is pumped at a rate of 300 L/min and O_2 addition is carried out (1) with simple aeration (2) with pure oxygen sparging, and (3) by the addition of H_2O_2 at a concentration of 250 mg/L.

As shown in Table 2.6, 2.26 g O_2 are required for the mineralization of 1 g toluene. Consequently, 11,300 kg O_2 are required for the biodegradation of the 5000 kg of toluene.

1. In the case of simple aeration, the rate of oxygen supply in the aquifer will be

   300 L/min × 8 g/L × 1440 min/d = 3.5 kg/d

   With this mode of O_2 supply, almost 9 years are needed for the complete degradation of the toluene.
2. In the case of pure oxygen sparging, the rate of oxygen supply in the aquifer will be

   300 L/min × 40 g/L × 1440 min/d = 17.3 kg/d

   The duration of the treatment in this case is 1.8 years.
3. According to reaction 2.23, 2 mol H_2O_2 produces 1 mol O_2; i.e., $2 \times 34 = 68$ g H_2O_2 produces 32 g O_2. Assuming that 250 mg/L H_2O_2 is added in the pumped waters, the rate of oxygen supply will be

   300 L/min × (250 × 32/68) mg/L × 1440 min/d = 50.8 kg/d

With the addition of H_2O_2, the required treatment time is 7.4 months.

It is obvious from the above calculations that realistic remediation times can be obtained using either pure oxygen or H_2O_2. The final selection will be based on the overall environmental and

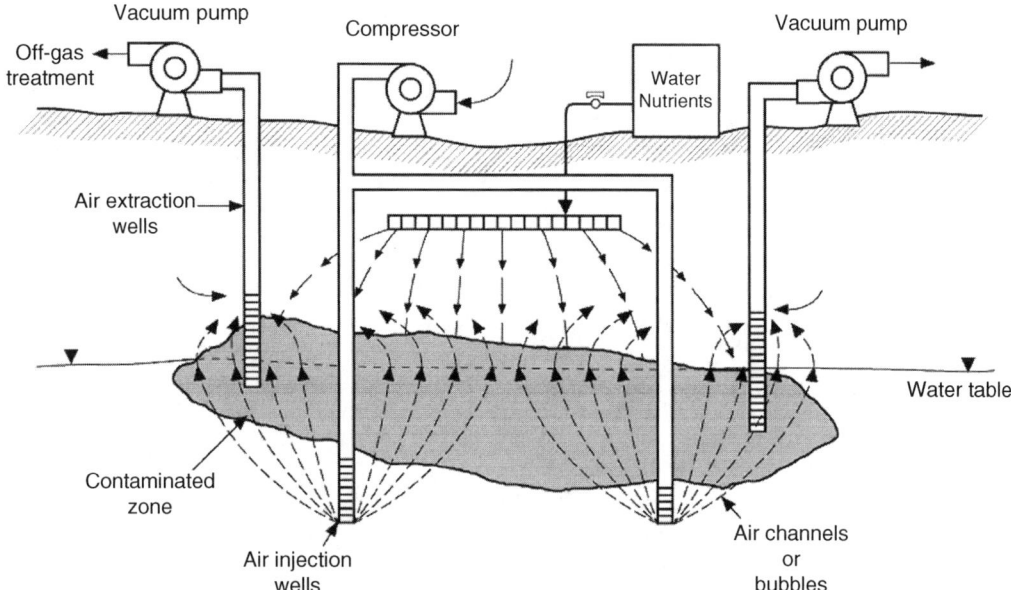

FIGURE 2.6 Air sparging system.

techno-economic evaluation of each alternative, taking into consideration the available timeframe and the financial resources available for the remediation of the area of concern.

Air sparging systems
An important innovation in bioremediation technology has been the use of air sparging to oxygenate groundwater. A typical system is presented in Figure 2.6. Using this technique, the removal of contaminants is achieved using two simultaneous mechanisms:

1. Volatilization of the dissolved volatile contaminants through the air–water interface
2. Biodegradation as a result of the enrichment of the groundwater with oxygen

A continuous oxygen supply is thus achieved, which is limited only by the mass transport phenomena between the gas and aqueous phases; this is the main advantage of air sparging over the alternative of water circulation systems.

Which mechanism will be dominant during air sparging (volatilization or biodegradation) depends on the relative volatility and biodegradability of the specific contaminants. The volatility of dissolved contaminants is usually characterized using the Henry's Law constant. In contrast with volatility, biodegradability cannot be estimated with a simple physical constant. Many data, however, have been published describing the biodegradation of organic contaminants under various laboratory and field conditions. In most cases, biodegradability is expressed in terms of half-life $t_{1/2}$, representing the time required to biodegrade half of the initial amount of contaminant under particular laboratory or field conditions. Available biodegradability data for a long list of contaminants are compiled in some environmental engineering textbooks and handbooks[9,40] and such compilations constitute a highly valuable information source for conducting initial biofeasibility studies. The Henry's Law constants K_H and aerobic biodegradation data for some characteristic organic compounds are presented in Table 2.7. The first four aromatic compounds, known as BTEX, are removed easily with both volatilization and biodegradation. Chlorinated hydrocarbons, such as trichloroethylene and trichloroethane, usually have low biodegradability and high volatility. These compounds are therefore removed mainly through volatilization. Some compounds are highly soluble in the

TABLE 2.7
Henry's Law Constants and Biodegradation Data for Some Characteristic Organic Compounds

Compound	Henry Constant (atm.m³/mol)	Aerobic Biodegradation in Soils Half-life (h)	
		From	To
Benzene	5.5×10^{-3}	120	384
Toluene	6.6×10^{-3}	96	528
m-Xylene	6.3×10^{-3}	168	672
Ethyl-benzene	8.7×10^{-3}	72	240
Trichloroethylene	9.1×10^{-3}	4320	8640
1,1,1-Trichloroethane	16.2×10^{-3}	3360	6552
Acetonitrile	3.5×10^{-6}	168	672
Phenol	4.0×10^{-7}	24	240

Source: Suthersan S.S., *Remediation Engineering: Design Concepts*, CRC Press, Boca Raton, FL, 1997. With permission.

aqueous phase, e.g., acetonitrile and phenol, and are easily biodegraded, but their volatility from the aqueous phase is low. In such cases, the main removal mechanism is biodegradation.

Design parameters

The main aim in the design of air sparging systems is to achieve the maximum possible interface between air and groundwater. A large interface is necessary not only for the volatilization of contaminants (transfer from the aqueous phase to the air), but also for the oxygenation of groundwater (oxygen transfer from the air to the aqueous phase). The dispersion and movement of air in the water-saturated zone are very complex phenomena that are not yet fully understood. For instance, two different approaches are used to describe the upward movement of air. The first approach suggests that air travels in the form of discrete air channels. In the second, the air travels in the form of air bubbles. Owing to the complexity of the process and the absence of simple and reliable mathematical models, the design of air sparging systems is based mainly on experience and *in situ* tests. The most important design parameters are the following[9]:

1. *Zone of influence*. Whereas in SVE systems the zone of influence around each extraction well can be described as a cylinder of a particular radius, in air sparging systems it is not possible to define a radius of influence. Air sparging is usually carried out through one or more injection wells that are installed in such a way that their end is located below the contaminated area. Air bubbles emerging from the end of each well are transferred upwards, in the shape of an inverted cone. The width of the cone depends mainly on the permeability and the homogeneity of the soil. Permeable and homogeneous soils usually form narrow cones. Low-permeability soils or soils containing low-permeability zones form broader cones. The zone of influence is usually determined with *in situ* tests.

2. *Depth of air injection*. The end of the well from which air sparging is conducted is usually located 30 to 60 cm below the contaminated area.

3. *Air injection pressure and flow rate*. Air pressure must be greater than the hydrostatic pressure of the overlying water column at the depth of injection. Additional overpressure is also required to overcome the capillary forces inhibiting the penetration of air into the porous medium. The required overpressure depends on the permeability of the aquifer. A high overpressure, in the range of 0.3 to 3 m of H_2O, is usually applied in fine-grained soils with

low permeability. For permeable coarse soils, an overpressure between 3 and 30 cm of H_2O is sufficient. The typical values of volumetric flow rates per well range from 25 to 400 L/min.

2.3.5 EX SITU BIOLOGICAL TREATMENT

In cases where *in situ* biological treatment cannot be applied, the contaminated soil is excavated and transferred to specially prepared areas where bioremediation can be carried out under well-controlled conditions. Some common *ex situ* biological methods are the landfarming technique and the biopile or biopit treatment options.[32]

2.3.5.1 Landfarming

In this treatment method, the soil is spread on a wide flat surface, creating a layer of thickness between 45 and 60 cm. At regular time intervals, the soil is plowed using classical tilling equipment to obtain good aeration and provide the oxygen necessary for the biological actions. Water and nutrients are also added as required using garden-type sprinkling equipment, which must be easily moved so that tilling can be conducted without destroying it. A schematic diagram of the landfarming treatment system is presented in Figure 2.7. The underlying surface is constructed with a slight slope (0.5 to 1%) towards a drainage collection point and is covered with an impermeable liner to obtain efficient recovery of the drainage and prevent eventual contamination of the subsoil. A permeable layer of sand is placed over the liner to protect it from the tilling equipment and promote the drainage of excess water. The contaminated soil is placed on top of the sand layer. This method is easily carried out and presents no technical difficulties. The main prerequisite is the availability of a large amount of surface, because the thickness of the soil layer cannot exceed 60 cm. This limit corresponds to the maximum plow depth of the available tilling machines.

2.3.5.2 Treatment in Biopile or Biopit

When available land space is insufficient for land farming, soil treatment can be carried out in piles or pits. Typical biopile and biopit constructions are presented in Figure 2.8 and Figure 2.9, respectively. When the soil has relatively low permeability, the pile can be constructed with sequential "lifts" of soil, approximately 60 cm in thickness, separated by permeable sand layers. These layers are connected with a vacuum pump or blower that is used to produce airflow though the soil pile. Water and nutrients are sprinkled on the top of the pile.

Treatment in a pit (Figure 2.9) can be carried out in the same area from which the soil was excavated, following isolation of the area with an impermeable liner. In this case, the upper surface of the pit can be covered with asphalt and rendered for use before the completion of the bioremediation project. If the upper surface is covered, an appropriate venting system must be installed to

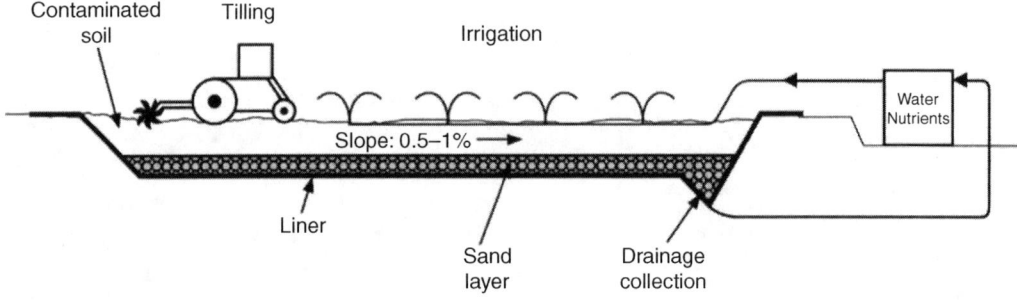

FIGURE 2.7 Landfarming treatment system.

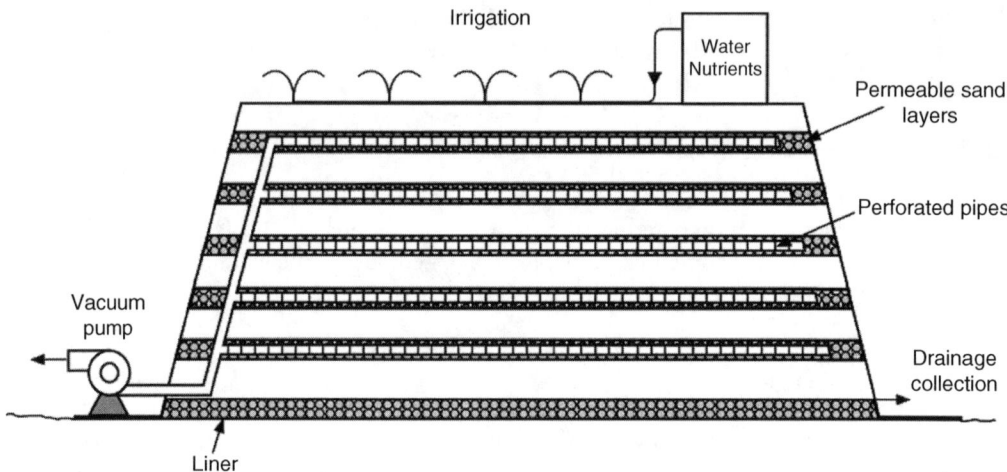

FIGURE 2.8   Treatment in a biopile.

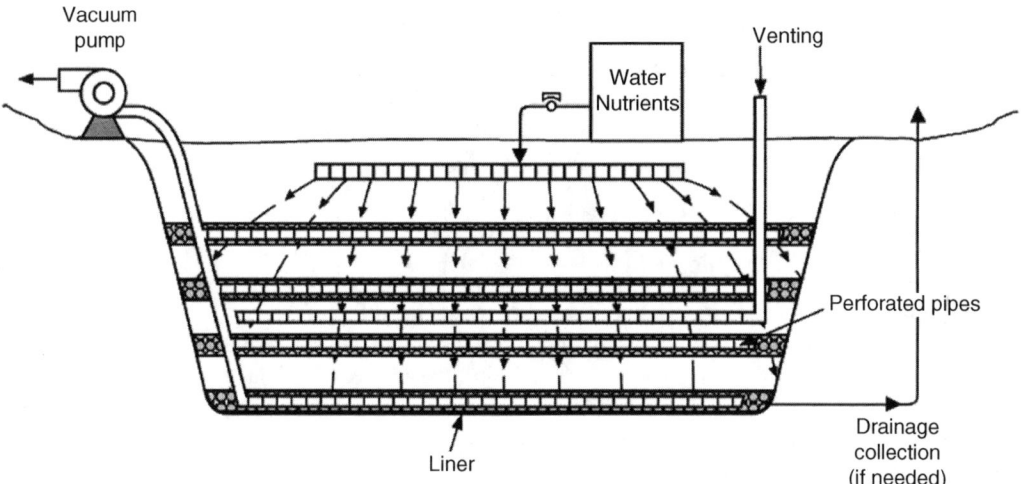

FIGURE 2.9   Treatment in a biopit.

allow the infiltration of fresh air inside the pit. Biopit is also a treatment option well adapted for anaerobic processes. With this configuration, the soil can be easily kept under water-saturated anaerobic conditions, and an impermeable cover on the surface will exclude any contact with atmospheric oxygen.

2.4   PHYTOREMEDIATION

2.4.1   GENERAL DESCRIPTION

Phytoremediation (also called green remediation, botano-remediation, agroremediation, and vegetative remediation) is the name given to a set of technologies that use living green plants and their associated microorganisms for *in situ* (in-place or on-site) partial or substantial remediation of contaminated soils, sludges, sediments, and groundwater. Both organic and inorganic contaminants can be addressed by applying phytoremediation technologies. Typical contaminants include petroleum hydrocarbons, crude oil, chlorinated compounds, pesticides, explosives, heavy metals, metalloids, and radioactive materials.

Plants aid remediation of polluted sites by means of several mechanisms. Some plants may withstand relatively high concentrations of organic chemicals without toxic effects, and in some cases can take up and quickly convert chemicals to less toxic metabolites. In addition, they stimulate the degradation of organic chemicals in the rhizosphere by releasing root exudates, enzymes and the buildup of organic carbon in the soil. Other plants, called hyperaccumulators, absorb unusually large amounts of metals. Growing these plants on contaminated soil and harvesting at certain times may result in decontamination of the soil. Still other plants may immobilize contaminants in the soil through absorption and accumulation into the roots, adsorption onto the roots, or precipitation or immobilization within the root zone. Therefore, based on the outcome for the contaminants, phytoremediation may be classified as a degradation, extraction, or containment technique. Another way of categorizing phytoremediation is based on the mechanisms involved. Such mechanisms include the following:

1. Extraction of contaminants from the soil and accumulation in the plant tissue for removal (phytoextraction)
2. Degradation of organic contaminants in the root zone by microorganisms (rhizodegradation)
3. Uptake of contaminants from the soil and metabolism above or below ground, within the root, stem, or leaves (phytodegradation)
4. Volatilization or transportation of volatile contaminants from the plants to the air (phytovolatilization)
5. Immobilization of contaminants in the root zone (phytostabilization)
6. Adsorption of contaminants on roots for containment or removal (rhizofiltration)

Phytoremediation is considered a low-cost remediation alternative for low-depth contamination, offering a permanent solution and improving the aesthetics of the polluted site. It is well-suited for use in the following situations:

1. At very large field sites where other methods of remediation are not cost-effective or practicable
2. At sites with low concentrations of contaminants where only a "polishing treatment" is required over long periods of time
3. In conjunction with other technologies where vegetation is applied as a final cap and closure of the site

Limitations need to be carefully considered before selecting this method for site remediation. These include the depth of contamination, the total length of time required for cleanup to below accepted limits, potential contamination of vegetation and the food chain, and difficulty in establishing and maintaining vegetation at some polluted sites.[8]

2.4.2 PHYTOREMEDIATION MECHANISMS

Phytoremediation takes advantage of the natural processes of plants (Figure 2.10). These processes include water and chemical uptake, metabolism within the plant, release of inorganic and organic compounds (exudates) into the soil, and the physical and biochemical impact of plant roots.[8,41] Plants require 13 essential inorganic plant nutrients (N, P, K, Ca, Mg, S, Fe, Cl, Zn, Mn, Cu, B, and Mo) for growth; these are taken up by the root system. In addition to these essential nutrients, other nonessential inorganics (such as various common contaminants like Pb, Cd, and As) or organics can be taken up. For uptake into a plant, a chemical must be in solution, either in the groundwater or in

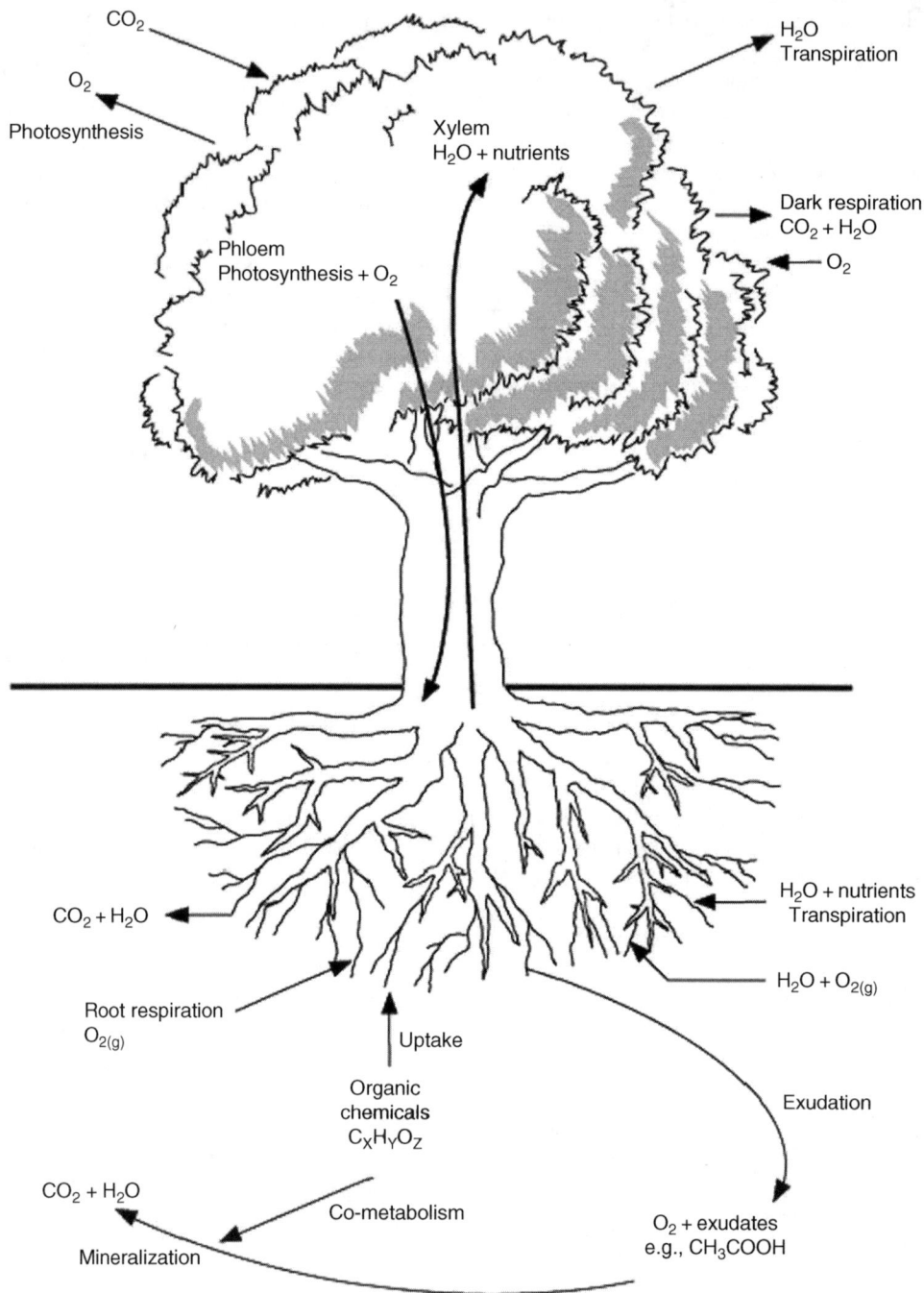

FIGURE 2.10   Oxygen, water, and chemical flows through a woody tree.

the soil solution. Water is absorbed from the soil solution into the outer tissue of the root and contaminants in the water can move to different parts of plants where they can be absorbed, bound, or metabolized.

Factors that affect the accessibility of chemicals to plant roots include hydrophobicity, polarity, sorption properties and solubility. In order to apply phytoremediation techniques to soils polluted by organic contaminants, the contaminant must come into contact with the plant roots and be dissolved in the soil water. One chemical characteristic that influences the uptake of organics into a plant is the octanol–water partition coefficient, log K_{ow}. Chemicals that are able to enter the plant have log K_{ow} values[8] between 1 and 3.5. Hydrophobic chemicals presenting log K_{ow} values greater than 3.5 are generally not sufficiently soluble in water or are bound so strongly to the surface of the roots that they cannot be easily translocated into the plant. On the other hand, chemicals that are highly polar and very water soluble (log $K_{ow} < 1.0$) are not sufficiently absorbed by the roots nor are they actively transported through plant membranes due to their high polarity.[42] Most benzene, toluene, ethylbenzene and xylene (BTEX) chemicals, chlorinated solvents, and short-chain aliphatic chemicals fall within the log K_{ow} range that allows them to be susceptible to phytoremediation.[8,41,42]

Plant roots cause changes at the soil–root interface as they release inorganic and organic compounds (root exudates) into the area of soil immediately surrounding the roots (the rhizosphere). Root exudates affect the number and activity of microorganisms, the aggregation and stability of the soil particles around the root, and the availability of elements. Root exudates can increase (mobilize) or decrease (immobilize), directly or indirectly, the availability of elements in the rhizosphere. Mobilization or immobilization of elements in the rhizosphere can be caused by changes in soil pH, the release of complexing substances such as metal-chelating molecules, changes in oxidation–reduction potential, and increases in microbial activity.

Different forms of phytoremediation may require different types of plants and be relevant for specific types of contaminants (Table 2.8). In the following section, each remediation form is presented separately.

2.4.2.1   Phytoextraction

Phytoextraction (also called phytoaccumulation, phytosequestration, phytoabsorption, and phytomining) refers to the use of certain plants to transport metals from the soil and concentrate them into the roots and aboveground shoots. One or a combination of these plants can be selected and planted at a site based on the type of metals present and other site conditions. After the plants have been allowed to grow for several weeks or months, they are harvested and either incinerated or recycled as metal ore. This procedure may be repeated as necessary to lower soil contaminant levels to allowable limits. Phytoextraction may be applied to metals (e.g., Ag, Cd, Co, Cr, Cu, Hg, Mn, No, Ni, Pb, and Zn), metalloids (e.g., As and Se), radionuclides (e.g., ^{90}Sr, ^{137}Cs, ^{234}U, and ^{238}U), and nonmetals (e.g., B). It has generally not been considered for organic or nutrient contaminants, as these can be metabolized, changed, or volatilized by the plant, thus preventing accumulation.

Phytoextraction is mainly carried out by certain plants called hyperaccumulators, which absorb unusually large amounts of metals compared to other plants. A hyperaccumulator is a plant species capable of accumulating 100 times more metal than a common nonaccumulating plant. Therefore, a hyperaccumulator will concentrate more than 1000 mg/kg or 0.1% (dry weight) of Co, Cu, Cr, or Pb, or 10,000 mg/kg (1%) of Zn and Ni (dry matter).[43,44] Similarly, halophytes are plants that can tolerate and, in many cases, accumulate large amounts of salt (typically sodium chloride but also Ca and Mg chlorides). Hyperaccumulators and halophytes may be selected and planted at a site based on the type of metals or salts present, the concentrations of these constituents, and other site conditions.

Almost all known metal-hyperaccumulating species were discovered on metal-rich soils and they are endemic to such soils, suggesting that hyperaccumulation is an important ecophysiological adaptation to metal stress and one of the manifestations of resistance to metals. These plants are generally rare and found only in localized areas around the world, with fewer than 400 identified species for eight heavy metals.[41] Phytoextraction occurs in the root zone of plants. The root zone typically may be relatively shallow, with the bulk of the roots at shallower rather than greater depths. This is a potential limitation of phytoextraction. One type of plant may take up different metals to different degrees. Experimental studies[45] have indicated that the phytoextraction coefficients (the ratio of the metal concentration in the shoot to the metal concentration in the soil) for different metals taken up by Indian mustard vary significantly (as shown in Table 2.9).

TABLE 2.8

Typical Plants Used in Various Phytoremediation Applications

Mechanism	Media	Typical Contaminants	Plant Types
Phytodegradation	Soils, groundwater, landfill leachate, land application of wastewater	Herbicides (atrazine, alachlor) Aromatics (BTEX) Chlorinated aliphatics (TCE) Nutrients (NO_3^-, NH_4^+, PO_4^{3-}) Ammunition wastes (TNT, RDX)	Phreatophyte trees (poplar, willow, cottonwood, aspen) Grasses (rye, Bermuda, sorghum, fescue) Legumes (clover, alfalfa, cowpeas)
Rhizodegradation	Soils, sediments, land application of wastewater	Organic compounds (TPH, PAHs, BTEX, pesticides, chlorinated solvents, PCBs)	Phenolics releasers (mulberry, apple, osage orange) Grasses with fibrous roots (rye, fescue, Bermuda) for contaminants 0–3 ft deep Phreatophyte trees for 0–10 ft Aquatic plants for sediments
Phytostabilization	Soils, sediments	Metals and metalloids (As, Cd, Cr, Cu, Pb, Zn, U, Se) Hydrophobic organics (PAHs, PCBs, dioxins, furans, pentachlorophenol, DDT, dieldrin)	Phreatophyte trees to transpire large amounts of water for hydraulic control Grasses with fibrous roots to stabilize soil erosion Dense root systems are needed to sorb/bind contaminants
Phytoextraction	Soils, sediments	Metals (Ag, Au, Cd, Co, Cr, Cu, Hg, Mn, Mo, Ni, Pb, Zn) Radionuclides (^{90}Sr, ^{137}Cs, ^{239}Pu, ^{234}U, ^{238}U)	Sunflowers Indian mustard Rape seed plants Barley, hops Crucifers Serpentine plants Nettles, dandelions
Phytovolatilization	Soils, sediments, sludges, groundwater	Chlorinated solvents, MTBE, some inorganics (Se, Hg, As)	Herbaceous species Trees Wetland species
Rhizofiltration	Groundwater, water and wastewater in lagoons or created wetlands	Metals (Pb, Cd, Zn, Ni, Cu) Radionuclides (^{137}Cs, ^{90}Sr, ^{234}U, ^{238}U) Hydrophobic organics	*Aquatic plants*: Emergents (bullrush, cattail, coontail, pondweed, arrowroot, duckweed) Submergents (algae, stonewort, parrotfeather, Eurasian water milfoil, Hydrilla)

Source: Schnoor, J.L., Phytoremediation. Technology Evaluation Report TE-98-01, Ground-Water Remediation Technologies Analysis Center, Pittsburgh, PA, 1997; Interstate Technology and Regulatory Cooperation Work Group (ITRC), Phytotechnologies Work Team, Technical/Regulatory Guidelines. Phytotechnology Technical and Regulatory Guidance Document, 2001, www.itrcweb.org/Documents/PHYTO-2.pdf. With permission.

Higher phytoextraction coefficients indicate higher metal uptake. The effectiveness of phytoextraction can be limited by the sorption of metals to soil particles and the low solubility of the metals; however, metals can be solubilized through the addition of acids or chelating agents and so allow uptake of the contaminant by the plant. Ethylene diamine tetra-acetic acid (EDTA), citric acid, and ammonium nitrate have been reported to help in the solubilization of lead, uranium, and cesium

TABLE 2.9
Phytoextraction Coefficients

Metal	Phytoextraction Coefficient
Cr^{6+}	58
Cd^{2+}	52
Ni^{2+}	31
Cu^{2+}	7
Pb^{2+}	1.7
Cr^{3+}	0.1
Zn^{2+}	17

Source: Kumar, P.B.A.N., Dushenkov, V., Motto, H. and Raskin, I., Phytoextraction: The use of plants to remove heavy metals from soils, *Environ. Sci. Technol.*, 29, 1232–1238, 1995. With permission.

137, respectively.[46] However, the potential adverse impact of such chemicals on groundwater, plant growth, or other elements' solubility must be considered before use.

Phytoextraction has several advantages. The contaminants are permanently removed from the soil and the quantity of the waste material produced is substantially decreased. In some cases, the contaminant can be recycled from the contaminated biomass. However, the use of hyperaccumulating plants is limited by their slow growth, shallow root systems, and small biomass production. In order for this remediation scheme to be feasible, plants must tolerate high metal concentrations, extract large concentrations of heavy metals into their roots, translocate them into the surface biomass, and produce a large quantity of plant biomass.

2.4.2.2 Rhizodegradation

Rhizodegradation (also called phytostimulation, rhizosphere biodegradation, or plant-assisted bioremediation/degradation) is the breakdown of contaminants in the soil through the enhanced bioactivity existing in the rhizosphere. Typical compounds exuded by plant roots in the rhizosphere include sugars, amino acids, organic acids, fatty acids, sterols, growth factors, nucleotides, flavanones, and enzymes.[47] Root exudates provide sufficient carbon to support large numbers of microbes (approximately 1×10^8 to 1×10^9 vegetative microbes per gram of soil in the rhizosphere). Because of these exudates, microbial populations and activities between 5 and 100 times greater in the rhizosphere than in bulk soil. This plant-induced enhancement of the microbial population is referred to as the "rhizosphere effect."[48,49] The increased microbial populations and activity in the rhizosphere can increase contaminant biodegradation in the soil, and degradation of the exudates can stimulate co-metabolism of contaminants in the rhizosphere.

Organic contaminants such as petroleum hydrocarbons or chlorinated solvents can be directly metabolized by proteins and enzymes, leading to the degradation, metabolism, or mineralization of the contaminants. Furthermore, many of these contaminants can be broken down into harmless products or converted into a source of food and energy for the plants or soil organisms.[50]

Rhizodegradation is a symbiotic relationship that has evolved between plants and soil microbes. The plants provide the nutrients necessary for the microbes to thrive, and the microbes provide a healthier soil environment in which the plant roots can proliferate.

2.4.2.3 Phytodegradation

Phytodegradation (also known as phytotransformation) is the uptake, metabolizing, and degradation of contaminants within the plant, or the degradation of contaminants external to the plant through the effect of compounds such as enzymes produced and released by the plant. Phytodegradation is not dependent on the microorganisms associated with the rhizosphere. For the type of phytodegradation that occurs within a plant, the plant must be able to take up the contaminant. Therefore, only moderately hydrophobic compounds, with an octanol–water partition coefiecient log K_{ow} between 1 and 3.5, are susceptible to phytodegradation.[8] The direct uptake of a chemical into a plant through its roots depends on the uptake efficiency, transpiration rate, and the concentration of the contaminant in the soil water.[51] Once an organic chemical is translocated, the plant may store the chemical and its fragments into new plant structures through lignification (covalent bonding of chemical or its fragments into lignin of the plant), or it can volatilize, metabolize, or mineralize the chemical completely to carbon dioxide and water.

2.4.2.4   Phytovolatilization

Phytovolatilization involves using plants to take up volatile or nonvolatile contaminants from the soil, transforming them into volatile forms and transpiring them into the atmosphere. Phytovolatilization is primarily a contaminant extraction process. However, metabolic processes within the plant might alter the initial form of the contaminant, and in some cases transform it to less toxic forms. Phytovolatilization may be applied to both organic and inorganic contamination. An example of phytovolatilization of inorganic contaminants is the transformation of the highly toxic mercuric ion to the less toxic elemental mercury. A disadvantage of this technique is that mercury released to the atmosphere is likely to be recycled by precipitation. Because phytovolatilization involves the transfer of contaminants to the atmosphere, a risk analysis of the impact of this transfer on the ecosystem and human health may be necessary.

2.4.2.5   Phytostabilization

Phytostabilization (also known as in-place inactivation or phytoimmobilization) is the use of certain plant species to immobilize contaminants in the soil through absorption and accumulation by roots, adsorption onto roots, or precipitation, complexation, and metal valence reduction within the root zone. The following three mechanisms determine the fate of the contaminants within the phytostabilization process[46]:

1. *Phytoremediation in the root zone.* Proteins and enzymes produced by the plant can be exuded by the roots into the rhizosphere. These plant products target contaminants in the surrounding soil, leading to precipitation or immobilization in the root zone. This mechanism within phytostabilization may reduce the fraction of the contaminant in the soil that is bioavailable.
2. *Phytostabilization on the root membranes.* Proteins and enzymes directly associated with the root cell walls can bind and stabilize the contaminant on the exterior surfaces of the root membranes. This prevents the contaminant from entering the plant.
3. *Phytostabilization in the root cells.* Proteins and enzymes present on the root cell walls can facilitate the transport of contaminants across the root membranes. Upon uptake, these contaminants can be sequestered into the vacuole of the root cells, preventing further translocation to the shoots.

Phytostabilization has generally focused on metal contamination, with lead, chromium, and mercury being identified as the top potential candidates for phytostabilization.[44,52] However, there is potential for phytostabilization of organic pollutants, because some organic contaminants or metabolic byproducts of these contaminants can be attached to or incorporated into plant components.[41] Very hydrophobic organic compounds with log K_{ow} values greater than 3.5 are candidates for phytostabilization.[8]

Effective phytostabilization requires a thorough understanding of the chemistry of the root zone, root exudates, contaminants, and fertilizers or soil amendments to avoid unintended effects that might increase contaminant solubility and leaching. It has been suggested that phytostabilization might be most appropriate for heavy-textured soils and soils with high organic matter contents.[53]

Phytostabilizing plants should be able to tolerate high levels of contamination (i.e., metal-tolerant plants for heavy-metal-contaminated soils), with roots growing into the zone of contamination, and should be able to alter the biological, chemical, and physical conditions in the soil. Furthermore, contaminants should not be accumulated in the plant tissues in order to eliminate the possibility that the harvested plants might themselves become hazardous wastes. Most research on phytostabilization deals with mining wastes or soils polluted by mining activities. Following field applications conducted in Liverpool, England, varieties of three grasses were made available for phytostabilization (*Agrostis tenuis, cv Parys* for copper wastes, *Agrosas tnuis, cv Coginan* for acid lead and zinc wastes, and *Festuca rubra, cv Merlin* for calcareous lead and zinc wastes.[54] Laboratory studies have indicated that other plants such as Indian mustard also have the potential for effective phytostabilization of Pb and Cr(VI).[55,56] Furthermore, poplar trees are being studied for possible use in phytostabilization, as they may be able to form roots up to the maximum depth of contamination.

Advantages associated with this technology include the fact that the disposal of hazardous material or biomass is not required, soil removal is unnecessary, the application cost is low, and the degree of disruption to site activities may be less than with other more vigorous remedial technologies. The technique is very effective when rapid immobilization is needed to preserve ground and surface waters. The presence of plants also reduces soil erosion and decreases the amount of water available in the system. The main disadvantage of phytostabilization is that the contaminants remain in the soil, and so the future release of contaminants should be prevented. Therefore, long-term maintenance of the vegetation or verification that the vegetation will be self-sustaining should be secured.

2.4.2.6   Rhizofiltration

Rhizofiltration (also known as phytofiltration) is adsorption or precipitation onto plant roots or absorption into the roots of contaminants that are in solution surrounding the root zone. The contaminant may remain on the root, within the root, or be taken up and translocated into other portions of the plant, depending on the contaminant, its concentration, and the plant species.[41] Applications of rhizofiltration are currently at the pilot-scale stage. It is intended to be generally applicable to the treatment of large volumes of water with low contaminant concentrations (in the ppb range). It is to be used in the treatment of metals, radionuclides, or mixed wastes, but it is also suitable for ammunition wastes. Rhizofiltration is effective in areas where wetlands can be created and all of the contaminated water may be allowed to come into contact with the roots.

2.4.3   DESIGN CONSIDERATIONS

The design of a phytoremediation system is determined by several factors associated with the contaminants (type, concentration, and depth), the conditions at the site, the plants, the level of cleanup required and the available time. Extraction techniques have different design requirements than immobilization or degradation methods. Nevertheless, it is possible to specify a few design factors that are a part of most phytoremediation efforts.

2.4.3.1   Root System

Remediation with plants requires that the contaminants be in contact with the root zone of the plants. Therefore, root morphology and depth directly affect the depth of soil that can be remediated or the depth of groundwater that can be influenced. A fibrous root system such as that found in grasses has numerous fine roots spread throughout the soil and provides maximum contact with the soil because of the high surface area of the roots. A tap root system (such as in alfalfa) is dominated

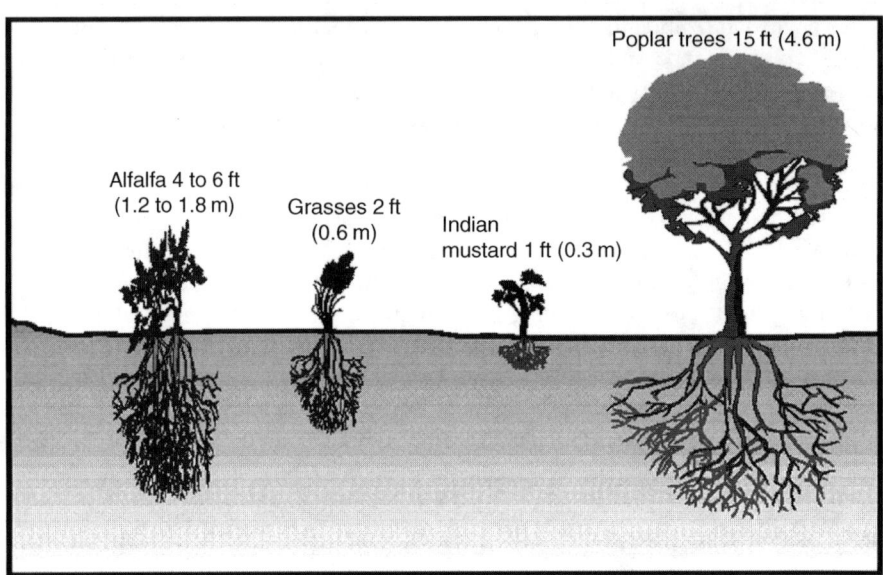

FIGURE 2.11   Root depths for different plants.

by one central root. Many hyperaccumulators have a tap root system, which limits root contact to relatively small volumes of soil. As shown in Figure 2.11, the effective root depth of plants varies by species and depends on soil and climate condition. The root depth ranges provided in the following represent maximum depths[57]:

1. *Legumes.* Alfalfa roots can extend down to about 9.1 m (30 ft), given the proper conditions.
2. *Grasses.* Some grass fibrous root systems can extend 2.4 to 3 m (8 to 10 ft) deep. The roots of major prairie grasses can extend to about 1.8 to 3 m (6 to 10 ft).
3. *Shrubs.* The roots of phreatophytic shrubs can extend to about 6 m (20 ft).
4. *Trees.* Phreatophyte roots will tend to extend deeper than other tree roots. Phreatophytic tree roots can reach as deep as 24 m (80 ft). Two examples are mesquite tap roots, which range from 12 to 30 m (40 to 100 ft), and river birch tap roots, which go to a depth of 27 to 30 m (90 to 100 ft).
5. *Other plants.* Indian mustard roots are generally about 15 to 22 cm (6 to 9 in.) deep.

These maximum depths are not likely to occur in most cases. The effective depth for phytoremediation using most nonwoody plant species is likely to be only 30 or 61 cm (1 or 2 ft). Most accumulators have root zones limited to the top foot of soil, which restricts the use of phytoextraction to shallow soils. The effective depth of tree roots is likely to be in the few tens of feet or less, with one optimistic estimate that trees will be useful for extraction of groundwater up to 9 m (30 ft) deep.[41,58]

2.4.3.2   Plant Growth Rate

The time required to clean up a site through phytoremediation may be longer than is acceptable for some redevelopment objectives. Phytoremediation is limited by the natural growth rate of plants and the length of the growing season. Several growing seasons may be required before phytoremediation systems become effective, and traditional methods may require a few weeks or months. Low removal rates may therefore prohibit the use of phytoremediation in cases where the time period available for cleanup is limited and is a key criterion in selecting a technology.[59] Growth rates can be defined differently for different forms of phytoremediation. For rhizodegradation, rhizofiltration, and phytostabilization, it is desirable to have fast growth in terms of root depth, density, volume, surface area, and lateral extension. For phytoextraction, a fast growth rate of aboveground plant mass is desirable.

A large root mass and large biomass are desired for an increased mass of accumulated contaminants, for greater transpiration of water, greater assimilation and metabolism of contaminants, or for production of a greater amount of exudates and enzymes. A fast growth rate will minimize the time required to reach a large biomass. For the phytoextraction of metals, the metals concentration in the biomass and the amount of biomass produced must both be considered. Metal hyperaccumulators are able to concentrate a very high level of some metals; however, their generally low biomass and slow growth rate means that the total mass of metals removed will tend to be low. A plant that extracts a lower concentration of metals, but that has a much greater biomass than many hyperaccumulators, is more desirable than the hyperaccumulator because the total mass of metals removed will be greater.[57]

The growth rate of a plant species will have a direct effect on its potential for use at a particular site. Fast-growing grasses will begin treating soil contamination more quickly than a tree, which must first establish deeper roots to treat target contaminants. As plants (particularly trees used in phytoremediation) mature their root structures deepen and their capacity to treat deeper levels of contamination improves. Phytoremediation can provide a number of benefits during the course of vegetation maturation. Plantings during initial stages can provide a cover that minimizes water infiltration. As the tree roots mature, phytoremediation processes take place to treat contaminants at increasing depths below the surface. Poplars, which have been widely used in phytoremediation applications, present high growth rates, varying between 2.7 and 4.6 m/yr (9 and 15 ft/yr).[57]

2.4.3.3 Plant Selection

Plant selection is probably one of the most important factors determining the success or failure of a phytotechnology project. Careful selection of the plant and plant variety is critical—first to ensure that the plant is appropriate for the climatic and soil conditions at the site, and second for the effectiveness of the phytoremediation. Once growing conditions at the site have been identified, the next goal of the plant selection process is to choose plants with appropriate characteristics for growth under site-specific conditions that also meet the objectives of the phytotechnology project. A screening test or knowledge from the literature of plant attributes will aid the design team in the selection of plants. Typical information needed for plant selection includes the species name (common and scientific), various tolerances (temperature, moisture, diseases, pests, etc.), growth habit (annual, perennial, biennial, evergreen vs. deciduous), climate zone, and general form (grass, leafy plant, shrub, tree, etc.). Another consideration in plant selection is the decision whether to use a monoculture or several plant species. In general, the use of a mixed variety of vegetation is preferred over monostands due to the following several advantages[46]:

1. Monostands can be susceptible to diseases that can destroy the entire phytotechnology system, while mixed stands may only lose one or two species.
2. Mixed stands support more diverse microbial communities (promoting potentially more complete rhizodegradation by further breaking down byproducts).
3. Synergistic effects such as nutrient cycling can be obtained in mixed stands.
4. Mixed stands have a more naturalized appearance.
5. Mixed stands promote biodiversity and potential habitat restoration qualities.

Plants are selected according to application needs and the contaminants concerned. For phytodegradation of organics, the design requirements are that vegetation is fast growing and hardy, easy to plant and maintain, utilizes a large quantity of water by evapotranspiration (if groundwater is an issue), and transforms the contaminants of concern to nontoxic or less toxic products. In temperate climates, phreatophytes (e.g., hybrid poplar, willow, cottonwood, aspen) are often selected because of their fast growth, deep rooting ability down to the surface of groundwater, large transpiration rates, and the fact that they are native throughout most of the U.S.[8] Indian mustard is a fast-growing accumulator plant with a relatively high biomass, which has the ability to take up and accumulate metals and radionuclides. Sunflower (*Helianthus annuus*) can accumulate metals and has about the same biomass as Indian mustard. Examples of metal hyperaccumulators that have been investigated include *Thlaspi caerulescens* (Alpine pennycress), but which is slow-growing and has a low biomass; *Thlaspi rotundifolium* spp. *cepaeifolium*, the only known hyperaccumulator of Pb; and other *Thlaspi* species that can hyperaccumulate cadmium, nickel, or zinc.[41]

Grasses are often planted in tandem with trees at sites with organic contaminants or even as the primary remediation method. They provide a tremendous amount of fine roots in the surface soil, which is effective at binding and transforming hydrophobic contaminants such as TPH, BTEX, and PAHs.[8] Grasses are often planted between rows of trees to provide for soil stabilization and protection against the wind-blown dust that can move contaminants off site. Some grasses, such as *Festuca ovina*, can take up metals but are not hyperaccumulators. Alfalfa has been investigated because of its deep root system, its ability to fix nitrogen, and the fact that there is a large knowledge base about this plant. These plants have been popular for research to date, but future screening studies will undoubtedly add many more candidates, some of which may prove to be much more effective for phytoremediation.

2.4.3.4 Treatability Studies

Treatability studies are recommended and may be required for all phytoremediation projects unless adequate site-specific information is available indicating a probable successful outcome. These studies may take the form of laboratory-scale germination tests, greenhouse-scale fate and transport studies and/or mass balances, or field-scale (up to 15×15 m) tests to examine site-specific survivability and treatment efficacy under existing site conditions.[8,46,57] Treatability tests should be carried out in real time, because plant growth cannot be accelerated and should continue for at least one growth cycle, including dormancy. Toxicity and transformation data are obtained in treatability studies. Regulators may require total mass balance information, which necessitates use of radio-labeled compounds in the laboratory-scale tests.

2.4.3.5 Plant Density and Patterns

Planting density depends on the application. For hybrid poplar trees, between 400 and 800 trees per hectare (1000 to 2000 trees per acre) are typically planted with a conventional tree planter at 30 to 46 cm (12 to 18 in.) depth or in trenched rows 0.30 to 1.8 m (1 to 6 ft) deep.[8] If a row conformation is used, the trees may be spaced with 0.6 m (2 ft) between trees and 3 m (10 ft) between rows. Several phreatophytes, such as willow and cottonwood, can be planted in a similar manner. Hardwood trees and evergreens may require a lower initial planting density. A high initial planting density assures a significant amount of evapotranspiration in the first year, which is normally desirable, but the trees will naturally thin themselves by competition to 240 to 320 trees per hectare (600 to 800 trees per acre) over the first six years. If desired, hybrid poplars can be harvested on a six-year rotation and sold for fuelwood or pulp and paper, and the trees will grow back from the cut-stump (coppicing trait). The dense, deep root system stays in place to sustain growth for the next year. The lifetime of hybrid poplars is on the order of 30 years, which is usually sufficient as the design life of the project. Grasses are usually drilled or broadcast for planting at contaminated sites. Biomass densities (above ground) of 200 to 600 g/m^2 are achieved by the second crop, with 1 to 3 crops per year depending on climate and water availability.[8]

2.4.3.6 Groundwater Capture and Transpiration

For the estimation of the contaminants' uptake rate and consequently the time required for the phytoremediation of a contaminated site, single mathematical models may be applied. In the following paragraphs, analysis as well as the examples given in *Groundwater Remediation Technologies Analysis Center Technology Evaluation Report on Phytoremediation*, by Schnoor, is presented without modification.[8]

A simple capture zone calculation[60] can be used to estimate whether the phytoremediation "pump" can be effective at entraining the plume of contaminants. Trees can be grouped for consideration as average withdrawal points. The goal of such a phytoremediation effort is to create a water table depression where contaminants will flow to the vegetation for uptake and treatment. Organic contaminants are not taken up at the same concentration as in the soil or groundwater; rather, there is a transpiration stream concentration factor (a fractional efficiency of uptake) that accounts for the partial uptake of the contaminant (due to membrane barriers at the root surface). The uptake rate is given by the following equation:

$$U = \text{TSCF} \times T \times C \tag{2.24}$$

where U = uptake rate of contaminant (mg/d), TSCF = transpiration stream concentration factor (dimensionless), T = transpiration rate of vegetation (L/d), and C = aqueous phase concentration in soil water or groundwater (mg/L).

If the plants do not take up the dissolved contaminant, the plume that emerges will be concentrated (i.e., the mass of contaminant in the plume will be the same, but the concentration remaining will actually be greater due to the reduction in water volume caused by the vegetation). This is a potential concern for phytoremediation of groundwater plumes or in created wetlands, where a relatively hydrophilic contaminant can be concentrated on the downstream side of the phytotechnology system.

A method for estimating the TSCF for equation 2.24 is given in Table 2.10. The root concentration factor is also defined in Table 2.10 as the ratio of the contaminant in the roots to the concentration dissolved in the soil water (μg/kg root per μg/L). This is important in estimating the mass of contaminant sorbed to roots in phytoremediation systems. The values of TSCF and RCF for metals depend on the metals' redox states and chemical speciation in soil and groundwater.

Mature phreatophyte trees (poplar, willow, cottonwood, aspen, ash, alder, eucalyptus, mesquite, bald cypress, birch, and river cedar) typically can transpire 3700 to 6167 m^3 (3 to 5 acre-ft) of water per year. This is equivalent to about 2 to 3.8 m^3 (600 to 1000 gal) of water per tree per year for a mature species planted at a density of 600 trees per hectare (1500 trees per acre). Transpiration rates in the first two years would be somewhat less, about 0.75 m^3 per tree per year (200 gal per tree per year), and hardwood trees would transpire about half the water of a phreatophyte. Two meters of water per year is a practical maximum for transpiration in a system with complete canopy coverage (a theoretical maximum would be 4 m/yr based on the solar energy supplied at latitude 40°N on a clear day).

If evapotranspiration of the system exceeds precipitation, it is possible to capture water that is moving vertically through soil. Areas that receive precipitation in the wintertime (the dormant season for deciduous trees) must be modeled to determine if the soil will be sufficiently dry to hold water for the next spring's growth period.

2.4.3.7 Contaminant Uptake Rate and Cleanup Time

From equation 2.24 it is possible to estimate the uptake rate of the contaminant(s). First-order kinetics can be assumed as an approximation for the time duration needed to achieve remediation goals. The uptake rate should be divided by the mass of contaminant remaining in the soil:

$$k = U/M_o \tag{2.25}$$

TABLE 2.10

Estimating the Transpiration Stream Concentration Factor (TSCF) and Root Concentration Factor (RCF) for Some Typical Contaminants (8)

Chemical	Log K_{ow}[a]	Solubility[a], $-\text{Log } C_w^{sat}$ @25°C (mol/L)	Henry's Constant[a] kH @25°C (dimensionless)	Vapor Pressure $-\text{Log } P_o$@ 25°C (atm)	TSCF[b]	RCF[c] (1/kg)
Benzene	2.13	1.64	0.2250	0.90	0.71	3.6
Toluene	2.69	2.25	0.2760	1.42	0.74	4.5
Ethylbenzene	3.15	2.80	0.3240	1.90	0.63	6.0
m-Xylene	3.20	2.77	0.2520	1.98	0.61	6.2
TCE	2.33	2.04	0.4370	1.01	0.74	3.9
Aniline	0.90	0.41	2.2×10^{-5}	2.89	0.26	3.1
Nitrobenzene	1.83	1.77	0.0025[d]	3.68	0.62	3.4
Phenol	1.45	0.20	$>1.0 \times 10^{-5}$	3.59	0.47	3.2
Pentachlorophenol	5.04	4.27	$1.5 \times 10^{-4 d}$	6.75[d]	0.07	54
Atrazine	2.69	3.81	$1.0 \times 10^{-7 d}$	9.40[d]	0.74	4.5
1,2,4-Trichlorobenzene	4.25	3.65	0.1130	3.21	0.21	19
RDX	0.87	4.57	—	—	0.25	3.1

[a]Physical chemical properties[61] unless otherwise noted.
[b]TSCF = 0.75 exp{−[(log K_{ow}−2.50)²/2.4]} (ref. 62).
[c]RCF = 3.0 + exp(1.497 log K_{ow}−3.615) (ref. 62).
[d]*Source*: Schnoor, J.L., *Environmental Modeling—Fate and Transport of Pollutants in Water, Air, and Soil*, John Wiley & Sons, New York, 1996. With permission.
Vic, E.A. and Bardos, P., Remediation of Contaminated Land. Technology Implementation in Europe, Federal Environmental Agency, Austria. CLARINET Report, available at www.clarinet.at, 2002. With permission.

where k = the first-order rate constant for uptake (yr⁻¹), U = the contaminant uptake rate (kg/yr), And M_o = the initial mass of contaminant (kg). Then, an estimate for mass remaining at any time is expressed by equation 2.26:

$$M = M_o e^{-kt} \tag{2.26}$$

where M = mass remaining (kg) and t = time (yr).

Solving for the time required to achieve cleanup of a known action level:

$$t = -(\ln M/M_o)/k \tag{2.27}$$

where t = time required for cleanup to action level (yr), M = mass allowed at action level (kg), and M_o = initial mass of contaminant (kg).

2.4.3.8 Examples

Equations 2.24 to 2.27 can be applied to most sites where soil cleanup regulations are known for metals or organic contaminants. Two examples follow, one for TCE treatment by phytotransformation and another for lead removal by phytoextraction, which demonstrate the use of the design equations.

Example 1: Organics
TCE residuals have been discovered in an unsaturated soil profile at a depth of 3 m. From lysimeter samples, the soil water concentration is approximately 100 mg/L. Long cuttings of hybrid poplar

trees will be planted through the waste at a density of 600 trees per hectare (1500 trees per acre) for uptake and phytotransformation of the TCE waste. By the second or third year, the trees are expected to transpire $3700 \, m^3/yr$ (3 acre-ft/yr) of water or about $2.27 \, m^3/tree$ (600 gal/tree) per year. Estimate the time required for cleanup if the mass of TCE per hectare is estimated to be 400 kg/hectare (1000 kg/acre), and the cleanup standard has been set at 40 kg/hectare (100 kg/acre) (90% cleanup).

The uptake rate of TCE can be determined by equation 2.24:

$$U = \text{TSCF} \times T \times C,$$

where TSCF = 0.74 (from Table 2.10), $T = (2.27 \, m^3/\text{tree-yr})(600 \, \text{tree/hectare})(1000 \, L/m^3) = 1.362 \times 10^6$ L/hectare-yr, and $C = 100$ mg/L (given). Therefore,

$$U = \text{TSCF} \times T \times C = 0.74 \times (1.362 \times 10^6 \, \text{L/hectare-yr}) \times (100 \, \text{mg/L})$$
$$= 1.00788 \times 10^8 \, \text{mg/hectare-yr} = 100.788 \, \text{kg/hectare-yr}$$

The coefficient k can be determined from equation 2.25:

$$k = U/M_o = (100.788 \, \text{kg/hectare-yr})/(400 \, \text{kg/hectare}) = 0.259 \, \text{yr}^{-1}$$

Therefore, the time required to achieve the remediation goal is calculated from equation 2.27:

$$t = -(\ln M/M_o)/k = -(\ln 40/400)/(0.259 \, \text{yr}^{-1}) = 8.9 \, \text{yr}$$

Most of the TCE that is taken up by the poplars is expected to volatilize slowly to the atmosphere. A portion will be metabolized by the leaves and woody tissue of the trees.

Example 2: Metals
Lead at a lightly contaminated brownfield site has a concentration in soil of 600 mg/kg to a depth of 1 ft. The cleanup standard has been set at 400 mg/kg. Indian mustard, *Brassica juncea*, will be planted, fertilized, and harvested three times each year for phytoextraction. Using small doses of EDTA, it is possible to achieve concentrations in the plant of 5000 mg/kg (dry weight basis), and harvestable densities of 2.72 t (3 short tons) dry matter per crop. Estimate the time required for cleanup:

$$U = \text{uptake rate} = (5000 \, \text{mg/kg}) \times (3 \times 2.72 \, \text{t/hectare-yr}) \times (1000 \, \text{kg/t})$$
$$= 4.09 \times 10^7 \, \text{mg/hectare-yr} = 40.9 \, \text{kg/hectare-yr}$$
M_o = mass of Pb in soil at a dry bulk density of 1.5 kg/L
$M_o = (600 \, \text{mg/kg}) \times (1.5 \, \text{kg/L}) \times (1233 \, m^3) \times (1000 \, L/m^3) \times (10^{-6} \, \text{mg/kg})$
$M_o = 1110$ kg/hectare (initial mass in soil)
$M = 740$ kg/hectare (cleanup standard of 400 mg/kg)

Zero-order kinetics is assumed (constant rate of Pb uptake each year), because EDTA will make the lead continue to be bioavailable to the sunflowers, so

$$t = (M - M_o)/U = 9.0 \, \text{yr}$$

The time to cleanup may actually be somewhat less than 9 years if Pb migrates down in the soil profile with the addition of EDTA, or if tillage practices serve to "smooth out" the hot spots. Regulatory cleanup levels are usually based on a limit that cannot be exceeded, such as 400 mg/kg, and soil concentrations would need to be analyzed to ensure compliance at the end of each year.

2.5   SOIL WASHING

2.5.1   GENERAL DESCRIPTION

Soil washing is a physical and/or chemical separation technology in which excavated soil is washed with fluids to remove contaminants. It is considered feasible for the treatment of a wide range of inorganic and organic contaminants including heavy metals, radionuclides, cyanides, polynuclear aromatic compounds, pesticides, and PCBs. Soil washing removes contaminants from soils by (1) concentrating them into a smaller volume of soil through mineral-processing techniques and (2) by dissolving or suspending them in the wash solution.

In the first technique, clean and contaminated soil particles are separated by taking advantage of their physical properties, such as selective adsorption of contaminants onto fine clay particles of soil,[64,65] variations in specific gravity, magnetic,[66] and surface properties of clean and contaminated soil particles.[67,68,84] Research studies have shown that a large percentage of soil contamination, especially organic, is sometimes associated with, or bound to, very small (silt and clay) soil particles. In these situations, a physical separation of the large soil particles (sand and gravel) from the silt, clay, and humic material effectively concentrates the contaminants in the fine fraction.

The second soil-washing technique involves chemical treatment using water or chemical agents and aims at the selective leaching of contaminants from soil particles, or the total dissolution of contaminated particles. Chemical treatment is mainly applied for the removal of heavy metals using leaching reagents such as inorganic acids (hydrochloric acid, sulfuric acid, pH < 2), organic acids (acetic, lactic, citric acid, pH ≥ 4), complexing reagents such as EDTA, and nitrilotriacetic acid (NTA), and combinations of these reagents.[69–72] In the case of organic contaminants the use of surfactants (surface active agents) or co-solvents may be considered in order to increase their solubility in aqueous solutions.

The procedure of soil washing involves three main operations:

1. Intensive mixing of contaminated soil with washing fluid
2. Separation of clean soil particles
3. Treatment of the supernatant solution containing the dissolved or suspended contaminants[73–76]

A general flow diagram of a soil-washing treatment is shown in Figure 2.12. Initially, the contaminated soil is sieved to remove large objects such as pieces of wood, plant roots, stones, etc. The maximum size of particles allowed in the feedstock varies with the equipment used, ranging from 10 to 50 mm.[77]

The main soil-washing stage involves mixing, washing, rinsing, and size separation steps. In the soil-washing stage, two main mechanisms are involved. The first is the dispersion of fine contaminated particles, which either occur as aggregates or cover the surface of soil particles. For better dispersion, sodium hydroxide and surface active reagents such as lye are used. The second mechanism is the dissolution of contaminants in the aqueous solution; this can be enhanced by the addition of appropriate chemical reagents such as inorganic or organic acids, complexing agents, and surfactants or co-solvents, depending on the types of contaminants. Intensive contact between the soil grains and the wash fluid causes the soil contaminants to be dissolved and dispersed into the water. Energy is introduced into the mixture by high-pressure water jets, vibration devices, and other means.

After mixing for an appropriate time, clean soil particles and wash water containing the dissolved and suspended contaminants are separated. Separation techniques in soil-washing systems are similar to those applied in the mineral-processing industry.[78] The most common separation techniques are as follows:

1. *Hydrocyclones.* Particle separation in hydrocyclones uses the centrifugal force as the means of separation. The slurry, consisting of clean soil and contaminated particles, is separated

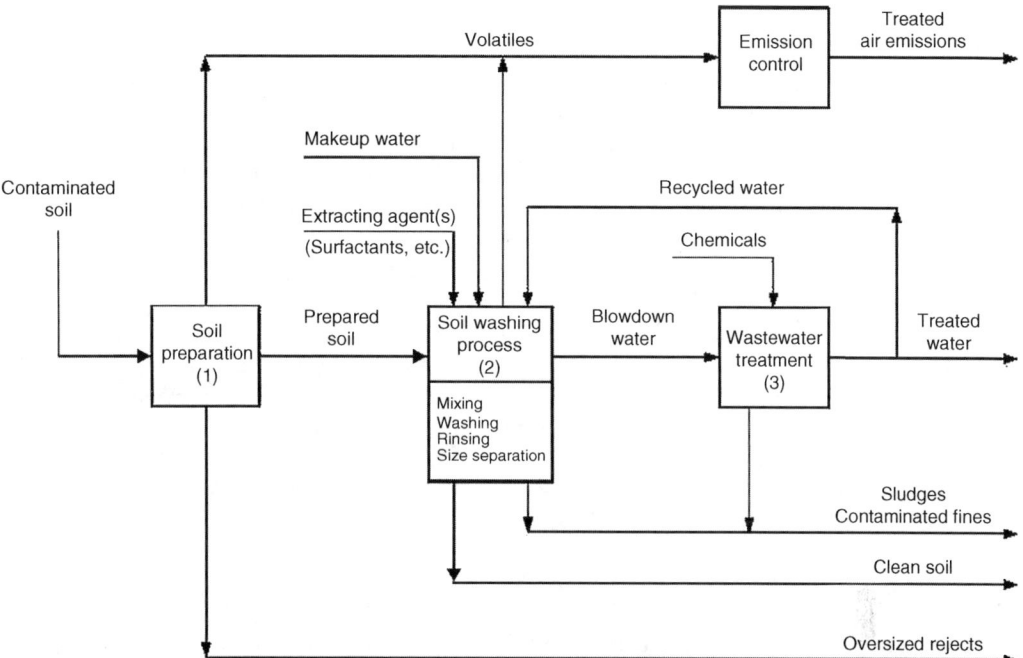

FIGURE 2.12 Typical soil-wash flow diagram for the treatment of contaminated soils.

into an underflow stream (where coarse particles are collected) and an overflow stream (containing the fine particles). To make operation more effective, multiple hydrocyclones may be placed in series. Selective separation is possible for particles with grain sizes greater than 10 to 20 μm.

2. *Fluidized bed separation system.* This separation system is based on the difference in gravimetric settling velocity of fine contaminated particles and coarse clean soil particles. The pulp containing the contaminated and clean soil particles is fed from the upper part of a vertical column countercurrent to the leaching solution. The rate of solution injection inside the column is adjusted so that sinking of the coarse soil particles is possible. The wash fluid containing the contaminated particles is removed from the upper part of the column. With this system, a selective separation of particles with grain size greater than 50 μm is achieved.

3. *Gravimetric separation systems.* These include jigs, shaking tables, Humphrey-type spiral concentrators, and so on.

4. *Flotation.* In many cases, contaminants adsorbed on the surface of clay particles, or contaminants occurring in soil as discriminate particles, have different surface properties to clean soil particles. By adding special chemical substances, the formation of a hydrophobic surface on the contaminated particles is possible. Pulp aeration results in the attachment of hydrophobic contaminated particles to the surface of the small bubbles that are formed. In this way, selective flotation of these particles is achieved. Contrary to the gravimetric separation methods, flotation offers the possibility to separate contaminated and noncontaminated particles of the same grain size and density but with different surface properties.

After the separation stage, the coarse soil fraction is rinsed with clean water to remove residual contaminants and any fine soil particles that may adhere to the coarse particles. Soil washing is not usually a stand-alone technology. Typically, both the fine soil fraction (silts and clays) recovered after washing and the spent wash water are subject to further specific treatment and disposal techniques, as appropriate, to complete cleanup. For the stripping of the washing solution, which contains dissolved contaminants and fine contaminated soil particles, a large number of physical and chemical systems are available. Dissolved contaminants can be removed by applying chemical methods such as neutralization, precipitation, ion exchange, and so on, whereas suspended particles originating either from the contaminated soil or produced during solution treatment can be removed by applying physical techniques such as flocculation, thickening, and filtering. Part or all of the purified washing solution is recycled back to the soil-washing stage.

The sludge produced from the wash water stripping stage can be dewatered with a centrifuge, filter press, or sieve belt press. The amount of sludge is a determinant factor for the cost-effectiveness of the soil-washing technique. This sludge mainly consists of clay soil particles. Although the fraction of contaminated compounds is relatively small, total contaminants concentration is rather high, so the sludge is usually characterized as a hazardous waste. Its final management may involve either disposal in a hazardous waste landfill or further treatment using thermal as well as stabilization techniques.

The soil-washing method already described may generate sidestreams, such as air emissions, spent solvents, and exhausted resins, which must also be properly managed.

2.5.2 Design Considerations

The main parameters that affect the cost-effectiveness of soil washing include the physicochemical parameters of the soil (grain size distribution, cation exchange capacity, percentage of silt, clay, or organic matter), and the type and concentration of contaminants.

Soils with relatively high percentages of sand and gravel respond better to soil washing than fine-grained soils. High percentages of clay and silt (i.e., fine particles with size <0.25 mm) reduce the efficiency of contaminant removal. Practically, soil washing is most appropriate for soils that contain at least 50% sand or gravel, i.e., coastal sandy soils and soils with glacial deposits. Soils rich in clay and silt tend to be poor candidates for soil washing. Modifications of soil washing with a view to being applied to predominantly silt and clay soils have been investigated at the laboratory scale, but it is not known whether they have yet been applied on an industrial scale. Figure 2.13 presents different difficulty levels in the application of soil-washing techniques for different particle size distributions of contaminated soil according to the evaluation of U.S. EPA.[77]

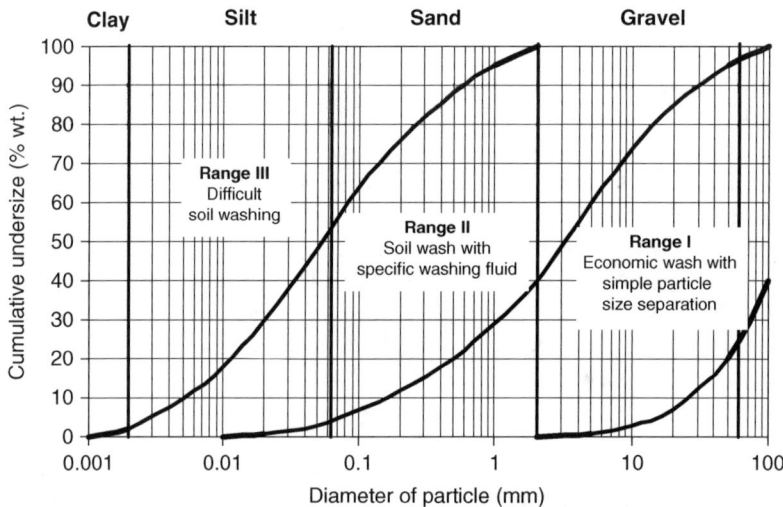

FIGURE 2.13   Ranges of soil-washing difficulty, based on the particle size distribution in the soil.

Cation exchange capacity (CEC) measures the tendency of the soil to exchange weakly held cations in the soil particles for cations in the wash solution. Soils with relatively low CEC values (<50 to 100 meq/kg) respond better to soil washing than soils with higher CEC values. Furthermore, high humic content in the soil makes separation of contaminants very difficult. Humus consists of decomposed plant and animal residues and offers binding sites for the accumulation of both organics and metals. Early characterization of these parameters and their variability throughout the site provides valuable information for the initial screening of soil washing as an alternative treatment technology.

Chemical and physical properties of the contaminant should also be investigated. Solubility in water (or other washing fluids) is one of the most important physical characteristics. Hydrophobic contaminants can be difficult to separate from the soil particles and into the aqueous washing fluid. Reactivity with wash fluids may, in some cases, be another important characteristic to consider. Other contaminant characteristics such as volatility and density may be important for the design of remedy screening studies and related residuals treatment systems. Speciation is important in metal-contaminated sites.

Complex mixtures of contaminants in the soil, such as a mixture of metals, nonvolatile organics, semivolatile organics, and so on, make it difficult to formulate a single suitable washing fluid that will remove all the different types of contaminants from the soil. Sequential washing steps, using different additives, may be needed. In fact, each type of contaminated soil requires a special treatment procedure, which is determined through laboratory or pre-industrial tests, so that system modifications and optimum operative conditions are specified.

Frequent changes in contaminant type and concentration in the feed soil can disrupt the efficiency of the soil-washing process. To accommodate changes in the chemical or physical composition of the feed soil, modifications to the wash fluid formulation and the operating settings may be required. Alternatively, additional feedstock preparation steps, such as blending soils to provide a consistent feedstock may be appropriate.[77,83]

Additives such as surfactants may be required to improve removal efficiencies. However, larger volumes of washing fluid may be needed when additives are used. Chelating agents, surfactants, solvents, and other additives are often difficult and expensive to recover from the spent washing fluid and recycle in the soil-washing process. The presence of additives may make the spent washing fluid difficult to treat by conventional treatment processes such as settling, chemical precipitation, or activated carbon. Furthermore, the presence of additives in the contaminated soil and treatment sludge residuals may increase difficulty in disposing of these residuals.

2.6 *IN SITU* SOIL FLUSHING

2.6.1 GENERAL DESCRIPTION

In situ flushing is the injection or infiltration of an aqueous solution into a contaminated zone, followed by downgradient extraction of groundwater and elutriate (flushing solution mixed with the contaminants) and aboveground treatment and discharge or reinjection. A schematic representation of *in situ* soil flushing is given in Figure 2.14. The goal of *in situ* flushing is to enhance conventional pump and treat methods of remediation by enhancing the solubility or mobility of contaminants, thus accelerating the remediation process.

Introduction of the flushing solution may occur within the vadose zone, the saturated zone, or both. Flushing solutions may consist of plain water, or surfactants, co-solvents, acids, bases, oxidants, chelants, and solvents. The infiltrating flushing solution percolates through the soil and soluble compounds present in the soil are dissolved. The elutriate is pumped from the bottom of the contaminated zone into a water treatment system to remove pollutants. The process is carried out until the residual concentrations of contaminants in the soil satisfy given limits.

Any variety of configurations of injection wells, horizontal wells, trenches, infiltration galleries, aboveground sprayers or leach fields, and extraction wells, open ditches, or subsurface collection

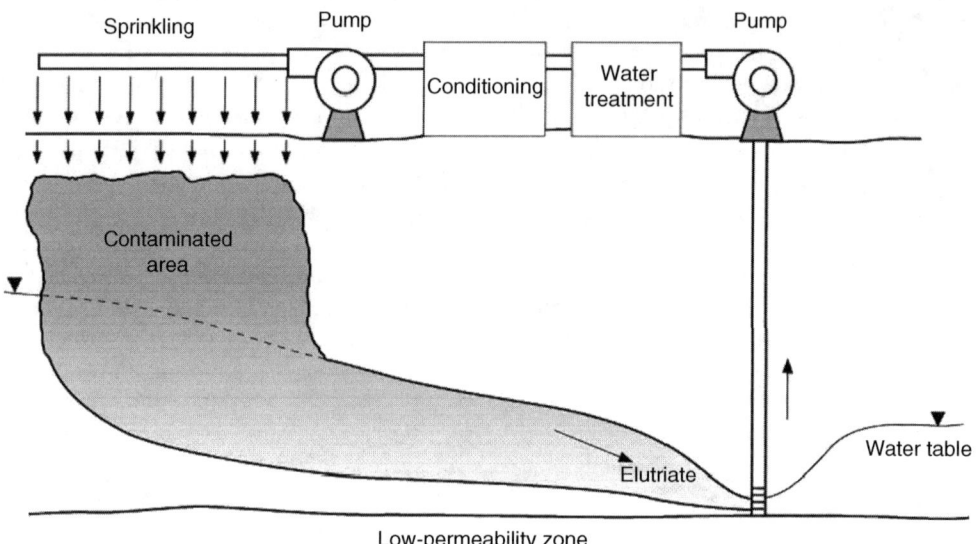

FIGURE 2.14 Typical soil-flushing system.

drains or trenches can be used to put the flushing solution in contact with the contaminated zone and collect elutriate.[76,79–81]

The *in situ* flashing technique is applicable to a variety of organic and inorganic contaminants. Organics, such as nonaqueous phase liquid (NAPL), volatile organic compounds (VOCs), semivolatile organic compounds (SVOCs), polychlorinated biphenyls (PCBs), halogenated pesticides, dioxins/furans, and corrosives, and inorganics including volatile and nonvolatile metals, cyanides, and radioactive contaminants may potentially be removed using *in situ* flushing. Removal efficiencies of contaminants depend on the contaminant as well as the soil type. Halogenated volatiles, nonhalogenated semivolatiles and nonvolatile metals are amongst the classes of chemical compounds treated successfully by *in situ* flashing.[80]

2.6.2 Design Considerations

There are a number of critical parameters to be considered for the application of *in situ* soil flushing, including the hydrogeological conditions of the contaminated site, the type and properties of contaminants, the properties of the additives, and the treatability of the flushing solution.[76,80]

2.6.2.1 Site Conditions

Regarding the hydraulic properties of contaminated soil, high permeability is a prerequisite for the application of this method. Hydraulic conductivity values should ideally be greater than 1.0×10^{-3} cm/sec to allow flushing solutions to pass through the geologic matrix in a reasonable period of time. This means that contaminated sandy soils are susceptible to *in situ* flushing. Less permeable materials, with hydraulic conductivity values ranging from 1×10^{-5} to 1×10^{-3} cm/sec may also be considered for *in situ* flushing. The presence of impermeable clay layers does not exclude the application of *in situ* soil flushing providing that special techniques for the effective injection of the flashing agent through the impermeable layer are applied. It is also important to ensure the percolation of the flushing agent through the entire area of contamination. Therefore, as well as vertical variability, lateral variability of permeability should also be taken into account. Hydraulic conductivity should be measured at several locations within the potential treatment zone, and the total number of measurements should be consistent with the size of the potential treatment area and the potential for heterogeneity.

Containment of the flushed contaminants and spent flushing solutions is essential to successful application of *in situ* flushing. This happens when the treatment zone is bounded geologically by materials with relative low hydraulic conductivity. Depth to the contaminated zone is a limiting factor because of the higher injection and extraction costs that are required compared with more shallow contaminated zones. Contaminants can be easily removed when the flushing solution follows the same channels as the pollutant. Also, possible mechanical disturbance of the surface layer of the contaminated area may render the contaminants inaccessible.

2.6.2.2   Flushing Solution

Flushing with plain water is effective only in cases where contaminants are soluble inorganic salts or hydrophilic organics, i.e., those presenting log K_{ow} values <1, such as lower molecular weight alcohols, phenols, and carboxylic acids. For medium- or low-solubility contaminants, additives should be introduced into the flushing solution to enhance mobilization of the contaminants. Surfactants such as detergents or emulsifiers may be used for low-solubility (hydrophobic) organics, such as chlorinated pesticides, PCBs, SVOCs (chlorinated benzenes and PAHs), petroleum products, aromatic solvents (BTEX), and chlorinated solvents (such as TCE). Co-solvents consisting of reagents such as alcohol may assist in enhancing the solubility of hydrophobic contaminants.[80]

The application of the *in situ* soil flushing technique for the removal of heavy metals is possible only when their solubility in the leaching solution is relatively high. It is possible to increase heavy metals extraction rates by adding appropriate chemical reagents that favor desorption or dissolution of these metals. The following reagents have been evaluated at the laboratory scale as additives to flushing solutions: inorganic and organic acids, sodium hydroxide, complexing reagents such as EDTA and NTA, chloride solutions, as well as oxidative or reductive media.[70–72,82]

2.6.2.3   Injection and Extraction Systems

The selection of a soil flushing solution delivery technique depends mainly on the hydraulic properties of the treatment zone, which can generally be classified as gravity-driven and pressure-driven. In gravity-driven delivery systems, sprinklers, trenches, or infiltration galleries are used, and the flushing solution infiltrates into the treatment zone as a result of natural hydraulic gradients. In pressure-driven delivery systems, the flushing solution is injected into the treatment zone through vertical or horizontal injection wells.

NOMENCLATURE

A_{contam} Surface area corresponding to the contaminated zone (m^2)
C_a Vapor concentration in pore air (mg/L)
$C_{a,equil}$ Concentration calculated from the vapor pressure data (mg/L)
C_{or} Concentration of the contaminant in the organic phase (mg/L)
C_s Adsorbed concentration of contaminant in the soil particle (mg/kg)
C_t Total quantity of contaminant per unit soil volume (mg/L)
C_w Dissolved concentration in pore water (mg/L)
C_w^o Water solubility (mg/L)
f_{oc} Fraction of organic carbon in soil
g Gravity acceleration constant (981 cm/sec^2)
H Thickness of the vadose zone (m)
K Intrinsic permeability of soil (cm^2)
k_a Air permeability (cm/sec)

k	First-order rate constant for contaminant uptake (yr^{-1})
K_H	Henry's constant (dimensionless)
K_{oc}	Organic carbon partitioning coefficient (L/kg)
k_w	Hydraulic conductivity (cm/sec)
m	Mass fraction of contaminant in NAPL mixture
M	Mass of contaminant remaining (kg)
M_o	Initial mass of contaminant (kg)
M_{final}	Residual acceptable quantity according to the cleanup objectives (t)
M_{rem}	Amount of contaminant that must be removed from the soil (t)
M_{spill}	Initial total amount of spill (t)
MW	Molecular weight (g/mol)
N_{wells}	Number of extraction wells
P_{atm}	Ambient atmospheric pressure (atm)
P^o	Vapor pressure (mmHg, atm)
P_I	Pressure in the subsurface at a distance corresponding to the radius of influence R_I of the SVE well (atm)
P_r	Pressure in the subsurface at a radial distance r from the SVE well (atm)
P_w	Pressure in the SVE well (atm)
Q_w	Volumetric flow rate (m^3/min)
Q_w^*	Volumetric flow rate in standard conditions (m^3/min)
r	Radial distance from SVE well (m)
R	Ideal Gas Law constant (L·atm/mol·K)
R_I	Radius of influence of SVE well (m)
R_{rem}	Contaminants removal rate (g/min)
R_w	Radius of SVE well
T	Absolute temperature (K)
T_{clean}	Required cleanup time (min)
TSCF	Transpiration stream concentration factor (dimensionless)
u	Air velocity (m/sec)
u_r	Air velocity in the radial r direction (m/sec)
u_w	Air velocity at the wellbore (m/sec)
U	Uptake rate of contaminant (mg/d)
X_i	Mol fraction of constituent i in the NAPL mixture

GREEK

γ_i	Activity coefficient of contaminant i in the NAPL mixture
η	Removal effectiveness factor
θ_a	Pore volume occupied by the gas phase (L/L)
θ_{or}	Pore volume occupied by NAPL (L/L)
θ_t	Total porosity of the soil (L/L)
θ_w	Pore volume occupied by water (L/L)
λ	Molar heat of vaporization (L·atm/mol)
μ	Viscosity (g/cm·sec)
μ_a	Air viscosity (g/cm·sec)
μ_w	Water viscosity (g/cm·sec)
ρ	Density (kg/L)
ρ_a	Air density (kg/L)
ρ_b	Soil bulk density (kg/L)
ρ_{or}	Density of the NAPL mixture (kg/L)
ρ_w	Water density (kg/L)

REFERENCES

1. NATO/CCMS, Evaluation of Demonstrated and Emerging Technologies for the Treatment and Clean Up of Contaminated Land and Groundwater, NATO CCMS Pilot Study, Phase III, 1999 Annual Report, EPA 542/R-99/007, no. 235, 1999.
2. Ferguson, C.C., Assessing risks from contaminated sites: Policy and practice in 16 European countries, *Land Contam. Reclamat.*, 7, 33–54, 1999.
3. Ferguson, C. and Kasamas, H., Eds., *Risk Assessment for Contaminated Sites in Europe, Vol. 2. Policy Frameworks*, LGM Press, Nottingham, UK, 1999.
4. Steffens, K., Integration of Technologies, in Final Report NATO CCMS Pilot Study, Phase II, Evaluation of Demonstrated and Emerging Technologies for the Treatment and Clean Up of Contaminated Land and Groundwater, EPA 542–R-98-001a, no. 219, 1998, pp. 10.1–10.19.
5. Vic, E.A., Bardos, P., Borgan, J., Gondi, F., Henrysson, T., Jensen, B.K., Jorge, C., Mariotti, C., Nathanail, P. and Papassiopi, N., Towards a framework for selecting remediation technologies for contaminated sites, *Land Contam. Reclamat.*, 9, 119–127, 2001.
6. Vic, E.A. and Bardos, P., Remediation of Contaminated Land. Technology Implementation in Europe, Federal Environmental Agency, Austria. CLARINET Report, available at www.clarinet.at, 2002.
7. Martin, I. and Bardos, P., A Review of Full Scale Treatment Technologies for the remediation of Contaminated Soil, Report for the Royal Commission on Environmental Pollution, EPP Publications, UK, 1996.
8. Schnoor, J.L., Phytoremediation. Technology Evaluation Report TE-98-01, Ground-Water Remediation Technologies Analysis Center, Pittsburgh, PA, 1997.
9. Suthersan S.S., *Remediation Engineering: Design Concepts*, CRC Press, Boca Raton, FL, 1997.
10. LaGrega, M.D., Buckingham, P.L. and Evans, J.C., *Hazardous Waste Management*, McGraw-Hill, New York, 1994.
11. Grasso, D., *Hazardous Waste Site Remediation: Source Control*, Lewis Publishers, Boca Raton, FL, 1993.
12. Kuo, J., *Practical Design Calculations for Groundwater and Soil Remediation*. Lewis Publishers, Boca Raton, FL, 1999.
13. Wilson, D.J., Clarke, A.N. and Clarke, J.H., Soil cleanup by in situ aeration, *Sep. Sci. Technol.*, 23, 991–1037, 1988.
14. Kirtland, B.C. and Aelion, C.M., Petroleum mass removal from low permeability sediment using air sparging/soil vapor extraction: impact of continuous or pulsed operation, *J. Contam. Hydrol.*, 41, 367–383, 2000.
15. Washington, J.B. and Ong, S.K., Air sparging effectiveness: laboratory characterization of air-channel mass transfer zone for VOC volatilization, *J. Hazard. Mater.*, 87, 241–258, 2001.
16. Johnston, C.D., Rayner, J.L. and Briegel, D., Effectiveness of in situ air sparging for removing NAPL gasoline from a sandy aquifer near Perth, Western Australia, *J. Contam. Hydrol.*, 59, 87–111, 2002.
17. Johnston, C.D., Rayner, J.L., Patterson, B.M. and Davis, G.B., Volatilisation and biodegradation during air sparging of dissolved BTEX-contaminated groundwater, *J. Contam. Hydrol.*, 33, 377–404, 1998.
18. Mindevalli, O. and Pedram, E.O., Soil bioventing and vapor extraction (in-situ and ex-situ) and groundwater treatment and reinjection at a diesel and gasoline contaminated site, *Proc. HAZMACON 94*, Bursztynsky, T., Ed., 1994, pp. 386–398.
19. Malina, G., Grotenhuis, J.T.C. and Rulkens, W.H., Vapor extraction/bioventing sequential treatment of soil contaminated with volatile and semivolatile hydrocarbon mixtures, *Bioremed. J.*, 6, 159–176, 2002.
20. Poppendieck, D.G., Loehr, R.C. and Webster, M.T., Predicting hydrocarbon removal from thermally enhanced soil vapor extraction systems. 2. Field study, *J. Hazard. Mater.*, 69, 95–109, 1999.
21. Lowe, D.F., Oubre, C.L. and Ward, C.H., Eds., Soil vapor extraction using radio frequency heating, in *Resource Manual and Technology Demonstration*, Lewis Publishers, Boca Raton, FL, 1999.
22. Falta, R.W., Pruess, K. and Chestnut, D.A., Modeling advective contaminant transport during soil vapor extraction, *Ground Water*, 31, 1011–1020, 1993.
23. Roy, W.R. and Griffin, R.A., An analytical model for in-situ extraction of organic vapors, *J. Hazard. Mater.*, 26, 301–317, 1991.
24. Lyman, W.J. and Hirsch. R.D., *Handbook of Chemical Property Estimation Methods*, McGraw-Hill, New York, 1982.
25. Lide, D.R., *CRC Handbook of Chemistry and Physics*, 87th ed., Taylor & Francis, CRC Press, Boca Raton, FL, 2006.
26. Verschueren, K., *Handbook of Environmental Data for Organic Chemicals*, 2nd ed., Van Nostrand Reinhold, New York, 1983.

27. Perry, R.H. and Green, D., *Chemical Engineers' Handbook*, 6th ed., McGraw-Hill, New York, 1984.

28. Wilson, D.J., Gomez-Lahoz, C. and Rodriquez-Maroto, J.M., Soil cleanup by in-situ aeration. XVI. Solution and diffusion in mass-transport-limited operation and calculation of Darcy's constants. *Sep. Sci. Technol.*, 29, 1133–1163, 1994.

29. Campagnolo, J.F. and Akgerman, A., Modeling of soil vapor extraction (SVE) systems—Part I, *Waste Management*, 15, 379–389, 1995.

30. Fischer, U., Hinz, C., Schulin, R. and Stauffer, F., Assessment of nonequilibrium in gas-water mass transfer during advective gas-phase transport in soils, *J. Contam. Hydrol.*, 33, 133–148, 1998.

31. Rathfelder, K.M., Lang, J.R. and Abriola, L.M., A numerical model (MISER) for the simulation of coupled physical, chemical and biological processes in soil vapor extraction and bioventing systems, *J. Contam. Hydrol.*, 43, 239–270, 2000.

32. King, R.B., Long, G.M. and Sheldon, J.K., *Practical Environmental Bioremediation: The Field Guide*, CRC Press, Boca Raton, FL, 1998.

33. U.S. National Academy of Science, *In Situ Bioremediation. When Does It Work?*, National Academy Press, Washington, 1993, available at http://www.nap.edu/openbook/03090489666/ html/R1.html

34. Wilson, J.T. and Wilson, J.M., Biotransformation of trichloroethylene in soil, *Appl. Environ. Microbiol.*, 49, 242–243, 1985.

35. Hollinger, C., The anaerobic microbiology and biotreatment of chlorinated ethens, *Curr. Biol.*, 6, 347–351, 1995.

36. Lovely, D.R. and Coates, J.D., Bioremediation of metal contamination, *Curr. Opin. Biotechnol.*, 8, 285–289, 1997.

37. Lovely, D.R., Phillips, E.J.P., Gorby, Y.A. and Landa, E.R., Biological reduction of uranium, *Nature*, 350, 413–416, 1991.

38. Stolz, J.F. and Oremland, R.S., Bacterial respiration of selenium and arsenic, *FEMS Rev.*, 23, 615–627, 1999.

39. Michel, C., Brugna, M., Aubert, C., Bernadac, A. and Bruschi, M., Enzymatic reduction of chromate: comparative studies using sulfate-reducing bacteria. Key role of polyheme cytochroms c and hydrogenases. *Appl. Microbiol. Biotechnol.*, 55, 95–100, 2001.

40. Howard, P.H., Boethling, R.S., Jarvis, W.F., Meylan, W.M. and Michalenko, E.M., *Handbook of Environmental Degradation Rates*, Lewis Publishers, Chelsea, MI, 1991.

41. Pivetz, B.E., Ground Water Issue. Phytoremediation of Contaminated Soil and Ground Water at Hazardous Waste Sites, EPA/540/S-01/500, Superfund Technology Support Center for Ground Water, OK, 2001.

42. Briggs, G.G., Bromilow, R.H. and Evans, A.A., Relationships between lipophilicity and root uptake and translocation of non-ionized chemicals by barley, *Pestic. Sci.*, 13, 495–504, 1982.

43. Baker, A.J.M., Brooks, R.R. and Reeves, R.D., Growing for gold, and copper and zinc, *New Science*, 177, 44–48, 1998.

44. Henry, J.R., An Overview of the Phytoremediation of Lead and Mercury, National Network of Environmental Management Studies (NNEMS), U.S. EPA, Technology Innovation Office, Washington, 2000.

45. Nanda Kumar, P.B.A., Dushenkov, V., Motto, H. and Raskin, I., Phytoextraction: The use of plants to remove heavy metals from soils, *Environ. Sci. Technol.*, 29, 1232–1238, 1995.

46. Interstate Technology and Regulatory Cooperation Work Group (ITRC), Phytotechnologies Work Team, Technical/Regulatory Guidelines. **Phytotechnology Technical and Regulatory Guidance Document**, 2001, available at http://www.itrcweb.org/Documents/PHYTO-2.pdf.

47. Shimp, J.F., Tracy, J.C., Davis, L.C., Lee, E., Huang, W., Erickson, L.E. and Schnoor, J.L., Beneficial effects of plants in the remediation of soil and groundwater contaminated with organic materials, *Crit. Rev. Environ. Sci. Technol.*, 23, 41–77, 1993.

48. Atlas, R.M. and Bartha, R., *Microbial Ecology: Fundamentals and Applications*, Benjamin/Cummings Publishing Company, Don Mills, ON, 1998.

49. Frick, C.M., Farrel, R.E. and Germida, J.J., *Assessment of Phytoremediation as an In-Situ Technique for Cleaning Oil-Contaminated Sites*, Petroleum Technology Alliance of Canada, Calgary, AB, 1999.

50. Donnelly, P.K. and Fletcher, J.S., Potential use of mycorrhizal fungi as bioremediation agents, in *Bioremediation Through Rhizosphere Technology*, Anderson, T.A. and Coats, J.R., Eds., American Chemical Society, Washington, 1994.

51. Burken, J.G. and Schnoor, J.L., Uptake and metabolism of atrazine by poplar trees, *Environ. Sci. Technol.*, 31, 1399–1406, 1997.

52. U.S. EPA, Recent Developments for In Situ Treatment of Metal Contaminated Soils, EPA-542-R-97-004, Technology Innovation Office, Washington, 1997.

53. Cunningham, S.D., Berti, W.R. and Huang, J.W., Phytoremediation of contaminated soils, *Trends Biotechnol.*, 13, 393–397, 1995.

54. Smith, R.A.H. and Bradshaw, A.D., The use of metal tolerant plant populations for the reclamation of metalliferrous wastes, *J. Appl. Ecol.*, 16, 596–612, 1979.

55. Salt, D.E., Blaylock, M., Nanda Kumar, P.B.A., Dushenkov, V., Ensley, B.D., Chet, I. and Raskin, I., Phytoremediation: A novel strategy for the removal of toxic metals from the environment using plants, *Biotechnology*, 13, 468–474, 1995.

56. Dushenkov, V., Nanda Kumar, P.B.A., Motto, H. and Raskin, I., Rhizofiltration: The use of plants to remove heavy metals from aqueous streams, *Environ. Sci. Technol.*, 29, 1239–1245, 1995.

57. U.S. EPA, Introduction to Phytoremediation, EPA/600/R-99/107, National Risk Management Research Laboratory, Office of Research and Development, Cincinnati, OH, 1999.

58. Gatliff, E.G., Vegetative remediation processes offers advantages over traditional pump-and-treat technologies, *Remediation*, 4, 343–352, 1994.

59. U.S. EPA, Brownfields Technology Primer: Selecting and Using Phytoremediation for Site Cleanup, EPA 542-R-01-006, Office of Solid Waste and Emergency Response, Technology Innovation Office, Washington, 2001.

60. Domenico, P.A. and Schwartz, F.W., *Physical and Chemical Hydrogeology*, John Wiley, New York, 1997.

61. Schwarzenbach, R.P., Gschwend, P.M. and Imboden, D.M., *Environmental Organic Chemistry*, John Wiley & Sons, New York, 1993.

62. Burken, J.G. and Schnoor, J.L., Predictive relationships for uptake of organic compounds following uptake by hybrid poplars, *Environ. Sci. Technol.*, 32, 3379–3385, 1998.

63. Schnoor, J.L., *Environmental Modeling—Fate and Transport of Pollutants in Water, Air, and Soil*, John Wiley & Sons, New York, 1996.

64. Marot, F., Ringeling, R.H.P. and Thomas, A.G., Application of the hydrocyclone in soil washing, *Recycling Derelict Land*, 3, 51–55, 1998.

65. Anderson, R., Rasor, E. and Van Ryn, F., Particle size separation via soil washing to obtain volume reduction, *J. Hazard. Mater.*, 66, 89–98, 1999.

66. Rikers, R.A., Rem, P. and Dalmijn, W.L., Improved method for prediction of heavy metal recoveries from soil using high intensity magnetic separation (HIMS), *Int. J. Miner. Process.*, 54, 165–182, 1998.

67. Gotlieb, I., Bozzelli, J.W. and Gotlieb, E., Soil and water decontamination by extraction with surfactants, *Sep. Sci. Technol.* 28, 793–804, 1993.

68. Deshpande, S., Shiau, B.J., Wade, D., Sabatini, D.A. and Harwell, J.H., Surfactant selection for enhancing ex-situ soil washing, *Water Res.*, 33, 351–360, 1999.

69. Ried, M., Heavy Metal Removal from Sewage Sludge: Practical Experiences with Acid Treatment, in Proceedings of the 3rd Gothenburg Symposium, Pretreatment in Chemical Water and Wastewater Treatment, Hann, H.H. and Klute, R., Eds., Springer-Verlag, Heidelberg, 1988, pp. 327–334.

70. Neale, C.N., Bricka, R.M. and Chao, A.C., Evaluating acids and chelating agents for removing heavy metals from contaminated soils, *Environ. Progr.*, 16, 274–280, 1997.

71. Peters, R.W., Chelant extraction of heavy metals from contaminated soils, *J. Hazard. Mater.*, 66, 151–210, 1999.

72. Pichtel, J. and Pichtel, T.M., Comparison of solvents for ex-situ removal of chromium and lead from contaminated soil, *Environ. Eng. Sci.*, 14, 97–104, 1997.

73. Hoogendoorn, D., Review of the Development of Remedial Action Techniques for Soil Contamination in the Netherlands, 5th National Conference on Management of Uncontrolled Hazardous Waste Sites, Washington, DC, 1984, pp. 569–575.

74. Rulkens, W.H. and Assink, J.W., Extraction as a Method for Cleaning Contaminated Soil: Possibilities, Problems and Research, 5th National Conference on Management of Uncontrolled Hazardous Waste Sites, Washington, DC, 1984, pp. 576–583.

75. Raghavan, A., Coles, E. and Dietz, D., Cleaning excavated soil using extraction agents: a state-of-the art review, *J. Hazard. Mater.*, 26, 81–87, 1991.

76. Rulkens, W.H., Grotenhuis, J.T.C. and Tichy, R., Methods for cleaning contaminated soils and sediments, in *Heavy Metals, Problems and Solutions*, Salomons, W., Forstner, U. and Mader, P., Eds., Springer, New York, 1998, pp. 165–191.

77. U.S. EPA, Soil Washing Treatment, EPA/540/2-90/017, U.S. EPA Office of Solid Waste and Emergency Response, Washington, DC, 1990.

78. Morizot, G.J.P., Mineral processing technology applied to the remediation of contaminated soils, in *Mineral Processing and the Environment*, Gallios, G.P. and Matis, K.A., Eds., Kluwer Academic Publications, the Netherlands, 1998, pp. 313–334.

79. U.S. EPA, Recent developments for in-situ treatment of metal contaminated soils, U.S. EPA, Office of Solid Waste and Emergency Response, Technology Innovation Office, EPA Contract No. 68-W5-006, 1997, pp. 1–47.

80. Roote, D.S., In-Situ Flushing, Groundwater Remediation Technologies Analysis Center (GWRTAC), Technology Overview Report, O Series: TO-97-02, 1997.

81. Yin, Y. and Allen, H.E., In-Situ Chemical Treatment, Groundwater Remediation Technologies Analysis Center (GWRTAC), Technology Evaluation Report, E Series: TE-99-01, 1999.

82. Garcia-Delgado, R.A. Rodriguez-Maroto, J.M. Gomez-Lahoz, C., Vereda-Alonso, C. and Garcia-Herruzo, F. Soil flushing with EDTA solutions: a model channelled flow, *Sep. Sci. Technol.*, 33, 867–886, 1998.

83. De-qing, S., Jian, Z., Zhao-long, G., Jian, D., Tian-li, W., Murygina, V. and Kalyuzhnyi, S., Bioremediation of oil sludge in Shengli Oilfield, *Water, Air, Soil Pollut.*, 185, 177–184, 2007.

84. Kulkarni, M. and Chaudhari, A., Microbial remediation of nitro-aromatic compounds: An overview, *J. Environ. Manage.*, 85, 496–512, 2007.

3 Remediation of Soils Contaminated with Metals

Nazih K. Shammas

CONTENTS

3.1 INTRODUCTION

Metals account for much of the contamination found at hazardous waste sites. They are present in the soil and groundwater at approximately 65% of the Superfund or CERCLA (Comprehensive Environmental Response, Compensation, and Liability Act)[1] sites for which the U.S. Environmental Protection Agency (U.S. EPA) has signed records of decisions (RODs).[2] The metals most frequently identified are lead, arsenic, chromium, cadmium, nickel, and zinc. Other metals often identified as contaminants include copper and mercury. In addition to the Superfund program, metals make up a significant portion of the contamination requiring remediation under the Resource Conservation and Recovery Act (RCRA)[3] and contamination present at federal facilities, notably those that are the responsibility of the Department of Defense (DOD) and the Department of Energy (DOE).

This chapter provides remedial project managers, engineers, on-scene coordinators, contractors, and other state or private remediation managers and their technical support personnel with information to facilitate the selection of appropriate remedial alternatives for soil contaminated with arsenic (As), cadmium (Cd), chromium (Cr), mercury (Hg), and lead (Pb).[4–6]

Common compounds, transport, and fate are discussed for each of these five elements. A general description of metal-contaminated Superfund soils is provided. The technologies covered are containment (immobilization), solidification–stabilization, vitrification, soil washing, soil flushing, pyrometallurgy, electrokinetics and phytoremediation. Use of treatment trains and remediation costs are also addressed.

It is assumed that users of this chapter will, as necessary, familiarize themselves with (1) the applicable or relevant and appropriate regulations pertinent to the site of interest, (2) applicable health and safety regulations and practices relevant to the metals and compounds discussed, and (3) relevant sampling, analysis, and data interpretation methods. Information on Pb battery (Pb, As), wood preserving (As, Cr), pesticide (Pb, As, Hg), and mining sites have been addressed in U.S. EPA Superfund documents.[7–12] The greatest emphasis is on remediation of inorganic forms of the metals of interest. Organometallic compounds, organic–metal mixtures, and multimetal mixtures are briefly addressed.

3.2 OVERVIEW OF METALS AND THEIR COMPOUNDS

This section provides a brief, qualitative overview of the physical characteristics and mineral origins of the five metals, and factors affecting their mobility. More comprehensive and quantitative reviews of the behavior of these five metals in soil can be found in readily available U.S. EPA Superfund documents.[4,13,14]

3.2.1 OVERVIEW OF PHYSICAL CHARACTERISTICS AND MINERAL ORIGINS

Arsenic is a semimetallic element or metalloid that has several allotropic forms. The most stable allotrope is a silver-gray, brittle, crystalline solid that tarnishes in air. Arsenic compounds, mainly As_2O_3, can be recovered as a byproduct of processing complex ores mined mainly for Cu, Pb, Zn, Au, and Ag. Arsenic occurs in a wide variety of mineral forms, including arsenopyrite ($FeAsS_4$), which is the main commercial ore of As worldwide.

Cadmium is a bluish-white, soft, ductile metal. Pure Cd compounds are rarely found in nature, although occurrences of greenockite (CdS) and otavite ($CdCO_3$) are known. The main sources of Cd are sulfide ores of lead, zinc, and copper. Cd is recovered as a byproduct when these ores are processed.

Chromium is a lustrous, silver-gray metal. It is one of the less common elements in the Earth's crust, and occurs only in compounds. The chief commercial source of Cr is the mineral chromite ($FeCr_2O_4$). Cr is mined as a primary product and is not recovered as a byproduct of any other mining operation. There are no chromite ore reserves, nor is there primary production of chromite in the U.S.

Mercury is a silvery, liquid metal. The primary source of Hg is cinnabar (HgS), a sulfide ore. In a few cases, Hg occurs as the principal ore product, but it is more commonly obtained as the byproduct of processing complex ores that contain mixed sulfides, oxides, and chloride minerals (these are usually associated with base and precious metals, particularly gold). Native or metallic Hg is found in very small quantities in some ore sites. The current demand for Hg is met by secondary production (i.e., recycling and recovery).

Lead is a bluish-white, silvery, or gray metal that is highly lustrous when freshly cut, but tarnishes when exposed to air. It is very soft and malleable, has a high density (11.35 g/cm^3) and low melting point ($327.4°C$), and can be cast, rolled, and extruded. The most important Pb ore is galena (PbS). Recovery of Pb from the ore typically involves grinding, flotation, roasting, and smelting. Less common forms of the mineral are cerussite ($PbCO_3$), anglesite ($PbSO_4$), and crocoite ($PbCrO_4$).

3.2.2 OVERVIEW OF BEHAVIOR OF AS, CD, CR, PB, AND HG

As metals cannot be destroyed, remediation of metal-contaminated soil consists primarily of manipulating (i.e., exploiting, increasing, decreasing, or maintaining) the mobility of metal contaminant(s) to produce a treated soil that has an acceptable total or leachable metal content. Metal mobility depends upon numerous factors. Metal mobility in soil-waste systems is determined by the following factors[13]:

1. The type and quantity of soil surfaces present
2. The concentration of the metal of interest
3. The concentration and type of competing ions and complexing ligands, both organic and inorganic
4. pH
5. Redox status

McLean and Bledsoe[13] state that

Generalization can only serve as rough guides of the expected behavior of metals in such systems. Use of literature or laboratory data that do not mimic the specific site soil and waste system will not be adequate to describe or predict the behavior of the metal. Data must be site specific. Long term effects must also be considered. As organic constituents of the waste matrix degrade, or as pH or redox conditions change, either through natural processes of weathering or human manipulation, the potential mobility of the metal will change as soil conditions change.

Cd, Cr(III), and Pb are present in cationic forms under natural environmental conditions.[13] These cationic metals are generally not mobile in the environment and tend to remain relatively close to the point of initial deposition. The capacity of soil to adsorb cationic metals increases with increasing pH, cation exchange capacity, and organic carbon content. Under the neutral to basic conditions typical of most soils, cationic metals are strongly adsorbed on the clay fraction of soils and can be adsorbed by the hydrous oxides of Fe, Al, or Mn present in soil minerals. Cationic metals will precipitate as hydroxides, carbonates, or phosphates. In acidic, sandy soils, the cationic metals are more mobile. Under conditions that are atypical of natural soils (e.g., pH < 5 or pH > 9; elevated concentrations of oxidizers or reducers, high concentrations of soluble organic or inorganic complexing or colloidal substances), but may be encountered as a result of waste disposal or remedial processes, the mobility of these metals may be substantially increased. Also, competitive adsorption between various metals has been observed in experiments involving a number of solids with oxide surfaces (γFeOOH, α-SiO$_2$, and γ-Al$_2$O$_3$). In several experiments, Cd adsorption was decreased by the addition of Pb or Cu for all three of these solids. The addition of Zn resulted in the greatest decrease of Cd adsorption. Competition for surface sites occurred when only a few percent of all surface sites were occupied.[15]

The behavior of As, Cr(VI), and Hg differs considerably from that of Cd, Cr(III), and Pb. Typically, As and Cr(VI) exist in anionic forms under environmental conditions. Hg, although it is a cationic metal, has unusual properties (e.g., liquid at room temperature, easily transforms among several possible valence states).

In most As-contaminated sites, As appears as As$_2$O$_3$ or as anionic As species leached from As$_2$O$_3$, oxidized to As(V), and then sorbed onto iron-bearing minerals in the soil. It may also be present in organometallic forms, such as methylarsenic acid (H$_2$AsO$_3$CH$_3$), and dimethylarsenic acid [(CH$_3$)$_2$AsO$_2$H], which are active ingredients in many pesticides, as well as the volatile compounds arsine (AsH$_3$) and its methyl derivatives [i.e., dimethylarsine HAs(CH$_3$)$_2$ and trimethylarsine, As(CH$_3$)$_3$]. These As forms illustrate the various oxidation states that As commonly exhibits (–III, 0, III, and V) and the resulting complexity of its chemistry in the environment.

As(V) is less mobile and less toxic than As(III). As(V) exhibits anionic behavior in the presence of water, and hence its aqueous solubility increases with increasing pH, and it does not complex or precipitate with other anions. As(V) can form low solubility metal arsenates. Calcium arsenate [Ca$_3$(AsO$_4$)$_2$] is the most stable metal arsenate in well-oxidized and alkaline environments, but it is unstable in acidic environments. Even under initially oxidizing and alkaline conditions, absorption of CO$_2$ from the air will result in the formation of CaCO$_3$ and the release of arsenate. In sodic soils, sufficient sodium is available such that the mobile compound Na$_3$AsO$_4$ can form. The slightly less stable manganese arsenate [Mn$_2$(AsO$_4$)$_2$] forms in both acidic and alkaline environments, and iron arsenate is stable under acidic soil conditions. In aerobic environments, HAsO$_4$ predominates at pH < 2 and is replaced by H$_2$AsO$_4^-$, HAsO$_4^{2-}$, and AsO$_4^{3-}$ as pH increases to about 2, 7, and 11.5, respectively. Under mildly reducing conditions, H$_3$AsO$_3$ is a predominant species at low pH, but is replaced by H$_2$AsO$_3^-$, HAsO$_3^{2-}$, and AsO$_3^{3-}$ as pH increases. Under still more reducing conditions and in the presence of sulfide, As$_2$S$_3$ can form. As$_2$S$_3$ is a low-solubility, stable solid. AsS$_2$ and AsS$_2^-$ are thermodynamically unstable with respect to As$_2$S$_3$ (ref. 16). Under extreme reducing conditions, elemental As and volatile arsine (AsH$_3$) can occur. Just as competition between cationic metals affects mobility in soil, competition between anionic species (chromate, arsenate, phosphate, sulfate, etc.) affects anionic fixation processes and may increase mobility.

The most common valence states of Cr in the Earth's surface and near-surface environment are +3 [trivalent or Cr(III)] and +6 [hexavalent or Cr (VI)]. The trivalent Cr (discussed above) is the most thermodynamically stable form under common environmental conditions. Except in leather tanning, industrial applications of Cr generally use the Cr(VI) form. Owing to kinetic limitations, Cr(VI) does not always readily reduce to Cr(III) and can remain present over an extended period of time.

Cr(VI) is present as the chromate (CrO_4^{2-}) or dichromate ($Cr_2O_7^{2-}$) anion, depending on pH and concentration. Cr(VI) anions are less likely to be adsorbed onto solid surfaces than Cr(III). Most solids in soils carry negative charges that inhibit Cr(VI) adsorption. Although clays have a high capacity to adsorb cationic metals, they interact little with Cr(VI) because of the similar charges carried by the anion and clay in the common pH range of soil and groundwater. The only common soil solid that adsorbs Cr(VI) is iron oxyhydroxide. Generally, a major portion of Cr(VI) and other anions adsorbed in soils can be attributed to the presence of iron oxyhydroxide. The quantity of Cr(VI) adsorbed onto the iron solids increases with decreasing pH.

At metal-contaminated sites, Hg can be present in mercuric form (Hg^{2+}), mercurous form (Hg_2^{2+}), elemental form (Hg), or alkylated form (e.g., methyl and ethyl Hg). Hg_2^{2+} and Hg^{2+} are more stable under oxidizing conditions. Under mildly reducing conditions, both organically bound Hg and inorganic Hg compounds can convert to elemental Hg, which can then be readily converted to methyl or ethyl Hg by biotic and abiotic processes. Methyl and ethyl Hg are mobile and toxic forms.

Hg is moderately mobile, regardless of the soil. Both the mercurous and mercuric cations are adsorbed by clay minerals, oxides, and organic matter. Adsorption of cationic forms of Hg increases with increasing pH. Mercurous and mercuric Hg are also immobilized by forming various precipitates. Mercurous Hg precipitates with chloride, phosphate, carbonate, and hydroxide. At concentrations of Hg commonly found in soil, only the phosphate precipitate is stable. In alkaline soils, mercuric Hg precipitates with carbonate and hydroxide to form a stable (but not exceptionally insoluble) solid phase. At lower pH and high chloride concentration, soluble $HgCl_2$ is formed. Mercuric Hg also forms complexes with soluble organic matter, chlorides, and hydroxides, which may contribute to its mobility.[13] In strong reducing conditions, HgS, a very low solubility compound, is formed.

3.3 DESCRIPTION OF SUPERFUND SOILS CONTAMINATED WITH METALS

Soils can become contaminated with metals from direct contact with industrial plant waste discharges, fugitive emissions, or leachate from waste piles, landfills, or sludge deposits. The specific type of metal contaminant expected at a particular Superfund site would obviously be directly related to the type of operation that had occurred there. Table 3.1 lists the types of operations that are directly associated with each of the five metal contaminants.[5]

Wastes at CERCLA sites are frequently heterogeneous on a macro- and micro-scale. Contaminant concentration and the physical and chemical forms of the contaminant and matrix are usually complex and variable. Waste disposal sites collect a wide variety of waste types; therefore, concentration profiles can vary by orders of magnitude through a pit or pile. Limited volumes of high-concentration "hot spots" may develop due to variations in the historical waste disposal patterns or local transport mechanisms. Similar radical variations frequently occur on the scale of particle size too. The waste often consists of a physical mixture of very different solids, for example, paint chips in spent abrasive.

Industrial processes may result in a variety of solid metal-bearing waste materials, including slags, fumes, mold sand, fly ash, abrasive wastes, spent catalysts, spent activated carbon, and refractory bricks.[17] These process solids may be found above ground as waste piles or below ground in landfills. Solid-phase wastes can be dispersed by well-intended but poorly controlled reuse projects. Waste piles can be exposed to natural disasters or accidents, causing further dispersion.

3.4 SOIL CLEANUP GOALS AND TECHNOLOGIES FOR REMEDIATION

Table 3.2 provides an overview of cleanup goals (actual and potential) for both total and leachable metals. Based on an inspection of the total metals cleanup goals, one can see that they vary considerably both within the same metal and between metals.

TABLE 3.1
Principal Sources of As, Cd, Cr, Hg, and Pb Contaminated Soils

Contaminant	Principal Sources
As	Wood preserving
	As-waste disposal
	Pesticide production and application
	Mining
Cd	Plating
	Ni–Cd battery manufacturing
	Cd-waste disposal
Cr	Plating
	Textile manufacturing
	Leather tanning
	Pigment manufacturing
	Wood preserving
	Cr-waste disposal
Hg	Chloralkali manufacturing
	Weapons production
	Copper and zinc smelting
	Gas line manometer spills
	Paint application
	Hg-waste disposal
Pb	Ferrous/nonferrous smelting
	Pb-acid battery breaking
	Ammunition production
	Leaded paint waste
	Pb-waste disposal
	Secondary metals production
	Waste oil recycling
	Firing ranges
	Ink manufacturing
	Mining
	Pb-acid battery manufacturing
	Leaded glass production
	Tetraethyl Pb production
	Chemical manufacturing

Source: U.S. EPA, Technology Alternatives for the Remediation of Soils Contaminated with AS, Cd, Cr, Hg, and Pb, EPA/540/S-97/500, U.S. Environmental Protection Agency, Cincinnati, OH, August 1997.

Similar variation is observed in the actual or potential leachate goals. The observed variation in cleanup goals has at least two implications with regard to technology alternative evaluation and selection. First, the importance of identifying the target metal(s), contaminant state (leachable vs. total metal), the specific type of test and conditions, and the numerical cleanup goals early in the remedy evaluation process is made apparent. Depending on which cleanup goal is selected, the required removal or leachate reduction efficiency of the overall remediation can vary by several orders of magnitude.[5,18] Second, the degree of variation in goals both within and between the metals, plus the many factors that affect the mobility of the metals, suggest that generalizations about effectiveness of a technology for meeting total or leachable treatment goals should be viewed with some caution.

TABLE 3.2

Cleanup Goals (Actual and Potential) for Total and Leachable Metals

Description	As	Cd	Cr (Total)	Hg	Pb
Total metals goals (mg/kg)					
Background (mean)	5	0.06	100	0.03	10
Background (range)	1–50	0.01–0.70	1–1000	0.01–0.30	2–200
Superfund site goals from TRD	5–65	3–20	6.7–375	1–21	200–500
Theoretical minimum total metals to ensure TCLP Leachate < threshold (i.e., TCLP × 20)	100	20	100	4	100
California total threshold limit concentration	500	100	500	20	1000
Leachable metals (µg/L)					
TCLP threshold for RCRA waste	5000	1000	5000	200	5000
Extraction procedure toxicity test	5000	1000	5000	200	5000
Synthetic precipitate leachate	—[b]	—	—	—	—
Multiple extraction procedure	—	—	—	—	—
California soluble threshold leachate concentration	5000	1000	5000	200	5000
Maximum contaminant level[a]	50	5	100	2	15
Superfund site goals from TRD	50	—	50	0.05–2	50

TRD, Technical Report Data.

[a]Maximum contaminant level = the maximum permissible level of contaminant in water delivered to any user of a public system.

[b]— indicates no specified level and no example cases identified.

Source: U.S. EPA, Technology Alternatives for the Remediation of Soils Contaminated with As, Cd, Cr, Hg, and Pb, EPA/540/S-97/500, U.S. Environmental Protection Agency, Cincinnati, OH, August 1997.

Technologies potentially applicable for the remediation of soils contaminated with the five metals or their inorganic compounds are listed in Table 3.3.[2,5]

The best demonstrated available technology (BDAT) status refers to the determination under the RCRA of the BDAT for various industry-generated hazardous wastes that contain the metals of interest. Whether the characteristics of a Superfund metal-contaminated soil (or fractions derived from it) are similar enough to the RCRA waste to justify serious evaluation of the BDAT for a specific Superfund soil must be made on a site-specific basis. Other limitations relevant to BDATs include the following:

1. The regulatory basis for BDAT standards focus BDATs on proven, commercially available technologies at the time of the BDAT determination.
2. A BDAT may be identified, but that does not necessarily preclude the use of other technologies.
3. A technology identified as a BDAT may not necessarily be the current technology of choice in the RCRA hazardous waste treatment industry.

The U.S. EPA's Superfund Innovative Technology Evaluation (SITE) program evaluates many emerging and demonstrated technologies in order to promote the development and use of innovative technologies to clean up Superfund sites across the country. The major focus of SITE is the Demonstration Program, which is designed to provide engineering and cost data for selected technologies.

Cost is not discussed in each technology narrative here; however, a summary table is provided at the end of Section 3.14 that illustrates technology cost ranges and treatment train options.

TABLE 3.3
Technologies Potentially Applicable for the
Remediation of Contaminated Soils

Technology Class	Specific Technology
Containment	Caps
	Vertical barriers
	Horizontal barriers
Solidification–stabilization	Cement-based
	Polymer microencapsulation
	Vitrification
Separation–concentration	Soil washing
	Soil flushing
	Pyrometallurgy
	Electrokinetics
	Phytoremediation

Source: U.S. EPA, Technology Alternatives for the Remediation of Soils
Contaminated with As, Cd, Cr, Hg, and Pb, EPA/540/S-97/500,
U.S. Environmental Protection Agency, Cincinnati, OH, August 1997;
U.S. EPA, Recent Developments for In Situ Treatment of Metal
Contaminated Soils, Contract no. 68-W5-0055 U.S. Environmental
Protection Agency, Washington, DC, 1997.

3.5   CONTAINMENT

Containment technologies for application at Superfund sites include landfill covers (caps), vertical barriers, and horizontal barriers.[4] For metal remediation, containment is considered an established technology except for *in situ* installation of horizontal barriers.

3.5.1   PROCESS DESCRIPTION

Containment ranges from a surface cap (which limits infiltration of uncontaminated surface water) to subsurface vertical or horizontal barriers (which restrict lateral or vertical migration of contaminated groundwater). The material provided here is primarily from U.S. EPA references.[5,9]

3.5.1.1   Caps

Capping systems reduce surface water infiltration, control gas and odor emissions, improve aesthetics, and provide a stable surface over the waste. Caps can range from a simple native soil cover to a full RCRA Subtitle C composite cover.

Cap construction costs depend on the number of components in the final cap system (i.e., costs increase with the addition of barrier and drainage components). Additionally, cost escalates as a function of topographic relief. Side slopes steeper than 3 horizontal to 1 vertical can cause stability and equipment problems that dramatically increase the unit cost.[4,19]

3.5.1.2   Vertical Barriers

Vertical barriers minimize the movement of contaminated groundwater off site or limit the flow of uncontaminated groundwater on site. Common vertical barriers include slurry walls in excavated trenches, grout curtains formed by injecting grout into soil borings, vertically injected, cement–bentonite grout-filled borings or holes formed by withdrawing beams driven into the ground, and sheet-pile walls formed of driven steel.

Certain compounds can affect cement–bentonite barriers. The impermeability of bentonite may significantly decrease when it is exposed to high concentrations of creosote, water-soluble salts (Cu, Cr, As), or fire-retardant salts (borates, phosphates, and ammonia). The specific gravity of salt solutions must be greater than 1.2 to impact bentonite.[20,21] In general, soil–bentonite blends resist chemical attack best if they contain only 1% bentonite and between 30 and 40% natural soil fines. Treatability tests should evaluate the chemical stability of the barrier if adverse conditions are suspected.

Carbon steel used in pile walls quickly corrodes in dilute acids, slowly corrodes in brines or salt water, and remains mostly unaffected by organic chemicals or water. Salts and fire retardants can reduce the service life of a steel sheet pile; corrosion-resistant coatings can extend their anticipated life. Major steel suppliers will provide site-specific recommendations for cathodic protection of piling.

Construction costs for vertical barriers are influenced by the soil profile of the barrier material used and by the method of placing it. The most economical shallow vertical barriers are soil–bentonite trenches excavated with conventional backhoes; the most economical deep vertical barriers consist of a cement–bentonite wall placed by a vibrating beam.

3.5.1.3   Horizontal Barriers

In situ horizontal barriers can underlie a sector of contaminated materials on site without removing the hazardous waste or soil. Established technologies use grouting techniques to reduce the permeability of underlying soil layers. Studies performed by the U.S. Army Corps of Engineers[22] indicate that conventional grout technology cannot produce an impermeable horizontal barrier because it cannot ensure uniform lateral growth of the grout. These same studies found greater success with jet grouting techniques in soils that contain fines sufficient to prevent collapse of the wash hole and that present no large stones or boulders that could deflect the cutting jet.

Few *in situ* horizontal barriers have been constructed, so accurate costs have not yet been established. Work performed by the Corps of Engineers for U.S. EPA has shown that it is very difficult to form effective horizontal barriers. The most efficient barrier installation used a jet wash to create a cavity in sandy soils into which cement–bentonite grouting was injected. The costs relate to the number of borings required and each boring takes at least one day to drill.

3.5.2   Site Requirements

In general, the site must be suitable for a variety of heavy construction equipment including bulldozers, graders, backhoes, multishaft drill rigs, various rollers, vibratory compactors, forklifts, and seaming devices.[23,24] When capping systems are being utilized, on-site storage areas are necessary for the materials to be used in the cover. If site soils are adequate for use in the cover, a borrow area needs to be identified and the soil tested and characterized. If site soils are not suitable, it may be necessary to truck in other low-permeability soils.[23] In addition, an adequate supply of water may also be needed in order to achieve the optimum soil density.

The construction of vertical containment barriers, such as slurry walls, requires knowledge of the site, the local soil and hydrogeologic conditions, and the presence of underground utilities.[25] Preparation of the slurry requires batch mixers, hydration ponds, pumps, hoses, and an adequate supply of water. Therefore, on-site water storage tanks and electricity are necessary. In addition, areas adjacent to the trench need to be available for the storage of trench spoils (which could potentially be contaminated) and the mixing of backfill. If excavated soils are not acceptable for use as backfill, suitable backfill must be trucked onto the site.[25]

3.5.3 APPLICABILITY

Containment is most likely to be applicable to the following[5]:

1. Wastes that are low-hazard (e.g., low toxicity or low concentration) or immobile
2. Wastes that have been treated to produce low-hazard or low-mobility wastes for on-site disposal
3. Wastes whose mobility must be reduced as a temporary measure to mitigate risk until a permanent remedy can be tested and implemented

Situations where containment would not be applicable include the following:

1. Wastes for which there is a more permanent and protective remedy that is cost-effective
2. Where effective placement of horizontal barriers below existing contamination is difficult
3. Where drinking water sources will be adversely affected if containment fails, and if there is inadequate confidence in the ability to predict, detect, or control harmful releases due to containment failure

Containment has the following important advantages[5]:

1. Surface caps and vertical barriers are relatively simple and rapid to implement at low cost, and can be more economical than the excavation and removal of waste.
2. Caps and vertical barriers can be applied to large areas or volumes of waste.
3. Engineering control (containment) is achieved, and may be a final action if metals are well immobilized and potential receptors are distant.
4. A variety of barrier materials are available commercially.
5. In some cases it may be possible to create a land surface that can support vegetation or be applicable for other purposes.

Containment also has the following disadvantages[5]:

1. Design life is uncertain.
2. Contamination remains on site, available to migrate should containment fail.
3. Long-term inspection, maintenance, and monitoring are required.
4. The site must be amenable to effective monitoring.
5. The placement of horizontal barriers below existing waste is difficult to implement successfully.

3.5.4 PERFORMANCE AND BDAT STATUS

Containment is widely accepted as a means of controlling the spread of contamination and preventing the future migration of waste constituents. Table 3.4 presents a list of selected sites where containment has been selected for remediating metal-contaminated solids.

The performance of capping systems, once installed, may be difficult to evaluate.[23] Monitoring well systems or infiltration monitoring systems can provide some information, but it is often not possible to determine whether the water or leachate originated as surface water or groundwater.

With regard to slurry walls and other vertical containment barriers, performance may be affected by a number of variables including geographic region, topography, and material availability. A thorough characterization of the site and a compatibility study are highly recommended.[25]

Containment technologies are not considered "treatment technologies" and hence no BDATs involving containment have been established.

TABLE 3.4

Containment Applications at Selected Superfund Sites with Metal Contamination

Site Name and State	Specific Technology	Key Metal Contaminants	Associated Technology
Ninth Avenue Dump, IN	Containment—slurry wall	Pb	Slurry wall/capping
Industrial Waste Control, AK	Containment—slurry wall	As, Cd, Cr, Pb	Capping/French drain
E.H. Shilling Landfill, OH	Containment—slurry wall	As	Capping/clay berm
Chemtronic, NC	Capping	Cr, Pb	Capping
Ordnance Works Disposal, WV	Capping	As, Pb	Capping
Industriplex, MA	Capping	As, Pb, Cr	Capping

Source: U.S. EPA, Technology Alternatives for the Remediation of Soils Contaminated with As, Cd, Cr, Hg, and Pb, EPA/540/S-97/500, U.S. Environmental Protection Agency, Cincinnati, OH, August 1997.

3.5.5 SITE PROGRAM DEMONSTRATION PROJECTS

Ongoing SITE demonstrations applicable to soils contaminated with the metals of interest include the following:

1. Morrison Knudsen Corporation (high clay grouting technology)
2. RKK, Ltd (frozen soil barriers)

3.6 SOLIDIFICATION–STABILIZATION TECHNOLOGIES

The term "solidification–stabilization" refers to a general category of processes that are used to treat a wide variety of wastes, including solids and liquids. Solidification and stabilization are each distinct technologies, as described below.[26]

Solidification refers to processes that encapsulate a waste to form a solid material and to restrict contaminant migration by decreasing the surface area exposed to leaching or by coating the waste with low-permeability materials. Solidification can be accomplished by a chemical reaction between a waste and binding (solidifying) reagents or by mechanical processes. Solidification of fine waste particles is referred to as microencapsulation, and solidification of a large block or container of waste is referred to as macroencapsulation.

Stabilization refers to processes that involve chemical reactions that reduce the leachability of a waste. Stabilization chemically immobilizes hazardous materials (such as heavy metals) or reduces their solubility through a chemical reaction. The physical nature of the waste may or may not be changed by this process.

Solidification–stabilization (S/S) aims to accomplish one or more of the following objectives[4]:

1. To improve the physical characteristics of the waste by producing a solid from liquid or semiliquid wastes
2. To reduce contaminant solubility by formation of sorbed species or insoluble precipitates (e.g., hydroxides, carbonates, silicates, phosphates, sulfates, or sulfides)
3. To decrease the exposed surface area across which mass transfer loss of contaminants may occur by the formation of a crystalline, glassy, or polymeric framework that surrounds the waste particles
4. To limit the contact between transport fluids and contaminants by reducing the material's permeability

S/S technology is usually applied by mixing contaminated soils or treatment residuals with a physical binding agent to form a crystalline, glassy, or polymeric framework surrounding the waste particles. In addition to microencapsulation, some chemical fixation mechanisms may improve the waste's leach resistance. Other forms of S/S treatment rely on macroencapsulation, where the waste is unaltered but macroscopic particles are encased in a relatively impermeable coating,[27] or on specific chemical fixation, where the contaminant is converted to a solid compound resistant to leaching. S/S treatment can be accomplished primarily through the use of either inorganic binders (e.g., cement, fly ash, or blast furnace slag) or by organic binders such as bitumen.[4] Additives may be used, for example, to convert the metal to a less mobile form or to counteract adverse effects of the contaminated soil on the S/S mixture (e.g., accelerated or retarded setting times, and low physical strength). The form of the final product from S/S treatment can range from a crumbly, soil-like mixture to a monolithic block. S/S is more commonly done as an *ex situ* process, but an *in situ* option is available. The full range of inorganic binders, organic binders, and additives is too broad, so the emphasis in this chapter is on *ex situ*, cement-based S/S, which is widely used, *in situ*, cement-based S/S, which has been applied to metals at full scale, and polymer microencapsulation, which appears applicable to certain wastes that are difficult to treat with cement-based S/S.

Additional information and references on solidification–stabilization of metals can be found in U.S. EPA documents.[4,28–30] Innovative S/S technologies (e.g., sorption and surfactant processes, bituminization, emulsified asphalt, modified sulfur cement, polyethylene extrusion, soluble silicate, slag, lime, and soluble phosphates) are addressed in U.S. EPA reports.[31–36]

3.6.1 PROCESS DESCRIPTION

3.6.1.1 *Ex Situ* Cement-Based S/S

Ex situ cement-based S/S is performed on contaminated soil that has been excavated and classified to reject oversize. Cement-based S/S involves mixing contaminated materials with an appropriate ratio of cement or similar binder/stabilizer, and possibly water and other additives. A system is also necessary for delivering the treated wastes to molds, surface trenches, or subsurface injection. Off-gas treatment (if volatiles or dust are present) may be necessary. The fundamental materials used to perform this technology are Portland-type cements and pozzolanic materials. Portland cements are typically composed of calcium silicates, aluminates, aluminoferrites, and sulfates. Pozzolans are very small spheroidal particles that are formed in the combustion of coal (fly ash) and in lime and cement kilns, for example. Pozzolans of high silica content are found to have cement-like properties when mixed with water. Cement-based S/S treatment may involve using only Portland cement, only pozzolanic materials, or blends of both. The composition of the cement and pozzolan, together with the amount of water, aggregate, and other additives, determines the set time, cure time, pour characteristics, and material properties (e.g., pore size, compressive strength) of the resulting treated waste. The composition of cements and pozzolans, including those commonly used in S/S applications, are classified according to American Society for Testing and Materials (ASTM) standards. S/S treatment usually results in an increase (>50% in some cases) in the treated waste volume. *Ex situ* treatment provides high throughput (100 to 200 m^3/d/mixer).

Cement-based S/S reduces the mobility of inorganic compounds by formation of insoluble hydroxides, carbonates, or silicates, substitution of the metal into a mineral structure, sorption, physical encapsulation, and perhaps other mechanisms. Cement-based S/S involves a complex series of reactions, and there are many potential interferences (e.g., coating of particles by organics, excessive acceleration or retardation of set times by various soluble metal and inorganic compounds; excessive heat of hydration; pH conditions that solubilize anionic species of metal compounds) that can prevent attainment of S/S treatment objectives for physical strength and leachability. Although there are many potential interferences, Portland cement is widely used and studied, and a knowledgeable vendor may be able to identify, and confirm through treatability studies, approaches to counteract adverse effects by use of appropriate additives or other changes in formulation.

3.6.1.2   *In Situ* Cement-Based S/S

In situ cement-based S/S has only two steps: mixing and off-gas treatment. The processing rate for *in situ* S/S is typically considerably lower than for *ex situ* processing. *In situ* S/S has been demonstrated to depths of 10 m and may be able to extend to 50 m. The most significant challenge in applying S/S *in situ* for contaminated soils is achieving complete and uniform mixing of the binder with the contaminated matrix.[37] Three basic approaches are used for *in situ* mixing of the binder with the matrix[5]:

1. Vertical auger mixing.
2. In-place mixing of binder reagents with waste by conventional earthmoving equipment, such as draglines, backhoes, or clamshell buckets.
3. Injection grouting, which involves forcing a binder containing dissolved or suspended treatment agents into the subsurface, allowing it to permeate the soil. Grout injection can be applied to contaminated formations lying well below the ground surface. The injected grout cures in place to produce an *in situ* treated mass.

3.6.1.3   Polymer Microencapsulation S/S

Polymer microencapsulation S/S can include the application of thermoplastic or thermosetting resins. Thermoplastic materials are the most commonly used organic-based S/S treatment materials. Potential candidate resins for thermoplastic encapsulation include bitumen, polyethylene and other polyolefins, paraffins, waxes, and sulfur cement. Of these candidate thermoplastic resins, bitumen (asphalt) is the least expensive and by far the most commonly used.[38] The process of thermoplastic encapsulation involves heating and mixing the waste material and the resin at elevated temperature, typically 130 to 230°C in an extrusion machine. Any water or volatile organics in the waste boil off during extrusion and are collected for treatment or disposal. Because the final product is a stiff, yet plastic resin, the treated material typically is discharged from the extruder into a drum or other container.

S/S process quality control requires information on the range of contaminant concentrations, potential interferences in waste batches awaiting treatment, and treated product properties such as compressive strength, permeability, leachability, and in some instances, toxicity.[28]

3.6.2   SITE REQUIREMENTS

The site must be prepared for the construction, operation, maintenance, decontamination, and decommissioning of the equipment. The size of the area required for the process equipment depends on several factors, including the type of S/S process involved, the required treatment capacity of the system, and site characteristics, especially soil topography and load-bearing capacity. A small mobile *ex situ* unit occupies space for two, standard flatbed trailers. An *in situ* system requires a larger area to accommodate a drilling rig as well as a larger area for auger decontamination.

3.6.3   APPLICABILITY

This section addresses expected applicability based on the chemistry of the metal and the S/S binders. The soil–contaminant–binder equilibrium and kinetics are complicated, and many factors influence metal mobility, so there may be exceptions to the generalizations presented below.

3.6.3.1   Cement-Based S/S

For cement-based S/S, if a single metal is the predominant contaminant in the soil, then Cd and Pb are the most amenable to cement-based S/S. The predominant mechanism for immobilization of metals in Portland and similar cements is precipitation of hydroxides, carbonates, and silicates.

Both Pb and Cd tend to form insoluble precipitates in the pH ranges found in cured cement. They may resolubilize, however, if the pH is not carefully controlled. For example, Pb in aqueous solutions tends to resolubilize as $Pb(OH)^{3-}$ around pH 10 and above. Hg, although it is a cationic metal like Pb and cadmium, does not form low-solubility precipitates in cement, so it is difficult to stabilize reliably by cement-based processes, and this difficulty would be expected to be greater with increasing Hg concentration and with organomercury compounds. Owing to its formation of anionic species, As also does not form insoluble precipitates in the high-pH cement environment, and cement-based solidification is generally not expected to be successful. Cr(VI) is difficult to stabilize in cement due to the formation of anions that are soluble at high pH. However, Cr(VI) can be reduced to Cr(III), which does form insoluble hydroxides. Although Hg, As(III), and As(V) are particularly difficult candidates for cement-based S/S, this should not necessarily eliminate S/S (even cement-based) from consideration for the following reasons:

1. As with Cr(VI), it may be possible to devise a multistep process that will produce an acceptable product for cement-based S/S.
2. A non-cement-based S/S process (e.g., lime and sulfide for Hg; oxidation to As(V) and co-precipitation with Fe) may be applicable.
3. The leachable concentration of the contaminant may be sufficiently low that a highly efficient S/S process may not be required to meet treatment goals.

The discussion of applicability above also applies to *in situ*, cement-based S/S. If *in situ* treatment introduces chemical agents into the ground, this chemical addition may cause a pollution problem in itself, and may be subject to additional requirements under the Land Disposal Restrictions.

3.6.3.2   Polymer Microencapsulation

Polymer microencapsulation has been mainly used to treat low-level radioactive wastes. However, organic binders have been tested or applied to wastes containing chemical contaminants such as As, metals, inorganic salts, polychlorinated biphenyls (PCBs), and dioxins.[38] Polymer microencapsulation is particularly well suited to treating water-soluble salts such as chlorides or sulfates that are generally difficult to immobilize in a cement-based system.[39] Characteristics of the organic binder and extrusion system impose compatibility requirements on the waste material. The elevated operating temperatures place a limit on the quantity of water and volatile organic chemicals (VOCs) in the waste feed. Low-volatility organics will be retained in the bitumen but may act as solvents, causing the treated product to be too fluid. The bitumen is a potential fuel source so the waste should not contain oxidizers such as nitrates, chlorates, or perchlorates. Oxidants present the potential for rapid oxidation, causing immediate safety concerns, as well as slow oxidation, which results in waste form degradation.

Cement-based S/S of multiple metal wastes is particularly difficult if a set of treatment and disposal conditions cannot be found that simultaneously produces low-mobility species for all the metals of concern. For example, the relatively high pH conditions that favor Pb immobilization would tend to increase the mobility of As. On the other hand, the various metal species in a multiple metal waste may interact (e.g., the formation of low-solubility compounds by the combination of Pb and arsenate) to produce a low-mobility compound.

Organic contaminants are often present with inorganic contaminants at metal-contaminated sites. S/S treatment of organic-contaminated waste with cement-based binders is more complex than treatment of inorganics alone. This is particularly true with VOCs, where the mixing process and heat generated by cement hydration reactions can increase vapor losses.[40–43] However, S/S can be applied to wastes that contain lower levels of organics, particularly when inorganics are present or the organics are semivolatile or nonvolatile. Also, recent studies indicate that the addition of silicates or modified clays to the binder system may improve S/S performance with organics.[27]

3.6.4 Performance and BDAT Status

Information in 2000 about the use of S/S at Superfund remedial sites has indicated that S/S has been used at 167 sites since FY 1982.[34] Figure 3.1 shows the number of projects by status for the following stages: predesign/design, design completed/being installed, operational, and completed. Data are shown for *in situ* and *ex situ* S/S projects. In addition, information about all source control technologies is provided. With respect to S/S projects, the majority of *in situ* and *ex situ* projects (62%) are completed, followed by projects in the predesign/design stage (21%). Overall, completed S/S projects represent 30% of all completed Superfund projects in which treatment technologies have been used for source control.

Figure 3.2 shows the types of binder materials used for S/S projects at Superfund remedial sites, including inorganic binders, organic binders, and combination organic and inorganic binders. Many of the binders used include one or more proprietary additives. Examples of inorganic binders include cement, fly ash, lime, soluble silicates, and sulfur-based binders; organic binders on the other hand include asphalt, epoxide, polyesters, and polyethylene. More than 90% of the S/S projects used inorganic binders. In general, inorganic binders are less expensive and easier to use than organic binders. Organic binders are generally used to solidify radioactive wastes or specific hazardous organic compounds.

Figure 3.3 shows the types of contaminant groups and combination of contaminant groups treated by S/S at Superfund remedial sites. S/S was used to treat metals only in 56% of the projects, and used to treat metals alone or in combination with organics or radioactive metals at approximately 90% of the sites. S/S was used to treat organics only at 6% of the sites.[34] Figure 3.4 provides a further breakdown of the metals treated by S/S at Superfund remedial sites. The top five metals treated by S/S are Pb, Cr, As, Cd, and Cu.

S/S with cement-based and pozzolan binders is a commercially available, established technology.[5] Table 3.5 presents a list of sites where S/S has been selected for remediating metal-contaminated

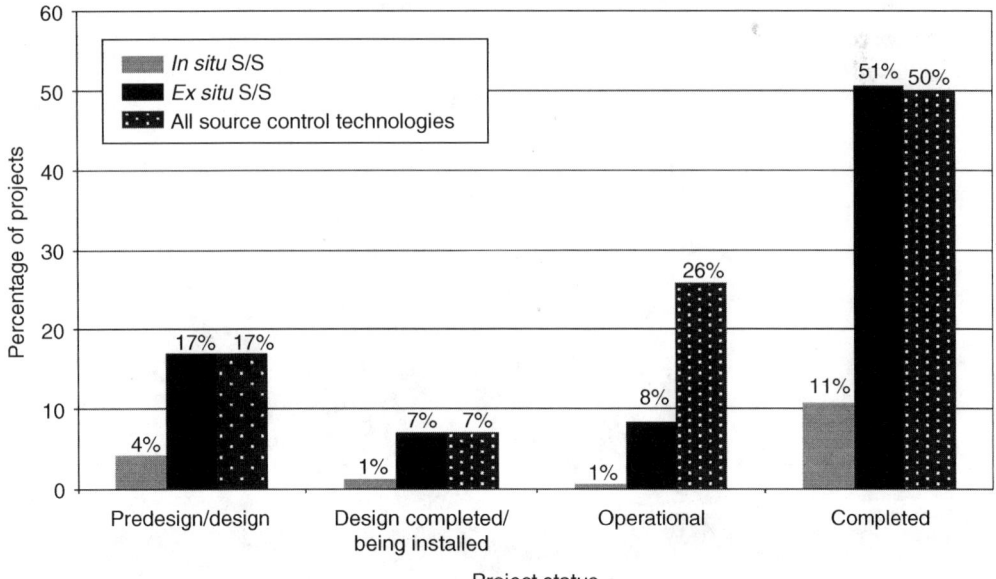

FIGURE 3.1 Percentage of Superfund remedial projects by status. Number of projects: source control, 682; *ex situ* solidification–stabilization (S/S), 139; *in situ* S/S, 28. (From U.S. EPA, Solidification/Stabilization Use at Superfund Sites, EPA-542-R-00-010, U.S. Environmental Protection Agency, Washington, DC, September 2000.)

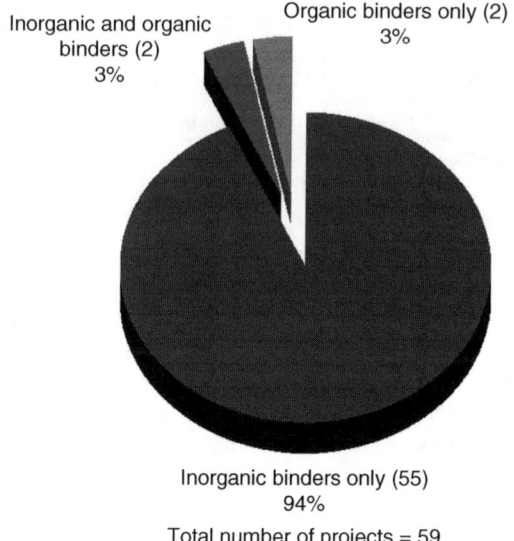

Total number of projects = 59

FIGURE 3.2 Binder materials used for solidification–stabilization projects. (From U.S. EPA, Solidification/ Stabilization Use at Superfund Sites, EPA-542-R-00-010, U.S. Environmental Protection Agency, Washington, DC, September 2000.)

solids. Note that S/S has been used to treat all five metals (Cr, Pb, As, Hg, and Cd). Although it would not generally be expected that cement-based S/S would be applied to As- and Hg-contaminated soils, it was beyond the scope of the project to examine in detail the characterization data, S/S formulations, and performance data upon which the selections were based, so the selection/implementation data are presented without further comment.

Applications of polymer microencapsulation have been limited to special cases where the specific performance features are required for the waste matrix, and contaminants allow reuse of the treated waste as a construction material.[44]

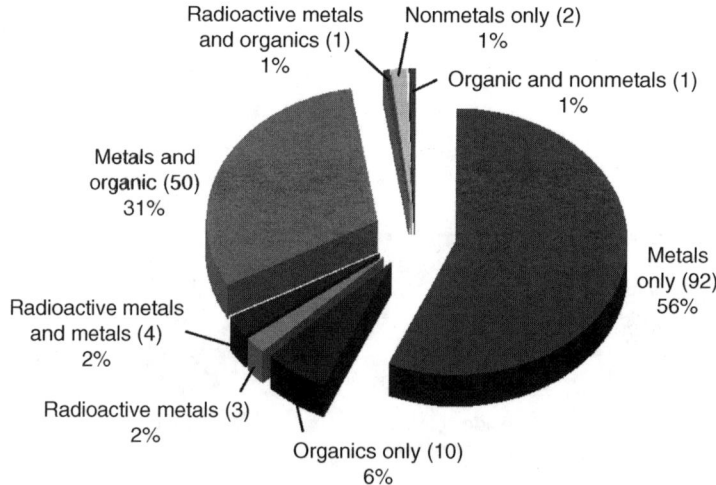

Total number of projects = 163

FIGURE 3.3 Contaminant types treated by solidification–stabilization. (From U.S. EPA, Solidification/ Stabilization Use at Superfund Sites, EPA-542-R-00-010, U.S. Environmental Protection Agency, Washington, DC, September 2000.)

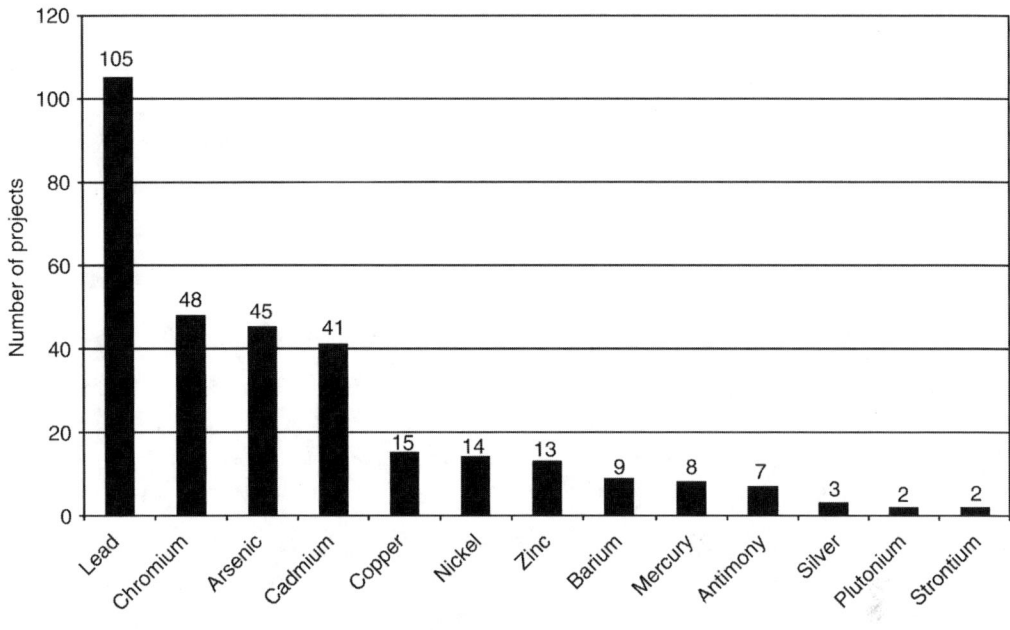

FIGURE 3.4   Number of solidification–stabilization projects treating specific metals. (From U.S. EPA, Solidification/Stabilization Use at Superfund Sites, EPA-542-R-00-010, U.S. Environmental Protection Agency, Washington, DC, September 2000.)

S/S is a BDAT for the following waste types[5]:

1. Cd nonwastewaters other than Cd-containing batteries
2. Cr nonwastewaters following reduction to Cr(III)
3. Pb nonwastewaters
4. Wastes containing low concentrations (<260 mg/kg) of elemental Hg-sulfide precipitation
5. Plating wastes and steel-making wastes

Although vitrification, not S/S, has been selected as BDAT for RCRA As-containing nonwaste-waters, U.S. EPA does not preclude the use of S/S for treatment of As (particularly inorganic As) wastes, but recommends that its use be determined on a case-by-case basis. A variety of stabilization techniques including cement, silicate, pozzolan, and ferric co-precipitation were evaluated as candidate BDATs for As. Because of concerns about long-term stability and increase in waste volume, particularly with ferric co-precipitation, stabilization was not accepted as BDAT.

3.6.5   SITE PROGRAM DEMONSTRATION PROJECTS

Completed SITE demonstrations applicable to soils contaminated with the metals of interest include the following[5]:

1. Advanced Remediation Mixing, Inc. (*ex situ* S/S)
2. Funderburk and Associates (*ex situ* S/S)
3. Geo-Con, Inc. (*in situ* S/S)
4. Soliditech, Inc. (*ex situ* S/S)
5. STC Omega, Inc. (*ex situ* S/S)

TABLE 3.5

Solidification–Stabilization (S/S) Applications at Superfund Sites with Metal Contamination

Site Name and State	Specific Technology	Key Metal Contaminants	Associated Technology
DeRewal Chemical, NJ	Solidification	Cr, Cd, Pb	GW pump and treatment
Marathon Battery Co., NY	Chemical fixation	Cd, Ni	Dredging, off-site disposal
Nascolite, Millville, NJ	Stabilization of wetland soils	Pb	On-site disposal of stabilized soils; excavation and off-site disposal of wetland soils
Roebling Steel, NJ	S/S	As, Cr, Pb	Capping
Waldick Aerospace, NJ	S/S	Cd, Cr	Off-site disposal
Aladdin Plating, PA	Stabilization	Cr	Off-site disposal
Palmerton Zinc, PA	Stabilization, fly ash, lime, potash	Cd, Pb	—
Tonolli Corp., PA	S/S	As, Pb	*In situ* chemical limestone barrier
Whitmoyer Laboratories, PA	Oxidation/fixation	As	GW pump and treatment, capping, grading, and revegetation
Bypass 601, NC	S/S	Cr, Pb	Capping, regrading, revegetation, GW pump and treatment
Flowood, MS	S/S	Pb	Capping
Independent Nail, SC	S/S	Cd, Cr	Capping
Pepper's Steel and Alloys, FL	S/S	As, Pb	On-site disposal
Gurley Pit, AR	*In situ* S/S	Pb	
Pesses Chemical, TX	Stabilization	Cd	Concrete capping
E.I. Dupont de Nemours, IA	S/S	Cd, Cr, Pb	Capping, regrading, and revegetation
Shaw Avenue Dump, IA	S/S	As, Cd	Capping, groundwater monitoring
Frontier Hard Chrome, WA	Stabilization	Cr	
Gould Site, OR	S/S	Pb	Capping, regrading, and revegetation

GW, groundwater.

Source: U.S. EPA, Technology Alternatives for the Remediation of Soils Contaminated with As, Cd, Cr, Hg, and Pb, EPA/540/S-97/500, U.S. Environmental Protection Agency, Cincinnati, OH, August 1997.

6. WASTECH Inc. (*ex situ* S/S)
7. Separation and Recovery Systems, Inc. (*ex situ* S/S)
8. Wheelabrator Technologies Inc. (*ex situ* S/S)

3.6.6 COST OF S/S

Information about the cost of using S/S to treat wastes at Superfund remedial sites was reported by U.S. EPA for 29 completed projects in 2000.[34] Total costs[45] in terms of 2007 USD for S/S projects ranged from USD 86,000 to USD 18,000,000, including the cost of excavation, treatment, and disposal (if *ex situ*). The cost ranged from 12 USD/m^3 to approximately 1,800 USD/m^3. The average cost for these projects was 396 USD/m^3, including two projects with relatively high costs (approximately 1,800 USD/m^3). Excluding those two projects, the average cost per cubic meter was USD 291.[34]

3.7 VITRIFICATION

Vitrification applies a high-temperature treatment aimed primarily at reducing the mobility of metals by their incorporation into a chemically durable, leach-resistant, vitreous mass. Vitrification can be carried out on excavated soils as well as *in situ*.

3.7.1 PROCESS DESCRIPTION

During the vitrification process, organic wastes are pyrolyzed (*in situ*) or oxidized (*ex situ*) by the melt front, whereas inorganics, including metals, are incorporated into the vitreous mass. Off-gases released during the melting process, containing volatile components and products of combustion and pyrolysis, must be collected and treated.[4,46,47] Vitrification converts contaminated soils to a stable glass and crystalline monolith.[47] With the addition of low-cost materials such as sand, clay, or native soil, the process can be adjusted to produce products with specific characteristics, such as chemical durability. Waste vitrification may be able to transform the waste into useful, recyclable products such as clean fill, aggregate, or higher valued materials such as erosion-control blocks, paving blocks, and road dividers.

3.7.1.1 *Ex Situ* Vitrification

Ex situ vitrification (ESV) technologies apply heat to a melter through a variety of sources such as combustion of fossil fuels (coal, natural gas, and oil) or input of electric energy by direct joule heat, arcs, plasma torches, and microwaves. Combustion or oxidation of the organic portion of the waste can contribute significant energy to the melting process, thus reducing energy costs. The particle size of the waste may need to be controlled for some of the melting technologies. For wastes containing refractory compounds that melt above the unit's nominal processing temperature, such as quartz or alumina, size reduction may be required to achieve acceptable throughputs and a homogeneous melt. For high-temperature processes using arcing or plasma technologies, size reduction is not a major factor. For intense melters using concurrent gas-phase melting or mechanical agitation, size reduction is needed for feeding the system and for achieving a homogeneous melt.

3.7.1.2 *In Situ* Vitrification

In situ vitrification (ISV) technology is based on electric melter technology, and the principle of operation is joule heating, which occurs when an electrical current is passed through a region that behaves as a resistive heating element. Electrical current is passed through the soil by means of an array of electrodes inserted vertically into the surface of the contaminated soil zone. Because dry soil is not conductive, a starter path of flaked graphite and glass frit is placed in a small trench between the electrodes to act as the initial flow path for electricity. Resistance heating in the starter path transfers heat to the soil, which then begins to melt. Once molten, the soil becomes conductive. The melt grows outward and downward as power is gradually increased to the full constant operating power level. A single melt can treat a region of up to 1000 T. The maximum treatment depth has been demonstrated to be about 6 m. Large contaminated areas are treated in multiple settings, and fuse the blocks together to form one large monolith.[4] Further information on *in situ* vitrification can be found in references 48 to 51.

3.7.2 SITE REQUIREMENTS

The site must be prepared for the mobilization, operation, maintenance, and demobilization of the equipment. Site activities such as clearing vegetation, removing overburden, and acquiring backfill material are often necessary for ESV and ISV. *Ex situ* processes will require areas for storage of excavated, treated, and possibly pretreated materials. The components of one ISV system are contained in three transportable trailers: an off-gas and process control trailer, a support trailer, and an electrical trailer. The trailers are mounted on wheels sufficient for transportation to and over a compacted ground surface.[52]

The field-scale ISV system evaluated in the SITE Program required three-phase electrical power at either 12,500 or 13,800 V, which is usually taken from a utility distribution system.[53] Alternatively, the power may be generated on site by means of a diesel generator. Typical applications require 800 kWh/T to 1000 kWh/T.[48]

3.7.3 APPLICABILITY

Setting cost and implementability aside, vitrification should be most applicable where nonvolatile metal contaminants have glass solubilities exceeding the level of contamination in the soil. Cr-contaminated soil should pose the least difficulties for vitrification, because it has low volatility, and glass solubility between 1% and 3%. Vitrification may or may not be applicable for Pb, As, and Cd, depending on the level of difficulty encountered in retaining the metals in the melt, and controlling and treating any volatile emissions that may occur. Hg clearly poses problems for vitrification due to its high volatility and low glass solubility (<0.1%), but may be allowable at very low concentrations.

Chlorides present in the waste in excess of about 0.5% by weight (wt%) typically will not be incorporated into and discharged with the glass but will fume off and enter the off-gas treatment system. If chlorides are excessively concentrated, salts of alkali, alkaline earths, and heavy metals will accumulate in solid residues collected by off-gas treatment. Separation of the chloride salts from the other residuals may be required before or during the return of residuals to the melter. When excess chlorides are present, there is also a possibility that dioxins and furans may form and enter the off-gas treatment system.

Waste matrix composition affects the durability of the treated waste. Sufficient glass-forming materials, SiO_2 (>30 wt%), and combined alkali (Na + K; >1.4 wt%), are required for vitrification of wastes. If these conditions are not met, frit or flux additives typically are needed. Vitrification is also potentially applicable to soils contaminated with mixed metals and metal–organic wastes.

Specific situations where ESV would not be applicable or would face additional implementation problems include those that involve the following[5]:

1. Wastes containing >25% moisture content, which can cause excessive fuel consumption
2. Wastes where size reduction and classification are difficult or expensive
3. Volatile metals, particularly Cd and Hg, which will vaporize and must be captured and treated separately
4. Arsenic-containing wastes, which may require pretreatment to produce less volatile forms
5. Metal concentrations in soil that exceed their solubility in glass
6. Sites where commercial capacity is not adequate or transportation cost to a fixed facility is unacceptable

Specific situations, in addition to those cited above, where ISV would not be applicable or would face additional implementation problems include the following[5]:

1. Metal-contaminated soil where a less costly and adequately protective remedy exists
2. Projects that cannot be undertaken because of limited commercial availability
3. Contaminated soil <2 m or >6 m below the ground surface
4. The presence of an aquifer with high hydraulic conductivity (e.g., soil permeability >1 × 10^{-5} cm/sec) limits economic feasibility due to excessive energy requirements
5. Contaminated soil mixed with buried metal, which can result in a conductive path causing short circuiting of the electrodes
6. Contaminated soil mixed with loosely packed rubbish or buried coal, which can start underground fires and overwhelm the off-gas collection and treatment system
7. Volatile heavy metals near the surface, which can be entrained in combustion product gases and not retained in the melt
8. Sites where a surface slope >5% may cause melt to flow
9. *In situ* voids >150 m^3, which can interrupt conduction and heat transfer
10. Underground structures and utilities <6 m from the melt zone that must be protected from heat or avoided

Where it can be successfully applied, the advantages of vitrification include the following[5]:

1. The vitrified product is an inert, impermeable solid that should reduce leaching for long periods of time.
2. The volume of the vitrified product will typically be smaller than the initial waste volume.
3. The vitrified product may be usable.
4. A wide range of inorganic and organic wastes can be treated.
5. There is both an *ex situ* and an *in situ* option available.

A particular advantage of *ex situ* treatment is better control of processing parameters. Also, fuel costs may be reduced for *ex situ* vitrification by the use of combustible waste materials. This fuel cost-saving option is not directly applicable for *in situ* vitrification, because combustibles would increase the design and operating requirements for gas capture and treatment.

3.7.4 PERFORMANCE AND BDAT STATUS

ISV has been implemented at metal-contaminated Superfund sites and has been evaluated under the SITE Program.[54] Some improvements are needed with regard to melt containment and air emission control systems. ISV has been operated at a large scale on many occasions, including two demonstrations on radioactively contaminated sites at the DOE's Hanford Nuclear Reservation.[46,55] Pilot-scale tests have been conducted at Oak Ridge National Laboratory, Idaho National Engineering Laboratory, and Arnold Engineering Development Center. More than 150 tests and demonstrations at various scales have been performed on a broad range of waste types in soils and sludges. The technology has been selected as a preferred remedy at ten private, Superfund, and Department of Defense (DoD) sites.[56] Table 3.6 provides a summary of ISV technology selection/application at metal-contaminated Superfund sites. A number of ESV systems are under development. The technical resource document[27] identified one full-scale *ex situ* melter that was reported to be operating on RCRA organics and inorganics. Vitrification is also a BDAT for As-containing wastes.

3.7.5 SITE PROGRAM DEMONSTRATION PROJECTS

Completed SITE demonstrations applicable to soils contaminated with the metals of interest include the following[5]:

1. Babcock & Wilcox Co. (cyclone furnace—ESV)
2. Retech, Inc. (Plasma arc—ESV)

TABLE 3.6
***In Situ* Vitrification Applications at Superfund Sites with Metal Contamination**

Site Name and State	Key Metal Contaminants
Parsons Chemical, MI	Hg (low)
Rocky Mountain Arsenal, CO	As, Hg

Source: U.S. EPA, Technology Alternatives for the Remediation of Soils Contaminated with As, Cd, Cr, Hg, and Pb, EPA/540/S-97/500, U.S. Environmental Protection Agency, Cincinnati, OH, August 1997.

3. Geosafe Corporation (ISV)
4. Vortec Corporation (*ex situ* oxidation and vitrification process)

3.8 SOIL WASHING

Soil washing is an *ex situ* remediation technology that uses a combination of physical separation and aqueous-based separation unit operations to reduce contaminant concentrations to site-specific remedial goals.[57] Although soil washing is sometimes used as a stand-alone treatment technology, more often it is combined with other technologies to complete site remediation. Soil-washing technologies have successfully remediated sites contaminated with organic, inorganic, and radioactive contaminants.[57] The technology does not detoxify or significantly alter the contaminant, but transfers the contaminant from the soil into the washing fluid or mechanically concentrates the contaminants into a much smaller soil mass[58] for subsequent treatment (Figure 3.5).

Further information on soil washing can be found in U.S. EPA innovative technology reports and programs.[59,60]

3.8.1 PROCESS DESCRIPTION

Soil-washing systems are quite flexible in terms of the number, type, and order of processes involved. Soil washing is performed on excavated soil and may involve some or all of the following, depending on the contaminant–soil matrix characteristics, cleanup goals, and specific process employed[5,58]:

1. Mechanical screening to remove various oversize materials
2. Crushing to reduce applicable oversize to suitable dimensions for treatment
3. Physical processes (e.g., soaking, spraying, tumbling, and attrition scrubbing) to liberate weakly bound agglomerates (e.g., silts and clays bound to sand and gravel) followed by size classification to generate coarse-grained and fine-grained soil fraction(s) for further treatment
4. Treatment of the coarse-grained soil fraction(s)
5. Treatment of the fine-grained fraction(s)
6. Management of the generated residuals

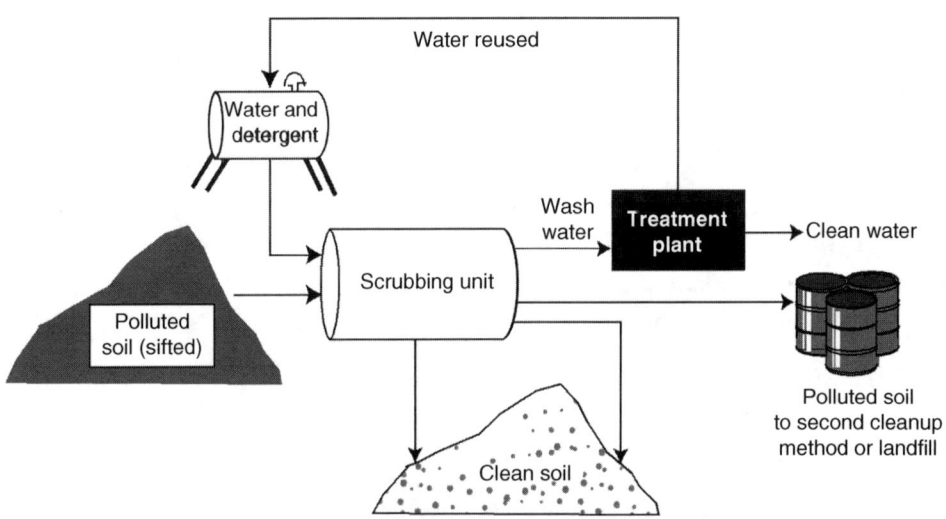

FIGURE 3.5 Soil-washing operation. (From U.S. EPA, A Citizen's Guide to Soil Washing, EPA 542-F-01-008, U.S. Environmental Protection Agency, Washington, DC, May 2001.)

Treatment of the coarse-grained soil fraction typically involves additional application of physical separation techniques and possibly aqueous-based leaching techniques. Physical separation techniques (e.g., sorting, screening, elutriation, hydrocyclones, spiral concentrators, and flotation) exploit physical differences (e.g., size, density, shape, color, and wetability) between contaminated particles and soil particles in order to produce a clean (or nearly clean) coarse fraction and one or more metal-concentrated streams. Many of the physical separation processes listed above involve the use of water as a transport medium, and if the metal contaminant has significant water solubility, then some of the coarse-grained soil cleaning will occur as a result of transfer to the aqueous phase. If the combination of physical separation and unaided transfer to the aqueous phase cannot produce the desired reduction in the soil's metal content, which is frequently the case for metal contaminants, then solubility enhancement is an option for meeting cleanup goals for the coarse fraction. Solubility enhancement can be accomplished in several ways[5,61,62]:

1. Converting the contaminant into a more soluble form (e.g., oxidation/reduction, conversion to soluble metal salts)
2. Using an aqueous-based leaching solution (e.g., acidic, alkaline, oxidizing, reducing) in which the contaminant has enhanced solubility
3. Incorporating a specific leaching process into the system to promote increased solubilization via increased mixing, elevated temperatures, higher solution/soil ratios, efficient solution/soil separation, multiple stage treatment, etc.
4. A combination of the above

After the leaching process is completed on the coarse-grained fraction, it will be necessary to separate the leaching solution and the coarse-grained fraction by settling. A soil rinsing step may be necessary to reduce the residual leachate in the soil to an acceptable level. It may also be necessary to readjust soil parameters such as pH or redox potential before replacement of the soil on the site. The metal-bearing leaching agent must also be treated further to remove the metal contaminant and permit reuse in the process or discharge, and this topic is discussed below under management of residuals.

Treatment of fine-grained soils is similar in concept to the treatment of the coarse-grained soils, but the production rate would be expected to be lower and hence more costly than for the coarse-grained soil fraction. The reduced production rate arises from factors including (1) the tendency of clays to agglomerate, thus requiring time, energy, and high water/clay ratios to produce leachable slurry and (2) slow settling velocities that require additional time or capital equipment to produce acceptable soil/water separation for multibatch or countercurrent treatment, or at the end of treatment. A site-specific determination needs to be made whether the fines should be treated to produce clean fines or whether they should be handled as a residual wastestream.

Management of generated residuals is an important aspect of soil washing. The effectiveness, implementability, and cost of treating each residual stream are important to the overall success of soil washing for the site. Perhaps the most important of the residual streams is the metal-loaded leachant that is generated, particularly if the leaching process recycles the leaching solution. Furthermore, it is often critical to the economic feasibility of the project that the leaching solution be recycled. For these closed- or semiclosed-loop leaching processes, successful treatment of the metal-loaded leachant is imperative to the successful cleaning of the soil. The leachant must (1) have adequate solubility for the metal so that the metal reduction goals can be met without using excessive volumes of leaching solution and (2) be readily, economically, and repeatedly adjustable (e.g., pH adjustment) to a form in which the metal contaminant has very low solubility so that the recycled aqueous phase retains a favorable concentration gradient compared to the contaminated soil. Also, efficient soil–water separation is important prior to recovering metal from the metal-loaded leachant in order to minimize contamination of the metal concentrate. Recycling the leachant reduces logistical requirements and costs associated with makeup water, storage, permitting, compliance analyses, and leaching agents. It also reduces external coordination requirements and eliminates the dependence of the remediation on the ability to meet publicly owned treatment works (POTW) discharge requirements.

Other residual streams that may be generated and require proper handling include the following[5]:

1. Untreatable, uncrushable oversize
2. Recyclable metal-bearing particulates, concentrates, or sludges from physical separation or leachate treatment
3. Nonrecyclable metal-bearing particulates, concentrates, soils, sludges, or organic debris that fail toxicity characteristic leaching procedure (TCLP) thresholds for RCRA hazardous waste
4. Soils or sludges that are not RCRA hazardous wastes but are also not sufficiently clean to permit return to the site
5. Metal-loaded leachant from systems where leachant is not recycled
6. Rinsate from treated soil

3.8.2 Site Requirements

The area required for a unit at a site will depend on the vendor system selected, the amount of soil storage space, or the number of tanks or ponds needed for wash water preparation and wastewater storage and treatment. Typical utilities required are water, electricity, steam, and compressed air; the quantity of each is vendor- and site-specific. It may be desirable to control the moisture content of the contaminated soil for consistent handling and treatment by covering the excavation, storage, and treatment areas. Climatic conditions such as annual or seasonal precipitation cause surface runoff and water infiltration; therefore, runoff control measures may be required. As soil washing is an aqueous-based process, cold weather effects include freezing as well as potential effects on leaching rates.

3.8.3 Applicability

Soil washing is potentially applicable to soils contaminated with all five metals of interest. Conditions that particularly favor soil washing include the following[5]:

1. There is just a single principal contaminant metal, which occurs in dense, insoluble particles that report to a specific, small mass fraction(s) of the soil.
2. There is just a single contaminant metal and species, which is very water or aqueous leachant soluble and has a low soil/water partition coefficient.
3. The soil contains a high proportion (e.g., >80%) of soil particles >2 mm, which is desirable for efficient contaminant–soil and soil–water separation.

Conditions that clearly do not favor soil washing include the following[5]:

1. Soils with a high (i.e., >40%) silt and clay fraction
2. Soils that vary widely and frequently in significant characteristics such as soil type, contaminant type and concentration, and where blending for homogeneity is not feasible
3. Complex mixtures (e.g., multicomponent, solid mixtures where access of leaching solutions to contaminant is restricted; mixed anionic and cationic metals where pH of solubility maximums are not close)
4. High clay content, cation exchange capacity, or humic acid content, which would tend to interfere with contaminant desorption

5. The presence of substances that interfere with the leaching solution (e.g., carbonaceous soils will neutralize extracting acids; similarly, high humic acid content will interfere with an alkaline extraction)
6. Metal contaminants in a very low-solubility, stable form (e.g., PbS), which may require long contact times and excessive amounts of reagent to solubilize

3.8.4 PERFORMANCE AND BDAT STATUS

Soil washing has been used at waste sites in Europe, in particular in Germany, the Netherlands, and Belgium.[63] Table 3.7 lists selected Superfund sites where soil washing has been selected or implemented. Acid leaching, which is a form of soil washing, is also the BDAT for Hg.

3.8.5 SITE DEMONSTRATIONS AND EMERGING TECHNOLOGIES PROGRAM PROJECTS

SITE demonstrations applicable to soils contaminated with the metals of interest include the following[5]:

1. Bergmann USA (physical separation/leaching) BioGenesis[SM] (physical separation/leaching)
2. Biotrol, Inc. (physical separation)
3. Brice Environmental Services Corporation (physical separation)
4. COGNIS, Inc. (leaching)
5. Toronto Harbor Commission (physical separation/leaching)

Four SITE Emerging Technologies Program projects have been completed that are applicable to soils contaminated with the metals of interest.

3.9 SOIL FLUSHING

Soil flushing is the *in situ* extraction of contaminants from the soil via an appropriate washing solution. Water or an aqueous solution is injected into or sprayed onto the area of contamination, and the contaminated elutriate is collected and pumped to the surface for removal, recirculation,

---

TABLE 3.7
Soil-Washing Applications at Selected Superfund Sites with Metal Contamination

Site Name and State	Specific Technology	Key Metal Contaminants	Associated Technology
Ewan Property, NJ	Water washing	As, Cr, Cu, Pb	Pretreatment by solvent extraction to remove organics
GE Wiring Devices, PR	Water with KI solution additive	Hg	Treated residues disposed on site and covered with clean soil
King of Prussia, NJ	Water with washing agent additives	Ag, Cr, Cu	Sludges to be land disposed
Zanesville Well Field, OH	Soil washing	Hg, Pb	SVE to remove organics
Twin Cities Army Ammunition Plant, MN	Soil washing	Cd, Cr, Cu, Hg, Pb	Soil leaching
Sacramento Army Depot Sacramento, CA	Soil washing	Cr, Pb	Off-site disposal of wash liquid

Source: U.S. EPA, Technology Alternatives for the Remediation of Soils Contaminated with As, Cd, Cr, Hg, and Pb, EPA/540/S-97/500, U.S. Environmental Protection Agency, Cincinnati, OH, August 1997.

or on-site treatment and reinjection. The technology is applicable to both organic and inorganic contaminants, and metals in particular[4]. For the purpose of metals remediation, soil flushing has been operated at full scale, but for a small number of sites.

3.9.1 Process Description

Soil flushing uses water, a solution of chemicals in water, or an organic extractant to recover contaminants from the *in situ* material. The contaminants are mobilized by solubilization, formation of emulsions, or a chemical reaction with the flushing solutions. After passing through the contamination zone, the contaminant-bearing fluid is collected by strategically placed wells or trenches and brought to the surface for disposal, recirculation, or on-site treatment and reinjection. During elutriation, the flushing solution mobilizes the sorbed contaminants by dissolution or emulsification.

One key to the efficient operation of a soil-flushing system is the ability to reuse the flushing solution, which is recovered along with groundwater. Various water-treatment techniques can be applied to remove the recovered metals and render the extraction fluid suitable for reuse. Recovered flushing fluids may need treatment to meet appropriate discharge standards prior to release to a POTW or receiving waters. The separation of surfactants from recovered flushing fluid, for reuse in the process, is a major factor in the cost of soil flushing. Treatment of the flushing fluid results in process sludges and residual solids, such as spent carbon and spent ion exchange resin, which must be appropriately treated before disposal. Air emissions of volatile contaminants from recovered flushing fluids should be collected and treated, as appropriate, to meet applicable regulatory standards. Residual flushing additives in the soil may be a concern and should be evaluated on a site-specific basis.[64] Subsurface containment barriers can be used in conjunction with soil-flushing technology to help control the flow of flushing fluids.

Further information on soil flushing can be found in references 59 and 64 to 66.

3.9.2 Site Requirements

Stationary or mobile soil-flushing systems are located on site. The exact area required will depend on the vendor system selected and the number of tanks or ponds needed for wash water preparation and wastewater treatment. Certain permits may be required for operation, depending on the system being utilized. Slurry walls or other containment structures may be needed along with hydraulic controls to ensure capture of contaminants and flushing additives. Impermeable membranes may be necessary to limit infiltration of precipitation, which could cause dilution of the flushing solution and loss of hydraulic control. Cold weather freezing must also be considered for shallow infiltration galleries and aboveground sprayers.[67]

3.9.3 Applicability

Soil flushing may be easy or difficult to apply, depending on the ability to wet the soil with the flushing solution and to install collection wells or subsurface drains to recover all the applied liquids. The achievable level of treatment varies and depends on the contact of the flushing solution with the contaminants and the appropriateness of the solution for contaminants, and the hydraulic conductivity of the soil. Soil flushing is most applicable to contaminants that are relatively soluble in the extracting fluid, and that will not tend to sorb onto soil as the metal-laden flushing fluid proceeds through the soil to the extraction point. Based on the earlier discussion of metal behavior, some potentially promising scenarios for soil flushing would include Cr(VI), As(III), or As(V) in permeable soil with low iron oxide, low clay, and high pH; Cd in permeable soil with low clay, low cation exchange capacity, and moderately acidic pH; and Pb in acid sands. A single target metal would be preferable to multiple metals, due to the added complexity of selecting a flushing fluid that would be reasonably efficient for all contaminants. Also, the flushing fluid must be compatible with not only the contaminant, but also the soil. Soils that counteract the acidity or alkalinity of the flushing solution will decrease its effectiveness. If precipitants occur due to interaction between the soil and the flushing fluid, then this could obstruct the soil pore structure and inhibit flow to and through sectors of the contaminated soil. It may take long periods of time for soil flushing to achieve cleanup standards.

A key advantage of soil flushing is that the contaminant is removed from the soil. Recovery and reuse of the metal from the extraction fluid may be possible in some cases, although the value of the recovered metal would not be expected to fully offset the costs of recovery. The equipment used for the technology is relatively easy to construct and operate. It does not involve excavation, treatment, and disposal of the soil, which avoids the expense and hazards associated with these activities.

3.9.4 PERFORMANCE AND BDAT STATUS

Table 3.8 lists the Superfund sites where soil flushing has been selected or implemented. Soil flushing has a more established history for removal of organics, but has been used for Cr removal (e.g., United Chrome Products Superfund Site, near Corvallis, OR). *In situ* technologies, such as soil flushing, are not considered RCRA BDAT for any of the five metals.[5]

Soil-flushing techniques for mobilizing contaminants can be classified as conventional or unconventional. Conventional applications employ water only as the flushing solution. Unconventional applications that are currently being researched include the enhancement of the flushing water with additives, such as acids, bases, and chelating agents to aid in the desorption/dissolution of the target contaminants from the soil matrix to which they are bound.

Researchers are also investigating the effects of numerous soil factors on heavy metal sorption and migration in the subsurface. Such factors include pH, soil type, soil horizon, particle size, permeability, specific metal type and concentration, and type and concentrations of organic and inorganic compounds in solutions. Generally, as the soil pH decreases, cationic metal solubility and mobility increase. In most cases, metal mobility and sorption are likely to be controlled by the organic fraction in topsoils and the clay content in the subsoils.

3.9.5 SITE DEMONSTRATION AND EMERGING TECHNOLOGIES PROGRAM PROJECTS

There are no *in situ* soil-flushing projects reported to be completed either as SITE demonstration or Emerging Technologies Program projects.[67]

TABLE 3.8
Soil-Flushing Applications at Selected Superfund Sites with Metal Contamination

Site Name and State	Specific Technology	Key Metal Contaminants	Associated Technology
Lipari Landfill, NJ	Soil flushing of soil and wastes contained by slurry wall and cap; excavation from impacted wetlands	Cr, Hg, Pb	Slurry wall and cap
United Chrome Products, OR		Cr	Electrokinetic pilot test; considering *in situ* reduction

Source: U.S. EPA, Technology Alternatives for the Remediation of Soils Contaminated with As, Cd, Cr, Hg, and Pb, EPA/540/S-97/500, U.S. Environmental Protection Agency, Cincinnati, OH, August 1997.

3.10   PYROMETALLURGY

Pyrometallurgy is used here as a broad term encompassing elevated temperature techniques for extraction and processing of metals for use or disposal. High-temperature processing increases the rate of reaction and often makes the reaction equilibrium more favorable, lowering the required reactor volume per unit output.[4] Some processes that clearly involve both metal extraction and recovery include roasting, retorting, or smelting. Although these processes typically produce a metal-bearing waste slag, metal is also recovered for reuse. A second class of pyrometallurgical technologies included here is a combination of high-temperature extraction and immobilization. These processes use thermal means to cause volatile metals to separate from the soil and report to the fly ash, but the metal in the fly ash is then immobilized, instead of recovered, and there is no metal recovered for reuse. A third class of technologies includes those that are primarily incinerators for mixed organic–inorganic wastes, but which have the capability of processing wastes containing the metals of interest by either capturing volatile metals in the exhaust gases or immobilizing the nonvolatile metals in the bottom ash or slag. Some of these systems may have applicability to some cases where metals contamination is the primary concern, so a few technologies of this type are noted that are in the SITE Program. Vitrification has already been addressed in a previous section. It is not considered pyrometallurgical treatment as there is typically neither a metal-extraction nor a metal-recovery component in the process.

3.10.1   PROCESS DESCRIPTION

Pyrometallurgical processing is usually preceded by physical treatment[5] to produce a uniform feed material and upgrade the metal content.

Solids treatment in a high-temperature furnace requires efficient heat transfer between the gas and solid phases while minimizing particulate in the off-gas. The particle size range that meets these objectives is limited and is specific to the design of the process. The presence of large clumps or debris slows heat transfer, so pretreatment to either remove or pulverize oversize material is normally required. Fine particles are also undesirable because they become entrained in the gas flow, increasing the volume of dust to be removed from the flue gas. The feed material is sometimes pelletized to give a uniform size. In many cases a reducing agent and flux may be mixed in prior to pelletization to ensure good contact between the treatment agents and the contaminated material and to improve gas flow in the reactor.[4]

Owing to its relatively low boiling point (357°C) and ready conversion at elevated temperature to its metallic form, Hg is commonly recovered through roasting and retorting at much lower temperatures than the other metals. Pyrometallurgical processing to convert compounds of the other four metals to elemental metal requires a reducing agent, fluxing agents to facilitate melting and to slag off impurities, and a heat source. The fluid mass is often called a melt, but the operating temperature, although quite high, is often still below the melting points of the refractory compounds being processed. The fluid forms as a lower-melting-point material due to the presence of a fluxing agent such as calcium. Depending on processing temperatures, volatile metals such as Cd and Pb may fume off and be recovered from the off-gas as oxides. Nonvolatile metals, such as Cr or nickel, are tapped from the furnace as molten metal. Impurities are scavenged by the formation of slag.[4] The effluents and solid products generated by pyrometallurgical technologies typically include solid, liquid, and gaseous residuals. Solid products include debris, oversized rejects, dust, ash, and the treated medium. Dust collected from particulate control devices may be combined with the treated medium or, depending on analyses for carryover contamination, recycled through the treatment unit.

3.10.2   SITE REQUIREMENTS

Few pyrometallurgical systems are available in mobile or transportable configurations. This is typically an off-site technology, so the distance of the site from the processing facility has an important influence on transportation costs. Off-site treatment must comply with U.S. EPA's off-site treatment policies and procedures. The off-site facility's environmental compliance status must be acceptable, and the waste must be of a type allowable under their operating permits. In order for pyrometallurgical processing to be technically feasible, it must be possible to generate a concentrate from the contaminated soil that will be acceptable to the processor. The processing rate of the off-site facility must be adequate to treat the contaminated material in a reasonable amount of time. Storage requirements and responsibilities must be determined. The need for air discharge and other permits must be determined on a site-specific basis.

3.10.3 APPLICABILITY

With the possible exception of Hg, or a highly contaminated soil, pyrometallurgical processing where metal recovery is the goal would not be applied directly to the contaminated soil, but rather to a concentrate generated via soil washing. Pyrometallurgical processing in conventional rotary kilns, rotary furnaces, or arc furnaces is most likely to be applicable to large volumes of material containing metal concentrations (particularly, Pb, Cd, or Cr) higher than 5 to 20%. Unless a very concentrated feed stream can be generated (e.g., approximately 60% for Pb), there will be a charge, in addition to transportation, for processing the concentrate. Lower metal concentrations can be acceptable if the metal is particularly easy to reduce and vaporize (e.g., Hg) or is particularly valuable (e.g., Au or Pt). Arsenic is the weakest candidate for pyrometallurgical recovery, because there is almost no recycling of As in the U.S. It is also the least valuable of the metals. The price ranges for the five metals[4] are reported here in terms of 2007 USD/T[45]:

As: 300 to 600 (as As trioxide)
Cd: 7320
Cr: 9630
Pb: 860 to 950
Hg: 6500 to 11,000

3.10.4 PERFORMANCE AND **BDAT** STATUS

The U.S. EPA technical document of reference 4 contains a list of approximately 35 facilities/addresses/contacts that may accept concentrates of the five metals of interest for pyrometallurgical processing. Sixteen of the 35 facilities are Pb recycling operations, seven facilities recover Hg, and the remainder address a range of RCRA wastes that contain the metals of interest. Owing to the large volume of electric arc furnace emission control waste, extensive processing capability has been developed to recover Cd, Pb, and Zn from solid waste matrices. The available process technologies include the following[5]:

1. Waelz kiln process (Horsehead Resource Development Company, Inc.)
2. Waelz kiln and calcination process (Horsehead Resource Development Company, Inc.)
3. Flame reactor process (Horsehead Resource Development Company, Inc.)
4. Inclined rotary kiln (Zia Technology)

Plasma arc furnaces are successfully treating waste at two steel plants. These are site-dedicated units that do not accept outside material for processing.

Pyrometallurgical recovery is a BDAT for the following waste types[5]:

1. Cd-containing batteries
2. Pb nonwastewaters in the noncalcium sulfate subcategory
3. Hg wastes prior to retorting

4. Pb acid batteries
5. Zinc nonwastewaters
6. Hg from wastewater treatment sludge

3.10.5 SITE DEMONSTRATION AND EMERGING TECHNOLOGIES PROGRAM PROJECTS

SITE demonstrations applicable to soils contaminated with the metals of interest include the following[5]:

1. RUST Remedial Services, Inc. (X-Trax Thermal Desorption)
2. Horsehead Resource Development Company, Inc. (Flame Reactor)

3.11 ELECTROKINETICS

Electrokinetic remediation relies on the application of low-intensity direct current between electrodes placed in the soil. Contaminants are mobilized in the form of charged species, particles, or ions.[2] Attempts to leach metals from soils by electro-osmosis date back to the 1930s. In the past, research focused on removing unwanted salts from agricultural soils. Electrokinetics has been used for dewatering of soils and sludges since the first recorded use in the field in 1939.[68] Electrokinetic extraction has been used in the former Soviet Union since the early 1970s to concentrate metals and to explore for minerals in deep soils. By 1979, research had shown that the content of soluble ions increased substantially in electro-osmotic consolidation of polluted dredgings, and metals were not found in the effluent.[69] By the mid-1980s, numerous researchers had realized independently that electrokinetic separation of metals from soils was a potential solution to contamination.[70]

Several organizations are developing technologies for the enhanced removal of metals by transporting contaminants to electrodes, where they are removed and subsequently treated above ground. A variation of the technique involves treatment without removal by transporting contaminants through specially designed treatment zones that are created between electrodes. Electrokinetics can also be used to slow or prevent migration of contaminants by configuring cathodes and anodes in a manner that causes contaminants to flow toward the center of a contaminated area of soil. Performance data illustrate the potential for achieving removals greater than 90% for some metals.[2]

The range of potential metals is broad. Commercial applications in Europe have treated Cu, Pb, Zn, As, Cd, Cr, and Ni. There is also potential applicability for radionuclides and some types of organic compounds. The electrode spacing and duration of remediation is site-specific. The process requires adequate soil moisture in the vadose zone, so the addition of a conducting pore fluid may be required (particularly as there is a tendency for soil drying near the anode). Specially designed pore fluids are also added to enhance the migration of target contaminants. The pore fluids are added at either the anode or cathode, depending on the desired effects.

Table 3.9 presents an overview of two variations of electrokinetic remediation technology. Geokinetics International, Inc., Battelle Memorial Institute, Electrokinetics, Inc., and Isotron Corporation are all developing variations of technologies categorized under Approach 1, "enhanced removal." The consortium of Monsanto, E.I. du Pont de Nemours and Company, General Electric, DOE, and the U.S. EPA Office of Research and Development is developing the Lasagna Process, which is categorized under Approach 2, "treatment without removal."[2]

3.11.1 PROCESS DESCRIPTION

Electrokinetic remediation, also referred to as electrokinetic soil processing, electromigration, electrochemical decontamination, or electroreclamation, can be used to extract radionuclides, metals, and some types of organic wastes from saturated or unsaturated soils, slurries, and sediments.[71]

TABLE 3.9

Overview of Electrokinetic Remediation Technology

General characteristics
- The depth of soil that is amenable to treatment depends on electrode placement.
- It is best used in homogeneous soils with high moisture content and high permeability.

Approach 1 Enhanced Removal	**Approach 2 Treatment Without Removal**
Description	*Description*
Electrokinetic transport of contaminants toward the polarized electrodes to concentrate the contaminants for subsequent removal and *ex situ* treatment.	Electro-osmotic transport of contaminants through treatment zones placed between the electrodes. The polarity of the electrodes is reversed periodically, which reverses the direction of the contaminants back and forth through treatment zones. The frequency with which electrode polarity is reversed is determined by the rate of transport of contaminants through the soil.
Status	*Status*
Demonstration projects using full-scale equipment are reported in Europe. Bench- and pilot-scale laboratory studies are reported in the U.S. and at least two full-scale field studies are ongoing in the U.S.	Demonstrations are ongoing.
Applicability	*Applicability*
Pilot scale: Pb, As, Ni, Hg, Cu, Zn. Laboratory scale: Pb, Cd, Cr, Hg, Zn, Fe, Mg, U, Th, Ra.	Technology developed for organic species and metals.
Comments	*Comments*
Field studies are under evaluation by U.S. EPA, DOE, DOD, and the Electric Power Research Institute (EPRI). The technique primarily would require the addition of water to maintain the electric current and facilitate migration; however, there is ongoing work in application of the technology in partially saturated soils.	This technology is being developed for deep clay formations.

Source: U.S. EPA, Recent Developments for In Situ Treatment of Metal Contaminated Soils, Contract no. 68-W5-0055 U.S. Environmental Protection Agency, Washington, DC, 1997.

This *in situ* soil-processing technology is primarily a separation and removal technique for extracting contaminants from soils.

The principle of electrokinetic remediation relies upon the application of a low-intensity direct current through the soil between two or more electrodes. Most soils contain water in the pores between the soil particles and have an inherent electrical conductivity that results from salts present in the soil.[72] The current mobilizes charged species, particles, and ions in the soil by the following processes[73]:

1. Electromigration (transport of charged chemical species under an electric gradient)
2. Electro-osmosis (transport of pore fluid under an electric gradient)
3. Electrophoresis (movement of charged particles under an electric gradient)
4. Electrolysis (chemical reactions associated with the electric field)

Figure 3.6 presents a schematic diagram of a typical conceptual electrokinetic remediation application.

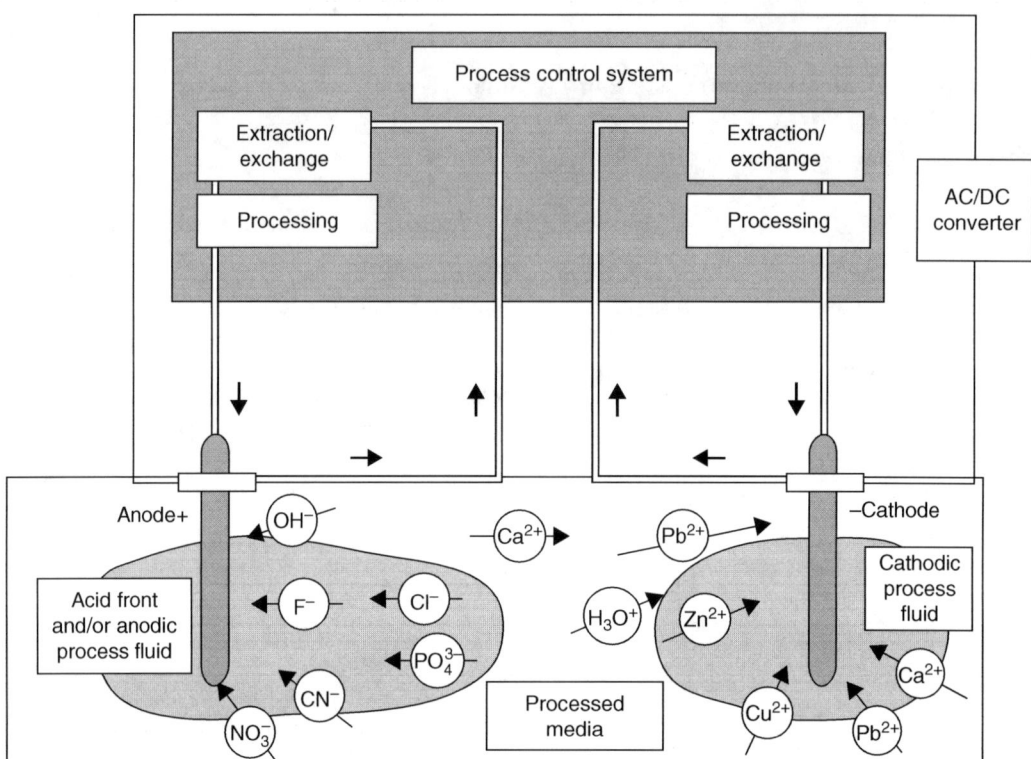

FIGURE 3.6 Diagram of one electrode configuration used in the field implementation of electrokinetics. (From U.S. EPA, Recent Developments for In Situ Treatment of Metal Contaminated Soils, Contract no. 68-W5-0055 U.S. Environmental Protection Agency, Washington, DC, 1997.)

Electrokinetics can be efficient in extracting contaminants from fine-grained, high-permeability soils. A number of factors determine the direction and extent of the migration of the contaminant. Such factors include the type and concentration of the contaminant, the type and structure of the soil, and the interfacial chemistry of the system.[74] Water or some other suitable salt solution may be added to the system to enhance the mobility of the contaminant and increase the effectiveness of the technology (e.g., buffer solutions may change or stabilize pore fluid pH). Contaminants arriving at the electrodes may be removed by any of several methods, including electroplating at the electrode, precipitation or co-precipitation at the electrode, pumping of water near the electrode, or complexing with ion-exchange resins.[74]

Electrochemistry associated with this process involves an acid front that is generated at the anode if water is the primary pore fluid present. The variation of pH at the electrodes results from the electrolysis of the water. The solution becomes acidic at the anode because hydrogen ions are produced and oxygen gas is released, and the solution becomes basic at the cathode, where hydroxyl ions are generated and hydrogen gas is released.[75] At the anode, the pH could drop to below 2, and it could increase at the cathode to above 12, depending on the total current applied. The acid front eventually migrates from the anode to the cathode. Movement of the acid front by migration and advection results in the desorption of contaminants from the soil.[71] The process leads to temporary acidification of the treated soil, and there are no established procedures for determining the length of time needed to reestablish equilibrium. Studies have indicated that metallic electrodes may dissolve as a result of electrolysis and introduce corrosion products into the soil mass. However, if inert electrodes, such as carbon, graphite, or platinum, are used, no residue will be introduced in the treated soil mass as a result of the process.[2]

3.11.2 Site Requirements

Before electrokinetic remediation is undertaken at a site, a number of different field and laboratory screening tests must be conducted to determine whether the particular site is amenable to the treatment technique:

1. *Field conductivity surveys.* The natural geologic spatial variability should be delineated, because buried metallic or insulating material can induce variability in the electrical conductivity of the soil and, therefore, the voltage gradient. In addition, it is important to assess whether there are deposits that exhibit very high electrical conductivity, at which the technique may be inefficient.
2. *Chemical analysis of water.* The pore water should be analyzed for dissolved major anions and cations, as well as for the predicted concentration of the contaminant(s). In addition, the electrical conductivity and pH of the pore water should be measured.
3. *Chemical analysis of soil.* The buffering capacity and geochemistry of the soil should be determined at each site.
4. *pH effects.* The pH values of the pore water and the soil should be determined because they have a great effect on the valence, solubility, and sorption of contaminant ions.
5. *Bench-scale test.* The dominant mechanism of transport, removal rates, and amounts of contamination left behind can be examined for different removal scenarios by conducting bench-scale tests. Because many of these physical and chemical reactions are interrelated, it may be necessary to conduct bench-scale tests to predict the performance of electrokinetics remediation at the field scale.[70,71]

3.11.3 Applicability and Demonstration Projects

Various methods, developed by combining electrokinetics with other techniques, are being applied for remediation. This section describes different types of electrokinetic remediation methods for use at contaminated sites. The methods discussed were developed by Electrokinetics, Inc., Geokinetics International, Inc., Isotron Corporation, Battelle Memorial Institute, a consortium effort, and P&P Geotechnik GmbH.[2]

3.11.3.1 Electrokinetics, Inc.

Electrokinetics, Inc. operates under a licensing agreement with Louisiana State University. The technology is patented by and assigned to Louisiana State University[76] and a complementing process patent is assigned to Electrokinetics, Inc.[77] As depicted in Figure 3.5, groundwater and/or a processing fluid (supplied externally through the boreholes that contain the electrodes) serve as the conductive medium. The additives in the processing fluid, the products of electrolysis reactions at the electrodes, and the dissolved chemical entities in the contaminated soil are transported across the contaminated soil by conduction under the influence of electric fields. This transport, when coupled with sorption, precipitation/dissolution, and volatilization/complexation, provides the fundamental mechanism that can affect the electrokinetic remediation process. Electrokinetics, Inc. accomplishes extraction and removal by electrodeposition, evaporation/condensation, precipitation, or ion exchange, either at the electrodes or in a treatment unit that is built into the system that pumps the processing fluid to and from the contaminated soil. Pilot-scale testing was carried out with support from the U.S. EPA, which also developed a design and analysis package for the process.[78]

3.11.3.2 Geokinetics International, Inc.

Geokinetics International, Inc. (GII) has obtained a patent for an electroreclamation process. The key claims in the patent are the use of electrode wells for both anodes and cathodes and the management of the pH and electrolyte levels in the electrolyte streams of the anode and the cathode. The patent also includes claims for the use of additives to dissolve different types of contaminants.[79] Fluor Daniel is licensed to operate GII's metal removal process in the U.S.

GII has developed and patented an electrically conductive ceramic material (EBONEX®) that has an extremely high resistance to corrosion. It has a lifetime in soil of at least 45 years and is self-cleaning. GII has also developed a batch electrokinetic remediation (BEK®) process. The process, which incorporates electrokinetic technology, normally requires 24 to 48 h for complete remediation of the substrate. BEK® is a mobile unit that remediates *ex situ* soils on site. GII also has developed a solution treatment technology (EIX®) that allows removal of contamination from the anode and the cathode solutions up to a thousand times faster than can be achieved through conventional means.[2]

3.11.3.3 Isotron Corporation

Isotron Corporation participated in a pilot-scale demonstration of electrokinetic extraction supported by DOE's Office of Technology Development. The demonstration took place at the Oak Ridge K-25 facility in Tennessee. Completed laboratory tests showed that the Isotron process could effect the movement and capture of uranium present in soil from the Oak Ridge site.[80]

Isotron Corporation was also involved with Westinghouse Savannah River Company in a demonstration of electrokinetic remediation. The demonstration, supported by DOE's Office of Technology Development, took place at the old TNX basin at the Savannah River site in South Carolina. Isotron used the Electrosorb® process with a patented cylinder to control buffering conditions *in situ*. An ion-exchange polymer matrix called Isolock® was used to trap metal ions. The process was tested for the removal of Pb and Cr.[80]

3.11.3.4 Battelle Memorial Institute

Another method that uses electrokinetic technology is electroacoustical soil decontamination. This technology combines electrokinetics with sonic vibration. Through the application of mechanical vibratory energy in the form of sonic or ultrasonic energy, the properties of a liquid contaminant in soil can be altered in a way that increases the level of removal of the contaminant. Battelle Memorial Institute of Columbus, OH, developed an *in situ* treatment process that uses both electrical and acoustical forces to remove floating contaminants, and possibly metals, from subsurface zones of contamination. The process was selected for U.S. EPA's SITE Program.[81]

3.11.3.5 Consortium Process

Monsanto Company coined the name "Lasagna" to identify its products and services that are based on the integrated *in situ* remediation process that has been developed by a consortium. The proposed technology combines electro-osmosis with treatment zones that are installed directly in the contaminated soils to form an integrated *in situ* remedial process, as shown in Figure 3.7. The consortium consists of Monsanto, E.I. du Pont de Nemours and Company (DuPont), and General Electric (GE), with participation by the U.S. EPA Office of Research and Development and DOE.

The *in situ* decontamination process is carried out as follows[2]:

1. Highly permeable zones are created in close proximity, sectioned through the contaminated soil region, and turned into sorption–degradation zones by introducing appropriate materials (sorbents, catalytic agents, microbes, oxidants, buffers, and others).
2. Electro-osmosis is used as a liquid pump to flush contaminants from the soil into the treatment zones of degradation.

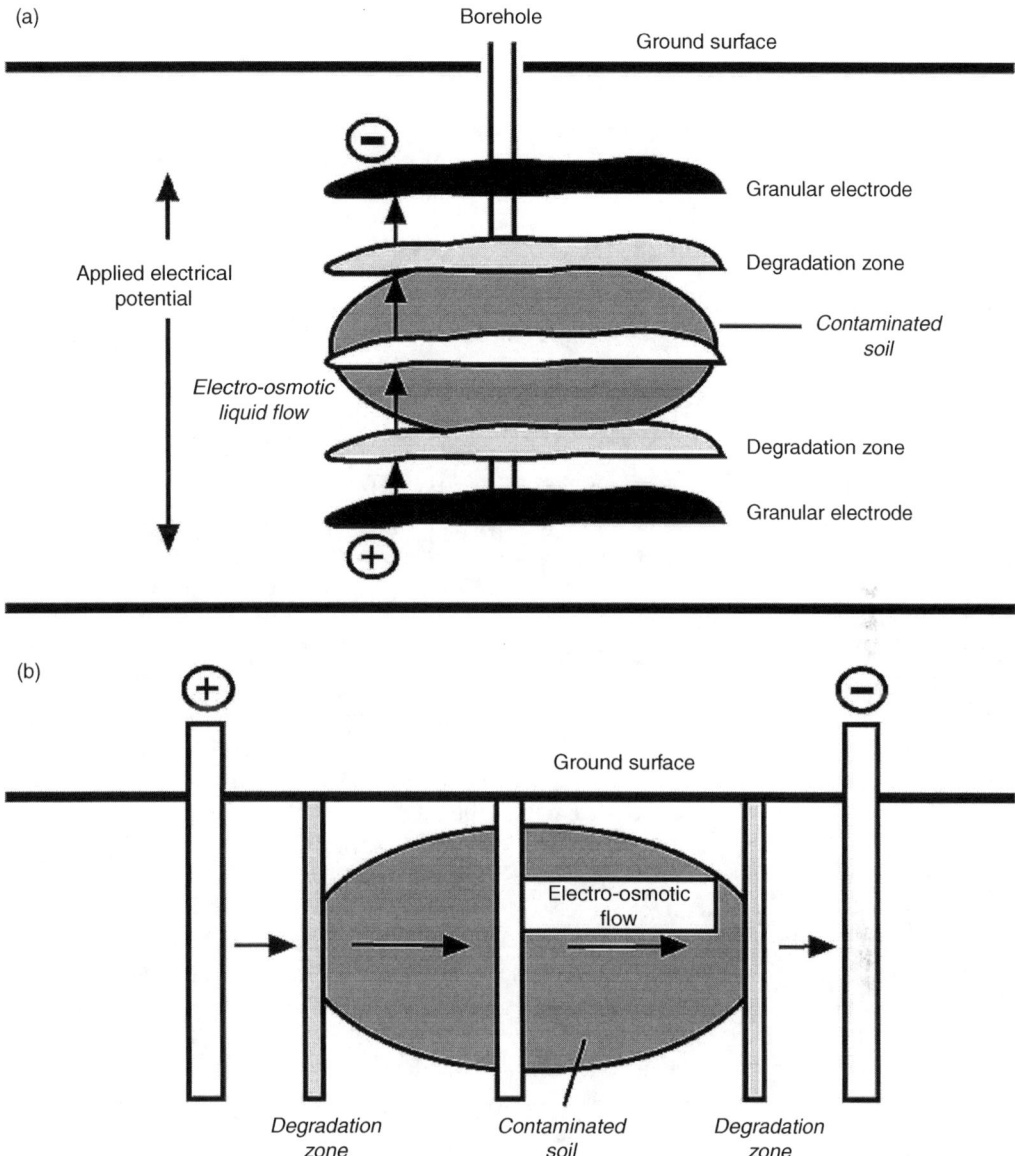

FIGURE 3.7 Schematic diagram of the Lasagna™ Process. (a) Horizontal configuration and (b) vertical configuration. *Note*: Electro-osmotic flow is a reversed upon switching electrical polarity. (From U.S. EPA, Recent Developments for In Situ Treatment of Metal Contaminated Soils, Contract no. 68-W5-0055 U.S. Environmental Protection Agency, Washington, DC, 1997.)

3. Liquid flow is reversed, if desired, by switching the electrical polarity, a mode that increases the efficiency with which contaminants are removed from the soil; this allows repeated passes through the treatment zones for complete sorption.

Initial field tests of the consortium process were conducted at DOE's gaseous diffusion plant in Paducah, KY. The experiment tested the combination of electro-osmosis and *in situ* sorption in treatment zones. Technology development for the degradation processes and their integration into the overall treatment scheme were carried out at bench and pilot scales, followed by field experiments of the full-scale process.[82]

3.11.4 PERFORMANCE AND COST

Work sponsored by U.S. EPA, DOE, the National Science Foundation, and private industry, when coupled with the efforts of researchers from academic and public institutions, have demonstrated the feasibility of moving electrokinetics remediation to pilot-scale testing and demonstration stages.[71]

This section describes testing and cost summary results reported by Louisiana State University, Electrokinetics, Inc., GII, Battelle Memorial Institute, and the consortium.[2]

3.11.4.1 Louisiana State University–Electrokinetics, Inc.

The Louisiana State University (LSU)–Electrokinetics, Inc. Group has conducted bench-scale testing on radionuclides and on organic compounds. Test results have been reported for Pb, Cd, Cr, Hg, Zn, Fe, and Mg. Radionuclides tested include U, Th, and Ra.

In collaboration with U.S. EPA, the LSU–Electrokinetics, Inc. Group has completed pilot-scale studies of electrokinetic soil processing in the laboratory. Electrokinetics, Inc. carried out a site-specific pilot-scale study of the Electro-Klean™ electrical separation process. Pilot field studies have also been reported in the Netherlands on soils contaminated with Pb, As, Ni, Hg, Cu, and Zn.

A pilot-scale laboratory study investigating the removal of 2000 mg/kg of lead loaded onto kaolinite has been completed. Removal efficiencies of 90 to 95% were obtained. The electrodes were placed 1 in. apart in a 2 T kaolinite specimen for four months, at a total energy cost of about (2007) 22 USD/T.[81]

With the support of DOD, Electrokinetics, Inc. carried out a comprehensive demonstration study of Pb extraction from a creek bed at a U.S. Army firing range in Louisiana. U.S. EPA took part in an independent assessments of the results of that demonstration study under the SITE Program. The soils were contaminated with levels as high as 4500 mg/kg of Pb, and pilot-scale studies have demonstrated that the concentrations of Pb decreased to less than 300 mg/kg in 30 weeks of processing. The TCLP values dropped from more than 300 mg/L to less than 40 mg/L within the same period. At the site of the demonstration study, Electrokinetics, Inc. used the CADEX™ electrode system, which promotes the transport of species into the cathode compartment where they are precipitated or electrodeposited directly. Electrokinetics, Inc. used a special electrode material that is cost-effective and does not corrode. Under the supervision and support of the Electric Power Research Institute and power companies in the southern U.S., a treatability and a pilot-scale field testing study of soils in sites contaminated with As was performed, in a collaborative effort between Southern Company Services Engineers and Electrokinetics, Inc.[2]

The processing cost of a system designed and installed by Electrokinetics, Inc. consists of energy cost, conditioning cost, and fixed costs associated with installation of the system. Power consumption is related directly to the conductivity of the soil across the electrodes. Electrical conductivity of soils can span orders of magnitude, from 30 mhos/cm to more than 3000 µmhos/cm, with higher values in saturated, high-plasticity clays. A mean conductivity value is 500 µmhos/cm. The voltage gradient is held to approximately 1 V/cm in an attempt to prevent adverse effects of temperature increases and for other practical reasons.[71] It may be cost-prohibitive to attempt to remediate high-plasticity soils that have high electrical conductivities. However, for most deposits having conductivities of 500 µmhos/cm, the daily energy consumption will be approximately 12 kWh/m^3/d or about 1.20 USD/m^3/d (0.10 USD/kWh) and 36 USD/m^3/month. The processing time will depend upon several factors, including the spacing of the electrodes and the type of conditioning scheme that will be used. If an electrode spacing of 4 m is selected, it may be necessary to process the site over several months.

Pilot-scale studies using "real-world" soils indicate that the energy expenditures in the extraction of metals from soils may be 500 kWh/m^3 or more at an electrode spacing of between 1.0 m and 1.5 m.[78] The vendor estimates that the direct cost of about 50 USD/m^3 (0.10 USD/kWh) suggested for this energy expenditure, together with the cost of enhancement, could result in direct costs of 100 USD/m^3. If no other efficient *in situ* technology is available to remediate fine-grained and heterogeneous subsurface deposits contaminated with metals, this technique would remain potentially competitive.

3.11.4.2   Geokinetics International, Inc.

GII has successfully demonstrated *in situ* electrochemical remediation of metal-contaminated soils at several sites in Europe. Geokinetics, a sister company of GII, has also been involved in the electrokinetics arena in Europe. Table 3.10 summarizes the physical characteristics of five of the sites, including the size, the contaminant(s) present, and the overall performance of the technology at each site. GII estimates its typical costs for "turn key" remediation projects are in the range of 160–260 USD/m^3 (2007 USD).[2]

3.11.4.3   Battelle Memorial Institute

The technology demonstration through the SITE Program has been completed,[81] and the results indicate that the electroacoustical technology is technically feasible for the removal of inorganic species from clay soils.[83]

3.11.4.4   Consortium Process

The Phase I field test of the Lasagna™ process has been completed. Scale-up from laboratory units was successfully achieved with respect to electrical parameters and electro-osmotic flow. Soil samples taken throughout the test site before and after the test indicate a 98% removal of trichloroethylene (TCE) from a tight clay soil (i.e., hydraulic conductivity less than 1×10^{-7} cm/sec). TCE soil levels were reduced from the 100 to 500 mg/kg range to an average concentration of 1 mg/kg.[84] Various treatment processes are being investigated in the laboratory to address other types of contaminants, including heavy metals.[84]

3.11.5   SUMMARY OF ELECTROKINETIC REMEDIATION

Electrokinetic remediation may be applied to both saturated and partially saturated soils. One problem to overcome when applying electrokinetic remediation to the vadose zone is the drying of

---

TABLE 3.10
Performance of Electrochemical Soil Remediation Applied at Five Field Sites in Europe

Site Description	Soil Volume (m³)	Soil Type	Contaminant	Initial Concentration (mg/kg)	Final Concentration (mg/kg)
Former paint factory	230	Peat/clay soil	Cu	1220	<200
			Pb	>3780	<280
Operational galvanizing plant	40	Clay soil	Zn	>1400	600
Former timber plant	190	Heavy clay soil	As	>250	<30
Temporary landfill	5440	Argillaceous sand	Cd	>180	<40
Military air base	1900	Clay	Cd	660	47
			Cr	7300	755
			Cu	770	98
			Ni	860	80
			Pb	730	108
			Zn	2600	289

Source: U.S. EPA, Recent Developments for In Situ Treatment of Metal Contaminated Soils, Contract no. 68-W5-0055 U.S. Environmental Protection Agency, Washington, DC, 1997.

soil near the anode. When an electric current is applied to soil, water will flow by electro-osmosis in the soil pores, usually toward the cathode. The movement of the water will deplete soil moisture adjacent to the anode, and moisture will collect near the cathode. However, processing fluids may be circulated at the electrodes. The fluids can serve both as a conducting medium and as a means to extract or exchange the species and introduce other species. Another use of processing fluids is to control, depolarize, or modify either or both electrode reactions. The advance of the process fluid (acid or the conditioning fluid) across the electrodes assists in desorption of species and dissolution of carbonates and hydroxides. Electro-osmotic advection and ionic migration lead to the transport and subsequent removal of the contaminants. The contaminated fluid is then recovered at the cathode.

Spacing of the electrode will depend upon the type and level of contamination and the selected current voltage regime. When higher voltage gradients are generated, the efficiency of the process might decrease because of increases in temperature. A spacing that will generate a potential gradient in the order of 1 V/cm is preferred. The spacing of electrodes generally will be as much as 3 m. The duration of the remediation will be site-specific. The remediation process should be continued until the desired removal is achieved. However, it should be recognized that, in cases in which the duration of treatment is reduced by increasing the electrical potential gradient, the efficiency of the process will decrease.[85,86]

The advantage of the technology is its potential for cost-effective use for both *in situ* and *ex situ* applications. The fact that the technique requires the presence of a conducting pore fluid in a soil mass may have site-specific implications. Also, heterogeneities or anomalies found at sites (such as submerged foundations, rubble, large quantities of iron or iron oxides, large rocks, or gravel) or submerged cover material, such as seashells, are expected to reduce removal efficiencies.[71]

3.12 PHYTOREMEDIATION

This technology is in the stage of commercialization for treatment of soils contaminated with metals, and in the future may provide a low-cost option under specific circumstances. At the current stage of development, this process is best suited for sites with widely dispersed contamination at low concentrations where only treatment of soils at the surface (in other words, within the depth of the root zone) is required.[2]

Phytoremediation is the use of plants to remove, contain, or render harmless environmental contaminants. This definition applies to all biological, chemical, and physical processes that are influenced by plants and that aid in the cleanup of contaminated substances.[87] Plants can be used in site remediation, both to mineralize and immobilize toxic organic compounds at the root zone and to accumulate and concentrate metals and other inorganic compounds from soil into aboveground shoots.[88] Although phytoremediation is a relatively new concept in the waste management community, techniques, skills, and theories developed through the application of well-established agroeconomic technologies are easily transferable. The development of plants for restoring sites contaminated with metals will require the multidisciplinary research efforts of agronomists, toxicologists, biochemists, microbiologists, pest-management specialists, engineers, and other specialists.[87,88] Table 3.11 presents an overview of phytoremediation technology.

Two basic approaches for metals remediation include phytoextraction and phytostabilization. Phytoextraction relies on the uptake of contaminants from the soil and their translocation into aboveground plant tissue, which is harvested and treated. Although hyperaccumulating trees, shrubs, herbs, grasses, and crops have potential, crops seem to be most promising because of their greater biomass production. Ni and Zn appear to be the most easily absorbed, although tests with Cu and Cd are encouraging.[2] Significant uptake of Pb, a commonly occurring contaminant, has not been demonstrated on a large scale. However, some researchers are experimenting with soil amendments that would facilitate uptake of Pb by the plants.

TABLE 3.11

Overview of Phytoremediation Technology

General characteristics
- It is best used at sites with low to moderate disperse metals content and with soil media that will support plant growth.
- Applications are limited to the depth of the root zone.
- Longer times are required for remediation compared with other technologies.
- Different species have been identified to treat different metals.

Approach 1: Phytoextraction (Harvest)	**Approach 2: Phytostabilization (Root-Fixing)**
Description	*Description*
Uptake of contaminants from soil into aboveground plant tissue, which is periodically harvested and treated.	Production of chemical compounds by the plant to immobilize contaminants at the interface of roots and soil. Additional stabilization can occur by raising the pH level in the soil.
Status	*Status*
Field testing for effectiveness on radioactive metals is ongoing in the vicinity of the damaged nuclear reactor in Chernobyl, Ukraine. Field testing is also being conducted in Trenton, NJ, and Butte, MT, and by the Idaho National Engineering Laboratory (INEL) in Fernald, OH.	Research is ongoing.
Applicability	*Applicability*
Potentially applicable for many metals. Ni and Zn appear to be most easily absorbed. Preliminary results for absorption of Cu and Cd are encouraging.	Potentially applicable for many metals, especially Pb, Cr, and Hg.
Comments	*Comments*
Cost is affected by the volume of biomass produced that may require treatment before disposal. Cost is also affected by the concentration and depth of contamination and the number of harvests required.	Long-term maintenance is required.

Source: U.S. EPA, Recent Developments for In Situ Treatment of Metal Contaminated Soils, Contract no. 68-W5-0055 U.S. Environmental Protection Agency, Washington, DC, 1997.

3.12.1 PROCESS DESCRIPTION

Metals considered essential for at least some forms of life include vanadium (V), Cr, Mn, Fe, Co, Ni, Cu, Zn, and Mo.[88] Because many metals are toxic in concentrations above minute levels, an organism must regulate the cellular concentrations of such metals. Consequently, organisms have evolved transport systems to regulate the uptake and distribution of metals. Plants have remarkable metabolic and absorption capabilities, as well as transport systems that can take up ions selectively from the soil. Plants have evolved a great diversity of genetic adaptations to handle potentially toxic levels of metals and other pollutants that occur in the environment. In plants, uptake of metals occurs primarily through the root system, in which the majority of mechanisms to prevent metal toxicity are found.[89] The root system provides an enormous surface area that absorbs and accumulates the water and nutrients essential for growth. In many ways, living plants can be compared to solar-powered pumps that can extract and concentrate certain elements from the environment.[90]

Plant roots cause changes at the soil–root interface as they release inorganic and organic compounds (root exudates) in the area of the soil immediately surrounding the roots (the rhizosphere).[91] Root exudates affect the number and activity of microorganisms, the aggregation and stability of soil particles around the root, and the availability of elements. Root exudates can increase (mobilize) or decrease (immobilize) directly or indirectly the availability of elements in the rhizosphere. Mobilization and immobilization of elements in the rhizosphere can be caused by the following events[92,93]:

1. Changes in soil pH
2. The release of complexing substances, such as metal-chelating molecules
3. Changes in oxidation–reduction potential
4. An increase in microbial activity

Phytoremediation technologies can be developed for different applications in environmental cleanup and are classified into three types:

1. Phytoextraction
2. Phytostabilization
3. Rhizofiltration

3.12.1.1 Phytoextraction

Phytoextraction technologies use hyperaccumulating plants to transport metals from the soil and concentrate them into the roots and aboveground shoots, which can then be harvested.[87,88,91] A plant containing more than 0.1% of Ni, Co, Cu, Cr, or 1% Zn and Mn in its leaves on a dry weight basis is called a hyperaccumulator, regardless of the concentration of metals in the soil.[88,94,95]

Almost all metal-hyperaccumulating species known today were discovered on metal-rich soils, either natural or artificial, often growing in communities with metal excluders.[88,96] In fact, almost all metal-hyperaccumulating plants are endemic to such soils, suggesting that hyperaccumulation is an important ecophysiological adaptation to metal stress and one of the manifestations of resistance to metals. The majority of hyperaccumulating species discovered so far are restricted to a few specific geographical locations.[88,94] For example, Ni hyperaccumulators are found in New Caledonia, the Philippines, Brazil, and Cuba. Ni and Zn hyperaccumulators are found in southern and central Europe and Asia Minor.

Dried or composted plant residues or plant ashes that are highly enriched with metals can be isolated as hazardous waste or recycled as metal ore.[98] The goal of phytoextraction is to recycle as "bio-ores" metals reclaimed from plant ash in the feed stream of smelting processes. Even if the plant ashes do not have enough concentration of metal to be useful in smelting processes, phytoextraction remains beneficial because it reduces by as much as 95% the amount of hazardous waste to be landfilled.[2] Several research efforts in the use of trees, grasses, and crop plants are being pursued to develop phytoremediation as a cleanup technology. The following paragraphs briefly discuss these three phytoextraction techniques.

The use of trees can result in the extraction of significant amounts of metal because of their high biomass production. However, the use of trees in phytoremediation requires long-term treatment and may create additional environmental concerns about falling leaves. When leaves containing metals fall or blow away, recirculation of metals to the contaminated site and migration off site by wind transport or through leaching can occur.[2]

Some grasses accumulate surprisingly high levels of metals in their shoots without exhibiting toxic effects. However, their low biomass production results in relatively low yield of metals. Genetic breeding of hyperaccumulating plants that produce relatively large amounts of biomass could make the extraction process highly effective.[99]

It is known that many crop plants can accumulate metals in their roots and aboveground shoots, potentially threatening the food chain. For example, in May 1980 regulations proposed under RCRA for hazardous waste include limits on the amounts of Cd and other metals that can be applied to crops. Recently, however, the potential use of crop plants for environmental remediation has been under investigation. Using crop plants to extract metals from the soil seems practical because of their high biomass production and relatively fast rate of growth. Other benefits of using crop plants are that they are easy to cultivate and they exhibit genetic stability.[97]

3.12.1.2 Phytostabilization

Phytostabilization uses plants to limit the mobility and bioavailability of metals in soils. Ideally, phytostabilizing plants should be able to tolerate high levels of metals and to immobilize them in the soil by sorption, precipitation, complexation, or the reduction of metal valences. Phytostabilizing plants should also exhibit low levels of accumulation of metals in shoots to eliminate the possibility that residues in harvested shoots might become hazardous wastes.[90] In addition to stabilizing the metals present in the soil, phytostabilizing plants can also stabilize the soil matrix to minimize erosion and migration of sediment. Dr. Gary Pierzynski of Kansas State University is studying phytostabilization in poplar trees, which were selected for the study because they can be deep-planted and may be able to form roots below the zone of maximum contamination.[2]

Because most sites contaminated with metals lack established vegetation, metal-tolerant plants are used to revegetate such sites to prevent erosion and leaching.[100] However, that approach is a containment rather than a remediation technology. Some researchers consider phytostabilization to be an interim measure to be applied until phytoextraction becomes fully developed. However, other researchers are developing phytostabilization as a standard protocol of metal remediation technology, especially for sites at which removal of metals does not seem to be economically feasible. After field applications conducted by a group in Liverpool, England, varieties of three grasses were made commercially available for phytostabilization[90]:

1. *Agrostis tenuis, cv Parys* for Cu wastes
2. *Agrosas tenuis, cv Coginan* for acid Pb and Zn wastes
3. *Festuca rubra, cv Merlin* for calcareous Pb and Zn wastes

3.12.1.3 Rhizofiltration

One type of rhizofiltration uses plant roots to absorb, concentrate, and precipitate metals from wastewater,[90] which may include leachate from soil. Rhizofiltration uses terrestrial plants instead of aquatic plants because the terrestrial plants develop much longer, fibrous root systems covered with root hairs that have extremely large surface areas. This variation of phytoremediation uses plants that remove metals by sorption, which does not involve biological processes. Use of plants to translocate metals to the shoots is a slower process than phytoextraction.[100]

Another type of rhizofiltration, which is more fully developed, involves construction of wetlands or reed beds for the treatment of contaminated wastewater or leachate. The technology is cost-effective for the treatment of large volumes of wastewater that have low concentrations of metals.[100] Because rhizofiltration focuses on the treatment of contaminated water, it is not discussed further in this chapter.

Table 3.12 presents the advantages and disadvantages of each of the types of phytoremediation currently being researched that are categorized as either phytoextraction on phytostabilization.[90]

3.12.1.4 Future Development

Faster uptake of metals and higher yields of metals in harvested plants may become possible through the application of genetic engineering or selective breeding techniques. Recent laboratory-scale testing has revealed that a genetically altered species of mustard weed can uptake mercuric ions from the soil and convert them to metallic mercury, which is transpired through the leaves.[2] Improvements in phytoremediation may be attained through research and a better understanding of

TABLE 3.12

Types of Phytoremediation Technology: Advantages and Disadvantages

Type of Phytoremediation	Advantages	Disadvantages
Phytoextraction by trees	High biomass production	Potential for off-site migration and leaf transportation of metals to surface.
		Metals are concentrated in plant biomass and must be disposed of eventually.
Phytoextraction by grasses	High accumulation	Low biomass production and slow growth rate.
		Metals are concentrated in plant biomass and must be disposed of eventually.
Phytoextraction by crops	High biomass and increased growth rate	Potential threat to the food chain through ingestion by herbivores.
		Metals are concentrated in plant biomass and must be disposed of eventually.
Phytostabilization	No disposal of contaminated biomass required	Remaining liability issues, including maintenance for an indefinite period of time (containment rather than removal).
Rhizofiltration	Readily absorbs metals	Applicable for treatment of water only.
		Metals are concentrated in plant biomass and must be disposed of eventually.

Source: U.S. EPA, Recent Developments for In Situ Treatment of Metal Contaminated Soils, Contract no. 68-W5-0055 U.S. Environmental Protection Agency, Washington, DC, 1997.

the principles governing the processes by which plants affect the geochemistry of their soils. In addition, future testing of plants and microflora may lead to the identification of plants that have metal accumulation qualities that are far superior to those currently known.

3.12.2 APPLICABILITY

Plants have been used to treat wastewater for more than 300 years, and plant-based remediation methods for slurries of dredged material and soils contaminated with metals have been proposed since the mid-1970s.[87,101] Reports of successful remediation of soils contaminated with metals are rare, but the suggestion of such applications is more than two decades old, and progress is being made at a number of pilot test sites.[96] Successful phytoremediation must meet cleanup standards in order to be approved by regulatory agencies.

No full-scale applications of phytoremediation have been reported. One vendor, Phytotech, Inc., is developing phytostabilization for soil remediation applications. Phytotech also has patented strategies for phytoextraction and is conducting several field tests in Trenton, NJ, and in Chernobyl, Ukraine.[97] Also, as was previously mentioned, a group in Liverpool, England, has made three grasses commercially available for the stabilization of Pb, Cu, and Zn wastes.[90]

3.12.3 PERFORMANCE AND COST

A variety of new research approaches and tools are expanding understanding of the molecular and cellular processes that can be employed through phytoremediation.[102]

3.12.3.1 Performance

The potential for phytoremediation (phytoextraction) can be assessed by comparing the concentration of contaminants and volume of soil to be treated with the particular plant's seasonal productivity of

TABLE 3.13

Examples of Metal Hyperaccumulators

Metal	Plant Species	Percentage of Metal in Dry Weight of Leaves (%)	Native Location
Zn	*Thlaspi calaminare*	<3	Germany
	Viola species	1	Europe
Cu	*Aeolanthus biformifolius*	1	Zaire
Ni	*Phyllanthus serpentinus*	3.8	New Caledonia
	Alyssum bertoloni and 50 other species of alyssum	>3	Southern Europe and Turkey
	Sebertia acuminata	25 (in latex)	New Caledonia
	Stackhousia tryonii	4.1	Australia
Pb	*Brassuca juncea*	<3.5	India
Co	*Haumaniastrum robertii*	1	Zaire

Source: U.S. EPA, Recent Developments for In Situ Treatment of Metal Contaminated Soils, Contract no. 68-W5-0055 U.S. Environmental Protection Agency, Washington, DC, 1997.

biomass and ability to accumulate contaminants. Table 3.13 lists selected examples of plants identified as metal hyperaccumulators and their native countries.[2,103] If plants are to be effective remediation systems, one ton of plant biomass, costing from several hundred to a few thousand dollars to produce, must be able to treat large volumes of contaminated soil. For metals that are removed from the soil and accumulated in aboveground biomass, the total amount of biomass per hectare required for soil cleanup is determined by dividing the total weight of metal per hectare to be remediated by the accumulation factor, which is the ratio of the accumulated weight of the metal to the weight of the biomass containing the metal. The total biomass per hectare (T/ha) then can be divided by the productivity of the plant (t/ha/yr) to determine the number of years (yr) required to achieve cleanup standards—a major determinant of the overall cost and feasibility of phytoremediation.[102]

As discussed earlier, the amount of biomass is one of the factors that determine the practicality of phytoremediation. Under the best climatic conditions, with irrigation, fertilization, and other factors, total biomass productivity can approach 100 T/ha/yr. One unresolved issue is the tradeoff between accumulation of toxic elements and productivity.[104] In practice, a maximum harvest biomass yield of 10 to 20 T/ha/yr is likely, particularly for plants that accumulate metals.

These values for productivity of biomass and the metal content of the soil would limit annual capacity for removal of metals to approximately 10 to 400 kg/ha/yr, depending on the pollutant, species of plant, climate, and other factors. For a target soil depth of 30 cm (4000 T/ha), this capacity amounts to an annual reduction of 2.5 to 100 mg/kg of soil contaminants. This rate of removal of contamination often is acceptable, allowing total remediation of a site over a period of a few years to several decades.[102]

3.12.3.2 Cost

The practical objective of phytoremediation is to achieve major reductions in the cost of cleanup of hazardous sites. Salt[90] and others have noted the cost-effectiveness of phytoremediation with an example: Using phytoremediation to clean up one acre of sandy loam soil to a depth of 50 cm will typically cost USD 60,000 to USD 100,000, compared with a cost of at least USD 400,000 for excavation and disposal storage without treatment.[90] One objective of field tests is to use commercially available agricultural equipment and supplies for phytoremediation in order to reduce costs. Therefore, in addition to their remediation qualities, the agronomic characteristics of the plants must be evaluated.

The processing and ultimate disposal of the biomass generated is likely to be a major percentage of overall costs, particularly when highly toxic metals and radionuclides are present at a site. Analysis of the costs of phytoremediation must include the entire cycle of the process, from the growing and harvesting of the plants to the final processing and disposal of the biomass. It is difficult to predict costs of phytoremediation, compared with overall cleanup costs at a site. Phytoremediation may also be used as a follow-up technique after areas having high concentrations of pollutants have been mitigated or in conjunction with other remediation technologies, making cost analysis more difficult.

3.12.3.3 Future Directions

Because metal hyperaccumulators generally produce small quantities of biomass, they are unsuited agronomically for phytoremediation. Nevertheless, such plants are a valuable store of genetic and physiologic material and data.[87] To provide effective cleanup of contaminated soils, it is essential to find, breed, or engineer plants that absorb, translocate, and tolerate levels of metals in the range of 0.1 to 1.0%. It also is necessary to develop a methodology for selecting plants that are native to the area.

Three grasses are commercially available for the stabilization of Pb, Cu, and Zn wastes.[90] An integrated approach that involves basic and applied research, along with consideration of safety, legal, and policy issues, will be necessary to establish phytoremediation as a practicable cleanup technology.[87]

According to a DOE report, three broad areas of research and development can be identified for the *in situ* treatment of soil contaminated with metals[102]:

1. *Mechanisms of uptake, transport, and accumulation.* Research is needed to develop a better understanding of the use of physiological, biochemical, and genetic processes in plants. Research on the uptake and transport mechanisms is providing improved knowledge about the adaptability of those systems and how they might be used in phytoremediation.
2. *Genetic evaluation of hyperaccumulators.* Research is being conducted to collect plants growing in soils that contain high levels of metals and screen them for specific traits useful in phytoremediation. Plants that tolerate and colonize environments polluted with metals are a valuable resource, both as candidates for use in phytoremediation and as sources of genes for classical plant breeding and molecular genetic engineering.
3. *Field evaluation and validation.* Research is being carried out to use early and frequent field testing to accelerate implementation of phytoremediation technologies and to provide data to research programs. Standardization of field-test protocols and subsequent application of test results to real problems are also needed.

Research in these areas is expected to grow as many of the current engineering technologies for cleaning surface soil of metals are costly and physically disruptive. Phytoremediation, when fully developed, could result in significant cost savings and in the restoration of numerous sites by a relatively noninvasive, solar-driven, *in situ* method that, in some forms, can be aesthetically pleasing.[87]

3.12.4 Summary of Phytoremediation Technology

Phytoremediation is in the early stage of development and is being field tested at various sites in the U.S. and overseas for its effectiveness in capturing or stabilizing metals, including radioactive wastes. Limited cost and performance data are currently available. Phytoremediation has the potential to develop into a practicable remediation option at sites at which contaminants are near the surface, are relatively nonleachable, and pose little imminent threat to human health or the environment.[87] The efficiency of phytoremediation depends on the characteristics of the soil and the contaminants; these factors are summarized in the sections that follow.

3.12.4.1 Site Conditions

The effectiveness of phytoremediation is generally restricted to surface soils within the rooting zone. The most important limitation to phytoremediation is rooting depth, which can be 20, 50, or even 100 cm, depending on the plant and soil type. Therefore, one of the favorable site conditions for phytoremediation is contamination with metals that is located at the surface.[102]

The type of soil, as well as the rooting structure of the plant relative to the location of the contaminants, can have a strong influence on the uptake of any metal substance by the plant. Amendment of soils to change soil pH, nutrient compositions, or microbial activities must be selected in treatability studies to govern the efficiency of phytoremediation. Certain generalizations can be made about such cases; however, much work is needed in this area.[87] Because the amount of biomass that can be produced is one of the limiting factors affecting phytoremediation, optimal climatic conditions, with irrigation and fertilization of the site, should be considered to promote increased productivity of the best plants for the site.[102]

3.12.4.2 Waste Characteristics

Sites that have low to moderate contamination with metals might be suitable for growing hyperaccumulating plants, although the most heavily contaminated soils do not allow plant growth without the addition of soil amendments. Unfortunately, one of the most difficult metal cations for plants to translocate is Pb, which is present at numerous sites in need of remediation. Although significant uptake of Pb has not yet been demonstrated, one researcher is experimenting with soil amendments that make lead more available for uptake.[90]

Capabilities to accumulate Pb and other metals are dependent on the chemistry of the soil in which the plants are growing. Most metals, and Pb in particular, occur in numerous forms in the soil, not all of which are equally available for uptake by plants.[87,105] Maximum removal of Pb requires a balance between the nutritional requirements of plants for biomass production and the bioavailability of Pb for uptake by plants. Maximizing the availability of Pb requires low pH and low levels of available phosphate and sulfate. However, limiting the fertility of the soil in such a manner directly affects the health and vigor of plants.[87]

3.13 USE OF TREATMENT TRAINS

Several of the metal remediation technologies discussed are often enhanced through the use of treatment trains. Treatment trains use two or more remedial options applied sequentially to the contaminated soil and often increase the effectiveness while decreasing the cost of remediation. Processes involved in treatment trains include soil pretreatment, physical separation designed to decrease the amount of soil requiring treatment, additional treatment of process residuals or off-gases, and a variety of other physical and chemical techniques, which can greatly improve the performance of the remediation technology. Table 3.14 provides examples of treatment trains used to enhance each of the proved and commercialized metal remediation technologies.[5]

3.14 COST RANGES OF REMEDIAL TECHNOLOGIES

Estimated cost ranges for the basic operation of the technology are presented in Table 3.15. The reader is cautioned that the cost estimates generally do not include pretreatment, site preparation, regulatory compliance costs, costs for additional treatment of process residuals (e.g., stabilization of incinerator ash or disposal of metals concentrated by solvent extraction), or profit.[5,106] Since the actual cost of employing a remedial technology at a specific site may be significantly different than these estimates, data are best used for order-of-magnitude cost evaluations only.

TABLE 3.14
Typical Treatment Trains

	Containment	S/S	Vitrification	Soil Washing	Pyromet-allurgical	Soil Flushing
Pretreatment						
Excavation	×	E,P	I,E	×	×	
Debris removal		E,P	E	×	×	
Oversize reduction		E,P	E	×	×	
Adjust pH	×	I,E,P				
Reduction [e.g., Cr(VI) to Cr(III)]	×	I,E				
Oxidation [e.g., As(III) to As (V)]	×	I,E				
Treatment to remove or destroy organics		I,E				
Physical separation of rich and lean fractions		I,E,P	E	×	×	
Dewatering and drying for wet sludge	×	P	E		×	
Conversion of metals to less volatile forms [e.g., As_2O_3 to $Ca_3(AsO_4)_2$]			E			
Addition of high-temperature reductants					×	
Pelletizing					×	
Flushing fluid delivery and extraction system						×
Containment barriers	×	I,E,P	I	×		×
Post-treatment/residuals management						
Disposal of treated solid residuals (preferably below the frost line and above the water table)		I,E,P	E		×	
Containment barriers		I,E,P	I,E			×
Off-gas treatment		I,E,P	I,E		×	
Reuse for on-site paving		P				
Metal recovery from extraction fluid by aqueous processing (ion exchange, electrowinning, etc.)				×		
Pyrometallurgical recovery of metal from sludge				×		
Processing and reuse of leaching solution				×	×	
S/S treatment of leached residual				×		
Disposal of solid process residuals (preferably below the frost line and above the water table)				×		
Disposal of liquid process residuals				×		×
S/S treatment of slag or fly ash					×	
Reuse of slag/vitreous product as construction material			E	×		
Reuse of metal or metal compound					×	
Further processing of metal or metal compound					×	
Flushing liquid/groundwater treatment/disposal						×

Technology has been divided into the following categories: I = *in situ* process; E = *ex situ* process; P = polymer microencapsulation (*ex situ*).

Source: U.S. EPA, Technology Alternatives for the Remediation of Soils Contaminated with As, Cd, Cr, Hg, and Pb, EPA/540/S-97/500, U.S. Environmental Protection Agency, Cincinnati, OH, August 1997.

TABLE 3.15

Estimated Cost Ranges of Metals Remediation Technologies

Type of Remediation	Cost Range (2007 USD/T)
Containment[a]	13–120
Solidification–stabilization	80–380
Vitrification	520–1140
Soil washing	80–320
Soil flushing[b]	80–215
Pyrometallurgical	330–730
Electrokinetics[b]	60–160
Phytoremediation[c]	30–50

[a] Includes landfill caps and slurry walls. A slurry wall depth of 6 m is assumed.
[b] Costs reported in USD/m³, assuming soil specific gravity of 1.6.
[c] Costs reported per acre for a soil depth of 0.50 m.

Source: U.S. EPA, Recent Developments for In Situ Treatment of Metal Contaminated Soils, Contract no. 68-W5-0055 U.S. Environmental Protection Agency, Washington, DC, 1997; U.S. EPA, Technology Alternatives for the Remediation of Soils Contaminated with As, Cd, Cr, Hg, and Pb, EPA/540/S-97/500, U.S. Environmental Protection Agency, Cincinnati, OH, August 1997.

REFERENCES

1. Federal Register, Comprehensive Environmental Response, Compensation, and Liability Act (CERCLA or Superfund) 42 U.S.C. s/s 9601 et seq., United States Government, Public Laws, 1980. Available at www.access.gpo.gov/uscode/title42/chapter103_.html.
2. U.S. EPA, Recent Developments for In Situ Treatment of Metal Contaminated Soils, Contract no. 68-W5-0055 U.S. Environmental Protection Agency, Washington, DC, 1997.
3. Federal Register, Resource Conservation and Recovery Act (RCRA), 42 US Code s/s 6901 et seq., 1976, U.S. Government, Public Laws. Available at www.access.gpo.gov/uscode/title42/chapter82_.html.
4. U.S. EPA, Contaminants and Remedial Options at Selected Metal-Contaminated Sites, EPA/540/R-95/512, U.S. Environmental Protection Agency, Washington, DC, July 1995.
5. U.S. EPA, Technology Alternatives for the Remediation of Soils Contaminated with As, Cd, Cr, Hg, and Pb, EPA/540/S-97/500, U.S. Environmental Protection Agency, Cincinnati, OH, August 1997.
6. U.S. EPA, In Situ Technologies for the Remediation of Soils Contaminated with Metals—Status Report. U.S. Environmental Protection Agency, Cincinnati, OH, July 1996.
7. U.S. EPA, Selection of Control Technologies for Remediation of Lead Battery Recycling Sites, EPA/540/2-91/014, U.S. Environmental Protection Agency, Cincinnati, OH, 1991.
8. U.S. EPA, Engineering Bulletin: Selection of Control Technologies for Remediation of Lead Battery Recycling Site, EPA/540/S-92/011, U.S. Environmental Protection Agency, Cincinnati, OH, 1992.
9. U.S. EPA, Contaminants and Remedial Options at Wood Preserving Sites, EPA 600/R-92/182, U.S. Environmental Protection Agency, Washington, DC, 1992.
10. U.S. EPA, Presumptive Remedies for Soils, Sediments, and Sludges at Wood Treater Sites, EPA/540/R-95/128, U.S. Environmental Protection Agency, Washington, DC, 1995.
11. U.S. EPA, Contaminants and Remedial Options at Pesticide Sites, EPA/600/R-94/202, U.S. Environmental Protection Agency, Washington, DC, 1994.
12. U.S. EPA, Separation/Concentration Technology Alternatives for the Remediation of Pesticide-Contaminated Soil, EPA/540/S-97/503, U.S. Environmental Protection Agency, Washington, DC, 1997.
13. McLean, J.E. and Bledsoe, B.E., Behavior of Metals in Soils, EPA/540/S-92/018, U.S. Environmental Protection Agency, Washington, DC, 1992.
14. Palmer, C.D. and Puls, R.W., Natural Attenuation of Hexavalent Chromium in Ground Water and Soils, EPA/540/S-94/505, U.S. Environmental Protection Agency, Washington, DC, 1994.

15. Benjamin, M.M. and Leckie, J.D., Adsorption of metals at oxide interfaces: Effects of the concentrations of adsorbate and competing metals, in *Contaminant sand Sediments, Volume 2: Analysis, Chemistry, Biology*, Baker R.A., Ed., Ann Arbor Science Publishers, Ann Arbor, MI, 1980, chap. 16.

16. Wagemann, R., Some theoretical aspects of stability and solubility of inorganic As in the freshwater environment, *Water Res.*, 12, 139–145, 1978.

17. Zimmerman, L. and Coles, C., Cement industry solutions to waste management—The utilization of processed waste by-products for cement manufacturing, in Proceedings of the 1st International Conference for Cement Industry Solutions to Waste Management, Calgary, Alberta, Canada, 1992, pp. 533–545.

18. Earth Platform, *Contaminated Soil Remediation*, 2007. Available at http://www.earthplatform. com/contaminated/soil/remediation.

19. Sharma, H.D. and Reddy, K.R., *Geoenvironmental Engineering: Site Remediation, Waste Containment, and Emerging Waste Management Technologies*, John Wiley & Sons, Hoboken, NJ, 2004.

20. Weston, R.F., Installation Restoration General Environmental Technology Development Guidelines for In-Place Closure of Dry Lagoons, U.S. Army Toxic and Hazardous Materials, May 1985.

21. U.S. EPA, Slurry Trench Construction for Pollution Migration Control, EPA/540/2-84/001, U.S. Environmental Protection Agency, Washington, DC, February 1984.

22. U.S. EPA, Grouting Techniques in Bottom Sealing of Hazardous Waste Sites, EPA/600/2-86/020, U.S. Environmental Protection Agency, Washington, DC, 1986.

23. U.S. EPA, Engineering Bulletin: Landfill Covers, EPA/540/S-93/500, U.S. Environmental Protection Agency, Cincinnati, OH, February 1993.

24. FRTR, *Physical Barriers*, Remediation Technologies Screening Matrix and Reference Guide, 2007. Available at http://www.frtr.gov/matrix2/section4/4-53.html.

25. U.S. EPA, Engineering Bulletin: Slurry Walls, EPA/540/S-92/008, U.S. Environmental Protection Agency, Cincinnati, OH, October 1992.

26. U.S. EPA, Solidification/Stabilization Use at Superfund Sites, EPA-542-R-00-010, U.S. Environmental Protection Agency, Washington, DC, September 2000.

27. U.S. EPA, Technical Resource Document: Solidification/Stabilization and Its Application to Waste Materials, EPA/530/R-93/012, U.S. Environmental Protection Agency, Cincinnati, OH, June 1993.

28. U.S. EPA, Engineering Bulletin: Solidification/Stabilization of Organics and Inorganics, EPA/540/S-92/015, U.S. Environmental Protection Agency, Cincinnati, OH, 1992.

29. Conner, J.R., Chemical Fixation and Solidification of Hazardous Wastes, Van Nostrand Reinhold, New York, 1990.

30. U.S. EPA, Solidification/Stabilization and Its Application to Waste Materials, EPA/530/R-93/012, U.S. Environmental Protection Agency, Washington, DC, June 1993.

31. Anderson, W.C., Ed., Innovative site remediation technology: solidification/stabilization, in *Innovative Site Remediation Technology: Phase I (Process Descriptions and Limitations)*, Vol. 4, EPA/542-B-94-001, June 1994.

32. WASTECH, *Solidification/Stabilization*, Waste Technology, American Academy of Environmental Engineers, EPA, printed under license EPA/542-B-94-001, June 1994.

33. U.S. EPA, A Citizen's Guide to Solidification/Stabilization, EPA 542-F-01-024, U.S. Environmental Protection Agency, Washington, DC, December 2001.

34. U.S. EPA, Solidification/Stabilization Use at Superfund Sites, EPA-542-R-00-010, U.S. Environmental Protection Agency, Washington, DC, September 2000.

35. U.S. ACE, Solidification/Stabilization of Contaminated Material, Unified Facility Guide Specification, UFGS-02160a, U.S. Army Corps of Engineers, October 2000.

36. ANL. Fact Sheet—Solidification/Stabilization. Drilling Waste Management Information System, Argonne National Laboratory, 2007. Available at http://web.ead.anl.gov/dwm/techdesc/solid/index.cfm.

37. U.S. EPA, Handbook on In Situ Treatment of Hazardous Waste-Contaminated Soils, EPA/540/2-90/002, U.S. Environmental Protection Agency, Cincinnati, OH, 1990.

38. Arniella, E.F. and Blythe, L.J., Solidifying traps hazardous waste, *Chem. Eng.*, 97, 92–102, 1990.

39. Kalb, P.D., Burns, H.H. and Meyer, M., Thermo-plastic encapsulation treatability study for a mixed waste incinerator off-gas scrubbing solution, in *Third International Symposium on Stabilization/Solidification of Hazardous, Radioactive, and Mixed Wastes*, Gilliam, T.M., Ed., ASTM STP 1240, American Society for Testing and Materials, Philadelphia, PA, 1993.

40. Ponder, T.G. and Schmitt, D., Field assessment of air emission from hazardous waste stabilization operation, in Proceedings of the 17th Annual Hazardous Waste Research Symposium, EPA/600/9-91/002, Cincinnati, OH, 1991.

41. Shukla, S.S., Shukla, A.S. and Lee, K.C., Solidification/stabilization study for the disposal of pentachlorophenol, *J. Hazard. Mater.*, 30, 317–331, 1992.

42. U.S. EPA, Evaluation of Solidification/Stabilization as a Best Demonstrated Available Technology for Contaminated Soils, EPA/600/2-89/013, U.S. Environmental Protection Agency, Cincinnati, OH, 1989.

43. Weitzman, L. and Hamel, L.E., Volatile emissions from stabilized waste, in Proceedings of the 15th Annual Research Symposium, EPA/600/9-90/006, U.S. Environmental Protection Agency, Cincinnati, OH, 1990.

44. Means, J.L., Nehring, K.W. and Heath, J.C., Abrasive blast material utilization in asphalt roadbed material, in *Third International Symposium on Stabilization/Solidification of Hazardous, Radioactive, and Mixed Wastes*, ASTM STP 1240, American Society for Testing and Materials, Philadelphia, PA, 1993.

45. U.S. ACE, Yearly average cost index for utilities, in Civil Works Construction Cost Index System Manual, 110-2-1304, U.S. Army Corps of Engineers, Washington, DC, 2007, p. 44. Available at http://www.nww.usace.army.mil/cost.

46. Buelt, J.L., Timmerman, C.L., Oma, K.H., FitzPatrick, V.F. and Carter, J.G., *In Situ Vitrification of Transuranic Waste: An Updated Systems Evaluation and Applications Assessment*, PNL-4800, Pacific Northwest Laboratory, Richland, WA, 1987.

47. U.S. EPA, Vitrification Technologies for Treatment of Hazardous and Radioactive Waste, EPA/625/R-92/002, U.S. Environmental Protection Agency, Cincinnati, OH, May 1992.

48. U.S. EPA, Engineering Bulletin—In Situ Vitrification Treatment, EPA/540/S-94/504, U.S. Environmental Protection Agency, Cincinnati, OH, October 1994.

49. U.S. EPA, Engineering Bulletin: In Situ Vitrification Treatment, EPA/540/S-94/504, U.S. Environmental Protection Agency, Washington, DC, revised May 2002.

50. FRTR, Solidification/Stabilization—In Situ Soil Remediation Technology, Remediation Technologies Screening Matrix and Reference Guide, 2007. Available at http://www.frtr.gov/matrix2/section4/4-8.html.

51. U.S. EPA, Geosafe Corporation In Situ Vitrification Innovative Technology Evaluation Report, EPA/540/R-94/520, U.S. Environmental Protection Agency, Washington, DC, March 1995.

52. FitzPatrick, V.F., Timmerman, C.L. and Buelt, J.L., In situ vitrification: An innovative thermal treatment technology, in Proceedings of the Second International Conference on New Frontiers for Hazardous Waste Management, 1987, pp. 305–322, EPA/600/9-87/018F, U.S. Environmental Protection Agency, EPA printed under license EPA/542-B-94-001, June 1994.

53. Timmerman, C.L., In Situ Vitrification of PCB Contaminated Soils, EPRI CS-4839, Electric Power Research Institute, Palo Alto, CA, 1986.

54. U.S. EPA, The Superfund Innovative Technology Evaluation Program: Technology Profiles, 4th ed., EPA/540/5-91/008, U.S. Environmental Protection Agency, Washington, DC, 1991.

55. Luey, J., Koegler, S.S., Kuhn, W.L., Lowery, P.S. and Winkelman, R.G., In Situ Vitrification of a Mixed-Waste Contaminated Soil Site, The 116-B-6A Crib at Hanford, PNL-8281. Pacific Northwest Laboratory, Richland, WA, 1992.

56. Hansen, J.E. and FitzPatrick, V.F., *In Situ Vitrification Applications*, Geosafe Corporation, Richland, WA, 1991.

57. U.S. EPA, Engineering Bulletin: Soil Washing Treatment, EPA/540/2-90/017, U.S. Environmental Protection Agency, Cincinnati, OH, 1996.

58. U.S. EPA, A Citizen's Guide to Soil Washing, EPA 542-F-01-008, U.S. Environmental Protection Agency, Washington, DC, May 2001.

59. William, C.A., Ed., *Innovative Site Remediation Technology: Soil Washing/Flushing*, Vol. 3, American Academy of Environmental Engineers (published by EPA under EPA 542-B-93-012), November 1993.

60. U.S. EPA, *Technology Focus—Soil Washing*, Technology Innovation Program, U.S. Environmental Protection Agency, Washington, DC, 2007. Available at http://clu-in.org/techfocus/default.focus/sec/Soil_Washing/cat/Overview.

61. Ehsan, S., Prasher, S.O. and Marshall, W.D., A washing procedure to mobilize mixed contaminants from soil. II. Heavy metals, *J. Environ. Qual.*, 35, 2084–2091, 2006.

62. Fischer, K. and Bipp, H.P., Removal of heavy metals from soil components and soils by natural chelating agents. Part II. Soil extraction by sugar acids, *Water, Air, Soil Pollut.*, 38, 271–288, 2002.

63. U.S. EPA, *Citizens Guide to Soil Washing*, EPA/542/F-92/003, U.S. Environmental Protection Agency, Washington, DC, March 1992.

64. U.S. EPA, Engineering Bulletin: In Situ Soil Flushing, EPA/540/2-91/021, U.S. Environmental Protection Agency, Cincinnati, OH, October 1991.

65. FRTR, Soil Flushing—In Situ Soil Remediation Technology, Remediation Technologies Screening Matrix and Reference Guide, 2007. Available at http://www.frtr.gov/matrix2/section4/4-6.html.

66. CPEO, *Soil Flushing*, Center for Public Environmental Oversight (CPEO), San Francisco, CA, 2007. Available at http://www.cpeo.org/techtree/ttdescript/soilflus.htm.

67. U.S. EPA, Superfund Innovative Technology Evaluation Program: Technology Profiles, 7th ed., EPA/540/R-94/526, U.S. Environmental Protection Agency, Washington, DC, November 1994.

68. Pamukcu, S. and Wittle, J.K., Electrokinetic removal of selected metals from soil, *Environ. Prog.*, II, 241–250, 1992.

69. Acar, Y.B., Electrokinetic cleanups, *Civil Eng.*, October, 58–60, 1992.

70. Mattson, E.D. and Lindgren, E.R., Electrokinetics: an innovative technology for in situ remediation of metals, in Proceedings, National Groundwater Association, Outdoor Acnon Conference, Minneapolis, MN, May 1994.

71. Acar, Y.B. and Gale, R.J. Electrokinetic remediation: basics and technology status, *J. Hazard. Mater.*, 40, 117–137, 1995.

72. Will, F., Removing toxic substances from the soil using electrochemistry, *Chem. Ind.*, May 15, 376–379, 1995.

73. Rodsand, T. and Acar, Y.B., Electrokinetic extraction of lead from spiked Norwegian marine clay, *Geoenvironment 2000*, 2, 1518–1534, 1995.

74. Lindgren, E.R., Kozak, M.W. and Mattson, E.D., Electrokinetic remediation of contaminated soils: an update, *Waste Management 92*, Tuscon, AZ, 1992, p. 1309.

75. Jacobs, R.A. and Sengun, M.Z., Model of experiences on soil remediation by electric fields, *J. Environ. Sci. Health*, 29A, 9, 1994.

76. Acar, Y.B. and Gale, R.J., Electrochemical Decontamination of Soils and Slurries, U.S. Patent 5,137,608, August 15, 1992.

77. Marks, R., Acar, Y.B. and Gale, R.J., *In situ* Bioelectrokinetic Remediation of Contaminated Soils Containing Hazardous Mixed Wastes, U.S. Patent 5,458,747, October 17, 1995.

78. Acar, Y.B. and Alshawabkeh, A.N., Electrokinetic remediation: I. Pilot-scale tests with lead spiked kaolinite, II. Theoretical model, *J. Geotech. Eng.*, 122, 173–196, 1996.

79. Pool, W., Process for the Electroreclamation of Soil Material, U.S. Patent 5,433,829, July 18, 1995.

80. U.S. EPA, *In Situ Remediation Technology Status Report: Electrokinetics*, EPA 542-K-94-007, U.S. Environmental Protection Agency, Washington, DC, 1995.

81. Editorial, Innovative in situ cleanup processes, *The Hazardous Waste Consultant*, September/October, 1992.

82. DOE, Development of an Integrated In-Situ Remediation Technology, Technology Development Data Sheet, DE-AR21-94MC31185, U.S. Department of Energy, 1995.

83. U.S. EPA, Superfund Innovative Technology Evaluation Program Technology Profiles, 7th ed., EPA 540-R-94-526, U.S. Environmental Protection Agency, Washington, DC, 1994.

84. U.S. EPA, Lasagna™ Public–Private Partnership, EPA 542-F-96-010A, U.S. Environmental Protection Agency, Washington, DC, 1996.

85. Szpyrkowicz, L., Radaelli, M., Bertini, S., Daniele, S. and Casarin, F., Simultaneous removal of metals and organic compounds from a heavily polluted soil, *Electrochim. Acta*, 52, 3386–3392, 2007.

86. CPEO, *Electrokinetics*, Center for Public Environmental Oversight (CPEO), San Francisco, CA, 2007. Available at http://www.cpeo.org/techtree/ttdescript/elctro.htm.

87. Cunningham, S.D. and Berti, W.R., Remediation of contaminated soils with green plants: an overview, *In Vitro Cell. Dev. Biol.*, 29, 207–212, 1993.

88. Raskin, I., Bioconcentration of metals by plants, *Environ. Biotechnol.*, 5, 285–290, 1994.

89. Goldsbrough, P., Phytochelatins and metallothioneins: complementary mechanisms for metal tolerance, in Fourteenth Annual Symposium 1995 in Current Topics in Plant Biochemistry, Physiology and Molecular Biology, Columbia, MO, 1995.

90. Salt, D.E., Phytoremediation: A novel strategy for the removal of toxic metals from the environment using plants, *Biotechnology*, 13, 468–474, 1995.

91. Kumar, P.B.A., Phytoextraction: the use of plants to remove metals from soils, *Environ. Sci. Technol.*, 29, 1232–1238, 1995.

92. Durham, S., *Using Plants to Clean Up Soil*, U.S. Department of Agriculture (USDA), 2007. Available at http://www.ars.usda.gov/is/pr/2007/070123.htm.

93. Morel, I.L., Root exudates and metal mobilization, in Fourteenth Annual Symposium 1995 in Current Topics in Plant Biochemistry, Physiology and Molecular Biology, Columbia, MO, 1995.

94. Baker, A.J.M. and Brooks, R.R., Terrestrial higher plants which hyperaccumulate metallic elements— a review of their distribution, ecology, and phytochemistry, *Biorecovery*, 1, 81–126, 1989.

95. Hyperaccumulation in the genus *Alyssum*, in Fourteenth Annual Symposium 1995 in Current Topics in Plant Biochemistry, Physiology and Molecular Biology, Columbia, MO, 1995.

96. Baker, A.J.M., Metal hyperaccumulation by plants: our present knowledge of ecophysiological phenomenon, in Fourteenth Annual Symposium 1995 in Current Topics in Plant Biochemistry, Physiology and Molecular Biology, Columbia, MO, 1995.

97. King Communications Group, Inc., Promise of heavy metal harvest lures venture funds, *The Bioremediation Report*, 4, 1, Washington, DC, 1995.

98. Greger, M. and Landberg, M.T., Improving removal of metals from soil by *Salix*, in Proceedings of the 7th International Conference on the Biogeochemistry of Trace Elements, Uppsala, Sweden, June 15–19, 2003.

99. Chaney, R.L, Malik, M., Li, Y.M., Brown, S.L., Angle, J.S. and Baker, A.J.M., Phytoremediation of soil metals, *Curr. Opin. Biotechnol.*, 8, 279–284, 1997.

100. Ensley, B.D., Will plants have a role in bioremediation?, in Fourteenth Annual Symposium 1995 in Current Topics in Plant Biochemistry, Physiology and Molecular Biology, Columbia, MO, 1995.

101. Cunningham, S.D. and Lee, C.R., Phytoremediation: Plant-based remediation of contaminated soil and sediments, in Proceedings of a Symposium of the Soil Science Society of America, Chicago, IL, November 1994.

102. DOE, Summary Report of a Workshop on Phytoremediation Research Needs, U.S. Department of Energy, Santa Rosa, CA, July 24–26, 1994.

103. Baker, A.J.M., Brooks, R.R. and Reeves, R.D., Growing for gold ... and copper ... and zinc, *New Scientist*, 1603, 44–48, 1989.

104. Parry, I., Plants absorb metals, *Pollut. Eng.*, February, 40–41, 1995.

105. USDA, Acidifying soil helps plant remove cadmium, zinc metals, Agricultural Research Service, *Science Daily*, 2007. Available at http://www.sciencedaily.com/releases/2005/06/050619192657.htm.

106. Hyman, M. and Dupont, R.R., *Groundwater and Soil Remediation: Process Design and Cost Estimating of Proven Technologies*, ASCE Publications, Reston, VA, 2001, p. 534.

4 Treatment of Wastes from Metal Finishing Industry

Nazih K. Shammas and Lawrence K. Wang

CONTENTS

4.1 INDUSTRY DESCRIPTION

The metal finishing industry is one of many industries subject to regulation under the Resource Conservation and Recovery Act (RCRA)[1,2] and the Hazardous and Solid Waste Amendments (HSWA).[3] The metal finishing industry has also been subject to extensive regulation under the Clean Water Act (CWA).[4] Compliance with these regulations requires highly coordinated regulatory, scientific, and engineering analyses to minimize costs.[5]

4.1.1 GENERAL DESCRIPTION

The metal finishing industry comprises 44 unit operations involving the machining, fabrication, and finishing of metal products (Standard Industrial Classification (SIC) groups 34 through 39). There are approximately 160,000 manufacturing facilities in the U.S. that are classified as being part of the metal finishing industry.[6] These facilities are engaged in the manufacturing of a variety of products constructed primarily by using metals. The operations performed usually begin with a raw stock in the form of rods, bars, sheets, castings, forgings, and so on, and can progress to sophisticated surface-finishing operations. The facilities vary in size from small job shops employing fewer than ten people to large plants employing thousands of production workers. Wide variations also exist in the age of the facilities and the number and type of operations performed within facilities. Because of the differences in size and processes, production facilities are custom-tailored to the specific needs of each plant. The possible variations in unit operations within the metal finishing industry are extensive. Some complex products could require the use of nearly all 44 possible unit operations, but a simple product might require only a single operation. Each of the 44 individual unit operations is listed with a brief description in the following[7]:

1. *Electroplating* is the production of a thin coating of one metal upon another by electrodeposition.
2. *Electroless plating* is a chemical reduction process that depends upon the catalytic reduction of a metallic ion in an aqueous solution containing a reducing agent and the subsequent deposition of metal without the use of external electric energy.
3. *Anodizing* is an electrolytic oxidation process that converts the surface of the metal to an insoluble oxide.
4. *Chemical conversion coatings* are applied to previously deposited metal or basis material for increased corrosion protection, lubricity, preparation of the surface for additional coatings, or formulation of a special surface appearance. This operation includes chromating, phosphating, metal coloring, and passivating.
5. *Etching and chemical milling* are used to produce specific design configurations and tolerances on parts by controlled dissolution with chemical reagents or etchants.
6. *Cleaning* involves the removal of oil, grease, and dirt from the surface of the basis material using water with or without a detergent or other dispersing material.
7. *Machining* is the general process of removing stock from a workpiece by forcing a cutting tool through the workpiece, removing a chip of basis material. Machining operations such as turning, milling, drilling, boring, tapping, planing, broaching, sawing and cutoff, shaving, threading, reaming, shaping, slotting, hobbing, filing, and chamfering are included in this definition.
8. *Grinding* is the process of removing stock from a workpiece by the use of a tool consisting of abrasive grains held by a rigid or semirigid binder. The processes included in this unit operation are sanding (or cleaning to remove rough edges or excess material), surface finishing, and separating (as in cutoff or slicing operations).
9. *Polishing* is an abrading operation used to remove or smooth out surface defects (scratches, pits, tool marks, and so on) that adversely affect the appearance or function of a part. The operation usually referred to as buffing is included in the polishing operation.

10. *Barrel finishing* or tumbling is a controlled method of processing parts to remove burrs, scale, flash, and oxides, as well as to improve surface finish.
11. *Burnishing* is the process of finish sizing or smooth finishing a workpiece (previously machined or ground) by displacement, rather than removal, of minute surface irregularities. It is accomplished with a smooth point or line-contact and fixed or rotating tools.
12. *Impact deformation* is the process of applying an impact force to a workpiece such that the workpiece is permanently deformed or shaped. Impact deformation operations include shot peening, forging, high-energy forming, heading, and stamping.
13. *Pressure deformation* is the process of applying force (at a slower rate than an impact force) to permanently deform or shape a workpiece. Pressure deformation includes operations such as roiling, drawing, bending, embossing, coining, swaging, sizing, extruding, squeezing, spinning, seaming, staking, piercing, necking, reducing, forming, crimping, coiling, twisting, winding, flaring, or weaving.
14. *Shearing* is the process of severing or cutting a workpiece by forcing a sharp edge or opposed sharp edges into the workpiece, stressing the material to the point of shear failure and separation.
15. *Heat treating* is the modification of the physical properties of a workpiece through the application of controlled heating and cooling cycles. Such operations as tempering, carburizing, cyaniding, nitriding, annealing, normalizing, austenizing, quenching, austempering, siliconizing, martempering, and malleabilizing are included in this definition.
16. *Thermal cutting* is the process of cutting, slotting, or piercing a workpiece using an oxyacetylene oxygen lance or electric arc cutting tool.
17. *Welding* is the process of joining two or more pieces of material by applying heat, pressure, or both, with or without filler material, to produce a localized union through fusion or recrystallization across the interface. Included in this process are gas welding, resistance welding, arc welding, cold welding, electron beam welding, and laser beam welding.
18. *Brazing* is the process of joining metals by flowing a thin, capillary thickness layer of non-ferrous filler metal into the space between them. Bonding results from the intimate contact produced by the dissolution of a small amount of base metal in the molten filler metal, without fusion of the base metal. The term brazing is used where the temperature exceeds 425°C (800°F).
19. *Soldering* is the process of joining metals by flowing a thin, capillary thickness layer of nonferrous filler metal into the space between them. Bonding results from the intimate contact produced by the dissolution of a small amount of base metal in the molten filler metal, without fusion of the base metal. The term soldering is used where the temperature range falls below 425°C (800°F).
20. *Flame spraying* is the process of applying a metallic coating to a workpiece using finely powdered fragments of wire and suitable fluxes, which are projected together through a cone of flame onto the workpiece.
21. *Sand blasting* is the process of removing stock, including surface films, from a workpiece by the use of abrasive grains pneumatically impinged against the workpiece. The abrasive grains used include sand, metal shot, slag, silica, pumice, or natural materials such as walnut shells.
22. *Abrasive jet machining* is a mechanical process for cutting hard, brittle materials. It is similar to sand blasting but uses much finer abrasives carried at high velocities (150 to 910 m/s [500 to 3000 ft/sec]) by a liquid or gas stream. Uses include frosting glass, removing metal oxides, deburring, and drilling and cutting thin sections of metal.
23. *Electrical discharge machining* is a process that can remove metal with good dimensional control from any metal. It cannot be used for machining glass, ceramics, or other non-conducting materials. Electrical discharge machining is also known as spark machining or

electronic erosion. The operation was developed primarily for machining carbides, hard nonferrous alloys, and other hard-to-machine materials.

24. *Electrochemical machining* is a process based on the same principles used in electroplating, except the workpiece is the anode and the tool is the cathode. Electrolyte is pumped between the electrodes and a potential applied, resulting in rapid removal of metal.

25. *Electron beam machining* is a thermoelectric process in which heat is generated by high-velocity electrons impinging the workpiece, converting the beam into thermal energy. At the point where the energy of the electrons is focused, the beam has sufficient thermal energy to vaporize the material locally. The process is generally carried out in a vacuum. The process results in X-ray emission, so the work area needs to be shielded to absorb radiation. At present the process is used for drilling holes as small as 0.05 mm (0.002 in.) in any known material, cutting slots, shaping small parts, and machining sapphire jewel bearings.

26. *Laser beam machining* is the process of using a highly focused, monochromatic collimated beam of light to remove material at the point of impingement on a workpiece. Laser beam machining is a thermoelectric process, and material removal is largely accomplished by evaporation, although some material is removed in the liquid state at high velocity. Because the metal removal rate is very small, this process is used for such jobs as drilling microscopic holes in carbides or diamond wire drawing dies, and for removing metal in the balancing of high-speed rotating machinery.

27. *Plasma arc machining* is the process of material removal or shaping of a workpiece by a high-velocity jet of high-temperature ionized gas. A gas (nitrogen, argon, or hydrogen) is passed through an electric arc, causing it to become ionized and raising its temperature in excess of 16,000°C (30,000°F). The relatively narrow plasma jet melts and displaces the workpiece material in its path.

28. *Ultrasonic machining* is a mechanical process designed to remove material by the use of abrasive grains, which are carried in a liquid between the tool and the work, and which bombard the work surface at high velocity. This action gradually chips away minute particles of material in a pattern controlled by the tool shape and contour. Operations that can be performed include drilling, tapping, coining, and the making of openings in all types of dies.

29. *Sintering* is the process of forming a mechanical part from a powdered metal by fusing the particles together under pressure and heat. The temperature is maintained below the melting point of the basis metal.

30. *Laminating* is the process of adhesive bonding of layers of metal, plastic, or wood to form a part.

31. *Hot dip coating* is the process of coating a metallic workpiece with another metal by immersion in a molten bath to provide a protective film. Galvanizing (hot dip zinc) is the most common hot dip coating.

32. *Sputtering* is the process of covering a metallic or nonmetallic workpiece with thin films of metal. The surface to be coated is bombarded with positive ions in a gas discharge tube, which is evacuated to a low pressure.

33. *Vapor plating* is the process of decomposing a metal or compound on a heated surface by reduction or decomposition of a volatile compound at a temperature below the melting point of either the deposit or the basis material.

34. *Thermal infusion* is the process of applying a fused zinc, cadmium, or other metal coating to a ferrous workpiece by imbuing the surface of the workpiece with metal powder or dust in the presence of heat.

35. *Salt bath descaling* is the process of removing surface oxides or scale from a workpiece by immersion of the workpiece in a molten salt bath or a hot salt solution. The work is immersed in the molten salt (temperatures range from 400 to 540°C [750 to 1000°F]),

quenched with water, and then dipped in acid. Oxidizing, reducing, and electrolytic baths are available, and the particular type needed depends on the oxide to be removed.

36. *Solvent degreasing* is a process for removing oils and grease from the surfaces of a workpiece by the use of organic solvents, such as aliphatic petroleum, aromatics, oxygenated hydrocarbons, halogenated hydrocarbons, and combinations of these classes of solvents. However, ultrasonic vibration is sometimes used with liquid solvent to decrease the required immersion time for complex shapes. Solvent cleaning is often used as a precleaning operation, for example, prior to the alkaline cleaning that precedes plating, as a final cleaning of precision parts, or as a surface preparation for some painting operations.

37. *Paint stripping* is the process of removing an organic coating from a workpiece. The stripping of such coatings is usually performed with caustic, acid, solvent, or molten salt.

38. *Painting* is the process of applying an organic coating to a workpiece. This process includes the application of coatings such as paint, varnish, lacquer, shellac, and plastics by methods such as spraying, dipping, brushing, roll coating, lithographing, and wiping. Other processes included in this unit operation are printing, silk screening, and stenciling.

39. *Electrostatic painting* is the application of electrostatically charged paint particles to an oppositely charged workpiece followed by thermal fusing of the paint particles to form a cohesive paint film. Both water-borne and solvent-borne coatings can be sprayed electrostatically.

40. *Electropainting* is the process of coating a workpiece by either making it anodic or cathodic in a bath that is generally an aqueous emulsion of the coating material. The electrodeposition bath contains stabilized resin, dispersed pigment, surfactants, and sometimes organic solvents in water.

41. *Vacuum metalizing* is the process of coating a workpiece with metal by flash-heating metal vapor in a high- vacuum chamber containing the workpiece. The vapor condenses on all exposed surfaces.

42. *Assembly* is the fitting together of previously manufactured parts or components into a complete machine, unit of a machine, or structure.

43. *Calibration* is the application of thermal, electrical, or mechanical energy to set or establish reference points for a component or complete assembly.

44. *Testing* is the application of thermal, electrical, or mechanical energy to determine the suitability or functionality of a component or complete assembly.

Table 4.1 presents an industry summary for the metal finishing industry, including the total number of subcategories, number of subcategories studied, and the type and number of dischargers.

TABLE 4.1
Metal Finishing Industry Summary

Item	Number
Total subcategories	51
Subcategories studied	28
Discharges in industry	98,418
Direct	20,632
Indirect	77,586
Zero discharge	200

Source: From U.S. EPA, Treatability Manual, Vol. II, Industrial Descriptions, Report EPA-600/2-82-001b, U.S. EPA, Washington, DC, September 1981.

4.1.2 SUBCATEGORY DESCRIPTIONS

The primary purpose of subcategorization is to establish groupings within the metal-finishing industry such that each subcategory has a uniform set of quantifiable effluent limitations. Several bases were considered in establishing subcategories within the metal finishing industry. These included the following:

1. Raw waste characteristics
2. Manufacturing processes
3. Raw materials
4. Product type or production volume
5. Size and age of facility
6. Number of employees
7. Water usage
8. Individual plant characteristics

After these subcategorization bases were evaluated, raw waste characterization was selected as the basis for subcategorization. The raw waste characterization is divided into two components, inorganic and organic wastes. These components are further subdivided into the specific types of wastes that occur within the components. Inorganics include common metals, precious metals, complexed metals, hexavalent chromium, and cyanide. Organics include oils and solvents.

Table 4.2 lists the unit operations associated with each of the seven industry subcategories (raw waste characteristics). Common metals are found in the raw waste of all 44 unit operations. Precious metals are found in only seven unit operations; cornplexed metals are found in three unit operations; hexavalent chromium is found in seven unit operations; and cyanide is found in eight unit operations. Within the organics, oils are found in 22 unit operations and solvents are found in nine unit operations. A unit operation will often be found in more than one subcategory.

4.2 WASTEWATER CHARACTERIZATION

In this section, the uses of water in the metal finishing industry are presented, and the waste constituents are identified and quantified.

Water is used for rinsing workpieces, washing away spills, air scrubbing, process fluid replenishment, cooling and lubrication, washing of equipment and workpieces, quenching, spray booths, and assembly and testing. Unit operations with significant water usage include electroplating, electroless plating, anodizing, conversion coating, etching, cleaning, machining, grinding, tumbling, heat treating, welding, sand blasting, salt bath descaling, paint stripping, painting, electrostatic painting, electroplating, and testing. Unit operations with zero discharge include electron beam machining, laser beam machining, plasma arc machining, ultrasonic machining, sintering, sputtering, vapor plating, thermal infusion, vacuum metalizing, and calibration.[7]

Table 4.3 displays the ranges of flows in the metal finishing industry. Approximately 81% of the plants have flows of between 1.9 and 57 m^3/h (67 to 2000 ft^3/h). For those plants with common metals wastestreams, the average contribution of these streams to the total wastewater flow within a particular plant is 62.4% (range, 0.007 to 100%). All of the plants have a wastestream requiring common metals treatment.

Of the plants, 4.8% have production processes that generate precious metals wastewater. The average precious metals wastewater flow is 21.5% of total plant flow.

The average contribution of the complexed metal streams to total plant flow is 22.2%. The percentage was computed from data for plants whose complexed metal streams could be segregated from the total stream.

Of the plants, 42.5% have segregated hexavalent chromium wastestreams. The average flow contribution of these wastestreams to the total wastewater stream is 28.7%. At those plants with cyanide

TABLE 4.2
Subcharacterization of Unit Operations

Industry Subcategory (Raw Waste Characteristics)	Unit Operations	
Common metals		
All 44 unit operations		
Precious metals		
Electroplating	Etching	Burnishing
Electroless plating	Cleaning	
Conversion coating	Polishing	
Complexed metals		
Electroless plating		
Etching		
Cleaning		
Hexavalent chromium		
Electroplating	Etching	Electrostatic painting
Anodizing	Cleaning	
Conversion coating	Tumbling	
Cyanide		
Electroplating	Cleaning	Heat treating
Electroless plating	Tumbling	Electrochemical machining
Conversion coating	Burnishing	
Oils		
Cleaning	Pressure deformation	Solvent degreasing
Machining	Shearing	Paint stripping
Grinding	Heat treating	Painting
Polishing	Other abrasive jet machining	Assembly
Tumbling	Electrostatic painting	Calibration
Burnishing	Electrical discharge machining	Testing
Impact deformation	Electrochemical machining	
Solvents		
Cleaning	Solvent degreasing	Electrostatic painting
Heat treating	Paint stripping	Electropainting
Electrochemical machining	Painting	Assembly

Source: From U.S. EPA, Treatability Manual, Vol. II, Industrial Descriptions, Report EPA-600/2-82-001b, U.S. EPA, Washington, DC, September 1981.

wastes, the average contribution of the cyanide-bearing stream to the total wastewater generated is 28.8% (range, 0.1 to 100%). Of the plants, 31.2% have segregated cyanide-bearing wastes.

Segregated oily wastewater is defined as oil waste collected from machine sumps and process tanks. The water is segregated from other wastewaters until it has been treated by an oily waste removal system. Of the plants, 12.4% are known to segregate their oily wastes. The average contribution of these wastes to the total plant wastewater flow is 6.6% (range, ca. 0.0 to 55.4%).

In order to characterize the wastestreams in each subcategory, raw waste data were collected. Discrete samples of raw wastes were taken for each subcategory and analyses were performed on the samples. The results of these analyses are presented for each subcategory in Tables 4.4 to 4.9. In each table, data are presented on the number of detections of a pollutant, the number of samples analyzed, the median concentration, the range in concentrations, and the mean concentration of

TABLE 4.3
Wastewater Flow Characterization of the Metal Finishing Industry

Flow of Plants (m³/h)	Percentage of Plants Represented by This Flow
<0.38	2.8
0.38–1.9	5.0
1.9–3.8	13
3.8–9.5	17
9.5–19	20.7
19–28	10.7
28–38	10.7
38–57	9.1
57–95	5.0
95–190	3.8
190–380	0.7
>380	1.5

Source: From U.S. EPA, Treatability Manual, Vol. II, Industrial Descriptions, Report EPA-600/2-82-001b, U.S. EPA, Washington, DC, September 1981.

those samples detected. The minimum detection limit for the toxic pollutants in the sampling program was 1 µg/L and any value below this is listed in the six tables as BDL, indicating "below detection limit."

4.2.1 "COMMON METALS" SUBCATEGORY

Pollutant parameters found in the "common metals" subcategory of raw wastestream from sampled plants are shown in Table 4.4. The major constituents shown are parameters that originate in process solutions (such as from plating or galvanizing) and enter wastewaters by drag-out to rinses. These metals appear in wastestreams in widely varying concentrations.

4.2.2 "PRECIOUS METALS" SUBCATEGORY

Table 4.5 shows the concentrations of pollutant parameters found in the "precious metals" subcategory of raw wastestreams. The major constituents are silver and gold, which are much more commonly used in metal finishing industry operations than palladium and rhodium. Because of their high cost, precious metals are of special interest to metal finishers.

4.2.3 "COMPLEXED METALS" SUBCATEGORY

The concentrations of metals found in the "complexed metals" subcategory of raw wastestreams are presented in Table 4.6. Complexed metals may occur in a number of unit operations, but come primarily from electroless and immersion plating. The most commonly used metals in these operations are copper, nickel, and tin. Wastewaters containing complexing agents must be segregated and treated independently of other wastes in order to prevent further complexing of free metals in the other streams.

4.2.4 "CYANIDE" SUBCATEGORY

Cyanide has been used extensively in the surface-finishing industry for many years; however, it is a hazardous substance that must be handled with caution. The use of cyanide in plating and stripping solutions stems from its ability to weakly complex many metals typically used in plating. Metal

TABLE 4.4
Concentrations of Pollutants Found in the "Common Metals" Subcategory of Raw Wastewater

Pollutant	Number of Samples	Number of Detections	Range of Detections	Median of Detections	Mean of Detections
Toxic pollutants (concentrations shown in µg/L)					
Metals and inorganics					
Antimony	106	22	1–430	6	34
Arsenic	105	31	2–64	10	16
Beryllium	27	23	1–44	5	9
Cadmium	108	60	BDL–19,000	8	1,000
Chromium	105	89	3–35,000	180	16,000
Copper	108	105	3–500,000	180	16,000
Lead	108	73	3–42,000	120	1,400
Mercury	99	32	BDL–400	10	18
Nickel	108	88	4–420,000	200	24,000
Selenium	26	21	1–60	5	9
Thallium	26	21	1–62	3	10
Zinc	108	107	9–330,000	290	19,000
Phthalates					
Bis(2-ethylhexyl)phthalate	93	91	BDL–1,900	6	57
Butyl benzyl phthalate	65	38	BDL–10	BDL	1
Di-n-butyl phthalate	89	79	BDL–10	BDL	BDL
Di-n-octyl phthalate	65	25	BDL–10	BDL	BDL
Diethyl phthalate	83	66	BDL–240	5	31
Dimethyl phthalate	65	7	BDL–10	BDL	2
Nitrogen compounds					
3,3-dichlorobenzidene	4	1	BDL		
N-nitroso-di-n-propylamine	4	1	570		
Phenols					
2-Nitrophenol	4	1	24		
Phenol	23	15	BDL–1,000	45	240
Aromatics					
Benzene	6	4	BDL–16	7	8
Ethylbenzene	37	9	BDL–1,200	250	340
Toluene	39	17	2–690	77	140
Polycyclic aromatic hydrocarbons					
Fluoranthene	4	1	74		
Isophorone	4	4	13–310	180	170
Napthalene	89	61	BDL–2,000	1	83
Anthracene	82	56	BDL–30	1	2
Fluorene	2	2	BDL–160		80
Phenanthrene	71	55	BDL–30	1	2
Pyrene	4	1	190		
Halogenated aliphatics					
Carbon tetrachloride	57	37	BDL–1	BDL	BDL
1,2-Dichloroethane	4	1	3		
1,1,1-Trichloroethane	57	43	BDL–550	BDL	18
1,1,2-Trichloroethane	57	21	BDL–3	BDL	BDL
Chloroform	65	48	BDL–140	BDL	5
1,1-Dichloroethylene	58	4	BDL–110	BDL	20

Continued

TABLE 4.4 (continued)

Pollutant	Number of Samples	Number of Detections	Range of Detections	Median of Detections	Mean of Detections
Halogenated aliphatics					
1,2-*Trans*-dichloroethylene	5	3	1–5	2	3
1,2-Dichloropropylene	4	1	2		
Methylene chloride	80	27	BDL–570	BDL	53
Methyl chloride	74	3	BDL–60	3	21
Methyl bromide	4	1	2		
Dichlorobromomethane	5	2	3–8		
Chlorodibromomethane	4	1	8		
Tetrachloroethylene	59	23	BDL–66	BDL	6
Trichloroethylene	77	49	BDL–480	BDL	22
Pesticides and metabolites					
Dieldrin	4	1	BDL		
α-Endosulfan	4	1	9		
Endrin aldehyde	4	1	BDL		
α-BHC	4	1	BDL		
β-BHC	4	1	4		
δ-BHC	4	1	BDL		
Classical pollutants (concentrations shown in mg/L)					
TSS	107	104	0.56–11,000	63	520
Aluminum	8	6	0.03–200	0.29	62
Barium	4	3	0.027–0.071	0.03	0.043
Calcium	3	3	25–76	52	51
Cobalt	4	4	0.009–0.023	0.02	0.017
Fluorides	7	3	0.021–36	1.1	5.3
Iron	85	76	0.035–490	1.9	28
Magnesium	88	87	5.6–31	14	16
Manganese	4	4	0.059–0.5	0.085	0.22
Molybdenum	7	7	0.031–0.3	0.27	0.2
Phosphorous	4	3	0.007–77	3	7.9
Sodium	4	3	17–310	140	160
Tin	4	4	0.002–15	0.86	3.7
Titanium	5	2	0.006–0.08	0.03	0.039
Vanadium	7	3	0.01–0.22	0.036	0.087
Yttrium	4	3	0.002–0.02	0.018	0.013

BDL, below detection limit; TSS, total suspended solids; BHC, a chemical that is the sum of isomers of 1,2,3,4,5,6,-hexachlorocyclohexane, such as lindane $C_6H_6Cl_6$.

Source: From U.S. EPA, Development Document for Effluent Limitations Guidelines and Standards for the Metal Finishing Point Source Category, Report EPA-440/ 1-80/091, U.S. EPA, Washington, DC, 1980.

deposits produced from cyanide plating solutions are finer grained than those plated from an acidic solution. In addition, cyanide-based plating solutions tend to be more tolerant of impurities than other solutions, offering preferred finishes over a wide range of conditions. In particular, cyanide is used in the following applications:

1. Cyanide-based strippers are used to selectively remove plated deposits from the base metal without attacking the substrate.
2. Cyanide-based electrolytic alkaline descalers are used to remove heavy scale from steel.
3. Cyanide-based dips are often used before plating or after stripping processes to remove metallic smuts on the surface of parts.

TABLE 4.5

Concentrations of Pollutants Found in the "Precious Metals" Subcategory of Raw Wastewater

Pollutant	Number of Samples	Number of Detections	Range of Detections	Median of Detections	Mean of Detections
Classical pollutants (concentrations shown in mg/L)					
Silver	15	12	0.033–600	0.38	86
Gold	15	9	0.56–43	0.86	15
Palladium	13	3	0.09–0.12	0.09	0.10
Rhodium	12	1	0.22		

Source: U.S. EPA, Development Document for Effluent Limitations Guidelines and Standards for the Metal Finishing Point Source Category, Report EPA-440/ 1-80/091, U.S. EPA, Washington, DC, 1980.

Cyanide-based metal finishing solutions usually operate at basic pH levels to avoid decomposition of the complexed cyanide and the formation of highly toxic hydrogen cyanide gas.

The cyanide concentrations found in the "cyanide" subcategory of raw wastestreams are shown in Table 4.7. The levels of cyanide range from 0.045 to 500 mg/L. Streams with high cyanide concentrations normally originate in electroplating and heat-treating processes. Cyanide-bearing wastestreams should be segregated and treated before being combined with other raw wastestreams.

TABLE 4.6

Concentrations of Pollutants Found in the "Complexed Metals" Subcategory of Raw Wastewater

Pollutant	Number of Samples	Number of Detections	Range of Detections	Median of Detections	Mean of Detections
Toxic pollutants (concentrations shown in μg/L)					
Cadmium	31	9	1–3,600	67	850
Copper	31	28	10–63,000	6,700	11,000
Lead	31	10	2–3,600	420	1,200
Nickel	31	25	26–290,000	3,200	28,000
Zinc	31	31	23–18,000	210	3,000
Classical pollutants (concentrations shown in mg/L)					
Aluminum	1	1	0.1		
Calcium	1	1	17		
Iron	31	31	0.038–99	0.74	9.9
Magnesium	1	1	2		
Manganese	1	1	0.1		
Phosphorus	31	31	0.023–100	8.2	23
Sodium	1	1	110		
Tin	31	10	0.013–6	0.68	1.6

Source: U.S. EPA, Development Document for Effluent Limitations Guidelines and Standards for the Metal Finishing Point Source Category, Report EPA-440/ 1-80/091, U.S. EPA, Washington, DC, 1980.

TABLE 4.7

Concentrations of Pollutants Found in the "Cyanide" Subcategory of Raw Wastewater

Pollutant	Number of Samples	Number of Detections	Range of Detections	Median of Detections	Mean of Detections
Toxic pollutants (concentrations shown in µg/L)					
Cyanide	20	20	45–500,000	45,000	110,000
Cyanide, amenable to chlorination	19	18	5–460,000	4,500	86,000

Source: U.S. EPA, Development Document for Effluent Limitations Guidelines and Standards for the Metal Finishing Point Source Category, Report EPA-440/ 1-80/091, U.S. EPA, Washington, DC, 1980.

4.2.5 "Hexavalent Chromium" Subcategory

Concentrations of hexavalent chromium from metal finishing raw wastes are shown in Table 4.8. Hexavalent chromium enters wastewater as a result of many unit operations and can be very concentrated. Because of its high toxicity, it requires separate treatment so that it can be efficiently removed from the wastewater.

4.2.6 "Oils" Subcategory

Pollutant parameters and their concentrations found in the "oily waste" subcategory streams are shown in Table 4.9. The oily waste subcategory for the metal finishing industry is characterized by both concentrated and dilute oily wastestreams that consist of a mixture of free oils, emulsified oils, greases, and other assorted organics. The appropriate treatment for oily wastestreams is dependent on the concentration levels of the wastes, but oily wastes normally receive specific treatment for oil removal prior to solids removal waste treatment.

The majority of the pollutants listed in Table 4.9 are priority organics that are used either as solvents or as oil additives to extend the useful life of the oils. Organic priority pollutants, such as solvents, should be segregated and disposed of or reclaimed separately. However, when they are present in wastewater streams, they are most often at the highest concentration in the oily wastestream, because organic pollutants generally have a higher solubility in hydrocarbons than in water. Oily wastes will normally receive treatment for oil removal before being directed to waste treatment for solids removal.

4.2.7 "Solvent" Subcategory

The "solvent" subcategory of raw wastes includes solvents generated in the metal finishing industry by the dumping of spent solvents from degreasing equipment (including sumps, water traps, and stills).

TABLE 4.8

Concentrations of Pollutants Found in the "Hexavalent Chromium" Subcategory of Raw Wastewater

Pollutant	Number of Samples	Number of Detections	Range of Detections	Median of Detections	Mean of Detections
Toxic pollutants (concentrations shown in µg/L)					
Chromium, hexavalent	49	41	5–13,000,000	20,000	420,000

Source: U.S. EPA, Development Document for Effluent Limitations Guidelines and Standards for the Metal Finishing Point Source Category, Report EPA-440/ 1-80/091, U.S. EPA, Washington, DC, 1980.

TABLE 4.9

Concentrations of Pollutants Found in the "Oils" Subcategory of Raw Wastewater

Toxic Pollutants (Concentrations Shown in μg/L)	Number of Samples	Number of Detections	Range of Detection	Median of Detections	Mean of Detections
Phthalates					
Bis(2-ethylhexyl) phthalate	37	20	2–9,300	73	820
Butyl benzyl phthalate	37	9	1–10,000	130	1,600
Di-n-butyl phthalate	37	19	1–3100	16	270
Di-n-octyl phthalate	37	3	4–120	—	62
Diethyl phthalate	37	9	1–1,900	40	420
Dimethyl phthalate	37	34	1–1,200	1	400
Ethers					
Bis(chloromethyl)ether	37	1	9	—	—
Bis(2-chloroethyl)ether	37	2	4–10	—	7
Bis(2-chloroisopropyl)ether	37	1	4	—	—
Bis(2-chloroethoxy)methane	37	1	3	—	—
Nitrogen compounds					
1,2-Diphenylhydrazine	37	2	5–12	—	8
Phenols					
2,4,6-Trichlorophenol	37	3	10–1,000	10	610
Perachlorometacresol	37	8	4–800,000	2,300	100,000
2-Chlorophenol	37	2	76–620	—	350
2,4-Dichlorophenol	37	2	10–68	—	39
2,4-Dimethylphenol	37	6	1–31,000	10	5,200
2-Nitrophenol	37	3	10–320	35	120
4-Nitrophenol	37	1	10	—	—
2,4-Dinitrophenol	37	3	10–10,000	13	3,300
N-Nitrosodiphenylamine	37	5	4–900	750	490
Pentachlorophenol	37	3	10–50,000	5,200	18,000
Phenol	27	3	3–6,600	440	1,700
4,6-Dinitro-o-cresol	37	2	10–5,700	—	2,800
Aromatics					
Benzene	37	18	1–110	8	12
Chlorobenzene	37	2	11–610	—	310
Nitrobenzene	37	2	1–10	—	5
Toluene	37	25	1–37,000	33	1,800
Ethylbenzene	37	16	1–5,500	12	380
Polynuclear aromatic hydrocarbons					
Acenaphthane	37	2	57–5,700	—	2,900
2-Chloronaphthalene	37	1	130	—	—
Fluoranthene	37	8	1–55,000	110	8,300
Naphthalene	37	10	1–260	100	36
Benzo (a) pyrene	37	1	10	—	—
Chrysene	37	3	1–73	2	25
Acenaphthalene	37	3	77–1,000	140	410
Anthracene	43	7	3–2,000	34	360
Fluorine	37	7	1–760	75	180
Phenanthrene	37	8	2–2,000	28	400
Pyrene	37	5	31–150	75	79

Continued

TABLE 4.9   (continued)

Toxic Pollutants (Concentrations Shown in µg/L)	Number of Samples	Number of Detections	Range of Detection	Median of Detections	Mean of Detections
Halogenated hydrocarbons					
Carbon tetrachloride	37	5	1–10,000	97	2,600
1,2-Dichloroethane	37	6	9–2,100	1,400	1,100
1,1,1-Trichloroethane	37	18	1–1,300,000	260	75,000
1,1-Dichloroethane	37	11	2–1,100	600	460
1,1,2-Trichloroethane	37	4	6–1,300	10	330
1,1,2,2-Tetrachloroethane	37	2	6–570	—	290
Chloroform	37	19	2–690	10	58
1,1-Dichloroethylene	37	12	2–10,000	200	1,500
1,2-Trans-dichloroethylene	43	9	8–1,700	88	510
Methylene chloride	37	29	5–7,600	92	600
Methyl chloride	37	4	1–4,700	9	1,200
Bromoform	37	1	10	—	—
Dichlorobromomethane	37	2	1–10	—	5
Trichlorofluoromethane	37	2	260–290	—	280
Chlorodibromomethane	37	3	1–10	2	4
Tetrachloroethylene	37	18	1–110,000	10	8,900
Trichloroethylene	37	11	1–130,000	110	23,000
Pesticides and metabolites					
Aldrin	37	2	4–11	—	7
Dialdrene	37	1	3	—	—
Chlordane	37	2	1–13	—	7
4,4-DDT (DichloroDiphenyl Trichloroethane)	37	2	2–10	—	6
4,4-DDE (Dichlorodiphenyl DichloroEthylene)	37	4	BDL–53	2	14
4,4-DDD (DichloroDiphenyl Dichloroethane)	37	3	1–10	4	5
α-Endosulfan	37	2	8–28	—	18
β-Endosulfan	37	2	BDL–6	—	3
Endosulfan sulfate	37	4	1–16	11	10
Endrin	37	2	7–10	—	8
Endrin aldehyde	37	2	10–14	—	12
Heptachlor	37	1	BDL	—	—
Heptachlor epoxide	37	1	BDL	—	—
α-BHC (lindane)	37	3	4–18	13	12
β-BHC (lindane)	37	3	1–9	7	6
δ-BHC (lindane)	37	2	4–11	—	7
Polychlorinated biphenyls					
Aroclor 1254	37	2	76–1,100	—	590
Aroclor 1248	37	2	160–1,800	—	580
Classical pollutants (concentrations shown in mg/L)					
Ammonia	37	10	0.46–270	7.9	46
Biochemical oxygen demand (BOD)	37	21	10–17,000	1,400	3,200
Chemical oxygen demand (COD)	37	16	310–1,500,000	12,000	120,000
Oil and grease	37	37	65–800,000	6,100	41,000
Phenols, total	37	34	0.002–49	0.24	2.5

Continued

TABLE 4.9   (continued)

Toxic Pollutants (Concentrations Shown in µg/L)	Number of Samples	Number of Detections	Range of Detection	Median of Detections	Mean of Detections
Total dissolved solids, TDS	37	9	250–4,900	1,600	2,000
Total organic carbon, TOC	37	37	3–560,000	1,600	28,000
Total suspended solids, TSS	37	35	35–18,000	680	2,700
BDL, below detection limit.					

Source: U.S. EPA, Development Document for Effluent Limitations Guidelines and Standards for the Metal Finishing Point Source Category, Report EPA-440/ 1-80/091, U.S. EPA, Washington, DC, 1980.

These solvents are predominately composed of compounds classified by the U.S. EPA as toxic pollutants. Spent solvents should be segregated, hauled for disposal or reclamation, or reclaimed on site. Solvents that are mixed with other wastewaters tend to appear in the common metals or oily wastestreams.

4.3   SOURCE REDUCTION

It is not currently feasible to achieve a zero discharge of chemical pollutants from metal finishing operations. However, substantial reductions in the type and volume of hazardous chemicals wasted from most metal finishing operations are possible.[8] Because end-of-pipe waste detoxification is costly for small- and medium-sized metal finishers, and the cost and liability of residuals disposal have increased for all metal finishers, management and production personnel may be more willing to consider production process modifications to reduce the amount of chemicals lost to waste.

This section provides guidance for reducing water-borne wastes from metal finishing operations in order to avoid or reduce the need for waste detoxification and the subsequent off-site disposal of detoxification residuals. Waste reduction practices may take the form of the following[5]:

1. Chemical substitution
2. Waste segregation
3. Process modifications to reduce dragout loss
4. Capture/concentration techniques

4.3.1   CHEMICAL SUBSTITUTION

The incentive for substituting process chemicals containing nonpolluting materials has only been present in recent years with the advent of pollution control regulations. Chemical manufacturers are gradually introducing such substitutes. By eliminating polluting process materials such as hexavalent chromium and cyanide-bearing cleaners and deoxidizers, the treatments required to detoxify these wastes are also eliminated. It is particularly desirable to eliminate processes using hexavalent chromium and cyanide, because special equipment is needed to detoxify them.

Substituting nonpolluting cleaners for cyanide cleaners can avoid cyanide treatment entirely. For a 7.6 L/min rinsewater flow, this means a savings of about USD 18,400 in equipment costs and USD 10 per kilogram of cyanide treatment chemical costs. In this case, treatment chemical costs are about four times the cost of the raw sodium cyanide cleaner.

There can be disadvantages in using nonpolluting chemicals. Before making a decision the following questions should be asked of the chemical supplier[5]:

• Are substitutes available and practical?
• Will substitution solve one problem but create another?

- Will tighter chemical controls be required of the bath?
- Will product quality or production rate be affected?
- Will the change involve any cost increases or decreases?

Based on a survey of chemical suppliers and electroplaters who use nonpolluting chemicals, some commonly used chemical substitutes are summarized in Table 4.10.

The chemical supplier can also identify any regulated pollutants in the facility's treatment chemicals and offer available substitutes. The federally regulated pollutants are cyanide, chromium, copper, nickel, zinc, lead, cadmium, and silver. Local or state authorities may regulate other substances, such as tin, ammonia, and phosphate. The current status of cyanide and noncyanide substitute plating processes is shown in Table 4.11.

4.3.2 WASTE SEGREGATION

After eliminating as many pollutants as possible, the next step is for polluting streams to be segregated from nonpolluting streams. Nonpolluting streams can go directly to the sewer, although pH adjustment may be necessary. The segregation process will likely require some physical re-layout or re-piping of the shop. These potentially nonpolluting rinse streams represent about one-third of all plating process water. Caution must be exercised to make certain that so-called nonpolluting baths contain no dissolved metal. The cost savings in segregating polluting from nonpolluting

TABLE 4.10
Chemical Substitutes

Polluting	Substitute	Comments
Fire dip (NaCN)	Muriatic acid with additives	Slower acting than + H_2O_2 traditional fire dip
Heavy copper cyanide plating bath	Copper sulfate	Excellent throwing power with a bright, smooth, rapid finish
		A copper cyanide strike may still be necessary for steel, zinc, or tin–lead base metals
		Requires good pre-plate cleaning
		Noncyanide process eliminates carbonate buildup in tanks
Chromic acid pickles, deoxidizers and bright dips	Sulfuric acid and hydrogen perioxide	Nonchrome substitute
		Nonfuming
Chrome-based antitarnish	Benzotriazole (0.1–1.0% solution in methanol) or water-based proprietaries	Nonchrome substitute
		Extremely reactive, requires ventilation
Cyanide cleaner	Trisodium-phosphate or ammonia	Noncyanide cleaner
		Good degreasing when hot and in an ultrasonic bath
		Highly basic
		May complex with soluble metals if used as an intermediate rinse between plating baths where metal ion may be dragged into the cleaner and cause wastewater treatment problems
Tin cyanide	Acid tin chloride	Works faster and better

Source: U.S. EPA, Meeting Hazardous Waste Requirements for Metal Finishers, Report EPA/625/4-87/018, U.S. EPA, Cincinnati, OH, 1987.

TABLE 4.11
Cyanide and Noncyanide Plating Processes

Metal	Cyanide	Noncyanide
Brass	Proven	No
Bronze	Proven	No
Cadmium	Proven	Yes
Copper	Proven	Proven
Gold	Proven	Developing
Indium	Proven	Yes
Silver	Proven	Developing
Zinc	Proven	Proven

Source: U.S. EPA, Managing Cyanide in Metal Finishing, Capsule Report
EPA 625/R-99/009, U.S. EPA, Cincinnati, OH, December 2000.

streams is realized through wastewater treatment equipment and operating costs. The remaining polluting sources, which require some form of control, include all dumped spent solutions (e.g., tumble finishing and burnishing washes), cyanide cleaner rinses, plating rinses, rinses after "bright dips," and aggressive cleaning solutions.

4.3.3 PROCESS MODIFICATIONS TO REDUCE DRAG-OUT LOSS

Plating solution that is wasted by being carried over into the rinsewater as a workpiece emerges from the plating bath is known as dragout, and is the largest-volume source of chemical pollutant in the electroplating shop. Numerous techniques have been developed to control dragout; the effectiveness of each method varies as a function of the plating process, operator cooperation, racking, barrel design, transfer dwell time, and plated part configuration.

Wetting agents and longer workpiece withdrawal/drainage times are two techniques that significantly control drag-out. These and other techniques are discussed below.

4.3.3.1 Wetting Agents

Wetting agents lower the surface tension of process baths. To remove plating solution dragged out with the plated part, gravity-induced drainage must overcome the adhesive force between the solution and the metal surface. The drainage time required for racked parts is a function of the surface tension of the solution, part configuration, and orientation. Lowering the surface tension reduces the drainage time and also minimizes the edge effect (the bead of liquid adhering to the part edge), leading to less drag-out. Plating baths such as nickel and heavy copper cyanide also use wetting agents to maintain grain quality and provide improved coverage. The chemical supplier should be asked if the baths he supplies contain wetting agents and, if not, whether wetting agents can be added. In some baths the use of wetting agents has the potential to reduce drag-out by 50%.

4.3.3.2 Longer Drain Times

By using slower withdrawal rates or longer drain times, the drag-out of process solutions can be reduced by up to 50%. Where high-temperature plating solutions are used, slow withdrawal of the rack may also be necessary to prevent evaporative "freezing," which can actually increase drag-out. In the extreme case, too rapid a withdrawal rate causes "sheeting," where huge volumes of drag-out are lost to waste. Figure 4.1 shows the drainage rates for plain and bent pieces. Drainage for all shapes is almost complete within 15 sec of withdrawal, indicating that this is an optimum drain time for most pieces.

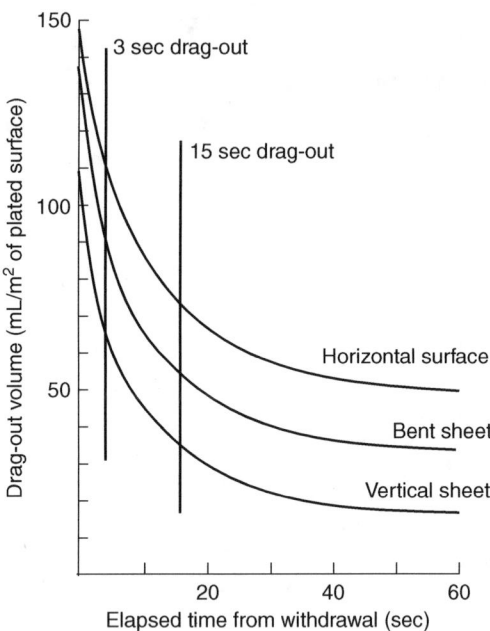

FIGURE 4.1 Typical drag-out drainage rates. (From U.S. EPA, Meeting Hazardous Waste Requirements for Metal Finishers, Report EPA/625/4-87/018, U.S. EPA, Cincinnati, OH, 1987.)

One of the best ways to control drag-out loss from rack plating on hand lines is to provide drain bars over the tank, from which the rack can be hung to drain for a brief period. Hanging and removing the racks from the drain bars ensures an adequate drain time. Slightly jostling the racks helps shake off adhering solution.

In barrel plating, the barrel should be rotated for a time just above the plating tank in order to reduce the volume of dragged-out chemical. Holes in the barrels should be as large as possible to improve solution drainage while still containing the pieces. A fog spray directed at the barrel or its contents can also help drag out drainage. Deionized water is recommended to minimize bath contamination.

The combined application of wetting agents and longer withdrawal/drainage times can significantly reduce the amount of drag-out for many cleaning or plating processes. For example, a typical nickel drag-out can be reduced from 1 L/h to 1/4 L/h by these techniques.

4.3.3.3 Other Drag-Out Reduction Techniques

Rinse elimination
The rinse between a soak cleaner and an electrocleaner may be eliminated if the two baths are compatible.

Low-concentration plating solutions
Low-concentration plating solutions reduce the total mass of chemicals being dragged out. The mass of chemicals removed from a bath is a function of the solution concentration and the volume of solution carried from the bath. Traditionally, bath concentration is maintained at a midpoint within a range of operating conditions. In contrast with the high cost of replacement, treatment, and disposal of dragged-out chemicals, the economics of low-concentration baths are favorable.

As an illustration, a typical nickel plating operation with five nickel tanks has an annual nickel drag-out of about 10,000 L. Assuming the nickel baths are maintained at the midpoint operating

TABLE 4.12
Standard Nickel Solution Concentration Limits

Chemical	Concentration Range (g/L)	Midpoint Operating Condition (g/L)	Modified Operating Condition (g/L)
Nickel sulfate			
$NiSO_4 \cdot 6H_2O$	300–375	338	308
as $NiSO_4$	—	200	182
Nickel chloride			
$NiCl_2 \cdot 6H_2O$	60–90	75	64
as $NiCl_2$	—	41	35
Boric acid, H_3BO_3	45–49	47	46

Source: U.S. EPA, Meeting Hazardous Waste Requirements for Metal Finishers, Report EPA/625/4-87/018, U.S. EPA, Cincinnati, OH, 1987.

concentration, as shown in Table 4.12, the annual cost of chemical replacement, treatment, and disposal are about USD 20,700 (in 2007 dollars). If the bath is converted to the modified operating condition as shown in the table, the annual cost of chemical replacement, treatment, and disposal are approximately USD 18,700, a saving of about USD 2,000 per year. Generally, any percent decrease in bath chemical concentration results in the same percent reduction in the mass of chemicals lost in the drag-out. The disadvantage of low-concentration baths may be lowered plating efficiencies, which may require higher current densities and closer process control in order to compensate. The reduction in plating chemical replacement, treatment, and disposal costs could be partially offset by the added labor and power costs associated with the use of the lower concentration baths.

Clean plating baths
Contaminated plating baths, for example through carbonate buildup in cyanide baths, can increase drag-out by as much as 50% by increasing the viscosity of the bath. Excessive impurities also make the application of recovery technology difficult, if not impossible.

Low-viscosity conducting salts
Bath viscosity indexes are available from chemical suppliers. As bath viscosity increases, drag-out volume also increases.

High-temperature baths
These reduce surface tension and viscosity, thus decreasing drag-out volume. Disadvantages to be considered are more rapid solution decomposition, higher energy consumption, and possible dry-on pattern on the workpiece.

No unnecessary components
Additional bath components (chemicals) tend to increase both viscosity and drag-out.

Fog sprays or air knives
These may be used over the bath to remove drag-out from pieces as they are withdrawn. The spray of deionized water or air removes plating solution from the part and returns as much as 75% of the drag-out back to the plating tank. Fog sprays, located just above the plating bath surface, dilute and drain the adhering drag-out solution, thus reducing the concentration and mass of chemicals lost. Fog sprays are best when tank evaporation rates are sufficient to accommodate the added volume of spray water. Air knives, also located just above the plating bath surface, reduce the volume of drag-out by mechanically scouring the adhering liquid from the workpiece. The drag-out concentration remains constant, but the mass of chemicals lost is reduced. Air knives are best when the surface

evaporation rates of the bath are too low to allow additional spray water. In some cases, use of supplementary atmospheric evaporators may be justified by economic considerations.

Air knives can be installed for about USD 750 to 800 per bath if an oil-free, compressed air source is available. Fog sprays can be installed for the same amount per bath if a deionized water source is available. The spray should be actuated only when work is in the spraying position. Properly designed spray nozzles distribute the water evenly over the work, control the volume of water used, and avoid snagging workpieces as they are withdrawn from the tank.

Proper racking
Every piece has at least one racking position in which drag-out will be at a minimum. In general, to minimize drag-out, the following should be considered:

1. Parts should be racked with major surfaces vertically oriented.
2. Parts should not be racked directly over one another.
3. Parts should be oriented so that the smallest surface area of the piece leaves the bath surface last.

The optimum orientation will provide faster drainage and less drag-out per piece. However, in some cases this may reduce the number of pieces on a rack, or the optimum draining configuration may not be the optimum plating configuration. In addition, the user should maintain rack coatings, replace rack contacts when broken, strip racks before plating buildup becomes excessive, and ensure that all holes on racks are covered or filled.

4.3.3.4   Capture/Concentration Techniques

Capture/concentration with full reuse of drag-out
The pioneer in simple, low-cost methods of reducing waste in the plating shop was Dr. Joseph B. Kushner.[29] In *Water and Waste Control for the Plating Shop* (1994), he describes a "simple waste recovery system" that captures drag-out in a static tank or tanks for return to the plating bath. The drag-out tanks are followed by a rinse tank, which flows to the sewer with only trace amounts of polluting salts and is often in compliance with sewer discharge standards. A simplified diagram of this reuse system is shown in Figure 4.2. It is not difficult to automate the direct drag-out recovery process, and commercial units are available.

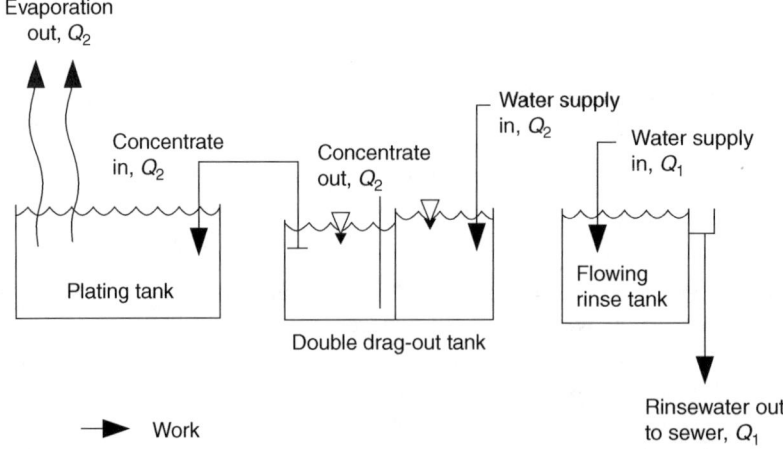

FIGURE 4.2   Kushner method of double drag-out for full reuse. (From U.S. EPA, Meeting Hazardous Waste Requirements for Metal Finishers, Report EPA/625/4-87/018, U.S. EPA, Cincinnati, OH, 1987.)

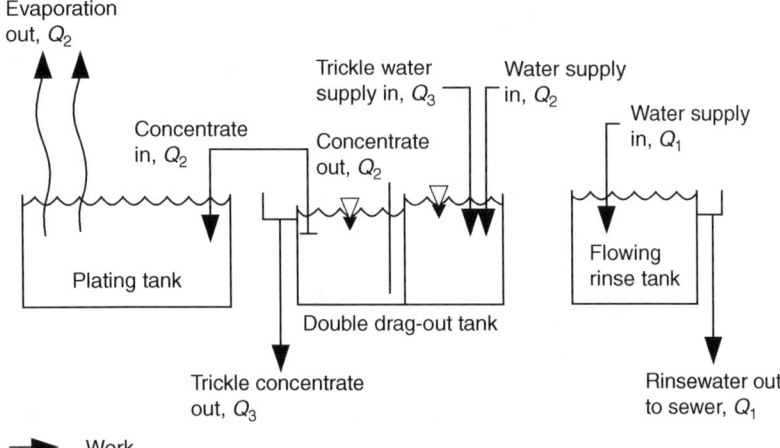

FIGURE 4.3 Modified method of double drag-out for partial reuse. (From U.S. EPA, Meeting Hazardous Waste Requirements for Metal Finishers, Report EPA/625/4-87/018, U.S. EPA, Cincinnati, OH, 1987.)

The Kushner concept is easily applicable to hot plating baths where the bath evaporation rate equals or exceeds the pour-back rate, Q_2. The drag-out concentration depends on the bath drag-out rate, the number of drag-out tanks, the rinsewater flow rate, Q_2, the plating bath evaporation rate, and drag-out return rate. The number of drag-out tanks must be based on the available space. The higher the number of counterflowed drag-out tanks, the smaller will be the return rate necessary to obtain good rinsing. The Kushner multiple drag-outs are not feasible if there is no room for the required drag-out tanks. If there is little or no evaporation from the bath, supplementary evaporation should be considered. Bath contamination must be minimized by using reverse osmosis (RO) purified water for Q_2.

Capture/concentration with partial reuse of drag-out
By adding a trickling water supply and drain, Q_3, to the drag-out tank, the application of Kushner's concept can be extended to other metal finishing processes that may not be amenable to full reuse but can allow partial reuse. Figure 4.3 depicts the partial reuse scheme. The trickle concentrate can also be batch-treated in a small volume on site, recycled at a central facility, or mixed with Q_1 for discharge, if the combined metal content is below sewer discharge standards.

4.3.4 Waste Reduction Costs and Benefits

The benefits of waste reduction in the metal finishing shop include the following:

1. Reduced chemical cost
2. Reduced water cost
3. Reduced volume of "hazardous" residuals
4. Reduced pretreatment cost

The benefits of saving valuable chemicals and water and reducing sludge disposal costs can best be illustrated by an example. An electroplating operation discharges 98,400 L/d of wastewater containing 0.91 kg copper, 1.14 kg nickel, and 0.91 kg cyanide. The shop can reduce its generation of cyanide and copper waste by about 50% by eliminating cyanide cleaners and utilizing pour-back of copper cyanide solution. The generation of nickel waste can be reduced by 90% by pour-back of the nickel solution. Reducing wasted salts also allows a reduced rinsewater flow rate, thus saving water and sewer use fees. The chemical costs of treatment are given in Table 4.13 and the annual replacement

TABLE 4.13
Chemical Costs of Treatment and Disposal

	Chemical Cost (2007 USD/kg)[a]	
Pollutant	Treatment[b]	Disposal[c]
Nickel	2.73	6.70
Copper	2.73	6.70
Cyanide	17.63	NA

[a]Costs were converted from 1979 USD to 2007 USD using the U.S. ACE Yearly Average Cost Index for Utilities.
[b]Cost of NaOH @ USD 1.00/kg and NaOCl @ USD 2.35/kg.
[c]Cost of disposal @ USD 1.84/kg of sludge (USD 400/drum) @ 30% solids content.
Source: U.S. EPA, Meeting Hazardous Waste Requirements for Metal Finishers, Report EPA/625/4-87/018, U.S. EPA, Cincinnati, OH, 1987.

costs of chemicals are given in Figure 4.4. Calculations of the annual dollar savings are shown in Table 4.14. All costs have been converted into 2007 USD using the U.S. ACE Yearly Average Cost Index for Utilities.[9]

4.4 POLLUTANT REMOVABILTY

This section reviews the technologies currently available used to remove or recover pollutants from the wastewater generated in the metal-finishing industry.[5–7,10] Treatment options are presented for each subcategory within the metal finishing industry. Table 4.15 lists the treatment techniques available for treating wastes from each subcategory.

4.4.1 COMMON METALS

The treatment methods used to treat wastes within the "common metals" subcategory fall into two groups:

1. Recovery techniques
2. Solids removal techniques

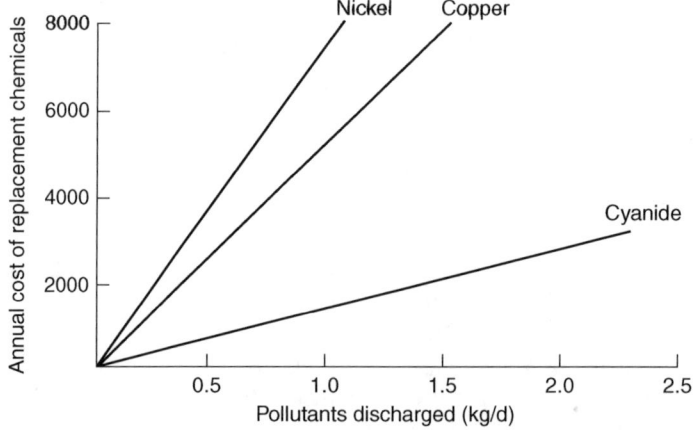

FIGURE 4.4 Annual replacement cost of chemicals in 2007 USD. (From U.S. EPA, Meeting Hazardous Waste Requirements for Metal Finishers, Report EPA/625/4-87/018, U.S. EPA, Cincinnati, OH, 1987.)

TABLE 4.14
Illustration of Annual Cost Savings for Waste Reduction

Item	Cost Saving[a] 2007 USD
Process chemical savings[b]	
Copper	2,425
Cyanide	485
Nickel	7,760
Treatment chemical saving[c]	
Copper	310
Cyanide	2,000
Nickel	700
Reduced treatment sludge disposal[c]	
Copper	760
Cyanide	0
Nickel	1,700
Water and sewer use fee reduction[d]	4,360
Total annual savings	20,500

[a] Costs were converted from 1979 USD to 2007 USD using the U.S. ACE Yearly Average
Cost Index for Utilities.
[b] From Figure 4.4.
[c] From Table 4.12 and Figure 4.4.
[d] USD 0.77/m^3.
Source: U.S. EPA, Meeting Hazardous Waste Requirements for Metal Finishers, Report
   EPA/625/4-87/018, U.S. EPA, Cincinnati, OH, 1987.

TABLE 4.15
Treatment Methods in Current Use or Available for Use in the Metal Finishing Industry

Subcategory/Technology	Number of Plants
Common metals	
Hydroxide followed by sedimentation	103
Hydroxide followed by sedimentation and filtration	30
Evaporation (metal recovery, bath concentrates, rinse waters)	41
Ion exchange	63
Electrolytic recovery	11
Electrodialysis	3
Reverse osmosis	8
Post-adsorption	0
Insoluble starch xanthate	2
Sulfide precipitation	3
Flotation	29
Membrane flotation	7
Precious metals	
Evaporation	1
Ion exchange	NR
Electrolytic recovery	NR
Complexed metals	
High-pH precipitation with sedimentation	NR
High-pH precipitation with sedimentation	NR

Continued

TABLE 4.15   (continued)

Subcategory/Technology	Number of Plants
Hexavalent chromium	
Chemical chrome reduction	343
Electrochemical chromium reduction	2
Electrochemical chromium regeneration	0
Advanced electrodialysis	NR
Evaporation	1
Ion exchange	1
Cyanide	
Oxidation by chlorine	201
Oxidation by ozone	2
Oxidation by ozone with UV radiation	NR
Oxidation by hydrogen peroxide	3
Electrochemical cyanide oxidation	4
Chemical precipitation	3
Reverse osmosis	NR
Evaporation	NR
Oils (segregated)	
Emulsion breaking	28
Skimming	94
Emulsion breaking and skimming	NR
Ultrafiltration	20
Reverse osmosis	3
Carbon adsorption	10
Coalescing	3
Flotation	29
Centrifugation	5
Integrated adsorption	0
Resin adsorption	0
Ozonation	0
Chemical oxidation	0
Aerobic decomposition	14
Thermal emulsion breaking	0
Solvent waste	
Segregation	NR
Contract handling	NR
Sludges	
Gravity thickening	78
Pressure filtration	66
Vacuum filtration	68
Centrifugation	55
Sludge bed drying	77
In-process control	
Flow reduction	NR

NR, not reported.

Source: U.S. EPA, Treatability Manual, Vol. II, Industrial Descriptions, Report EPA-600/2-82-001b, U.S. EPA, Washington, DC, September 1981.

Recovery techniques are treatment methods used for the purpose of recovering or regenerating process constituents that would otherwise be discarded. Included in this group are the following[5–7]:

1. Evaporation
2. Ion exchange
3. Electrolytic recovery
4. Electrodialysis
5. Reverse osmosis

Solids removal techniques are used to remove metals and other pollutants from process wastewaters to make these waters suitable for reuse or discharge. These methods include the following[5–7]:

1. Hydroxide and sulfide precipitation
2. Sedimentation
3. Diatomaceous earth filtration
4. Membrane filtration
5. Granular bed filtration
6. Peat adsorption
7. Insoluble starch xanthate treatment
8. Flotation

Three treatment options are used in treating common metals wastes:

1. The *Option 1* system consists of hydroxide precipitation[11] followed by sedimentation.[12] This system accomplishes end-of-pipe metals removal from all common metals-bearing wastewater streams that are present at a facility. The recovery of precious metals, the reduction of hexavalent chromium, the removal of oily wastes, and the destruction of cyanide must be accomplished prior to common metals removal.
2. The *Option 2* system is identical to the Option 1 treatment system but with the addition of filtration devices[13] after the primary solids removal devices. The purpose of these filtration units is to remove suspended solids such as metal hydroxides that do not settle out in the clarifiers. The filters also act as a safeguard against pollutant discharge should an upset occur in the sedimentation device. Filtration techniques applicable to Option 2 systems are diatomaceous earth and granular bed filtration.[14,15]
3. The *Option 3* treatment system for common metals wastes consists of the Option 2 end-of-pipe treatment system plus the addition of in-plant controls for lead and cadmium. In-plant controls would include evaporative recovery, ion exchange, and recovery rinses.[15]

In addition to these three treatments, there are several alternative treatment technologies applicable to the treatment of common metals wastes. These technologies include electrolytic recovery, electrodialysis, reverse osmosis, peat adsorption, insoluble starch xanthate treatment, sulfide precipitation, flotation, and membrane filtration.[14,15]

4.4.2 Precious Metals

Precious metal wastes can be treated using the same treatment alternatives as those described for the treatment of common metals wastes. However, due to the intrinsic value of precious metals, every effort should be made to recover them. The treatment alternatives recommended for precious metal wastes are the recovery techniques of evaporation, ion exchange, and electrolytic recovery.

4.4.3 Complexed Metal Wastes

Complexed metal wastes within the metal finishing industry are a product of electroless plating, immersion plating, etching, and the manufacture of printed circuit boards. The metals in these wastestreams are tied up or complexed by particular complexing agents whose function is to prevent metals from coming out of solution. This counteracts the technique used by most conventional solids removal methods. Therefore, segregated treatment of these wastes is necessary. The treatment method most suited to treating complexed metal wastes is high-pH precipitation. An alternative method is membrane filtration[16], which is primarily used in place of sedimentation for solids removal.

4.4.4 Hexavalent Chromium

Hexavalent chromium-bearing wastewaters are produced in the metal finishing industry in chromium electroplating, in chromate conversion coatings, in etching with chromic acid, and in metal finishing operations carried out on chromium as a basis material.

The selected treatment option involves the reduction of hexavalent chromium to trivalent chromium either chemically or electrochemically. The reduced chromium can then be removed using a conventional precipitation-solids removal system. Alternative hexavalent chromium treatment techniques include chromium regeneration, electrodialysis, evaporation, and ion exchange.[15]

4.4.5 Cyanide

Cyanides are introduced as metal salts for plating and conversion coating or are active components in plating and cleaning baths. Cyanide is generally destroyed by oxidation. Chlorine, in either elemental or hypochlorate form, is the primary oxidation agent used in industrial waste treatment to destroy cyanide. Alternative treatment techniques for the destruction of cyanide include oxidation by ozone, ozone with ultraviolet radiation (oxyphotolysis), hydrogen peroxide, and electrolytic oxidation.[17] Treatment techniques that remove cyanide but do not destroy it include chemical precipitation, reverse osmosis, and evaporation.[15,17]

4.4.6 Oils

Oily wastes and toxic organics that combine with the oils during manufacturing include process coolants and lubricants, wastes from cleaning operations, wastes from painting processes, and machinery lubricants. Oily wastes are generally of three types: free oils, emulsified or water-soluble oils, and greases. Oil removal techniques commonly employed in the metal finishing industry include skimming, coalescing, emulsion breaking, flotation, centrifugation, ultrafiltration, reverse osmosis, carbon adsorption, and aerobic decomposition.[17–19]

Because emulsified oils and processes that emulsify oils are used extensively in the metal finishing industry, the exclusive occurrence of free oils is nearly nonexistent.

Treatment of oily wastes can be carried out most efficiently if oils are segregated from other wastes and treated separately. Segregated oily wastes originate in the manufacturing areas and are collected in holding tanks and sumps. Systems for treating segregated oily wastes consist of the separation of oily wastes from the water. If oily wastes are emulsified, techniques such as emulsion breaking or dissolved air flotation (DAF)[20] with the addition of chemicals are necessary to remove the oil. Once the oil–water emulsion is broken, the oily waste is physically separated from the water by decantation or skimming. Following oil–water separation, the water is sent to the precipitation/sedimentation unit used for metals removal. There are three options for oily waste removal:

1. The *Option 1* system involves the emulsion breaking process followed by surface skimming (gravity separation is adequate if only free oils are present).
2. The *Option 2* system consists of the Option 1 system followed by ultrafiltration.

3. The *Option 3* treatment system consists of the Option 2 system with the addition of either carbon adsorption or reverse osmosis.

In addition to these three treatment options, several alternative technologies are applicable to the treatment of oily wastewater. These include coalescing, flotation, centrifugation, integrated adsorption, resin adsorption, ozonation, chemical oxidation, aerobic decomposition, and thermal emulsion breaking.[17–19]

4.4.7 SOLVENTS

Spent degreasing solvents should be segregated from other process fluids to maximize the value of the solvents, to preclude contamination of other segregated wastes, and to prevent the discharge of priority pollutants to any wastewaters. This segregation may be accomplished by providing and identifying the necessary storage containers, establishing clear disposal procedures, training personnel in the use of these techniques, and checking periodically to ensure that proper segregation is occurring. Segregated waste solvents are appropriate for on-site solvent recovery or may be contract hauled for disposal or reclamation.

Alkaline cleaning is the most feasible substitute for solvent degreasing. The major advantage of alkaline cleaning over solvent degreasing is the elimination or reduction in the quantity of priority pollutants being discharged. Major disadvantages include high energy consumption and the tendency to dilute oils removed and to discharge these oils as well as the cleaning additive.

4.5 TREATMENT TECHNOLOGIES

4.5.1 NEUTRALIZATION

One technique that is used in a number of facilities that utilize molten salt for metal surface treatment prior to pickling takes advantage of the alkaline values generated in the molten salt bath in treating other wastes generated in the plant. When the bath is determined to be spent, it is in many instances manifested, hauled off site, and land disposed. One technique is to take the solidified spent molten salt (molten salt is solid at ambient temperatures) and circulate acidic wastes generated in the facility over the material prior to entry to the waste-treatment system. This in effect neutralizes the acid wastes and eliminates the requirements of manifesting and land disposal.

4.5.2 CYANIDE-CONTAINING WASTES

There are eight methods applicable to the treatment of cyanide wastes for metal finishing[5,21]:

1. Alkaline chlorination
2. Electrolytic decomposition
3. Ozonation
4. UV/ozonation
5. Hydrogen peroxide
6. Thermal oxidation
7. Acidification and acid hydrolysis
8. Ferrous sulfate precipitation

Alkaline chlorination is the most widely applied in the metal finishing industry. A schematic for cyanide reduction via alkaline chlorination is provided in Figure 4.5. This technology is generally applicable to wastes containing less than 1% cyanide, generally present as free cyanide. It is conducted in two stages. The first stage is operated at a pH greater than 10 and the second stage with a pH in the range of 7.5 to 8. Alkaline chlorination is performed using sodium hypochlorite and chlorine.

$$NaCN + Cl_2 \rightarrow CNCl + NaCl$$

$$CNCl + 2\ NaOH \rightarrow NaCNO + H_2O$$

$$2\ NaCNO + 3\ Cl_2 + 4\ NaOH \rightarrow 2\ CO_2 + N_2 + 6\ NaCl + 2\ H_2O$$

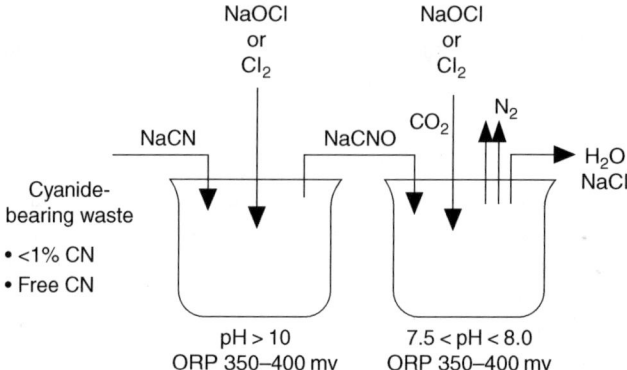

FIGURE 4.5 Cyanide reduction via alkaline chlorination. (From U.S. EPA, Meeting Hazardous Waste Requirements for Metal Finishers, Report EPA/625/4-87/018, U.S. EPA, Cincinnati, OH, 1987.) ORP, oxidation-reduction potential; mv, millivolt.

Electrolytic decomposition technology was applied to cyanide-containing wastes in the early part of this century, but it fell from favor as alkaline chlorination came into use at large-scale facilities. However, as wastes become more concentrated, because this technology is applicable to wastes containing cyanide in excess of 1%, it may find more widespread applications in the future. The basis of this technology is electrolytic decomposition of the cyanide compounds at an elevated temperature (200 °F) to yield nitrogen, CO_2, ammonia, and amines (see Figure 4.6).

Ozonation treatment can be used to oxidize cyanide, thereby reducing the concentration of cyanide in wastewater. Ozone, with an electrode potential of +1.24 V in alkaline solutions, is one of the most powerful oxidizing agents known. Cyanide oxidation with ozone is a two-step reaction similar to alkaline chlorination.[21] Cyanide is oxidized to cyanate, and the ozone is reduced to oxygen according to the following equation:

$$CN^- + O_3 \rightarrow CNO^- + O_2 \tag{4.1}$$

The cyanate is then hydrolyzed in the presence of excess ozone to bicarbonate and nitrogen, and oxidized according to the following reaction:

$$2\ CNO^- + 3\ O_3 + H_2O \rightarrow N_2 + 2\ HCO_3^- + 3\ O_2 \tag{4.2}$$

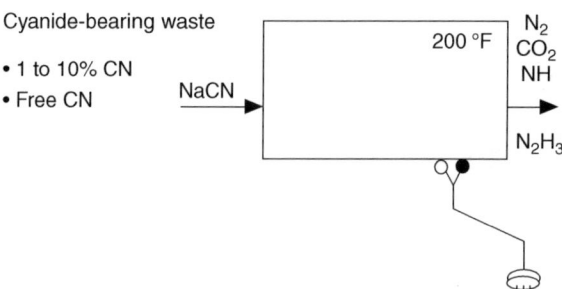

FIGURE 4.6 Cyanide reduction via electrolytic decomposition. (From U.S. EPA, Meeting Hazardous Waste Requirements for Metal Finishers, Report EPA/625/4-87/018, U.S. EPA, Cincinnati, OH, 1987.)

The reaction time for complete cyanide oxidation is rapid in a reactor system, with retention times of 10 to 30 min being typical. The second-stage reaction is much slower than the first-stage reaction. The reaction is typically carried out in the pH range 10 to 12, where the reaction rate is relatively constant. Temperature does not influence the reaction rate significantly.

One interesting variation on ozonation technology is augmentation with UV radiation. This is a technology that has been applied on wastes in the coke byproduct manufacturing industry. A significant development has been made that has resulted in a significant reduction in ozone consumption—the use of UV radiation. UV absorption has the following effects:

1. Ozone and cyanide are raised to a higher energy status.
2. Free radicals are formed.
3. There is more rapid reaction.
4. Less ozone is required.

Cyanide reduction with hydrogen peroxide is effective in reducing cyanide. It has been applied on a less frequent basis within this industry, due to the fact that there are high operating costs associated with the generation of hydrogen peroxide. The reduction of cyanide with peroxide occurs in two steps and yields CO_2 and ammonia:

$$NaCN + H_2O_2 \longrightarrow NaCNO + H_2O \tag{4.3}$$

$$NaCNO + 2\,H_2O \longrightarrow CO_2 + NH_3 + NaOH \tag{4.4}$$

Thermal oxidation is another alternative for destroying cyanide. Thermal destruction of cyanide can be accomplished through either high-temperature hydrolysis or combustion. At temperatures between $140\,^\circ C$ and $200\,^\circ C$ and pH 8, cyanide hydrolyzes quite rapidly to produce formate and ammonia.[22,23] Pressures up to 100 bar are required, but the process can effectively treat wastestreams over a wide concentration range and is applicable to both rinsewater and concentrated solutions.[21] The process involves the following reaction:

$$CN^- + 2\,H_2O \longrightarrow HCOO^- + NH_3 \tag{4.5}$$

In the presence of nitrites, formate and ammonia can be destroyed in another reactor at 150°C, according to the following equations:

$$NH_4^+ + NO_2^- \longrightarrow N_2 + 2\,H_2O \tag{4.6}$$

$$3\,HCOOH + 2\,NO_2^- + 2\,H^+ \longrightarrow 3\,CO_2 + 4\,H_2O + N_2 \tag{4.7}$$

Direct acidification of cyanide wastestreams was once a relatively common treatment. Cyanide is acidified in a sealed reactor that is vented to the atmosphere through an air emission control system. Cyanide is converted to gaseous hydrogen cyanide, treated, vented, and dispersed.

Acid hydrolysis of cyanates is still commonly used, following a first-stage cyanide oxidation process. At pH 2 the reaction proceeds rapidly, but at pH 7 cyanate may remain stable for weeks. This treatment process requires specially designed reactors to ensure that HCN is properly vented and controlled. The hydrolysis mechanisms are as follows.

In an acid medium:

$$H_2O + HOCN + H^+ \longrightarrow NH_4^+ + CO_2 \text{ (rapid)} \tag{4.8}$$

$$HOCN + H_2O \longrightarrow NH_3 + CO_2 \text{ (slow)} \tag{4.9}$$

In strongly alkaline medium:

$$NCO^- + 2\,H_2O \longrightarrow NH_3 + HCO_3^- \text{ (very slow)} \tag{4.10}$$

Each of the technologies described above is effective in treating wastes containing free cyanides; that is, cyanides present as CN in solution. There are instances in metal finishing facilities where complex cyanides are present in wastes. The most common are complexes of iron, nickel, and zinc. A technology that has been applied to remove complex cyanides from aqueous wastes is ferrous sulfate precipitation. The technology involves a two-stage operation in which ferrous sulfate is first added at pH 9 to complex any trace amounts of free cyanide. In the second stage, the complex cyanides are precipitated through the addition of ferrous sulfate or ferric chloride in a pH range of 2 to 4.[5]

4.5.3 CHROMIUM-CONTAINING WASTES

There are three treatment methods applicable to wastes containing hexavalent chromium. Wastes containing trivalent chromium can be treated using chemical precipitation and sedimentation, which is discussed below. The three methods applicable to the treatment of hexavalent chromium use the following:

1. Sulfur dioxide
2. Sodium metabisulfite
3. Ferrous sulfate

Hexavalent chromium reduction through the use of sulfur dioxide and sodium metabisulfite has found the widest application in the metal finishing industry. It is not truly a treatment step, but a conversion process in which the hexavalent chromium is converted to trivalent chromium. The hexavalent chromium is reduced through the addition of the reductant at a pH in the range 2.5 to 3 with a retention time of approximately 30 to 40 min (see Figure 4.7).

Ferrous sulfate has not been as widely applied. However, it is particularly applicable in facilities where ferrous sulfate is produced as part of the process, or is readily available. The basis for this technology is that the hexavalent chromium is reduced to trivalent chromium and the ferrous iron is oxidized to ferric iron.

4.5.4 ARSENIC- AND SELENIUM-CONTAINING WASTES

It may be necessary to segregate wastestreams containing elevated concentrations of arsenic and selenium, especially wastestreams with concentrations in excess of 1 mg/L for these pollutants.

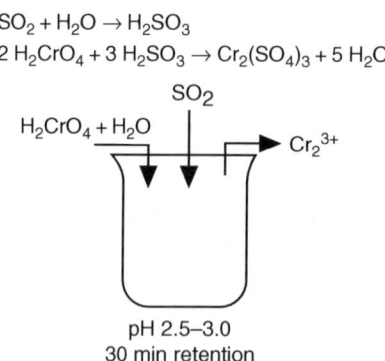

$$SO_2 + H_2O \rightarrow H_2SO_3$$
$$2\,H_2CrO_4 + 3\,H_2SO_3 \rightarrow Cr_2(SO_4)_3 + 5\,H_2O$$

pH 2.5–3.0
30 min retention

FIGURE 4.7 Hexavalent chromium reduction. (From U.S. EPA, Meeting Hazardous Waste Requirements for Metal Finishers, Report EPA/625/4-87/018, U.S. EPA, Cincinnati, OH, 1987.)

Arsenic and selenium form anionic acids in solution (most other metals act as cations) and require special preliminary treatment prior to conventional metals treatment. Lime, a source of calcium ions, is effective in reducing arsenic and selenium concentrations when the initial concentration is below 1 mg/L. However, preliminary treatment with sodium sulfide[21] at a low pH (i.e., 1 to 3) may be required for wastestreams with concentrations in excess of 1 mg/L. The sulfide reacts with the anionic acids to form insoluble sulfides, which are readily separated by means of filtration.

4.5.4.1 Chemical Precipitation and Sedimentation

The most important technology in metals treatment is chemical precipitation and sedimentation. It is accomplished through the addition of a chemical reagent to form metal precipitants, which are then removed as solids in a sedimentation step. The options available to a facility as precipitation reagents are lime [$Ca(OH)_2$], caustic (NaOH), carbonate ($CaCO_3$ and Na_2CO_3), sulfide (NaHS and FeS), and sodium borohydride ($NaBH_4$). The advantages and disadvantages of these reagents are summarized in the following[21]:

1. *Lime*
   a) It is the least expensive precipitation reagent.
   b) It generates the highest sludge volume.
   c) The sludges generally cannot be sold to smelters or refiners.
2. *Caustic*
   a) It is more expensive than lime.
   b) It generates a smaller volume of sludge.
   c) The sludges can be sold to smelter and refiners.
3. *Carbonates*
   a) These may be used for metals where solubility within a pH range is not sufficient to meet treatment standards.

Lime is the least expensive reagent, but it generates the highest volume of residue. It also generates a residue that cannot be resold to smelters and refiners for reclaiming because of the presence of the calcium ion. Caustic is more expensive than lime, but it generates a smaller volume of residue. One key advantage to caustic is that the resulting residues can be readily reclaimed. Carbonates are particularly appropriate for metals where solubility within a pH range is not sufficient to meet a given set of treatment standards. The sulfides offer the benefit of achieving effective treatment at lower concentrations due to the lower solubilities of the metal sulfides. Sodium borohydride has application where small volumes of sludge that are suitable for reclamation are desired.

It is appropriate to look at reagent use in the context of the current regulatory framework under HSWA. Historically, lime has been the reagent of choice. It was relatively inexpensive and simple to handle. The phrase "lime and settle" refers to the application of lime precipitation and sedimentation technology. In the 1970s, new designs made use of caustic as the precipitation reagent because of the reduction in residue volume realized and the possibility of reclamation. In the 1980s, a return to lime and the use of combined reagent techniques came into use.

One obvious question is "Why return to lime as a treatment reagent, given that caustic results in a smaller residue volume and a waste that can undergo reclamation?" The answer lies in the three points that result from the implementation of the HSWA hierarchy. As source reduction and material reuse and recovery techniques are applied, facilities will be generating wastes with the following characteristics:

- Greater concentration
- A varied array of constituents
- A greater degree of complexation

4.5.4.2 Complexation

Complexation is a phenomenon that involves a coordinate bond between a central atom (the metal) and a ligand (the anions). In a coordinate bond, the electron pair is shared between the metal and the ligand. A complex containing one coordinate bond is referred to as a monodentate complex. Multiple coordinate bonds are characteristic of polydentate complexes. Polydentate complexes are also referred to as chelates. An example of a monodentate-forming ligand is ammonia. Examples of chelates are oxylates (bidentates) and EDTA (hexadentates).

The reason for the return to lime is the calcium ion present in lime. The calcium ion that is present in solution on the addition of lime is very effective in competing with the ligand for the metal ion. The sodium ion contributed by caustic is not effective. As such, lime dramatically reduces complexation and is therefore more effective in treating complexed wastes. The term "high lime treatment" is used in cases where excess calcium ions are introduced into solution. This is accomplished through the addition of lime to raise the pH to ca. 11.5 or through the addition of calcium chloride (which has a greater solubility than lime).

The use of combinations of precipitation reagents has been most effective in taking advantage of the attributes of caustic as well as the advantages of lime. As an example, a system may use caustic in a first stage to make a coarse pH adjustment, followed by the addition of lime to make a fine adjustment. This achieves an overall reduction in the sludge volume through the use of the caustic, and more effective metal removal through the use of lime. Sulfide reagents are used in a similar fashion in combination with caustic or lime to provide additional metal removal, taking advantage of the lower solubility of the metal sulfides. Sulfides are also applicable to wastes containing elevated concentrations (i.e., in excess of 2 mg/L) of selenium and arsenic compounds.[21]

4.5.5 OTHER METALS WASTES

There are three techniques applicable to managing solids generated in metal finishing:

1. Dewatering
2. Stabilization
3. Incineration

There are four dewatering techniques (centrifugation, vacuum filtration, belt filtration, and evaporation/drying) that have been applied in metal processing. The most widely used are vacuum and belt filtration.[24,30] They have a higher relative capital cost, but generally have a lower relative operating cost. Plate and frame filter presses have experienced less widespread application. Belt filters generally have a lower relative capital cost and have higher relative operating costs in comparison with other dewatering techniques. The higher operating costs are due to the fact that the units are more labor-intensive to operate. Centrifuges[24] have been applied in specific instances, but are more difficult to operate when a widely varying mix of wastes is treated.

Experience has shown that companies are most successful in applying a dewatering technique that they have successfully designed and operated in similar applications within the company. As an example, many companies operate plate and frame filter presses as a part of metal manufacturing operations. The knowledge gained in metal processing had been successfully transferred to treatment of metal finishing wastes.

There are many stabilization techniques currently available; however, only two of these have found widespread application. These are cementation and stabilization through the addition of lime and fly ash.[24,25,30] Developmental work is currently being undertaken to make use of bitumen, paraffin, and polymeric materials to reduce the degree to which metals can be taken into solution. Encapsulation with inert materials is also under development.

4.6 COSTS

The investment, operating and maintenance,[26,27] and energy costs for the application of control technologies to the wastewaters of the metal finishing industry have been analyzed. These costs were developed to reflect the conventional use of technologies in this industry. A detailed presentation of the cost methodology and cost data is available in a U.S. EPA publication.[6] The available industry-specific cost information is characterized in the following.

4.6.1 TYPICAL TREATMENT OPTIONS

Several unit operation/unit process configurations have been analyzed for their cost of application to the wastewater of this industry. The components included in these configurations are as follows:

1. *Option 1* includes emulsion breaking and oil separation by skimming, cyanide oxidation, chromium reduction, chemical precipitation and sedimentation, and sludge drying beds.
2. *Option 2* includes all of Option 1, plus multimedia filtration.
3. *Option 3* includes all of Option 2, plus ultrafiltration and carbon adsorption for oily waste, and achieving zero discharge of any processes using either cadmium or lead by using an evaporative system.

A flow diagram for suggested Option 1 is shown in Figure 4.8. Flow diagram for the other options would be similar.

4.6.2 COSTS

Cost estimates for the treatment technologies commonly used in this industry are described briefly in the following. More details on the factors considered in the cost analyses are available from the source document.[6]

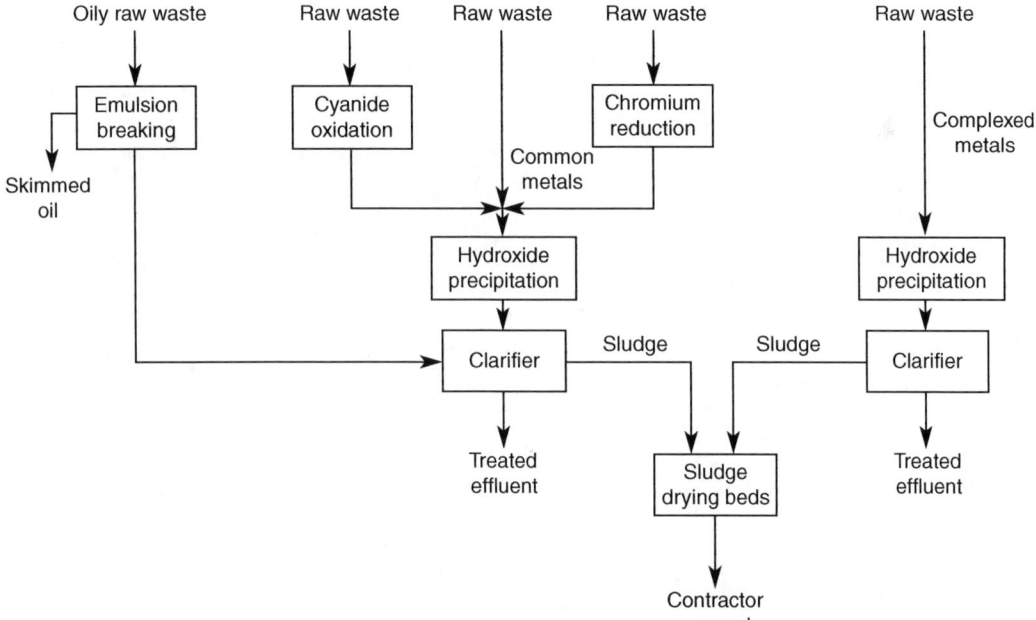

FIGURE 4.8 Metal finishing wastewater treatment flow diagram. (From U.S. EPA, Treatability Manual, Vol. II, Industrial Descriptions, Report EPA-600/2-82-001b, U.S. EPA, Washington, DC, September 1981.)

4.6.2.1   Emulsion Breaking and Oil Separation

1. *Method.* The emulsion is broken by mixing oily waste with alum in a chemical emulsion breaker, followed by gravity oil separation in a tank.
2. *System components.* These comprise a small mixing tank, two chemical feed tanks, a mixer, and a large tank equipped with an oil skimmer and a sludge pump. The mixing tank has a retention time of 15 min and the oil skimming tank a retention time of 2.5 h.

4.6.2.2   Cyanide Oxidation

1. *Method.* Cyanide is destroyed by reaction with sodium hypochlorite under alkaline conditions.
2. *System components.* These comprise reaction tanks, a reagent storage and feed system, mixers, sensors, and controls. Two identical reaction tanks sized as above ground cylindrical tank with a retention time of 4 h. The chemical storage consists of covered concrete tanks able to store a 60 d supply of sodium hypochlorite and 90 d supply of sodium hydroxide.

4.6.2.3   Chromium Reduction

1. *Method.* This involves the chemical reduction of hexavalent chromium by sulfur dioxide under acid conditions for continuous operating systems and by sodium bisulfite under acid conditions for batch operating systems. The reduced trivalent form of chromium is subsequently removed by precipitation as the hydroxide.
2. *System components.* These comprise reaction tanks, a reagent storage and feed system, mixers, sensors, and controls for continuous chromium reduction. A single above-ground concrete tank with retention time of 45 min is provided. For batch operation, dual above-ground concrete tanks with 4 h retention time are provided.

4.6.2.4   Chemical Coagulation, Precipitation and Clarification

1. *Method.* This involves the chemical coagulation/precipitation of dissolved and complexed metals by reaction with lime alum and polyelectrolyte and subsequent removal of the precipitated solids by gravity settling or dissolved air flotation (DAF) in a clarifier.
2. *System components.* This is a continuous treatment system including reagent storage and feed equipment, a mix tank for reagent feed addition, sensors and controls, and a clarification basin with associated sludge rakes and pumps. Lime is fed as 30% lime slurry prepared by using hydrated lime. The mix tank is sized for a retention time of 45 min, and the setting clarifier is sized for hydraulic loading of 1360 L/m^2 and a retention time of 4 h. Batch setting treatment includes dual reaction-settling tanks sized for 8 h retention time and sludge pumps. The retention time of a DAF clarifier is in the range of 20–60 min.[14,20]

4.6.2.5   Sludge Drying Beds

1. *Method.* Sludge is dewatered by means of gravity drainage and natural evaporation.
2. *System components.* Beds of highly permeable gravel and sand with underlying drain pipes.[28]

4.6.2.6   Multimedia Filter

1. *Method.* This involves a polishing treatment after chemical precipitation and sedimentation by filtration through a bed of particles of several distinct size ranges.
2. *System components.* These comprise filter beds, media, backwash mechanism, pumps, and controls. Filter beds are sized for hydraulic loading of 81 L/min/m^2 (2 gpm/ft^2).

TABLE 4.16
Total Annual Unit Cost (USD/m³ in 2007 Dollars)[a]

Flow (m³/h)	Option 1		Option 2		Option 3	
	Continuous	Batch	Continuous	Batch	Continuous	Batch
2.36	—	14.28	—	23.94	—	28.35
11.81	6.09	5.04	9.66	8.4	11.34	10.29
59.07	2.52	—	4.62	—	5.25	—
118.16	2.10	2.10	3.57	3.78	4.20	4.41

[a]Costs were converted from 1979 USD to 2007 USD using the U.S. ACE Yearly Average Cost Index for Utilities.
Source: From U.S. EPA, Treatability Manual, Vol. II, Industrial Descriptions, Report EPA-600/2-82-001b, U.S. EPA, Washington, DC, September 1981.

4.6.2.7   Ultrafiltration

1. *Method.* This process is used for oily wastestreams after emulsion breaking–gravity oil separation.
2. *System components.* These comprise filter modules sized on the basis of a hydraulic loading of 1 L/min/m².

4.6.2.8   Carbon Adsorption

1. *Method.* This is a packed-bed throwaway system to remove organic pollutants from oily wastestreams.
2. *System component.* These comprise a contactor system, and a pump station designed for a contact time of 30 min and hydraulic loading of 162 L/min/m² (4 gpm/ft²).

Unit costs shown in Table 4.16 are for the complete treatment options described previously. Unit costs are computed for a model plant where flows have contributions from several wastestreams:

1. 30% oily wastestream
2. 4% cyanide wastestream
3. 9% chromium wastestream
4. 52.5% common metals stream
5. 4.5% complex metal stream

REFERENCES

1. Federal Register, Resource Conservation and Recovery Act (RCRA), 42 U.S. Code s/s 6901 et seq. 1976, United States Government, Public Laws, January 2004. Available at www.access.gpo.gov/uscode/title42/chapter82_.html.
2. U.S. EPA, Resource Conservation and Recovery Act (RCRA)—Orientation Manual, U.S. EPA, Report EPA530-R-02-016, Washington, DC, January 2003.
3. U.S. EPA, Federal Hazardous and Solid Wastes Amendments (HSWA), U.S. EPA, Washington, DC, November 1984. Available at http://www.epa.gov/osw/laws-reg.htm.
4. Federal Register, Clean Water Act (CWA), 33 U.S.C. ss/1251 et seq. (1977), U.S. Government, Public Laws, May 2002. Available at www.access.gpo.gov/uscode/title33/chapter26_.html.
5. U.S. EPA, Meeting Hazardous Waste Requirements for Metal Finishers, Report EPA/625/4-87/018, U.S. EPA, Cincinnati, OH, 1987.

6. U.S. EPA, Development Document for Effluent Limitations Guidelines and Standards for the Metal Finishing Point Source Category, Report EPA-440/1-80/091, U.S. EPA, Washington, DC, 1980.

7. U.S. EPA, Treatability Manual, Vol. II, Industrial Descriptions, Report EPA-600/2-82-001b, U.S. EPA, Washington, DC, September 1981.

8. PRC Environmental Management, *Hazardous Waste Reduction in the Metal Finishing Industry*, Noyes Data Corporation, Park Ridge, NJ, 1989.

9. U.S. ACE, Yearly average cost index for utilities, in *Civil Works Construction Cost Index System Manual*, 110-2-1304, U.S. Army Corps of Engineers, Washington, DC, 2007. Available at http://www.nww.usace.army.mil/cost.

10. Patterson, J.W., *Industrial Wastewater Treatment Technology*, 2nd ed., Butterworths, Boston, MA, 1985.

11. Wang, L.K., Vaccari, D.A., Li, Y., and Shammas, N.K., Chemical precipitation, in *Physicochemical Treatment Processes*, Wang, L.K., Hung, Y.T., and Shammas, N.K., Eds., Humana Press, Totowa, NJ, 2005, pp. 141–198.

12. Shammas, N.K., Kumar, I.J., Chang, S.Y., and Shammas, N.K., Sedimentation, in *Physicochemical Treatment Processes*, Wang, L.K., Hung, Y.T., and Shammas, N.K., Eds., Humana Press, Totowa, NJ, 2005, pp. 379–430.

13. Chen, J.P., Chang, S.Y., Huang, J.Y.C., Baumann, E.R., and Hung, Y.T., Gravity filtration, in *Physicochemical Treatment Processes*, Wang, L.K., Hung, Y.T., and Shammas, N.K., Eds., Humana Press, Totowa, NJ, 2005, pp. 501–544.

14. Wang, L.K., Hung, Y.T., and Shammas, N.K., Eds., *Physicochemical Treatment Processes*, Humana Press, Totowa, NJ, 2005, 723 p.

15. Wang, L.K., Hung, Y.T., and Shammas, N.K., Eds., *Advanced Physicochemical Treatment Processes*, Humana Press, Totowa, NJ, 2006, 690 p.

16. Chen, J.P., Mou, H., Wang, L.K., and Matsuura, T., Membrane filtration, in *Advanced Physicochemical Treatment Technologies*, Humana Press, Totowa, NJ, 2007, pp. 203–260.

17. Wang, L.K., Hung, Y.T., and Shammas, N.K., Eds., *Advanced Physicochemical Treatment Technologies*, Humana Press, Totowa, NJ, 2007, 710 p.

18. Wang, L.K., Pereira, N., Hung, Y.T., Eds., and Shammas, N.K., Consulting Ed., *Biological Treatment Processes*, Humana Press, Totowa, NJ, 2008.

19. Wang, L.K., Shammas, N.K., and Hung, Y.T., Eds., *Advanced Biological Treatment Processes*, Humana Press, Totowa, NJ, 2008.

20. Wang, L.K., Fahey, E.M., and Wu, Z., Dissolved air flotation, in *Physicochemical Treatment Processes*, Wang, L.K., Hung, Y.T., and Shammas, N.K., Eds., Humana Press, Totowa, NJ, 2005, pp. 431–500.

21. U.S. EPA, Managing Cyanide in Metal Finishing, Capsule Report EPA 625/R-99/009, U.S. EPA, Cincinnati, OH, December 2000.

22. Hartinger, L., *Handbook of Effluent Treatment and Recycling for the Metal Finishing Industry*, 2nd ed., Finishing Publications, Stevenage, UK, 1994.

23. Eilbeck, W.J. and Mattock, G., *Chemical Processes in Wastewater Treatment*, Ellis Horwood Ltd., New York, 1987.

24. Wang, L.K., Shammas, N.K., and Hung, Y.T., Eds., *Biosolids Treatment Processes*, Humana Press, Totowa, NJ, 2007, 820 p.

25. Singh, I.B., Chaturvedi, K., Singh, D.R., and Yegneswaran, A.H., Thermal stabilization of metal finishing waste with clay, *Environ. Technol.*, 26, 877–884, 2005.

26. Roy, C.H., *Operation and Maintenance of Surface Finishing Wastewater Treatment Systems*, American Electroplaters and Surface Finishers Society, Washington, DC, 1988.

27. Altmayer, F., *Plating and Surface Finishing, Advice & Council*, AESF, Orlando, FL, 1997.

28. Wang, L.K., Li, Y., Shammas, N.K., and Sakellaropoulos, G.P., Drying beds, in *Biosolids Treatment Processes*, Wang, L.K., Shammas, N.K., and Hung, Y.T., Eds., Humana Press, Totowa, NJ, 2007.

29. Kushner, J.B. *Water and Waste Control for the Plating Shop*, 3rd ed., Gardner Publications Inc., Cincinnati, OH, 1994.

30. Wang, L.K., Shammas, N.K., Hung, Y.T., Eds., *Biosolids Engineering and Management*, Humana Press, Totowa, NJ, 2008, 788 p.

5 Leachate Treatment Using Bioremediation

Azni Idris, Katayon Saed, and Yung-Tse Hung

CONTENTS

5.1 INTRODUCTION

More than 90% of municipal solid waste is directly disposed of on land, the vast majority of it in an unsatisfactory manner. Open and burning dumps are common in many developing countries; these contribute to water and air pollution and provide food and breeding grounds for birds, rats, insects, and other carriers of disease. The presence of these dumps often reduces the property value of nearby land and residences.

Sanitary landfilling is an acceptable and recommended method for ultimately disposing of solid wastes. This method has sometimes been confused with waste disposal on open and burning dump sites, but this is a misconception. The sanitary landfill is an engineered landfill that requires sound and detailed planning and specification, careful construction, and efficient operation. In essence, modern landfilling involves spreading the wastes in thin layers, compacting them to the smallest practical volume, and covering them with daily earth cover in a manner that minimizes adverse environmental pollution.

The sanitary landfill, the most acceptable alternative to the present poor practices of land disposal, involves the long-term planning and application of sound engineering principles and construction techniques. By definition, no burning of solid waste will ever occur at a sanitary landfill. A sanitary landfill is not only an acceptable and economic method of solid waste disposal, it is also an excellent way to make otherwise unsuitable or marginal land valuable.[1]

All landfills produce a liquid stream called leachate, which is a highly complex and polluted wastewater. Leachate pollution is a concern for many local authorities as it directly degrades river water quality. Many researchers continue to search for ways to treat leachate effectively using different biological processes. To secure long-term dewatering of landfills and reduce the increasing treatment costs, it is therefore necessary to control leachate quantity and quality. This is often difficult, as increasing water quality standards make the requirements on leachate treatment ever more stringent.

Treatment procedures must consider the highly varying flow and complex composition of the leachate; this often results in special operational problems. The following chapters give an overview of leachate generation and the development of leachate control and treatment applicable to many landfills.

5.2   SANITARY LANDFILL

A sanitary landfill is defined as a land disposal site that applies an engineered method of disposing of solid wastes on land in a manner that minimizes environmental hazards by spreading the solid wastes to the smallest practical volume, and applying and compacting cover material at the end of each day.[2]

Landfills are the physical facilities used for the ultimate disposal of residual solid wastes in the ground. In the past, the term sanitary landfill was used to denote a landfill in which the wastes were placed in the landfill and then covered at the end of daily operation. Today, sanitary landfill refers to an engineered facility for the disposal of municipal solid waste (MSW), designed and operated to minimize public health and environmental impact.

Solid wastes deposited in a landfill undergo slow degradation to produce residual solid, liquid, and gaseous products. Ferrous and other metals are oxidized and organic and inorganic wastes are utilized by microorganisms through aerobic and anaerobic processes. Organic acids, which are produced as a result of microbial degradation, increase chemical activity within the fill. Food wastes degrade quite readily, but other materials, such as plastics, rubber, glass, and some demolition wastes, are highly resistant to decomposition.

The degree of degradation of organic waste in landfills is very much dependent on the organic content of the waste. Wastes in Asian countries are reported to have a larger organic fraction, which leads to more problems in leachate generation. Waste data from Indonesia and China show that the organic fraction comprised 70.2% and 67.3%, respectively.[3]

Landfill methods are considered the most economical and environmentally acceptable way of disposing of solid wastes throughout the world. Even with the implementation of waste reduction, recycling, and transformation technologies, disposal of residual solid waste in landfill will still remain an important component of an integrated solid waste management strategy.[4]

In engineering terms, a sanitary landfill is also sometimes identified as a bioreactor due to the presence of anaerobic activities in the wastes. As such, landfilling sites need the incoming waste stream top be monitored, as well as placement and compaction of the waste, and installation of landfill environmental monitoring and control facilities. Gas vent and leachate collection pipes are important features of a modern landfill.

5.3 LEACHATE

The harmful liquid that collects at the bottom of a landfill is known as leachate. The generation of leachate is a result of uncontrolled runoff, and percolation of precipitation and irrigation water into the landfill. Leachate can also include the moisture content initially contained in the waste, as well as infiltrating groundwater. Leachate contains a variety of chemical constituents derived from the solubilization of the materials deposited in the landfill and from the products of the chemical and biochemical reactions occurring within the landfill under the anaerobic conditions.

An estimation of leachate generation in a landfill can be carried out by calculating the infiltration through a landfill cover using a water budget model such as the Hydrologic Evaluation of Landfill Performance (HELP).[5] The model uses conservation of mass to predict water movement, which enables the volumetric flux of water infiltrating into the waste to be calculated on a time-varying basis.

The generated leachate can cause significant environmental damage, becoming a major pollution hazard when it comes into contact with the surrounding soil, ground, or surface waters. One such problem is caused by infiltrating rainwater and the subsequent movement of liquid or leachate out of the fill into the surrounding soil. This leachate often contains a high concentration of organic matter and inorganic ions, including ammoniacal nitrogen and heavy metals. Therefore, in order to avoid environmental damage, landfill leachate must be collected and appropriately treated before being discharged into any water body.

5.4 COMPOSITION AND CHARACTERISTICS OF LEACHATE

Leachate tends to percolate downward through solid waste, continuing to extract dissolved or suspended materials. In most landfills, leachate seeps through the landfill from external sources, such as surface drainage, rainfall, groundwater, and water from underground springs, as well as from the liquid produced from the decomposition of the wastes, if any.[3]

Many factors influence the production and composition of leachate. One major factor is the climate of the landfill. For example, where the climate is prone to higher levels of precipitation, there will be more water entering the landfill and therefore more leachate generated. Another factor is the site topography of the landfill, which influences the runoff patterns and again the water balance within the site.

The composition of leachate is important in determining its potential effects on the quality of nearby surface water and groundwater. Contaminants carried in leachate are dependent on solid waste composition and on the simultaneously occurring physical, chemical, and biological activities within the landfill. The quantity of contaminants in leachate from a completed landfill where no more waste is being disposed of can be expected to decrease with time, but it will take several years to stabilize.

5.4.1 LEACHATE OF DIFFERENT AGE

The decomposition of solid urban waste in landfills is essentially a result of microbiological processes and, therefore, the production of biogas and leachate are both directly related to the activity of microorganisms. It has been demonstrated that large variations in leachate quality exist for different landfills, but also at different locations within the same landfill.[6]

New landfills generate more organic pollutants than older landfills. The BOD:COD (biochemical oxygen demand : chemical oxygen demand) ratio in young leachate is typically in the range of 0.5 to 0.7, which indicates higher biodegradability than that of mature landfills, which produce leachate with a BOD:COD ratio of less than 0.4.

5.4.2 LEACHATE IN DIFFERENT COUNTRIES

It is expected that leachate characteristics will vary by country. This is because the soil under a landfill site, the composition of disposed waste, the climate, sampling and landfill management vary among countries.[7,8]

5.5   LEACHATE TREATMENT

Many landfills pollute water bodies by discharging untreated leachate. When leachate percolates through the ground, it entrains landfill components such as decaying organic matter, micro-organisms, metals, and inorganic compounds into the underlying groundwater, causing serious contamination.

Landfill leachates are commonly classified as a high-strength wastewater containing dissolved and entrained landfill components.[9] Freshly produced landfill leachates are characterized by low pH values, high BOD_5 and COD values, as well as by the presence of several other toxic/hazardous compounds.[10] Several treatment options have been utilized for leachate treatment, with varying degrees of efficiency. The main applicable methods are biological, chemical, membrane separation, and thermal treatment processes.[11]

Physico-chemical processes are generally considered to incur high operating costs and sometimes lower effectiveness. A biological process is normally preferred, such as a conventional activated sludge process, which has been proven to be effective for the removal of organic carbon and nutrient content. Nevertheless, the problem of poor sludge settleability has usually been encountered, as well as the need for longer aeration times, for settling tanks of larger volume, and for total biomass recycling.

Some landfills practice leachate recycling in the fill area, where leachate percolates through the waste cell and undergoes further degradation. The treatment process or processes selected will depend to a large extent on the contaminants to be removed.[4]

Biological processes have been increasingly used in the treatment of leachate in combination with physical and chemical processes. Selected microorganisms are introduced in the aerobic treatment to achieve a better process efficiency. However, because of the variation in leachate composition from site to site, the remedial process train will generally be tailored to the site and consist of several unit operations. The following section discusses applications of bioremediation processes to landfill leachates. It is important to remember that characterization of leachate plumes through groundwater modeling, analysis of leachate physical and chemical characteristics, and development of leachate recovery systems are all important in selecting a leachate treatment system.[12]

5.6   BIOREMEDIATION METHODS

Bioremediation is defined as the use of microorganisms or microbial processes to degrade environmental contaminants. Bioremediation has numerous applications, including cleanup of groundwater, soils, lagoons, sludge, and process waste streams.

In general terms, bioremediation involves multiphase but heterogeneous environments, such as soils in which the contaminant is present in association with the soil particles, dissolved in soil liquids, and in the soil atmosphere. Because of these complexities, successful bioremediation depends on an interdisciplinary approach involving microbiology, biochemistry, and engineering.

For leachate treatment, the bioremediation method may be carried out either on or off site. Both methods have their advantages and disadvantages, depending on the site condition. Two factors favor the treatment of leachates on site: the expense of off-site transportation and the reluctance of communities nationwide to permit transportation routes or treatment facilities within their jurisdictions. The desirability of on-site leachate treatment should encourage the development of small-scale technology requiring low capital investment. Biological processes are well suited to on-site leachate treatment for the removal of organic compounds.[9]

5.6.1   *In Situ* and *Ex Situ* Methods

Depending on the situation, a bioremediation method can be either *ex situ* or *in situ*. *Ex situ* treatments are treatments that involve the physical removal of the contaminated material in order to undergo the treatment process. *In situ* techniques involve treatment of the contaminated material in place. Examples of *in situ* and *ex situ* bioremediation are listed in the following:

1. *Land farming.* This is a solid-phase treatment system for contaminated soils; it may be carried out *in situ* or *ex situ.*
2. *Bioreactors.* Biodegradation is carried out in a container or reactor; it may be used to treat liquids or slurries.
3. *Composting.* This is an aerobic, thermophilic treatment process in which contaminated material is mixed with a bulking agent; it can be carried out using static piles or aerated piles.
4. *Bioventing.* This is a method of treating contaminated soils by drawing air or oxygen through the soil to stimulate microbial activity.
5. *Biofilters.* Microbial stripping columns are used to treat air or liquid emissions.
6. *Bioaugmentation.* Bacterial cultures are added to a contaminated medium; this is frequently used in both *in situ* and *ex situ* systems.
7. *Biostimulation.* Indigenous microbial populations in soils or groundwater are stimulated by providing the necessary nutrients.
8. *Intrinsic bioremediation.* This is the unassisted bioremediation of the contaminant; the only process carried out is regular monitoring.

5.6.2 ADVANTAGES AND DISADVANTAGES OF BIOREMEDIATION

Successful bioremediation requires microbes and suitable environmental factors for degradation to occur. The most suitable microbes are bacteria or fungi that have the physiological and metabolic capabilities to degrade the pollutants.

Bioremediation offers several advantages over conventional methods of waste treatment such as landfilling or incineration. Bioremediation can be done on site, it is often less expensive, involves minimal site disruption, eliminates waste permanently, eliminates long-term liability, has greater public acceptance with regulatory encouragement, and can be coupled with other physical or chemical treatment methods.

Bioremediation also has its limitations. Some chemicals are not amenable to biodegradation, for instance, heavy metals, radionuclides, and some chlorinated compounds. In some cases, the microbial metabolism of the contaminants may produce toxic metabolites. Bioremediation is a scientifically intensive procedure that must be tailored to site-specific conditions, and usually requires treatability studies to be conducted on a small scale before the actual cleanup of a site.[13] The treatability procedure is important, as it establishes the extent of degradation and evaluates the potential use of a selected microorganism for bioremediation. A precise estimate on vessel size or area involved, speed of reaction, and economics can therefore be determined at the laboratory stage.

5.6.3 PHYSIOLOGY OF BIODEGRADATIVE MICROBES

Bioremediation is based on the activities of aerobic or anaerobic heterotrophic microorganisms. Microbial activity is affected by a number of physicochemical environmental parameters. Factors that directly affect bioremediation are energy sources (electron donors), electron acceptors, nutrients, pH, temperature, and inhibitory substrates or metabolites. One of the primary distinctions between surface soils, subsurface soils, and groundwater sediments is the organic material content. Surface soils, which typically receive regular inputs of organic material from plants, will undoubtedly have higher organic matter content.

High organic matter content is typically associated with high microbial numbers and a great diversity of microbial populations. Organic matter serves as a wardrobe of carbon and energy as well as a source of other macronutrients such as nitrogen, phosphorous, and sulfur. Subsurface soils and groundwater sediments have lower levels of organic matter and thus lower microbial numbers and population diversity than surface soils.[14] Bacteria become more dominant in the microbial community with increasing depth in the soil profile, because the numbers of other organisms such as fungi or actinomycetes decrease. This is attributed to the ability of bacteria to use alternative electron acceptors to oxygen. Other factors that control microbial populations are moisture content, dissolved oxygen, nutrient, and temperature.[13]

5.6.4 METABOLIC PROCESSES

The primary metabolism of an organic compound uses a substrate as a source of carbon and energy. For the microorganism, this substrate serves as an electron donor, which results in the growth of the microbial cell. The application of co-metabolism for bioremediation of a xenobiotic is necessary because the compound cannot serve as a source of carbon and energy due to the nature of the molecular structure, which does not induce the required catabolic enzymes. Co-metabolism has been defined as the metabolism of a compound that does not serve as a source of carbon and energy or as an essential nutrient, and can be achieved only in the presence of a primary (enzyme-inducing) substrate.

Two conditions favor metabolic activities: aerobic and anaerobic environments. Aerobic processes are characterized by metabolic activities involving oxygen as a reactant. Dioxygenases and monooxygenases are two primary enzymes used by aerobic organisms during the transformation and mineralization of xenobiotics. Anaerobic microbes take advantage of a range of electron acceptors, including, depending on their availability and the prevailing redox conditions, nitrate, iron, manganese, sulfate, and carbon dioxide.

5.6.5 FACTORS AFFECTING BIOREMEDIATION

5.6.5.1 Energy Sources

The primary factor that affects the activity of bacteria is the ability and availability of reduced organic material to serve as an energy source. Whether a contaminant will serve as an effective energy source for an aerobic heterotrophic organism is a function of the average oxidation state of the carbon in the material. Each degradation process depends on microbial (biomass concentration, population diversity, and enzyme activities), substrate (physico-chemical characteristics, molecular structure, and concentration), and a range of environmental (pH, temperature, moisture content, availability of electron acceptors, and carbon and energy sources) factors. These parameters affect the acclimation period of the microbes to the substrate. Molecular structure and contaminant concentration have been shown to strongly affect the feasibility of bioremediation and the type of microbial transformation occurring, as well as whether the compound will serve as a primary, secondary, or co-metabolic substrate.

5.6.5.2 Bioavailability

The rate at which microbial cells can convert contaminants during bioremediation depends on the rate of contaminant uptake and metabolism and the rate of transfer to the cell (mass transfer). Increased microbial conversion capacities do not lead to higher biotransformation rates when mass transfer is a limiting factor.[15] This appears to be the case in most contaminated soils and sediments. For example, contaminating explosives in soil did not undergo biodegradation even after 50 years. Treatments involving rigorous mixing of the soil and breaking up of the larger soil particles stimulated biodegradation drastically.[16] The bioavailability of a contaminant is controlled by a number of physico-chemical processes such as sorption and desorption, diffusion, and dissolution. Slow mass transfer causes a reduced bioavailability of the contaminants in the soil to the degrading microbes. Contaminants become unavailable when the rate of mass transfer is zero. The decrease of bioavailability over the course of time is often referred to as aging or weathering. It may result from the following:

1. Chemical oxidation reactions incorporating contaminants into natural organic matter
2. Slow diffusion into very small pores and absorption into organic matter
3. The formation of semirigid films around nonaqueous-phase liquids (NAPL) with a high resistance to NAPL–water mass transfer

These bioavailability problems may be overcome by the use of food-grade surfactants,[17] which increase the availability of contaminants for microbial degradation.

5.6.5.3 Bioactivity

Bioactivity refers to the operating state of microbiological processes. Improving bioactivity implies that system conditions are adjusted to optimize biodegradation.[18] For example, if the use of bioremediation requires meeting a certain minimum rate, adjusting the conditions to improve biodegradation becomes important and a bioremediation configuration that makes this control possible has an advantage over one that does not.

In nature, the ability of organisms to convert contaminants to both simpler and more complex molecules is very diverse. In light of our current limited ability to measure and control biochemical pathways in complex environments, favorable or unfavorable biochemical conversions are evaluated in terms of whether individual or groups of parent compounds are removed, whether increased toxicity is a result of the bioremediation process, and sometimes whether the elements in the parent compound are converted to measurable metabolites. These biochemical activities can be controlled in an *in situ* operation when one can control and optimize the conditions to achieve a desirable result.

5.7 BIOREMEDIATION OF LANDFILL LEACHATE

Bioremediation is the treatment of choice for mineralizing most organic compounds in landfill leachate.[19] Mineralization is carried out by microorganisms, which can degrade organic compounds to carbon dioxide under aerobic conditions and to a mixture of carbon dioxide and methane under anaerobic conditions. Microorganisms are also capable of changing the oxidation state of metals and inorganic compounds and can concentrate heavy metals and hydrophobic compounds through ingestion or adsorption. Microorganisms are ubiquitous, self-replicating, adaptable to a variety of leachate compositions, and active at moderate reaction conditions. In addition, biodegradation benefits from a long process history in the treatment of domestic sewage.

Leachate that comes from mixed landfills, that is, those with municipal waste combined with industrial wastes, may contain a host of xenobiotics (synthetic or unnatural) compounds. A number of these xenobiotics are normally classified as hazardous waste. A vast majority of organic hazardous wastes can be degraded if the proper microbial communities are established, maintained, and controlled.[20] Degradation is not necessarily growth-associated,[21] as organic compounds may be transformed to microbial storage polysaccharides under nitrogen-limiting conditions rather than being mineralized to carbon dioxide. Research regarding the mechanisms controlling xenobiotic degradation is important in understanding the capabilities and limitations of biological leachate treatment.[22]

An important element in xenobiotic biodegradation is the broad specificity of some microbial enzymes, which permits an enzyme-catalyzed reaction to occur without providing energy or carbon for cell replication. This phenomenon is divided into two categories: fortuitous metabolism, in which a growth co-substrate is not obligate, and co-metabolism, in which the growth co-substrate is obligate.[23] One of the most thoroughly characterized examples of broad enzyme specificity is the ability of the methane mono-oxygenase enzyme (MMO) to oxygenate hydrocarbons other than methane, its natural substrate. The oxygenated hydrocarbons then accumulate stoichiometrically in the reactor.[24] MMO-catalyzed reactions are co-metabolic, because energy from a co-substrate is required to supply reducing power for the reaction.

Fortuitous or co-metabolic biodegradation may account for a significant portion of the removal of xenobiotics in the environment.[24] Numerous examples of co-metabolic activity have been described for pure substrates,[22] but co-metabolism has been very difficult to demonstrate in mixed-substrate, mixed-culture systems, because products of the co-metabolic reactions of one species may be degraded by another.[24] To encourage co-metabolism, easily degradable co-substrates should be included in the leachate prior to biological treatment. Fatty acids, which often occur in landfill leachates, may fulfill this requirement.

In the case of industrial landfill leachate, it is unlikely that the microbial enzymatic machinery would be sufficient to degrade all the compounds present,[25] especially if a single microbial species is used. Furthermore, the adaptability of a single microbial species is limited and the mutation rate is too slow to make single-species adaptation practical. In order to increase the diversity of degradative enzymes it is common to use a mixed microbial population, also known as a microbial consortium or mixed culture. Mixed cultures have two advantages over pure cultures in the degradation of complex substrates. First, the product of an incomplete mineralization by one microbe, such as from a co-metabolic transformation, may serve as a substrate for another microbe. Second, the transfer of genetic information between species may enhance the degradability of the culture.[26] It has been demonstrated that DDT (dichloro diphenyl trichloroethane) can be co-metabolized to pentachlorophenol-induced periplasmic protein (PCPA) by one species and that PCPA can be mineralized by another species. A combined culture of the two species results in the complete mineralization of DDT.[27] Stable mixed cultures degrading xenobiotics have been isolated in which the microbial consortia can degrade a substrate better than the individual species.[22]

Many strains of microorganism have been isolated that can degrade xenobiotics or families of xenobiotics.[28] For example, a white rot fungus studied for its lignin-degrading potential has been shown in laboratory studies to mineralize a number of recalcitrant organics, such as a tetrachlorodibenzo-paradioxin (TCDD) and DDT.[29] Degradation is carried out by extracellular enzymes whose production is stimulated by nitrogen limitation. Because of the requirements of nitrogen limitation and an acidic environment, the fungus is incompatible with many activated-sludge-derived organisms. Whether such organisms will be useful for degrading mixtures of compounds or will be active in a full-scale process has yet to be demonstrated.

Gross genetic changes brought about by the interspecies transfer of genetic material may be important in the microbial degradation of xenobiotics. Although there are several mechanisms for such transfers, the most important is thought to be conjugation. In this process, loops of extra-chromosomal DNA mediate their own replication from host to recipient microorganisms. Conjugative plasmids, as these DNA loops are known, carry coding for a variety of proteins, which, although not required for reproduction, may confer a selective environmental advantage such as heavy metal resistance or extended substrate range.[30] In some cases, nonconjugative plasmids can link to conjugative plasmids and "piggy-back" from organism to organism.[23] Once a plasmid is transferred, DNA sequences called transposons may play a role in the integration of portions of the plasmid DNA into the genome of the new host. The rapid spread of antibiotic resistance among various classes of microorganisms is an example of the transfer of plasmid-born information.

The key issues in developing an effective biological landfill leachate treatment process are the following:

1. Process configuration
2. Microbial culture selection and development
3. Substrate modification
4. Process control

Due to the complex and varying nature of landfill leachate, these factors must be evaluated for each site. Chemical species thought to be biologically recalcitrant may be biodegradable given the proper acclimation. The principal mechanisms of acclimation are macromolecule modification, population selection, and genetic transfer. Modification of cellular components, for example, enzyme induction or increased membrane permeability, occurs when a substrate interacts with biological molecules of the cell. The time frame for such interactions ranges from minutes to hours.[30] Population selection, or shifts in the representation of preexisting species, occurs because some species or mutants within a species may be better adapted to a new environment. The time frame depends on growth rates and may range from hours to days for aerobic cultures and from days to weeks for anaerobic cultures.[31] Favorable genetic adaptation, alteration of the microbial DNA, may occur over periods ranging from months or years.[32]

Carbon limiting is also used to encourage enzyme induction, place the population under selective pressure for degradation of recalcitrant substrates, and favor the simultaneous rather than sequential metabolism of a mixed carbon source.[33] Carbon-limiting conditions can be achieved either through continuous culture (chemostat) or through a fed batch reaction.

To facilitate biodegradation, the leachate may require modification through pH adjustment, removal or addition of oxygen, amendment with nutrients, or dilution or removal of toxic species. Microbial nutrition is complex and is better understood for aerobes than for anaerobes.[34] Biological processes typically favor a pH near 7. Pretreatment processes to remove inhibitory components include coagulation and precipitation, carbon adsorption, and possibly ozonation.

A variety of biological processes options may be used to treat leachate.[35] The basic decision is whether to treat a particular leachate aerobically or anaerobically. Both aerobic and anaerobic processes can degrade a wide range of xenobiotics.[36] Aerobic processes are generally superior in mineralizing aromatic compounds; anaerobic processes are superior for short-chain aliphatic groups.[27] Aerobic processes have the advantage of speed and ease of control and acclimation. However, aerobic processes accumulate large quantities of microbial sludge that may contain adsorbed organics and heavy metals, and may strip volatile compounds. Anaerobic processes produce less sludge and can provide energy through methane production. They also reduce sulfate to sulfide, which is a powerful precipitator of heavy metals. However, because of their low reproduction rates, anaerobes require a long start-up time and are sensitive to toxic shocks.[37] Both aerobic and anaerobic processes have been shown to be capable of degrading landfill leachate.[38] However, many landfill leachate treatments have been found to be insufficient if the anaerobic process is used alone without the aerobic. Systems comprising combined anaerobic–aerobic treatment are therefore recommended to achieve effective treatment at landfill.

The rate of mineralization of organic carbon in a biological process depends on the concentration of active cell mass. The maximum cell growth in a process will depend on nutrient availability, gas transfer, and toxicity of the leachate. In aerobic and anaerobic treatment lagoons, no provision is made for concentrating the suspended cells. Therefore, lagoons must have a large surface area to facilitate effective organic destruction. The advantage of lagoons is that very low maintenance is needed except for a periodic desludging of the microbial sludge.[20]

The reduction in organic carbon achievable by microorganisms is limited to some extent by the minimum concentration required to maintain cellular metabolic processes.[39] Microbial species known as oligotrophs can operate at low substrate concentrations, but they may not be able to reduce contaminant concentrations below water quality standards. There are methods to circumvent the biological maintenance barrier to leachate degradation. A well-known approach involves the use of activated carbon to enhance the biodegradation reaction.[40] There are three known beneficial effects of adsorbent addition: organic carbon is concentrated for microbial attack in the microenvironment around the adsorbent particle; the concentration of potentially inhibitory organic compounds in the bulk solution is lowered; and the carbon particles serve as a surface for microbial growth.[41]

Leachate can also be degraded biologically *in situ* at the landfill site. Conditions within the landfill are controlled to encourage microbial activity, and leachate is recirculated through the landfill. Recirculated leachate may require nutrient amendment, neutralization, or heavy metal removal. Aerobic microbial activity occurs at the landfill surface, and anaerobic activity occurs in the landfill interior. Recirculation, combined with anaerobic activity, may stabilize heavy metals through the precipitation of heavy metal sulfides.[42] Aerobic biodegradation is faster and better understood, and methods for encouraging aerobic activity within a landfill by the addition of hydrogen peroxide or air microbubbles have been investigated.[43] Subsurface aeration wells have also been used to encourage *in situ* degradation.

Biodegradation is considered the first option for the primary removal of organic compounds from landfill leachate. However, some organic compounds are resistant to biological attack. In addition, biological sludge resulting from biological processes may become a disposal problem, particularly because of its capacity to store adsorbed undegraded hydrophobic organic species and heavy metals. No biological leachate treatment processes have yet to take advantage of microbial transformations, nor has adsorption of heavy metals though suitable microorganisms been studied in the laboratory.[44,48] Bioremediation processes are still relatively unsophisticated and the potential exists for combining various types of microbial process schemes for selective component removal.[9]

5.8   CASE STUDIES

5.8.1   CASE 1: ANAEROBIC/AEROBIC TREATMENT OF MUNICIPAL LANDFILL LEACHATE IN SEQUENTIAL TWO-STAGE UP-FLOW ANAEROBIC SLUDGE BLANKET REACTOR (UASB)/AEROBIC COMPLETELY STIRRED TANK REACTOR (CSTR) SYSTEMS

A project was conducted to study the treatability of leachate produced from a laboratory-scale simulated reactor treating food wastes using a two-stage sequential up-flow anaerobic sludge blanket reactor (UASB)/aerobic completely stirred tank reactor (CSTR).[45] Experiments were performed in two UASB reactors and a CSTR reactor having effective volumes of 2.5 and 9 L, respectively. The hydraulic retention times in the anaerobic and aerobic stages were 1.25 and 4.5 d, respectively. Following the startup period, the COD concentration of the leachate steadily increased from 5400 to 20,000 mg/L. The organic loading rate (OLR) was increased from 4.3 to 16 kg/m^3/d by increasing the COD concentrations from 5400 to 20,000 mg/L.

As reported, the effluent of the first anaerobic UASB reactor (Run1) was used as the influent of the second UASB reactor (Run2), and the effluent of the second UASB reactor was used for the influent of the aerobic CSTR reactor (Run3). COD removal efficiencies for the first UASB reactor and in the whole system (two-step UASB + CSTR) were 58%, 62%, 65%, 72%, 74%, 79% and 96%, 96.8%, 97.3%, 98%, 98%, and 98%, respectively. As the OLR increased from 4.3 to 16 kg/m^3/d, the COD removal efficiency reached a maximum of 80%. NH$_4$–N removal efficiency was ca. 99.6% after the aerobic stage. The maximum methane percentages of the first and second UASB reactors were 64% and 43%, respectively.

The study used two continuously fed stainless steel anaerobic UASB (2.5 L) reactors and an aerobic CSTR reactor (9 L). The UASB was operated at 37 to 42 °C using an electronic heater located in the central part of the reactor. The system was provided with a settling compartment (with an effective volume of 1.32 L). The dissolved oxygen concentration was maintained above 2 mg/L in the CSTR reactor. Partially granulated anaerobic sludge taken from the methanogenic reactor of the Pakmaya Yeast Baker Factory in Izmir was used as seed in the UASB reactor. The activated sludge culture was obtained from the DYO Dye Industry in Izmir and was used as seed for the aerobic CSTR reactor.

In this study, anaerobic and aerobic processes using sequential two-step UASB/CSTR reactors were found to form a feasible process for treating the leachate from food solid waste. COD removal efficiencies for the first and second anaerobic, aerobic and total system processes were 79%, 42%, 89%, and 98%, respectively. The COD loading rate used ranged from 4.3 to 16 kg/m^3/d.

The methane content of the first UASB reactor was ca. 60%. The NH_4-N removal efficiency of the total system was 99.6%. Ammonium nitrogen was converted to nitrate in the aerobic system via nitrification. The BOD_5/COD value obtained at the final stage was in the range 0.12 to 0.15.

5.8.2 CASE 2: COMPARISON OF TWO BIOLOGICAL TREATMENT PROCESSES USING ATTACHED-GROWTH BIOMASS FOR SANITARY LANDFILL LEACHATE TREATMENT

Two biological systems were compared using attached-growth biomass for the treatment of leachate generated from a municipal waste sanitary landfill. A moving-bed biofilm process, which is a relatively new type of biological treatment system, was used.[46] The process was based on the use of small, free-floating polymeric (polyurethane) elements, and biomass was grown and attached as biofilm on the surface of these porous carriers. For comparison, a granular activated carbon (GAC) moving-bed biofilm process was also tested. This method offered the advantages of combining both physico-chemical and biological removal mechanisms for the removal of pollutants. The presence of GAC in the reaction tank provided a porous surface able to adsorb both organic matter and ammonia, as well as to provide an appropriate surface onto which biomass could grow. A laboratory-scale sequencing batch reactor (SBR) was used for examination of both carriers. The effects of different operation strategies on the efficiency of these biological treatment processes were studied in order to optimize their performance, especially for the removal of nitrogen compounds and biodegradable organic matter. It was found that these processes were able to remove nitrogen content almost completely, and the removal of organic matter such as BOD_5 and COD was acceptable.

The SBR reactor used was constructed from cylindrical Plexiglas® with a working capacity of 8 L (as shown in Figure 5.1).[47] The contents of the reactor were mixed with a magnetic stirrer, and

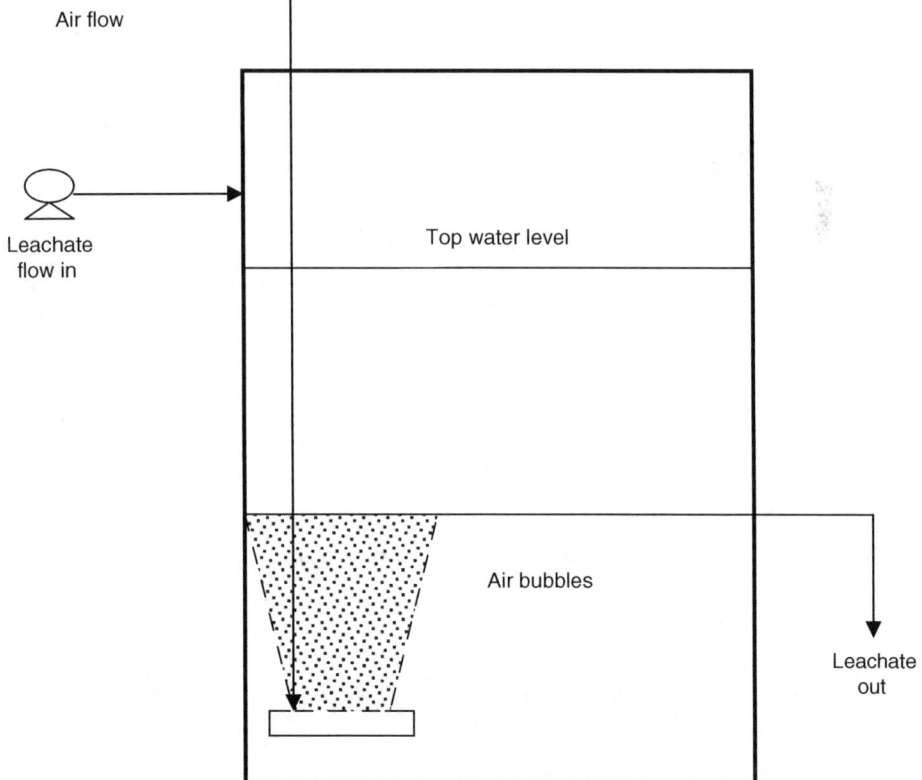

FIGURE 5.1 Schematic diagram of a laboratory-scale, sequencing batch (bio) reactor (SBR).

a ceramic diffuser was used for aeration. A peristaltic pump was used to feed leachate directly into the SBR, as well as to remove the treated effluent.

The study consisted of two separate treatment cycles using a suspended-carrier attached-biofilm process. During the first cycle, the SBR was filled up to 50% of its empty volume with cube-shaped waste polyurethane particles (total dry weight 30 g). The density of the carrier media was slightly below 1 g/cm^3, so the waste particles could easily follow the water flow pattern, circulating in the filled reactor. The continuous motion eliminated problems with clogging and dead space, which can often decrease the efficiency of fixed-bed biofilm systems. The cubes (having an approximate dimension of 1 cm) present high porosity (20 to 40 pores/cm^2). During the second operational cycle, GAC (type F400, supplied by Chemviron Co., Belgium) was added to the reactor (90 g total), with a specific surface area of 1100 m^2/g and density of 1.2 g/cm^3. The main parameters studied during this investigation included the following:

1. The addition of alkalinity, phosphorus, and methanol (different concentrations and rates were evaluated)
2. An increase in the hydraulic retention time
3. A replacement sequence for used carrier media
4. The application of intermittent aeration, i.e., operation with alternate aerobic and anoxic conditions

Table 5.1 summarizes process efficiency during the first operation cycle of the SBR, and Table 5.2 shows the treatment results for the second operation cycle (GAC).

This study demonstrated that the suspended carrier–biofilm treatment method can offer an alternative option to the conventional activated sludge process for the effective removal of carbon and nitrogen in sanitary landfill leachates. Although raw leachate is very difficult to treat, complete nitrification and a high degree of organic carbon removal were achieved using the moving-bed biofilm SBR process.

The study reported some problems regarding the data for the biofilm from the media after 3 weeks of operation, and also sludge accumulation at the bottom of the bioreactor. It was also found that an external carbon source, such as methanol, was necessary for controlling the denitrification stage.

An alternative moving-bed biofilm SBR process using GAC has also been proven to be an effective treatment method for the removal of nitrogen from landfill leachates. This method can remove biodegradable organic carbon (BOD$_5$) and COD. However, the main disadvantage of this process is the buildup of a large amount of residual suspended solids, hence increasing sludge disposal costs. An overall comparison between the two attached biomass biological treatment processes showed an advantage for the process that used porous polyurethane as its carrier material.[46]

TABLE 5.1

Average Treatment Results during First Operation Cycle (SBR)

Parameters	Influent Concentrations	Effluent Concentrations	Total Removal (%)
Ortho-P (mg/L)	3.2	1.1	66
Total-P (mg/L)	8.3	3.2	62
Cl^{-1} (mg/L)	4,640	3,062	34
Alkalinity (mg/L CaCO$_3$)	7,800	2,890	63
Conductivity (mS/cm)	24	14.3	40
TDS (mg/L)	14,000	7,000	50

SBR, sequencing batch reactor; TDS, total dissolved solids.

TABLE 5.2
Treatment Results for the Second Operation Cycle (GAC)

Parameters	Influent	Effluent	Total (%)
BOD_5 (mg/L)	1,292	114	91
Ortho-P (mg/L)	3.8	0.5	88
Total-P (mg/L)	9.3	2.2	73
Cl^{-1} (mg/L)	5,050	3,396	32.8
Alkalinity (mg/L $CaCO_3$)	733	1,395	81
Turbidity (NTU)	142	173	—
Conductivity (mS/cm)	25.4	12.5	50.7
TDS (mg/L)	14,900	6,900	54

BOD_5, 5d-biochemical oxygen demand; GAC, granular activated carbon; TDS, total dissolved solids; NTU, normal turbidity units.

5.8.3 Case 3: Leachate Treatment Using an Aerobic Biofilm Reactor

In this project, leachate was treated using an innovative aerobic biofilter utilizing special plastic media. Aerobic biofilters have been shown to be very effective in many treatments for removing organic pollutants and also their nutrient content. This study focused on leachate treatment using an attached growth biofilm reactor, which contains a packing of 80 mm diameter plastic media called "Cosmo balls."[47] Figure 5.2 shows how the experiment was set up. The selected parameters for the study include COD, ammonia nitrogen, pH, and BOD. The results showed that the COD removal percentages were above 90% for COD but declined to 70% at very high loading. The ammonia nitrogen removal achieved in the study was above 85%.

The use of an attached growth aerobic biofilm reactor to treat leachate is relatively new. Past studies on anaerobic biofilters showed excellent organic removal up to 90%, and the retention time needed to treat high-strength effluent was between 3 and 5d. The use of aerobic biofilters using

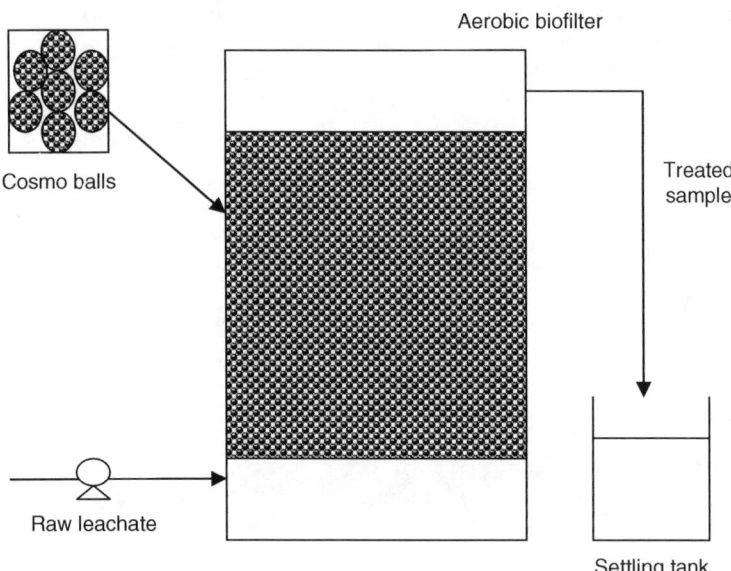

FIGURE 5.2 Schematic diagram of leachate treatment using an attached growth biofilm reactor.

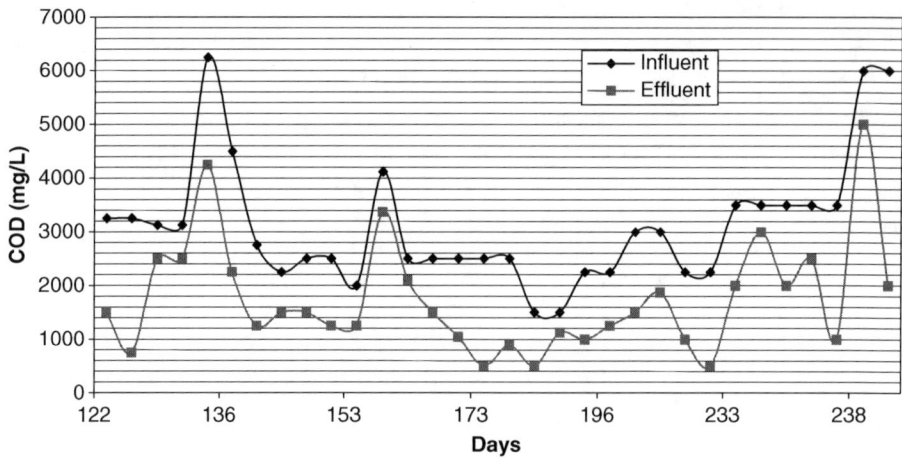

FIGURE 5.3 Relationship of COD content (mg/L) of influent and effluent over time (d).

Cosmo balls has been successful in treating sewage effluent with a short hydraulic retention time of only 4 h. This study was carried out to evaluate process efficiency using Cosmo balls with a hydraulic retention time of 5 d.

The aerobic biofilter used in the study had a capacity of 10 L. The reactor was packed to 60% of the empty bed volume with Cosmo ball media. The biofilter was seeded with active innoculum taken from an active aerated lagoon of a nearby landfill leachate treatment. Fresh raw leachate was used as feed to the reactor at a rate of 5 L/d over 24 h. The loading rates applied to the bioreactor were between 1.6 and 22.2 kg COD/m^3 d. Initial studies were conducted as a batch process lasting for a period of 24 d. Thereafter, the biofilter was fed continuously for a total period of 240 d.

Figure 5.3 shows that percentages of COD removed in the biofilter decreased with increasing feed COD concentration. The value of the influent fluctuated, indicating that leachate characteristics were never uniform. The aerobic bioreactor was shown to be capable of treating leachate with about 80% COD removal using the designed hydraulic retention time of 5 d. Figure 5.4 shows that the ammonia nitrogen levels in the treated effluent were fluctuating and that the percentage of ammonia nitrogen removed declined very slightly at increased ammonia loading. Ammonia nitrogen removal showed very good results, with more than 80% destruction achieved in this study.

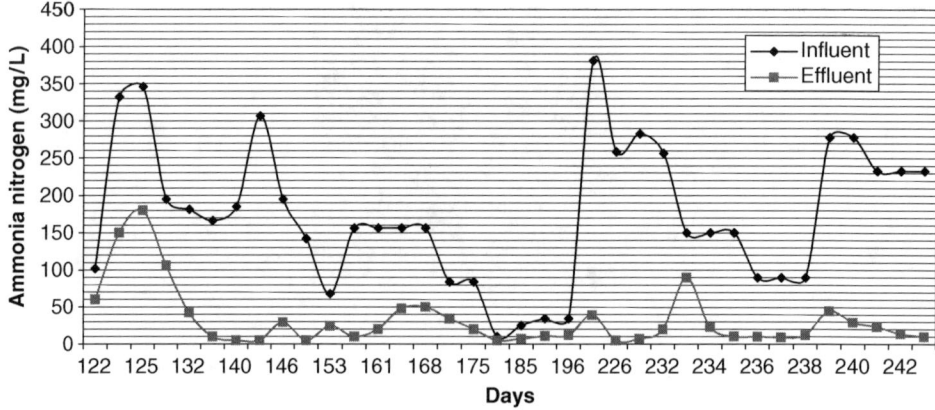

FIGURE 5.4 Relationship of ammonia nitrogen level (mg/L) in the influent and effluent over time (d).

It was observed that the factors contributing to the variation of leachate data are solid waste characteristics, for example, the composition and size of the waste and degree of compaction, the moisture content and degree of rainwater infiltration, temperature, sampling, and analytical methods.[47]

NOMENCLATURE

BOD	Bochemical oxygen demand
BOD_5	Five-day biochemical oxygen demand
COD	Chemical oxygen demand
CSTR	Completely stirred tank reactor
DDT	Dichloro diphenyl trichloroethane
DNA	Deoxyribonucleic acid
GAC	Granular activated carbon
MBAS	Methyl blue active substances
MMO	Methane mono-oxygenase enzyme
MSW	Municipal solid waste
OLR	Organic loading rate
PCPA	Pentachlorophenol-induced periplasmic protein
SBR	Sequencing batch reactor
TCDD	Tetrachlorodibenzo-paradioxin
TOC	Total organic carbon
TDS	Total dissolved solids
UASB	Up-flow anaerobic sludge blanket reactor
VDS	Volatile dissolved solids
VSS	Volatile suspended solids

REFERENCES

1. Weiss, S., *Sanitary Landfill Technology*, Noyes Data Corporation, NJ, 1974.
2. Davis, M.L. and Cornwell, D.A., *Introduction to Environmental Engineering*, 3rd ed., McGraw-Hill, New York, 1998.
3. Idris, A., Inanc, B., and Hassan, M.N., Overview of waste disposal and landfills/dump in Asian countries, *J. Mater.Cycles Waste Manage.*, 6, 104–110, 2004.
4. Tchobanoglous, G., Theisen, H., and Vigil, S., *Integrated Solid Waste Management; Engineering Principles and Management Issues*, McGraw-Hill, New York, 1993.
5. Schweder, P.R., Gibson, A.C., and Smolen, M.D., The Hydrologic Evaluation of Landfill Performance (HELP) Model, EPA/830-SW-84-010, 1984.
6. Robinson, H.D. and Luo, M.M.H., Characterization and treatment of leachates from Hong Kong landfill sites, *J. Inst. Water Environ. Manage.*, 5, 326–334, 1991.
7. Lea Rastas, A Literature Review; Typical Leachate—Does It Exist, SBA199 Senior Design Project, Lulea Tekniska Universitet, Finland, 2002.
8. Idris, A., Hassan, M.N., and Chong, T.L., Overview of Municipal Solid Wastes Landfill Sites in Malaysia, Proceeding of 2nd Workshop on Material Cycles and Waste Management in Asia, NIES Tsukuba, Japan, December 2–3, 2003.
9. Enzminger, J.D., Robertson, D., Ahlert, R.C., and Kosson, D.S., *Treatment of Landfill Leachates*, Elsevier Science Publishers, Amsterdam, 1986.
10. Ehrig, H.J., Leachate Quality. In: Christensen, T.H., Cossu, R., Eds., *Sanitary Landfilling: Process, Technology and Environmental Impact*, Academic Press, London, 1989, pp. 213–229.
11. Forgie, D.J.L., Selection of the most appropriate leachate treatment methods. Part 1: A review of potential biological leachate treatment methods, *Water Pollut. Res. J. Can.*, 23, 308–329, 1988.
12. Evans, R.B. and Schweiter, G.E., Hazardous waste assessment, *Environ. Sci. Technol.*, 18, 330a–340a, 1984.
13. Boopathy, R., Factors limiting bioremediation technologies, *Biores. Technol.*, 74, 63–67, 2000.

14. Adrians, P. and Hickey, W.J., in *Biotechnology for the Treatment of Hazardous Waste*, Stone, D.L., Ed., Lewis Publishers, Ann Arbor, 1993, pp. 97–120.

15. Boopathy, R. and Manning, J., A laboratory study of the bioremediation of 2,4,6-trinitrotoluene-contaminated soil using aerobic anaerobic soil slurry reactor, *Water Environ. Res.*, 70, 80–86, 1998.

16. Manning, J., Boopathy, R., and Kulpa, C.F., A Laboratory Study in Support of the Pilot Demonstration of a Biological Soil Slurry Reactor, Report no. SFIM-AEC-TS-CR-94038, U.S. Army Environmental Center, Aberdeen Proving Ground, MD, 1995.

17. Boopathy, R. and Manning, J., Surfactant-enhanced bioremediation of soil contaminated with 2,4,6-trinitrotoluene in soil slurry reactors, *Water Environ. Res.*, 71, 119–124, 1999.

18. Blackburn, J.W. and Hafker, W.R., The impact of biochemistry, bioavailability, and bioactivity on the selection of bioremediation technologies, *TIB Tech.*, 11, 328–333, 1993.

19. Ghassemi, M., Quintara, S., and BachMare, J., Characteristics of leachates from hazardous waste landfills, *J. Environ. Sci. Health. A*, 19, 679–620, 1984.

20. Michael, L.G., Philip, L.B., and Jeffrey, C.E., *Hazardous Waste Management*, 2nd ed., Biological Sciences Series, McGraw-Hill, New York, 2001.

21. Rozich, A.F. and Lowe, W.L., Oxidative assimilation treatment of a nitrogen-deficient toxic waste, *Biotechnol. Bioeng.*, 26, 613–619, 1984.

22. Johnston, J.B. and Robinson, S.G., Genetic engineering and the development of new pollution control technologies, EPA-600/2-84-037 (PB84-148972), 1984.

23. Dalton, H. and Stirling, D.I., Co-metabolism, *Philos. Trans. R. Soc. London, Ser. B*, 297, 481–496, 1982.

24. Venkataramani, E.S. and Ahlert, R.C., Role of co-metabolism in biological oxidation of synthetic compounds, *Biotechnol. Bioeng.*, 27, 1306–1311, 1985.

25. Grady, Jr, C.P.L., Biodegradation: its measurement and microbiological basis, *Biotechnol. Bioeng.*, 27660–27674, 1985.

26. Reineke, W. and Knackmuss, H.J., Construction of haloaromatic utilizing bacteria, *Nature*, 277, 285–286, 1979.

27. KoybayashI, H. and Rittman, B.E., Microbial removal of hazardous organic compounds, *Environ. Sci. Technol.*, 16, 170a–183a, 1982.

28. Slater, J.H. and Bull, A.T., Environmental microbiology: biodegradation, *Philos. Trans. R. Soc. London, Ser. B*, 297, 676–697, 1982.

29. Bumpus, J.A., Tien, M., Wright, D., and Aust, S.D., Oxidation of persistent environmental pollutants, *Science*, 228, 1434–1436, 1985.

30. Williams, P.A., Genetic interactions between mixed microbial populations, *Philos. Trans. R. Soc. London, Ser. B*, 297, 481–496, 1982.

31. Varma, M.M., Wan, L.W., and Prasad, C., Acclimation of wastewater bacteria by induction or mutation selection, *J. Water Pollut. Control Fed.*, 48, 832–835, 1976.

32. Kellogg, S.T., Chatterjee, D.K., and Chakrabarty, A.M., Plasmid-assisted molecular breeding: New techniques for enhanced biodegradation of persistent toxic chemicals, *Science*, 214, 1133–1135, 1981.

33. Harder, W. and Dijkhuizen, L., Strategies of mixed substrate utilization in microorganisms, *Philos. Trans. R. Soc. London, Ser. B*, 297, 469–480, 1982.

34. Speece, R.E., Anaerobic technologies for industrial wastewater, *Environ. Sci. Technol.*, 17, 416a–427a, 1983.

35. Peyton, T.O., Biological isposal of hazardous waste, *Enz. Microbiol. Technol.*, 6, 146, 1984.

36. Patterson, J.W. and Kodukala, P.S., Biodegradation of hazardous organic pollutants, *Chem. Eng. Prog.*, 77, 48–55, 1981.

37. Archer, D.B., The microbial basis of process control in methanogenic fermentation of soluble waste, *Enz. Microbial. Technol.*, 6, 162–170, 1983.

38. Corbo, P. and Ahlert, R.C., Anaerobic treatment of concentrated industrial wastewater, *Environ. Prog.*, 4, 22–26, 1986.

39. Alexander, M., Biodegradation of organic chemicals, *Environ. Sci. Technol.*, 19, 106–111, 1985.

40. Andrews, G.F. and Trapasso, R., A novel adsorbing bioreactor for wastewater treatment, *Environ. Prog.*, 3, 57–63, 1984.

41. Suidan, M.T., Siekerka, G.L., Kao, S.-W., and Pfeffer, J.T., Anaerobic filters for the treatment of coal gasification wastewater, *Biotechnol. Bioeng.*, 25, 1581–1596, 1983.

42. Pohland, F.G. and Gould, J.P., Containment of Heavy Metals in Landfills with Leachate Recycle, Proc. 7th Annual Symposium Disposal of Hazardous Waste, EPA 600/g-81-002a, pp. 171–189, 1981.

43. Michelson, D.L., Wallis, D.A., and Sebba, F., In-situ biological oxidation of hazardous organics, *Environ. Prog.*, 3, 103–107, 1984.

44. Norberg, A.B. and Persson, H., Accumulation of heavy-metal ions by *zooglea ramigere*, *Biotechnol. Bioeng.*, 26, 239–246, 1984.

45. Osman, N.A, and Sponza, D.T., Anaerobic/aerobic treatment of municipal landfill leachate in sequential two-stage up-flow anaerobic sludge blanket reactor (UASB)/completely stirred tank reactor (CSTR) system, *Proc. Biochem.*, 40, 895–902, 2005.

46. Loukidou, M.X. and Zouboulis, A.I., Comparison of two biological treatment processes using attached-growth biomass for sanitary landfill leachate treatment, *Environ. Pollut.*, 111, 273–281, 2000.

47. Idris, A. and Abdullah, A.G.L., Leachate Treatment Using Aerobic Biofilm Reactor, Proc. JSPS-VCC Seminar on Water Resource Management, Kota Kinabalu, Malaysia, 2001.

48. Haarstad K. and Mæhlum T., Electrical conductivity and chloride reduction in leachate treatment systems, *J. Environ. Eng.*, 133, 659, 2007.

6 Remediation of Sites Contaminated by Hazardous Wastes

Lawrence K. Wang, Nazih K. Shammas, Ping Wang, and Robert LaFleur

CONTENTS

6.1   INTRODUCTION

Site contamination generally results from leakage, spillage, or disposal of industrial wastes, and can arise from the past uncontrolled disposal of chemical wastes or any recent negligence. Contaminated sites are a threat to human beings by the following means of contamination:

1. Contact with contaminated soil
2. Inhalation of evaporated toxic gases
3. Drinking of contaminated groundwater
4. Consumption/intake of a secondary contaminant, for example, by eating contaminated crops or livestocks fed in the contaminated area

The Comprehensive Environmental Response, Compensation, and Liability Act (CERCLA)[1-3] and the Superfund Amendments and Reauthorization Act (SARA)[4] protect the public from the risks created by past and recent chemical disposal practices. Cleanup of contaminated sites is needed in order to protect human and natural resources, as defined by the Clean Air Act,[5] the Clean Water Act,[6] the Safe Drinking Water Act,[7] and the Resource Conservation and Recovery Act (RCRA).[8,9]

This chapter presents a regulatory overview of on-site remediation, remedial investigations (RI), feasibility studies (FS), remedial technologies, and a simulated case study. The discussion of remedial investigations and feasibility studies also includes the development and selection of remedial technologies. The case study outlines a remedial investigation and feasibility study, as well as the selection of remedial technologies.

6.2   LEGISLATIVE AND REGULATORY OVERVIEW

6.2.1   COMPREHENSIVE ENVIRONMENTAL RESPONSE, COMPENSATION, AND LIABILITY ACT

In 1980, the U.S. Congress enacted the Comprehensive Environmental Response, Compensation and Liability Act (CERCLA), the first comprehensive federal law addressing the protection of the environment from the threat of hazardous substances. The primary goal of CERCLA is to establish an organized cost-effective mechanism for response to abandoned or uncontrolled hazardous waste sites that pose a serious threat to human health and the environment.[8,9] To accomplish this goal, two types of response capabilities are mandated by CERCLA[1-3]:

1. An emergency response action for handling major chemical spills or incidents requiring immediate action, usually only at the surface of a site (e.g., to avert an explosion, to clean up a hazardous waste spill, or to stabilize a site until a permanent remedy can be found); these action are limited to 12 months or USD 2 million in expenditure, although in certain cases these limits may be extended.
2. A remedial response capability for undertaking the long-term cleanup of abandoned hazardous waste disposal sites. These remedial actions represent the final remedy for a site and are generally more expensive and of a longer duration than emergency removals. The U.S. Environmental Protection Agency (U.S. EPA) deals only with remedial actions for hazardous waste sites that are on the National Priorities List (NPL).

Both removal and remedial actions may be carried out at the same site. To accomplish these tasks, CERCLA has given cleanup authority to U.S. EPA, has established the Hazardous Substance Response Trust Fund (Superfund) to finance the remedial actions at CERCLA sites, has initiated a procedure for the emergency response to accidental spills, and has imposed cleanup liability on those responsible. The National Contingency Plan (NCP) was developed in 1982 and in 1985 as the regulatory framework to guide these responses.

Preliminary assessments have been conducted at more than 31,000 sites reported as possible sources of contamination. In 1990 there were over 1100 sites (presenting the greatest health risk and hence eligible for Superfund reimbursement) on the NPL.[8,9] The NCP has outlined the level of cleanup necessary at Superfund sites and established the basic procedures that have to be followed for the discovery, notification, response, and remediation of the hazardous waste sites.[10]

6.2.2 SUPERFUND AMENDMENTS AND REAUTHORIZATION ACT (SARA)

SARA has added several important new dimensions to CERCLA, including an increased emphasis on health assessments and the consideration of air releases.[11]

It should be noted that early remedial actions for contaminated soil consisted primarily of excavation and removal of the contaminated soil from the site and its disposal at a landfill. SARA strongly recommends on-site treatment that permanently and significantly reduces the volume, toxicity, or mobility of hazardous substances, and utilizes cost-effective permanent solutions. The legislation prohibits land disposal of hazardous wastes unless U.S. EPA determines otherwise (as in the Hazardous and Solid Waste Amendments, HSWA).

SARA requires that remedial actions meet all applicable or relevant federal standards or any more stringent state standards. Nine criteria that need to be met are set by CERCLA as amended by SARA for a complete assessment of treatment alternatives applicable for a site remedial action[12]:

1. The overall protection of human health and the environment by permanently and significantly reducing the volume, toxicity, or mobility of hazardous substance, pollutants, and contaminants
2. Compliance with applicable, relevant, and appropriate requirements (ARARs)
3. Long-term effectiveness and permanence
4. Reduction of toxicity, mobility, or volume
5. Short-term effectiveness
6. Implementability
7. Cost
8. State acceptance
9. Community acceptance

The CERCLA reauthorization regards off-site transport and disposal without treatment as the least favored alternative where practicable treatment technologies are available. It also favors the use of permanent solutions and alternative treatment technologies or resource recovery technologies and using them to the maximum extent practicable.

6.2.3 RESOURCE CONSERVATION AND RECOVERY ACT (RCRA)

RCRA has a regularity focus (in contrast to CERCLA, which has a response focus), and authorizes control over the management of wastes from the moment of generation until final disposal, including transportation, storage, and other processes.

6.3 OVERVIEW OF REMEDIAL STRATEGIES AND PHASES

The remedial strategies include the following:

1. Site selection from the NPL
2. Scoping

3. Remedial investigation including site characterization and a treatability study
4. Feasibility study including analysis and selection of alternatives
5. Remedial design and action (see Figure 6.1)[12]

The remedial strategies of concern focus on how to select a remedial method and how to complete the remediation at the most effective cost.

6.3.1 SCOPING

Scoping is the prework for RI and FS study. The task of scoping consists mainly of site data collection. As this is required for the RI phase, some investigators have regarded scoping as an early subphase of RI. However, scoping also involves project planning and other prework for FS, so it is to be regarded as a separate phase that precedes both RI and FS.

6.3.2 REMEDIAL INVESTIGATION/FEASIBILITY STUDY (RI/FS)

In accordance with §105 of CERCLA, U.S. EPA has established a process for locating releases, evaluating remedies, determining the appropriate extent of response, and ensuring that the remedies selected are cost-effective. This process is commonly referred to as the RI/PS process. The overall purpose of the RI/PS process represents the methodology that the Superfund program has established for characterizing the nature and extent of the risks posed by uncontrolled hazardous waste sites and for evaluating their potential remedial options.

The NCP requires that a detailed RI/PS be conducted for every site that is targeted for remedial response action under §104 of CERCLA.

Figure 6.2 outlines the major tasks carried out in the RI/FS process under CERCLA guidance.[13] The components of RI comprise the following:

1. Collecting data to characterize site conditions
2. Determining the nature of the waste

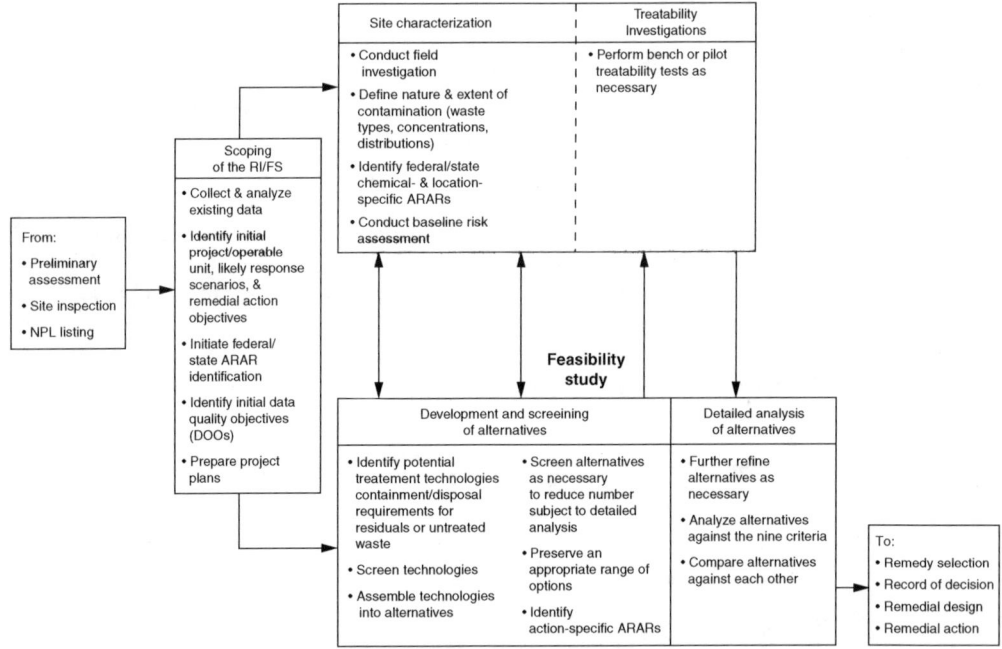

FIGURE 6.1 Phased remedial investigation process.

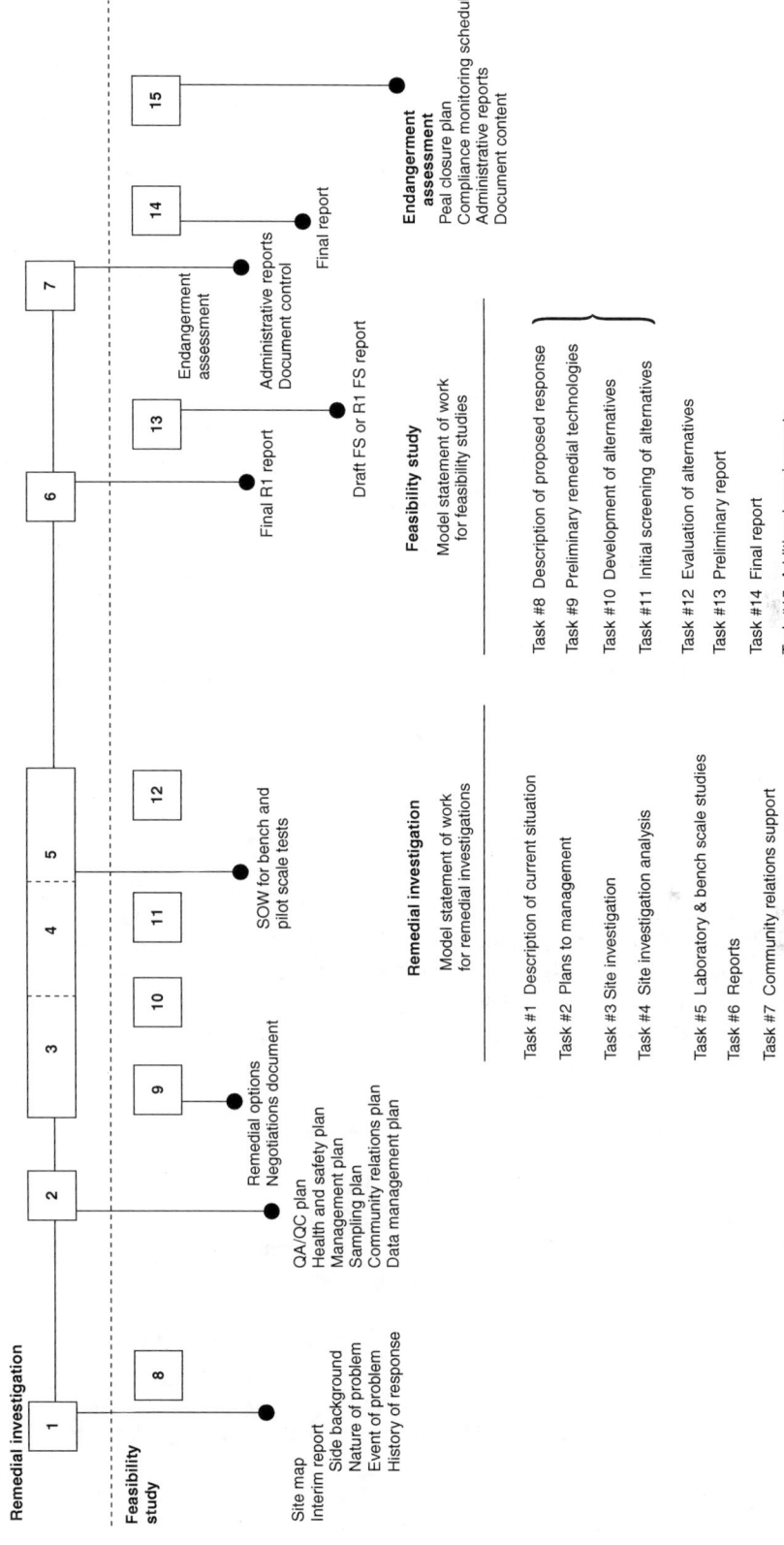

FIGURE 6.2 RI/FS process.

3. Assessing risk to human health and the environment
4. Conducting treatability testing as necessary to evaluate the potential performance and cost of the treatment technologies that are being considered[12,13]

The components of FS comprise developing, screening, and evaluating alternative remedial actions.

RI and FS are interdependent processes and are generally performed concurrently rather than sequentially, although the FS uses the data provided by the RI. This approach should be viewed as a dynamic, flexible process that can and should be tailored to specific circumstances of individual sites. It is not a rigid step-by-step approach that must be conducted identically at every site. Figure 6.3 illustrates a generic timeline of the phasing of RI/PS activities.

6.4   SCOPING THE REMEDIAL INVESTIGATION AND FEASIBILITY STUDY

Scoping is the initial planning phase of site remediation and is a part of the funding allocation and planning process.[12] Scoping of the RI/FS comprises the following steps:

1. Evaluating existing data
2. Developing the conceptual site model
3. Identifying the initial project/operable unit, likely response scenarios, and remedial action objects
4. Initiating potential federal/state ARARs identification
5. Identifying initial data quality objectives
6. Preparing project plans

6.4.1   PROJECT PLANNING

There are 12 tasks involved in project planning:

1. *Conducting project meetings.* This includes meeting with the lead agency, the support agency, and contractor personnel to discuss site issues and assign responsibilities for RI/FS activities.
2. *Collecting and analyzing existing data.* Existing data (Table 6.1) are collected and analyzed to develop a conceptual site model that can be used to assess both the nature and the extent of contamination and to identify potential exposure pathways and potential human health or environmental receptors.
3. *Describing the current situation.*
4. *Developing a conceptual site model.* An example of this is presented in Figure 6.4.[12]
5. *Developing preliminary remedial action alternatives.* This involves initiating limited field investigations if available data are inadequate to develop a conceptual site model and adequately scope the project, and identifying preliminary remedial action objectives and likely response actions for the specific project.
6. *Evaluating the need for a treatability study.* The requirement and schedule for treatability studies so as to better evaluate potential remedial alternatives are identified. If remedial actions involving treatment have been identified for a site, then the need for treatability studies should be evaluated as early as possible in the RI/FS process. This is because many treatability studies may take several months or longer to complete.
7. *Beginning preliminary identification of ARARs and "to be considered" (TBC) information.* This preliminarily identifies the ARARs that are expected to apply to site characterization and site remediation activities.
8. *Identifying data needs.* Data requirements and the level of analytical and sampling certainty required for additional data if currently available data are inadequate to conduct the FS is

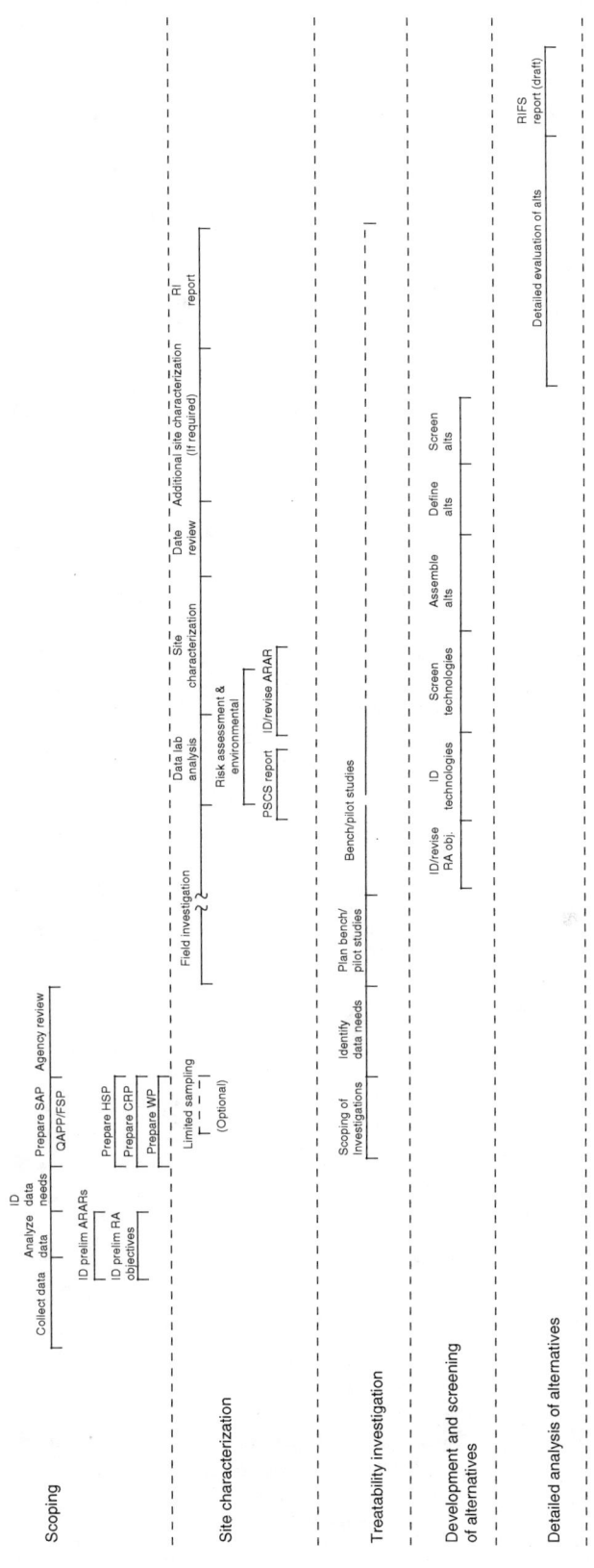

FIGURE 6.3　Generic phased RI/FS timeline.

TABLE 6.1
Data Collection Information

Information Source	Hazardous Waste Sources	Migration Pathways Subsurface	Surface	Air	Receptors
U.S. EPA files	×	×	×	×	×
U.S. Geological Survey		×	×		
U.S. DOA, Soil Conservation Service		×	×		
U.S. DOA, Agricultural Stabilization and Conservation Service		×	×		
U.S. DOA, Forest Service			×		×
U.S. DOI, Fish and Wildlife Agencies					×
U.S. DOI, Bureau of Reclamation	×	×	×		
U.S. Army Corps of Engineers	×				
Federal Emergency Management Agency			×		
U.S. Census Bureau					×
National Oceanic and Atmospheric Administration				×	
State Environmental Protection or Public Health Agencies	×	×	×	×	×
State Geological Survey		×	×		
State Fish and Wildlife Agencies					×
Local Planning Boards		×	×	×	×
County or City Health Departments	×	×	×	×	×
Town Engineer or Town Hall	×				×
Local Chamber of Commerce	×				×
Local airport				×	
Local library		×			×
Local well drillers		×			
Sewage treatment plants	×	×	×		
Local water authorities		×			×
City fire departments	×	×	×	×	
Regional geologic and hydrologic publications		×	×		
Court records of legal action	×				
Department of Justice files	×				
State Attorney General files	×				
Facility records	×				
Facility owners and employees	×	×			×
Citizens residing near site	×	×	×	×	×
Waste haulers and generators	×		×		
Site visit reports	×		×	×	×
Photographs	×		×		×
Preliminary assessment report	×	×	×	×	×
Field investigation analytical data	×	×	×	×	
FIT/TAT reports	×	×	×	×	×
Site inspection report	×	×	×	×	×
HRS scoring package	×	×	×	×	×
EMSL/EPIC	×		×		×

EMSL, Environmental Monitoring Support Laboratory; EPIC, Environmental Photographic Information Center; DOA, Department of Agriculture; DOI, Department of Interior; FIT, Field Investigation Team; TAT, Technical Assistance Team.
Source: From U.S. EPA, Guidance for Conducting Remedial Investigations and Feasibility Studies under CERCLA, EPA/540/G-89/004, October, U.S. EPA, Washington, DC, 1988.

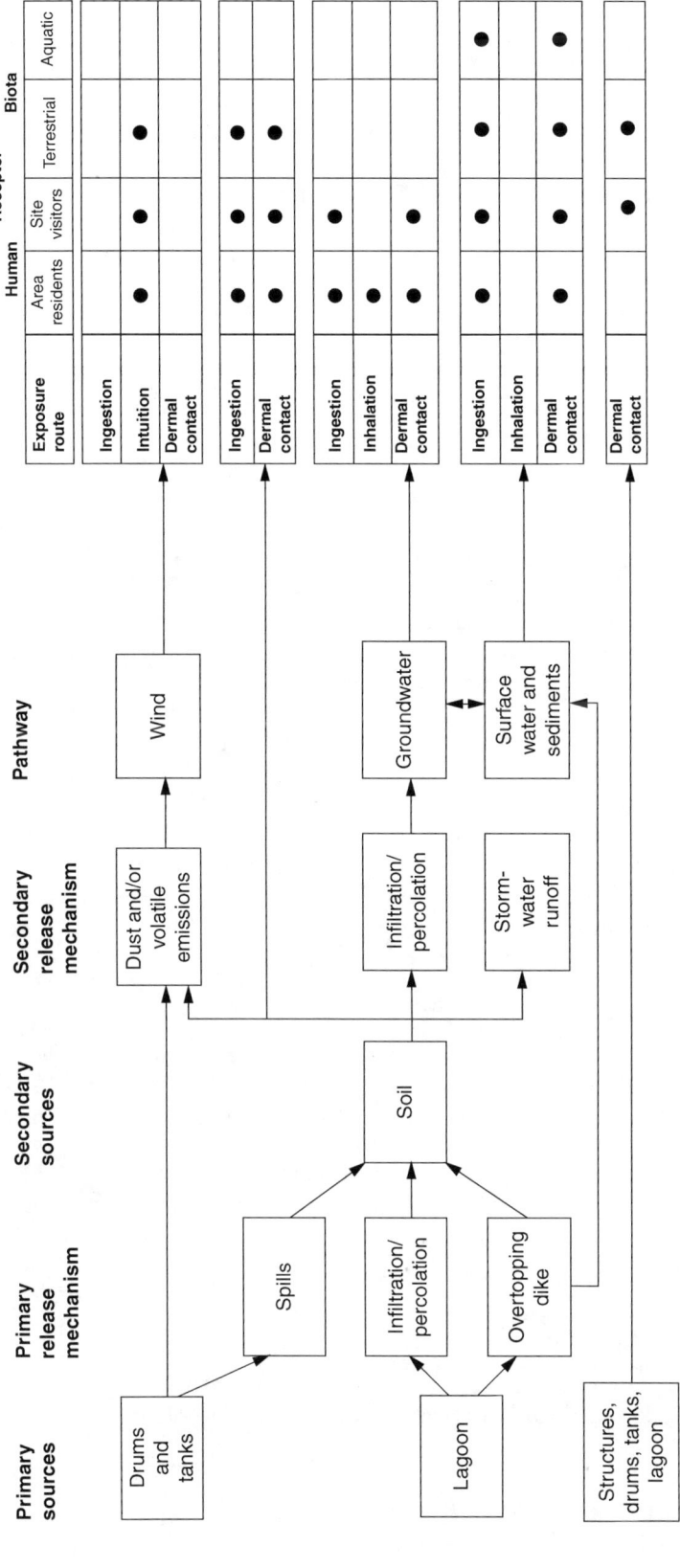

FIGURE 6.4 Example of a conceptual site model. (From U.S. EPA, Guidance for Conducting Remedial Investigations and Feasibility Studies under CERCLA, EPA/540/G-89/004, U.S. EPA, Washington, October 1988.)

identified, as well as possible uses of the data, including monitoring during implementation, health and safety planning, site characterization, risk assessment, evaluating alternatives, determining the potential responsible party (PRP), and the design of alternatives.

9. *Designing a data collection program.* A data collection program is designed to describe the selection of sampling approaches and analytic options, to establish the level of confidence required for the data, and to develop strategies for sampling and analysis.

10. *Developing a work plan.* A work plan is established that documents the scoping process and presents anticipated future tasks.

11. *Identifying health and safety protocols.* In this stage, health and safety protocols required during field investigations are identified and documented, and a site health and safety plan is prepared to support the field effort and conform to the firm's or agency's health and safety program.

12. *Conducting community interviews.* Community interviews are carried out to obtain information that can be used to develop a site-specific community relations plan that documents the objectives and approaches of the community relations program.

The identification of sampling requirements involves specifying the sampling design, the sampling method, sample numbers, types, and locations, and the level of sampling quality control. Data quality requirements include precision, accuracy, representativeness, completeness, and comparability.

The purpose of a sampling and analysis plan (SAP) is to ensure that sampling data collection activities will be comparable to and compatible with previous data collection activities performed at the site, while providing a mechanism for planning and approving field activities. The plan also serves as a basis for estimating costs of field efforts for inclusion in the work plan.

The SAP consists of the field sampling plan (FSP) and the quality assurance project plan (QAPP) elements. The QAPP describes the policy, organization, functional activities, and quality assurance and quality control protocols necessary to achieve data quality objects dictated by the intended use of the data. The FSP provides guidance for all fieldwork by defining in detail the sampling and data-gathering methods to be used on a project, including plan preparation and responsibilities (timing, preparation and review, field sampling plan, and so on). Table 6.2 lists the format for the FSP and QAPP.

6.4.2 Deliverables and Communications

There are several points during the scoping process when communication is required between the lead agency and its contractor or the support agency (Table 6.3). It is especially important that discussion and information exchange occur if interim actions or limited field investigations are considered necessary.

Deliverables required for all RI/FSs in which field investigations are planned consist of a work plan, an SAP, a health and safety plan (HSP), and a community relations plan (CRP).

6.5 REMEDIAL INVESTIGATION

6.5.1 Site Characterization

Site characterization is necessary in order to determine to what extent a site poses a threat to human health or the environment.[12] Site characterization is the core of RI, and includes the following stages:

1. Conducting field investigations as appropriate
2. Analyzing field samples in the laboratory

TABLE 6.2

Suggested Format for the Sampling and Analysis Plan (Comprising the Field Sampling Plan and Quality Assurance Project Plan)

FSP

Site background
Sampling objectives
Sample location and frequency
Sample designation
Sampling equipment and procedures
Sample handling and analysis

QAPP

Title page
Table of contents
Project description
Project organization and responsibilities
QA objectives for measurement
Sampling procedures
Sample custody
Calibration procedures
Analytical procedures
Data reduction, validation, and reporting
Internal quality control
Performance and systems audits
Preventative maintenance
Data assessment procedures
Corrective actions
Quality assurance reports

Source: U.S. EPA, Guidance for Conducting Remedial Investigations and Feasibility Studies under CERCLA, EPA/540/G-89/004, October, U.S. EPA, Washington, DC, 1988.

3. Evaluating the results of data analysis to characterize the site and develop a baseline risk assessment
4. Determining if data are sufficient for developing and evaluating potential remedial alternatives

6.5.1.1 Field Investigation

The major components of field investigation are air, biota, close support laboratories, RI-derived waste disposal, soil, gas, support, well logging, mapping and survey, geophysical characteristics, well installation, groundwater, source testing, and surface water. A complete field investigation includes at least prefield work, site physical characteristics investigation, contamination sources identification, and contamination determination.

Prefield work

The following prefield work is often needed before beginning an official field work:

1. Analyzing the collected existing data, including site characteristics, history of site (including disposal practices, disposal locations, disposed waste condition, waste degradations, storage of raw materials)

TABLE 6.3
Communications and Deliverables during Scoping

Information Needed	Purpose	Potential Methods of Information Exchange
Interim actions (if necessary)	For lead agency and contractor to identify actions that will abate immediate threat to public health or prevent further degradation of the environment; to obtain concurrence of support agency	Meeting Tech memo Other
Limited field investigations (if necessary)	For lead agency and contractor to improve focus of RI and reduce time and cost; to obtain concurrence of support agency	Meeting Tech memo Other
Summary of existing data; field studies conducted prior to FS; identification of preliminary remedial action alternatives	For lead agency and contractor to confirm need for FS; for lead agency and contractor to plan data collection; to obtain support agency review and concurrence	Meeting Tech memo Other
Documentation of QA and field sampling procedures	For contractor to obtain lead agency review and approval; for lead agency to obtain support agency review and comment	SAP (FSP, QAPP)
Documentation of health and safety procedures	For contractor to obtain lead agency agreement that OSHA safety requirements are met	Health and safety plan
Documentation of all RI/FS tasks	For contractor to obtain lead agency review and approval; for lead agency to obtain support agency concurrence	Work plan

QA, quality assurance; RI, remedial investigation; FS, feasibility study; SAP, sampling and analysis plan; FSP, field sampling plan; QAPP, quality assurance project plan; OSHA, Occupational Safety and Health Administration.

Source: U.S. EPA, Guidance for Conducting Remedial Investigations and Feasibility Studies under CERCLA, EPA/540/
G-89/004, October, U.S. EPA, Washington, DC, 1988.

2. Ensuring that access to the site and any other areas to be investigated has been obtained
3. Procuring equipment protective ensembles, air monitoring devices, sampling equipment, decontamination apparatus, and supplies (disposables, tape, notebook, and so on)
4. Coordinating with analytical laboratories, including sample scheduling, sample bottle acquisition reporting, chain-of-custody records, and procurement of close support laboratories or other in-field analytical capabilities
5. Procuring on-site facilities for office and laboratory space, decontamination equipment, and vehicle maintenance and repair, and sample storage, as well as on-site water, electric, telephone, and sanitary utilities
6. Providing for storage or disposal of contaminated materials (e.g., decontamination solutions, disposable equipment, drilling muds and cuttings, well-development fluids, well-purging water, and spill-contaminated materials)
7. Preparing field work, including time table, health, instrument, container, RCRA, equipment, and sample aspects

Site physical characteristics investigation
A site physical characteristics investigation examines the following[12]:

1. Surface features, including facility dimensions and locations, surface disposal areas, fencing, property lines and utility lines, roadways and railways, drainage ditches, leachate

springs, surface water bodies, vegetation, topography, and residence and commercial buildings

2. Geology information, including the geology of unconsolidated overburden and soil deposits (thickness and areal extent of units, petrology, mineralogy, particle size and sorting, and porosity) and the geology of the bedrock (type of bedrock, petrology, structure and texture, discontinuities such as joints, fractures, and foliation, and unusual features such as dikes, lavas, and karsts)

3. Soils and vadose zone information, including soil characteristics (type, holding capacity, temperature, biological activity, and engineering properties), soil chemical characteristics (solubility, ion specification, adsorption, leachability, cation exchange capacity, mineral partition coefficient, and chemical and sorptive properties), and vadose zone characteristics (permeability, variability, porosity, moisture content, chemical characteristics, and extent of contamination)

4. Surface water information, including drainage patterns (overland flow, topography, channel flow pattern, tributary relationships, soil erosion, and sediment transport and deposition), surface water bodies (flow, stream widths and depths, channel elevations, flooding tendencies, and physical dimensions of surface water impoundments; structures; surface water/ groundwater relationships), and surface water quality (pH, temperature, total suspended solid, salinity, and specific contaminant concentrations)

5. Hydrogeology information, including geologic aspects (type of water-bearing unit or aquifer; thickness, areal extent of water-bearing units and aquifers; type of porosity; presence or absence of impermeable units or confining layers; depths to water table; thickness of vadose zone), hydraulic aspects (hydraulic properties of water-bearing unit or aquifer, such as hydraulic transmissivity, storativity, porosity, and dispersivity; pressure conditions such as confined, unconfined, or leaky confined), groundwater flow directions (hydraulic gradients horizontally and vertically, specific discharge, rate; recharge and discharge area; groundwater or surface water interactions; areas of groundwater discharge to surface water; seasonal variations of groundwater conditions), and groundwater use aspects (existing or potential aquifers; determination of existing near-site use of groundwater)

6. Atmospheric information, including local climate (precipitation, temperature, wind speed and direction, presence of inversion layers), weather extremes (storms, floods, winds), release characteristics (direction and speed of plume movement, rate, amount, and temperature of release, relative densities), and types of atmospheric hazards and hazards assessment

7. Human populations and land use

8. Ecological information, including information needed for public health evaluation (land use characteristics, water use characteristics) and information needed for environmental evaluation (ecosystem components and characteristics, critical habitats and biocontamination)

Contamination sources identification

The sources of contamination are usually those hazardous materials that are contained in drums, tanks, surface impoundments, waste piles, and landfills, as well as heavily contaminated media (such as soil) affected by the original leaking or spilling source. The purpose of defining sources of contamination is to help to identify the source location, potential releases, and engineering characteristics that are important in the evaluation of remedial actions, as well as waste characteristics, such as the type and quantity of contaminants that may be contained in or released to the environment, and the physical or chemical characteristics of the hazardous wastes present in the source.

Contamination determination

The targets for the determination of the nature and extent of contamination are groundwater, soil, surface water, sediments, and air.

6.5.1.2   Laboratory Analysis

Laboratory analysis provides data that will be used as the basis for decision-making. The data require that the analysis of samples in laboratories meets specific quality assurance and quality control (QA/QC) requirements.

6.5.1.3   Data Analysis

Data analysis should focus on the development or refinement of the conceptual site model by analyzing data on source characteristics, the nature and extent of contamination, the contaminants transport pathways and fate, and the effects on human health and the environment. All field activities, sample management and tracking, and document control and inventory should be well managed and documented to ensure their quality, validity, and consistency.

6.5.1.4   Community Relations Activities

Community relations should be properly maintained throughout the RI, including site characterization.

6.5.1.5   Reporting and Communication

During site characterization, communication is required between the lead and support agencies. The information is mainly on identifying ARARs, and includes a description of the contaminants of concern, the affected media, and any physical features. This information may be supplied by the preliminary site characterization summary or by a letter or other document.

A draft RI report should be produced for review by the support agency and submitted to the Agency for Toxic Substances and Disease Registry (ATSDR) for its use in preparing a health assessment and also to serve as documentation of data collection and analysis in support of the FS. The draft RI report can be prepared any time between the completion of the baseline risk assessment and the completion of the draft FS. Therefore, the draft RI report should not delay the initiation or execution of the FS.

6.5.2   Treatability Study

The objectives of the treatability study are primarily to achieve the following:

1. To provide sufficient data to allow treatment alternatives to be fully developed and evaluated during the detailed analyses, and to support the remedial design of a selected alternative
2. To reduce cost and performance uncertainties for treatment alternatives to acceptable levels so that a remedy can be selected

Figure 6.5 shows a decision process for treatability studies.[12]

Certain technologies have been sufficiently demonstrated so that the site-specific information collected during site characterization is adequate to evaluate and cost those technologies without conducting treatability testing.

A treatability study performed during an RI/FS is used to adequately evaluate a specific technology, including evaluating performance, determining process sizing, and estimating costs in sufficient detail to support the remedy selection processes. In general, treatability studies include the following steps:

1. Preparing a work plan (or modifying the existing work plan) for the bench or pilot studies
2. Performing field sampling, and/or bench testing, and/or pilot testing

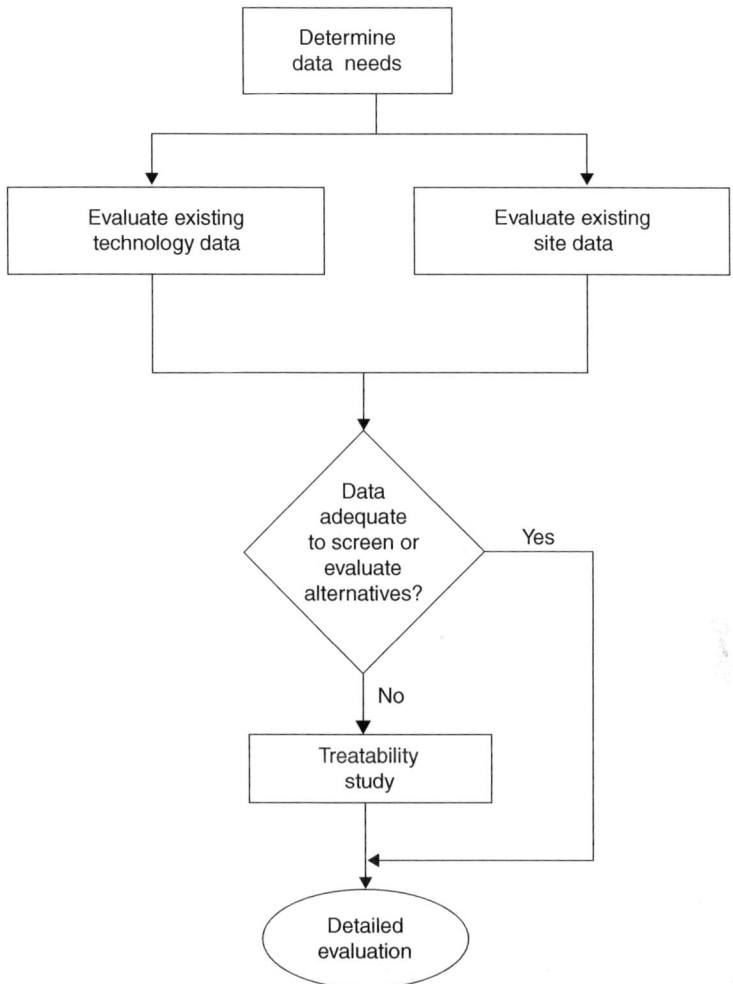

FIGURE 6.5 Treatability investigations.

3. Evaluating data from field studies, and/or bench testing, and/or pilot testing
4. Preparing a brief report documenting the results of testing

A treatability study can be performed by using bench-scale or pilot-scale techniques. Bench study is usually performed in a laboratory, in which comparatively small volumes of waste are tested for the individual parameters of a treatment technology to determine effectiveness of the treatment alternative on the waste, differences in performance between competing manufacturers, differences in performance between alternative chemicals, sizing requirements for pilot-scale studies, feasible technologies to be pilot tested, sizing of those treatment units that would sufficiently affect the cost of implementing the technology, and compatibility of materials with the waste.

Pilot testing is intended to simulate the physical, biological, and chemical parameters of a full-scale process; therefore, the treatment unit size and the volume of waste to be processed in pilot systems greatly increase over those of bench-scale testing. As such, pilot tests are intended to bridge the gap between bench-level analyses and full-scale operation, and are intended to more accurately simulate the performance of a selected full-scale process.

Once a decision is made to perform treatability studies, the type of treatability testing (bench or pilot scale) should be decided. The choice of bench versus pilot testing is affected by the level of development of the technology. For a technology that is well developed and tested, bench studies are often sufficient to evaluate performance on new wastes. For innovative technologies, however, pilot tests may be required as information necessary to conduct full-scale test is either limited or nonexistent.

6.6 FEASIBILITY STUDY

The feasibility study (FS) utilizes the data on site characterization and remedial technology screening to establish remedial alternatives, in turn, to select the cost-effective remedial actions. The FS may be viewed as occurring in three phases:

1. The development of alternatives
2. The screening of alternatives
3. The detailed analysis of alternatives

In practice, the specific point at which the first phase ends and the second phase begins is not so distinct. Therefore, the development and screening of alternatives are discussed together to better reflect the interrelation of these efforts. Furthermore, in many instances, there is only a limited number of available options and it may not be necessary to screen alternatives prior to detailed analysis.

6.6.1 DEVELOPMENT AND SCREENING OF ALTERNATIVES

The primary objective is to develop an appropriate range of waste management options to be analyzed more fully in the detailed analysis phase of the FS.[12] Appropriate waste management ensures the protection of human health and the environment. It may involve, depending on site-specific circumstances, complete elimination or destruction of hazardous substances at the site, significant reduction of concentrations of hazardous substances to acceptable health-based levels, and prevention of exposure to hazardous substances via engineering or institutional controls, or some combination of the above.

Alternatives are typically developed concurrently with the RI site characterization, with the results of one influencing the other in a methodology of iteration. Alternatives for remediation are developed by assembling combinations of technologies, including the media to which they would be applied, into alternatives that address contamination on a site-wide basis or for an identified operable unit. The methodology of development and screening of alternatives consists of six general steps[12]:

1. Developing remedial action objectives specifying the contaminant and media of interest, exposure pathways, and preliminarily remediation goals that permit a range of treatment and containment alternatives to be developed on the basis of chemical-specific ARARs when available, other available information, and site-specific, risk-related factors
2. Developing general response actions for each medium of interest defining containment, treatment, excavation, pumping, or other actions, singly or in combination, that may be taken to satisfy the remedial action objectives for the site
3. Identifying volumes or areas of media to which general response actions might be applied, taking into account the requirements for protectiveness as identified in the remedial action objectives and the chemical and physical characterization of the site
4. Identifying and screening the technologies applicable for each general response action to eliminate those that cannot be implemented technically at the site and to specify remedial technology types

5. Identifying and evaluating technology process options to select a representative process for each technology type retained for consideration, alternative development and evaluation, with an intention to represent the broader range of process options within a general technology type
6. Assembling the selected representative technologies into alternatives representing a range of treatment and containment combinations, as appropriate

6.6.2 Detailed Analysis of Alternatives

Analysis and presentation of the relevant information are needed to allow decision-makers to select a site remedy.

6.6.2.1 Overview of Detailed Analysis of Alternatives

A detailed analysis of alternatives consists of the following:

1. Further definition of each alternative, if necessary, with respect to the volumes or areas of contaminated media to be addressed, the technologies to be used, and any performance requirements associated with those technologies
2. An assessment and a summary profile of each alternative against the evaluation criteria
3. A comparative analysis among the alternatives to assess the relative performance of each alternative with respect to each evaluation criterion

6.6.2.2 Criteria for Detailed Analysis of Alternatives

During the detailed analysis, each alternative is assessed against the evaluation criteria. The results provide decision-makers with sufficient information to adequately compare the alternatives, select an appropriate remedy for a site, and demonstrate satisfaction of the CERCLA remedy selection requirements in the record of decision:

1. *Overall protection of human health and the environment.* This is the overall aim of the process.
2. *Compliance with ARARs.* It is considered how each alternative will comply with ARARs, or if a waiver is required and how it is justified.
3. *Long-term effectiveness.* The long-term effectiveness of alternatives in maintaining protection of human health and the environment after response objectives have been met is investigated.
4. *Reduction of toxicity, mobility, and volume through treatment.* The anticipated performance of the specific treatment technologies an alternative may employ is evaluated.
5. *Short-term effectiveness.* This is an examination of the effectiveness of alternatives in protecting human health and the environment during the construction and implementation of a remedy until response objectives have been met.
6. *Implementability.* This is an evaluation of the technical and administrative feasibility of alternatives and the availability of the required goods and services.
7. *Cost.* Capital and operation and maintenance costs of each alternative are evaluated.

   The overall criteria include cost-effectiveness, utilization of permanent solutions and alternative treatment technologies or resource recovery technologies to the maximum extent practicable, and satisfaction of the preference for treatment to reduce toxicity, mobility, or volume as a principal element, or the provision of an explanation if this preference is not met.[12] This is needed in order to attain acceptance by the support agency and the community.

6.6.2.3 Factors Affecting Potentially Applicable Remedial Technologies

The following factors may affect the potentially applicable remedial technologies:

1. *Site characteristics*, which may limit or promote the use of certain remedial technologies
2. *Waste characteristics*, which may limit the effectiveness or feasibility of the remedial technologies: quantity/concentration, chemical composition, acute toxicity, persistence, biodegradability, radioactivity, ignitability, reactivity/corrosivity, infectiousness, solubility, volatility, density, partition coefficient, compatibility with chemicals, and treatability
3. *Technology limitations*, including level of technology development, performance record, inherent construction, operation, and maintenance problems

6.6.2.4 Procedure for Detailed Analysis of Alternatives

The procedure for a detailed analysis of alternatives can be generalized into the following steps[14]:

1. Data analyzing
2. Modeling, such as groundwater modeling
3. Defining the objectives of remedial actions
4. Identifying technologies
5. Posing alternatives—preliminary screening
6. Scrutinizing selected alternatives, including technical analysis, regulation compliance, public health and environmental analysis, and cost analysis
7. Recording the feasibility study
8. Selecting the remedial alternative

6.7 REMEDIAL TECHNOLOGIES

This section will cover site control for waste movement, site cleanup technologies, and point-of-entry protection. The main focus will be on site cleanup technologies including remediation for contaminated groundwater, soil, and sediments. The technologies involve *in situ* treatment, which converts contaminants to less hazardous materials, and *ex situ* methodologies, which use soil excavation or groundwater pumping to remove contaminants from the site, and then treat them.[14–102]

The techniques to remove the free product of nonaqueous phase liquids (NAPL) will not be included in this chapter, because NAPL is one of the main releases from leaking underground storage tanks and is covered in Chapter 8, "Remediation of Sites Contaminated by Underground Storage Tank Releases," which addresses remediation techniques for organic contaminants, especially volatile organic compounds (VOCs) in soil and groundwater.

6.7.1 SURFACE SITE CONTROL OF WASTE MOVEMENT

The purpose of site control is to achieve the following:

1. To prevent waste movement (in air, surface water, and groundwater)
2. To contain wastes in a limited area
3. To reduce and eliminate impact on the environment
4. To lower the overall remedial cost

Gas may be formed by microbiological degradation of organics, evaporation and volatilization of volatile materials, or chemical reactions. The high combustibility of methane—a major component of landfill-generated gas—is a potential hazard. The emission of gas can be accelerated by elevated temperatures and venting conditions. Air pollution, which may result from gaseous emissions and fugitive dusts, should be controlled at uncontrolled waste sites. The main tasks include control of air contamination associated with natural forces, control of air contamination associated with remedial actions, and monitoring air pollution.

6.7.1.1 Control of Hazardous Gas Emission

According to U.S. EPA, the techniques that are used to control air pollution include the following[15]:

1. Covering surface impoundments
2. Passive perimeter gas control systems
3. Active perimeter gas control systems
4. Active interior gas collection/recovery systems

Covering surface impoundments
Covering surface impoundments is important for the control of hazardous gases emission. A common covering method places a barrier at the water–air interface to reduce gaseous emissions. The technology available includes lagoon covers, floating immiscible liquids, and floating (polypropylene) spheres.

Covers provide temporary methods for reducing volatile emission from surface impoundment. Floating lagoon covers function as both a surface water control mechanism and a mechanism for controlling gaseous emissions. They are suitable in situations where more than a year will elapse before final closure of a lagoon. They are not suitable for lagoons with weak berms or for lagoons located in areas that cannot support heavy construction equipment.

Floating immiscible liquids are suitable for controlling emissions of water-soluble organics. However, the effectiveness is temporary, estimated to be between one and two weeks. Some chemicals in water may prevent the formation of a monolayer, and wave action can destroy the monolayer effectiveness.

Floating polyethylene spheres are capable of reducing volatile emissions by up to 90%. Polyethylene spheres are compatible with a broad range of compounds including inorganic acids and bases and most aromatic and aliphatic organic compounds.[16]

Passive perimeter gas control systems
Passive gas control systems control gas movement by altering the paths of flow without the use of mechanical components. There are generally two types, high-permeability and low-permeability.

High-permeability passive perimeter gas control systems entail the installation of highly permeable (relative to the surrounding soil) trenches or wells between the hazardous waste site and the area to be protected (Figure 6.6). The permeable material offers conditions more conductive to gas flow than the surrounding soil, and provides paths of flow to the points of release. High-permeability systems usually take the form of trenches or wells excavated outside the site, then backfilled with a highly permeable medium such as coarse crushed stone.

Low-permeability passive perimeter gas control systems (Figure 6.7) effectively block gas flow into the areas of concern by using barriers (such as synthetic membranes or natural clays) between the contaminated site and the area to be protected. In the low-permeability system, gases are not collected and therefore cannot be conveyed to a point of controlled release or treatment. The low-permeability system can also alter the paths of convective flow.

High-permeability and low-permeability passive perimeter gas control systems are often combined to provide controlled venting of gases and blockage of available paths for gas migration.[15]

The applications and limitations of passive gas control systems must also be understood. They can be used at virtually any site where there is the capability to trench or drill and excavate to at least the same depth as the landfill. Limiting factors could include the presence of a perched water table or rock strata. Passive vents should generally be expected to be less effective in areas of high rainfall or prolonged freezing temperatures.

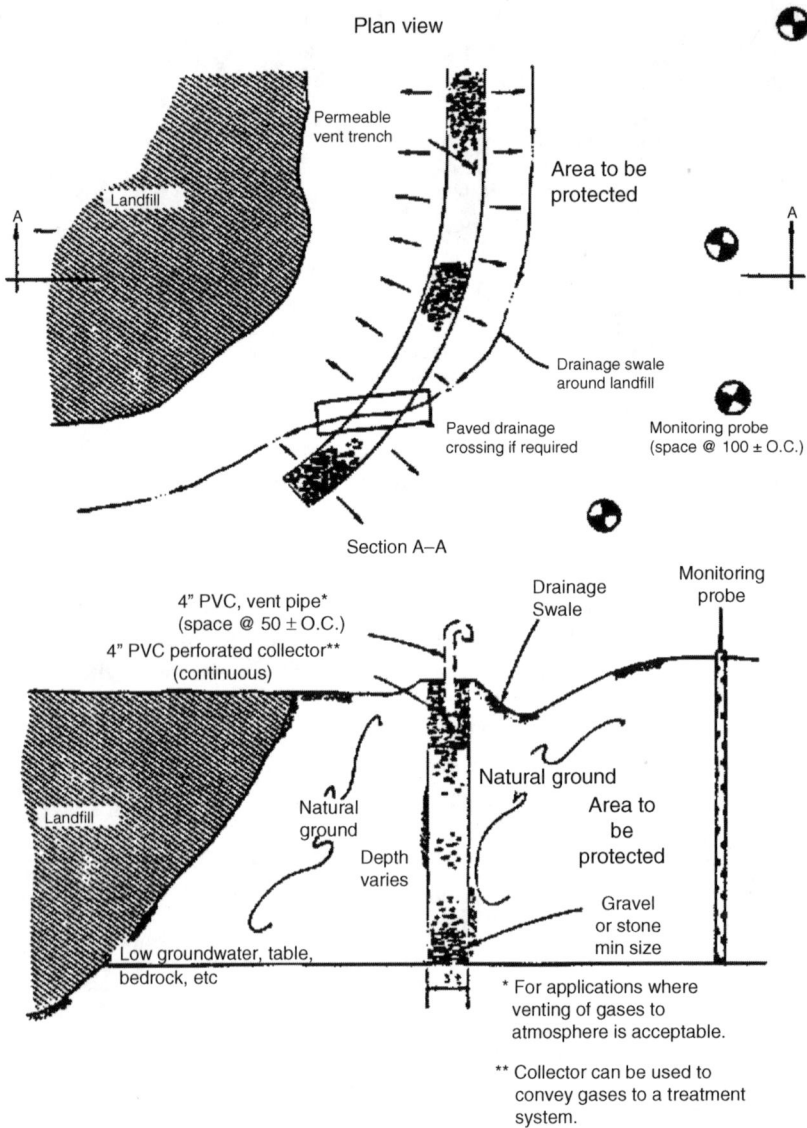

FIGURE 6.6 Passive gas control using a permeable trench.

The cost of passive gas control systems is low. The "passive" concept has virtually no operating or maintenance costs. However, it is recommended that periodic inspections be made and that the surface gas be periodically monitored in the area being protected to ensure that the systems are performing their intended functions.

Active perimeter gas control systems
Active perimeter gas control systems control off-site gas migration with the use of an active control system to alter pressure gradients and paths of gas movement by mechanical means. Three or four major components are required in active perimeter gas control systems:

1. Gas extraction wells
2. Gas collection headers

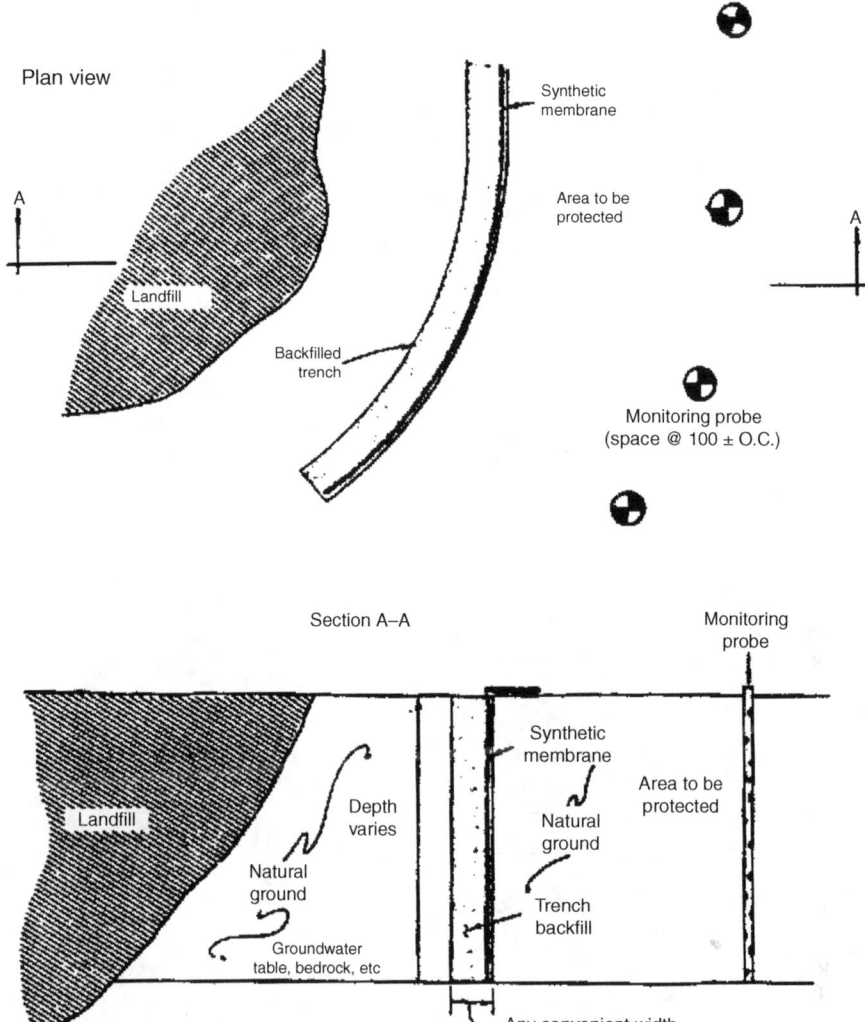

FIGURE 6.7　Passive gas control synthetic membrane.

3. Vacuum blowers or compressors
4. Gas treatment or utilization systems

Figure 6.8 shows an active perimeter gas extraction system. Active systems can be used at virtually any site where there is the capability to drill and excavate through the materials in the action area to the required depth. Limiting factors of active systems include the presence of free-standing leachate (i.e., saturation) or impenetrable materials. Active perimeter gas control systems are not sensitive to freezing or saturation of the surface or cover soils.

Centrifugal blowers create a vacuum through the collection headers and wells to the wastes and ground surrounding the wells. A pressure gradient is thereby established, inducing flow from the landfill (which is normally under positive pressure) to the blower (creating a negative, or vacuum, pressure). Subsurface gases flow in the direction of decreasing pressure gradient (through the wells, the header, and the blower) and are released directly to the atmosphere, treated and released to the atmosphere, or recovered for use as fuel.[15]

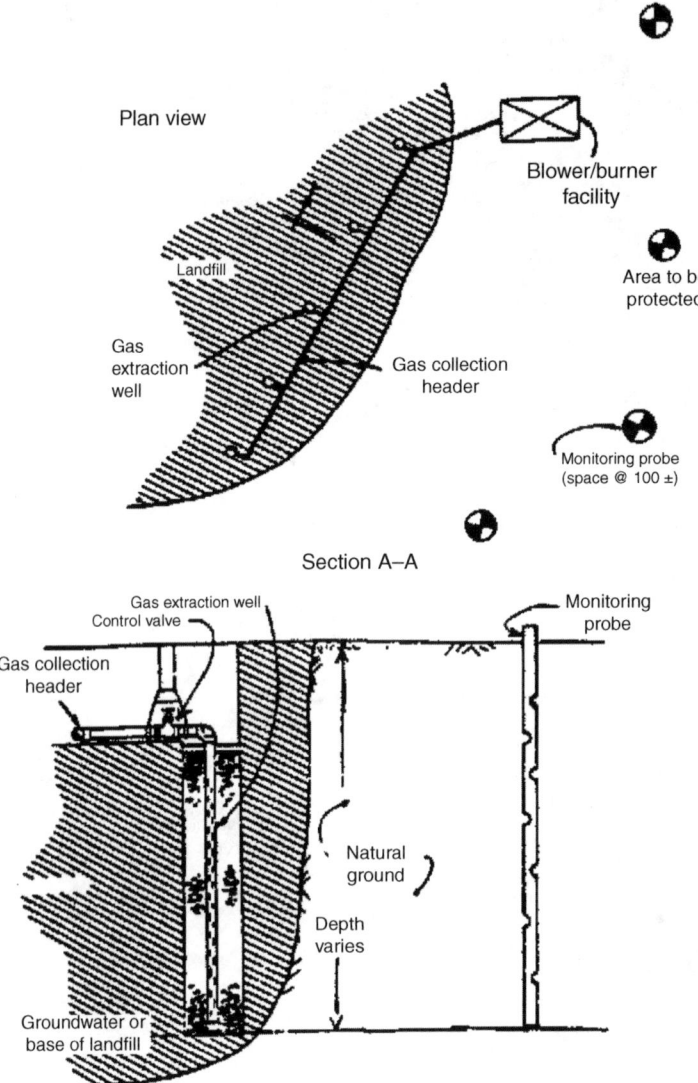

FIGURE 6.8   Active gas extraction.

Active interior gas collection/recovery system

Similar to the active perimeter gas control system, an active interior gas collection/recovery system consists of gas extraction wells, gas collection headers, vacuum blowers or compressors, and a treatment system. However, it is used to directly remove the hazardous gases from the site (beneath a landfill), instead of off-site removal. Figure 6.9 shows a schematic view of such a system.

Applications and limitations of the active interior gas collection/recovery system are similar to those of the active parameter gas control system. The active interior gas collection/recovery systems can be used at virtually any site where there is the capability to drill and excavate through landfilled material to the required depth. Limiting factors of the active interior gas collection/ recovery systems include the presence of free-standing leachate or impenetrable materials within the landfill.

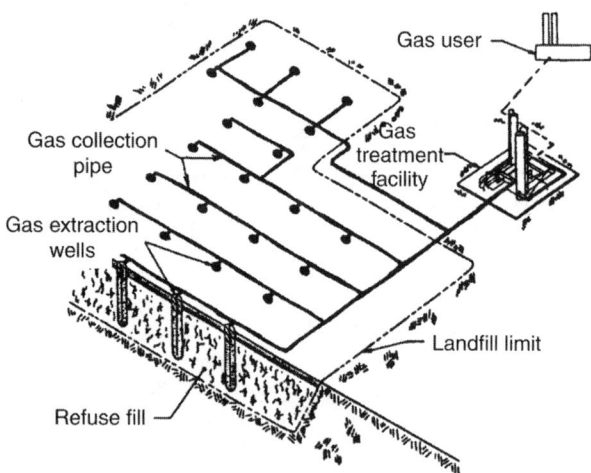

FIGURE 6.9   Gas collection and recovery system.

6.7.1.2   Control of Fugitive Dusts

Fugitive dusts are caused by wind erosion on waste sites, by vehicular traffic, and by excavation of waste during remedial action. The most commonly used control methods include the following:

1. Dust suppressants
2. Wind fences/screens
3. Water spraying

The dust suppressant method uses chemicals to (temporarily) strengthen bonds between soil particles and reduce fugitive dust emissions from inactive waste piles. Dust suppressant is expected to be 100% effective for a period of one to four weeks if the use of chemical is appropriate and undisturbed. Dust suppressants can also be used to control dust from work areas; however, it is less effective and requires frequent reapplications.

The wind fences/screens method uses screens, which take up or deflect a sufficient amount of wind so that the wind velocity is lowered below the threshold required for initiation of soil movement. The maximum reduction of wind velocity is expected for a distance of one to five fence heights downstream. Tests have shown that wind screens can achieve up to 60% efficiency in controlling inhalable particulates and 75% of total suspended particulates at wind speeds of about 10 to 13 mi/h.

The water spraying method is most commonly used to reduce fugitive dusts emission by spraying water onto the exposed surface area, for example, along active travel paths, excavation areas, and truck boxes loaded with soils.

6.7.1.3   Treatment of Emitted Gases

The gaseous phase of organic and inorganic contaminants that are collected from gaseous waste-streams can be treated. The most common methods are carbon adsorption and scrubbing with water or chemicals.

A mobile gaseous waste treatment unit developed by QUAD Environmental Technologies Corporation[17] utilizes atomizing nozzles within the scrubber chamber to disperse droplets of a controlled chemical solution, resulting in 85 to 100% removal (for benzene, toluene, phenol, and so on). Very small droplet sizes (less than 10 μm) and long retention times allow the use of a "once-through system" that generates low volumes of liquid residuals. This technology is best suited for

VOCs, although it is claimed to treat gaseous wastestreams containing a wide variety of organic or inorganic contaminants.

6.7.1.4 Surface Water Control through Control of Run-On and Runoff

Surface water control is necessary to minimize contamination of surface waters, to prevent surface water infiltration, and to prevent off-site transport of surface waters that have been contaminated. Control of run-on and runoff will accomplish the following:

1. It will prevent surface runoff, which carries contaminants to rivers and to places where the contaminants will infiltrate and percolate into soil and groundwater.
2. It will prevent surface water runoff from entering contaminated areas and in turn migrate into the contaminated plume. The methodology used involves dikes, terraces, diversion channels, floodwalls, grading, and revegetation, for example, using bench, terrace, or grading to divert or intercept surface water.

6.7.1.5 Surface Water Control through Prevention of Infiltration

Capping
Capping is a process used to cover buried waste materials to prevent them coming into contact with the land surface. Hence, capping on landfill can prevent infiltration of surface water to ensure minimum liquid migration through the waste. The materials used for capping usually have a permeability lower than or equal to the underlying liner system or natural soils, and high resistance to damage by settling or subsidence. Capping requires low cover maintenance and increases the efficiency of site drainage.

Capping is necessary whenever contaminated materials are to be buried or left in place at a site. Capping is often performed together with groundwater extraction or containment technologies to reduce further plume development, thus reducing the time needed to complete groundwater cleanup operations. In addition, groundwater monitoring wells are often used to detect any unexpected migration of capped wastes. A gas collection system should always be incorporated into a cap when wastes may generate gases. Capping is also associated with other surface water control technologies as discussed later. The main disadvantages of capping are the need for long-term maintenance and uncertain design life. A final cap should be inspected on a regular basis for signs of erosion, settlement, ponding liquid, invasion of deep-rooted vegetation, or subsidence, especially in the first six months when problems are most likely to appear. However, the long-term maintenance requirements are usually considerably more economical than excavation and removal of the wastes. Another disadvantage is the high cost of proper soil and drainage materials in certain areas of the country.

Caps can be single-layered or multilayered depending on the cap materials used. For construction and implementation considerations the reader can refer to U.S. EPA[15] and Matrecon, Inc.[18]

Grading
Grading is the technique used to reshape the surface in order to minimize infiltration by maximizing the amount of water thath will run off without causing significant erosion. Grading is often performed in conjunction with surface sealing practices and revegetation as part of an integrated landfill closure plan.

Grading is a relatively inexpensive remedial action component when suitable cover materials are available on site or close to the disposal site. Surface grading serves several functions:

1. It reduces ponding, which minimizes infiltration and reduces subsequent differential settling
2. It reduces runoff velocities and do reduces soil erosion
3. It roughens and loosens soils in preparation for revegetation
4. It is a factor in reducing or eliminating leaching of wastes

It is important upon completion of grading to establish vegetation cover as quickly as possible. This cover is essential to help prevent drying and erosion.

Revegetation

Revegetation is a cost-effective method to stabilize the surface of hazardous waste disposal sites, especially when preceded by capping and grading. Revegetation decreases erosion by wind and water and contributes to the development of a naturally fertile and stable surface environment. It may be part of a long-term site reclamation project, or it may be used on a temporary or seasonal basis to stabilize intermediate cover surfaces at waste disposal sites.

A systematic revegetation plan includes the following steps:

1. Selection of suitable plant species
2. Seedbed preparation
3. Seeding/planting
4. Mulching and/or chemical stabilization
5. Fertilization
6. Maintenance

Revegetation may not be feasible at disposal sites with high cover soil concentrations of phytotoxic chemicals, unless these sites are properly sealed and vented and then recovered with suitable topsoil. In some cases, clays or synthetic barriers below supporting topsoil in poorly drained areas may cause swamping of the cover soil and subsequent anaerobic conditions. A cover soil that is too thin may dry excessively in arid seasons and irrigation may be necessary. Improperly vented gases and soluble phytotoxic waste components may kill or damage vegetation. The roots of shrubs or trees may penetrate the waste cover and cause leaks of water infiltration and gas exfiltration. Also, periodic maintenance of revegetation areas (liming, fertilizing, mowing, replanting, or regarding eroded slopes) will add to the costs associated with this remedial technique.

Although vegetation cover requires frequent maintenance, it prevents the more costly maintenance that would result from erosion of surface soils.

6.7.1.6   Surface Water Control through Control of Erosion

Control of erosion is usually implemented through reducing slope length (using interception dikes, diversion channels, and terraces), slope steepness (using proper grading), or improving soil management, as well as controlling infiltration or erosion (using grading and revegetation). Most of these technologies have been addressed above (e.g., grading and revegetation) or will be addressed later (e.g., dikes and channels).

6.7.1.7   Surface Water Control through Collection and Transfer of Water

The purpose of the collection and transfer of water is to collect water that has been diverted away from the site or been prevented from infiltrating, and discharging or transferring the collected water to storage or treatment.[15] Surface water control can be carried out using dikes and berms, channels, chutes, and downpipes.

Dikes and berms are well-compacted earthen ridges or ledges located immediately upslope from or along the perimeter of a disturbed area (e.g., disposal sites). They can prevent excessive erosion of newly constructed slopes until more permanent drainage structures are installed or until the slope is stabilized with vegetation, and are widely used to provide temporary isolation of wastes until they can be removed or effectively contained, particularly during excavation and removal operations, to prevent runoff and mixing of incompatible wastes. For cost estimates of various technologies used to prevent infiltration one can refer to the U.S. EPA publication "Remedial Action at Waste Disposal Sites."[15]

Dikes and berms usually provide short-term protection of critical areas by intercepting storm runoff and diverting the flow to natural or manmade drainage ways, to stabilized outlets, or to sediment traps. These can only handle relatively small amounts of runoff and are not recommended for drainage areas larger than five acres.[19] Channels are wide and shallow excavated ditches used to intercept or divert water as well as collect and transfer the diverted water elsewhere. Chutes (or flumes) and downpipes are used to carry surface runoff from one level to a lower level without erosive damage and to enable the transfer of water away from diversion structures. They provide temporary erosion control while slopes are being stabilized with vegetative growth. Chutes are limited to head-drops of about 5.5 m (18 ft) or less, and downpipes are limited to drainage areas five acres in size.

6.7.1.8 Surface Water Control through Protection from Flooding

Flood control dikes (or embankment), levees, and floodwalls are the most common flood protection structures. They are used in areas subject to inundation from tidal flow or riverine flooding, but not for areas directly within open floodways. Levees create a barrier to confine floodwaters to a floodway and to protect structures behind the barrier. Floodwalls perform much the same function as levees, but are constructed from concrete.

6.7.1.9 Surface Water Control through Storage and Discharge of Water

Sedimentation basins can be used to collect and store surface water flow and to settle suspended solid particles. Seepage basins and ditches can be used to discharge uncontaminated or treated water downgradient of the site. It is important to separate clean surface runoff from contaminated water and store and treat them separately. Table 6.4 summarizes the surface water control methods.

6.7.1.10 Control of Waste Movement at Roads and Residential Areas

Site control at roads and residential areas will include at least the following activities:

1. Clearing the road, or, alternatively, building a detour route
2. Establishing signs at dangerous areas

TABLE 6.4
Primary Functions of Various Surface Water Control

Technology	Prevent or Intercept Run-on/Runoff	Prevent or Minimize Infiltration	Reduce Erosion	Collect and Transfer Water	Protection from Flooding	Discharge Water
Capping		×				
Lagoon covers		×				
Grading	×		×			
Revegetation	×	×	×			
Dikes and berms	×		×		×	
Channels and waterways			×	×		
Terraces and benches	×		×			
Chutes and downpipes			×	×		
Seepage basins and ditches						×
Sedimentation basins and ponds	×					×
Levees and floodwalls	×				×	

Source: U.S. EPA, Remedial Action at Waste Disposal Sites, EPA/625/6-85/006, U.S. EPA, Washington, DC, 1985.

3. Preventing fire associated with low ignition point volatile organics
4. Evacuating residents and protecting the area, or providing a facility for treatment of drinking water and cleanup of the site
5. Providing subsurface control of migration of contaminants

6.7.2 Subsurface Site Control of Waste Movement

6.7.2.1 Controls of Groundwater

The purpose of groundwater control includes the following aspects:

1. To contain a plume
2. To prevent migration of contaminated groundwater that may enlarge the size of the contaminated area and lead to the contamination of clean groundwater
3. To prevent clean groundwater from moving into the contaminated site, which may cause further migration and enlargement of the contaminated area
4. To prevent leachate formation by lowering the water table beneath a source of contamination or by preventing infiltration
5. To pump out the contaminated groundwater or perform *in situ* treatment to halt the source of contamination

Groundwater pumping

Groundwater pumping can remove the contaminated plume directly, or reconfigure the migration of groundwater through the cone of depression, which can either prevent further migration of contaminants or prevent movement into clean groundwater, or lower the water table. Extraction wells or a combination of extraction and injection wells can be used for this purpose. Figure 6.10a shows how an extraction well controls the movement of groundwater through the cone of depression, thus ensuring that clean water will be withdrawn from the domestic well. Figure 6.10b shows the use of a line of extraction wells to protect a domestic well.

The cone of depression can be evaluated based on an expression that relates to the measured saturated thickness of the aquifer, the height of water at the well from the bottom of the aquifer, pumping rate, hydraulic conductivity, and the radius of the observation wells (Jacob and Theis methods). Note that the formulae for calculation of the cone of depression are different for different confining conditions, for example, unconfined, artesian, and leaking confined aquifers. Various computer models have been established for groundwater flow, or associated with particle transfer or with chemical reactions (such as MODFLOW, MODPATH, and MOC, developed by the U.S. Geological Survey). Graphical or computer-aided calculations are usually used for composite drawdowns by multiwells (extraction or injection).

Subsurface drains

Subsurface drains include any type of buried conduits that convey and collect aqueous discharges by gravity flow (Figure 6.11). Water collected in a storage tank or a collection sump is then pumped for further treatment. Filters are usually needed in drain systems to prevent fine particles from causing clogging.

Subsurface drains function like an infinite line of extraction wells, and can be used to contain and remove a plume or to lower the groundwater table (Figure 6.12). They are more cost-effective than pumping for shallow contamination problems at depths of less than 12 m (40 ft). Depths may be increased if the site is stable, if the soil has a low permeability, and if no rock excavations are encountered.

Subsurface barriers

Subsurface barriers, low-permeability cutoff walls or diversions below ground are used to contain, capture, or redirect groundwater flow. The most common method uses bentonite slurry

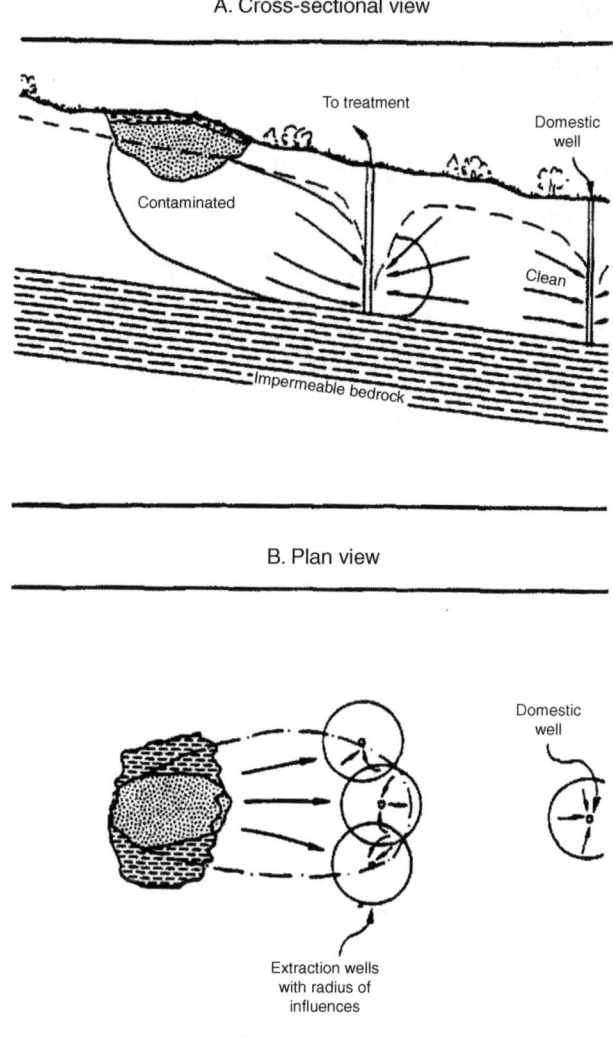

FIGURE 6.10 Containment using extraction wells: (a) Cross-sectional view; (b) plan view.

walls. Less common are other types of slurry walls (such as concrete), grouted barriers, and sheet piling cutoffs. The limiting factor for slurry walls is site topography, which may cause increasing engineering effort and cost. Also, slurry walls may not maintain good performance over a long period of time.

Grouted barriers use a variety of fluids injected into a rock or soil mass, which is set in place to reduce water flow and strengthen the formation. Grouted barriers are seldom used for containing groundwater flow in unconsolidated materials around hazardous waste sites because they cost more and have lower permeability than bentonite slurry walls. Nevertheless, they are suited to sealing voids in rock for waste sites remediation.

Sheet piling uses wood, precast concrete, or steel to form barriers for groundwater. They are seldom used because of high costs and unpredictable wall integrity, except for temporary dewatering for other construction or as erosion protection for other barriers. Bottom sealing is the technique used to place a horizontal barrier beneath an existing site to act as a floor and prevent downward migration of contaminants.

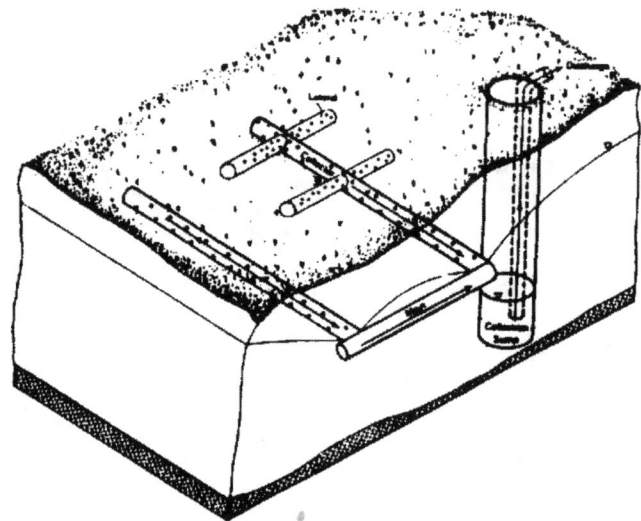

FIGURE 6.11 Subsurface drainage system components.

Control of sediments

Various technologies such as dikes, covers, and *in situ* grouting can be used for the control of migration of contaminants from contaminated sediments or for prevention of contamination of clean sediments.

6.7.3 *IN SITU* GROUNDWATER REMEDIATION

In situ groundwater treatment is an alternative to the conventional pump-and-treat methods. *In situ* treatment uses biological or chemical agents or physical manipulations that degrade, remove, or immobilize contaminants. *In situ* treatment technologies can usually treat both contaminated groundwater and soil. In many instances a combination of *in situ* and aboveground treatment will achieve the most cost-effective treatment at an uncontrolled waste site.

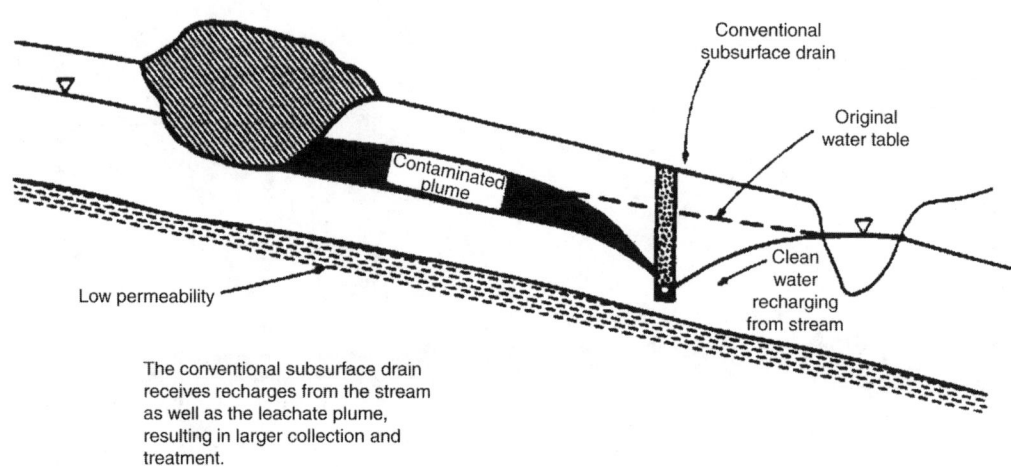

FIGURE 6.12 Use of a one-sided subsurface drain for reducing flow from uncontaminated sources.

6.7.3.1   Biological Treatment

Bioremediation is a technique for treating zones of contamination by microbial degradation, which involves altering the environmental conditions to enhance microbial catabolism or cometabolism of organic contaminants, resulting in the breakdown and detoxification of those contaminants.[15] According to microbial metabolic activity, bioremediation can be classified into three categories[20,21]:

1. Aerobic respiration, in which oxygen is required as a terminal electron acceptor
2. Anaerobic respiration, in which sulfate or nitrate serves as the terminal electron acceptor
3. Fermentation, in which the microorganism rids itself of excess electrons by exuding reduced organic compounds

The *in situ* biological treatment technique for organic contaminants is fully discussed in the Chapter 7. An example of a cost estimate for bioremediation is shown in Table 6.5. The data is based on a U.S. EPA study[15] of a project performed by Biocraft Laboratories, Waldwick, New Jersey.

6.7.3.2   Chemical Treatment

Chemical treatment of groundwater uses chemicals to immobilize or detoxify the organic or inorganic contaminants. Appropriate chemicals should be selected and pH or Eh are generally controlled. For example, for *in situ* treatment of inorganics, the most commonly used chemicals are sulfide, carbonate/hydroxide, and phosphate, which lead to the oxidation or reduction of contaminants or cause the precipitation of target materials;[22–24] for *in situ* treatment of organics, the methods of chemical oxidation and hydrolysis are used for detoxification, and polymerization is used to reduce the mobility of the contaminants. Generally, it is easier to control chemical processes in pumped

TABLE 6.5
Example—Summary of Project Costs[a] (Biocraft Laboratories, Waldwick, NJ)

Task	Actual Expenditure	Unit Cost	Period of Performance
Hydrogeological study: problem definition	$73,948	—	1976–1978
In-house process development (R&D)	$446,280	—	1978–1981
Groundwater collection/injection system total	$184,243	—	
Design	($61,490)		
Installation	($122,753)		1980–1981
Biostimulation plant design and construction total	$193,187	—	1981
Engineering design	($58,400)	—	1981
Masonry construction	($73,975)	—	1981
Equipment and miscellaneous installation	($60,812)	—	1981
Capital and R&D total	$926,158	—	
Operation and maintenance (O&M)			
Utilities	$47.40/d		
Electricity: 26.4 kW (24 h/d)	($195.25/d)	$7.396/kW	1983 rates
Steam: 72 lb (33 kg)/d & 90 psi	($61.92/d)	$0.86/lb	1981
Maintenance (see text)	$159.93/d		
O&M total	$226.53/d		
Total water treated	13,680 gal/d	$0.0165/gallon	
	(51,779 L/d)	($0.0044/L)	

[a]U.S. ACE (Cost Index for Utilities) may be used to convert costs into current USD.
Source: U.S. EPA, Remedial Action at Waste Disposal Sites, EPA/625/6-85/006, U.S. EPA, Washington, DC, 1985.

TABLE 6.6
Chemical Costs[a]

Category	Chemical	Cost/Unit
Acids	Hydrochloric acid, 20° Baume tanks	$55–105/t
	Nitric acid 36° to 42° Baume tanks	$195/t
	Sulfuric acid	
	Virgin, 100%	$61–95.9/t
	Smelter, 100%	$48–65/t
Bases	Caustic soda, liquid 50%, low iron	$255–285/t
Chelating agents	Ammonium chloride	$18/100 lb
	Citric acid	$0.81–$1.19/lb
Fertilizers	Ammonia, anhydrous, fertilizer	$140–$215/t
(microbial nutrients)	Ammonium chloride	$18/100 lb
	Ammonium sulfate	$73–79/t
	Sodium monophosphate	$55.75/100 lb
	Sodium diphosphate	$54.50/100 lb
	Phosphoric acid	
	75%, commercial grade	$27.5/100 lb
	52–54% a.p.a., agricultural grade	$3.10/unit-ton[b]
	Potassium–muriate, 60 to 62%, minimum	$0.82–0.92/unit-ton
	Potassium chloride	$105/t
	Potassium-magnesium sulfate	$59/t
Liming material	Agricultural limestone (dolomite)	$3.50–34/t
	Lime	$30.75–45/t
	Hydrated lime	$32.5–34.5/t
Oxidizing agents	Hydrogen peroxide, 35%	$0.24/1b
	Potassium permanganate	$1.03–1.06/lb
Reducing agents	Caustic soda, liquid 50%, low iron	$255–285/t
Precipitating agents	Ferrous sulfate	
	Heptahydrate	$130/t
	Monohydrate	$160/t
Surfactant		
Anionic	Witconate 605A	$0.65–0.85/lb
	Witconate P–1020BV (calcium sulfonates)	$0.70–0.88/lb
Nonionic	Adsee 799	$0.75–0.87/lb

[a]Use Appendix (USACE, Cost Index for Utilities) to convert costs into current USD.
[b]Unit-ton: 1% of 2000 lb of the basic constituent or other standard of the material. The percentage figure of the basic constituent multiplied by the unit-ton price gives the price of 2000 lb of the material.
Source: U.S. EPA, Remedial Action at Waste Disposal Sites, EPA/625/6-85/006, U.S. EPA, Washington, DC, 1985.

groundwater than in *in situ* groundwater. Costs of chemicals are listed in Table 6.6. Oxidizing agents, such as hydrogen peroxide, are commonly used for *in situ* groundwater remediation.[23,24,72]

6.7.3.3 Permeable Reactive Barrier

A permeable reactive barrier (PRB) is defined as an *in situ* method for remediating contaminated groundwater that combines a passive chemical or biological treatment zone with subsurface fluid flow management. Treatment media may include zero-valent iron, chelators, sorbents, and microbes to address a wide variety of groundwater contaminants, such as chlorinated solvents, other organics, metals, inorganics, and radionuclides. The contaminants are concentrated and either degraded or retained in the barrier material, which may need to be replaced periodically. There are approximately 100 PRBs operating in the U.S. and at least 25 internationally.

PRBs can be installed as permanent or semipermanent units. The most commonly used PRB configuration is that of a continuous trench in which the treatment material is backfilled. The trench is perpendicular to and intersects the groundwater plume. Another frequently used configuration is the funnel and gate, in which low-permeability walls (the funnel) direct the groundwater plume toward a permeable treatment zone (the gate). Some gates are *in situ* reactors, which are readily accessible so as to facilitate the removal and replacement of reactive media. These PRBs use collection trenches, funnels, or complete containment to capture the plume and pass the groundwater, by gravity or hydraulic head, through a vessel containing either a single treatment medium or sequential media. In circumstances where *in situ* treatment is found to be impracticable, reactive vessels have been located above ground.

Zero-valent iron has performed so successfully in PRB technology that it is now being applied directly for source zone treatment. Although this measure is not considered a PRB, examples of the technology will be included under the heading PRB because the reactive media and treatment mechanism are related. Pneumatic fracturing and injection, hydraulic fracturing, and injection via direct push rigs have been used successfully to introduce the reactive media to the groundwater or soil source area.[74–76]

6.7.3.4 Circulating Wells and In-Well Air Stripping Technologies

Circulating wells (CWs) provide a technique for subsurface remediation by creating a three-dimensional circulation pattern of the groundwater. Groundwater is drawn into a well through one screened section and is pumped through the well to a second screened section where it is reintroduced to the aquifer. The flow direction through the well can be specified as either upward or downward to accommodate site-specific conditions. Because groundwater is not pumped above ground, pumping costs and permitting issues are reduced and eliminated, respectively. Also, the problems associated with storage and discharge are removed. In addition to groundwater treatment, CW systems can provide simultaneous vadose zone treatment in the form of bioventing or soil vapor extraction.

CW systems can provide treatment inside the well, in the aquifer, or a combination of both. For effective in-well treatment, the contaminants must be adequately soluble and mobile so they can be transported by the circulating groundwater. Because CW systems provide a wide range of treatment options, they provide some degree of flexibility to a remediation effort.

In-well vapor stripping technology involves the creation of a groundwater circulation pattern and simultaneous aeration within the stripping well to volatilize VOCs from the circulating groundwater. Air-lift pumping is used to lift groundwater and strip it of contaminants. Contaminated vapors may be drawn off for aboveground treatment or released to the vadose zone for biodegradation. Partially treated groundwater is forced out of the well into the vadose zone, where it reinfiltrates to the water table. Untreated groundwater enters the well at its base, replacing the water lifted through pumping. Eventually, the partially treated water is cycled back through this process until contaminant concentration goals are met.

6.7.3.5 Air Sparging in Aquifers

Air sparging involves the injection of air or oxygen through a contaminated aquifer. Injected air traverses horizontally and vertically in channels through the soil column, creating an underground stripper that removes volatile and semivolatile organic contaminants by volatilization. The injected air helps to flush the contaminants into the unsaturated zone. Soil vapor extraction (SVE) is usually implemented in conjunction with air sparging to remove the generated vapor-phase contamination from the vadose zone. Oxygen added to the contaminated groundwater and vadose-zone soils can also enhance biodegradation of contaminants below and above the water table.[77]

6.7.3.6 Multiphase Extraction

Multiphase extraction uses a vacuum system to remove various combinations of contaminated groundwater, separate-phase petroleum product, and vapors from the subsurface. The system lowers the water table around the well, exposing more of the formation. Contaminants in the newly exposed vadose zone are then accessible to vapor extraction. Once above ground, the extracted vapors or liquid-phase organics and groundwater are separated and treated.

6.7.4 PUMP-AND-TREAT GROUNDWATER REMEDIATION

The pump-and-treat methodology is effective for groundwater remediation. It is also an effective way to prevent the further extension of a contaminated area. The cleanup involves two steps:

1. Pumping the contaminated groundwater out from the site
2. Treating the pumped contaminated water on ground so that it can be returned to the system

In order to effectively pump all contaminated water out of an aquifer (or soil) pore space, water injection is usually needed, and sometimes a chemical flushing agent.

The pump-and-treat method is comparable to soil flushing. In fact, the pump-and-treat method can treat both groundwater and aquifer soil at the same time, and can also be directly applied to unsaturated soil zones. The soil flushing method is mainly considered as a treatment in unsaturated zones.

6.7.4.1 Pumping Systems

There are three common methods for groundwater collection using pumping systems: a well point system, a gravity drain system, and in combination with injection wells.

Well point system
A well point system consists of several individual well points spaced at 0.6 m to 1.8 m intervals along a specified alignment. A well point is a well screen (length 0.5 to 1.0 m) with a conical steel drive point at bottom. Individual well points are attached to a riser pipe (diameter 2.5 to 7.5 cm) and connected to a header pipe (diameter 15 to 20 cm). At the midpoint, the header pipe is connected to a centrifugal pump. As yield at different well points may vary, a valve at the top of each riser pipe is used to control the drawdown so that the screen bottom is exposed. The pump provides 6 to 7.5 m of suction, but friction losses reduce the effective suction to 4.5 to 5.4 m.

The well point system is the most economical method of groundwater collection where the water table is less than 3 m and the contaminant is less 9 m below the surface.

Gravity drain system
A trench is excavated perpendicular to the flow of groundwater to a depth below the water table. A perforated pipe is placed in the trench and the remainder of the trench is backfilled with gravel. Groundwater is collected in a main collector pipe and flows to a sump, from which it is pumped to the surface for treatment.

The gravity drain system is most effective when circumstances are suitable to gravity flow, the water table is less than 3 m and the contaminant is less than 9 m below the surface.

Combination with injection wells
The main purpose of recharging water into an aquifer is to elevate the hydraulic gradient to promote the movement of groundwater towards the collection system, thus enhance the efficiency of pumping.

There are two general recharge systems, recharge basins and injection wells. The recharge of treated groundwater into the system provides a method for the discharge of treated groundwater. The recharging of water can also have other purposes, such as creating a hydraulic barrier to restrict the migration of a contaminant plume, and providing a method for introducing flushing solutions into the groundwater to flush the pollutants out of soil.

6.7.4.2   Treatment of Pumped Water

Gravity liquid separation

Gravity liquid separation uses gravity force to separate the liquid-phase contaminant from water (immiscible with the contaminant) by the force of gravity.

Gravity separators can take many shapes and arrangement, depending in part on the characteristics of the waste. Typical design configurations include horizontal cylindrical decanters, vertical cylindrical decanters, and cone-bottomed settlers.

Sedimentation

Sedimentation is commonly applied to the treatment of pumped groundwater containing high concentrations of suspended solids.[25] It can also be used to remove the suspended solids from collected surface runoff, leachate or landfill toe seepage, and dredge slurries as a pretreatment step for biological treatment or many chemical processes, including precipitation, carbon adsorption, ion exchange, stripping, reverse osmosis, and filtration.[22–24]

Chemical precipitation/coagulation, flocculation, and clarification

Chemical precipitation/coagulation methods transfer the target substances (mainly metals) in solution into a solid phase. Many heavy metal hydroxides and sulfides have very low solubility (within a certain pH range) and are therefore insoluble. The metal sulfides have significantly lower solubility than their hydroxide counterparts over a broad range of pH.[26] Precipitation/coagulation is also applicable for removing certain anionic species such as phosphate, sulfate, and fluoride.

Lime and sodium sulfide are the most common chemical agents added to contaminated water in a rapid mixing tank. Generally, flocculating agents (such as alum, lime, or iron salts) are added along with the precipitating agents.[27] Agglomerated particles are separated from the liquid phase by settling in a sedimentation clarifier, by floating in a dissolved air flotation (DAF) clarifier,[28,29,71–73] or by other physical processes such as filtration.[22] Figure 6.13 is a typical configuration for precipitation, flocculation, and sedimentation clarification,[15] in which the sedimentation clarifier may also be replaced by a DAF cell[28–30,71–73] for cost and space saving.

Certain physical or chemical characteristics of the wastestream may limit the application of precipitation. For example, some organic compounds (as well as cyanide or other ions) may form organometallic complexes with metals, decreasing the precipitation potential.

Wang and colleagues[71–73,100] have developed a physical–chemical sedimentation sequencing batch reactor (PCS-SBR) process and a physical–chemical flotation sequencing batch reactor (PCF-SBR) process for the treatment of contaminated groundwater, potable water, and wastewater. The reactor of a PCS-SBR process is similar to a conventional biological sequencing batch reactor (SBR), except that chemical flocs (instead of biological activated sludges) are used for water and wastewater treatment. A PCF-SBR is another physical–chemical SBR process in which flotation (instead of sedimentation) is used for the separation of chemical flocs from the flocculated water.

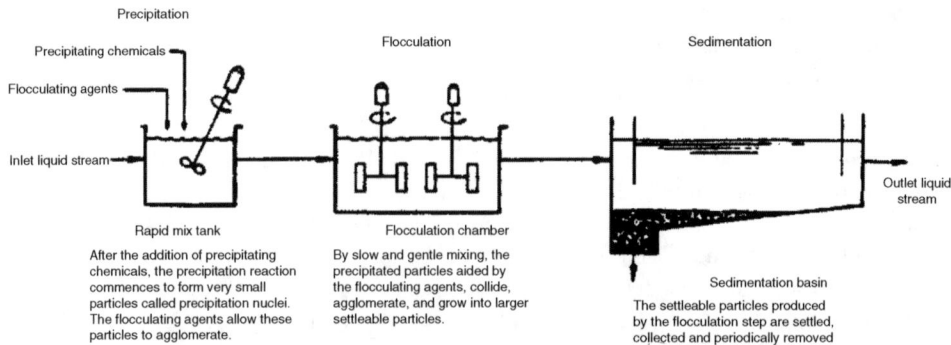

FIGURE 6.13   Representative configuration using precipitation, flocculation, and sedimentation.

Ion exchange

Ion exchange is a reversible interchange of ions between a liquid and a solid phase. The ions (contaminants) in a liquid wastestream and the ions on the surface of an ion-exchange resin are exchanged, purifying the wastestream while concentrating the waste constituent in the resin.[22–24] Mixed resins are sometimes effective in removing both cations and anions.

The ion-exchange process is applicable for removing a broad range of ionic species from water containing all metallic elements, inorganic anion such as halides, sulfates, nitrates, cyanides, organic acids such as carboxylics, sulfonics, some phenols at sufficiently alkaline pH conditions, and organic amines at sufficiently acidic conditions.

The upper concentration limit for ion exchange is about 2500 to 4000 mg/L. A high concentration of pollutants can result in the rapid exhaustion of the resin, resulting in high regeneration costs. The feed stream must be free of oxidants. Suspended soil material in the feed stream should be less than 50 mg/L to prevent plugging the resins. Recently, an ion-exchange sequencing batch reactor (IX-SBR) was developed by Wang and colleagues[71] for groundwater decontamination and industrial effluent treatment.[71,100]

Conventional filtration and automatic backwash filtration

Conventional filtration is widely used to remove suspended solids from solution by forcing the fluid through a porous medium. Filter media usually consist of a bed of granular particles, typically sand or sand with anthracite or coal. The filtrates are usually greater than 1 μm in diameter. The filtration is termed conventional, in order to distinguish it from other types of filtration such as membrane filtration (for particles less than 1 μm). As water passes through the filter bed, the particles become trapped on top of and within the filter bed, thus in time reducing the filtration rate. Therefore, backwash is periodically needed and filtration is often preceded by sedimentation[31] or flotation[28,32,33,71–73] to reduce the suspended solid load on the filter.

Membrane filtration processes

Membrane filtration processes have been successfully applied to the field of environmental engineering for air pollution control,[34] potable water purification,[22–24] groundwater decontamination,[35,36] industrial effluent treatment,[37] hazardous leachate treatment,[35,36] and site remediation,[36] mainly because membrane filtration can remove heavy metals and organics.

There are three major types of membrane processes, each with different physical means of operation: reverse osmosis (RO), ultrafiltration (UF), and microfiltration (MF). In addition, electrodialysis (ED) is also considered to be a membrane process.

In ED, cation-exchange membranes are alternated with anion-exchange membranes in a parallel manner to form compartments 0.5 to 1.0 mm thick. The entire membrane assembly is held between two electrodes. When an electrical potential is applied to the electrodes, all positive ions tend to travel towards the negative electrode, and all negative anions tend to move toward the positive electrode.

In the other three membrane processes, for example, in RO, a membrane is mounted in an apparatus so that a two-section compartment is formed. Contaminated water is pressurized and circulated through the high-pressure-solution compartment. Water permeates to the low-pressure side and is removed. The concentrated brine is removed separately.

The main difference between the UF, MF, and RO arrangements is membrane pore size, which allows different sizes of particles to pass through the membrane. All three processes allow certain solvent molecules to pass through, and impede certain sizes of particles. MF impedes the passage of large colloids and small particles, UF membranes impede the passage of molecules with a molecular weight of 100 or higher, and the membranes used in RO allow the passage of water, but impede the passage of salts and small molecules. of the three membrane filtration processes, RO requires the highest pressure.

The main advantages of membrane processes are their ability to separate impurities from water for recovery, low operation cost, and a requirement for only a small amount of space for installation. Their limitation lies in the possibility of deterioration of the membranes by certain kinds of water streams, for example, water containing certain strong oxidizing compounds or at high temperatures.

Recently, Wang[100] introduced a membrane sequencing batch reactor (membrane-SBR) process for groundwater decontamination, water purification, and industrial effluent treatment. A membrane-SBR is similar to conventional SBR except that membrane filtration is used (instead of sedimentation) for the separation of mixed liquor suspended solids (MLSS) from the mixed liquor.

Activated carbon adsorption

Activated carbon has high specific surface area with respect to its volume, and thus has high adsorption capacity. Activated carbon adsorption is considered to be one of the most versatile treatment technologies and can remove classical pollutants such as COD, TOC, BOD, and nitrogen, as well as toxic pollutants such as phenol, refractory organic compounds, VOCs, and soluble heavy metals.[38] Activated alumina and peat have also demonstrated similar abilities.

Once the micropore surfaces of activated carbon are saturated with target material, the spent carbon must be replaced or regenerated. Granular activated carbon (GAC) is favored over powder activated carbon (PAC) in most cases, because the former is considered to be capable of regeneration and sustainable to flow, although the costs of both carbons and the cost of regeneration are high.

Activated carbon adsorption is used to remove soluble organics, suspended solids, and refractory organics that cannot be biodegraded in groundwater. Because of its high cost and its ability to result in low pollutant concentration in effluents, activated carbon is usually used following biological treatment or granular media filtration in order to reduce the load on the carbon columns. PAC can be dosed into an SBR for facilitating physical–chemical or biological reactions for groundwater decontamination.[71–73]

Biological sorption

The biological sorption technique uses biogenetic materials for the adsorption of contaminants. The AlgaSorb sorption process developed by Bio-Recovery Systems, Inc., is designed to remove heavy metal ions from aqueous solution based on the mutual affinity of the cell walls of algae and heavy metal ions. The sorption medium comprises algal cells immobilized in a silica gel polymer. The system functions as a biological ion-exchange resin to bind both metallic cations (positively charged ions) and large metallic anions. Like ion-exchange resins, the algae–silica medium can be recycled. This technology is useful for removing metal ions from groundwater and surface leachate that contain high levels of dissolved solids.[23,24]

Solvent extraction

Solvent extraction is the separation of constituents from a liquid solution by contact with another immiscible liquid. It is mainly used for the recovery of organics from liquid solutions.[39] Specifically, solvent extraction uses an organic solvent as an extractant to separate organic and metal contaminants from soil. The organic solvent is mixed with contaminated soil in an extraction unit. The extracted solution is then passed through a separator, where the contaminants and extractant are separated from the soil. Organically bound metals may be extracted along with the target organic contaminants.[78]

From a process viewpoint, three steps are involved:

1. Actual extraction of the solvent by forced mixing or countercurrent flow
2. Solute removal from the extracting solvent
3. Solvent and extracted solute recovery

Significant energy consumption and other operating costs are expected. This method of treatment becomes cost-effective when material recovery is significant.

Chemical oxidation

Chemical oxidation typically involves reduction/oxidation (redox) reactions that chemically convert hazardous contaminants to nonhazardous or less toxic compounds that are more stable, less mobile, or inert. Redox reactions involve the transfer of electrons from one compound to another. Specifically, one reactant is oxidized (loses electrons) and one is reduced (gains electrons). The oxidizing agents most commonly used for the treatment of hazardous contaminants in soil are ozone, hydrogen peroxide, hypochlorites, chlorine, chlorine dioxide, potassium permanganate, and Fentons reagent (hydrogen peroxide and iron). Cyanide oxidation and dechlorination are examples of a chemical treatment. This method may be applied *in situ* or *ex situ*, to soils, sludges, sediments, and other solids, and may also be applied to the *in situ* treatment of groundwater.[22–24,79,80]

Chemical oxidation technology is primarily used for the detoxification of cyanide and other oxidizable organics such as aldehydes, mercaptans, phenols, unsaturated acids, and certain pesticides.[40]

For example, cyanide detoxification involved the following process[106]:

$$NaCN + KMnO_4 + H^+ \rightarrow MnO_2 + NaCNO + KOH$$

Oxidation can be an effective way of pretreating waste prior to biological treatment either by detoxification or by rendering refractory compounds to be more amenable to biological treatment.

Chemical Waste Management, Inc., has developed a technique that is a combination of evaporation and catalytic oxidation processes. Contaminated water is concentrated in an evaporator by boiling off most of the water and its volatile contaminants, both organic and inorganic. Air or oxygen is added to the vapor and the mixture is forced through a catalyst bed, where the organic and inorganic compounds are oxidized. This stream, composed mainly of steam, passes through a scrubber, if necessary, to remove any acid gases formed during oxidation. The stream is then condensed or vented to the atmosphere. This technique can be used to treat complex contaminated waters that contain volatile and nonVOCs, salts, soluble heavy metals, and volatile inorganic compounds.

The limitation for chemical oxidation is that oxidation is frequently not completed to the final products CO_2 and H_2O. This can be due to a number of factors, including oxidant concentration, pH, redox potential, or the formation of stable intermediate toxic oxidation products.

Chemical reduction

Chemical reduction is used to transform a toxic substance with a higher valence to a nontoxic or less-toxic substance with lower valence. The most promising application is the reduction of hexavalent chromium to trivalent chromium. This method is also applicable to other multivalent metals such as lead and mercury. Commonly used chemical agents for this purpose are sulfite salts, sulfur dioxide, and base metals (e.g., iron and aluminum).[22–24]

Biological treatment

Biological treatment technology, also known as bioremediation technology, is mainly used to treat organic contaminants (as terminal electron acceptor to bacteria). Bioremediation techniques include the use of two primary respiratory pathways: aerobic and anaerobic.[20,21] Each approach has advantages and limitations. To date, aerobic systems using naturally occurring microorganisms are most widely implemented. Aerobic systems tend to be more efficient when degrading petroleum-based organic contaminants such as benzene, toluene, ethylbenzene, xylenes (BTEX), and naphthalene. Research suggests that aerobic systems are not as effective for the treatment of highly chlorinated compounds such as PCBs (polychlorinated biphenyls). However, genetically engineered microbial system (GEMS) are increasingly used in research applications for recalcitrant compounds. Research scientists[41] have developed techniques to modify microbial DNA to enable organisms to degrade contaminants that are currently very recalcitrant (i.e., PCBs) or extremely toxic (i.e., dioxin). Some bacteria can use certain inorganics as the terminal electron acceptor, so biological decontamination of inorganic materials is feasible. The following presents an example on biological decontamination of inorganic materials.

Ehrlich[42] used biotechnology coupled with physicochemical extraction to remove chromium from contaminated soil including recovery and reuse. Ehrlich's biological treatment is based on an oxygen-insensitive bacterial respiration, with chromate as the terminal electron acceptor using intact cells and cell extracts. The bacterial strain used to reduce chromate is *Pseudomonas fluorescens* LB300, which has chromate resistance to more than 2000 mg/L of potassium chromate, although very slight resistance to potassium dichromate. In the Ehrlich process, the highly Cr-concentrated solution was recovered through ion exchange, and the low-concentration solution was then treated by reducing Cr(VI) to Cr(III) in a rotating biological contactor (RBC). The Cr(III) slurry was recovered through sedimentation and purification for reuse.

Wang and colleagues[43,71,100] have developed a biological flotation process for the treatment of contaminated groundwater. The process has a built-in air emission control device for the removal of toxic organics and inorganics from water without causing air pollution problems. One of the biological processes is a conventional biological SBR equipped with an enclosure on top for air emission control. Another new biological high rate process is the dissolved air flotation sequencing batch reactor (DAF-SBR), which is also equipped with an air pollution control enclosure on top, which is suitable for temporary groundwater decontamination in the field. The DAF-SBR process is similar to a conventional biological SBR process, except that DAF (instead of sedimentation) is used in the reactor for the separation of mixed liquor suspended solids (MLSS) from the mixed liquor.

Air and steam strippings

Stripping methods, including steam stripping and air stripping, are mainly used for the removal of volatile organics from contaminated water. The difference between steam stripping and air stripping is the stripping agent, the former obviously using steam and the latter air. Moreover, steam stripping is more like a distillation process, in which steam is used as both the heating medium and the driving force for removal of the volatile materials. After condensing the steam, the waste compounds are concentrated and separated from the water. Air stripping, on the other hand, is based on the distribution coefficients of volatile organics between the contaminated water and the stripping stream at a certain temperature.

Stripping can be integrated with vapor extraction for a better contamination removal. The stripping technology can also be combined with activated carbon adsorption to result in a higher removal efficiency. The conventional air stripping process can only remove VOCs from contaminated water while its gaseous effluent may pollute the air environment. A new stripping process developed by Wang and colleagues[32] and Hrycyk and colleagues[44] can remove VOCs, VICs (volatile inorganic compounds), and radioactive radon from water, without the creation of an air pollution problem.

6.7.5 IN SITU SOIL TREATMENT

6.7.5.1 In Situ Heating

In situ soil remediation with physical methods includes the *in situ* heating (*in situ* thermal treatment), ground-freezing, hydraulic fracturing, immobilization/stabilization, flushing, chemical detoxification, vapor extraction, steam extraction, biodegradation/bioremediation, electroosmosis/electrokinetic processes, etc.

In situ heating (*in situ* thermal treatment) uses thermal decomposition, vaporization, and distillation techniques to destroy or remove organic contaminants. The most common *in situ* heating methodologies include electrical resistance heating, radio frequency heating, hot air/water/steam injection, and thermal vitrification. These different methods or their combinations can be used to apply heat to polluted soil or groundwater *in situ*. The heat can destroy or volatilize organic chemicals. As the chemicals change into gases, their mobility increases, and the gases can be extracted via collection wells for capture and cleanup in an *ex situ* treatment unit. Thermal methods can be particularly useful for dense or light nonaqueous phase liquids (DNAPLs or LNAPLs). Heat can be introduced to the subsurface by electrical resistance heating, radio frequency heating, dynamic underground stripping, thermal conduction, or injection of hot water, hot air, or steam.

The main advantage of *in situ* thermal methods is that they allow soil to be treated without being excavated and transported, resulting in significant cost savings; however, *in situ* treatment generally requires longer time periods than *ex situ* treatment, and there is less certainty about the uniformity of treatment because of the variability in soil and aquifer characteristics and because the efficacy of the process is more difficult to verify.

Electrical resistance heating

Electrical resistance heating uses an electrical current to heat less permeable soils such as clays and fine-grained sediments so that water and contaminants trapped in these relatively conductive regions are vaporized and ready for vacuum extraction. Electrodes are placed directly into the less permeable soil matrix and activated so that electrical current passes through the soil, creating a resistance, which then heats the soil. The heat dries out the soil, causing it to fracture. These fractures make the soil more permeable, allowing the use of SVE to remove the contaminants. The heat created by electrical resistance heating also forces trapped liquids to vaporize and move to the steam zone for removal by SVE. Six-phase soil heating (SPSH) is a typical electrical resistance heating, and uses low-frequency electricity delivered to six electrodes in a circular array to heat the soil. With SPSH, the temperature of the soil and contaminant is increased, thereby increasing the contaminant's vapor pressure and its removal rate. SPSH also creates an *in situ* source of steam to strip contaminants from the soil. SPSH has been demonstrated, and all large-scale *in situ* projects utilize three-phase soil heating.

Radio frequency/electromagnetic heating

Radio frequency heating (RFH) is an *in situ* process that uses electromagnetic energy to heat soil and enhance SVE. The RFH technique heats a discrete volume of soil using rows of vertical electrodes embedded in the soil (or other media). Heated soil volumes are bounded by two rows of ground electrodes with energy applied to a third row midway between the ground rows. The three rows act as a buried triplet capacitor. When energy is applied to the electrode array, heating begins at the top center and proceeds vertically downward and laterally outward through the soil volume. The technique can heat soils to over 300°C.[45] RFH enhances SVE in four ways:

1. Contaminant vapor pressure and diffusivity are increased by heating
2. Soil permeability is increased by drying
3. There is an increase in the volatility of the contaminant from *in situ* steam stripping by the water vapor
4. There is a decrease in viscosity, which improves mobility

The technology is self-limiting; as the soil heats and dries, current will stop flowing. Extracted vapor can then be treated by a variety of existing technologies, such as GAC or incineration.

Hot air injection

Hot air, hot water, or hot steam are injected below the contaminated zone to heat the contaminated soil. The heating enhances the release of contaminants form the soil matrix. Some VOCs and semi-volatile organic compounds (SVOCs) are stripped from the contaminated zone and brought to the surface through soil vapor extraction. Hot air is introduced at high pressure through wells or soil fractures. In surface soils, hot air is usually applied in combination with soil mixing or tilling, either *in situ* or *ex situ*.

Hot water injection

Hot water injection via injection wells heats the soil and groundwater and enhances contaminant release. Hot water injection also displaces fluids (including LNAPL and DNAPL free product) and decreases contaminant viscosity in the subsurface to accelerate remediation through enhanced recovery.

Hot steam injection

Hot steam injection heats the soil and groundwater and enhances the release of contaminants from the soil matrix by decreasing viscosity and accelerating volatilization. Steam injection may also destroy some contaminants. As steam is injected through a series of wells within and around a source area, the steam zone grows radially around each injection well. The steam front drives the contamination to a system of groundwater pumping wells in the saturated zone and SVE wells in the vadose zone.[82,83] Figure 6.14 show the operation of a typical hot steam injection process.

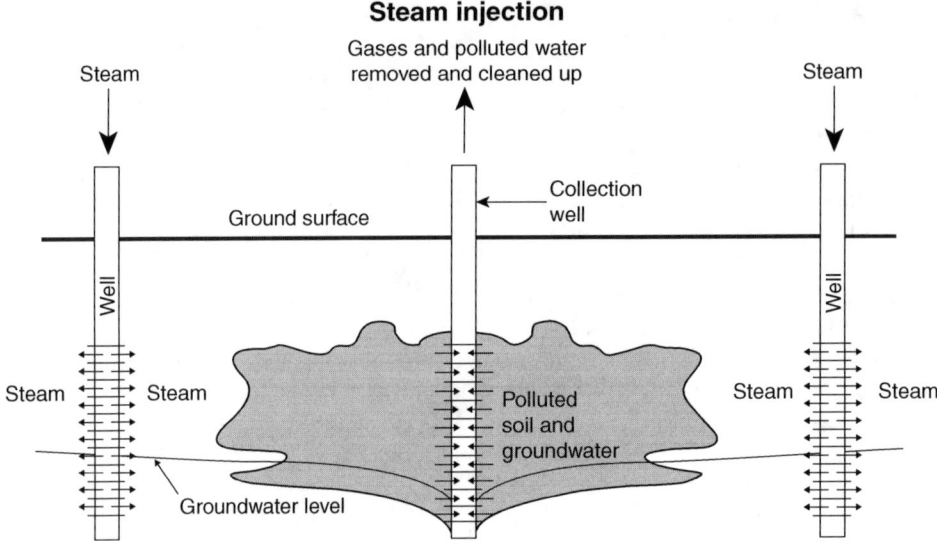

FIGURE 6.14 *In situ* thermal treatment by steam injection. (Taken from U.S. EPA, A Citizen's Guide to *In situ* Thermal Treatment Methods, Technical Report EPA-542-F-01-012, U.S. EPA, Washington, DC, 2001.)

In situ *thermal vitrification*

In situ thermal vitrification is based on electric melter technology. Contaminated soil is converted into durable glass and the waste is pyrolyzed or crystallized. Off-gases released during the melting process are trapped in an off-gas hood. The depth of the waste is a significant limiting factor for this application.[17,82] In essence, vitrification is a process that permanently traps harmful chemicals in a solid block of glass-like material. This keeps the chemicals from leaving the site. Vitrification can be done either in place (*in situ*) or above ground (*ex situ*). Specifically, vitrification uses electric power to create the heat needed to melt contaminated soil at elevated temperatures (1600 to 2000°C or 2900 to 3650°F). The high-temperature component of the process destroys or removes organic materials. Radionuclides and heavy metals are retained within the vitrified product.

Figure 6.15 shows that four rods (electrodes) are drilled in the polluted area. An electric current is passed between the electrodes, melting the soil between them. Melting starts near the ground surface and moves downward. As the soil melts, the electrodes sink further into the ground, causing deeper soil to melt. When the power is turned off, the melted soil cools and vitrifies, which means it

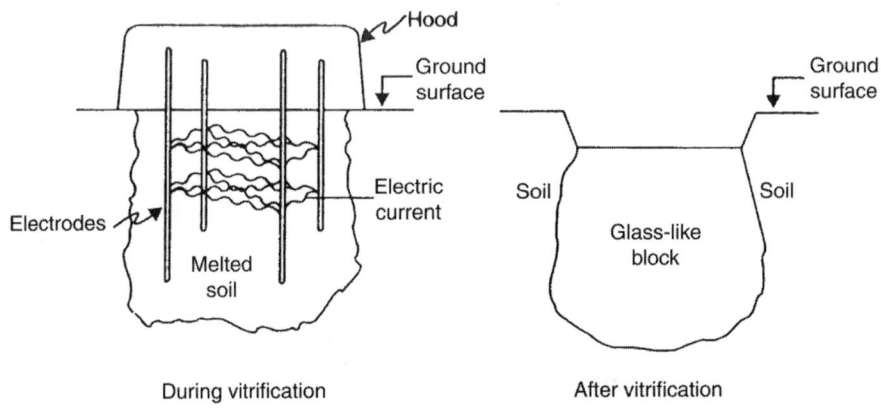

FIGURE 6.15 During the vitrification process and after vitrification. (Taken from U.S. EPA, A Citizen's Guide to Vitrification, Technical Report EPA-542-F-01-017, U.S. EPA, Washington, DC, 2001.)

turns into a solid block of glass-like material. The electrodes become part of the block. When vitrified, the original volume of soil shrinks. This causes the ground surface in the area to sink slightly. To level it, the sunken area is filled with clean soil. When used properly, vitrification can be quite safe. The gas hood must be large enough to cover the polluted area so it can capture all the chemicals released from the soil. Any wet soil must be dried first to prevent steam from forming. The release of steam can splash hot, melted soil above ground. The hood further prevents site workers from being splashed.[82] The vitrified block that is left in place is permanent and not harmful to people.

Thermal conduction

Thermal conduction (also referred to as electrical conductive heating or *in situ* thermal desorption) supplies heat to the soil through steel wells or with a blanket that covers the ground surface. As the polluted area is heated, the contaminants are destroyed or evaporated. Steel wells are used when the polluted soil is deep. The blanket is used where the polluted soil is shallow. Typically, a carrier gas or vacuum system transports the volatilized water and organics to a treatment system.

6.7.5.2   Artificial Ground Freezing

Artificial ground freezing involves the installation of freezing loops in the ground and a self-confined refrigeration system that pumps coolant around the freezing loop. This method is useful only as a temporary treatment approach because of the high thermal maintenance expense.

6.7.5.3   Fracturing

Fracturing is a way to crack rock or very dense soil, like clay, below ground. It is not necessarily a cleanup method in itself. Rather, fracturing is used to break up the ground to help other cleanup methods work better. The cracks, which are called fractures, create paths through which harmful chemicals can be removed or destroyed.[17,46,84]

Hydraulic fracturing

Hydraulic fracturing uses a liquid, usually water. The water is pumped under pressure into holes drilled in the ground. The force of the water causes the soil (or sometimes rock) to crack. It also causes existing fractures to grow larger. To fracture soil at greater depths, sand is pumped underground with the water. The sand helps prop the fractures open and keep them from closing under the weight of the soil.

Pneumatic fracturing

Pneumatic fracturing uses air to fracture the soil. It can also help to remove chemicals that evaporate or change to gases quickly when exposed to air. When air is forced into the soil, the chemicals evaporate and the gases are captured and treated above ground.

Air can be forced into the ground at different depths within a hole. When air is forced near the ground surface, the surface around the holes may rise by as much as an inch, but will settle back close to its original level. In both pneumatic and hydraulic fracturing, equipment placed underground directs the pressure to the particular zone of soil that needs to be fractured.

Blast-enhanced fracturing

Blast-enhanced fracturing uses explosives, such as dynamite, to fracture rock. The explosives are placed in holes and detonated. The main purpose is to create more pathways for polluted groundwater to reach wells drilled for pump-and-treat cleanup.

6.7.5.4   Immobilization and Stabilization

Immobilization and stabilization render contaminants insoluble and prevent leaching of the contaminants from the soil and their movement from the contamination area. The techniques used for immobilization are precipitation, chelation, redox reaction, and polymerization.

Precipitation is the most promising method for immobilizing dissolvable metals such as lead, cadmium, zinc, and iron.[15] Some forms of arsenic, chromium, mercury, and some fatty acids can also be treated by precipitation.[47] The common precipitating chemicals for metal cations are sulfide, phosphate, hydroxide, or carbonate. Among them, sulfide is the most promising, because sulfides have low solubility over a broad pH range. Precipitation is most applicable to sites with sand or coarse silt strata.

The use of chelating agents may also be a very effective means of immobilizing metals.

Redox reactions may cause mobile toxic ions to become either immobile or less toxic. Hexavalent chromium is mobile and highly toxic. It can be reduced to be rendered less toxic in the form of trivalent chromium sulfide by the addition of ferrous sulfate. Similarly, pentavalent (V) or trivalent (III) arsenic, arsenate or arsenite are more toxic and soluble forms. Arsenite (III) can be oxidized to As(IV). Arsenate (V) can be transformed to highly insoluble $FeAsO_4$ by the addition of ferrous sulfate.

Polymerization involves the injection of a catalyst into the groundwater plume to cause polymerization of organic monomers (e.g., vinyl chloride, isoprene, and methyl methacrylate), transforming the once fluid substance into a gel-like, nonmobile mass. It has been reported that 90% of an acrylate monomer leakage was polymerized by the injection of a catalyst, activator, and wetting agents.[48] *In situ* polymerization is suitable for groundwater cleanup following land spills or underground leaks of pure monomers. Applications for uncontrolled hazardous waste sites are very limited.

Various immobilization and stabilization methods can be applied to soils contaminated with heavy metals, petroleum products, PCB, peroxyacetyl nitrate (PAN), and so on.[17] The disadvantages of immobilization and stabilization methods include the following:

1. There is a requirement for numerous, closely spaced injection wells, even in coarse-grained deposits
2. Contaminants are not removed, and some of the chemical reactions could be reversed, producing monomers, which will again migrate with the groundwater
3. There is a possibility of the injection of a potential groundwater pollutant that in association with chemicals forms toxic byproducts
4. There is a potential for the clogging of soil pore spaces

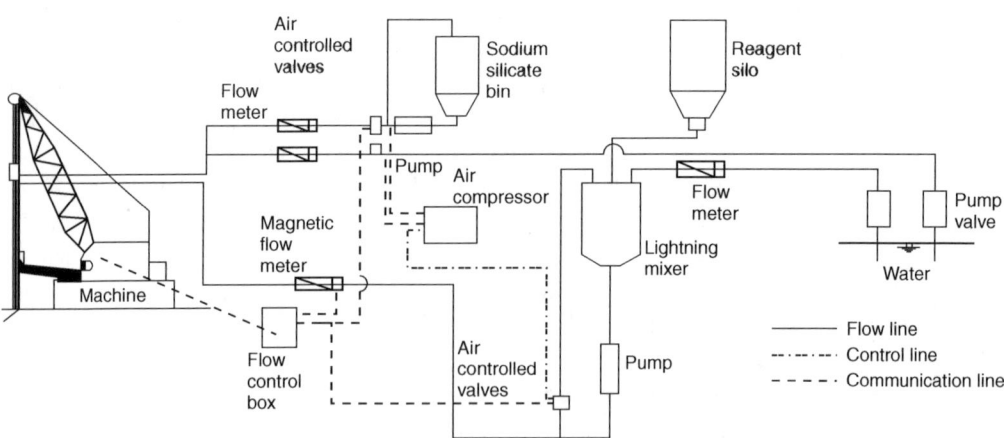

FIGURE 6.16 *In situ* solidification batch mixing plant process.

International Waste Technologies/Geo-Con, Inc., has used a deep soil mixing system to deliver and mix the chemicals with the soil *in situ*. The system involves mechanical mixing and injection, as shown in Figure 6.16.

ChemFix Technologies, Inc., has used silicates and silicate setting agents to stable polyvalent metal ions. Usually, there is a need to separate coarse and fine pollutants so as to crush coarse materials and reduce the material size required for the stabilization technology.

The soil–cement mixing wall technology developed by S. M. W. Seiko, Inc., involves the *in situ* fixation, solidification, and stabilization of contaminated soils by mixing soil, cement, and chemical grout, by including cutoff walls and soil stabilization, and by using hollow-stem augers to inject solidification and stabilization agents and blend them with contaminated soil *in situ*.

Table 6.7 and Table 6.8 summarize some promising *in situ* chemical treatment methods for organics and inorganics that can be applied to soil. Some of them can also be applied to groundwater.

TABLE 6.7
Summary of *In Situ* Chemical Treatment Methods for Organics

Method	Amenable to Treatment	Treatment Reagents	Process
Soil flushing Water flushing	Hydrophilic compounds (high solubility, low R_{ow})	Water	Contaminated soils are flooded with water or a water chemical mixture and the elutriated solution is collected
Water with surfactants	Hydrophobic compounds (low solubility, high K_{ow})	Aqueous solutions of surfactants	Contaminants are mobilized into solution by reason of solubility, formation of emulsion or reaction
Oxidation	Benzene and substituted benzenes	Ozone, hypochlorite, or hydrogen peroxide	Oxidation state of compounds is increased by loss of electrons
	Phenols		
	Halogenated phenols		
	Nitro aromatics		Contaminants are detoxified, mobility is increased or compounds are made more amenable to biological degradation
	PAHs		
	Heterocyclic nitrogen and oxygen compounds		
	Aldehydes and ketones		
	Sulfides, disulfides		
Hydrolysis (base-catalyzed)	Esters Amides	Water with lime or NaOH	Attack of nucleophile (e.g., water or hydroxyl ion) on an electrophile (e.g., carbon or phosphorus), resulting in bond cleavage and displacement of the leaving group
	Carbamates		
	Organophosphorus compounds		
	Certain pesticides (i.e., parathion, malathion, 2-4D esters, DDT)		
Polymerization	Aliphatic, aromatic and oxygenated monomers	Catalyst activation	Conversion of a compound to a larger chemical multiple of itself
	Vinyl chloride		
	Isoprene		
	Acrylonitrile		Reduces mobility of compound in soil

Source: U.S. EPA, Field Standard Operating Procedures for Decontamination of Response Personnel, FSOP7, U.S. EPA, Washington, DC, 1985.

TABLE 6.8
Summary of *In Situ* Chemical Treatment Methods for Inorganics

Method	Amenable to Treatment	Treatment Reagents	Process
Precipitation Sulfide	Heavy metals	Sodium or calcium sulfide	Formation of insoluble metal precipitate, thereby reducing the mobility of the metal
Carbonate/hydroxide	Heavy metals	Lime, calcium carbonate	
Phosphate	Heavy metals	Superphosphate fertilizer	
Soil flushing			
Acids/bases	Heavy metals	Dilute solutions of acids or bases	Involves solubilizing the metals followed by extraction of the metal ions
Chelates	Heavy metals	Chelating agents such as citric acid or EDTA	Formation of stable metal chelates; depending on chelating agent, metal chelate is either strongly sorbed to soil or is highly mobile and can be flumbed using water or dilute acid solutions.
Oxidation	Trivalent arsenic	Potassium permanganate	Oxidizes trivalent arsenic to pentavalent arsenic, and results in precipitation of arsenic-iron-manganese compounds.
Reduction	Hexavalent chromium	Ferrous sulfate	Reduces Cr(VI) to Cr(III)
	Hexavalent selenium	Ferrous sulfate	Reduces Se(VI) to Se(IV)

Source: U.S. EPA, Field Standard Operating Procedures for Decontamination of Response Personnel, FSOP7, U.S. EPA, Washington, DC, 1985.

6.7.5.5 Soil Flushing

For *in situ* soil flushing, large volumes of water, at times supplemented with surfactants, cosolvents, or treatment compounds, are applied to the soil or injected into the groundwater to raise the water table into the contaminated soil zone. Injected water and treatment agents are isolated within the underlying aquifer and recovered together with flushed contaminants.[50–52,85]

Water can be used to flush water-soluble or water-mobile organics and inorganics. The inorganics to which this can be applied include soluble salts such as the carbonates of nickel, zinc, and copper. The organics that it is feasible to remove from soil should have a certain degree of water affinity, in other words, they should have low soil–water partitioning coefficients ($P < 1000$; i.e., $k = \log P \leq 3$). Among them, the high-solubility organics ($k \leq 1$) include low-molecular-weight alcohols, phenols, and carboxylic acids, and the medium-solubility organics ($1 \leq k \leq 3$) include low- to medium-molecular-weight ketones, aldehydes, and aromatics, and lower-molecular-weight halogenated hydrocarbons such as TCE (tetrachloroethylene) and PCE (pentachloroethylene). It has been reported that an 18-month period of water flushing on soil for a PCE spill site in Germany removed 50% of the material.[49]

Adjusting pH to the optimum solubility of salt by adding dilute acid or base solution can enhance inorganic solubilization and removal. Week acids are recommended to avoid the high toxicity resulting from acidity. Sodium dihydrogen phosphate and acetic acid have low toxicity and are relatively stable. A stronger dilute acid such as sulfuric acid may be used for neutralizing soils containing sufficient alkalinity. Acidic solutions may also be used to flush some basic organics such amines, ethers, and anilines. Complexing and chelating agents (such as EDTA, DTPA, and acetic and citric acids) are also used to removal heavy metals.[50–52]

Some contaminants are adsorbed by iron and manganese oxides (which may exist as coatings on soil particles) in soil. By using acids or chelating agents (such as sodium dithionite/citrate), the iron and manganese coating can be dissolved, thus mobilizing the adsorbed contaminants.

Surfactant washing is among the most promising *in situ* chemical treatment methods. Surfactants can improve the solvent property of the flushing water, emulsify nonsoluble organics, and enhance the removal of hydrophobic organics sorbed onto soil particles.

In situ soil flushing should involve the design of a series of injection wells (for washing agents) and extraction wells. An economically feasible soil flushing method may involve the recycling of the elutrate through the contaminated material, with make-up solvent being added to the system while a fraction of the elutrate stream is routed to the portable wastewater treatment system. Soil flushing operations require soils with moderate to high permeability, and tend to work best for sandy soil conditions.[53]

6.7.5.6 Chemical Detoxification

Chemical detoxification uses oxidation, reduction, neutralization, and hydrolysis to reduce the toxicity of the contaminants. The basic theory is similar to that of treating pumped groundwater.

6.7.5.7 Soil Vapor Extraction

Soil vapor extraction (SVE) can be used to remove volatile contaminants and, when combined with another technology, to treat nonvolatile contaminants. If contamination has reached the aquifer, it is necessary to use SVE in combination with groundwater pumping and air stripping.

Soils with low air permeability are more difficult to treat. Heterogeneity can cause variable flow and desorption, making remediation more difficult. High organic carbon content causes a high sorption capacity for VOCs and is more difficult to remedy. Contaminants with low vapor pressure or high water solubilities become difficult to remove. The lower limit on vapor pressure is 1 mmHg absolute. The moisture in the soil hinders the removal of soluble compounds because water moisture acts as a sink for the compounds. Figure 6.17 shows how SVE works.[86]

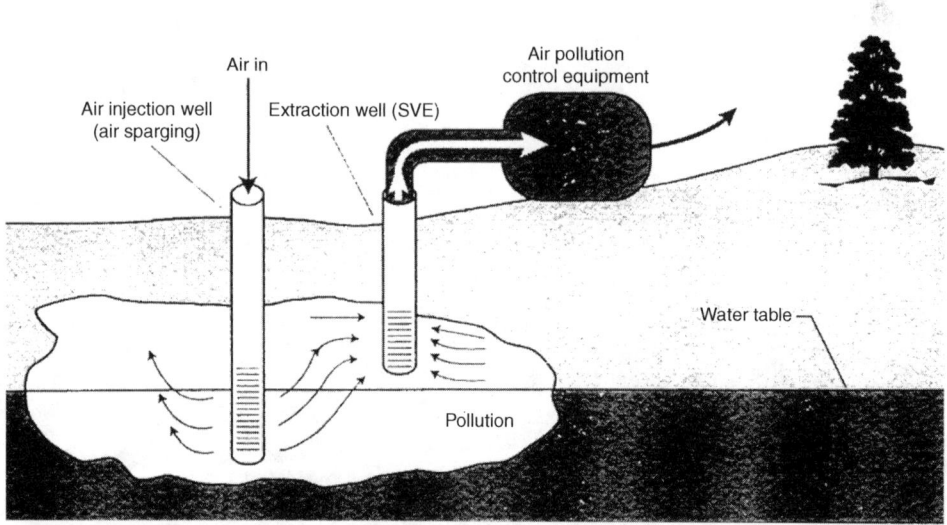

FIGURE 6.17   Soil vapor extraction and air sparging. (Taken from Rohayo, A.J., Cameron, R.J., Teters, B.B., Rossabi, J., Riha, R., and Downs, W., Passive Soil Vapor Extraction, Technical Report DE98051208, 20 p., U.S. Department of Commerce, National Technical Information Service, Springfield, VA, 1997. With permission.)

6.7.5.8 *In Situ* Steam Extraction

In situ steam extraction treatment is provided to effectively remove volatile and semivolatile soil contaminants, including volatile organic compounds, petroleum wastes, soluble inorganics (acids, bases, salts, and heavy metals). Steam is injected into the soil or groundwater and acts as a stripping agent, heating the soil/water and releasing the volatile contaminants. This produces both air and water streams that must be further treated.

Raising the temperature of the soil increases the vapor pressure of the contaminants, improving their ability to volatilize. Many semivolatile compounds will eventually be released as the temperature rises, although these compounds tend to need longer residence times.

Two types of *in situ* steam extraction systems, mobile and stationary, are available. The mobile system may have rotating cutter blades that release steam as they tunnel through the soil. This system treats small areas sequentially. The stationary system injects the steam into drilled wells, without disturbing the soil.

The mobile *in situ* steam extraction system has certain restrictions on its use. High silt and clay content may cause stability problems with respect to the support of the system (causing equipment to sink or tip), and may also require longer treatment times due to its lower permeability. The mobile *in situ* steam extraction system is also limited to a depth of 9 m (30 ft), and a height requirement of 9 m (30 ft) is needed for clearance. A slope of less than 1° is also required. Temperatures of –7 to 38°C (20 to 100°F) are desirable. Figure 6.18 shows a schematic illustration of the mobile unit system developed by Novaterra, Inc.[54] The boring unit contains two counter-rotating blades with nozzles that release steam and compressed air. The steam (at 400°F; 204°C) and air (at 275°F; 135°C) volatilize the organics, which are caught and collected. A blower provides suction to draw up the vapor and protect against leakage. The vapor is then separated into gas and water and treated. The mobile system can treat areas of 2.2 m × 1.2 m × 9 m (7 ft 4 in. × 4 ft × 30 ft).

The stationary *in situ* steam extraction system uses injection wells to introduce the steam, and recovery wells for removing it. Soil permeability is a major factor. Low-permeability soils require a far greater number of wells compared to high-permeable soil, driving up costs. To be effective

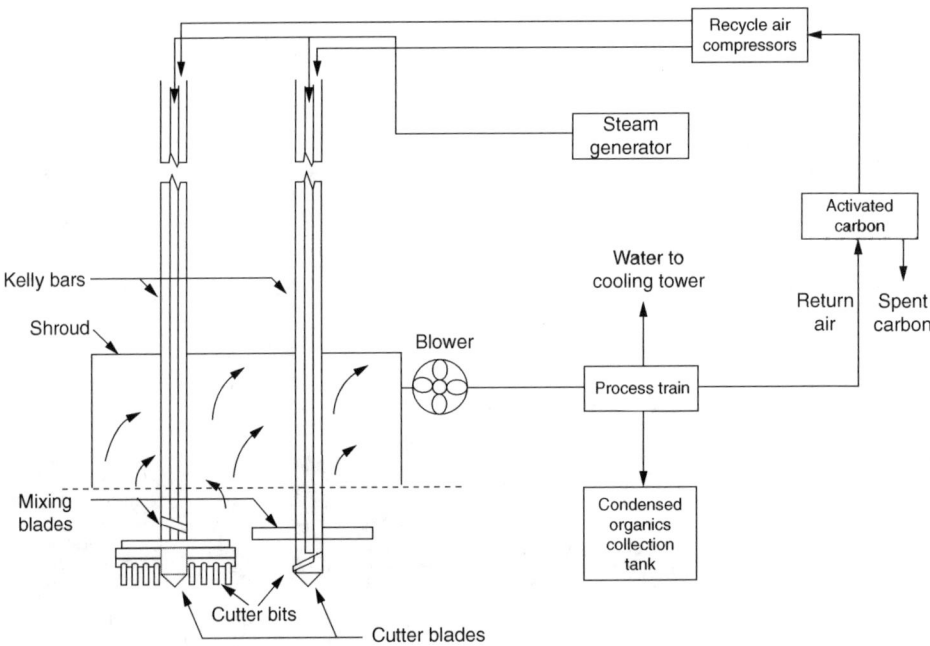

FIGURE 6.18 A mobile steam extraction system.

(85% contaminants removal), the stationary *in situ* steam extraction system requires homogeneous soils with high to medium permeability. Accordingly, further treatment may be necessary.

Steam extraction has been used for gasoline and diesel fuel. High-molecular-weight components of the diesel fuel cannot be removed easily, although a total removal of up to 91% is possible. When used to remove low-volatility compounds in a soil with a high percentage of clay, performance is expected to be ca. 85%. The mobile *in situ* steam extraction system can reduce VOCs in soils by more than 50% of their initial level. Based on pilot studies, the stationary steam extraction system is expected to have a 90% removal efficiency.[54]

6.7.5.9 *In Situ* Biodegradation/Bioremediation and Bioventing

Biodegradation or bioremediation has so far been developed for aerobic degradation of organic contaminants in soil.[41,87] Anaerobic bioremediation approaches have several limitations. For a strict anaerobic system to be effective, no oxygen should be present in the environment, because oxygen itself is toxic to strictly anaerobic microorganisms. This anaerobic condition is difficult to implement under field conditions, especially when a mechanical pumping system is used to extract groundwater. Also, anaerobic degradation of some contaminants can produce intermediate end products that may be less desirable than the target substance. For example, tetrachloroethylene (TCE) can be anaerobically degraded to vinyl chloride. This partial-breakdown end product does not undergo further anaerobic degradation. Vinyl chloride, a potent carcinogen, can accumulate in the environment. Furthermore, anaerobic degradation can produce unpleasant and potentially dangerous off-gases such as H_2S and CH_4. For these reasons, full-scale anaerobic bioremediation technologies have lagged behind aerobic approaches.

The bioventing system developed by the U.S. EPA Risk Reduction Engineering Laboratory[54] comprises mainly the injection of atmospheric air to treat contaminated soil *in situ* (Figure 6.19). This air provides a continuous oxygen source, which enhances the growth of microorganisms naturally present in the soil. The provided low-pressure air allows for an inflow of oxygen without volatilization of contaminants. Additional additives such as ozone or nutrients may also be required to stimulate microbial growth.[17]

6.7.5.10 Electroosmosis Remedial Technology

Electroosmotic soil processing is an *in situ* separation/removal technique for extracting heavy metals and organic contaminants from soils.[17,55,89] The fluid between the soil particles moves because a constant, low DC current is applied through electrodes inserted into the soil mass. The electroosmosis (EO) remedial method provides an advantage over conventional pumping techniques for *in situ* treatment of contaminated fine-grained soils and is more efficient in saturated conditions.

Electroosmosis is an electrokinetic effect, so a direct electric potential causes a movement of liquid through stationary particles. From primary electrode reactions,

$$2H_2O - 4e^- = O_2 + 4H^+ \quad \text{(anode)}$$

$$4H_2O + 4e^- = 2H_2 + 4OH^- \quad \text{(cathode)}$$

Electrokinetics relies upon the application of a low-intensity direct current through the soil between ceramic electrodes, which are divided into a cathode array and an anode array. This mobilizes charged species, causing ions and water to move toward the electrodes. Metal ions, ammonium ions, and positively charged organic compounds move toward the cathode. Anions such as chloride, cyanide, fluoride, nitrate, and negatively charged organic compounds move toward the anode. Removal of contaminants at the electrode may be accomplished by several means, among which are electroplating at the electrode, precipitation or coprecipitation at the electrode, pumping of water near the electrode, or complexing with ion-exchange resins.[89]

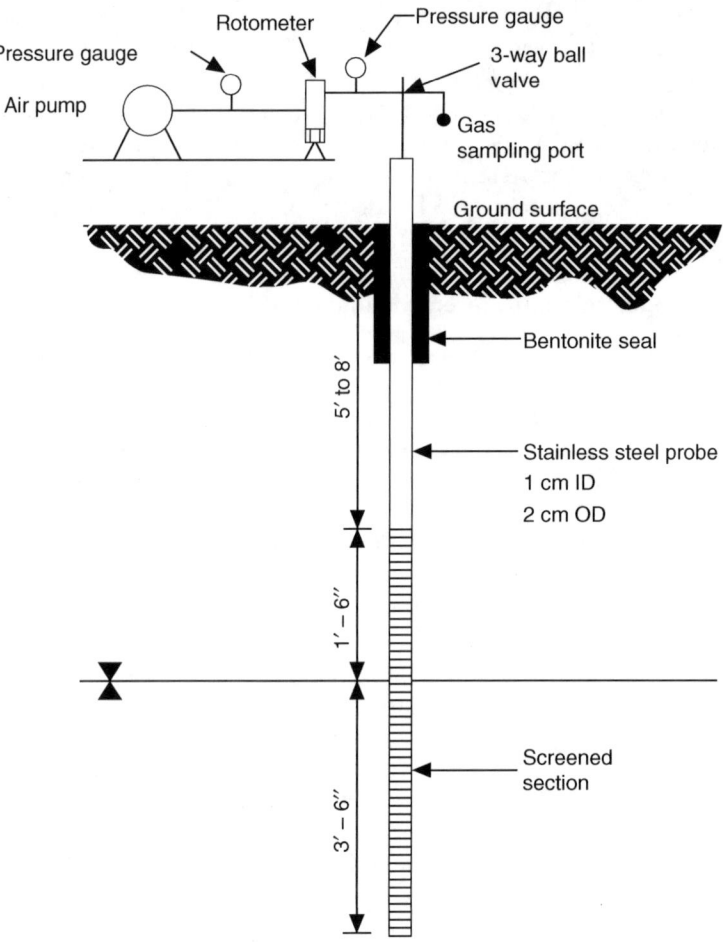

FIGURE 6.19 An air injection system.

For the same quantity of electricity, twice as many water molecules are electrolyzed at the cathode than at the anode, producing a chemical gradient of molecular water. As a consequence of the reactions, two supplemental ionic species, H^+ and OH^- (in addition to the migration of existing anions and cations in the pore fluid under the electrical field), are generated, and can have a significant influence on local conductance.

The electrical potential difference is initially distributed linearly across the specimen. The changing chemistry across the cell may result in variations in electrical potential difference in time and space. Specifically, an acid front is generated at the anode by the electrolysis reactions. A corresponding base front appears at the cathodes. This acid front generated at the anode advances toward the cathode in time under the influence of the imposed electrical, chemical, and hydraulic potential gradients. It is suggested that the movement of the acid front by migration (electrical potential), diffusion (chemical potential), and advection (hydraulic potential) will cause desorption of cations and other species from clay surfaces and facilitate their release into the pore fluid. With an open electrode configuration this front flushes through the specimen and reaches the cathode.[55] This phenomenon, together with the concurrent electroosmotic flow, would then constitute the mechanism for removing contaminants from soils.

Several factors influence the efficiency of removing contaminants from soils by EO. The first factor is the chemistry generated at the electrodes. Low-pH conditions generated at the anode cause desorption and ionization of most heavy metals and inorganic chemicals. However, a flux of high H^+ ion concentration results in an increase in conductivity and a decrease in electroosmotic flow. The pH conditions at the anode and the cathode should be strictly controlled and adjusted for continued flow. This adjustment depends on the cation-exchange capacity of the soil, the type and concentration of the chemicals in the soil, and the initial pH of the medium.

The second factor is the type and concentration of chemicals in soil. Soils with low initial ionic strengths favor high EO efficiencies. A lower initial ionic strength is responsible for a higher conductivity of the specimen, which in turn results in a decrease in the resistance offered to current flow, and hence the ion flow is governed more by diffusion and migration.

The third factor is the behavior of primary chemicals in the soil at different pH conditions. The chemistry in the system is governed by the pH gradients across the soil mass. Knowledge of the behavior of the primary chemicals in different pH environments is necessary for a better understanding of the efficiency and to enable a decision to be made on the required processing conditions and time.

The fourth factor is the current density. At an inert anode and for 100% Faradaic efficiency for water oxidation, the density of the current controls the flux of H^+ ions. The cathodic current density and the species available in its vicinity establish the efficiency of the reduction processes ($Pb^{2+} \rightarrow Pb$). These vary to a greater extent than the anode process, because the pH and the species reaching the cathode vary with processing time. Thus, control of the current density is critical to ensure optimal EO efficiency and contaminant removal.

The fifth factor is the water content of the soil. Electroosmotic flow is promoted at higher water contents. Therefore, high moisture content, and in particular saturated conditions are favored. However, the technique can be used in partially saturated deposits by supplying a pore fluid at the anode.

The sixth factor is conditioning. Similar to the changes in current density, the pore fluid at the anode and cathode compartments can be conditioned to a specific pH or chemistry to increase the efficiency of the process.

Preliminary laboratory data demonstrate the feasibility of removing Pb, Cr, Cd, Ni, Cu, Zn, As, TCE, BTEX compounds, and phenol from soils (clays and sandy clayey deposits, and dredged sediments) using EO technology. It has been demonstrated that 75 to 95% of Pb can be removed across the cell, in which a significant amount of the removed Pb can be electroplated at the cathode.

Metallic electrodes may dissolve as a result of electrolysis and may introduce corrosion products into the solid mass. However, if the electrodes are made of carbon or graphite, no residue will be introduced in the treated soil mass as a result of the process. The energy expenditure for Pb removal has been estimated to in the range 30 to 60 kWh/m^3 of soil. The EO method also provides an advantage over conventional pumping techniques for *in situ* treatment of contaminated fine-grained soils.

6.7.6 SOIL EXCAVATION AND *EX SITU* TREATMENT

If an *in situ* treatment method is not feasible, a soil excavation and treatment method should be conducted. The soil excavation and treatment method is usually more cost-effective for small sites and shallow contamination. Before excavation, planning is needed regarding the following steps of the treatment, among others:

1. Protecting fugitive gas accompanying the excavation
2. Pumping to remove liquids from the pounds and surface impoundments
3. Avoiding the mixing of clean soil with the excavated contaminated soil, and uncontrolled mixing of incompatible wastes
4. Covering excavated contaminated soils to prevent water leaching and fugitive dust production

6.7.6.1   Soil Washing Technology

For soil washing, contaminants sorbed onto fine soil particles are separated from bulk soil in a water-based system on the basis of particle size. The wash water may be augmented with a basic leaching agent, surfactant, or chelating agent, or by adjustment of pH to help remove organics and heavy metals. Soils and wash water are mixed *ex situ* in a tank or other treatment unit. The wash water and various soil fractions are usually separated using gravity settling.[90]

There are various agents that can be used to wash soil and drive its contaminants out, as discussed in the section on *in situ* soil flushing technology.

Washing with water may be used for the dissolution of soluble metallic ions and desorption of adsorbed metals and organics (such petroleum products), as long as the soil has high water affinity.

Surfactants or organic solvents are generally required for hydrophobic contaminants. When dealing with certain pesticides and metals that are insoluble in water, it may be necessary to add acids or chelating agents for their proper removal. However, these agents may create difficulties in wastewater treatment processes. If the soil contains a wide variety of contaminants, sequential washing steps may be needed along with adjustments in wash formulation or soil/wash-fluid ratios. A high percentage of silt and clay-sized particles in soil creates removal difficulties due to the contaminants being strongly adsorbed to these particles. Some sophisticated soil washing systems, such as the one developed by BioTrol, Inc., is claimed to be effective in washing contaminants (metals, PCB, pesticides, and petroleum products) concentrated in the fine-sized fraction of soil.[17]

6.7.6.2   Solvent Extraction

Solvent extraction has a similar procedure to the soil washing treatment. The difference is that solvent extraction uses organic chemicals as a solvent, whereas soil washing uses mainly water. Figure 6.20 illustrates the flow diagram of solvent extraction developed by CF System Corporation.[17,56] The waste and solvent are mixed, resulting in the organic contaminant dissolving into the solvent. The extracted organics are removed from the extractor with the solvent, which is transferred to a separator, where the pressure or temperature is changed, causing the organic contaminants to separate from the solvent. The solvent is recycled to the extractor and the concentrated contaminants are removed from the separator, disposed of, or reclaimed.

Solvent extraction shows effectiveness in the removal of organic wastes such as PCBs, VOCs, halogenated solvents, and petroleum wastes, but is less effective in removing inorganic compounds.[39] The removal of organic contaminants depends on the nature of the extracting solvent. Organic bound metals can become a constituent of the concentrated waste, which is undesirable because it can restrict both disposal and recycle options.

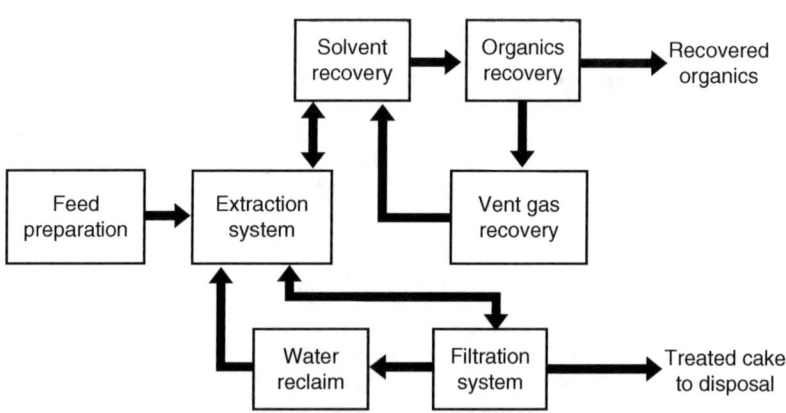

FIGURE 6.20   Solvent extraction remediation system.

Treated solids leave the extraction subsystem with trace amounts of extraction solvent, which usually volatilizes quickly. Ambient air should be monitored to determine if the volatilization of the solvent presents a problem.

6.7.6.3   Treatment of Washed Wastewater

The washed wastewater treatment techniques are basically the same as those used for pumped groundwater. Several integrated treatment technologies have been developed that can wash soil and treat washing water, such as that by BioTrol, Inc., in which the excavated soil is first screened, then washed, and finally the contaminated water is treated. As contaminants are difficult to wash from silt and clay, the clay and silt slurry contaminated with organics is treated in a bioslurry reactor.

In a technology developed by Excalibut Enterprises, Inc., a soil/liquid separator, such as a centrifuge or a cyclone, is used to separate the decontaminated soil from contaminated water.[17] Water is then treated with ozone and ultraviolet light, with ultrasound catalyzing the oxidation. This method is claimed to be able to treat soils contaminated with inorganics including cyanides, and organics such as PCBs, PCP, pesticides, herbicides, and dioxins.

Dissolved air flotation (DAF) technology, requiring a short detention time (less than 15 min) and a small space, combined with its mobility, is technologically and economically feasible for treatment of washed wastewater or contaminated groundwater.[57,58]

WasTech, Inc., has applied proprietary bonding reagents to a waste (soil or wastewater) containing organic and inorganic contaminants. The waste and reagent mixture is mixed with cementing materials that form a stabilizing matrix. The resultant material is a nonleaching, high-strength monolithic material that can be used to refill the excavated site.[17]

6.7.6.4   *Ex Situ* Thermal Desorption

In the thermal desorption technique excavated soil is heated to around 200 to 1000°F (93 to 538°C). Volatile and some semivolatile contaminants are vaporized and carried off by air, combustion gas, or inert gas. Off-gas is typically processed to remove particulates. Volatiles in the off-gas may be burned in an afterburner, collected on activated carbon, or recovered in condensation equipment. Thermal desorption systems are physical separation processes that are not designed to provide high levels of organic destruction, although some systems will result in localized oxidation or pyrolysis.

The thermal desorption process could be an excellent first step in soil treatment if used in conjunction with another *ex situ* treatment. Thermal desorption can remove TCE, most diesel fuel, and perhaps organically bound lead. Chemical Waste Management, Inc., has claimed that thermal desorption can reduce volatile organics to less than 1 mg/L and inorganics to less than 10 mg/L (sometimes even to less than 1 mg/L), and has shown a removal of 96 to 99+% of PCBs from soils containing 120 to 6000 mg/L of initial PCBs.[17,91]

6.7.6.5   Plasma Arc Verification

A plasma centrifugal furnace uses thermal heat transferred from arc plasma to create a molten bath that detoxifies the feed material. Organic contaminants are vaporized at temperatures of 2000 to 2500°F (1093 to 1371°C) to form innocuous products. Solids melt and are vitrified in the molten bath at 2800 to 3000°F (1540 to 1650°C). Metals are retained in this phase, which is a nonleachable, glassy residue. This method is applicable to soils contaminated with organic compounds and metals.

6.7.6.6   Direct Incineration

Direct incineration is mainly used for organically contaminated soil with sufficient concentration that no or little additional fuel is needed. Incineration of contaminated soil in a rotary kiln would result in virtually complete destruction of TCE and diesel fuel. The organic portion of lead dithiocarbonate would be destroyed, leaving lead and lead oxides in the soil. If lead cannot be removed from the soil subsequently, then it has to be disposed of as a hazardous waste.

Note that U.S. EPA regulations (under the RCRA) for hazardous waste incineration require that particulate emissions be no more than 180 mg/m^3 and that hydrogen chloride removal efficiency from the exhaust gas can be no less than 99%. Therefore, trial burns to determine the maximum ash and chlorine content that a waste can handle are needed prior to issuance of a permit.

6.7.6.7   Bioreactor Landfill

The Solid Waste Association of North America (SWANA) has defined a bioreactor landfill as "any permitted Subtitle D landfill or landfill cell where liquid or air is injected in a controlled fashion into the waste mass in order to accelerate or enhance biostabilization of the waste."

A bioreactor landfill operates to rapidly transform and degrade organic waste. The increase in waste degradation and stabilization is accomplished through the addition of liquid and air to enhance microbial processes. This bioreactor concept differs from the traditional "dry tomb" municipal landfill approach.[92,93] A bioreactor landfill is not just a single design and will vary to correspond to the operational process invoked. There are three different general types of bioreactor landfill configurations:

1. *Aerobic*. Leachate is removed from the bottom layer, piped to liquids storage tanks, and recirculated into the landfill in a controlled manner. Air is injected into the waste mass, using vertical or horizontal wells, to promote aerobic activity and accelerate waste stabilization.
2. *Anaerobic*. Moisture is added to the waste mass in the form of recirculated leachate and from other sources to obtain optimal moisture levels. Biodegradation occurs in the absence of oxygen (anaerobically) and produces landfill gas. Landfill gas, primarily methane, can be captured to minimize greenhouse gas emissions and for energy projects.
3. *Hybrid (aerobic–anaerobic)*. The hybrid bioreactor landfill accelerates waste degradation by employing a sequential aerobic–anaerobic treatment to rapidly degrade organics in the upper sections of the landfill and collect gas from lower sections. Operation as a hybrid results in an earlier onset of methanogenesis compared to aerobic landfills.

The bioreactor landfill is a remedial alternative that can be applied either on site or off site. However, landfilling is regarded as the least attractive alternative at a site cleanup action. Landfilling of hazardous materials is becoming increasingly difficult and more expensive due to steadily growing regulatory control.[92,93]

Bioreactor landfill operations should comply with RCRA landfill facility standards under 40 CFR Part 264. It should be noted that SARA strongly recommends on-site treatment that permanently and significantly reduces the volume, toxicity, or mobility of hazardous substances, and utilizes cost-effective permanent solutions. The legislation prohibits land disposal of hazardous wastes unless U.S. EPA determines otherwise. U.S. EPA guidance for CERCLA responses requires most on-site disposal actions to attain or exceed applicable and relevant standards of all Federal public health and environmental laws unless specific circumstances dictate otherwise.

The site conditions for an on-site landfill, such as location, geology, hydrogeology, physiography, climate, and so on, should also be suitable. Landfill should meet the minimum technology requirements and regulations for hazardous waste landfills such as double liners and leachate collection and removal systems, leak detection systems, closure procedures and final cover, and construction quality assurance.[59]

Off-site landfill is not desirable, because it faces more problems associated with off-site transportation. Other off-site treatment and disposal, such as incineration or other waste treatment methods performed off site, are also not attractive, because they are not the on-site permanent treatments as recommended by U.S. EPA. Off-site waste treatment should be used only if on-site applications are not possible.

The RCRA manifest requirements (40 CFR Parts 262 and 263) must be complied with for all wastes that are shipped off site. The regulations for transportation of hazardous wastes by the U.S. Department of Transportation, U.S. EPA, and states and local regulation agencies, should be complied with. A knowledge of RCRA regulations (40 CFR Parts 261–265) and other regulations developed by State Governments is required to determine the feasibility of off-site disposal.

6.7.7 SEDIMENTS REMEDIATION

Similar to soil remediation, *in situ* control and excavation-and-treat methods can be applied to sediment remediation.

6.7.7.1 *In Situ* Control and Containment

The aim of *in situ* control and containment is to reduce dispersion and leaching of a hazardous substance to other areas in the water body, in particular if removal of the substance is determined to be an unacceptable singular remedial response. The following briefly presents some common methods:

1. *Retaining dikes and berms.* Retaining dikes and berms include earthen embankments, earth-filled cellular and double-sheet pile walls, water inflated dams, and so on, which aim to minimize the transport of contaminated sediments.
2. *Cover methods.* Cover methods are used to cover contaminated sediments in order to minimize leaching of contaminants and prevent erosive transport of contaminated sediments.
3. *Surface sealing.* Surface sealing applies cement, quicklime, or other grouting materials to the surface or mixed with bottom sediments to create a seal.
4. *In situ grouting.* The *in situ* grouting method involves injecting grouting materials into sediments to stabilize the contaminated sediments. *In situ* containments can be either temporary or permanent. However, permanent containment of contaminated sediments has not been well demonstrated or widely used.

In situ methods have potential use as an interim or emergency measure until dredging can be undertaken or as a primary remedial action where it is determined to be more cost-effective than removal. The biggest advantages are that they are much less costly than dredging, eliminate the need for dredged material management, and minimize the resuspension of contained sediments.[15]

6.7.7.2 Sediment Removal

There are several methods used in the removal of contaminated sediments[60]:

1. *Mechanical dredging.* Mechanical dredging methods use mechanical excavation equipments such as backhoes, draglines, clamshells, and bucket ladder dredges.
2. *Hydraulic dredging.* Hydraulic dredging removes and transports sediment in a liquid slurry form.
3. *Pneumatic dredging.* The pneumatic dredging method utilizes pumping, operated using compressed air and hydrostatic pressure, to draw sediments to the collection head through transport piping. The dredged sediments are subsequently treated and disposed of. Other than the use of different instrumentation, the approaches used for soil remediation can be applied to sediments remediation.

6.7.8 Points-of-Entry Control and Alternative Methods

The protection of human health from the threat of contaminated sites is mainly relevant during site cleanup and site control. However, if residential well water is contaminated, then point-of-entry control has to be applied. Point-of-entry control is used to avoid contaminated well water from entering houses for drinking and other in-contact uses. It has been pointed out by U.S. EPA[61] that taking showers in contaminated groundwater (especially in the case of VOC contamination) probably leads to far greater exposure to the chemicals than drinking the same water. The following sections present some common methods used as point-of-entry control.

6.7.8.1 Aeration

Aeration can be applied to well water contaminated by VOCs. It has been reported that 95–99% reduction in high-level (>100 µg/L) VOCs can be obtained by aeration. However, it should be noted that aeration is less effective for VOCs removal at lower concentrations (<10 µg/L). Boiling can further enhance the reduction of VOCs.[60,101,102]

Aeration is also an efficient process for removing radioactive radon from contaminated well water.

6.7.8.2 Distillation

Distillation can cause the evaporation of compounds that have boiling temperatures lower than 100°C. Thus, distilled water will contain more of those compounds, but will have lower concentrations of heavy metals and other components that have high boiling temperatures. Although VOCs are also evaporated with the water, they mostly evaporate in the early phase and can be removed; the recondensation of water at high temperatures (less than 100°C but higher than the boiling temperatures of VOCs) allows the separation of water from its contaminants that have lower boiling points. The VOCs will, however, continue to pose a health hazard in the atmosphere.[22,23]

6.7.8.3 Chemical Precipitation

Chemical precipitation is commonly used to remove heavy metal cations through pH adjustment. However, it is not appropriate to adjust the pH far from neutral for household drinking water applications. Alum (which has only a mild pH effect) is commonly used for the removal of colloids and ions from water.[22,23]

6.7.8.4 Flotation

Both dissolved air flotation (DAF) and electroflotation have been successfully applied to the removal of contaminants from surface water as well as groundwater.[62–64]

The contaminants that can be removed by flotation include conventional pollutants such as BOD, COD, total suspended solids (TSS), phosphorus, phenols, oil and grease, as well as toxic pollutants including heavy metals, toxic organics, pathogenic microorganisms, and radioactive radon.[22,28,33,54,64,100–102]

6.7.8.5 Ion Exchange

Ion exchange is effective for the removal of cationic or anionic heavy metal contaminants. It can also be used for water softening. Ion-exchange resins are usually regenerable with salt.[65]

6.7.8.6 Activated Carbon Adsorption

Granular activated carbon (GAC) is commonly used to remove contaminants from drinking water. It has the ability to remove contaminants to very low concentrations. Brunotts and colleagues[66] have studied 11 chemical spills and 18 groundwater contamination cases, which have shown that most contaminants were removed to less than 1 µg/L. GAC can be used effectively to remove both inorganic and organic contaminants.[22–24,38]

Water pH, temperature, hardness, and type and concentration of other solutes can influence GAC adsorption capacity. Certain types of pathogenic bacteria that are frequently colonized in carbon treatment units can also be released to the treated water. As in most conventional water treatment applications the pH, hardness, and pathogens are controlled, carbon adsorption can work effectively in conjunction with conventional water treatment technologies. Research conducted by Wang and colleagues[67] indicates that activated carbon adsorption is extremely effective for removing radioactive radon from contaminated groundwater. Although GAC is regenerable, the completely exhausted GAC should be replaced.

6.7.8.7 Membrane Methods

Membrane methods such as RO, MF, UF, and so on, are effective for removing certain sizes of molecules from contaminated water. However, energy is required for this removal technique.

6.7.8.8 Alternative Water Supply

Providing an alternative water supply, instead of treatment of contaminated well water, can be an alternative method of point-of-entry control. The cost of alternative water supply varies widely depending on site locations.[100–102]

6.7.9 NATURAL ATTENUATION

Natural attenuation relies on natural processes to clean up or attenuate pollution in soil and groundwater. Natural attenuation occurs at most polluted sites. However, the right conditions must exist underground to clean sites properly. If not, cleanup will not be quick enough or complete enough. Scientists monitor or test these conditions to make sure natural attenuation is working. This is called monitored natural attenuation (MNA).[94–96]

When the environment is polluted with chemicals, nature can work in four ways to clean it up:

1. Tiny bugs or microbes that live in the soil and groundwater use some chemicals for food. When they completely digest the chemicals, they can change them into water and harmless gases.
2. Chemicals can stick or sorb to soil, which holds them in place. This does not clean up the chemicals, but it can keep them from polluting groundwater and leaving the site.
3. As pollution moves through soil and groundwater, it can mix with clean water. This reduces or dilutes the pollution.
4. Some chemicals, such as oil and solvents, can evaporate, which means they change from liquids to gases within the soil. If these gases escape to the air at the surface, sunlight may destroy them.

MNA works best where the source of pollution has been removed. For instance, buried waste must be dug up and disposed of properly. Or it can be removed using other available cleanup methods. After the source is removed, the natural processes get rid of the small amount of pollution that remains in the soil and groundwater. The soil and groundwater are monitored regularly to make sure they are cleaned up.

The U.S. EPA publishes natural attenuation reports[94–96] that provide a general description on approaches to clean up contaminated waste sites. One U.S. EPA report lists five questions about each cleanup approach:

1. What is it?
2. How does it work?

3. Is it safe?
4. How long will it take?
5. Why use it?

Other U.S. EPA reports explain what MNA means when the term is used to describe a potential strategy to remediate a contaminated site. They also describe the various physical, chemical, and biological processes of natural attenuation that may occur at a site. Other informational materials are in preparation and will provide more specific details and scientific depth for the evaluation of MNA as a remedy at specific sites.

Surampalli, Ong, Seagren, and Nuno compiled and edited a book by the American Society of Civil Engineers (ASCE) called *Natural Attenuation of Hazardous Wastes.*[97] In addition to a discussion of the regulatory framework, this book covers major pollutants and basic scientific principles on physical, chemical, and biological processes involved in natural attenuation. It also contains an extensive review of literature, examples of applications of natural attenuation, and site characterization and monitoring requirements and procedures.

6.7.10 PHYTOREMEDIATION

Phytoremediation is a process that uses plants to remove, transfer, stabilize, or destroy contaminants in soil, sediment, and groundwater. The phytoremediation process may be applied *in situ* or *ex situ*, to soils, sludges, sediments, other solids, or groundwater.[98]

The mechanisms of phytoremediation include enhanced rhizosphere biodegradation (takes place in soil or groundwater immediately surrounding plant roots), phytoextraction (also known as phytoaccumulation, the uptake of contaminants by plant roots and the translocation/accumulation of contaminants into plant shoots and leaves), phytodegradation (metabolism of contaminants within plant tissues), and phytostabilization (production of chemical compounds by plants to immobilize contaminants at the interface of the roots and soil). Phytoremediation applies to all biological, chemical, and physical processes that are influenced by plants (including the rhizosphere) and that aid in the cleanup of the contaminated substances. Plants can be used in site remediation, both through the mineralization of toxic organic compounds and through the accumulation and concentration of heavy metals and other inorganic compounds from soil into aboveground shoots.

6.7.11 REMEDIATION OPTIMIZATION

Remediation optimization uses defined approaches to improve the effectiveness and efficiency with which an environmental remedy reaches its stated goals. Optimization approaches might include third-party site-wide optimization evaluations conducted by expert teams, the use of mathematical tools to determine optimal operating parameters or monitoring networks, or the consideration of emerging technologies. Since 1999, U.S. EPA has promoted remediation optimization in the following manner:

1. It has commissioned over 40 third-party optimization evaluations known as Remediation System Evaluations (RSEs) at Superfund, RCRA, and Leaking Underground Storage Tank sites.
2. It has applied or demonstrated new mathematical tools for optimizing pumping strategies and monitoring networks.
3. It has developed fact sheets and training seminars to educate the remediation community about optimization and to convey lessons learned from U.S. EPA optimization projects.
4. It has worked on outreach efforts with many State and Federal partners to disseminate information on new optimization approaches for streamlining long-term remedial action.

Optimization efforts conducted by other organizations can be found at the Federal Remediation Technologies Roundtable (FRTR) optimization website.

6.8   CASE STUDY

This case example illustrates how the results of individual and comparative analyses of remedial alternatives may be presented in a feasibility study report. The study uses a U.S. EPA example[12] that focuses on a detailed analysis of the alternatives that had been selected after screening.

6.8.1   SITE BACKGROUND

The site is an old battery and cleaning solution storage facility located in a rural area. Improper handling and storage activities at the site during a ten-year period, from 1968 to 1978, resulted in soil and groundwater contamination.

Figure 6.21 presents a site map, showing the extent and types of contamination. Area 1 contains 19,110 m³ (25,000 yd³) of contaminated soil with concentrations of lead exceeding 200 mg/kg (reaching 500 mg/kg at several locations). Area 2 outlines a discrete area of approximately 15,290 m³ (20,000 yd³) of TCE-contaminated soil. A plume having TCE concentrations over 5 μg/L, the maximum contaminant level (MCL) (at certain points measured as high as 50 mg/L) is estimated to be moving in the direction of residential wells at an interstitial velocity of 19.82 m/yr (65 ft/yr). The large ruled area indicates the approximate location of groundwater contaminated with concentrations above the MCL. Analysis of soil samples from this area shows TCE concentrations up to 6% and slightly elevated levels of metals compared to background concentrations. Although the risk

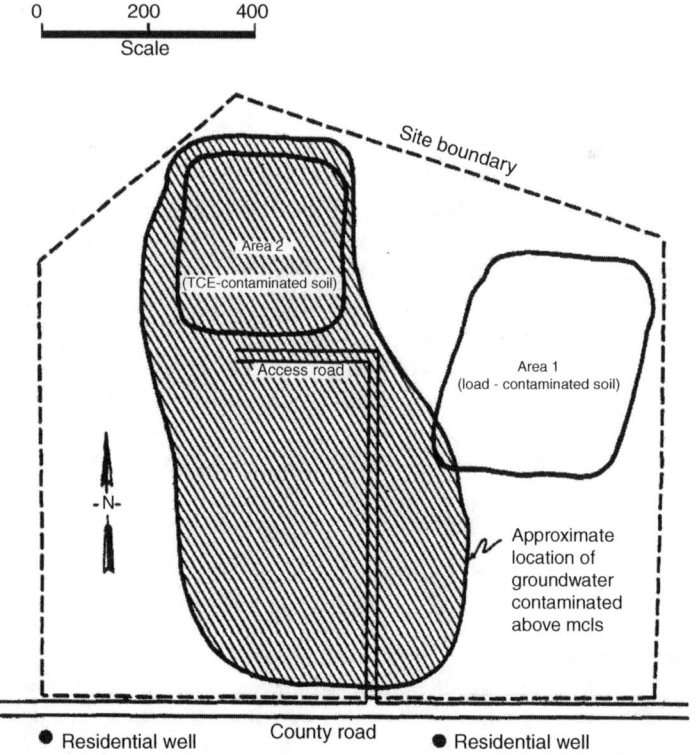

FIGURE 6.21   Site map: case example.

assessment did not identify a human health or environmental risk from these metals, there is a possibility that hot spots of metal contamination may have been missed. The soils of both Areas 1 and 2 are fairly permeable.

The affected aquifer that is used for drinking water is shallow. The water table lies approximately 3.66 m (12 ft) under the site. The aquifer consists of fractured bedrock, making groundwater containment difficult to implement. Groundwater extraction may also be difficult due to the fractured bedrock. The nearest residential well is 183 m (600 ft) from the site boundary, and the plume of contaminated groundwater is likely to reach the well in an estimated 1 to 3 yr at concentrations exceeding federal drinking water standards. Sampling conducted during the RI shows that no existing residential wells are currently contaminated.

The exposure pathways of concern identified during the baseline risk assessment include direct contact, with the possible ingestion of contaminated soil (1×10^{-3} associated excess cancer risk), and potential ingestion of contaminated groundwater in the future through existing or newly installed offsite wells (2×10^{-2} associated excess cancer risk).

The MCL for TCE (5 µg/L) has been determined to be a relevant and appropriate remediation level for the contaminated groundwater at this site because the groundwater is used as a source for drinking water. Based on the site-specific risk assessment, the MCL has been determined to be sufficiently protective as the aquifer remediation goal.

The risk assessment has also concluded that a level of 200 mg/kg for lead in the soil will be a protective level for expected site exposures along with an excess cancer risk level for TCE-contaminated soil (56 µg/L). Based on investigations of activities at the site, the TCE-contaminated soil has not been determined to be a listed RCRA hazardous waste, as the cleaning solution records indicate the solution contained less than 10% TCE. However, the lead-contaminated soil is an RCRA hazardous waste by characteristic in this instance due to extraction procedure (EP) toxicity. None of the waste is believed to have been disposed at the site after November 19, 1980 (the effective date for most of the RCRA treatment, storage, and disposal requirements).

6.8.2 THE LISTING OF ALTERNATIVES

Table 6.9 lists five remedial alternatives and their primary components. The nonaction alternative (Alternative 1) provides a baseline for comparison of other alternatives. Because no remedial activities will be implemented with the nonaction alternative, long-term human health and environmental risks for the site essentially will be the same as those identified in the baseline risk assessment. All other action alternatives with action have four common components:

1. *Fencing.* Fencing can be installed around the perimeter of a contaminated site to restrict public access. Signs warning of the presence and potential danger of hazardous materials can be posted on the fence to further discourage unauthorized access to the site.
2. *Institutional controls.* Many states are allowed by the current owner to place a deed restriction on the site that prohibits soil excavation and construction of buildings on any part of the site still containing hazardous materials upon completion of the remedy. In addition, a local groundwater well regulation requiring state review of all installation plans for groundwater wells can be used to prohibit the installation of drinking water supply wells in contaminated parts of the aquifer.
3. *Road reconstruction.* Some roads on the site can be restabilized and improved to allow construction activities and the movement of materials.
4. *Groundwater monitoring.* A selected number of new monitoring wells can be installed off site. Analytical results from new wells, some existing wells, and residential wells can be used to monitor future conditions and to assess the effectiveness of the final action. If the mean value of any compound at any facility boundary well is greater than the background concentration at the 0.05 significance level in two successive sampling rounds, appropriate investigative and remedial action(s) may be initiated as necessary.

TABLE 6.9
Alternative Component Case Example

	Alternative				
	1[a]	2[b]	3[c]	4[d]	5[e]
Groundwater					
Monitoring		●	●	●	●
Natural attenuation	N O A C T I O N	●			
Extraction wells			●	●	●
Onsite air stripping			●	●	●
Soil					
Soil/clay cap (Area 1)		●	●	●	●
Soil/clay cap (Area 2)		●			
Fixation (Area 1)				●	●
Soil vapor extraction (Area 2)			●	●	
Onsite incineration (Area 2)					●
Others					
Institutional controls		●	●	●	●
Road reconstruction		●	●	●	●
Fence		●	●	●	●

[a]Alternative 1—No action.
[b]Alternative 2—Cap and natural attenuation.
[c]Alternative 3—*In situ* soil vapor extraction, cap, and groundwater pump-and-treat.
[d]Alternative 4—*In situ* soil vapor extraction/soil fixation, cap, and groundwater pump-and-treat.
[e]Alternative 5—Incineration, *in situ* soil fixation, and groundwater pump-and-treat.

In fact, the nonaction alternative also requires groundwater monitoring and fencing. The following paragraphs describe the actions posed by the nonaction alternative and the four alternatives with actions, considering the site remediation case shown in Figure 6.21 and Table 6.9.

6.8.2.1 Alternative 1: No Action

The nonaction alternative (1) provides no control of exposure to the contaminated soil and no reduction in the risk to human health posed through the groundwater. It also allows for possible continued migration of the contaminated plume and further degradation of the groundwater.

6.8.2.2 Alternative 2: Cap and Natural Attenuation

The primary action in this alte rnative is capping of one or more contaminated areas (such as Areas 1 and 2 of Figure 6.21) and then natural attenuation of the contaminated groundwater. The cap would be consistent with the state RCRA (which is more stringent than the Federal requirement) landfill closure requirements.

A Geonet drainage layer can be chosen if the HELP model shows it to be more effective than sand in controlling leachate production and is comparable in cost. It is assumed that the HELP model predicts a 75 to 80% reduction in leachate production. A Geotextile layer would be laid on either side of the Geonet drain to prevent clogging. A minimum slope of 3% would be provided to meet state requirements.

Two assumptions about the surface have been made to determine the effect of natural attenuation on the contaminated groundwater. First, despite the fractured nature of the bedrock, it has been assumed that the subsurface is homogeneous so as to facilitate the evaluation. Second, the potential for reduction in TCE concentrations has been assessed using a hydrogeologic model in which the fact that the cap would reduce existing leachate production by 75% is taken into account. This model is assumed to predict that the concentration of TCE in the groundwater would be reduced to an excess cancer risk level of 28 μg/L in 60 yr and an excess cancer risk level of 5 μg/L, approximately equal to the MCL, in approximately 100 yr.

An alternative water supply would be included to provide a safe and reliable source of drinking water until the concentrations in the aquifer reach acceptable levels.

6.8.2.3   Alternative 3: *In Situ* SVE, Cap, and Groundwater Pump-and-Treat

This alternative consists of capping Area 1 (lead-contaminated soil), using *in situ* vapor extraction to treat the TCE-contaminated soil in Area 2, extracting the groundwater, and treating it on site through an air stripping system, and discharging it to the tributary of nearby receiving water.

It is demonstrated in the pilot tests that TCE can be removed by 99% for the direct contact exposure route within 3 to 5 yr using the vapor extraction system. The potential for fugitive losses of air contaminants would be minimal under good control conditions. A countercurrent packed tower air stripper (13.72 m tall and 1.22 m in diameter) would be used to treat the extracted groundwater to meet the performance goal of 5 μg/L TCE concentration. The exhaust air would be discharged through carbon beds for adsorption.

6.8.2.4   Alternative 4: *In Situ* SVE/Soil Fixation, Cap, and Groundwater Pump-and-Treat

For the site remediation case shown in Figure 6.21, this alternative consists of *in situ* SVE of TCE-contaminated soil (Area 2), *in situ* soil fixation of lead-contaminated soil (Area 1), cap (Area 1), and the groundwater pump-and-treat components of Alternative 3.

It is assumed that the moisture content of the soil has been determined to be approximately 50% under worst-case conditions. Using this information and the results from vendor tests, it has been determined that a minimum dose of one part solidification reagent to two parts soil is required for the migration control of lead. Testing has shown that the optimum solidification reagent mixture would comprise ca. 50% fly ash and ca. 50% kiln dust. Thus, ca. 7000 t (6364 T) each of fly ash and cement kiln dust would be required. The reagents would be added *in situ* with a backhoe. As one area of the soil is fixed, the equipment could be moved onto the fixed soil to blend the next section. It may be anticipated that the soil volume would expand by ca. 20% as a result of the fixation process. This additional volume would be used to achieve the required slope for the cap. An RCRA soil/clay cap placed over the solidified material is necessary to prevent infiltration and additional hydraulic stress on the fixed soil. It is estimated that the fixation would reduce lead migration by 40% and that the fixed soil may pass the U.S. EPA levels for lead.

6.8.2.5   Alternative 5: Incineration, *In Situ* Soil Fixation, and
Groundwater Pump-and-Treat

This alternative includes components of Alternatives 3 and 4 and introduces a thermal destruction component to address the TCE-contaminated soil. For the site remediation case shown in Figure 6.21, the lead-contaminated soil in Area 1 would be fixed and covered with a soil/clay cap, as described in Alternative 4. The groundwater would be addressed through pumping and treating, via an air stripper, as described in Alternatives 3 and 4. The TCE-contaminated soil in Area 2 would be excavated and treated on site by a thermal destruction unit comprisng a mobilized rotary kiln.

It is estimated that approximately 15,290 m³ (20,000 yd³) of contaminated soil would need to be excavated and treated. The incinerator would be operated continuously (24 h/d, 365 d/yr), although some downtime would be required (20%) for regular maintenance. It is assumed that the incinerator would be operated to achieve 99.8% TCE removal from the soil and destruction efficiency as required by RCRA. Specific operating practices would be enforced to meet performance objectives, including 99.99% destruction of stack emissions as dictated by subtitle O of RCRA.

The facility would use a dry scrubber system for emission control, which would eliminate the need for wastewater treatment. Any water from emission control and from decontamination procedures would be treated in the on-site groundwater treatment system. The residual soil and collected ash is assumed to be nonhazardous and can be disposed of in a solid waste disposal facility in compliance with subtitle D of RCRA. In the event that they cannot be delisted due to the presence of metals, the residuals will be managed as part of the closure of Area 2 shown in Figure 6.21 (lead-contaminated soil).

The groundwater model simulation indicates that the shallow aquifer could be restored to 5 μg/L (MCL) in 25 to 40 yr with soil remediation. Without soil remediation, between 60 and 100 yr would be required.

6.8.3 COMPARATIVE ANALYSIS OF ALTERNATIVES

The alternatives are evaluated in relation to one another for each of seven evaluation criteria. The purpose of this analysis is to identify the relative advantages and disadvantages of each alternative. Table 6.10 shows how these five alternatives comply with the seven major criteria:

1. Overall protection of human health and the environment
2. Compliance with ARARs
3. Long-term effectiveness and permanence
4. Reduction of toxicity, mobility, or volume through treatment
5. Short-term effectiveness
6. Implementability
7. Cost

6.8.3.1 Overall Protection of Human Health and the Environment

All of the alternatives, except Alternative 1, provide adequate protection of human health and the environment. Risk through direct contact and groundwater ingestion is reduced to cancer risk levels less that 1×10^{-6} through each pathway. Alternatives 3, 4, and 5 prevent further migration of the contaminated groundwater by extracting and treating the plume to health-based ARAR levels.

Alternative 2 achieves protection by preventing exposure through capping and natural attenuation of the contaminated groundwater. Alternative 3 combines treatment to reduce the risk from the TCE-contaminated soil and groundwater and capping of the lead area. Alternatives 4 and 5 reduce risks posed by all portions of the site through treatment.

There is some uncertainty about the potential presence of metal in the TCE-contaminated soil of Area 2. If metal concentrations of concern are present, only Alternatives 2 and 5 would protect against direct contact and further groundwater contamination through a cap and incineration, respectively. Incineration of metal-contaminated soil may result in a hazardous waste residue, which would have to be disposed of in a hazardous waste landfill. Alternatives 3 and 4 rely on vapor extraction and would not lower risks from metal to human health or the environment.

6.8.3.2 Compliance with Applicable, Relevant and Appropriate Requirements (ARARs)

The evaluation of the ability of the alternatives to comply with ARARs includes a review of chemical-specific and action-specific ARARs as listed in Table 6.10. All alternatives will meet all of their respective ARARs except the nonaction alternative.

TABLE 6.10

Individual Evaluation of Final Alternatives: Case Study

Evaluation	Alternative 1 (No Action)	Alternative 2 (Cap, Natural Attenuation)	Alternative 3 (*In Situ* Soil Vapor Extraction, Cap, Groundwater Pump-and-Treat)	Alternative 4 (*In Situ* Soil Vapor Extraction, *In Situ* Soil Fixation, Cap, Groundwater Pump-and-Treat)	Alternative 5 (*In Situ* Soil Fixation, Cap, Incineration, Groundwater Pump-and-Treat)
Overall protectiveness					
Human health protection					
Direct contact/soil ingestion	No significant reduction in risk; some reduction in access to risk through fence	Cap reduces direct contact risk and soil ingestion risk to less than 1×10^{-6}	Cap and vapor extraction reduce direct contact/soil ingestion risk to less than 1×10^{-6}	Cap, fixation, vapor extraction reduce direct contact/soil ingestion	Cap, fixation, incineration reduce direct contact/soil ingestion risk to less than 1×10^{-6}
Groundwater ingestion for existing users	No reduction in risk	Protect against existing risk by providing an alternative water supply	Reduces risk to less than 1×10^{-6} by pump-and-treat	See Alternative 3	See Alternative 3
Groundwater ingestion for future users	No reduction in risk	Institutional controls provide protection against risk from groundwater ingestion	Reduces risk to less than 1×10^{-6}	See Alternative 3	See Alternative 3
Environmental protection	Allows continued contamination of the groundwater	Continued contamination is curtailed by use of cap. Continued migration of contaminated groundwater is allowed	Continued contamination is curtailed by SVE and by cap; migration of contaminated groundwater is curtailed by pump-and-treat	Continued contamination is curtailed by SVE, soil fixation, and cap; migration of contaminated groundwater is curtailed by pump-and-treat	Continued contamination is curtailed by soil fixation and Incineration; migration of contaminated groundwater is curtailed by pump-and-treat
Compliance with ARARs					
Chemical-specific ARARS	Does not meet groundwater standards past the site boundary	Would meet MCLs at the waste boundary in over 50 years	Would meet MCLs at the waste boundary in 25–40 yr	See Alternative 3.	See Alternative 3
Location-specific ARARs	Not relevant; there are no location-specific ARARs	See Alternative 1	See Alternative 1	See Alternative 1	See Alternative 1
Action-specific ARARs	Would not meet any ARARs as there will be no action	Will meet RCRA landfill closure requirements	Would meet RCRA landfill closure requirements; would also meet air release standards from the vapor extraction system; would meet NPDES requirements	Would meet air release standards from air strippers and vapor extraction system; would meet NPDES requirements; would meet RCRA landfill closure requirements	Would meet regulations concerning incineration and air stripping; would meet NPDES requirements; would meet RCRA landfill closure requirements

Continued

Other criteria and guidance	Would allow ingestion of groundwater exceeding 1×10^{-4}; would not protect against Pb levels above 200 mg/kg in soil	Protects against soil ingestion 1×10^{-6} level and groundwater ingestion at 1×10^{-6} level; covers soil with Pb above 200 mg/kg	See Alternative 2	See Alternative 2	See Alternative 2
Long-term effectiveness and permanence					
Magnitude of residual risk					
Direct contact/soil ingestion	Source has not been removed; existing risk will remain	Risk is eliminated as long as cap is maintained; because source is only contained, inherent hazard of waste remains	Risk eliminated through vapor extraction and cap; some inherent hazard remains in the Pb material under the cap; risk from Pb would only occur if the cap were destroyed	Slight chance of future risk from fixed Pb-contaminated soil	See Alternative 4
Groundwater ingestion for existing users	Future risk greater as plume migrates to residents; eventually natural attenuation and dilution may decrease risk; risk significant for about 100 yr	Risk eliminated by providing alternative water supply; some risk would remain for over 100 yr if the groundwater is used	Risk eliminated by extracting groundwater exceeding 1×10^{-6} cancer risk levels; safe drinking water achieved in 25–40 yr with source control	See Alternative 3	See Alternative 3
Groundwater ingestion for future users	Risk greater as area of contamination increases; eventually natural attenuation and dilution may decrease risk; risk significant for about 100 yr	Institutional controls used to control use of contaminated groundwater; unauthorized use of groundwater would result in increased risk	Risk eliminated by extracting groundwater exceeding 1×10^{-6} cancer risk levels; safe drinking water achieved in 25–40 yr with source control	See Alternative 3	See Alternative 3
Adequacy and reliability of controls	No controls over remaining contamination; no reliability	Risk to groundwater controlled by alternative water supply and institutional controls; soil/clay cap controls contaminated soil; cap effective for Area 2 even if metals are present;	Soil/clay cap controls remaining contaminated soil in Area 1; would need additional controls for Area 2 if metals are present as SVE would not remove metals; groundwater extraction controls	See Alternative 3. Reliability of fixation with cap high, as are vapor extraction and groundwater pump-and-treat	Similar to Alternative 3; incinerator ash disposed in municipal landfill. If metals are present in Area 2, incinerator ash would be disposed in RCRA landfill. Incineration very reliable because material is

TABLE 6.10 (continued)

Evaluation	Alternative 1 (No Action)	Alternative 2 (Cap, Natural Attenuation)	Alternative 3 (*In Situ* Soil Vapor Extraction, Cap, Groundwater Pump-and-Treat)	Alternative 4 (*In Situ* Soil Vapor Extraction, *In Situ* Soil Fixation, Cap, Groundwater Pump-and-Treat)	Alternative 5 (*In Situ* Soil Fixation, Cap, Incineration, Groundwater Pump-and-Treat)
		institutional controls are limited in effectiveness; reliability of cap can be high if maintained; institutional controls to control use of groundwater not very reliable	contaminated groundwater; both are adequate; reliability of vapor extraction high because no long-term O&M is required; cap reliable if maintained; groundwater pump-and-treat is reliable		destroyed; fixation with cap and groundwater pump-and-treat are reliable
Need for 5 yr review	Review could be required to ensure that adequate protection of human health and the environment is maintained	See Alternative 1; TCE and Pb soil would remain on site	See Alternative 1; Pb-contaminated soil would remain on site	See Alternative 1; fixed Pb residuals would remain on site	See Alternative 1; fixed lead residuals would remain on site
Reduction of toxicity, mobility, or volume through treatment					
Treatment process used	None	None	Vapor extraction of soil and groundwater air stripping	Vapor extraction, soil fixation, and groundwater air stripping	Incineration, soil fixation, and groundwater air stripping
Amount destroyed or treated	None	None	9.99% of volatiles in soil and 96% volatiles in groundwater removed and destroyed by carbon regeneration	Same as Alternative 3 and 25,000 cy of contaminated soil is fixed	99.8% of volatiles in 20,000 cy of soil destroyed and 25,000 cy of contaminated soil is fixed
Reduction of toxicity, Mobility or volume	None	None	Reduced volume and toxicity of contaminated groundwater; toxicity of soil contamination reduced	Reduced volume and toxicity of contaminated groundwater; toxicity of soil contamination in Area 2 reduced by 97% mobility of contaminants in Area 1 reduced by 10% while volume increased by 20%	Incineration reduces volume of contaminated soil by 20,000 cy and reduces toxicity; mobility of contaminants in Area 1 is reduced; volume and toxicity of contaminated groundwater is reduced

Irreversible treatment	None	None	Vapor extraction and air stripping with irreversible regeneration of carbon used for air stream treatment	See Alternative 3	Incineration is irreversible; air stripping with subsequent gaseous carbon treatment and regeneration is irreversible
Type and quanity of residuals remaining after treatment	No residuals remain	None	No detectable residuals in Area 2 remain; carbon from gaseous treatment requires regeneration	No detectable residuals in Area 2 remain; 30,000 cy of fixed soils remain in Area 1	Incinerated soil (18,000 cy) and fixed soil (30,000 cy) remain; incinerated soil expected to nonhazardous; carbon from gaseous treatment remains requiring regeneration
Statutory preference for treatment	Does not satisfy	Does not satisfy	Satisfies	Satisfies	Satisfies
Community protection	Risk to community not increased by remedy implementation, but, contaminated water may reach the residents within 1–3 yr	Temporary increase in dust production through cap installation; contaminated soils remain undisturbed	Soil would remain uncovered during vapor extraction for 3–5 yr; Temporary increase in dust production during cap installation	Similar to Alternative 3; Fixation may result in dust and odor increase	Soil would remain uncovered during incineration (about 1 yr); excavation and fixation would release dust and odors to the atmosphere
Worker protection	No significant risk to workers	Protection required against dermal contact and inhalation of contaminated dust during cap construction	Protection required against dermal contact, vapor, or dust inhalation during construction and operation of vapor extraction system and air stripper	Protection required against dermal contact, vapor, or dust inhalation during construction and operation of vapor extraction system, fixation, and air stripper	Protection required against dermal contact and inhalation of volatiles and particulates as a result of excavation, fixing, and incinerating TCE soil
Environmental impacts	Continued impact from existing conditions	Would be some migration of contaminant pluma as part of attenuation process	Vapor extraction may affect air quality and odors although it will meet emission standards; would be aquifer drawdown during groundwater extraction	See Alternative 3; fixation may also affect air quality and produce odors	Incineration may affect air quality by producing odors, although if will meet emission standards
Time until action is not applicable	Complete	Cap installed in 6 months; risk from groundwater reduced within 3 months due to	SVE complete in 6 months; capping complete in 6 months; groundwater	Fixation and capping completed in 9 months; SVE complete in 3–5 yr;	Incineration complete in 2 yr from design completion; fixation and capping

Continued

TABLE 6.10   (continued)

Evaluation	Alternative 1 (No Action)	Alternative 2 (Cap, Natural Attenuation)	Alternative 3 (*In Situ* Soil Vapor Extraction, Cap, Groundwater Pump-and-Treat)	Alternative 4 (*In Situ* Soil Vapor Extraction, *In Situ* Soil Fixation, Cap, Groundwater Pump-and-Treat)	Alternative 5 (*In Situ* Soil Fixation, Cap, Incineration, Groundwater Pump-and-Treat)
		alternative water supply and institutional controls	remedial action complete in 25–40 yr	groundwater action complete in 25–40 yr	complete in 9 months; groundwater action complete in 25–40 yr
Implementability					
Ability to construct and operate	No construction or operation	Simple to operate and construct; would require materials handling of about 50,000 cy of soil and clay	Vapor extraction requires some operation; fairly straightforward to construct; cap construction would require materials handling of 25,000 cy of soil and clay; on-site groundwater treatment requires operation	Fixation with cap somewhat difficult to construct; otherwise similar to Alternative 3	Incineration is difficult to operate; fixation with cap is somewhat difficult to construct; similar to Alternative 3 with respect to groundwater
Ease of more action if needed	If monitoring indicates more action is necessary, may need to go through the FS/ROD process again	Simple to extend extraction system and cap; cap would be sufficient if metals were significant in Area 2; could implement groundwater treatment if necessary	Simple to extend groundwater extraction system, vapor extraction system, and cap; however, if significant metal concentration are present in Area 2, may need additional soil treatment or would need to extend cap	Fairly complete alternative; can increase volume of or modify all technologies; if significant metal concentrations are present in Area 2, could use fixation	Complete alternative; can handle varying volumes or concentrations

Ability to monitor effectiveness	No monitoring; failure to detect contamination means ingestion of contaminated groundwater	Proposed monitoring will give notice of failure before significant exposure occurs	See Alternative 2	See Alternative 2	See Alternative 2
Ability to obtain appropriate and coordinate with other agencies permit	No approval necessary	See Alternative 1	Need a NPDES permit. Should be easy to obtain	See Alternative 3	Need to demonstrate technical intent of incenerator permit. Need an NPDES permit
Availability of services and capacities	No services or capacities required	See Alternative 1	See Alternative 1	Need fixation services	Need fixation and incineration services
Availability of equipment, specialists, and materials	No services capacities required	No special equipment, material, or specialists required; cap materials available within 20 miles	Needs readily available specialists to install and monitor vapor extraction system; need treatment plant operators; cap materials available within 20 miles		Need a mobile incinerator and trained operators; need treatment plant operators; closest source of incinerator is 50 miles from site
Availability of technologies	None required	Cap technology readily available	Vapor extraction will need to be developed; require pilot testing	Vapor extraction and fixation well developed; will require pilot testing	Incineration and fixation well developed; will require pilot testing
Cost					
Capital cost ($)	0	$4,200,000	$3,300,000	$6,200,000	$13,000,000
First year annual O&M cost	0	$60,000	$440,000	$480,000	$1,200,000
Present worth cost	0	$4,800,000	$7,300,000	$10,200,000	$16,000,000

cy, cubic yard.

Source: U.S. EPA, Guidance for Conduction Remedial Investigations and Feasibility Studies under CERCLA, EPA/540/G-89/004, U.S. EPA, Washington, DC, October 1988.

6.8.3.3   Long-Term Effectiveness and Performance

Alternatives 4 and 5 afford the highest degrees of long-term effectiveness and permanence because both alternatives use treatment or fixation technologies to reduce the hazards posed by all known wastes at the site. Although some contaminated soil would remain after implementation of both alternatives, it would be fixed to reduce mobility. These two alternatives differ only in the technology used to treat the TCE-laden soil. Although incineration would destroy more TCE than SVE, both alternatives reduce risks posed by the waste to a 1×10^{-6} cancer risk level through both the groundwater and soil pathways.

Alternatives 4 and 5 would rely on a soil/clay cap to control infiltration for Area 1 (lead-contaminated) as well as treatment or fixation. Upon completion, some long-term maintenance of the cap and groundwater monitoring would be required until each alternative has met the health-based cleanup goals for groundwater. These alternatives would have almost no long-term reliance on institutional controls.

Alternative 3 eliminates the risk of exposure at the site to the same levels as Alternatives 4 and 5 in the short term; however, it relies solely on a cap for controlling the waste remaining in Area 1. Although capping is an effective and accepted approach for reducing risk from direct contact with wastes, it is less reliable in the long term than treatment, because the inherent hazard of the lead would remain.

Alternative 2 leaves all of the contaminated waste at the site and relies solely upon a cap and institutional controls to prevent exposure. Although the alternative water supply lowers the risk of ingesting contaminated groundwater from existing wells, the institutional controls would not be effective for more than 5 to 10 years in preventing the installation of new wells and the injection of contaminated groundwater.

Long-term groundwater monitoring and cap maintenance requirements are more critical for Alternative 2, because all of the waste remains at the site.

6.8.3.4   Reduction of Toxicity, Mobility, or Volume through Treatment

Alternatives 4 and 5 use treatment or fixation technologies to reduce the inherent hazards posed by all known waste at the site, posing more than a 1×10^{-6} excess cancer risk level by ingestion. However, neither alternative completely treats all of the soil at the site. Both alternatives produce 22,937 m^3 (30,000 yd^3) of fixed soil, and 13,762 to 15,291 m^3 (18,000 to 20,000 yd^3) of treated soil. Under Alternative 5, there would remain 13,762 m^3 (18,000 yd^3) of soil (with 99.8% TCE removal). Under Alternative 4, there would remain 15,291 m^3 (20,000 yd^3) of soil (with 99.9% TCE removal). These two alternatives would satisfy the statutory preference for treatment as a principal element.

Alternative 3 also treats soil and groundwater for TEC. However, ca. 19,114 m^3 (25,000 yd^3) of lead-contaminated soil would remain untreated on site, although the lead mobility would be very low.

Alternative 2 uses no treatment technologies. All contaminated soil and groundwater would remain; however, contaminates will in time attenuate naturally.

6.8.3.5   Short-Term Effectiveness

Alternative 2 is anticipated to have the greatest short-term effectiveness, and presents the least amount of risk to workers, the community, and the environment. The other alternatives could release volatiles during excavation activities or SVE.

The time required to achieve short-term protection would be shorter than for any other alternative. It is anticipated that only 6 months would be required to install a new cap and to provide an alternative water supply. Alternatives 3 and 4, involving vapor extraction, require 3 to 5 yr before the risk from direct soil contact and ingestion is controlled.

Alternative 4 would take longer to implement than Alternative 2 and has a greater potential of releasing volatiles to the atmosphere during excavation than Alternatives 3 and 4. However, implementation of Alternative 5 would take less time than Alternatives 3 and 4 because incineration would require less time than SVE to remediate the soil to safe levels. Alternative 5 has the disadvantage of requiring incineration equipment, which could increase the risk to workers in the event of a failure.

6.8.3.6 Implementability

Alternative 2 is the simplest system to construct and operate. Alternative 3 is fairly simple with regard to construction requirements but has more operational requirements than Alternatives 1 and 2 because of the adoption of the SVE system and the air stripper. Alternative 4 is more complex than Alternative 3 because of the inclusion of *in situ* soil fixation components.

Alternative 5 is the most complex alternative to construct and, during implementation, to operate. During operation of the incinerator, this alternative would require the most attention because incinerators require periodic sampling of the residue and modification of operating parameters. It is expected that the incinerator would operate for slightly more than a year, whereas the SVE system of Alternative 4 would operate for 3 to 5 yr.

6.8.3.7 Cost Analysis

Alternative 2 has a lower present worth and O&M cost than Alternative 3, but because of the additional cap required it has a higher capital cost (USD 11,200,000 versus USD 8,000,000). The cap is one of the most expensive components to construct. Alternative 4 has a higher capital, O&M, and present worth cost than Alternatives 2 and 3. Alternative 5 has the highest capital (USD 34,600,000), first-year O&M (USD 3,200,000), and present worth cost (USD 42,600,000) of all of the alternatives because of the incinerator component. All costs have been updated in terms of 2007 USD.[68]

6.9 REMEDIATION, DECONTAMINATION, AND SAFETY MANAGEMENT

6.9.1 SITE PREPARATION AND WORK ZONES

Several site control procedures can be implemented to reduce worker and public exposure to chemical, physical, biological, and safety hazards[69,70,102–105]:

1. Compiling a site map, showing topographic features, prevailing wind direction, drainage, and the location of buildings, containers, impoundments, pits, ponds, and tanks
2. Preparing the site for subsequent activities (see Table 6.11)
3. Establishing work zones
4. Using the buddy system when necessary
5. Establishing and strictly enforcing decontamination
6. Establishing site security measures
7. Setting up communication networks
8. Enforcing safe work practices

Time and effort must be spent in preparing a site for the cleanup activity to ensure that response operations go smoothly and that worker safety is protected. Site preparation can be as hazardous as site cleanup. Therefore, safety measures should be afforded the same level of care at this stage as during actual cleanup. Table 6.11 presents the major steps in site preparation prior to any cleanup activities.

To reduce the accidental spread of hazardous substances by workers from a contaminated area to a clean area, zones should be delineated on the site where different types of operations will occur, and the flow of personnel among the zones should be controlled. The establishment of work zones will help ensure that personnel are properly protected against the hazards present where they are working, that work activities and contamination are confined to the appropriate areas, and that personnel can be located and evacuated in an emergency.

TABLE 6.11

Site Preparation

Construct roadways to provide ease of access and a sound roadbed for heavy equipment and vehicles

Arrange traffic flow patterns to ensure safe and efficient operations

Eliminate physical hazards from the work area as much as possible, including:

    Ignition sources in flammable hazard areas

    Exposed or unground electrical wiring, and low overhead wiring that may entangle equipment

    Sharp or protruding edges, such as glass, nails, and torn metal, which can puncture protective clothing and
      equipment and inflict puncture wounds

    Debris, holes, loose steps or flooring, protruding objects, slippery surfaces, or unsecured railings, which can
      cause falls, slips, and trips

    Unsecured objects, such as bricks and gas cylinders, near the edges of elevated surfaces, such as catwalks, roof tops,
      and scaffolding, which may dislodge and fall on workers

    Debris and weeds that obstruct visibility

Install skid-resistant strips and other antiskid devices on slippery surfaces

Construct operation pads for mobile facilities and temporary structures

Construct loading docks, processing and staging areas, and decontamination pads

Provide adequate illumination for work activities. Equip temporary lights with guards to prevent accidental contact

Install all wiring and electrical equipment in accordance with the National Electric Code

Source: U.S. GPO, Occupational Safety and Health Guidance Manual for Hazardous Waste Site Activities, DHHS-NIOSH-
      85-115, U.S. Government Printing Office, Washington, DC, October, 1985. With permission.

Hazardous waste sites should be divided into as many different zones as needed to meet operational and safety objectives. For illustration, the following are three frequently used zones:

1. *Exclusion zone.* This is the contaminated area.
2. *Contamination reduction zone (CRZ).* This is the area where decontamination takes place.
3. *Support zone.* This is the uncontaminated area where workers should not be exposed to hazardous conditions (Table 6.12).

Delineation of these three zones should be based on sampling and monitoring results and on an evaluation of the potential routes and amount of contaminant dispersion in the event of a release. Movement of personnel and equipment among these zones should be minimized and restricted to specific access control points to prevent cross-contamination from contaminated areas to clean areas. A decision for evaluating health and safety aspects of decontamination methods is presented in Figure 6.22.[105]

To establish the hot lines, an environmental engineer will do the following:

1. Visually survey the immediate site
2. Determine the locations of (a) hazardous substances, (b) drainage, leachate, and spilled material, and (c) visible discolorations
3. Evaluate data from the initial site survey indicating the presence of (a) combustible gases, (b) organic and inorganic gases, particulates, or vapors, and (c) ionizing radiation
4. Evaluate the results of soil and water sampling
5. Consider the distances needed to prevent an explosion or fire from affecting personnel outside the exclusion zone
6. Consider the distances the personnel must travel to and from the exclusion zone
7. Consider the physical area necessary for site operation

8. Consider meteorological conditions and the potential for contaminants to be blown from the area
9. Secure or mark the hotline
10. Modify its location, if necessary, as more information becomes available

The support zone activities are briefly presented in Table 6.12.

TABLE 6.12
Support Zone Activities

Facility	Function
Command post	Supervision of all field operations and field teams
	Maintenance of communications, including emergency lines of communication
	Recordkeeping, including:
	– Accident reports
	– Chain-of-custody records
	– Daily logbooks
	– Manifest directories and orders
	– Personnel training records
	– Site inventories
	– Site safety map
	– Up-to-date site safety plans
	Providing access to up-to-date safety and health manuals and other reference materials
	Interfacing with the public: government agencies, local politicians, medical personnel, the media, and other interested parties
	Monitoring work schedules and weather changes
	Maintaining site security
	Sanitary facilities
Medical station	First-aid administration
	Medical emergency response
	Medical monitoring activities
	Sanitary facilities
Equipment and supply centers	Supply, maintenance, and repair of communications, respiratory, and sampling equipment
	Maintenance and repair of vehicles
	Replacement of expendable supplies
	Storage of monitoring equipment and supplies—storage may be here or in an on-site field laboratory
Administration	Sample shipment
	Interface with home office
	Maintenance of emergency telephone numbers, evacuation route maps, and vehicle keys
	Coordination with transporters, disposal sites, and appropriate federal, state, and local regulatory agencies
Field laboratory	Coordination and processing of environmental and hazardous waste samples; copies of the sampling plans and procedures should be available for quick reference in the laboratory
	Packaging of materials for analysis following the decontamination of the outsides of the sample containers, which should be done in the CRZ
	This packaging can also be done in a designated location in the CRZ
	Shipping papers and chain-of-custody files should be kept in the command post
	Maintenance and storage of laboratory notebooks in designated locations in the laboratory while in use, and in the command post when not in use

Source: U.S. EPA, Remedial Action at Waste Disposal Sites, EPA/625/6-85/006, U.S. EPA, Washington, DC, 1985.

6.9.2   HEALTH AND SAFETY HAZARDS

Although decontamination is performed to protect health and safety, it can pose hazards under certain circumstances. Decontamination methods may have the following characteristics:

1. They may be incompatible with the hazardous substances being removed (i.e., a decontamination method may react with contaminants to produce an explosion, heat, or toxic products)
2. They may be incompatible with the clothing or equipment being decontaminated (e.g., some organic solvents can permeate or degrade protective clothing)
3. They may pose a direct health hazard to workers (e.g., vapors from chemical decontamination solutions may be hazardous if inhaled or they may be flammable)

The chemical and physical compatibility of decontamination solutions or other decontamination materials must be determined before use. Any decontamination method that permeates, degrades, damages, or otherwise impairs the functioning of the personal protective equipment (PPE) is incompatible with such PPE and should not be used. If a decontamination method does pose a direct health hazard, measures must be taken to protect both decontamination personnel and the workers being decontaminated. Figure 6.22 presents a decision aid for the evaluation of health and safety aspects of decontamination methods.

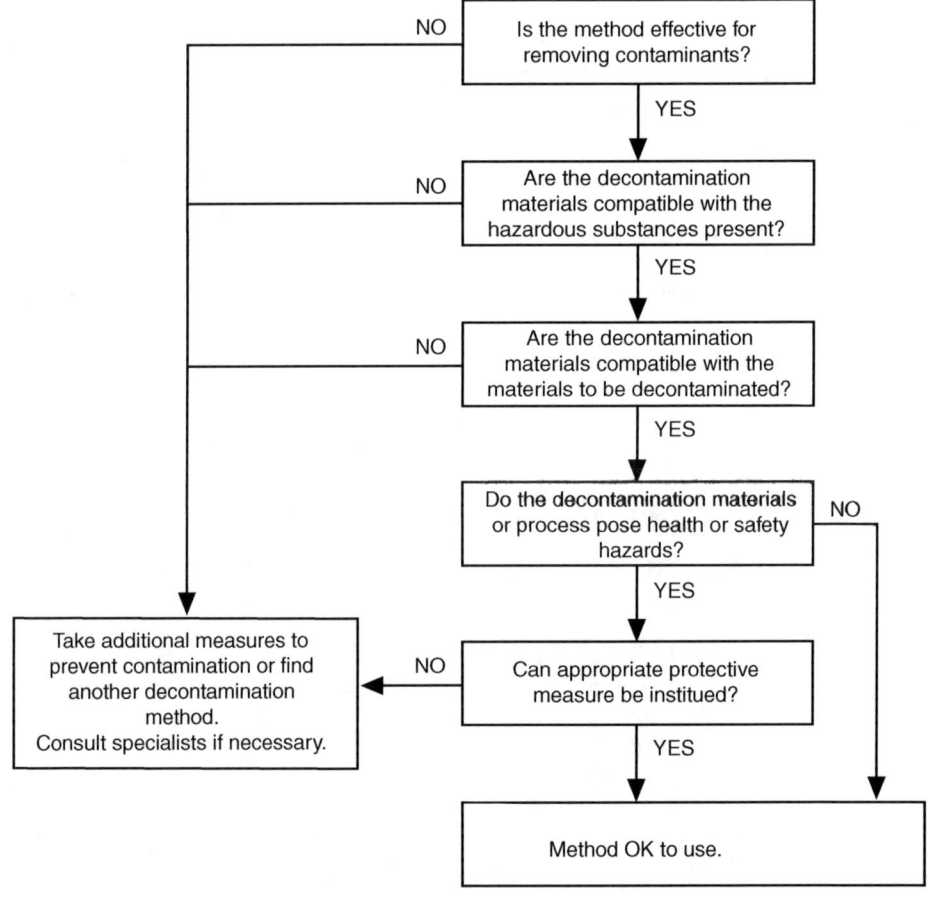

FIGURE 6.22   Decision aid for evaluating health and safety aspects of decontamination.

6.9.3 REMEDIATION/DECONTAMINATION FACILITY DESIGN

At a hazardous waste site, remediation and decontamination facilities should be located in the CRZ, that is, the area between the exclusion zone (the contaminated area) and the support zone (the clean area). The level and types of remediation and decontamination procedures required depend on several site-specific factors:

1. The chemical, physical, and toxicological properties of the wastes
2. The pathogenicity of infectious wastes
3. The amount, location, and containment of contaminants
4. The potential for, and location of, exposure based on assigned worker duties, activities, and functions
5. The potential for wastes to permeate, degrade, or penetrate materials used for personal protective clothing and equipment, vehicles, tools, buildings, and structures
6. The proximity of incompatible wastes
7. The movement of personnel or equipment among different zones
8. Emergencies
9. The methods available for protecting workers during decontamination
10. The impact of the decontamination process and compounds on worker safety and health

Decontamination procedures must provide an organized process by which levels of contamination are reduced. The decontamination process should consist of a series of procedures performed in a specific sequence. Each procedure should be performed at a separate station in order to prevent cross-contamination. The sequence of stations is called the decontamination line.

Stations should be separated physically to prevent cross-contamination and should be arranged in order of decreasing contamination, preferably in a straight line. Separate flow patterns and stations should be provided to isolate workers from different contamination zones containing incompatible wastes. Entry and exit points should be conspicuously marked, and the entry to the CRZ from the exclusion zone should be separate from the entry to the exclusion zone from the CRZ. Dressing stations for entry to the CRZ should be separate from redressing areas for exit from the CRZ. Personnel who wish to enter clean areas of the decontamination facility, such as locker rooms, should be completely decontaminated.

NOMENCLATURE

ARARs	Applicable, relevant, and appropriate requirements
ATSDR	Agency for Toxic Substances and Disease Registry
BOD	Biochemical oxygen demand
BTEX	Benzene, toluene, ethylbenzene, and xylenes (collectively)
CERCLA	Comprehensive Environmental Response, Compensation, and Liability Act
COD	Chemical oxygen demand
CRP	Community relations plan
CRZ	Contamination reduction zone
CW	Circulating well
DAF	Dissolved air flotation
DAF-SBR	Dissolved air flotation sequencing batch reactor
ED	Electrodialysis
EO	Electroosmosis
FS	Feasibility study
FSP	Field sampling plan
GAC	Granular activated carbon

GEMS	Genetically engineered microbial system
HSP	Health and safety plan
HSWA	Hazardous and Solid Waste Amendments
MF	Microfiltration
NAPL	Nonaqueous phase liquids
NCP	National Contingency Plan
NPL	National Priorities List
PAC	Powder activated carbon
PCB	Polychlorinated biphenyls
PCE	Pentachloroethylene
PCF-SBR	Physical-chemical flotation sequencing batch reactor
PCS-SBR	Physical-chemical sedimentation sequencing batch reactor
PPE	Personal protective equipment
PRB	Permeable reactive barrier
PRP	Potential responsible party
QA	Quality assurance
QAPP	Quality assurance project plan
RBC	Rotating biological contactor
RCRA	Resource Conservation and Recovery Act
Redox	Reduction/oxidation
RF	Radio frequency
RFH	Radio frequency heating
RI	Remedial investigation
RO	Reverse osmosis
SAP	Sampling and analysis plan
SARA	Superfund Amendments and Reauthorization Act
SBR	Sequencing batch reactor
SPSH	Six-phase soil heating
SVE	Soil vapor extraction
SVOC	Semi-volatile organic compounds
SWANA	Solid Waste Association of North America
TBC	To be considered
TCE	Tetrachloroethylene
TOC	Total organic carbon
UF	Ultrafiltration
USACE	United States Army Corps of Engineers
U.S. EPA	**United States Environmental Protection Agency**
U.S. GPO	United States Government Printing Office
USGS	United States Geological Services
VICs	Volatile inorganic compounds
VOCs	Volatile organic compounds

APPENDIX

U.S. Army Corps of Engineers Civil Works Construction Yearly Average Cost Index for Utilities

Year	Index	Year	Index
1967	100	1988	369.45
1968	104.83	1989	383.14
1969	112.17	1990	386.75
1970	119.75	1991	392.35
1971	131.73	1992	399.07
1972	141.94	1993	410.63
1973	149.36	1994	424.91
1974	170.45	1995	439.72
1975	190.49	1996	445.58
1976	202.61	1997	454.99
1977	215.84	1998	459.40
1978	235.78	1999	460.16
1979	257.20	2000	468.05
1980	277.60	2001	472.18
1981	302.25	2002	484.41
1982	320.13	2003	495.72
1983	330.82	2004	506.13
1984	341.06	2005	516.75
1985	346.12	2006	528.12
1986	347.33	2007	539.74
1987	353.35		

Source: U.S. ACE, Civil Works Construction Cost Index System Manual—Index for Utilities, 110-2-1304, U.S. Army Corps of Engineers, Washington, 2007, p. 44. With permission.

REFERENCES

1. Federal Register, Comprehensive Environmental Response, Compensation, and Liability Act (CERCLA or Superfund), 42 U.S.C. s/s 9601 et seq. (1980), U.S. Government Printing Office, Washington, 2007. Available at www.access.gpo.gov/uscode/title42/chapter103.html.
2. U.S. EPA, Superfund — Cleaning up the Nation's Hazardous Waste Sites, U.S. EPA, Washington, May 2006. Available at www.epa.gov/superfund.
3. U.S. EPA, Superfund, CERCLA Overview, U.S. EPA, Washington, 2007. Available at www.epa.gov/superfund/action/law/cercla.htm.
4. Federal Register, Superfund Amendments and Reauthorization Act (SARA), 42 U.S.C. 9601 et seq. (1986), U.S. Government Printing Office, Washington, 2004. Available at www.access.gpo.gov/uscode/title42/chapter103_html.
5. Federal Register, Clean Air Act (CAA), 42 U.S.C. s/s 7401 et seq. (1970), U.S. Government Printing Office, Washington, 2007. Available at www.epa.gov/airprogm/oar/caa/index.html.
6. Federal Register, Clean Water Act (CWA), 33 U.S.C. ss/1251 et seq. (1977), U.S. Government Printing Office, Washington, 2002. Available at www.access.gpo.gov/uscode/title33/chapter26_.html.
7. Federal Register, Safe Drinking Water Act (SDWA), 42 U.S.C. s/s 300f et seq. (1974), U.S. Government Printing Office, Washington, 2002. Available at http://frwebgate.access.gpo.gov/cgi-bin/getdoc.cgi?dbname=browse_usc&docid=Cite:+42USC300f.

8. Federal Register, Resource Conservation and Recovery Act (RCRA), 42 US Code s/s 6901 et seq., 1976, U.S. Government Printing Office, Washington, 2004. Available at www.access.gpo.gov/uscode/title42/chapter82_.html.

9. U.S. EPA, Resource Conservation and Recovery Act (RCRA)—Orientation Manual, Report EPA 530-R-02-016, U.S. EPA, Washington, 2003.

10. Hall Jr, R.M. and Bryson, N.S., Comprehensive Environmental Response, Compensation, and Liability Act (Superfund), in *Environmental Law Handbook*, 8th ed., Arbuckle, G.J., Ed., Government Institutes, Rockville, MD, 1985.

11. U.S. EPA, Handbook on *In situ* Treatment of Hazardous Waste-Contaminated Soils, EPA/540/2-90/002, U.S. EPA, Washington, 1990.

12. U.S. EPA, Guidance for Conducting Remedial Investigations and Feasibility Studies under CERCLA, EPA/540/G-89/004, U.S. EPA, Washington, October 1988.

13. U.S. EPA, Model Statement of Work for RI/PS under CERCLA Guidance, U.S. EPA, Washington, DC, February 1985.

14. O'Brien & Gere Engineerings, The engineering perspective, in *Hazardous Waste Site Remediation*, Bellandi, R., Ed., Van Nostrand Reinhold, New York, 1988, p. 422.

15. U.S. EPA, Remedial Action at Waste Disposal Sites, EPA/625/6-85/006, U.S. EPA, Washington, DC, 1985.

16. Vogel, G.A., Air emission control at hazardous waste management facilities, *J. Air Pollut. Control Assoc.*, 35, 550–566, 1985.

17. U.S. EPA, Superfund Innovative Technology Evaluation Program, 4th ed., EPA/540/5-91/008, U.S. EPA, Cincinnati, OH, 1991, p. 293.

18. Matrecon, *Lining of Waste Impoundments and Disposal Facilities*, Matrecon, Atlantic Ave, Alameda, CA, 1983.

19. Virginia SWCC, *Virginia Erosion and Sediment Control Handbook: Standards, Criteria and Guidelines*, Virginia Soil and Water Conservation Commission, Richmond, VA, 1988.

20. Wang, L.K., Pereira, N., Hung, Y.T., and Shammas, N.K., Eds., *Biological Treatment Processes*, The Humana Press, Totowa, NJ, 2008.

21. Wang, L.K., Shammas, N.K., and Hung, Y.T., Eds., *Advanced Biological Treatment Processes*, The Humana Press, Totowa, NJ, 2008.

22. Wang, L.K., Hung, Y.T., and Shammas, N.K., Eds., *Physicochemical Treatment Processes*, Humana Press, Totowa, NJ, 2005, pp. 526–671.

23. Wang, L.K., Hung, Y.T., and Shammas, N.K., Eds., *Advanced Physicochemical Treatment Processes*, Humana Press, Totowa, NJ, 2006, pp. 203–260.

24. Wang, L.K., Hung, Y.T., and Shammas, N.K., Eds., *Advanced Physicochemical Treatment Technologies*, Humana Press, Totowa, NJ, 2007, pp. 295–390.

25. Shammas, N.K., Kumar, I.J., Chang, S.Y., and Hung, Y.T., Sedimentation, in *Physicochemical Treatment Processes*, Wang, L.K., Hung, Y.T., and Shammas, N.K., Eds., The Humana Press, Totowa, NJ, 2005, pp. 379–429.

26. Wang, L.K., Vaccari, D.A., Li, Y., and Shammas, N.K., Chemical precipitation, in *Physicochemical Treatment Processes*, Wang, L.K., Hung, Y.T., and Shammas, N.K., Eds., The Humana Press, Totowa, NJ, 2005, pp. 141–197.

27. Shammas, N.K., Coagulation and flocculation, in *Physicochemical Treatment Processes*, Wang, L.K., Hung, Y.T., and Shammas, N.K., Eds., The Humana Press, Totowa, NJ, 2005, pp. 103–138.

28. Wang, L.K. and Wang, M.H.S., Decontamination of groundwater and hazardous industrial effluents by high-rate air flotation process, *Proc. Great Lakes 1990 Conference*, Hazardous Materials Control Research Institute, Silver Springs, MD, September 1990.

29. Wang, L.K., Fahey, E.M., and Wu, Z., Dissolved air flotation. in *Physicochemical Treatment Processes*, Wang, L.K., Hung, Y.T., and Shammas, N.K., Eds., Humana Press, Totowa, NJ, 2005, pp. 431–493.

30. Wang, L.K., Design and specification of Pittsfield Water Treatment System consisting of air flotation and sand filtration, *Water Treatment*, 6, 127–146, 1991.

31. DeRenzo, D., *Unit Operations for Treatment of Hazardous Industrial Wastes*, Noyes Data Corporation, Park Ridge, NJ, 1978.

32. Wang, L.K., Hrycyk, O., and Kurylko, L., Removal of Volatile Compounds and Surfactants from Liquid, U.S. Patent 5,122,165, June l6, 1992.

33. Wang, L.K., Wang, M.H.S., and Hoagland, F.M., Reduction of color, odor, humic acid and toxic substances by adsorption, flotation and filtration, *Water Treatment*, 7, 1–16, 1992.

34. Wang, L.K., Pereira, N.C., and Hung, Y.T., Eds., *Air Pollution Control Engineering*, Humana Press, Totowa, NJ, 2004, pp. 471–475.

35. Wang, L.K., Hung, Y.T., Lo, H.H., and Yapijakis, C., Eds., *Handbook of Industrial and Hazardous Wastes Treatment*, CRC Press, New York, 2004, pp. 515–618, 923–970, 945–946.

36. Wang, L.K., Hung, Y.T., Lo, H.H., and Yapijakis, C., Eds., *Hazardous Industrial Waste Treatment*, CRC Press, New York, 2007, pp. 241–288, 468–487.

37. Wang, L.K., Hung, Y.T., Lo, H.H., and Yapijakis, C., Eds., *Waste Treatment in the Process Industries*, CRC Press, New York, 2006, pp. 470–473.

38. Hung, Y.T., Lo, H.H., Wang, L.K., Taricska, J.R., and Li, K.H., Granular activated carbon adsorption, in *Physicochemical Treatment Processes*, Wang, L.K., Hung, Y.T., and Shammas, N.K., Eds., Humana Press, Totowa, NJ, 2005, pp. 573–630.

39. Scovazzo, P., Chen, W.Y., Wang, L.K., and Shammas, N.K., Solvent extraction, leaching, and supercritical extraction, in *Advanced Physicochemical Treatment Processes*, Wang, L.K., Hung, Y.T., and Shammas, N.K., Eds., Humana Press, Totowa, NJ, 2006, pp. 379–429.

40. Kiang, Y. and Metry, A.R., *Hazardous Waste Processing Technology*, Ann Arbor Science Publishers, Ann Arbor, MI, 1982.

41. Hicks, B.N. and Caplan, J.A., Bioremediation: a natural solution, *Pollut. Eng.*, 25, 2, 1993.

42. Ehrlich, H.L., Biological treatment of chromium, *Annual Hazardous Waste Management Conference*, Lake George, Rensselaer Polytechnic Institute, Troy, NY, 1991.

43. Wang, L.K., Kurylko, L., and Wang, M.H.S., Gas Dissolving and Releasing Liquid Treatment System, U.S. Patent 5,167,806, 1992.

44. Hrycyk, O., Kurylko, L., and Wang, L.K., *Removal* of Volatile Compounds and Surfactants from Liquid, U.S. Patent 5,122,166, June 16, 1992.

45. Dev, H., Bridges, J.E., and Sresty, G.C., Decontamination of hazardous waste substances from spills and uncontrolled waste sites by radio frequency *in situ* heating, in *Hazardous Material Spills Conference Proceedings*, Government Institutes, Rockville, MD, 1984.

46. Smith, S., Rock fracturing methods: their development and use, *Water Well J.*, February, 41–47, 1989.

47. Huibregtse, K.R. and Kastman, K.H., *Development of a System to Protect Groundwater Treatment by Hazardous Spills on Land*, U.S EPA, Edison, NJ, 1979.

48. Williams, E.G., Contaminant containment by *in situ* polymerization, in *Proc. Second National Symposium on Aquifer Restoration and Ground Water Monitoring*, National Water Well Association, Worthington, OH, 1982.

49. Stief, K., Remedial action for groundwater protection case studies within the Federal Republic of Germany, in *Hazardous Material Spills Conference Proceedings*, Government Institutes, Rockville, MD, 1984.

50. Rogoshewski, P.J. and Carstea, D.D., An Evaluation of Lime Precipitation as a Means of Treating Boiler Tube Cleaning Wastes, EPA/600/7-80/052, U.S EPA, Washington, 1980.

51. Gibson, M.J. and Farmer, J.G., Multi-step sequential chemical extraction of heavy metals from urban soils, *Environ. Poll. (Ser. B)*, 11, 117, 1986.

52. Linn, J.H. and Elliott, H.A., Mobilization of Cu and Zn in contaminated soil by nitrotriacetic acid, *Water Air Soil Poll.*, 37, 449, 1989.

53. U.S. EPA, Soil Washing Treatment, EPA/540/2-90/017, U.S. EPA, Washington, 1990.

54. U.S. EPA, *In Situ* Steam Extraction Treatment, EPA/540/2-91/005 U.S. EPA, Washington, DC, 1991.

55. Hamed, J., Acar, Y.B., and Gale, R.J., Pb(II) removal from kaolinite by electrokinetics, *J. Geotech. Eng.*, 117, 241–370, 1991.

56. U.S. EPA, Engineering Bulletin: Solvent Extraction Treatment, Hazardous Waste Engineering Laboratory, EPA/540/2-90/013, U.S. EPA, Washington, 1990.

57. Wang, L.K., Shammas, N.K., Selke, W.A., and Aulenbach, D.B., Eds., *Flotation Technology*, Humana Press, Totowa, NJ, 2009.

58. Wang, L.K., Shammas, N.K., Selke, W.A., and Aulenbach, D.B., Eds., *Flotation Engineering*, Humana Press, Totowa, NJ, 2009.

59. U.S. EPA, Requirements for Hazardous Waste Landfill Design, Construction, and Closure, EPA/625/4-89/022, U.S. EPA, Washington, 1989, p. 127.

60. Wang, L.K. Dredging operation and waste disposal, in *Water Resources and Natural Control Processes*, 1st ed., Wang, L.K. and Pereira, N.C., Eds., Humana Press, Totowa, NJ, 1986, pp. 447–492.

61. U.S. EPA, Cleanups of Releases from Petroleum USTs: Selected Technologies, EPA/530/UST-88/001, U.S. EPA, Washington, 1988.

62. Krofta, M. and Wang, L.K., Development of innovative flotation-filtration systems for water treatment, part A: first full-scale Sandfloat Plant in U.S., *Proc. American Water Works Association Water Reuse Symposium III*, San Diego, CA, 3, 1226–1237, l984.

63. Krofta, M. and Wang, L.K., Development of innovative flotation-filtration systems for water treatment, part B: dissolved-air flotation plants for small communities, *Proc. American Water Works Association Water Reuse Symposium III*, San Diego, CA, 3, 1238–1250, 1984.

64. Krofta, M. and Wang, L. K., Development of innovative flotation-filtration systems for water treatment, part: C electroflotation plant for single families and institutions, *Proc. American Water Works Association Water Reuse Symposium III*, San Diego, CA, 3, 1251–1264, 1984.

65. Chen, J.P., Yang, L., Ng, W.J., Wang, L.K., and Thong, S.L., Ion exchange, in *Advanced Physicochemical Treatment Processes*, Wang, L.K., Hung, Y.T., and Shammas, N.K., Eds., Humana Press, Totowa, NJ, 2006, pp. 261–290.

66. Brunotts, V.A., Cost-effective treatment of priority pollutant compounds with granular activated carbon, *National Conference on Management of Uncontrolled Hazardous Waste Sites*, Washington, 1983.

67. Wang, L.K., Kurylko, L., and Wang, M.H.S., *Method and Apparatus for Filtration with Plural Ultraviolet Treatment Stages*, U.S. Patent 5,236,595, 1993.

68. U.S. ACE, Civil Works Construction Cost Index System Manual—Index for Utilities, 110-2-1304, U.S. Army Corps of Engineers, Washington, 2007, p. 44. Available at http://www.nww.usace.army.mil/cost.

69. U.S. EPA, Field Standard Operating Procedures for Decontamination of Response Personnel, FSOP7, U.S. EPA, Washington, 1985.

70. U.S. GPO, *Occupational Safety and Health Guidance Manual for Hazardous Waste Site Activities*, DHHS-NIOSH-85-115, U.S. Government Printing Office, Washington, October 1985.

71. Wang, L.K., Kurylko, L., and Wang, M.H.S., *Sequencing Batch Liquid Treatment*, U.S. Patent 5,354,458, 1994.

72. Wang, L.K., Wang, P., and Clesceri, N.L., Development of a combined biological aeration and flotation system for groundwater decontamination, *Proc. Industrial Waste Conference*, Purdue University, West Lafayee, IN, 1993.

73. Wang, L.K., Wang, P., and Clesceri, N.L. Groundwater decontamination using sequencing batch process, *Water Treatment*, 10, 121–134, 1995.

74. U.S. EPA, A Citizen's Guide to Permeable Reactive Barriers, Technical Report EPA-542-F-01-005, U.S. EPA, Washington, 2001.

75. U.S. EPA, Permeable Reactive Subsurface Barriers for the Interception and Remediation of Chlorinated Hydrocarbon and Chromium (VI) Plumes in Ground Water, Technical Report EPA-600-F-97-008, U.S. EPA, Washington, 1997.

76. U.S. EPA, Treatment Technologies for Site Cleanup: Annual Status Report, 11th ed., Technical Report EPA-542-R-03-009, U.S. EPA, Washington, 2004.

77. U.S. EPA, A Citizen's Guide to Soil Vapor Extraction and Air Sparging, Technical Report EPA-542-F-01-006, U.S. EPA, Washington, 2001.

78. U.S. EPA, A Citizen's Guide to Solvent Extraction, Technical Report EPA-542-F-01-009, U.S. EPA, Washington, 2001.

79. U.S. EPA, A Citizen's Guide to Chemical Oxidation, Technical Report EPA-542-F-01-013, U.S. EPA, Washington, 2001.

80. U.S. EPA, Engineering Issue Paper: *In Situ* Chemical Oxidation, Technical Report EPA-600-R-06-072, U.S. EPA, Washington, 2006.

81. FRTR, Remediation Technologies Screening Matrix and Reference Guide, Federal Remediation Technologies Roundtable, Washington, DC, 2007. Available at http://clu-in.org/techfocus/default.focus/sec.

82. U.S. EPA, A Citizen's Guide to Vitrification, Technical Report EPA-542-F-01-017, U.S. EPA, Washington, 2001.

83. U.S. EPA, A Citizen's Guide to *In Situ* Thermal Treatment Methods, Technical Report EPA-542-F-01-012, U.S. EPA, Washington, 2001.

84. U.S. EPA, A Citizen's Guide to Fracturing, Technical Report EPA-542-F-01-015, U.S. EPA, Washington, 2001.

85. U.S. EPA, A Citizen's Guide to *In Situ* Flushing, Technical Report EPA-542-F-01-011, U.S. EPA, Washington, 2001.

86. Rohayo, A.J., Cameron, R.J., Teters, B.B., Rossabi, J., Riha, R., and Downs, W., Passive Soil Vapor Extraction, Technical Report DE98051208, 20 p., U.S. Department of Commerce, National Technical Information Service, Springfield, VA, 1997.

87. U.S. EPA, A Citizen's Guide to Bioremediation, Technical Report EPA-542-F-01-001, U.S. EPA, Washington, 2001.

88. Moretti, L., *In Situ Bioremediation of DNAPL Source Zones*, U.S. EPA, Washington, DC, 2005.

89. CPEO, Technology Overview Report: Electrokinetics, Center for Public Environmental Oversight, Washington, 1997, 36 p.

90. U.S. EPA, A Citizen's Guide to Soil Washing, Technical Report EPA-542-F-01-008, U.S. EPA, Washington, 2001.

91. U.S. EPA, A Citizen's Guide to Thermal Desorption, Technical Report EPA-542-F-01-003, U.S. EPA, Washington, 2001.

92. Hater, G., Green, G., and Vogt, G., Landfills as Bioreactors, Technical Report EPA-600-R-03-097, U.S. EPA, Washington, 2003, 406 p.

93. Modrak, M., Hashmonay, R.A., Varma, R., and Kagann, R., *Measurement of Fugitive Emissions at a Landfill Practicing Leachate Recirculation and Air Injection*, U.S. Department of Commerce, National Technical Information Service, Springfield, VA, 2005, 58 p.

94. U.S. EPA, A Citizen's Guide to Monitored Natural Attenuation, Technical Report EPA-542-F-01-004, U.S. EPA, Washington, 2001.

95. U.S. EPA, Monitored Natural Attenuation of Chlorinated Solvents, Technical Report EPA-600-F-98-022, U.S. EPA, Washington, 1998.

96. U.S. EPA, Monitored Natural Attenuation of Petroleum Hydrocarbons, Technical Report EPA-600-F-98-021, U.S. EPA, Washington, 1998.

97. ASCE, *Natural Attenuation of Hazardous Wastes*, American Society of Civil Engineers, Reston, VA, 2004, 256 p.

98. U.S. EPA, A Citizen's Guide to Phytoremediation, Technical Report EPA-542-F-01-002, U.S. EPA, Washington, 2001.

99. U.S. EPA, Radionuclide Biological Remediation Resource Guide, Technical Report EPA-905-B-04-001, U.S. EPA, Washington, 2004.

100. Wang, L.K., Innovative UV, Ion Exchange, Membrane and Flotation Technologies for Water and Waste Treatment, 2006 National Engineers Week Seminar, Practicing Institute of Engineers, Albany, NY, February 2006.

101. Wang, L.K., Kurylko, L., and Hrycyk, O., Biological Process for Groundwater and Wastewater Treatment, U.S. Patent 5,451,320, 1995.

102. Wang, L.K. and Kurylko, L., Site Remediation Technology, U.S. Patent 5,552,051, 1996.

103. Wang, L.K., Emerging Induced and Dissolved Gas Flotation Processes, 2007 National Engineers Week Seminar, Practicing Institute of Engineers, Albany, NY, February 2007.

104. Wang, L.K., Bubble Dynamics of Adsorptive Bubble Separation Processes, 2007 National Engineers Week Seminar, Practicing Institute of Engineers, Albany, NY, February 2007.

105. Wang, L.K., Hung, Y.T., Lo, H.H., and Yapijakis, C., *Hazardous Industrial Waste Treatment*, CRC Press, Boca Raton, FL, 2007, 516 p.

106. Wang, L.K. New Water and Wastewater Treatment Technologies: Chemical Oxidation and Emergency Response. Training Course. N S Dept. of Health No. ATC-232-2316-3645. N S Dept. of Environmental Conservation No. RTC-10234-07. Albany Water Treatment Plant, NY, May 18, 2007.

7 Enzymatic Removal of Aqueous Pentachlorophenol

Khim Hoong Chu, Eui Yong Kim, and Yung-Tse Hung

CONTENTS

7.1   INTRODUCTION

Pentachlorophenol (PCP) has been used extensively as a pesticide, herbicide, and wood-preserving agent at many wood treating sites. The chemical structure of PCP is shown in Figure 7.1. It is a probable human carcinogen and has been placed on the U.S. EPA priority pollutant list. Its presence in the environment is therefore of particular concern. In recent years many countries have banned the use of PCP. Unfortunately, past legal disposal practices coupled with the environmental stability of PCP have led to widespread contamination of soil, surface water, and groundwater aquifers. Many of the more than 700 wood preserving sites identified in the U.S. are currently being dealt with under federal, state, or voluntary cleanup programs.[1]

Various treatment methods can be used to remove PCP from contaminated environmental compartments, and the treatment of PCP-contaminated soil usually involves a combination of physical, chemical, and biological methods. An integrated system combining soil washing with a solvent, recovery of the spent solvent for reuse, and biodegradation of the desorbed PCP in aqueous solution has been proposed.[2–4] The biodegradation of aqueous PCP by microorganisms has several advantages over chemical and physical methods, including mild operating conditions and better environmental compatibility. Several species of bacteria and fungi can biodegrade PCP.[5–15] These organisms secrete

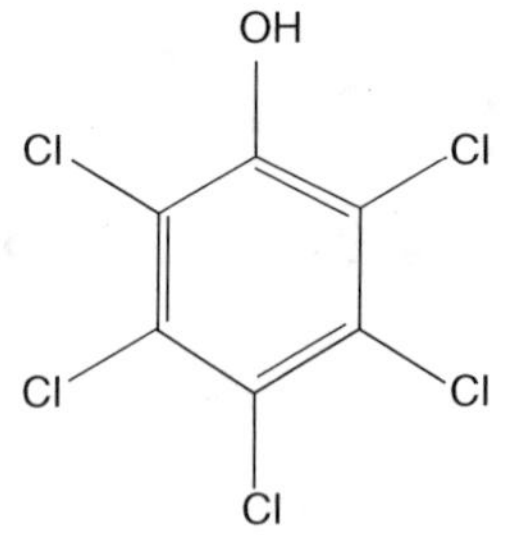

FIGURE 7.1   Chemical structure of pentachlorophenol (PCP).

a series of oxidative enzymes that are capable of catalyzing the oxidation of PCP. However, high concentrations of PCP can be inhibitory to the activity of the degrading organisms. Cho and colleagues[9] have shown that PCP concentrations higher than 50 mg/L inhibit the growth of some PCP-degrading white rot fungi such as *Gloeophyllum odoratum* and *Trametes versicolor* completely. As a result, the direct application of isolated enzymes has been proposed as an alternative method of removing PCP from aqueous solution. A number of reviews on the *in vitro* use of oxidative enzymes to catalyze the oxidation of phenolic substances including PCP are available in the literature.[16–18] For PCP oxidation, the enzymes that have been tested include horseradish peroxidase (HRP),[19–25] laccase,[26–28] ligninase,[29] and other extracellular peroxidases.[30]

Because HRP has been used extensively to transform a wide range of phenolic contaminants, this chapter focuses on the salient aspects of the HRP-catalyzed oxidation of PCP in the presence of hydrogen peroxide (H_2O_2). The oxidation process generates free aromatic radicals, which combine to form polymers of low solubility that eventually precipitate from solution. Thus, the enzyme-mediated removal process is also known as polymerization precipitation. The major product of the HRP-catalyzed oxidation of PCP over the pH range 4 to 7 is 2,3,4,5,6-pentachloro-4-pentachloro-phenoxy-2,5-cyclohexadienone (PPCHD).[31] PPCHD is formed by the coupling of two pentachloro-phenoxyl radicals, the expected products of one-electron oxidation reactions catalyzed by HRP and other peroxidases. The chemical structure of PPCHD is shown in Figure 7.2. Although the HRP-mediated oxidative coupling process has enormous potential for remediation of aqueous solutions contaminated by PCP, its application is hampered by the low operational stability of HRP as a result of inactivation by the enzyme's own substrate, H_2O_2.[32] The key area of interest reported in this chapter is the elucidation of the inhibitory effect of H_2O_2 on HRP activity. To this end, a theoretical model incorporating saturation kinetics and formation of a catalytically inactive form of HRP in the presence of excess H_2O_2 was developed to facilitate the quantitative evaluation of the oxidative inactivation of HRP.[20] It should be noted that HRP inactivation can occur via two other mechanisms: radical attack and sorption by precipitated products. The analysis of such mechanisms is beyond the scope of this chapter.

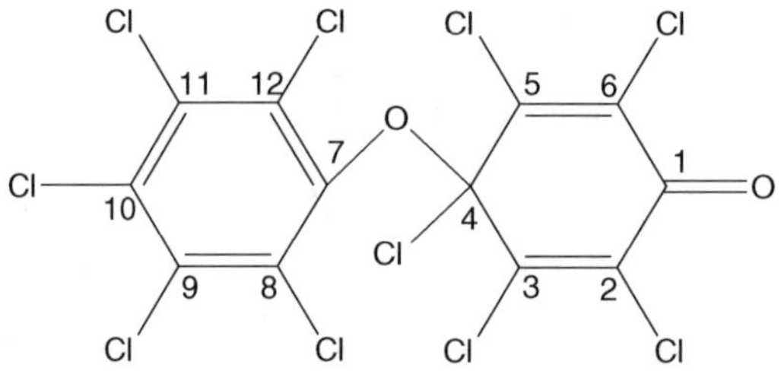

FIGURE 7.2   Chemical structure of 2,3,4,5,6-pentachloro-4-pentachlorophenoxy-2,5-cyclohexadienone (PPCHD).

7.2 DESCRIPTION OF HORSERADISH PEROXIDASE

As its name implies, HRP (EC 1.11.1.7) is isolated from the roots of horseradish (*Armoracia rusticana*). A comprehensive description of the structure, function, mechanism of action, and practical applications of HRP has recently been given by Veitch.[33] HRP exists in the form of several distinctive isoenzymes, with the C isoenzyme (HRP C) being the predominant form. It consists of 308 amino acid residues, a ferric heme prosthetic group, and 2 mol of calcium per mol of protein, adding up to a molecular weight of 34,520. It is glycosylated and contains four highly conserved disulfide bridges. Recently, there have been key advances in our understanding of HRP and some of these include X-ray crystallographic studies of the crystal structure[34] of HRP C as well as the intermediate species in the catalytic cycle of the enzyme.[35] HRP can accommodate a broad range of substrates in a variety of reactions. Although it is widely used in analytical diagnostics such as in enzyme immunoassays and biosensors, its low operational stability hampers its commercial applications in organic synthesis for the biotransformations of various drugs and chemicals and in the detoxification of aromatic contaminants.

7.3 MODEL DEVELOPMENT

7.3.1 CATALYTIC CYCLE AND INACTIVATION OF HRP

HRP catalyzes the oxidation of a variety of organic and inorganic substances, with H_2O_2 as electron acceptor. The global reaction catalyzed by HRP is described by Equation 7.1, in which an oxidant (H_2O_2) reacts with a reducing substrate (AH_2) to produce a radical product ($\cdot AH$) and H_2O:

$$H_2O_2 + 2\ AH_2 \xrightarrow{\text{HRP}} 2 \cdot AH + 2\ H_2O \tag{7.1}$$

The above reaction proceeds in three distinct steps. First, the native ferric enzyme reacts with the oxidizing substrate (H_2O_2). Following binding of H_2O_2 to the heme in the Fe(III) state, the heterolytic cleavage of the oxygen–oxygen bond of H_2O_2 results in the two-electron oxidation of the heme to form an intermediate (compound I) comprising a ferryl species (Fe(IV) = O) and a prophyrin radical cation, with the concomitant release of a water molecule. Compound I is a reactive intermediate with a higher formal oxidation state (+5 compared with +3 for the resting enzyme). Compound I is then converted back to the resting enzyme via successive single-electron transfers from two reducing substrate molecules (AH_2). The first reduction, of the prophyrin radical cation, yields a second enzyme intermediate, compound II, which retains the heme in the ferryl (Fe(IV) = O) state and the free radical $\cdot AH$. The second reduction regenerates the ferric heme resting state of the enzyme and delivers another free radical $\cdot AH$ and a water molecule. The catalytic cycle of HRP involving the oxidation and reduction of the heme group can be described by the following reaction scheme:

$$E + H_2O_2 \longrightarrow E_i + H_2O \tag{7.2}$$

$$E_i + AH_2 \longrightarrow E_{ii} + \cdot AH \tag{7.3}$$

$$E_{ii} + AH_2 \longrightarrow E + \cdot AH + H_2O \tag{7.4}$$

In these equations, E, E_i, and E_{ii} represent the resting enzyme, compound I, and compound II, respectively.

Numerous studies have shown that oxidation of a wide range of AH_2 by HRP in the presence of H_2O_2 is characterized by a loss of enzyme activity. It is now well established that HRP is inactivated by H_2O_2.[32] Because the final step (Equation 7.4), during which the oxidized ferryl intermediate is reduced, is very slow, inactivation of HRP is thought to occur by reaction of compound II with an additional molecule of H_2O_2:

$$E_{ii} + H_2O_2 \longrightarrow E_{iii} + H_2O \tag{7.5}$$

where E_{iii} is known as compound III, which is an inactive form of the enzyme. The degree of inactivation appears to depend on several factors, including the chosen electron donor AH_2, the amount of H_2O_2, and the concentration ratio of H_2O_2 and the electron donor. The above reaction scheme describes and summarizes the major catalytic and inactivation pathways that have been identified. Because the enzyme is a significant contributor to the cost of contaminant degradation, judicious control of H_2O_2 concentration to avoid enzyme inactivation will help to enhance the commercial viability of this approach. We describe here a mathematical model that can be used to predict the inhibitory effect of H_2O_2 on the catalytic behavior of HRP.

7.3.2 Proposed Reaction Mechanism

A kinetic model describing the HRP-catalyzed oxidation of PCP by H_2O_2 should account for the effects of the concentrations of HRP, PCP, and H_2O_2 on the reaction rate. To derive such an equation, a reaction mechanism involving saturation kinetics is proposed. Based on the reaction scheme described in Section 7.3.1, which implies that the catalytic cycle is irreversible, the three distinct reactions steps (Equations 7.2 to 7.4) are modified to include the formation of Michaelis–Menten complexes:

$$E + H_2O_2 \underset{k_{-1}}{\overset{k_1}{\rightleftharpoons}} E^* \overset{k_2}{\longrightarrow} E_i + H_2O \tag{7.6}$$

$$E_i + PCP \underset{k_{-3}}{\overset{k_3}{\rightleftharpoons}} E_i^* \overset{k_4}{\longrightarrow} E_{ii} + \cdot P \tag{7.7}$$

$$E_{ii} + PCP \underset{k_{-5}}{\overset{k_5}{\rightleftharpoons}} E_{ii}^* \overset{k_6}{\longrightarrow} E + \cdot P + H_2O \tag{7.8}$$

In these equations, E^*, E_i^*, and E_{ii}^* represent Michaelis–Menten complexes, $\cdot P$ is the PCP-derived radical, k_{-1}, k_{-3}, k_{-5} and k_1 to k_6 are the rate constants of the respective reactions. The existence of the Michaelis–Menten complexes between HRP and H_2O_2 (E^*) and between compound I or compound II and certain reducing substrates (E_i^* or E_{ii}^*) has been demonstrated by van Haandel and colleagues,[36] Baek and van Wart,[37] and Rodríguez-López and colleagues,[38] respectively. It should be noted that the radical generation steps in Equation 7.7 and Equation 7.8 have also been proven to be reversible.[39] The overall reaction is given by

$$H_2O_2 + 2\,PCP \xrightarrow{\text{HRP}} 2\cdot P + 2\,H_2O \tag{7.9}$$

The radical intermediates $\cdot P$ can couple with each other, leading to the formation of polymeric precipitates that can be readily removed from water (see Figure 7.2). The polymerization of the free radicals is known to be extremely fast, and it is therefore not included in the above reaction scheme.

7.3.3 Derivation of the Reaction Rate Equation

To derive a rate equation based on Equations 7.6, 7.7, and 7.8, the following assumptions are made. First, at the start of the reaction the concentrations of the products are assumed to be zero in comparison with those of the reactants. Thus, these equations can be considered to be essentially irreversible during the early stages of reaction. Second, Equation 7.8 is assumed to be the rate-limiting step, because under most steady-state conditions the reaction of HRP with H_2O_2 is very fast, and the reaction of compound II with the reducing substrate is at least 10 to 20 times slower than that of compound I.[40] The overall reaction rate V is thus given by

$$V = k_6[E_{ii}^*]$$ (7.10)

Applying steady-state assumptions, the rate equation for the reaction mechanism described by Equations 7.6, 7.7, and 7.8 can be obtained:

$$V = \cfrac{\cfrac{1}{\cfrac{1}{k_2} + \cfrac{1}{k_4} + \cfrac{1}{k_6}}[E_0][H_2O_2][PCP]}{\cfrac{k_{-1} + k_2}{k_1 k_2}}{\cfrac{1}{k_2} + \cfrac{1}{k_4} + \cfrac{1}{k_6}}[PCP] + \cfrac{\cfrac{k_{-3} + k_4}{k_3 k_4} + \cfrac{k_{-5} + k_6}{k_5 k_6}}{\cfrac{1}{k_2} + \cfrac{1}{k_4} + \cfrac{1}{k_6}}[H_2O_2] + [H_2O_2][PCP]}$$ (7.11)

where $[E_0]$ denotes the initial concentration of enzyme. Further details concerning the derivation of Equation 7.11 by the schematic method of King and Altman[41] are given in the Appendix. Equation 7.11 indicates that the reaction mechanism follows the well-known Ping–Pong Bi–Bi mechanism. This mechanism is characterized by the product of the enzyme's reaction with the first substrate (i.e., H_2O_2), being released before the reaction of the enzyme with the second substrate (i.e., PCP). The general form of the rate equation based on the Ping–Pong Bi–Bi mechanism is given by

$$V = \frac{K_{cat}[E_0][A][B]}{K_m^A[B] + K_m^B[A] + [A][B]}$$ (7.12)

where $[A]$ and $[B]$ denote the concentrations of two different substrates, and K_{cat}, K_m^B, and K_m^A are constants.

Recasting Equation 7.11 in the form of Equation 7.12 gives

$$V = \frac{K_{cat}[E_0][H_2O_2][PCP]}{K_m^{H_2O_2}[PCP] + K_m^{PCP}[H_2O_2] + [H_2O_2][PCP]}$$ (7.13)

where

$$K_{cat} = \cfrac{1}{\cfrac{1}{k_2} + \cfrac{1}{k_4} + \cfrac{1}{k_6}}; \quad K_m^{H_2O_2} = \cfrac{\cfrac{k_{-1} + k_2}{k_1 k_2}}{\cfrac{1}{k_2} + \cfrac{1}{k_4} + \cfrac{1}{k_6}}; \quad K_m^{PCP} = \cfrac{\cfrac{k_{-3} + k_4}{k_3 k_4} + \cfrac{k_{-5} + k_6}{k_5 k_6}}{\cfrac{1}{k_2} + \cfrac{1}{k_4} + \cfrac{1}{k_6}}$$

Equation 7.13 has been derived without taking account of HRP inactivation by H_2O_2, which is described in Equation 7.5. One simple way to remedy this situation is to introduce an inactivation constant into Equation 7.13:

$$V = \frac{K_{cat}[E_0][H_2O_2][PCP]}{K_m^{H_2O_2}[PCP] + (K_m^{PCP} + [PCP])\left(1 + \frac{[H_2O_2]}{K_i}\right)[H_2O_2]}$$

(7.14)

where K_i is an inactivation constant that describes the inhibitory effect of H_2O_2. Equation 7.14 may be used to predict the effects of enzyme, H_2O_2, and PCP concentrations as well as the inhibitory effect of H_2O_2 concentration on the reaction rate, provided that the four constants K_{cat}, $K_m^{H_2O_2}$, K_m^{PCP}, and K_i are known. In the next section, we describe how these constants may be estimated by fitting Equation 7.14 to experimental data.

7.4 PARAMETER ESTIMATION AND MODEL VALIDATION

7.4.1 EXPERIMENTAL DATA

To generate experimental data for parameter estimation, batch reaction experiments were conducted at 25°C using solutions containing equimolar concentrations of PCP and H_2O_2 (0.01 to 6 mM) in 100 mM sodium phosphate buffer (pH 6.5). The enzymatic reaction was initiated by adding a dose of HRP stock solution to the reaction mixture. Experiments were conducted at four different initial enzyme concentrations: 0.13, 0.148, 0.295, and 0.34 μM. Solution samples were taken at fixed time intervals and centrifuged to settle precipitated colloidal particles. The PCP concentration of the supernatant was determined using a UV spectrophotometer. Initial reaction rates were estimated from the initial slopes of PCP concentration versus time curves. Additional experiments were conducted to generate a new set of data for model validation. In these experiments, the initial enzyme and PCP concentrations were fixed at 0.72 μM and 1.5 mM, respectively, while the initial H_2O_2 concentration was varied in the range 0.01 to 12 mM.

7.4.2 PARAMETER ESTIMATION

The four constants in Equation 7.14 may be estimated by fitting the equation to the measured initial reaction rate data presented in Figure 7.3. Because equimolar concentrations of the two substrates, PCP and H_2O_2, were used in the experiments, Equation 7.14 may be simplified as follows:

$$V = \frac{K_{cat}[E_0][S]^2}{K_m^{H_2O_2}[S] + (K_m^{PCP} + [S])\left(1 + \frac{[S]}{K_i}\right)[S]}$$

$$= \frac{K_{cat}[E_0][S]}{K_m^{H_2O_2} + K_m^{PCP} + \left(1 + \frac{K_m^{PCP}}{K_i}\right)[S] + \frac{1}{K_i}[S]^2}$$

(7.15)

where $[S] = [PCP] = [H_2O_2]$. Fitting Equation 7.15 to the data shown in Figure 7.3 provides a simple way of estimating the four constants.

The best-fit values of the four constants were estimated by minimizing the error between experimental data and model calculations. The minimization algorithm is based on a genetic algorithm, which is a stochastic optimization technique patterned after the natural selection process taking place during biological evolution. It explores all regions of the solution space using a population of individuals. Each individual represents a set of the parameters to be optimized. Initially, a population of individuals is formed randomly. The fitness of each individual is evaluated using an objective

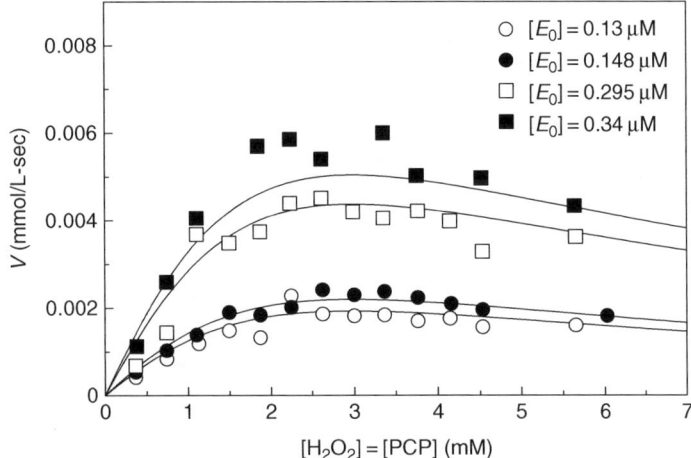

FIGURE 7.3 Experimental data (symbols) showing the variation of reaction rate V with equimolar substrate concentration ($[H_2O_2]$ = [PCP]) for different initial enzyme concentrations $[E_0]$. Also shown are the theoretical curves (lines) calculated according to Equation 7.15 with the constants of set A as given in Table 7.1.

function. Upon completion of the fitness evaluation, genetic operations such as mutation and crossover are applied to individuals selected according to their fitness to produce the next generation of individuals for fitness evaluation. This process continues until a near-optimum solution is found.

Because the genetic algorithm searches the entire input space in parallel, it is more robust than traditional deterministic methods and is more likely to converge to a unique global minimization. As with any artificial intelligence technique, the performance of a genetic algorithm is affected by a number of design parameters such as the initial population size, parent selection, crossover rate, mutation rate, and the number of generations. Some initial tests indicate that the genetic algorithm used in this work is robust to parameter variations, with the population size and the number of generations having the largest effect on performance. Using a population of 100, the solution successfully converged to the optimum values after 2000 to 3000 generations. All computations were conducted using the software package Matlab®. An excellent description of the implementation of genetic algorithms and their use as a problem-solving and optimization technique can be found in the book by Goldberg.[42]

Repetitive optimization runs reveal the existence of two distinct sets of best-fit values within the search space of 0 to 500 for each constant. These best-fit values are listed in Table 7.1. A comparison between the reaction rate profiles calculated from the two sets of constants (lines) and experimental data (symbols) is shown in Figure 7.3 and Figure 7.4. It is clear that there is generally good agreement, although at the highest $[E_0]$ examined the two theoretical curves underestimate the middle part of the experimentally measured reaction rate data. It is further observed that both sets of constants give congruent theoretical profiles. It can therefore be concluded that unique parameter estimates cannot be obtained for the simplified nonlinear model (Equation 7.15), because more than

TABLE 7.1

Best–Fits Values of K_{cat}, $K_m^{H_2O_2}$, K_m^{PCP}, and K_i

Set	K_{cat} (sec^{-1})	$K_m^{H_2O_2}$ (mM)	K_m^{PCP} (mM)	K_i (mM)
A	87.7	7.2	0.05	1.22
B	269.2	24.7	~0	0.34

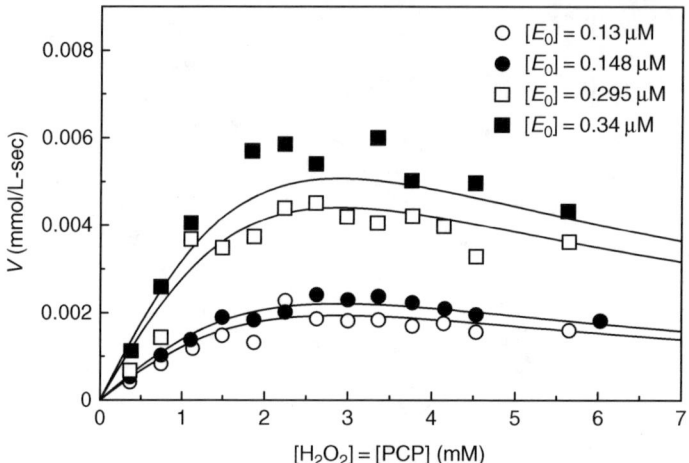

FIGURE 7.4 Experimental data (symbols) showing the variation of reaction rate V with equimolar substrate concentration ($[H_2O_2] = [PCP]$) for different initial enzyme concentrations $[E_0]$. Also shown are the theoretical curves (lines) calculated according to Equation 7.15 with the constants of set B as given in Table 7.1.

one combination of parameters can describe the same data set. In addition, the value of K_m^{PCP} identified by this multiparameter estimation routine is either zero or very close to zero, indicating that it is not a significant parameter. Simultaneous retrieval of unique estimates of the four constants may require fitting the original model equation (Equation 7.14) to data obtained from experiments with different combinations of $[E_0]$, $[PCP]$, and $[H_2O_2]$. As the fitted parameters are able to capture the general trends of the experimental data, as shown in Figure 7.3 and Figure 7.4, the best-fit constants of set A are used in the simulation studies reported in Section 7.5.

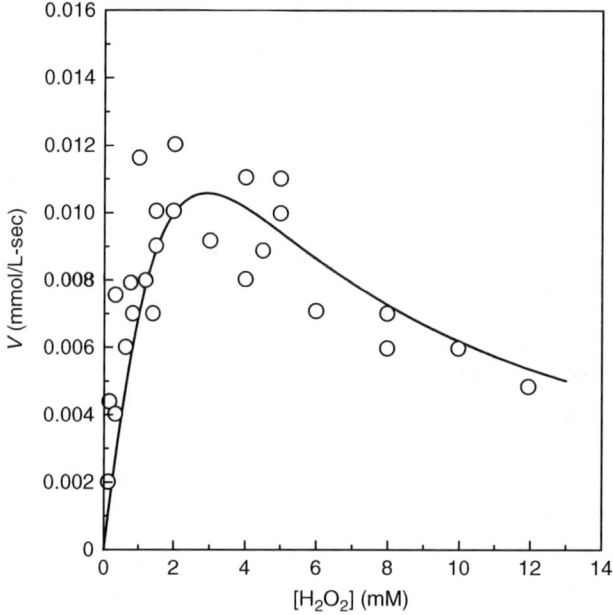

FIGURE 7.5 Experimental data (symbols) showing reaction rate V as a function of $[H_2O_2]$. The initial enzyme and PCP concentrations are $[E_0] = 0.72$ μM and $[PCP] = 1.5$ mM, respectively. The theoretical curve (line) calculated from Equation 7.14 with the constants of set A (Table 7.1) is shown for comparison.

7.4.3 MODEL VALIDATION

Because the constants identified by the parameter estimation approach described above are not unique, it is important to assess the predictive capability of the model equation before the model is used for simulation studies. The predictive capability of Equation 7.14 can be assessed by comparing its predictions with data obtained from experiments conducted at conditions that are different from those used to generate data for parameter estimation. A set of such data (symbols) is shown in Figure 7.5, together with the theoretical curve (line) calculated from Equation 7.14 with the constants of set A. As can be seen in Figure 7.5, although the simulation does not capture the measured reaction rate data accurately, it does predict the trend very well. Given that the experimentally measured data show some scatter, for all practical purposes the agreement achieved using the rate constants of set A is quite satisfactory. The results presented in Figure 7.5 clearly show that the reaction rate is inhibited when the H_2O_2 concentration is higher than ~3 to 4 mM. Having developed confidence in the theoretical model after matching the simulation results with the experimental observations, the model is used to examine the inhibitory effect of H_2O_2 in greater detail in the next section.

7.5 MODEL SIMULATION

7.5.1 DEPENDENCE OF THE REACTION RATE ON PCP CONCENTRATION

We first examine the dependence of the reaction rate V on PCP concentration. Figure 7.6 and Figure 7.7 show the effect of PCP concentration on reaction rate for different initial enzyme and H_2O_2 concentrations, respectively. The reaction rate profiles shown in these two figures are calculated from Equation 7.14 with the constants of set A. Both figures show highly rectangular reaction rate profiles, indicating that the reaction rate reaches its maximum value at very low PCP concentrations (~0.1 to 0.2 mM) for a given initial enzyme or H_2O_2 concentration. The plateau of the profiles gives the maximum rate of reaction. The profiles shown in Figure 7.6 indicate that the maximum reaction rate increases if more enzyme is added. This is of course a typical feature of enzyme kinetics. On the other hand, the profiles in Figure 7.7 do not show a monotonic rise in maximum reaction rate with

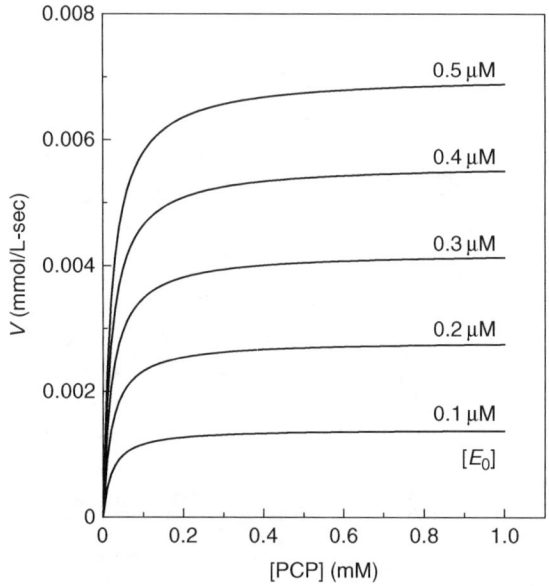

FIGURE 7.6 Theoretical profiles showing the variation of reaction rate V with [PCP] for different initial enzyme concentrations $[E_0]$. $[H_2O_2] = 2$ mM.

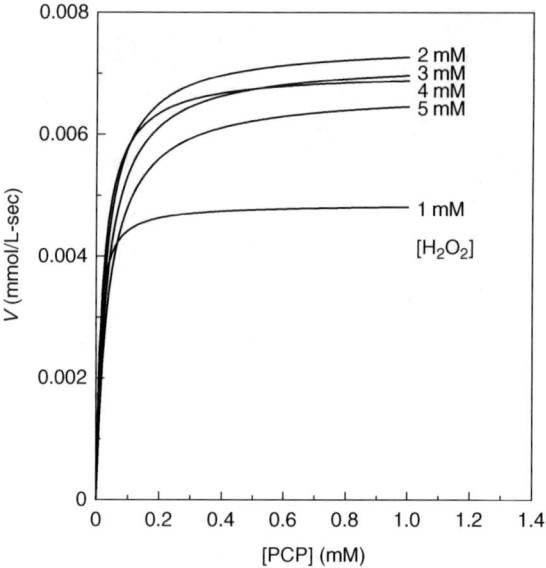

FIGURE 7.7 Theoretical profiles showing the variation of reaction rate V with [PCP] for different [H_2O_2]. The initial enzyme concentration is [E_0] = 0.5 µM.

H_2O_2 concentration. The simulation results suggest that the maximum reaction rate at first increases and then decreases with increasing H_2O_2 concentration, reflecting the inhibitory effect of H_2O_2.

7.5.2 DEPENDENCE OF THE REACTION RATE ON H_2O_2 CONCENTRATION

In this section, we describe the dependence of the reaction rate V on H_2O_2 concentration, as the reaction rate has been shown to be suppressed at high H_2O_2 concentrations (see Figure 7.5). Representative theoretical curves calculated according to Equation 7.14 with the constants of set A

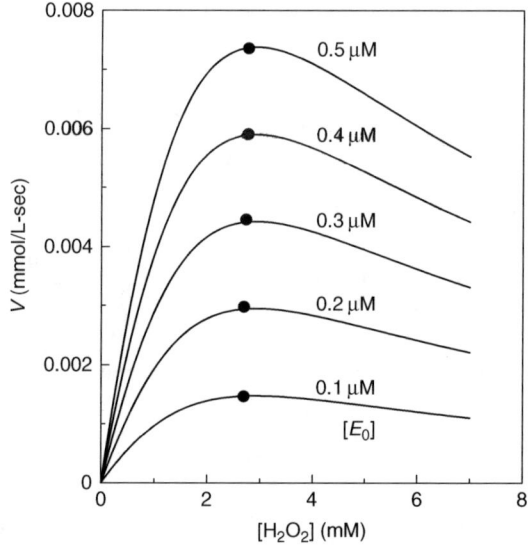

FIGURE 7.8 Theoretical profiles showing the variation of reaction rate V with [H_2O_2] for different initial enzyme concentrations [E_0]. [PCP] = 2 mM. The solid circles indicate the location of the maximum reaction rate.

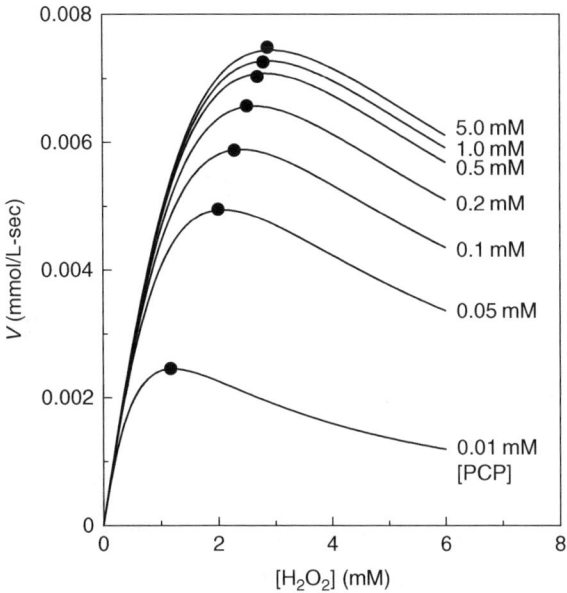

FIGURE 7.9 Theoretical profiles showing the variation of reaction rate V with $[H_2O_2]$ for different [PCP]. The initial enzyme concentration is $[E_0] = 0.5 \, \mu M$. The solid circles indicate the location of the maximum reaction rate.

are shown in Figure 7.8 and Figure 7.9. As expected, the figures show that the reaction rate increases with increasing initial enzyme or PCP concentration at a given H_2O_2 concentration. Also, the reaction rate profiles in both figures have almost the same shape, although they differ in absolute values; they initially increase with increasing H_2O_2 concentration, reaching a maximum value before declining with a further increase in H_2O_2 concentration. This type of curve is commonly observed for systems in which the substrate is inhibiting.

However, two major differences between the two sets of profiles are observed. First, the variation in the shape of the profiles in Figure 7.8 is directly proportional to the enzyme concentration, but although the variation in Figure 7.9 is quite pronounced in the low PCP concentration region, at higher values of PCP concentration this is no longer true. The curves lie quite close together when the PCP concentration is >0.5 mM. Second, Figure 7.8 shows that the maximum reaction rate for each profile, as indicated by the solid circles, does not vary with H_2O_2 concentration and it appears to occur at a H_2O_2 concentration of ~3 mM. By contrast, Figure 7.9 shows that the occurrence of the maximum reaction rate is governed by the H_2O_2 concentration when the PCP concentration is varied from 0.01 to 5 mM. The simulation results presented in Figure 7.9 suggest that for a given PCP concentration an optimum H_2O_2 concentration exists that gives the maximum reaction rate. As it is desirable to run the enzymatic reaction at the maximum possible reaction rate, knowledge of the relationship between the optimum H_2O_2 concentration and PCP concentration is of great practical interest. The next section describes how this relationship may be derived from the model equation.

7.5.3 Determination of Optimum H_2O_2 Concentration

The relationship between the optimum H_2O_2 concentration and PCP concentration may be derived from Equation 7.14. Differentiating V with respect to the H_2O_2 concentration and setting the derivative to zero ($dV/d[H_2O_2] = 0$) yields the following equation:

$$\left[H_2O_2\right]_{opt} = \sqrt{\frac{K_m^{H_2O_2} K_i [PCP]}{K_m^{PCP} + [PCP]}} \tag{7.16}$$

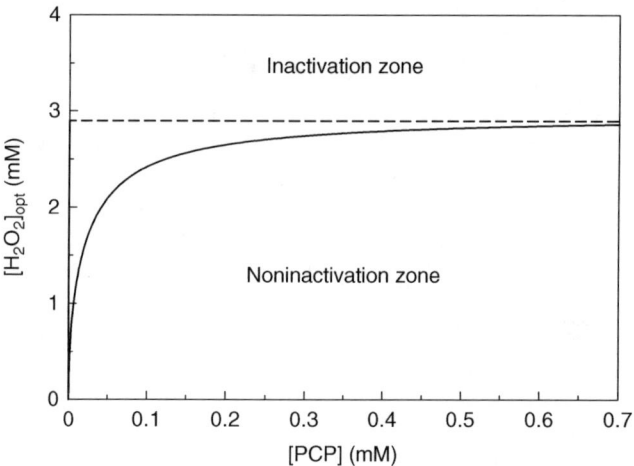

FIGURE 7.10 Theoretical profiles showing the variation of optimum [H$_2$O$_2$] with [PCP] according to Equation 7.16. The solid and broken lines are calculated from Equation 7.16 with the constants of sets A and B, respectively (Table 7.1).

where [H$_2$O$_2$]$_{opt}$ is the optimum H$_2$O$_2$ concentration. The solid line in Figure 7.10 is a plot of Equation 7.16 using the constants of set A. This curve gives the optimum H$_2$O$_2$ concentration at which the maximum reaction rate occurs for a given PCP concentration. As can be seen in Figure 7.10, when the PCP concentration is <0.5 mM, the optimum H$_2$O$_2$ concentration increases nonlinearly with increasing PCP concentration. When the PCP concentration is >0.5 mM the optimum H$_2$O$_2$ concentration approaches an asymptote and becomes independent of the PCP concentration. Figure 7.10 therefore serves as a useful guide for selecting combinations of H$_2$O$_2$ and PCP concentrations that would avoid enzyme inactivation by H$_2$O$_2$. For example, the noninactivation zone is designated by the area below the solid line, and the area above the solid line depicts the inactivation zone where the inactivated form of the enzyme, E_{iii}, is formed in the presence of excess H$_2$O$_2$, leading to reduced reaction rates.

The low operational stability of HRP as a result of inactivation by H$_2$O$_2$ seriously impedes commercial applications of the enzyme in detoxification of waste streams and industrial organic synthesis. One approach to improving the operational stability of HRP is to maintain the concentration of H$_2$O$_2$ at a low level. This can be achieved using an appropriate method of H$_2$O$_2$ addition or generation. Examples of these methods include the stepwise addition of H$_2$O$_2$, feed-on-demand addition of H$_2$O$_2$, and *in situ* generation of H$_2$O$_2$.[43]

A different curve is observed when Equation 7.16 is plotted using the constants of set B. The optimum H$_2$O$_2$ concentration curve goes straight up from the origin to a certain H$_2$O$_2$ concentration and then extends horizontally from that point, as depicted by the broken line in Figure 7.10. This curve overestimates the optimum H$_2$O$_2$ concentration when the PCP concentration is <0.5 mM. Such a limiting form is a consequence of K_m^{PCP} being set to 0. As a result, we can see from Equation 7.16 that the optimum H$_2$O$_2$ concentration becomes independent of the PCP concentration. Nevertheless, from Figure 7.10 it is clear that both curves predict a limiting optimum H$_2$O$_2$ concentration of ~2.9 mM, even though the values of $K_m^{H_2O_2}$ and K_i used in generating the two curves are quite different, as may be seen from Table 7.1. This is evidently due to the fact that for high values of PCP concentration Equation 7.16 will approach the asymptote $\sqrt{K_m^{H_2O_2} K_i}$. The practically important conclusion from this analysis is that the effective use of mathematical models for simulation studies requires the development of sound methodologies to identify the key model parameters. It is essential to know whether the measured data are sufficient for identifying the unknown parameters and the conditions under which they are identifiable. The development of robust parameter estimation methodologies is beyond the scope of this chapter.

NOMENCLATURE

AH_2	Reducing substrate
E	Free enzyme
E^*	Enzyme–substrate complex
E_i	Compound I
E_i^*	Enzyme–substrate complex
E_{ii}	Compound II
E_{ii}^*	Enzyme–substrate complex
E_{iii}	Compound III
HRP	Horseradish peroxidase
k_1	Forward rate constant, L/mol-sec
k_{-1}	Reverse rate constant, sec^{-1}
k_2	Rate constant, sec^{-1}
k_3	Forward rate constant, L/mol-sec
k_{-3}	Reverse rate constant, sec^{-1}
k_4	Rate constant, sec^{-1}
k_5	Forward rate constant, L/mol-sec
k_{-5}	Reverse rate constant, sec^{-1}
k_6	Rate constant, sec^{-1}
K_{cat}	Rate constant, sec^{-1}
K_i	Inactivation constant, mM
$K_m^{H_2O_2}$	Constant, mM
K_m^{PCP}	Constant, mM
PCP	Pentachlorophenol
PPCHD	2,3,4,5,6-Pentachloro-4-pentachlorophenoxy-2,5-cyclohexadienone
V	Reaction rate, mol/L-sec
$\cdot AH$	Radical product of AH_2
$\cdot P$	Radical product of PCP
$[A]$	Concentration of substrate, mM
$[B]$	Concentration of substrate, mM
$[E]$	Concentration of free enzyme, mM
$[E_0]$	Initial concentration of enzyme, mM
$[E^*]$	Concentration of enzyme–substrate complex, mM
$[E_i]$	Concentration of compound I, mM
$[E_i^*]$	Concentration of enzyme–substrate complex, mM
$[E_{ii}]$	Concentration of compound II, mM
$[E_{ii}^*]$	Concentration of enzyme–substrate complex, mM
$[E_{iii}]$	Concentration of compound III, mM
$[H_2O_2]$	Concentration of hydrogen peroxide, mM
$[H_2O_2]_{opt}$	Optimum concentration of hydrogen peroxide, mM
$[PCP]$	Concentration of pentachlorophenol, mM
$[S]$	Concentration of substrate, mM

APPENDIX

This appendix illustrates the steps involved in deriving the reaction rate equation (Equation 7.11) from the reaction scheme given in Section 7.3.2 using the King and Altman method.[41] This schematic method allows the derivation of a rate equation for essentially any enzyme mechanism in terms of the individual rate constants of the various steps in biocatalysis.

Step 1. An enzymatic reaction is considered as a cyclic process that displays all the interconversions among the various enzyme forms involved. For each step in the reaction a rate constant is defined in terms of the product of the actual rate constant for that step and the concentration of free substrate involved in the step. Hence, the cyclic form of the reaction scheme given in Equations 7.6, 7.7, and 7.8 is represented by

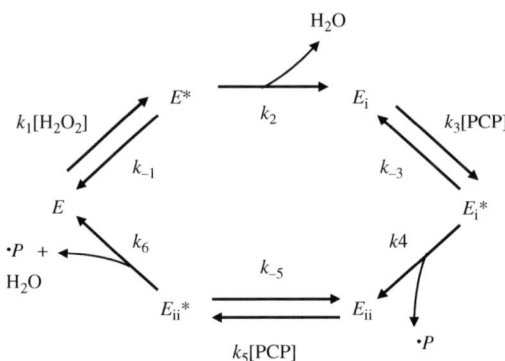

Because the enzyme serves as a catalyst and is not consumed, the conservation equation on the enzyme yields

$$[E_0] = [E] + [E^*] + [E_i] + [E_i^*] + [E_{ii}] + [E_{ii}^*]$$

(A7.1)

Step 2. Every reaction pathway in the reaction scheme involving five arrows, by which a particular enzyme species might be formed, is constructed. The concentration of a particular enzyme species is given by the sum of the rate constant products for that enzyme form. Consideration of the above cyclic reaction scheme yields the relationships given in Table A7.1.

Step 3. Equation 7.11 can now be derived from the overall reaction rate equation, Equation 7.10, using the expressions derived in Step 2 for the concentrations of the six enzyme species.

Dividing Equation 7.10 by $[E_0]$ gives

$$\frac{[E_0]}{V} = \frac{[E_0]}{k_6[E_{ii}^*]}$$

(A7.2)

Substituting the enzyme conservation Equation A7.1 in the left-hand side of Equation A7.2 yields

$$\frac{[E_0]}{V} = \frac{[E] + [E^*] + [E_i] + [E_i^*] + [E_{ii}] + [E_{ii}^*]}{k_6[E_{ii}^*]}$$

(A7.3)

Substituting the expressions derived in Step 2 for the six enzyme species into Equation A7.3 gives

$$\frac{[E_0]}{V} = \frac{1}{k_6}\left\{ \frac{(k_{-1} + k_2)k_6}{k_1 k_2 [H_2O_2]} + \frac{k_6}{k_2} + \frac{(k_{-3} + k_4)k_6}{k_3 k_4 [PCP]} + \frac{k_6}{k_4} + \frac{k_{-5} + k_6}{k_5 [PCP]} + 1 \right\}$$

(A7.4)

TABLE A7.1
King–Altman Relationships for the Various Enzyme Species

Enzyme Form	Pathways to Form	Sum of Rate Constant Products
	k_{-1}, $k_3[PCP]$, E, k_6, k_4, $k_5[PCP]$	$k_{-1}k_3k_4k_5k_6[PCP]^2$
E	k_2, $k_3[PCP]$, E, k_6, k_4, $k_5[PCP]$	$k_2k_3k_4k_5k_6[PCP]^2$
		$[E] = (k_{-1} + k_2)k_3k_4k_5k_6[PCP]^2$
E^*	$k_1[H_2O_2]$, E^*, $k_3[PCP]$, k_6, k_4, $k_5[PCP]$	$[E^*] = k_1k_3k_4k_5k_6[H_2O_2][PCP]^2$
	k_2, E_i, $k_1[H_2O_2]$, k_{-3}, k_6, $k_5[PCP]$	$k_{-3}k_1k_2k_5k_6[H_2O_2][PCP]$
E_i	k_2, E_i, $k_1[H_2O_2]$, k_6, k_4, $k_5[PCP]$	$k_1k_2k_4k_5k_6[H_2O_2][PCP]$
		$[E_i] = (k_{-3} + k_4)k_1k_2k_5k_6[H_2O_2][PCP]$
E_i^*	k_2, $k_3[PCP]$, $k_1[H_2O_2]$, E_i^*, k_6, $k_5[PCP]$	$[E_i^*] = k_1k_2k_3k_5k_6[H_2O_2][PCP]^2$

Continued

TABLE A7.1 (continued)

Enzyme Form	Pathways to Form	Sum of Rate Constant Products

E_{ii} $k_{-5}k_1k_2k_3k_4[H_2O_2][PCP]$

$k_1k_2k_3k_4k_6[H_2O_2][PCP]$

$[E_{ii}] = (k_{-5} + k_6)k_1k_2k_3k_4[H_2O_2][PCP]$

E_{ii}^* $[E_{ii}^*] = k_1k_2k_3k_4k_5[H_2O_2][PCP]^2$

Solving Equation A7.4 for V we find

$$V = \frac{k_6[E_0][H_2O_2][PCP]}{\frac{(k_{-1} + k_2)k_6}{k_1k_2}[PCP] + \left\{\frac{(k_{-3} + k_4)k_6}{k_3k_4} + \frac{k_{-5} + k_6}{k_5}\right\}[H_2O_2] + \left\{+ \frac{k_6}{k_2} + + \frac{k_6}{k_4} + 1\right\}[H_2O_2][PCP]} \quad (A7.5)$$

Rearranging the right-hand side of Equation A7.5 we obtain

$$V = \frac{\dfrac{1}{\dfrac{1}{k_2} + \dfrac{1}{k_4} + \dfrac{1}{k_6}}[E_0][H_2O_2][PCP]}{\dfrac{\dfrac{k_{-1} + k_2}{k_1k_2}}{\dfrac{1}{k_2} + \dfrac{1}{k_4} + \dfrac{1}{k_6}}[PCP] + \dfrac{\dfrac{k_{-3} + k_4}{k_3k_4} + \dfrac{k_{-5} + k_6}{k_5k_6}}{\dfrac{1}{k_2} + \dfrac{1}{k_4} + \dfrac{1}{k_6}}[H_2O_2] + [H_2O_2][PCP]} \quad (A7.6)$$

The above equation is formally identical to Equation 7.11.

REFERENCES

1. Evans, M.L., Porges, R.E., Coghlan, G.H., Eilers, R.G., and Glazer, A.G., Methodology and model for performance and cost comparison of innovative treatment technologies at wood preserving sites, *Remediation J.*, 12, 37–54, 2001.
2. Miller, K.M., Suidan, M.T., Sorial, G.A., Khodadoust, A.P., Acheson, C.M., and Brenner, R.C., Anaerobic treatment of soil wash fluids from a wood preserving site, *Water Sci. Technol.*, 38, 63–72, 1998.

3. Khodadoust, A.P., Sorial, G.A., Wilson, G.J., Suidan, M.T., Griffiths, R.A., and Brenner, R.C. Integrated system for remediation of contaminated soils, *J. Environ. Eng.*, 125, 1033–1041, 1999.

4. Koran, K.M., Suidan, M.T., Khodadoust, A.P., Sorial, G.A., and Brenner, R.C. Effectiveness of an anaerobic granular activated carbon fluidized-bed bioreactor to treat soil wash fluids: a proposed strategy for remediating PCP/PAH contaminated soils, *Water Res.*, 35, 2363–2370, 2001.

5. Pallerla, S. and Chambers, R.P., Reactor development for biodegradation of pentachlorophenol, *Catal. Today*, 40, 103–111, 1998.

6. Brandt, S., Zeng, A.P., and Deckwer, W.D., Adsorption capacity as a key parameter for enzyme induction and pentachlorophenol degradation in *Mycobacterium chlorophenolicum* PCP-1, *Biotechnol. Bioeng.*, 65, 93–99, 1999.

7. Cort, T. and Bielefeldt, A., Effects of surfactants and temperature on PCP biodegradation, *J. Environ. Eng.*, 126, 635–643, 2000.

8. Fahr, K., Wetzstein, H.G., Grey, R., and Schlosser, D., Degradation of 2,4-dichlorophenol and pentachlorophenol by two brown rot fungi, *FEMS Microb. Lett.*, 175, 127–132, 2001.

9. Cho, N.S., Nam, J.H., Park, J.M., Koo, C.D., Lee, S.S., Pashenova, N., Ohga, S., and Leonowicz, A., Transformation of chlorophenols by white-rot fungi and their laccase, *Holzforschung*, 55, 579–584, 2001.

10. Cort, T. and Bielefeldt, A., A kinetic model for surfactant inhibition of pentachlorophenol biodegradation, *Biotechnol. Bioeng.*, 78, 606–616, 2002.

11. Shim, S.S. and Kawamoto, K., Enzyme production activity of *Phanerochaete chrysosporium* and degradation of pentachlorophenol in a bioreactor, *Water Res.*, 36, 4445–4454, 2002.

12. Sedarati, M.R., Keshavarz, T., Leontievsky, A.A., and Evans, C.S., Transformation of high concentrations of chlorophenols by the white-rot basidiomycete *Trametes versicolor* immobilized on nylon mesh. *Electronic J. Biotechnol.* [online], 6(2), August 15, 2003.

13. Chen, S.T. and Berthouex, P.M., Use of an anaerobic sludge digestion process to treat pentachlorophenol (PCP)-contaminated soil, *J. Environ. Eng.*, 129, 1112–1119, 2003.

14. Kao, C.M., Chai, C.T., Liu, J.K., Yeh, T.Y., Chen, K.F., and Chen, S.C., Evaluation of natural and enhanced PCP biodegradation at a former pesticide manufacturing plant, *Water Res.*, 38, 663–672, 2004.

15. Walter, M., Boul, L., Chong, R., and Ford, C., Growth substrate selection and biodegradation of PCP by New Zealand white-rot fungi, *J. Environ. Manage.*, 71, 361–369, 2004.

16. Durán, N. and Esposito, E., Potential applications of oxidative enzymes and phenoloxidase-like compounds in wastewater and soil treatment: a review, *Appl. Catal. B: Environmental*, 28, 83–99, 2000.

17. Mester, T. and Tien, M., Oxidation mechanism of ligninolytic enzymes involved in the degradation of environmental pollutants, *Int. Biodeter. Biodegrad.*, 46, 51–59, 2000.

18. Torres, E., Bustos-Jaimes, I., and Le Borgne, S., Potential use of oxidative enzymes for the detoxification of organic pollutants, *Appl. Catal. B: Environmental*, 46, 1–15, 2003.

19. Tatsumi, K., Wada, S., and Ichikawa, H., Removal of chlorophenols from wastewater by immobilized horseradish peroxidase, *Biotechnol. Bioeng.*, 51, 126–130, 1996.

20. Choi, Y.J., Chae, H.J., and Kim, E.Y., Steady-state oxidation model by horseradish peroxidase for the estimation of the non-inactivation zone in the enzymatic removal of pentachlorophenol, *J. Biosci. Bioeng.*, 88, 368–373, 1999.

21. Zhang, G. and Nicell, J.A., Treatment of aqueous pentachlorophenol by horseradish peroxidase and hydrogen peroxide, *Water Res.*, 34, 1629–1637, 2000.

22. Song, H.Y., Liu, J.Z., Xiong, Y.H., Weng, L.P., and Ji, L.N., Treatment of aqueous chlorophenol by phthalic anhydride–modified horseradish peroxidase, *J. Mol. Catal. B: Enzymatic*, 22, 37–44, 2003.

23. Kim, E.Y., Choi, Y.J., Chae, H.J., and Chu, K.H., Removal of aqueous pentachlorophenol by horseradish peroxidase in the presence of surfactants, *Biotechnol. Bioprocess Eng.*, 11, 462–465, 2006.

24. Kim, E.Y., Chae, H.J., and Chu, K.H., Enzymatic oxidation of aqueous pentachlorophenol, *J. Environ. Sci.*, 19, 1032–1036, 2007.

25. Zhang, J., Ye, P., Chen, S., and Wang, W., Removal of pentachlorophenol by immobilized horseradish peroxidase, *Int. Biodeter. Biodegrad.*, 59, 307–314, 2007.

26. Ricotta, A., Unz, R.F., and Bollag, J.M., Role of a laccase in the degradation of pentachlorophenol, *Bull. Environ. Contam. Toxicol.*, 57, 560–567, 1996.

27. Cho, N.S., Rogalski, J., Jaszek, M., Luterek, J., Wojtas-Wasilewska, M., Malarczyk, E., Fink-Boots, M., and Leonowicz, A., Effect of coniferyl alcohol addition on removal of chlorophenols from water effluent by fungal laccase, *J. Wood Sci.*, 45, 174–178, 1999.

28. Ullah, M.A., Bedford, C.T., and Evans, C.S., Reactions of pentachlorophenol with laccase from *Coriolus versicolor*, *Appl. Microbiol. Biotechnol.*, 53, 230–234, 2000.

29. Wang, P., Woodward, C.A., and Kaufman, E.N., Poly(ethylene glycol)-modified ligninase enhances pentachlorophenol biodegradation in water–solvent mixtures, *Biotechnol. Bioeng.*, 64, 290–297, 1999.
30. Choi, S.H., Moon, S.H., and Gu, M.B., Biodegradation of chlorophenols using the cell-free culture broth of *Phanerochaete chrysosporium* immobilized in polyurethane foam, *J. Chem. Technol. Biotechnol.*, 77, 999–1004, 2002.
31. Kazunga, C., Aitken, M.D., and Gold, A., Primary product of the horseradish peroxidase–catalyzed oxidation of pentachlorophenol, *Environ. Sci. Technol.*, 33, 1408–1412, 1999.
32. Arno, M.B., Acosta, M., Del Rio, J.A., Varon, R., and García-Cánovas, F., A kinetic study on the suicide inactivation of peroxidase by hydrogen peroxide, *Biochim. Biophys. Acta*, 1041, 43–47, 1990.
33. Veitch, N.C., Horseradish peroxidase: a modern view of a classic enzyme, *Phytochemistry*, 65, 249–259, 2004.
34. Gajhede, M., Schuller, D.J., Henriksen, A., Smith, A.T., and Poulos, T.L., Crystal structure of horseradish peroxidase C at 2.15 Å resolution, *Nat. Struct. Biol.*, 4, 1032–1038, 1997.
35. Berglund, G.I., Carlsson, G.H., Smith, A.T., Szöke, H., Henriksen, A., and Hajdu, J., The catalytic pathway of horseradish peroxidase at high resolution, *Nature*, 417, 463–468, 2002.
36. Van Haandel, M.J.H., Primus, J.L., Teunis, C., Boersma, M.G., Osman, A.M., Veeger, C., and Rietjens, I.M.C.M., Reversible formation of high-valent-iron-oxo prophyrin intermediates in heme-based catalysis: revisiting the kinetic model for horseradish peroxidase, *Inorg. Chim. Acta*, 275/276, 98–105, 1998.
37. Baek, H.K. and van Wart, H.E., Elementary steps in the reaction of horseradish peroxidase with several peroxides: kinetics and thermodynamics of formation of compound 0 and compound I, *J. Am. Chem. Soc.*, 114, 718–725, 1992.
38. Rodríguez-López, J.N., Gilabert, M.A., Tudela, J., Thorneley, R.N.F., and García-Cánovas, F., Reactivity of horseradish peroxidase compound II toward substrates: kinetic evidence for a two-step mechanism, *Biochemistry*, 39, 13201–13209, 2000.
39. Taraban, M.B., Leshina, T.V., Anderson, M.A., and Grissom, C.B., Magnetic field dependence of electron transfer and the role of electron spin in heme enzymes: horseradish peroxidase, *J. Am. Chem. Soc.*, 119, 5768–5769, 1997.
40. Colonna, S., Gaggero, N., Richelmi, C., and Pasta, P., Recent biotechnological developments in the use of peroxidases, *Trends Biotechnol.*, 17, 163–168, 1999.
41. King, E.I. and Altman, C., A schematic method of deriving rate laws for enzyme-catalyzed reactions, *J. Phys. Chem.*, 60, 1375–1382, 1956.
42. Goldberg, D., *Genetic Algorithms in Search, Optimization, and Machine Learning*, Addison-Wesley, Reading, 1999.
43. Van de Velde, F., van Rantwijk, F., and Sheldon, R.A., Improving the catalytic performance of peroxidases in organic synthesis, *Trends Biotechnol.*, 19, 73–79, 2001.

8 Remediation of Sites Contaminated by Underground Storage Tank Releases

Lawrence K. Wang, Nazih K. Shammas,
Ping Wang, and Nicholas L. Clesceri

CONTENTS

8.1   INTRODUCTION

Underground storage tanks (USTs) comprise one or a combination of tanks (including the associated underground piping) that are used to contain substances regulated under the RCRA[1,2] (Resource Conservation and Recovery Act) or CERCLA[3,4] (Comprehensive Environmental Response, Compensation, and Liability Act—Superfund), the volume of which include 10% or more located below ground surface (bgs). Generally, this term does not encompass residential and farm tanks holding 4164 L (1100 gal) or less of motor fuel used for noncommercial purposes, tanks storing heating oil to be used on the premises where it is stored, tanks on or above the floor of an underground area, such as basements or tunnels, septic tanks, and systems for collecting wastewater and stormwater, flow-through process tanks, emergency spill and overfill tanks, and related pipeline facilities.[5–7]

When the UST program began, there were approximately 2.1 million regulated tanks in the U.S. Today there are far fewer, because many substandard UST systems have been closed.[8] According to the U.S. Environmental Protection Agency (U.S. EPA), less than 5% of the current number of UST tanks store hazardous substances.[6] The majority of these tanks are used to store petroleum products for retail and industrial purposes. of the regulated tanks, 80% are believed to be made of bare steel, which can quickly corrode, allowing the contaminants to seep into the ground, posing a significant threat to the environment. The greatest potential hazard from a leaking UST is that the petroleum or other hazardous substance may seep into the soil and contaminate groundwater, the source of drinking water for nearly half of all Americans.[8] A leaking UST can present other health and environmental risks, including the potential for fire and explosion.

Federal UST regulations[9,10] promulgated in September 1988 established the minimum requirements for the design, installation, operation, and testing of USTs in the U.S. Through the implementation of the Clean Water Act[11] (CWA) (including the regulations issued for oil pollution prevention) and the Occupational Safety and Health Administration[12] (OSHA) (incorporating underground motor fuel storage tanks in its regulations dealing with flammable and combustible liquids), the control of USTs has helped in the minimization of the adverse environmental impact caused by the leakage of products from underground tanks.

This chapter will discuss those USTs storing petroleum products, such as gasoline, fuel oil, kerosene, and crude oil, and the problems related to petroleum release. In this context, the term "oil" or "gasoline" will be used in the text. Accordingly, the sections on underground release and transport remedial technologies mainly deal with petroleum products. Most petroleum products are nonaqueous-phase liquids (NAPLs) that are immiscible with water and have a lower specific gravity. The remainder of NAPLs with specific gravities greater than water are called the dense nonaqueous-phase liquids (DNAPLs). DNAPLs constitute only a small percentage of the petroleum products stored in USTs.

8.2 LEGISLATIVE AND REGULATORY OVERVIEW

The consequences of the release of petroleum from leaking USTs include a loss of valuable fuel, contamination of drinking water supplies, and danger to human life, property, and the environment. The RCRA was enacted to regulate the generation, transportation, storage, treatment, and disposal of waste material that met the definition of hazardous waste. Subtitles I and J of the RCRA are specifically promulgated for the management of underground storage tanks.

8.2.1 SUBTITLE I OF THE RCRA

Subtitle I of RCRA was enacted to control and prevent leaks from underground storage tanks.[1,6] It regulates substances, including petroleum products and hazardous material. Tanks storing hazardous wastes, however, are regulated under Subtitle C, and are not the concern of this chapter.

On September 23, 1988, U.S. EPA issued the final technical performance standards and associated regulations for USTs.[13] On October 26, 1988, U.S. EPA issued the final regulations for financial responsibility for those USTs related mainly with petroleum products. The technical standards for USTs comprise eight components, as described in the following sections[13,14]:

8.2.1.1 Program Scope and Interim Prohibition

Both the program scope and the interim prohibition must be clearly identified and documented.

8.2.1.2 Design, Construction, Installation, and Notification Requirements

U.S. EPA has established standards for tanks and piping tightness tests. In lieu of the standards specified in the regulations, new USTs may be constructed using alternative standards as long as they are equally protective of human health and the environment. The cathodic protection systems of new USTs must be designed and installed in accordance with industry codes. Tank installation includes securing the tank, obtaining clean backfill, and ensuring that the substances to be stored are compatible with the tank system. Tanks must be properly installed following manufacturer specifications and certified by the state regulatory agency when installation is satisfactorily completed. USTs must also be fitted with equipment to prevent the spills and overfills that are the common causes of tank leakage. Existing USTs had to comply with all requirements for new tanks by December 22, 1998. Any UST systems that were unable to meet the deadline were closed.

8.2.1.3 General Operating Requirements

Four steps must be taken to meet the general operating requirements to prevent spills or overfills:

1. Ensuring that the capacity of the tank is greater than the volume of product to be transferred
2. Having someone present at all times during the transfer

3. Incorporating equipment that can prevent or severely limit spills, such as automatic shutoff devices that act when the tank is almost full
4. Following manufacturer recommendations regarding proper maintenance, including inspections, record keeping, periodic maintenance, and corrosion protection.

8.2.1.4 Release Detection

Release detection is one of the most important requirements of the UST program. The detection system should be capable of detecting a release from any part of the UST system. Detection methods will be discussed under Section 8.3.

8.2.1.5 Release Reporting, Investigation, and Confirmation

Any spill or overfill of over 95 L (25 gal) petroleum must be reported within 24 h. An amount less than 95 L (25 gal), that cannot be cleaned up within 24 h should also be reported.

8.2.1.6 Corrective Action Requirements

Following the immediate response activities (including release reporting, immediate containment, and monitoring of explosive hazards), the actions that the facility must implement as initial abatement measures include the following:

1. Further containing the regulated substance to prevent continued release
2. Preventing further migration of aboveground and underground release
3. Continuously monitoring and mitigating explosive hazards
4. Remedying hazards posed by excavated soils resulting from response activities
5. Performing a site check to evaluate the extent of the release
6. Determining the presence of free product on the water table
7. Compiling detailed corrective action plans if further corrective action is found to be required.

8.2.1.7 UST Closure

Unless permanently closed, all systems containing regulated substances must continue to comply with all the normal regulatory requirements. USTs closed for less than three months have no special requirements. USTs closed for between three and twelve months must leave vent lines open and cap all other lines. After 12 months out of service, USTs must be closed permanently. Before closing the UST system, the site must be assessed to ensure that no further release has occurred.

8.2.1.8 Financial Assurance

Under the new petroleum UST regulations, financial assurance (between ca. 0.5 and 1 million USD per occurrence or between 1 and 2 million USD for aggregate coverage) is required to cover both the cost of any required corrective action, and compensation for third-party liability from accidental release. State and federally owned facilities are exempt from these requirements.

As part of the amendments to Superfund, U.S. Congress created the Leaking Underground Storage Tank Trust Fund under RCRA Subtitle I. The Trust Fund is financed through a tax on gasoline, diesel, and aviation fuels and is used when the following conditions are met:

1. Cleanup costs exceed the coverage requirements of the financially responsible party.
2. The owner or operator refuses to comply with a corrective action order.
3. A solvent owner or operator cannot be found.

4. An emergency situation exists.
5. To cover the administrative and enforcement costs associated with a cleanup.

8.2.2 SUBTITLE J OF THE RCRA

In order to regulate USTs storing hazardous substances and to provide a second means of containing the substance should the tank fail, U.S. EPA revised Subtitle J of the RCRA, which regulates secondary containment systems. This secondary containment system would have the following features:

1. It will prevent waste or liquid from escaping to the soil or water for the life of the tank.
2. It will collect waste or leakage until the material is removed.
3. It will be constructed or lined with material compatible with the waste and with sufficient strength to prevent failure from pressure, climate, traffic, and daily use.
4. It will have an adequate base or foundation capable of resisting settlement compression and uplift.
5. It will have a system capable of detecting leaks within 24 h of occurrence.
6. It will have a slope or drain system to permit removal of leaks, spills, and precipitation, and contain provisions for such accumulation to be removed.
7. It will have 110% of the design capacity of the largest tank within the containment boundary.
8. It will prevent run-on or infiltration of precipitation unless the collection system has excess capacity (beyond the 110%) to hold precipitation consistent with the 25-yr, 24-h rainstorm prediction.

8.2.3 STATE AND LOCAL UST PROGRAMS

Several states already have, or are developing, regulatory programs for USTs. Subtitle I of the RCRA is designed to avoid interfering with those state programs and to encourage other states to press ahead with control programs.

According to the state program approval regulations (promulgated on September 23, 1989) U.S. EPA will evaluate various elements of the state program against the corresponding Federal requirements. U.S. EPA must determine that the state's requirements are "no less stringent" than the Federal program, and that there is provision for "adequate enforcement."

8.2.4 USTs CONTAINING OTHER HAZARDOUS CHEMICALS

The regulatory standards for leak detection in tank systems containing hazardous chemicals are more stringent than those for tanks containing petroleum motor fuels. Both above standards and those required in RCRA hazardous substances management should be met.

8.3 CAUSES OF LEAKS AND LEAK IDENTIFICATION METHODS

8.3.1 CAUSES OF TANK FAILURE

USTs release contaminants into the environment as a result of (1) corrosion, (2) faulty installation, (3) piping failure, and (4) overfills.[15–17]

Corrosion and poor installation are by far the most common causes of storage system leaks. The most common causes of release from bare-steel UST systems are galvanic corrosion and the breakdown of hard refined steel to its natural soft ore. Because older USTs are usually constructed from bare steel, corrosion is believed to be the leading factor contributing to release. The speed and severity of corrosion varies depending on site characteristics, such as soil conductivity, groundwater or soil water chemistry, and weather. Most commonly, part of a tank becomes negatively charged with respect to the surrounding area and acts as a battery. The negatively charged part of the steel UST starts to corrode at a rate proportional to the intensity of the current. Corrosion rate can be reduced significantly or eliminated if cathodic protection or other protection methods are used.

Faulty installation of USTs encompasses a wide variety of problems, for example, accidents from vehicles colliding with the storage system, or faulty installation arising from inadequate compaction of backfills and unsealing of joints. Therefore, precautions should be taken to ensure that poor construction or installation do not degrade the performance of the USTs.

Piping failure can be caused in several ways. A study by U.S. EPA[16] has shown that piping failure accounted for a substantial portion of releases at USTs. Spills and overfills are usually caused by human error. Repeated spill can also increase the corrosive nature of soils.

8.3.2   LEAK IDENTIFICATION METHODS

Three basic actions can be considered to identify leakage from USTs[5,18]: (1) direct observation (visual observation of losses or environmental and mechanical signs of leaks), (2) checking (inventory monitoring), and (3) testing (instrumental testing of tanks and piping for leaks). These are described in the following sections.

8.3.2.1   Visual Tank Inspection

Visual inspection may be carried out by entering the tank if it is large enough for a person to be able to enter and walk in the tank, or by inspection of the tank's outer walls following the removal of pads or backfill material.

8.3.2.2   Watching for Environmental Signs

There are at least five signs to look for:

1. The odor of motor fuel in the soil near the tank may be a sign of leakage.
2. The odor of motor fuel present in underground structures such as basements and sewers is also a sign of leakage.
3. Plants located on property near a UST may grow sluggishly, look sickly, or die.[5]
4. Motor fuels may be found in drinking water wells or rivers.
5. A higher than expected gain or loss of fuel in a tank may be caused by water infiltration or leakage of fuel through the tank wall.

8.3.2.3   Watching for Mechanical Signs

There are three phenomena to be monitored[5]:

1. Interruption in the delivery of motor fuel dispensed by the suction pump
2. A rattling sound and irregular fuel flow in the suction pumping system
3. Meter spin without motor fuel delivery

It should be noted that these can also be caused by other problems besides tank leakage, such as leaking valves, loose fittings, or other factors.

8.3.2.4   Checking Inventory

By carefully checking inventory records one is able to determine whether there is loss or gain of fuel in USTs. Inventory review is generally an inexpensive and relatively easy way to check for leakage.

This method is particularly useful for identifying large leaks, although small leaks may also be noticed, particularly in tanks with metered dispensing pumps. Interested readers can refer to U.S. EPA[5] and API (American Petroleum Institute)[20] for detailed procedures of inventory checking for tanks with metered or nonmetered pumps.

8.3.2.5   Environmental Tests with Instrumentation

Another method to examine tank leakage uses instrumentation. An instrumental test should be conducted if there is the suggestion of a leakage from various environmental or mechanical indicators or from an inventory review.

When the leaked motor fuel is at a deeper level or flows away from underground strata, there may be no visual sign, and instrumentation may be necessary to detect the leak. Such an instrumental test on the tank environment is called an external test, and is the counterpart of visual observation.

There are a number of methods for detecting the sign of leakage from external tanks. The most common method uses monitoring wells. Typically, the monitoring well reaches 2 ft below the bottom of the underground storage tank. Detection sticks are placed in the well, and indicate the existence of motor fuel within the well. Other methods include soil sampling, fuel vapor testing, ground penetrating radar, seismic methods, electromagnetic induction, resistivity, magnetometers, and X-ray fluorescence.

Tracer methods can also be used, in which tracers such as freon, fluorescent materials, and isotope-fuel are added to a tank, and are then detected externally. An analogy of tracer methods includes pressurizing the tank with a noble gas, then detecting the gas if it escapes from the tank through cracks or holes.

Some tanks are installed with permanent leak identification sensors, which can check for leaked fuel vapor or liquid as it comes into contact with the sensors.[21] However, these, as well as all the environmental sign tests (visual or instrumental) may be triggered by a spill instead of a leak. The success of external systems depends on the sensitivity of the sensor, the ability of the sensor to distinguish the stored chemical from other chemicals, the ambient background noise level of the stored chemical, the migration properties of the chemical, and the sampling network.

8.3.2.6   In-Tank Measurement with Instrumentation

In-tank measurement uses equipment that is placed inside the tank or pipes. Some tests can qualitatively determine whether a tank is leaking; others can establish the leakage rate. Most of the work can be performed within a time of 2 to 4h, excluding setup time.[5,21] A common method measures the changes in the amount of fuel in the tank by measuring the fuel level or pressure. These tests may be influenced by several factors, including evaporation, condensation, and changes in temperature, changes in the shape of the tank due to changes in the fuel load, temperature air packets, vibrations from traffic, groundwater, or soil moisture.

Other devices and methods can also be applied, such as laser interferometry, which measures the change in the height of fuel in the tank using lasers, or acoustics methods that measure the sound of fluid escaping or entering the tank.

8.3.2.7   Direct Tank Tests with Instrumentation

An instrument can be used to test tank walls directly, for example, by using acoustics or sound waves to identify holes or cracks in the tank walls.[18]

8.3.2.8   Release Detection Approaches for Modern Tank Systems

Release detection is an important aspect of the management of USTs. U.S. EPA regulations required an upgrade of release detection during the 10-yr period between 1988 and 1998. The external or internal detection systems should be in compliance with the requirements for modern tank systems.

There are three methods of release detection that are associated with modern tank systems.[18,22] The first approach is to conduct an annual tank or line tightness test to detect small releases and to use more frequent monitoring by another method to detect large releases. All tank and line tightness tests must be performed at least once a year and must be able to detect leaks of 0.38 L/h (0.1 gal/h). In all cases where annual tightness tests are used, the regulation requires an additional form of leak detection in which tests on tanks are conducted at least monthly and those on pressurized lines at least hourly; this ensures the detection of excessively large releases. For tanks, daily inventory records must be reconciled monthly. for pressurized lines, leaks of up to 11.4 L/h (3 gal/h) must be reliably detected.

The second approach is to install an automatic tank gage or automatic line leak detector that is capable of detecting leaks of 0.76 L/h (0.2 gal/h); all monitoring tests must be done at least once a month. This option also requires that there be a system for detecting large leaks. The tank gage can be used to satisfy inventory control requirements, and most automatic line leak detectors are designed so as to be able to satisfy the 11.4 L/h (3 gal/h) test for pressurized piping.

The third approach is to install an external monitoring system that can detect the presence of the stored chemical in or on the groundwater or in the backfill and soil surrounding the tank system. In many instances both internal and external methods are used in conjunction as a way to increase the liability of detection.

8.4 UNDERGROUND CONDITIONS AND FACTORS AFFECTING TRANSPORT OF LIQUIDS IN THE SUBSURFACE

8.4.1 UNDERGROUND FORMATIONS

Subsurface formations can be divided into the overburden (unconsolidated) and bedrock according to its solidarity. The upper subsurface can be further divided into the unsaturated zone and the saturated zone depending on pore structure and moisture saturation. The saturated zone is the zone in which the voids in the rock or soil are filled with water at a pressure greater than atmospheric. The water table is at the top of a saturated zone in an unconfined aquifer. The unsaturated zone is the zone between the land surface and the water table, and is also called the zone of aeration or the vadose zone. The pore spaces contain water at less than atmospheric pressure, air, and other gases. This zone is unsaturated except during periods of heavy infiltration.

In the lower region of the unsaturated zone, immediately above the water table, is the capillary fringe, where water is drawn upward by capillary attraction. Above the capillary fringe, moisture coats the solid surfaces of the soil or rock particles. If the liquid coating becomes too thick to be held by surface tension, a droplet will pull away and be drawn downward by gravity. The fluid can also evaporate and move through the air space in the pores as water vapor.

The moisture in the upper unsaturated zone can be affected by plant transpiration and atmospheric conditions. Some scholars classify the unsaturated zone into subzones such as the soil water zone and the intermediate zone.[23]

8.4.2 GRAVITATIONAL FORCE AFFECTING UNDERGROUND LIQUID MOVEMENT

Soil water, like other bodies in nature, has two principal forms of energy, kinetic and potential. Kinetic energy is proportional to the square of velocity. As the velocity of groundwater is quite slow, the kinetic energy is usually negligible. Potential energy, due to position or internal conditions, determines the movement of water from a higher energy level to a lower energy level in soil formations. Accordingly, there are three forces related to potential energy:

1. Gravity (the weight of the fluid)
2. External pressure (atmosphere pressure)
3. Molecular attraction (surface tension, adsorptive, diffusive, and osmotic forces)

The forces resisting groundwater flow are shearing stress and normal stress due to viscosity, collision, and turbulence.

Gravity force can be measured by means of the mass of the water. The direction of the force is, obviously, downwards toward the Earth's center. The gravitational potential of soil water at each position is determined by the elevation of the position relative to some reference level. If we only consider the elevation potential and the related velocity energy, then a water body at a higher elevation will flow to a lower elevation, decreasing the elevation potential but increasing its velocity.

8.4.3 ATMOSPHERE PRESSURE AFFECTING UNDERGROUND LIQUID MOVEMENT

Atmospheric pressure is not obvious, because it is balanced in opposite directions. The combination of atmospheric pressure and the weight of the overlying water create the total pressure in the saturation zone.[24]

8.4.4 SURFACE TENSION AND CAPILLARY POTENTIAL AFFECTING UNDERGROUND LIQUID MOVEMENT

Tension in the free surface of a liquid is the cause of the tendency of a liquid surface to assume the form having a minimum area, as manifested in the shape of a bubble or a drop of liquid.[25] The tendency to contract is a special case of the general principle that potential energy tends toward a minimum value.

8.4.4.1 Wetting and Nonwetting

When a drop of liquid is placed on a solid surface, it will displace the gas and spread over the surface. If the contact angle is <90°, the liquid wets the solid (wetting, Figure 8.1a); if the contact angle is >90°, the liquid does not wet the solid (nonwetting, Figure 8.1b).

Whether a liquid wets or does not wet a solid surface depends on the affinity between the liquid and the solid. In the case of wetting, the smaller contact angle enables the liquid to enlarge the liquid–solid interface area (which has a lower surface energy than the liquid surface energy) and shrinks the liquid surface area (which has a greater surface energy), thus reducing the total energy. In the case of nonwetting, the greater contact angle enables the reduction of the liquid–solid

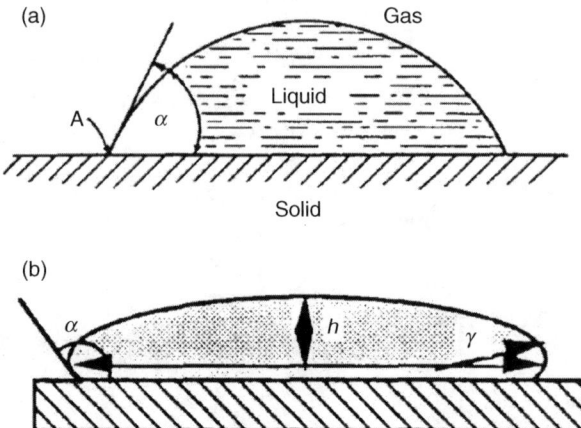

FIGURE 8.1   (a) Drop contact angle and (b) a sessile drop showing characteristic dimensions.

interface area (which has a higher surface energy than the liquid surface energy) and the enlarge-ment of the liquid surface area (which has a lower surface energy), thus bringing about a reduction in the total energy.

8.4.4.2 Capillary Potential

A liquid–solid contact angle away from 90° induces the formation of a meniscus on the free surface of the liquid in a vertical tube (the solid phase). In the nonwetting case, the meniscus concaves upwards to the air. The upwards meniscus is the result of a downward surface tension at the liquid–tube interface, causing a capillary depression. In the wetting case, the meniscus has a concave-downward configuration. The downwards meniscus is the result of an upward surface tension at the liquid–tube interface, causing a capillary rise.

A typical profile of the pressure potential of soil moisture tested by a tensometer across the free-water surface shows a negative pressure (lower than atmosphere pressure) in the capillary zone (Figure 8.2). The negative pressure in the capillary zone indicates that the capillary zone belongs to the unsaturated zone.

Surface tension is independent of tube size. However, the extent of capillary rise or depression by surface tension is dependent on tube size. This can be seen from Equation 8.1 in Section 8.4.6.1. For example, in the case of a capillary rise, the greater the tension, the higher the water rises above the free-water surface. For the same amount of water, the smaller the tube is, the higher the water rises.

8.4.4.3 Relative Soil Wettability of Two Liquid Phases

The predominant form of released petroleum products is a liquid that is immiscible with water; this is called the free product (in this section it will be referred to as oil). The behavior of water and oil in soil depends on the interaction of the three phases water, oil, and soil. The affinity of water or oil with the soil can be estimated by establishing the contact angle of oil/water/soil triple line.

Note that the contact angle of the fluid 1/fluid 2/solid triple line is still largely unpredictable, even though the material properties of the three phases, taken separately, are known. It is difficult to compare the wettabilities of a solid with respect to two fluids that wet it perfectly, or, in other terms, to measure the fluids' spreading parameters, even on an ideal surface. There are several methods used for wettability evaluation. The AMOTT-IFP (E. Amott – Institut Français du Pétrole) test is probably the most widely accepted one. Other advanced methods for measuring wettability include the computerized automated tomographic X-ray scanner, magnetic resonance imagery,

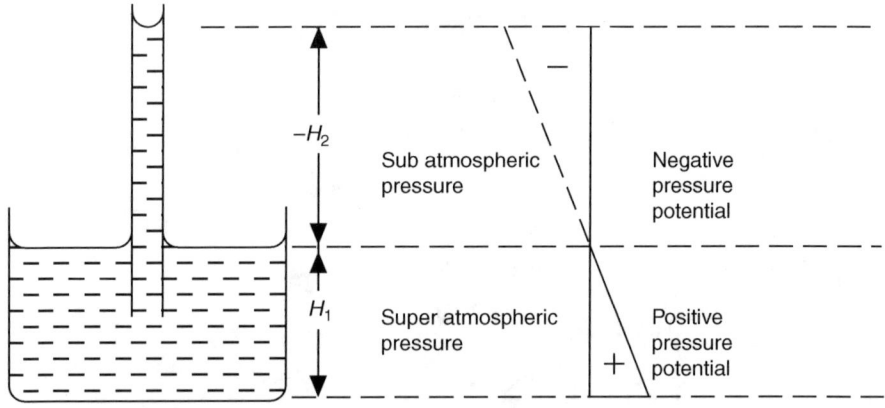

FIGURE 8.2 Superatmospheric and subatmospheric pressures below and above a free-water surface.

cryo-scanning electron microscopy, and nuclear magnetic resonance. For more information, see references 26 to 28.

Wettability measurements show that most soil constituents are water wettable or hydrophilic,[28] although calcium carbonates [calcite, $CaCO_3$, and dolomite, $CaMg(CO_3)_2$] are slightly hydrophobic; for example, the contact angle of water and heptane is 100 to 105°. Therefore, carbonaceous reservoirs are usually oil-wet.

In practice, evaluating the wettability of a soil is far more uncertain, because soil is a mixture of gravel, sand, silt, and clay particles, as well as other chemical precipitates. The mineral components of soil particles include quartz, feldspar, carbonates, and clay. These components have different wettability by water and oil. Therefore, the retention of oil or water in a soil matrix is heterogeneous and variable. The general wettability of soil or liquid retention in a soil is reported on a statistical basis.

The relative wettability of soil by oil and water determines the relative affinity of soil to oil and water, which in turn determines the level of retention of oil or water in the soil. A soil is hydrophilic (i.e., it has water affinity) if water has more affinity to the soil than the oil, although soil may also to a certain degree be somewhat wetted by oil. A soil is called hydrophobic if oil has more affinity to the soil than water.

When water-affinity soil that is originally saturated with oil is rinsed with water, most of the oil will be displaced by water; however, some oil will remain. This remaining amount of oil that can no longer be displaced by the flowing water is defined by the petroleum industry as the "residual oil saturation."[28] This term is used to measure the upper limit of the microscopic efficiency of the displacement of oil by water. There is no appreciable lower limit, as at a given pore level the wetting fluid should be spread over the mineral surface as a continuous wetting film, which might thin itself until the externally applied capillary pressure (or pressure difference across the wetting fluid/nonwetting fluid interface) is balanced by the thin film disjoining pressure.

If an open space of a water-affinity porous medium filled with oil is brought into contact with a reservoir of water, the oil will be spontaneously displaced by the water. Conceivably, a symmetric behavior can be observed if an oil-affinity medium at residual oil saturation is brought into contact with oil. The driving force for this spontaneous flow is the high initial capillary pressure (which equals the pressure in the nonwetting fluid minus the pressure in the wetting fluid) inside the medium compared to its value of zero outside, where the oil–water interface has no curvature at equilibrium. Capillary pressures, like relative permeabilities, are a function of saturations, the geometric properties of the porous medium, the fluid–fluid interfacial tension, and the wettability.[28] It is generally observed that if all other parameters of a system are maintained unchanged, but the wetting properties of the solid are changed, the nonwetting fluid will be displaced more easily by the wetting fluid than vice versa.

Surfactants are used to rinse oil from soil more effectively. Surfactants have higher soil affinity, and so reduce the interfacial tension between the oil and the soil. The replacement of the oil film by a surfactant solution is dependent on the contact angle between the oil–solution interface and the soil. As long as the contact angle (in the solution) is acute, the solution will tend to advance, displacing oil from the soil. Mechanical agitation would assist the spreading by compensating to some extent for the immobility of the molecules on the surface of the soil. Processes for the removal of surfactants in a contaminated liquid are described in works by Wang and colleagues[29] and Hrycyk and colleagues.[30]

8.4.5 Adsorptive Force Affecting Underground Liquid Movement

Adsorption results from bonding forces between the solute and soil particles. These forces are generally electrostatic, although entropy generation and magnetic forces can be involved. Bonding forces range from relatively weak to strong with respect to bond formation.[31]

Adsorption can be attributed to the following interactions: van der Waals–London interactions, charge transfer/hydrogen bonding, ligand exchanges, ion exchange, direct and induced ion–dipole and dipole–dipole forces, chemisorption, hydrophobic bonding, and magnetic bonding. These interactions are the result of electrostatic magnetic or entropy-generating forces. Most often, physical adsorption is due to the electrostatic interactions between atoms, ions, and molecules resulting from instantaneous dipoles. Van der Waals force, charge transfer/hydrogen bonding, ligand exchange, ion exchange, direct and induced ion–dipole and dipole–dipole force, and chemisorption interactions are all the result of electrostatic forces, variation in energy level, and method of interaction. Hydrophobic bonding is due to entropy-generating forces, and magnetic bonding is due to magnetic forces.[31]

Van der Waals–London interactions are due to fluctuations in electron distribution as the electrons circulate within their orbits. These instantaneous dipoles are usually weak, but are, regardless, the most common interaction resulting in adsorption.[31] Stronger interactions result from charge transfer.

8.4.6 COMBINATION OF CAPILLARY AND ADSORPTIVE FORCES AFFECTING UNDERGROUND LIQUID MOVEMENT

Both adsorptive and capillary forces play an important part in soil–liquid interaction (see Figure 8.3). This is very important for unsaturated soil. The total force (i.e., the sum of capillary force and adsorptive force) is termed the matrix potential, which has a negative gage pressure relative to the external gas pressure on the soil water (more often the gage pressure is referred to as the atmospheric pressure).

In fact, an unsaturated soil has no pressure potential, only a matrix potential (expressible as a negative pressure). The negative pressure causes water to move toward the soil with a higher suction potential, in contrast to the saturated flow where water moves from a high pressure potential to a low pressure potential. For soils with the same properties but with different saturation, the less saturated soil has more excessive suction force, causing water to move towards it.

The presence of water in films as well as under concave menisci is most important in clayey soil and at high suctions, because clay minerals have high specific surface area and often have a

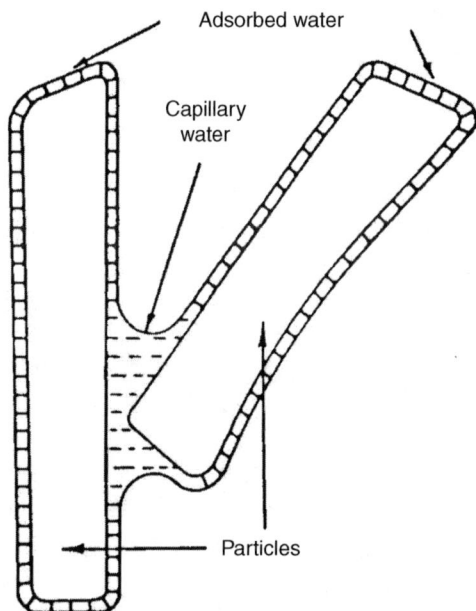

Adsorbed water

Capillary
water

Particles

FIGURE 8.3   Water in an unsaturated soil is subject to capillarity and adsorption, which combine to produce a matric suction.

high cation exchange capacity. In sandy soil, adsorption is relatively unimportant and the capillary effect predominates. A combination of capillary effects and adsorption results in negative pressure potential.

8.4.6.1 Viscosity and Shearing Stress

Viscosity is the property of a fluid that offers resistance to the relative motion of fluid molecules. The energy loss due to friction in a flowing fluid is due to its viscosity.

As a fluid moves, shearing stress develops in it. The magnitude of the shearing stress depends on the viscosity of the fluid. Shearing stress can be defined as the force required to slide one unit area layer of a substance over another. Considering a fluid moving along a fixed surface, the velocity is highest along the moving surface, and zero at the fixed surface. The shearing stress in a fluid is directly proportional to the velocity gradient:

$$S_s = \mu(dV_s/dy) \tag{8.1}$$

where S_s = shearing stress (M/LT^2), y = distance (L) between the moving and fixed surfaces, V_s = the velocity along the surfaces (L/T), dV_s/dy = velocity gradient across the surfaces, and μ = proportionality constant = dynamic viscosity (or absolute viscosity) of the fluid (M/LT).

Equation 8.1 is known as Newton's Law of Friction. In the SI system, the dynamic viscosity units are N-sec/m², Pa-sec, or kg/m-sec. Here 1 Pa-sec = 1 N-sec/m² = 1 kg/m-sec. The dynamic viscosity (or absolute viscosity) is also often expressed in the metric CGS (centimeter-gram-second) system as g/cm-sec, dyne-sec/cm², or poise (P) where 1 poise = 1 P = 1 dyne-sec/cm² = 1 g/cm-sec = 0.1 Pa-sec = 100 centipoises = 100 cP. Water at 20.2°C (68.4°F) has a dynamic viscosity of 1 cP.

Kinematic viscosity is the ratio of dynamic viscosity and density, and can be obtained by dividing the dynamic viscosity of a fluid with its mass density, as shown by Equation 8.2:

$$v = \mu/\rho \tag{8.2}$$

where v = kinematic viscosity (L^2/T), μ = dynamic viscosity (M/LT), and ρ = density (M/L^3).

In the SI system, the theoretical unit of v is m²/sec or the commonly used Stoke (St) where 1 St = 0.0001 m²/sec = 100 cSt = 100 centiStoke. Similarly, 1 centiStoke = 1 cSt = 0.000001 m²/sec = 0.01 Stoke = 0.01 st. The specific gravity of water at 20.2°C (68.4°F) is almost 1. The kinematic viscosity of water at 20.2°C (68.4°F) is for all practical purposes equal to 1 cSt. For a liquid, the kinematic viscosity will decrease with higher temperature. For a gas, the kinematic viscosity will increase with higher temperature. Another commonly used kinematic viscosity unit is Saybolt universal seconds (SUS), which is the efflux time required for 60 mL of petroleum product to flow through the calibrated orifice of a Saybolt universal viscometer, as described by ASTM-D88. Therefore, the relationship between dynamic viscosity and kinematic viscosity can be expressed as

$$v = 4.63 \, \mu/Sg \tag{8.3}$$

where v = kinematic viscosity (SUS), μ = dynamic or absolute viscosity (cP), and Sg = specific gravity (dimensionless).

The viscosities of water and gasoline increase with decreasing temperature. Gasoline has lower viscosity than water, and fuel and crude oil have a much higher viscosity that increases dramatically when temperature decreases.[32] The ease with which a fluid pours is an indication of its viscosity. It is observed that cold oil has a high viscosity and pours very slowly. The viscosity properties of various potential pollutants are discussed in Section 8.9.

8.4.6.2   Electrokinetic Effects

Flow movement also has a relationship with the electrokinetic phenomenon, which can promote or retard the motion of the fluid constituents. Electrokinetic effects can be described as when an electrical double layer exists at an interface between a mobile phase and a stationary phase. A relative movement of the two phases can be induced by applying an electric field and, conversely, an induced relative movement of the two will give rise to a measurable potential difference.[33]

Four phenomena are classified as electrokinetic effects:

1. Streaming potential
2. Electroosmosis
3. Sedimentation potential
4. Electrophoresis

The streaming potential (Dorn effect) relates to a movement of liquid that generates electric potential, and electroosmosis occurs when a direct electric potential causes movement of the liquid. The sedimentation potential relates to sedimentation (directed movement) of charged particles that generates electric potential, and electrophoresis occurs when a direct electric potential causes a movement of charged particles.

With regard to the movement of liquid versus particles under direct current, electrophoresis is the reverse of the effect of electroosmosis.[33] If particles move through a liquid that is stationary, this is called electrophoresis; conversely, if the liquid moves through particles that are stationary, that is called electroosmosis.

The potential governing these electrokinetic effects is clearly at the boundary (the face of shear) between the stationary phase (the fixed double layer) and the moving phase (the solution). This potential is called the electrokinetic potential or the zeta potential. An electrokinetic phenomenon in soil involves coupling between electrical, chemical, and hydraulic gradients.

Initially the electrical potential difference is distributed linearly across the specimen. The changing chemistry across the cell may result in variations in electrical potential difference in time and space.

It is suggested that the movement of the front by migration (electrical potential), diffusion (chemical potentials), and advection (hydraulic potentials) will cause desorption of cations and other species from clay surfaces and facilitate their release into the fluid.[34]

The relationship of electrokinetic phenomena and the movement of petroleum constituents is not of high importance; however, it can be important for the transport of some solutes related to a remedial technology such as electroosmosis remediation.

8.4.7   ENERGY CONSERVATION AFFECTING UNDERGROUND LIQUID MOVEMENT

Equation 8.4 describes the energy conservation when water moves between two points (1 and 2)

$$\frac{P_1}{\gamma} + Z_1 + \frac{v_1^2}{2g} + h_A - h_R - h_L = \frac{P_2}{\gamma} + Z_2 + \frac{v_2^2}{2g} \tag{8.4}$$

where P_1 = pressure at point 1 (M/L^2), P_2 = pressure at point 2 (M/L^2), P_1/γ = pressure head at point 1 (L), P_2/γ = pressure head at point 2 (L), γ = density of liquid (M/L^3), Z_1 = elevation of point 1 (L), Z_2 = elevation of point 2 (L), $v_1^2/2g$ = velocity head at point 1 (L), $v_2^2/2g$ = velocity head at point 2 (L), g = gravitational acceleration (L/T^2), h_A = the head added to the fluid with a mechanical device such as a pump (L), h_R = the head removed from the fluid with a mechanical device such as fluid motor (L), and h_L = the head losses from the system due to friction (L). Equation 8.4 reduces to the familiar Bernoulli's equation when there is no pump ($h_A = 0$), no motor ($h_R = 0$), and where the head loss (h_L) between the two points is negligible.

8.4.8   WATER MOVEMENT IN SATURATED ZONE OF SOIL FORMATION

Hydraulic conductivity is one of the characteristic properties of a soil relating to water flow. The movement of water in soil depends on the soil structure, in particular its porosity and pore size distribution. A soil containing more void space usually has a higher permeability. Most consolidated bedrocks are low in permeability. However, rock fractures could create a path for water movement.

Groundwater flowing through an aquifer is influenced by gravitational force, but the rate at which the groundwater moves can vary significantly. Depending on the permeability of an aquifer and the flow gradient, groundwater can move at a velocity varying from only a few meters per year to several meters per day.

The most important factor for movement in the saturated zone is the hydraulic gradient. The velocity head, which is generally more than ten orders of magnitude smaller than the pressure and gravitational head, may be neglected because of the slow water movement. Equation 8.4 can therefore be simplified to

$$\frac{P_1}{\gamma} + Z_1 + h_A - h_R - h_L = \frac{P_2}{\gamma} + Z_2 \qquad (8.5)$$

The relative importance of pressure and gravitational heads depends on whether the water formation is in a free water table condition or in a confined aquifer condition.

8.4.8.1   Water Table Condition

When considering two points on a water table 1 and 2, P_1 can be regarded as equal to P_2, because the external pressure is the same as the atmospheric pressure. If there is neither addition nor loss of head by mechanical devices (i.e., $h_A = 0$ and $h_R = 0$), then Equation 8.5 reduces to

$$Z_1 - h_L = Z_2 \qquad (8.6a)$$

and

$$Z_1 - Z_2 = h_L \qquad (8.6b)$$

The cause of flow between these points is the difference in elevation head between them, that is, $(Z_1 - Z_2)$, denoted as d_h, which is contributed by the gravitational potential. If d_l is the distance between the two points on the water table, then the d_h/d_l ratio is known as the hydraulic gradient.

Comparing the results from the above paragraphs, it is apparent that the elevation difference d_h causes the flow between the two points and the energy is lost by friction (i.e., h_L) during the movement.

8.4.8.2   Confined Aquifer

For the confined aquifer, the pressure head becomes more important than the elevation head. As can usually be seen in an artesian aquifer condition, the groundwater may flow from a lower elevation to a higher elevation if the water pressure at the lower elevation is higher.

8.4.9   WATER MOVEMENT IN UNSATURATED ZONE OF SOIL FORMATION

In an unsaturated zone, the capillary force becomes predominant, and the pressure gradient becomes a suction gradient. Hydraulic conductivity is no longer constant, but is a function of the water content or suction, which is greatest in value when the soil is saturated and decreases in value steeply when the soil water suction increases and the soil loses moisture.

If we consider the water transport between two points, water movement would increase when suction increases and moisture decreases. That is, water tends to move from higher moisture to lower moisture points, because the point with lower moisture has a higher suction force.

Both vapor and liquid movement can be important where appreciable temperature gradients occur.

8.5   PROPERTIES OF GASOLINE AND ITS MOVEMENT IN SOIL

8.5.1   Properties of Gasoline and the Forms of Release Underground

Gasoline is a mixture of different compounds. A typical blend contains nearly 200 different hydrocarbons and additives such as antioxidants and antiknock agents. Thirteen of the chemicals commonly found in gasoline (nine hydrocarbons and four additives) are regulated as hazardous substances under CERCLA. Table 8.1 lists the chemicals along with the values of toxicity, water solubility, vapor pressure, and biodegradability.[19]

In general, there are four major forms of released gasoline underground:

1. Free product
2. Solutes dissolved in groundwater
3. Gases in the vapor phase in the soil void
4. Adsorbates adsorbed by the soil matrix

TABLE 8.1

Physical and Chemical Properties of Toxic Gasoline Components

Compound	Mass		Prevalence	Fate and Transport			Toxicity
	% Volume in Gasoline	% Weight in Gasoline	% Gasolines Containing Chemical	Water Solubility at 20°C (mg/L)	Vapor Pressure (torr)[a]	Degree of Biogradability	Final RQ (kg)[b]
Benzene	1–2	0.81	>99	1,780	75.0	Some	4.54
Toluene	4.0	12.02	>99	515	22.0	Some	454.00
Xylene-M	5–8	3.83	>99	175	5.0	Some	454.00
Xylene-O	5–8	1.93	>99	162	6.0	Some	454.00
Xylene-P	5–8	1.58	>99	198	6.5	Some	454.00
Ethylbenzene	2–5	1.70	>99	152	7.0	Some	454.00
Naphthalene	0.7	0.10	>90	31.1	1.0	Readily	45.40
Phenol	—	—	>90	66,667	0.5	Readily	454.00
EDB	0.01	0.024	<40	4,310	11.0	Some	4.54
EDC	0.01	<0.024	<40	8,690	61.0	Some	45.40
Tetraethyl lead	—	—	<40	0.08	0.2	Some	4.54
Dimethylamine	—	—	—	—	1,345.0	Readily	454.00
Cyclohexane	<0.7	0.17	—	66.5	77.0	Some	454.00

EBD, ethylene dibromide; EDC, ethylene dichloride; RQ, reportable quantity.

[a] At 20°C.

[b] The lower the RQ value, the more toxic the chemical is in pure product form.

Source: U.S. EPA, Cleanups of Releases from Petroleum USTs: Selected Technologies, EPA/530/UST-88/001, U.S. EPA, Washington, 1988.

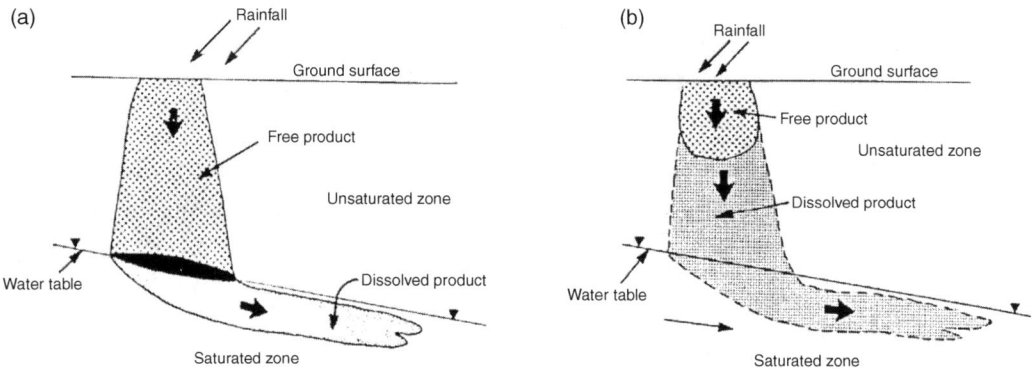

FIGURE 8.4 Schematic of contaminant plumes showing methods by which groundwater can be contaminated.

Figure 8.4 shows schematically the methods by which groundwater can be contaminated. Most of the gasoline components are immiscible with water—these are called the free product. The density of gasoline free product ranges from 0.72 to 0.78 g/mL with a viscosity less than that of water. Gasoline free product floats on and moves faster than groundwater. The density of crude oil and fuel oil ranges from 0.86 to 0.97 g/mL, with a viscosity greater than that of water.

There are many components of gasoline that readily dissolve in water and are transported as solutes in the groundwater. Most gasoline products are volatile and can release gas into the soil void in gaseous form, particularly in the unsaturated zone. Besides these three forms, gasoline components can be adsorbed by the soil matrix and exist in the soil as adsorbates.

Some gasoline constituents, particularly those that are highly volatile or soluble, are readily biodegraded in the presence of soil bacteria and oxygen. Gasoline constituents underground, specifically in the unsaturated zone, belong to the four forms or phases mentioned above. The released gasoline can be transported in the soil matrix in three forms: gas, liquid (free product), or solute. The distribution among these forms may change due to adsorption by soil, desorption from soil, and the extent of degradation.

8.5.2   FATES OF GASOLINE UNDERGROUND: ADSORPTION AND DEGRADATION OF GASOLINE AND THE EFFECT ON GASOLINE MOVEMENT

8.5.2.1   Adsorption of Gasoline by Soil

The forces associated with adsorption of gasoline by the soil are the same as those for adsorption of water by soil. The difference is in the adsorptive strengths of gasoline and water, because gasoline and water have different affinity to soil. Moreover, different gasoline constituents may also have different extents of adsorption by soil. For example, tetraethyl lead and naphthalene have relatively low mobility values and are likely to be adsorbed to the soil. Toluene, xylenes, benzene, and phenol have high mobility values and are therefore more likely to appear in either the dissolved or gaseous phases than being adsorbed. Table 8.2 lists the adsorption coefficients for common gasoline compounds.

The soil above the water table in a gasoline release site is most likely to have the highest concentration of adsorbed gasoline. The soil may be flushed by groundwater when the level of the water table fluctuates, or by infiltrating water, thus changing the adsorbed concentration. In gasoline movement, the gasoline constituent will transfer between the moving phases and the soil adsorptive sites. The extent of transfer depends on the concentration of gasoline in these phases and the distribution coefficient among these phases. Generally, in the release case, gasoline will be adsorbed more by the soil matrix when passing through a pristine soil. During remediation, the gasoline

TABLE 8.2
Adsorption Coefficients for Gasoline Compounds

Chemical	K_{oc} Value (mL/g)
Tetraethyl lead[a]	4900
(n) Heptane	2361
(n) Hexane	1097
Naphthalene[b]	976
(n) Pentane	568
Ethylbenzene[b]	565
Toluene[b]	339
1-Pentane	280
(o) Xylene[b]	255
Benzene[b]	50
Phenol[b]	50
Ethylene dibromide	44

[a] K_{oc} is a measure of the tendency for organic compounds to be adsorbed by soil. The higher the K_{oc} value for each compound, the lower the mobility and the higher the adsorption potential.
[b] Toxic compound.
Source: U.S. EPA, Cleanups of Releases from Petroleum USTs: Selected Technologies, EPA/530/ UST-88/00l, U.S. EPA, Washington, 1988.

constituents will be released from the soil, because the condition is manipulated to have a lower concentration than the previously partitioned concentration.

8.5.2.2 Degradation of Gasoline

Gasoline compounds are also subjected to chemical and biological processes.[19] Biodegradation and biotransformation are two basic biotic processes. Biodegradation is the decomposition of gasoline by microorganisms. The end products are water, carbon dioxide, and energy. Biotransformation is partial biodegradation. Gasoline compounds are partially degraded to simpler compounds that may be more or less soluble or toxic than the original compounds. Most of the biotic processes occur under aerobic conditions.

Abiotic chemical transformation is the reduction of chemical concentrations by degrading the chemicals into other products. The most important chemical transformations are hydrolysis and oxidation/reduction reactions.

Degradation is often the result of the combined effect of chemical transformation and biodegradation. For example, the oxidation/reduction of complex hydrocarbons can produce simple compounds such as peroxides, primary alcohols, and monocarbocylic acids. These compounds can then be further degraded by bacteria, leading to the formation of carbon dioxide, water, and new bacterial biomass.[19,35]

8.5.2.3 Movement of Gasoline Free Product

Most gasoline constituents are immiscible with water, and thus form free product of gasoline from water and usually float on groundwater.

The movement of free product is dependent on soil permeability and moisture. The released gasoline first infiltrates downward vertically, mainly governed by the gravity force, into and through the unsaturated zone, then reaches the water table. If there is an impermeable layer above the water table, the free product will be purged and may not reach the water table directly. In the unsaturated zone, gasoline can be retained by capillary forces and adsorbed onto soil particles. The capillary action in the unsaturated zone also enhances the extent of evaporation of both gasoline and groundwater.

As oil has a lighter specific gravity and lower viscosity than groundwater, the free product floats on the groundwater surface and moves at a faster rate than the groundwater. This horizontal movement is mainly governed by the hydraulic gradient. In the process, gasoline components are also partly adsorbed by the soil, evaporated into the soil void, and dissolved in the groundwater.

8.5.2.4   Movement of Gas-Phase Gasoline

Most gasoline constituents are volatile organics. Volatilization depends on the potential volatility of the compounds and on the soil and environmental conditions, which modify the vapor pressure of the chemicals. Factors affecting volatility are water content, clay content, surface area, temperature, surface wind speed, evaporation rate, and precipitation.

For vapor to move in the unsaturated zone, the soil formations must be sufficiently dry to permit the interconnection of air passages among the soil pores. Vapor concentration and vapor flow govern its movement. Vapor can move by diffusion from areas of higher concentration to areas of lower concentration and ultimately to the atmosphere. Therefore, the transportation of the vapor phase of gasoline components in the unsaturated zone can pose a significant health and safety threat because of inhalation and explosion potential.

Vapor can also move due to pressure gradient, as effected by a barometric-pumping-imposed pressure gradient, and due to density differences. If there is an impermeable layer above the rising vapors, such as a paved road, building, or a frozen ground surface, the vapors are able to move only by lateral underground travel; thus, migration can occur over relatively long distances.

The level of vapor movement in the unsaturated zone is much less important than transport in liquid form. However, this might not be true if the water content of the soil is very low or if there is a strong temperature gradient. The movement of vapor through the unsaturated zone is a function of temperature, humidity gradients, and molecular diffusion coefficients for water vapor in the soil.

8.5.2.5   Movement of Gasoline Solutes

Solubility causes gasoline compounds to be more mobile in association with the movement of groundwater. Dissolved gasoline compounds reach the saturated zone in several ways:

1. From groundwater flow that already has dissolved solute
2. From infiltrating water that has extracted solute from the soil or free product in its path due to the extraction of solute directly from soil adsorbates
3. From free product by the contacting groundwater

Dissolution of gasoline compounds to soil water is a function of each compound's solubility. A highly soluble gasoline substance often has a relatively low adsorption coefficient and also tends to be more readily degradable by microorganisms,[19] as shown in Table 8.1.

The most soluble gasoline compound is methyl tertiary-butyl ether (MTBE) (43,000 mg/L). In addition, MTBE in solution has a cosolvent effect, causing some of the other compounds in gasoline to solubilize at higher concentrations than they normally would in clean water.

8.5.3   Multiphase Movement of Gasoline Compounds

Because gasoline is composed of some highly volatile and soluble hydrocarbon fractions, its components can move in the subsurface in three states: vapor, solute, and liquid. The form of its components in the soil are vapor, solute, free product, and adsorbate. The multiphase flow of gasoline is further complicated due to the various characteristics of the undersurface formation. The partition coefficients of the gasoline constituents in the gasoline free product, groundwater, soil particles, and soil gas determine the transformation of the gasoline forms.

The fate of gasoline in the subsurface is dependent on its interaction with soil and groundwater, volatilization, chemical reaction, biodegradability, and its movement, which in turn depends on the properties of both gasoline and the underground structure.

Soil moisture may greatly affect the movement of gasoline constituents. The adsorptive sites in a soil saturated with moisture are less available than those in a less saturated soil, so an unsaturated condition may promote adsorption of gasoline and retard the movement of gasoline away from a drier soil. Water makes gasoline less able to "wet" the soil, thus promoting the movement of gasoline as long as the pore space of soil is not fully occupied with water.

The extent of soil adsorption and suction forces varies depending on soil components. For example, clay has a much greater adsorption capacity and suction force than sand. The depth of gasoline penetrating the subsurface depends on the volume release, and the adsorption capacity and permeability of the soil. Gravitational force causes downward vertical migration. Suction can cause both vertical movement and horizontal movement. A higher suction force may cause a wider dispersion of gasoline away from the contaminated area.

In different soil zones, the effect of the forces is different, so the movement of gasoline should be considered separately in each zone. Based on the above discussion, the reader should be able to determine the fate and movement of a gasoline compound in different soil zones. The following gives a brief summary:

1. In the saturated zone, the most important phase of gasoline is its free product above the groundwater, then the gasoline as adsorbate in the soil; the gasoline as solute in the groundwater is less important.
2. In the upper unsaturated zone (above the capillary fringe), multiphase movement and transformation are typical. Vapor-phase gasoline becomes more important; gasoline adsorption by soil, dissolution in pore water, and free product in the pore space can also be significant.
3. In the capillary fringe, movement by suction occurs in all directions. Transport in the capillary fringe is also governed by multiphase flow. The increased water content in the capillary zone affects the rates of volatilization and dissolution. As soil water content increases, volatilization and vapor transport generally decrease and dissolution and solute transport generally increase. Free product migration occurs on top of the water table; the free product continues to spread and is held by capillary forces in the soil matrix. When the free product is exhausted, migration stops and residual saturation is reached.

Note that the heterogeneity of underground conditions would favor the flow along the path of least resistance, which is another factor controlling flow besides control by the hydraulic or concentration gradients.

8.6 MANAGEMENT OF TANKS AND THE ENVIRONMENT AS REMEDIAL ACTIONS

Immediate response for release is required, including release reporting, immediate containment, monitoring of explosive hazards, performing a site check to evaluate the extent of release, determining the presence of free product on the water table, and remedying hazards posed by excavated soils. Further corrective actions may be required such as removing the released free product, soil gas, and contaminated groundwater and soils, as well as removal and replacement of tanks. Detailed correction action plans are required if such further corrective actions are needed.[36,37]

An underground storage system that is found to leak or likely to leak should be abandoned, repaired, or replaced. Removal and cleaning of the tank are usually carried out before repair.

8.6.1 TANK REMOVAL

Removal of a leaking storage tank can limit liability and environmental damage. The following steps may be followed[38,39]:

1. Analyzing the tank content according to the U.S. EPA hazardous waste characterization process to determine the proper disposal procedure for the contents
2. Emptying the tank
3. Cleaning the tank interior with high-pressure water, steam, or solvent
4. Purging vapors from the tank using air, carbon dioxide, or nitrogen
5. Removing the tank from the ground
6. Rendering the tank to ensure it will not be reused any further, then disposing of it
7. Examining soil around the excavation for contamination
8. Removing and disposing of obviously contaminated soil (note that groundwater analyses are usually not required when a tank is removed)
9. Obtaining soil samples in the cleaned area for analysis, and documenting the effectiveness of the cleanup effort
10. Backfilling the excavation
11. Documenting the removal and disposal of the tank and soils; filing a report with the controlling government agencies and with the tank's owner, if any spills occurred during the work

Tanks should be removed only by contractors familiar with pertinent government regulations and knowledgeable about the safeguards necessary to prevent environmental harm so as to limit potential liability to the owner of the storage system.

8.6.2 TANK REPAIR

Some tanks, after repair, may stay in service to store gasoline. Most steel tank repairs are done by lining the interior of the tank with epoxy-based resins or some other coating that is compatible with fuel products.[5] Before the tank can be repaired, all free products must be emptied, and all vapors must be removed completely. The tank should be cleaned thoroughly to ensure the lining material adheres to the interior surface of the tank. Before putting the tank back into service, the tank should be tested and examined to be sure that all leaks are repaired, and whether or not additional work needs to be done. For example, recoating the tank, reinforcing the tank area, and lining or relining can all extend a tank's life.

8.6.3 TANK REPLACEMENT

There are cases in which tanks should be replaced rather than repaired. For instance, the American Petroleum Institute (API) does not recommend the lining of a tank that has open seams more than 3 in. long, perforations larger than about 1.5 in. in diameter, more than five perforations per square foot of surface area, or more than 20 perforations per 500 square feet of surface area.[40] Some localities have certain restrictions on repairing tanks.[5] It is also recommended to replace an unsecured underground storage system with a new one.

Compared to earlier tanks, current underground storage systems have two advantages:

1. Minimization of leaks
2. Leak monitoring devices

Leaking is minimized in new tank systems by including corrosion protection and using a double-walled tank construction. Corrosion protection is achieved by coating, by using cathodic protection, or by using fiberglass-enforced plastic tanks. In double-walled construction, the outer wall protects the erosion of the inner wall and contains any leakage that may occur.

New tank systems are also equipped with leak monitoring devices that take advantage of the double-walled construction. Leakage can be reported in real time and more accurately using these detection devices, which include water- or product-sensitive probes, or pressure detection devices if the space between the two walls is designed to remain under vacuum.

8.6.4 ALTERNATIVES FOR TANK ABANDONMENT AND REPLACEMENT

There are two alternatives to tank abandonment and replacement:

1. Abandonment in place
2. Installation of an aboveground storage tank

Although it is more desirable to remove unreliable underground tanks, a tank may be abandoned in place, for example, when it is indoors, under a building, beneath a foundation, or barricaded with other constructions.

Before a tank is abandoned in place, the following measures should be taken into consideration[38,39]:

1. Assessment of the tank's integrity, knowing that a tank may be abandoned in place only if it has never leaked; otherwise, a broader remediation effort might be required if it has leaked and contaminated the soil and groundwater
2. Removal of all liquids
3. Removal and disposal of sludge and residues
4. Cleaning of the tank and disposal of the cleaning residue
5. Filling of the tank with inert material such as sand, gravel, or concrete
6. Disconnection of piping, and plugging it with concrete or nonshrinkable grout or removing all piping.

The second alternative is to construct an aboveground tank, whenever it is feasible, in order to avoid the liability of uncontrolled USTs. This alternative is being chosen by many tank owners for new storage. However, the aboveground storage system has the following disadvantages compared with underground storage:

1. There are more strict fire regulations.
2. Space is needed for installation.
3. It is more likely to be exposed to accidental damage.
4. It is more exposed to local building codes, which usually do not favor aboveground tank systems.

8.7 CONTROL OF CONTAMINANTS MIGRATION AS REMEDIAL ACTIONS

8.7.1 GAS CONTROL

Gas control is required, because the vapor phase of gasoline components in an unsaturated zone can pose a significant health and safety threat. The gas control and safety concern are discussed in another chapter. Some of the remedial technologies presented in subsequent sections of this chapter can also act as gas control measures.

8.7.2 CONTROL OF PLUME MIGRATION

Migration of the gasoline free product and the contaminated groundwater plumes should be controlled. The containment of a plume prevents its further migration and the enlargement of contaminated areas. The most effective method is to pump so as to cause a depression of the water

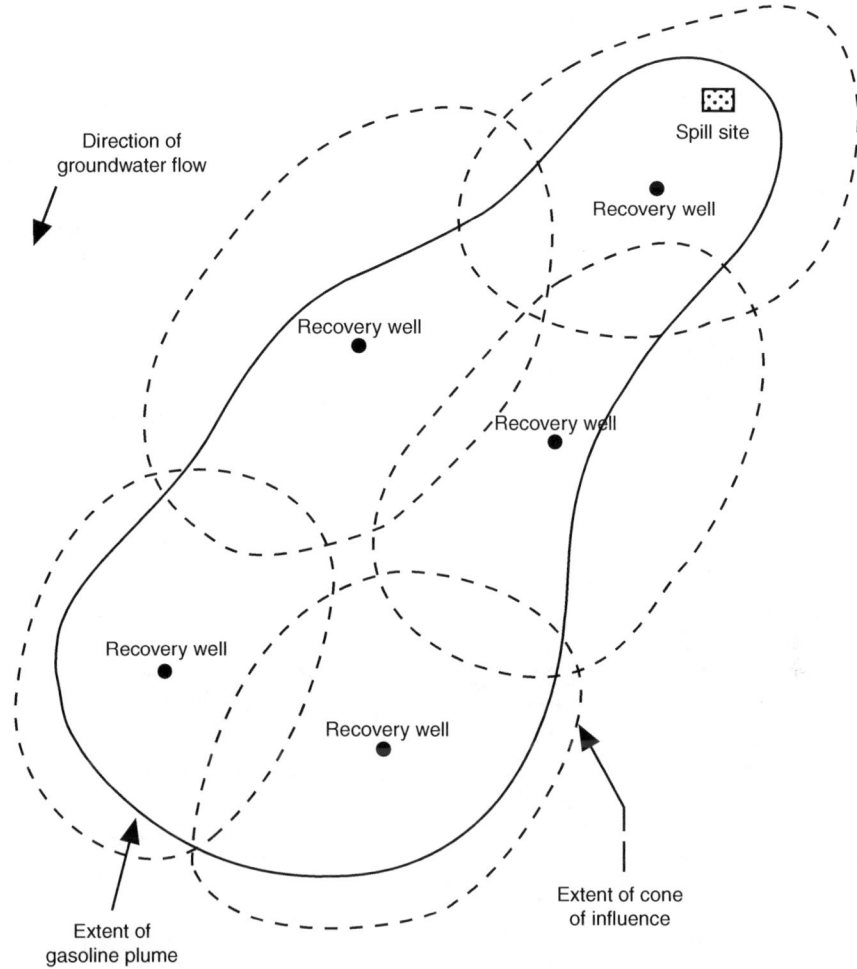

FIGURE 8.5 Using overlapping cones of influence to contain a gasoline plume.

table, which modifies and controls the flow direction of groundwater (Figure 8.5). The trench method can also intercept the plume and prevent it from further migration.

These two methods, which are used as an emergency action, can also be utilized for the cleaning of plumes. Containment methods can often be extended to plume treatment by using the trench or well pumping to recover the free product.

There are other methods for containment such as slurry walls and piling sheets, which are only used as methods for containment but not for treatment.

8.8 REMOVAL OF CONTAMINANTS AS REMEDIAL ACTIONS

8.8.1 REMOVAL AND RECOVERY OF FREE PRODUCT

Recovering free product comprises the following major steps[41]:

1. Establishing gasoline plume containment
2. Gathering and extracting (associated with gasoline/water separators) the contained plume from underground
3. Recovering gasoline

8.8.1.1   Gasoline Plume Containment and Extraction

Trench method

The trench method uses an excavator to dig a trench down to the water table to intercept the flow of the floating gasoline. The trench should be dug deep enough to pond groundwater and the floating gasoline. Pumping out the water in the trench can increase the hydraulic gradient and increase the movement of gasoline to the trench.

Groundwater flow direction should be predetermined. An impermeable membrane is placed on the downgradient side of the trench to ensure that the gasoline in the trench does not escape back into the soil. A better practice is to install an upgradient membrane that can allow more gasoline but less water to enter the trench, and a downgradient membrane to prevent gasoline from moving into the soil while allowing water to pass out to the soil on the downgradient side.

The ponded gasoline in the trench is removed separately from the water for recovery. Special equipment has been used for this purpose, including skimmers and filter separators that are automatically activated when gasoline is present in the trench to separate and remove the gasoline from the water.[19] It is inevitable that a gasoline and water mixture will be pumped out. Gasoline will then be recovered using methods (detailed later) for the treatment of pumped contaminated groundwater.

The trench method is applicable only when the water table is relatively shallow, less than 10 to 15 ft below the ground surface. For a deeper water table, the cost of the trench method becomes more expensive than other methods such as pump systems. Another limitation of the trench method is the soil structure. The soil above the water table has to be firm and well aggregated to allow for the trench to be self-supporting. Otherwise, embankment enforcement or screening would be needed. A third limitation is that continuous pumping and skimming is required to maintain a flow gradient towards the trench. Otherwise, the free product will move back and reenter the soil.

Pumping well method

The pumping well method is more suitable for a water table that is too deep for the trench method. Pumps draw water, forming a cone of depression in the water table to control the movement of floating gasoline. The gasoline is then pumped out. The pumps can be either single- or a dual-pump systems.

Groundwater models and other analytic techniques are available to assist in proper pump siting, choosing pump capacities, and calculating the movement of the contaminant plume. The characteristics of the aquifer, the flow of groundwater, and the size of the plume should be known.

In the single-pump system both gasoline and water are recovered through a single pipeline to aboveground storage tanks or oil/water separators (Figure 8.6a). There are two problems encountered with this single-pump system:

1. During pumping, gasoline and water are mixed, which complicates aboveground separation.
2. Large volumes of contaminated water must be stored, treated, and disposed of.

Therefore, the single-pump method is commonly used only for smaller spills when the gasoline–water recovery rates are relatively low (e.g., less than 1892 L/h or 500 gal/h).

The dual-pump system is used when a large amount of gasoline is to be recovered. Separate gasoline and water pumps are used. The dual-pump system significantly reduces the amount of water that must be treated. Water pumps are placed at a depth lower than the water table to be able to establish a cone of depression, and the gasoline pumps draw out the gasoline that floats into the depression on the top of distorted water table for product recovery (Figure 8.6b).

Dual-pump systems are better able to control a constant cone of depression than the single-pump system. It is important to maintain a nearly constant cone of depression to prevent the migration of the gasoline plume. If a constant cone of depression is not maintained, the water table and

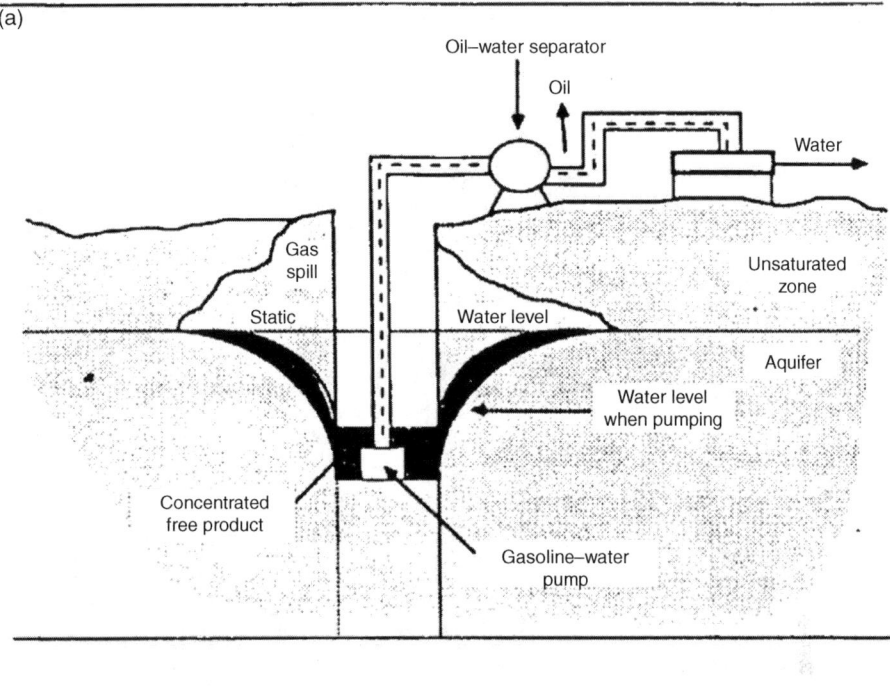

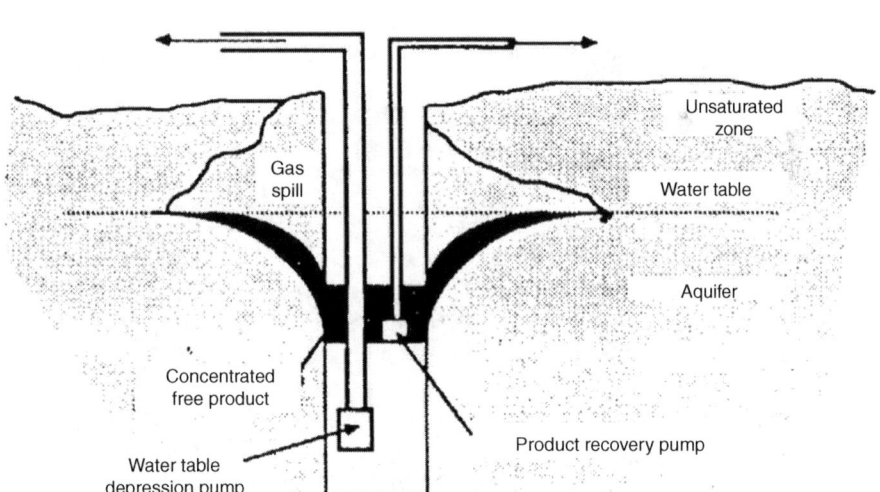

FIGURE 8.6 (a) Single-pump and (b) dual-pump gasoline recovery systems.

the gasoline plume will rise, and gasoline droplets may adhere to soil particles and consequently remain in the soil.

The cone of depression in a dual-pump system is controlled by a detection probe. Initially, the probe is set in the well at the depth of the proposed cone of depression of the gasoline–water interface. The water pump draws the water table down, reaching the pump probe. The water pump ceases when the pump probe detects gasoline. The depressed water table will rise slightly. As soon as the probe detects water again, the water pump resumes, thus maintaining a constant cone of depression. Gasoline will accumulate in the depression. The product pump, both inlet and probe of which are placed a few inches above the water probe, draws gasoline aboveground.

Installation of the pumping well is more time consuming than digging a trench. There is a lag period between the start of pumping, the formation of the depression cone, and containment of the plume. This limits its use as a rapid containment measure. The water table depression must be kept constant; otherwise, if the water table is allowed to fluctuate, gasoline droplets may adhere to soil particles and get trapped below the water table, especially when the depth of the cone of depression gets lower.

The pumped free product is usually accompanied by water. Hence, it is necessary to separate water from the oil, which is usually performed aboveground, although recently a subsurface recovery system has been developed.

8.8.1.2 Subsurface Gasoline Recovery

Subsurface gasoline recovery is analogous to *in situ* oil–water separation. The main advantage of this technique is that the pumped gasoline, which moves with the groundwater gradient, can be intercepted and recovered with minimum energy input.[19] The plume is trapped and directed to the separator influent nozzle. Other advantages are that it reduces the likelihood of water being frozen in the separator in cold weather, it eliminates the evaporation of potentially dangerous volatile organic compounds, and it saves aboveground space for other uses.

The disadvantages of subsurface gasoline recovery are as follows:

1. It is difficult to excavate a hole large enough and deep enough to install the separator at the water table.
2. Installation is time consuming and may not be completed quickly enough to contain the migration of a rapidly moving plume.
3. The separator effluent usually contains a residual dissolved gasoline concentration of 15 mg/L.
4. Treatment of separated gasoline is also needed if the reuse of gasoline is desired. In such a case, an aboveground advanced gasoline–water separator is needed.

8.8.1.3 Aboveground Gasoline Recovery

Aboveground separators are typically large tanks whose function is to slow down the flow of the incoming water; this allows gravity separation of the less dense gasoline emulsions.[19,41] Separators are composed of two or more chambers. The first chamber is used for the deposition of settleable solids, and the second is used for the separation of liquids of dissimilar specific gravities and the removal of the lighter liquid.

In the preseparation chamber, the less dense oil droplets rise, collide, and fuse with adjacent droplets. According to Stoke's law, the larger the diameter of a particle, the faster is its rate of rise. Thus, as small droplets coalesce to form larger droplets, their upward vertical velocity increases. Coalescing tubes or plates are designed to enhance the separation of oil–water emulsions. The emulsion free water is directed away from the tubes or plates and enters the separation section. Some separators are built with an outlet zone for the discharge of clarified water.

Under optimum conditions, an oil–water separator can reduce the hydrocarbon emulsion in water down to 15 mg/L. The separator is most effective when the gasoline plume is relatively small and the rate of water flow is slow enough to allow for complete separation.

If it is desirable to reuse the oil, then more efficient oil–water separators utilizing heating and nebulization techniques will be needed. U.S. patents issued to Weber and colleagues[42] and Wang and colleagues[43] make use of such techniques.

8.8.1.4   Recovered Gasoline

Recovered gasoline can either be disposed of by incineration or reused. If the gasoline is to be reused, it must be refined or mixed with other gasoline as it gets degraded while in the soil. There are three processes that affect the degradation of gasoline:

1. Aromatic hydrocarbons such as benzene, toluene, and xylene become oxidized in the presence of oxygen.
2. Gasoline constituents are metabolized by soil microbes.
3. Water particles may coalesce with the hydrocarbons.

8.8.1.5   Recovered Water

Recovered water that contains a small amount of floating free product and dissolved constituent is usually passed through an oleophilic–hydrophobic adsorbent filter to remove the remaining free product.[19]

If the remedial action involves the treatment of contaminated water (such as pump-treatment for groundwater recovery or soil-washing for soil recovery, which will be discussed in Section 8.8.2), then the preliminarily recovered water can be combined with a treatment stream for further treatment.

There are many options for the disposal of the filter-treated water and dissolved hydrocarbons:

1. The aquifer may be recharged with the recovered water in order to flush out the remaining pockets of free gasoline. A drawback to this technique is that the recharging water contains dissolved constituents.
2. The water may be discharged to a natural water course where dilution and exposure to oxygen will reduce the hazards of its dissolved gasoline constituents. In such a case, a National Pollutant Discharge Elimination System permit and a State Pollutant Discharge Elimination System permit must be obtained.
3. The water may be sent through a wastewater treatment plant where the remaining dissolved constituents can be removed.
4. The water may be treated with on-site air strippers and carbon adsorption filtration systems.

8.8.2   *In Situ* Biological Treatment of Groundwater Decontamination

Several methods are available to remove gasoline constituents from water, such as air stripping, biorestoration, activated carbon adsorption, reverse osmosis, ozonation, oxidation, resin adsorption, oxidation with hydrogen peroxide, ultraviolet irradiation, flotation, and land treatment.

Biological *in situ* treatment is based on the concept of stimulating microflora to decompose the contaminants in place, resulting in the breakdown and detoxification of those contaminants. Biological degradation or biological remediation is generally considered a cost-effective method for the removal of organic compounds, although it is site-specific for *in situ* biological degradation. For removing volatile organic compounds (VOCs), on the other hand, cost-efficiency may be achieved by using the technologies involving volatization (such as air-stripping), as well as other technologies. In fact, about 95% of cases that involve removing a gasoline plume dissolved in groundwater use air stripping and filtration through GAC.[19] Biological treatment is not widely applied in the field, although it is cost-effective and promising for coarse-grained soils.

8.8.2.1   Classification of Biological Treatment

Bacteria can grow in two main environments, aerobic and anaerobic. In aerobic treatment, aerobic and facultative bacteria use molecular oxygen as their terminal electron acceptor. The treatment occurs in the presence of a molecular oxygen supply. In anaerobic treatment, anaerobic and facultative bacteria use some other compound as their terminal electron acceptor, for example, carbon dioxide, sulfate, or nitrate, in the absence of molecular oxygen. In fact, there is another type of biological treatment called the fermentative and methanogenic process, which is carried out by what is referred to as a methanogenic consortium.[44,45]

So far, only aerobic processes have proved to be effective for *in situ* removal of organic waste in groundwater and soil.

8.8.2.2   Characteristics and Factors Affecting Aerobic Biological Treatment

In the aerobic process, organic contaminants such as gasoline releases are broken down by bacteria to produce new biomass (bacteria) and other byproducts:

$$\text{Bacteria + organics + oxygen + nutrients (N, P)} \longrightarrow \text{more bacteria + byproducts} \qquad (8.7)$$

The organics contaminants, whose concentration is usually expressed in terms of biochemical oxygen demand (BOD), are utilized as food for the bacteria. Besides oxygen, nutrients (nitrogen and phosphorus) are also needed by the bacteria for its metabolism. The concentrations of oxygen, bacteria, organic contaminants, and nutrients, as well as other factors, have an affect on the biological treatment rate.

Dissolved oxygen (DO) in a bioreactor should be maintained above a critical concentration in order to maintain good aerobic biological activity. The minimum required DO concentration ranges between 0.2 and 2.0 mg/L with 0.5 mg/L being the most reported value.

Significant and active microbial populations are usually found in the subsurface soil and groundwater. However, if there is a lack of required microorganisms, then bacteria can be injected *in situ*. An optimum food/microorganisms (F/M) ratio should be maintained for effective removal of organic contaminants.

An equally important factor is the biomass/oxygen ratio. If oxygen is deficient, then the biomass cannot be sustained under aerobic conditions. Thus, control of the oxygen supply becomes important. In fact, in bioremediation the most important part of the design is the provision of an appropriate level of oxygen supply to maintain an efficient process.

Another important factor is the food/nutrient ratio. Many of the necessary nutrients may already be present in the aquifer, such as K, Mg, Ca, S, Na, Mn, Fe, and trace elements; however, N and P may be deficient and need to be added. The optimum ratio of BOD : N : P is 100 : 5 : 1. It is not a good practice to inject a large quantity of nutrients in the aquifer at one go. They should be fed at the required usage rate throughout the cleanup process. Both the organic contaminants and the nutrients should be completely exhausted by the end of the *in situ* remediation of an aquifer.

pH should be maintained near neutral, between 6 and 8. Generally, the optimal value is slightly higher than 7.

The optimal temperature for bacterial growth is between 20 and 37°C. For every 10°C decrease in temperature, bacterial activity is approximately halved. Temperature in deep groundwater is rather constant. However, for shallow soil and water, in cold weather the rate of biodegradation becomes depressed compared to in warmer weather, and therefore warm water may need to be injected into the subsurface.

Other factors affecting performance include the presence of toxic material, the redox potential, salinity of the groundwater, light intensity, hydraulic conductivity of the soil, and osmotic potential. The rate of biological treatment is higher for more permeable soils or aquifers. Bioremediation is not applicable to soils with very low permeability, because it would take a long time for the cleanup process unless many more wells were installed, thus raising the cost.

Clogging of aquifers by the growth of biomass is an operational problem. The permeability of an aquifer could be reduced due to the precipitation of biomass sludges and chemicals, or due to clay dispersion.

8.8.2.3 Design of an *In Situ* Bioremediation System

The concentration of biomass is important for the degradation of organic contaminants. Designers can utilize the available microbial population in the soil and groundwater. However, the biomass grows slowly, and remediation requires an accelerated growth rate. This can be realized by a delivery and recovery system. The delivery directs oxygen and nutrients to the underground formations; the recovery stage recovers the spent treatment solution. Circulation of groundwater is very important. A complete delivery and recovery system will do the following:

1. Deliver a high concentration of oxygen and supply additional nutrients or commercially available bacteria if bacteria and nutrients are deficient
2. Provide adequate contact between the biomass and contaminants
3. Prevent the clogging of the soil voids to ensure a sufficient groundwater flow
4. Flush the groundwater
5. Provide hydrologic control of treatment agents and contaminants to prevent their migration beyond the treatment area
6. Provide for complete recovery of the spent treatment solution or contaminants where necessary

As bioremediation proceeds, the bacterial population increases due to the growth of the biomass. Thus, although bacteria may be deficient at the beginning they do not usually need to be added after the startup.

The following design example of an injection and extraction system (Figure 8.7) illustrates the bioremediation process. Both the soil and groundwater are contaminated. Groundwater is extracted downgradient and reinjected upgradient of the zone of contamination. Water is also injected to flush the soil.

There are two methods for the injection of oxygen: *in situ* and in line. In an *in situ* oxygen supply, oxygen is supplied directly from the aeration well to the contaminated plume. A mechanical aeration unit produces sufficient mixing of oxygen and bacteria with the leachate plumes. In an in-line oxygen supply, oxygen is added together with nutrients or bacteria to the mixing tanks (Figure 8.7b).

The most common sources as oxygen supply are air, pure oxygen, hydrogen peroxide, or possibly ozone. Table 8.3 summarizes the advantages and disadvantages of these oxygen supply alternatives.

Using air is economical. However, an in-line method using air may not provide adequate oxygen supply because the maximum oxygen supply is approximately 10 mg/L O_2, which is sufficient for the degradation of only about 5 mg/L of hydrocarbons. Even when using pressurized air or pure oxygen, an in-line supply of oxygen can only degrade low levels of contaminants, less than 5 to 25 mg/L of hydrocarbons.

Pure oxygen can also be used. The injection method can be the same as for air injection. The advantage of using pure oxygen over conventional aeration is that higher oxygen transfer to the biomass can be attained. The in-line injection of pure oxygen will provide sufficient dissolved oxygen to degrade 20 to 30 mg/L of organic material.

Using hydrogen peroxide (H_2O_2) has the following advantages:

1. Greater oxygen concentrations can be delivered to the subsurface.
2. Less equipment is required.
3. Hydrogen peroxide can be added in-line along with the nutrient solution, and aeration wells are not necessary.
4. Hydrogen peroxide keeps the well free of heavy biological growth, thus reducing clogging problems.

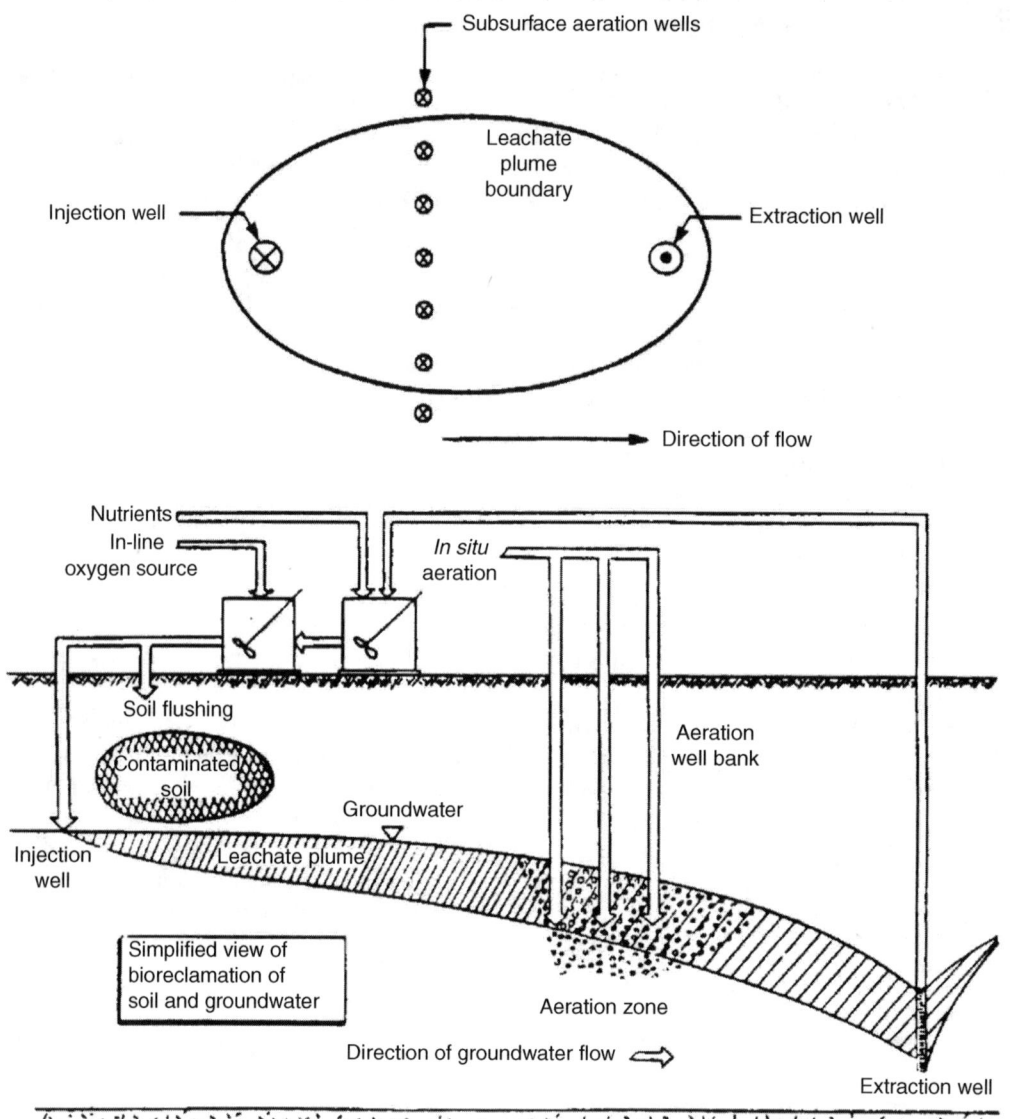

FIGURE 8.7 Simplified view of groundwater bioreclamation.

Ozone is not widely used, because of its high cost and the possibility of some toxicity to bacteria if used at high dosage for low BOD concentrations (higher than 1 mg/L of ozone per mg/L total organic carbon).

In situ oxygen supply requires aeration wells for the injection of oxygen. The criteria are that the aeration well zone must be wide enough to allow the total plume to pass through, and the flow of air must be sufficient to produce a substantial radius of aeration while small enough so as not to create an air barrier to groundwater flow. The required residence time t_r for aeration can be calculated from Darcy's law as a function of the groundwater head and hydraulic conductivity:

$$t_r = \frac{L_a^2}{K(h_1 - h_2)} \tag{8.8}$$

TABLE 8.3
Oxygen Supply Alternatives

Substance	Application Method	Advantages	Disadvantages
Air	In-line	Most economical	Not practical except for trace contamination <10 mg/L COD
	In situ wells	Constant supply of oxygen possible	Wells subject to blow out
Oxygen-enriched air or pure oxygen	In-line	Provides considerably higher O_2 solubility than does aeration	Not practical except for low levels of contamination <25 mg/L COD
	In situ wells	Constant supply of oxygen possible	Very expensive Wells subject to blow out
Hydrogen peroxide	In-line	Moderate cost Intimate mixing with groundwater Greater O_2 concentrations can be supplied to the subsurface (100 mg/L) H_2O_2 provides 50 mg/L O_2) Helps to keep wells free of heavy biogrowth	H_2O_2 decomposes rapidly upon contact with soil, and oxygen may bubble out prematurely unless properly stabilized
Ozone	In-line	Chemical oxidation will occur, rendering compounds more biodegradable	Ozone generation is expensive Toxic to microorganisms except at low concentrations May require additional aeration

Source: U.S. EPA, Cleanups of Releases from Petroleum USTs: Selected Technologies, EPA/530/UST-88/001, U.S. EPA, Washington, 1988.

where t_r = residence time (T), L_a = length of aerated zone (L), h_1 = groundwater elevation at beginning of aerated zone (L), h_2 = groundwater elevation at the end of the aerated zone (L), K = hydraulic conductivity (L/T).

The design conditions for the injection and extraction system are as follows:

1. The groundwater injection rate should be determined by a field testing program.
2. All injected groundwater and associated elements are to be kept within the site boundary to prevent the transport of contaminants to adjacent areas.
3. The distance between the injection-pumping wells should be such that approximately six injection-pumping cycles can be completed within a six-month period.
4. Aquifer flow rate should be sufficiently high so that the aquifer is flushed several times over the period of operation.
5. Flow and recycle rates should not be high enough to cause excessive pumping costs or loss of hydraulic containment efficiency due to turbulent conditions, corrosion, flooding, or well blow out.

8.8.2.4 Case History of *In Situ* Bioremediation

A bioremediation system described by U.S. EPA[19] consists of a downgradient dewatering trench and well, two mobile biological activating tanks, two mobile settling tanks, and two upgradient

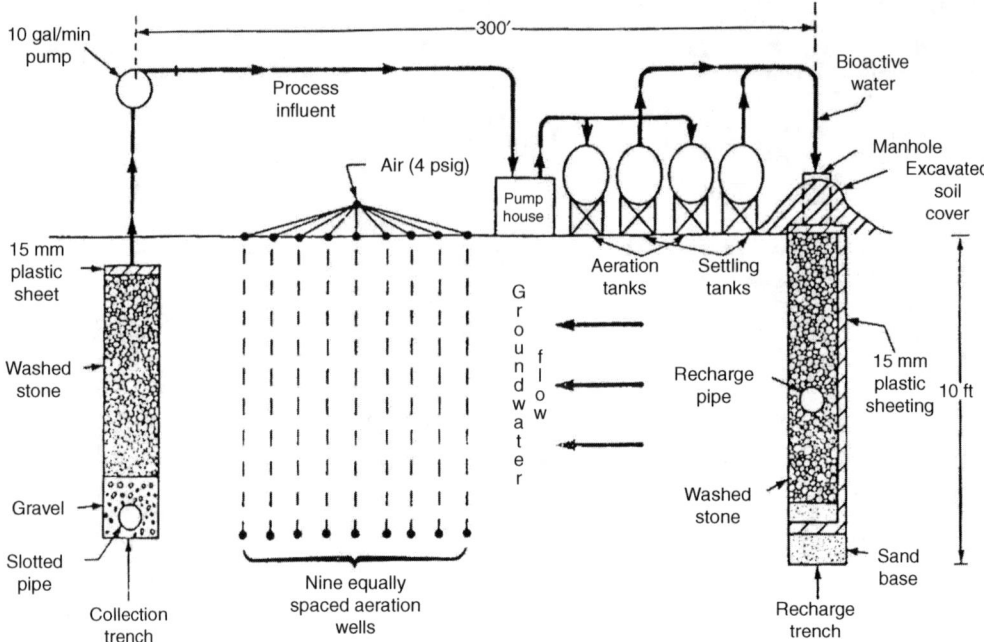

FIGURE 8.8 Flow diagram of Biocraft biorestoration.

reinjection trenches (Figure 8.8). The system was used to treat between 53,000 and 76,000 L/d (14,000 and 20,000 gal/d) of groundwater that had been contaminated with 114 m³ (30,000 gal) of organics that leaked from USTs. The reduction of contaminant mass ranged from 88 to 98% for methylene chloride, acetone, and n-butyl alcohol, and 64% removal for dimethylaniline. Most of the contaminants in the groundwater (over 95%) had been removed during its operation from 1981 to 1985.[46]

There are several advantages of using *in situ* bioremediation[47–49]:

1. Cost-effectiveness
2. Minimal disturbance to an existing site
3. On-site destruction of contaminants
4. Continuous treatment after shutdown of the project
5. Permanent solution
6. Possibility of simultaneous cleanup for both groundwater and soil

Most contaminations of aquifers are a result of material being released above the saturated zone. The contaminant pumping method is limited to the cleanup of the saturated zone. Contaminants in the unsaturated zone can still be a source of future contamination. *In situ* bioremediation techniques can also be designed to clean up the unsaturated zone simultaneously.

The limitations of *in situ* bioremediation are as follows:

1. It is not suitable for short-term projects (it usually needs two to eight weeks of startup period to have the bacteria grown to a sufficient concentration in order to effectively remove the contaminants).
2. It is not suitable for low-permeability and high-salinity areas, as well as areas with extreme pH levels.

3. It is not suitable for the removal of nonbiodegradable organics, toxic material, or material whose concentration is too high and thus toxic to bacteria.

4. It requires continuous operation (a biological treatment system cannot be turned on and off frequently).

8.8.3   PUMP-AND-TREAT PROCESSES FOR GROUNDWATER DECONTAMINATION

8.8.3.1   Air Stripping

Air stripping is an effective and widely used method to remove VOCs from water. It is the most cost-effective option for removal of gasoline from groundwater.[19]

The basic principle of air stripping is to provide contact between air and water to allow the volatile substances to diffuse from the liquid to the gaseous phase. Mass transfer occurs across the air–water interface. The theory of air stripping is related to Henry's law. At a given temperature, the partition of VOCs in the contacting air and water follows Henry's law:

$$P_a = HX_a \qquad (8.9)$$

where P_a = particle vapor pressure of VOC (atm), H = Henry's law constant (atm), and X_a = mole fraction of VOC in water (mol/mol).

The Henry's law constant can be regarded as the partitioning coefficient of VOCs between air and water. Molecules of VOCs can pass freely between gaseous and liquid phases. At equilibrium, the same numbers of molecules move in both directions through a unit area in a unit of time. Departure from equilibrium provides the driving force for mass transfer. This can be affected by a change of temperature or by driving the VOC out of the air phase. Air stripping can be regarded as a "controlled disequilibrium".[19,50] Removal of a VOC from the contacting air–water system leads to it being at a decreased concentration in the water. The eventual outcome is the removal of the VOC from water.

Types of air stripping facilities
There are many methods to introduce fresh air for air stripping, including diffused aeration, tray aerators, spray basins, and packed-towers methods.

In the air diffusion method, compressed air is injected into the water through diffusers or sparging devices that produce fine air bubbles.[51] Mass transfer occurs across the air–water interface of the bubbles. Consequently, contaminants are removed from the wastewater. Mass transfer rates can be improved by producing fine bubbles, increasing the air/water ratio, improving basin geometry, using a turbine to increase turbulence, or increasing the depth of the aeration tanks. Reported removal of organics by air diffusion is between 70 and 90%.[52]

The tray aeration method is a simple, low-maintenance method of aeration that does not use forced air.[19] Water is allowed to cascade through several layers of slat trays to increase the exposed surface area for contact with air (Figure 8.9). Tray aeration is capable of removing 10 to 90% of some VOCs, with a usual efficiency of between 40 and 60%.[53] This method cannot be used where low effluent concentrations are required, but could be a cost-effective method for reducing a certain amount of VOC concentration prior to activated carbon treatment.

The spray aeration method comprises a grid network of piping and nozzles over a pond or basin. Contaminated water is simply sprayed through the nozzles and into the air to form droplets. Mass transfer of the contaminant takes place across the air–water surface of the droplets. Mass transfer efficiency can be increased by multiple passing of the water through the nozzles. This method has three disadvantages:

1. A large land area is necessary for the spray pond.
2. Mist is formed, which could be carried into nearby residential areas.

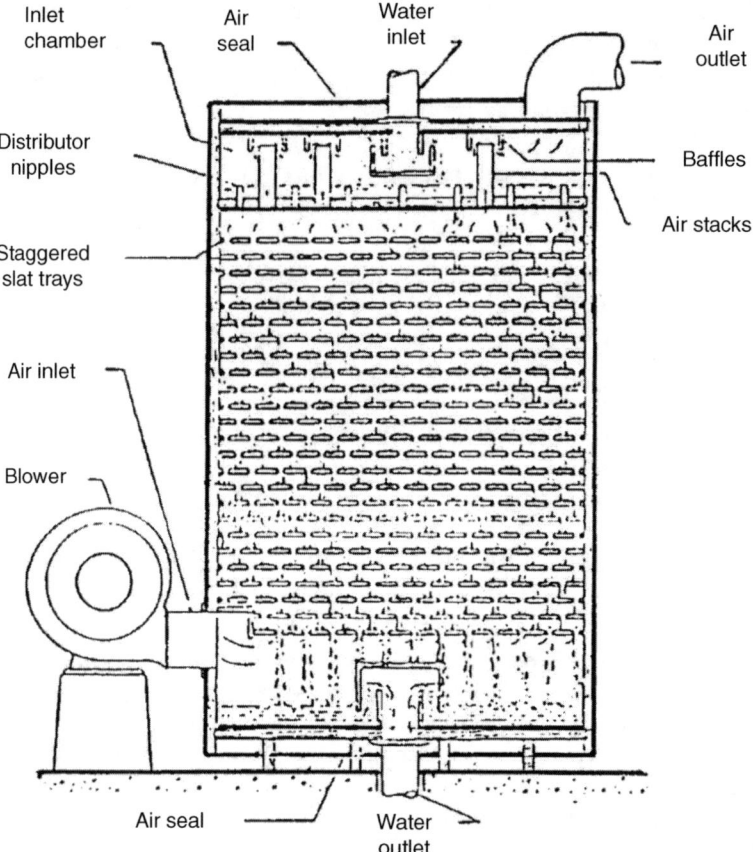

FIGURE 8.9   Schematic diagram of Redwood slatted tray aerator.

3. There is the possibility of ice formation, which lowers the usefulness of the technique in colder climates.

The packed-tower method involves passing water down through a column of packing material while pumping countercurrent air up through the packing (Figure 8.10). The packing material breaks the water into small droplets, causing a large surface area across which mass transfer takes place. The towers are very effective in removing VOCs. Typical removal efficiencies are between 90 and 99%, although 100% (i.e., down to nondetectable levels) removal has been reported. These countercurrent packed towers are the most common of the air-stripping methods. The air emission problems associated with air stripping units have been eliminated from the units developed in the early 1990s by Wang and colleagues[29] and Hrycyk and colleagues.[30]

Henry's law constant

Henry's law constant (H) is usually expressed as follows:

$$H = P_{atm}(M_w/M_c) \qquad (8.10)$$

where P_{atm} = pressure (atm) (here 1 atm = 760 mmHg), M_w = weight of water (mol), and M_c = weight of contaminant (mol).

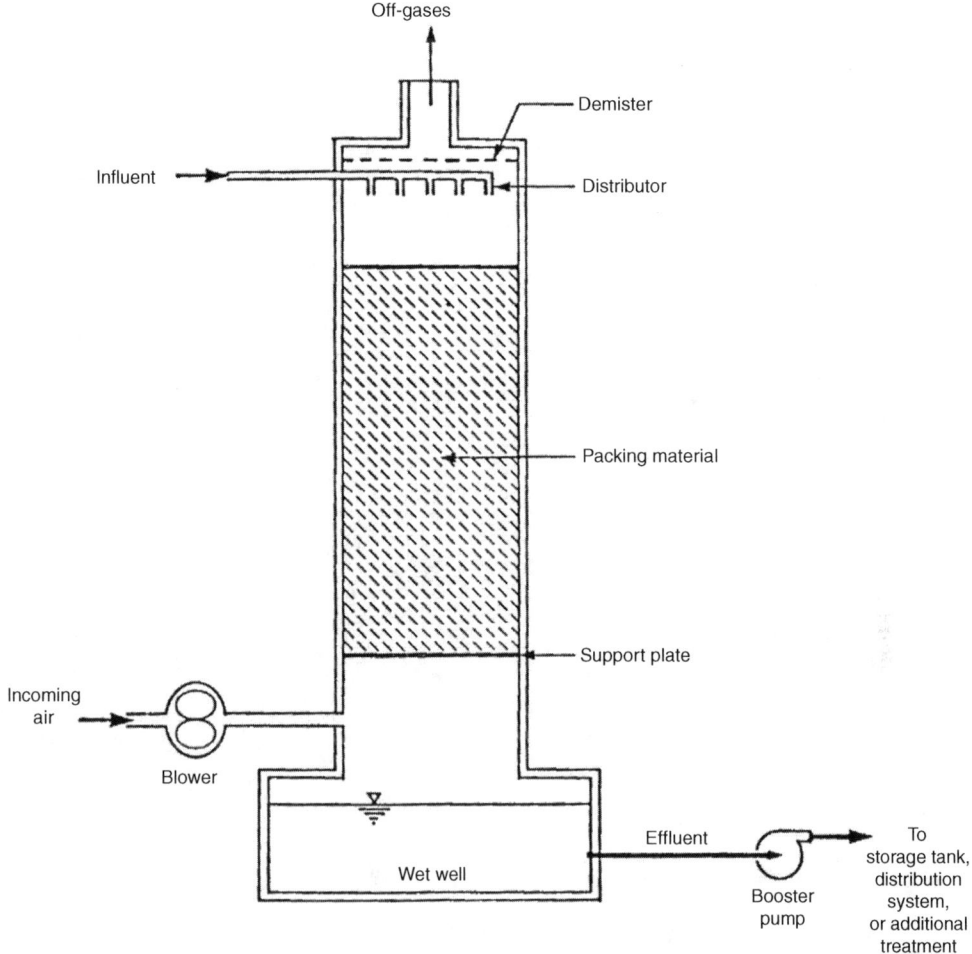

FIGURE 8.10   Schematic diagram of packed tower aerator.

Dimensionless units are also used that are valid only for systems that operate at standard pressure ($P_{atm} = 1$). The actual units are as follows:

$$H = P_{atm}(V_w/V_c) \qquad (8.11a)$$

and

$$H = (V_w/V_c) \qquad (8.11b)$$

where $P_{atm} = 1$ (standard pressure), V_w = volume of water (m³), and V_c = volume of contaminant (m³).

Typical values of H for gasoline components range between 20 and 500 atm (0.03 to 0.30 in dimensionless units at the standard condition [$P_{atm} = 1$]).

Henry's law constants for most of the compounds of interest can be found in the literature.[54] Figure 8.11 shows Henry's law constants for TCE, EDC and several gasoline compounds.[19] These data are derived from water solubility data and the equilibrium vapor pressure of pure liquids at certain temperatures, and may be extrapolated correctly to field design work. Temperature has a major effect on Henry's constant and on stripper performance. Each rise of 10°C in temperature

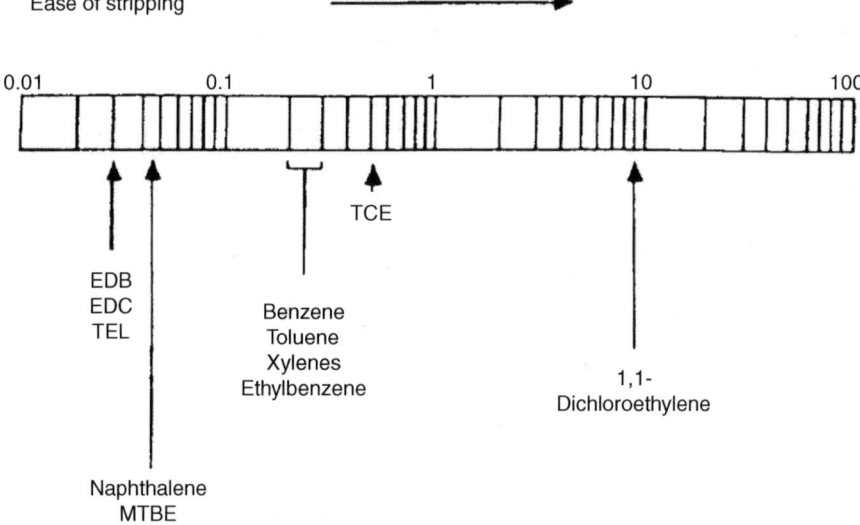

$$H = \frac{\text{Concentration in air, } \mu g/L}{\text{Concentration in water, } \mu g/L}$$

EDB = Ethylene dibromide (1,2-dibromoethane)
EDC = Ethylene dichloride (1,2-dichloroethane)
TEL = Tetraethyl lead
MTBE = Methyl tertiary butyl ether
TCE = Trichloroethylene

FIGURE 8.11   A comparison of stripping rates for TCE and gasoline compounds.

may cause an increase of Henry's constant by a factor of about 1.6.[55] Consequently, warmer temperatures can achieve higher rates of stripping.

Mass balance and air/water ratio
Contaminant mass transport in an air stripper is schematically shown in Figure 8.12. The removal process can be described mathematically by a mass balance for the contaminant assuming that there is no change in the accumulated contaminant in the stripper under steady-state conditions:

$$L(X_{in} - X_{out}) = G(Y_{out} - Y_{in}) \tag{8.12}$$

where L = volumetric rate of contaminated groundwater (L^3/T), G = volumetric rate of air (L^3/T), X_{in} = influent contaminant concentration in water (M/L^3), X_{out} = effluent contaminant concentration in water (M/L^3), Y_{in} = influent contaminant concentration in air (M/L^3), and Y_{out} = effluent contaminant concentration in air (M/L^3).

    For a further application of the mass balance equation to removal processes, four basic assumptions are made:

1. Influent air is free of VOCs (i.e., $Y_{in} = 0$).
2. Differential flow holds for air and water.
3. Changes of liquid and air volumes during mass transfer are negligible.
4. Henry's law holds for these conditions.

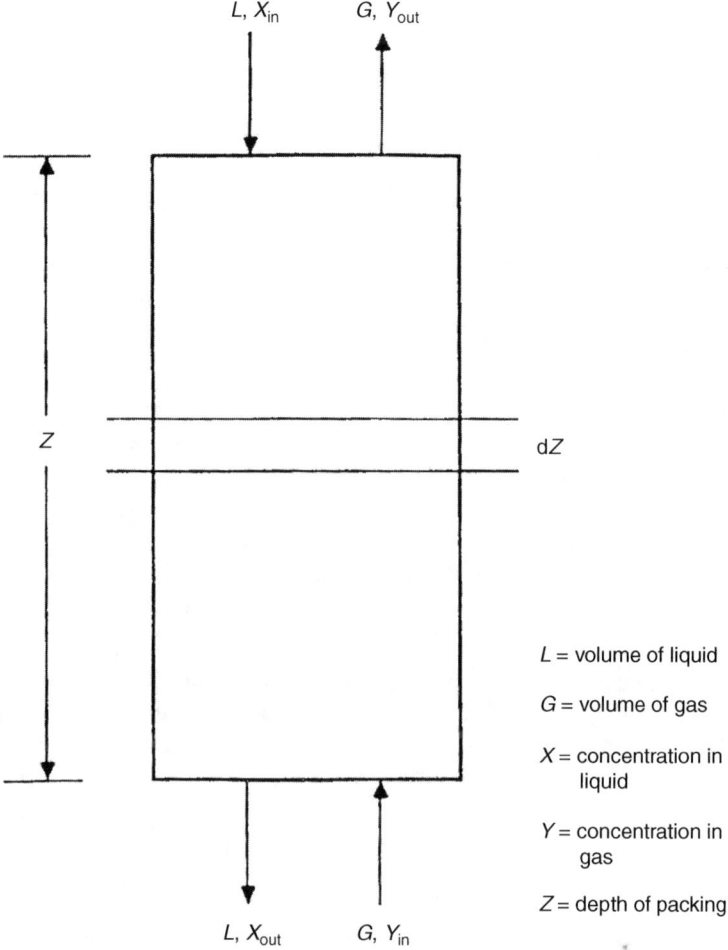

L, X_{in} G, Y_{out}

Z dZ

L = volume of liquid

G = volume of gas

X = concentration in
liquid

Y = concentration in
gas

Z = depth of packing

L, X_{out} G, Y_{in}

FIGURE 8.12   Differential element for an air stripping tower.

Applying the first assumption, $Y_{in} = 0$, Equation 8.12 can be rearranged as

$$G(Y_{out}) = L(X_{in} - X_{out}) \tag{8.13}$$

and

$$\frac{G}{L} = \frac{X_{in} - X_{out}}{Y_{out}} \tag{8.14}$$

Applying Henry's law at the point that air leaves the stripper (i.e., the contaminated water enters stripper) and assuming that equilibrium for mass transfer holds between air and water at that point, Equation 8.9 becomes

$$Y_{out} = HX_{in} \tag{8.15}$$

Substituting Equation 8.15 into Equation 8.14 yields

$$\frac{G}{L} = \frac{X_{in} - X_{out}}{HX_{in}} \tag{8.16}$$

where H = Henry's constant (dimensionless).

Note that $(X_{in} - X_{out})/X_{in}$ is the removal efficiency of a stripper, denoted as f. Then Equation 8.16 becomes

$$\frac{G}{L} = \frac{f}{H} \qquad (8.17a)$$

G/L in Equation 8.17a is the theoretical air/water ratio required for the removal efficiency f for a specific contaminant following Henry's law. In this context, the G/L is denoted $(G/L)_{theory}$, indicating the theoretical air/water ratio. This also means that a minimum amount of air must be brought into contact with the water for a certain length of detention time, the sparging size of the water droplets also affects the mass transfer, as does the air pressure.

Stripping factor

Theoretically, the required air/water ratio for a specific removal can be determined by the mass balance in the stripper:

$$\left(\frac{G}{L}\right)_{theory} = \frac{f}{H} \qquad (8.17b)$$

This air/water ratio is the theoretical or minimum air/water ratio for a given removal. However, in practice, the contaminant mass transfer is a long way from being at ideal equilibrium. A higher air/water ratio, denoted as $(G/L)_{actual}$, the actual air/water ratio, is required for that removal.

The stripping factor R is used to describe the ratio of the actual operating air/water ratio to the theoretical minimum ratio:

$$R = \frac{\left(\dfrac{G}{L}\right)_{actual}}{\left(\dfrac{G}{L}\right)_{theory}} \qquad (8.18)$$

The air stripping factor is directly related to the air/water ratio, and is in turn related to the gas pressure drop through the stripper.

Gas pressure drop

The gas pressure drop is a function of the gas and liquid flow rates and the size and type of packing. It relates to the overall cost of the air stripper and to its performance. The gas pressure drop through a stripping unit can be determined from the pressure drop curve.

A stripper operating at a high pressure drop will require a smaller volume than a similar stripper at a lower pressure drop. This reduces the capital cost for the tower, but increases the blower cost. Towers designed and built to operate at a low pressure drop have the flexibility to increase the gas flow rate and hence the air/water ratio, should the future influent concentration increase or the effluent limitation decrease. Towers designed for high pressure drops do not have this flexibility in operation and would need to decrease the liquid loading to increase the air/water ratio.

Applicability to air stripping

The removal effectiveness of air stripping depends on the following factors:

1. *Physical properties of the contaminants.* Physical properties, particularly Henry's constant, determine the ease of air stripping. The higher the value of Henry's constant, the higher is the removal efficiency.

2. *Temperature.* The temperature of contaminated water significantly influences the removal efficiency because Henry's constant increases with temperature.
3. *VOC concentration in water.* The higher the concentration of a target component in the contaminated water, the higher is its removal efficiency, because the driving force for the target compound to transfer from the contaminated water to the gas is greater when its concentration is higher than at equilibrium.
4. *Air/water ratio.* Increasing the air/water ratio will increase the removal efficiency.
5. *Packing material.* Packing materials are usually designed to be less susceptible to biological and mineral fouling in order to maintain a high surface area and a high void volume, both of which are necessary for maintaining a high operating efficiency.

There are several factors that may limit the use of conventional air stripping for the removal of dissolved gasoline from groundwater:

1. Applicability of air-stripping methods with respect to the type of groundwater contaminants is the most important factor. The major constituents of interest, such as benzene, toluene, xylene, and ethylbenzene, are all fairly volatile and thus are easily removed by this technique. Compounds with low volatility such as 1,2-dichloroethane (DCE) cannot be readily removed.
2. The air pollution impact of the stripping tower is significant, because the air-stripping treatment does not destroy the contaminant; it simply transfers it from the liquid to the gaseous phase. The stripper off-gas, after dilution in the tower, usually mixes with the ambient air in the atmosphere that would further lower the contaminant concentration to values below unsafe levels. Some states have regulated the limit of discharge of volatiles to the atmosphere. In New Jersey, the limit of discharge including benzene is 0.1 lb/h. Off-gas air pollution control is required if a stripper exceeds this limit. Most commonly, GAC adsorption is used to treat the vapor-phase contaminant.[56]
3. High concentrations of iron and magnesium or suspended solids in the influent will limit the efficiency of air stripping, because iron and manganese facilitate the growth of bacteria on the packing, causing decreased mass transfer rates and higher gas pressure drop (suspended solids can cause a similar problem if they are trapped by the packing).
4. High noise levels associated with tower operation may limit air stripping.

Air stripping processes[29,30] and air flotation process[57–59] introduced in the 1990s have solved some of the abovementioned problems.

8.8.3.2 Activated Carbon Adsorption

Applications
Many case studies[19,60,61] have demonstrated the ability of activated carbon to remove a variety of compounds in gasoline from contaminated water to nondetectable levels (99.99+% removal). GAC is more widely used than powdered activated carbon (PAC). Activated carbon adsorption, in general, is not cost-effective in removal of highly concentrated gasoline compounds in water, where the air-stripping method or biological treatment method may be applied. Thus, GAC is widely used for removing low concentrations of complex pollutants, in particular in polishing effluent or in point-of-entry treatment for drinking water.

The main limitation of GAC in removing gasoline compounds is its cost and the disposal of the generated spent carbon. However, the problem of spent carbon's regeneration has been solved, at least in part.[29]

The compounds MTBE and disopropyl ether (DIPE) are sometimes found as additives in gasoline. Both have very high carbon usage rates; thus, the costs of removing these compounds are prohibitive, especially if the influent concentrations are substantial. Therefore, the presence or absence of highly soluble compounds such as MTBE or DIPE or other additives may determine the appropriateness of using GAC for a particular gasoline spill.

Petroleum hydrocarbons (benzene, toluene, ethylbenzene, and xylenes—collectively BTEX), particularly benzene, are believed to pose significant health concerns, especially as they are contained in over 99% of all gasoline. However, additives such as MTBE and DIPE, which have high carbon usage rates, are not found in all gasolines and hence pose less significant health concerns. Thus, GAC is generally applicable for the removal of BTEX.

As mentioned, a major potential limitation of GAC use is the disposal of the spent carbon. The spent GAC can be regenerated or disposed of using sanitary landfills or incineration. GAC regeneration is possible and highly feasible by heating the carbon to very high temperatures (e.g., in a kiln) to remove the volatiles and incinerate them. However, on-site regeneration is economical only in very large projects, not in UST sites. Off-site regeneration, on the other hand, may be acceptable at a central regeneration facility. However, U.S. highways authorities consider any carbon with a flash point below 200°F to be hazardous and cannot therefore be transported on the highways. Under RCRA rules, many contaminant-laden carbons are considered hazardous materials, necessitating disposal in a permitted landfill.

Iron and manganese levels in the influent water may also limit the use of GAC. They will precipitate onto the carbon during treatment. If this happens, head losses will increase rapidly, the removal of organics will be hindered, and the carbon filter may eventually get clogged, making it ineffective and increasing cost substantially, or impractical due to space constraints. If these elements are present at concentration levels above 5 mg/L, they must be removed prior to GAC treatment.

Design of GAC systems for groundwater decontamination
An isotherm test can determine whether or not a particular contaminant can be adsorbed effectively by activated carbon. In very dilute solutions, such as contaminated groundwater, a logarithmic isotherm plot usually yields a straight line represented by the Freundlich equation[62,63]:

$$\log \frac{X}{W} = \log k + \frac{1}{n}\log C \tag{8.19}$$

where X = amount of contaminant adsorbed (M), W = weight of activated carbon (M), k = constant, n = constant, C = unabsorbed concentration of contaminant left in solution (M/L^3), and $1/n$ = represents the slope of the straight-line isotherm. The above equation also indicates the approximate capacity of activated carbon for groundwater decontamination and provides a rough estimate of the activated carbon dosage required.

For the design of a GAC system, the following interrelated parameters should be taken into consideration:

1. Influent flow
2. Carbon contact time
3. Dosage
4. Bed depth
5. Pretreatment requirements
6. Carbon breakthrough characteristics
7. Headloss characteristics
8. On-stream cycle time of carbon (i.e., the time between carbon regenerations)

In general, influent flow and contact time determine the carbon bed depth and size, which in turn determine the breakthrough characteristics of the carbon bed for the influent water, thereby deciding the actual carbon dosage. The carbon dosage determines the volume of influent water that can be treated, which sets the on-stream cycle time of the carbon. These operating variables are related by the following equations:

$$B_m = (Q_m)(t)(d_m) \tag{8.20a}$$

where B_m = carbon bed (kg), Q_m = influent rate (m³/min), t = contact time (min), and d_m = carbon density (kg/m³),

$$B = \frac{Q}{7.48} td \tag{8.20b}$$

where, B = carbon bed (lb), Q = influent rate (gal/min), t = contact time (min), and d = carbon density (lb/ft³),

$$B_m = Q_m(C_{cm}) (1440\ T) \tag{8.21a}$$

where C_{cm} = actual carbon dosage (kg/m³), T = on-stream cycle time (d), and Q_m = influent rate (m³/min),

$$B = Q\ \frac{C_c}{1000} (1440\ T) \tag{8.21b}$$

where C_c = actual carbon dosage (lb/1000 gal), T = on-stream cycle time (d), Q = influent rate (gal/min), and d = 25 (lb/ft³), and

$$P_{dm} = (55,922)\ \frac{\mu K_c Q_m B_{hm}}{D_p^2 D_{cm}} \tag{8.22a}$$

where P_{dm} = pressure drop (mmHg), μ = dynamic viscosity (centipoise), Q_m = influent flow rate (m³/min), B_{hm} = carbon bed depth (m), D_p = mean carbon particle diameter (mm), D_{cm} = carbon column diameter (cm), and K_c = carbon adsorption coefficient.

$$P_d = \frac{\mu K_c Q B_h}{D_p^2 D_c} \tag{8.22b}$$

where P_d = pressure drop (in.Hg), μ = dynamic viscosity (centipoise), Q = influent flow rate (gal/min), B_h = carbon bed depth (ft), D_p = mean carbon particle diameter (mm), D_c = carbon column diameter (in.), and K_c = carbon adsorption coefficient.

Gravity flow in downflow carbon beds is usually controlled at hydraulic loadings less than 9.78 m³/h/m² (4 gal/min/ft²). Upflow carbon beds with bed expansion should be considered when headloss is expected. It should be noted that TSS will break through an upflow carbon bed at about 10% bed expansion.

8.8.3.3 Air Stripping and Activated Carbon Combination

Activated carbon is more suitable for an influent with low VOC concentration, and air stripping is more suitable for treating high VOC concentrations, but yields a relatively higher effluent concentration in comparison to GAC treatment. The cost of air stripping may be doubled if one tries to yield an effluent concentration to be as low as in activated carbon treatment. This is because it would require a taller tower, a higher air/water ratio, or higher air pressure. The combination of air stripping and activated carbon can complement each other and avoid such a high cost.

Many cases have demonstrated that the combination of activated carbon adsorption and air stripping is one of the most common methods for the removal of dissolved gasoline compounds in groundwater. In this case, air stripping lowers the high concentration in the influent, and the GAC further polishes the effluent to a very low concentration. Generally, this method also reduces operation and maintenance costs. O'Brien and Stenzel[64] reported that, when using air stripping, a wastewater containing 1000 µg/L TEC was reduced to 200 µg/L. This 80% removal of TEC resulted in 58% reduction in the consumption of activated carbon.

Another process involving the use of both air stripping and activated carbon adsorption has been developed by Wang and colleagues.[29] This process purifies and recycles the emitted gas, thus not creating an air pollution problem. Also, the spent GAC can be automatically regenerated for reuse.

8.8.3.4   Integrated Vapor Extraction and Steam Vacuum Stripping

Integrated vapor extraction and steam vacuum stripping can simultaneously treat groundwater and soil contaminated with VOCs. The system developed by AWD Technologies consists of two basic processes: a vacuum stripping tower that uses low-pressure steam to treat contaminated groundwater; and a soil gas vapor extraction/reinjection process to treat contaminated soil. The two processes form a closed-loop system that provides simultaneous *in situ* remediation of contaminated groundwater and soil with no air emission.

The vacuum stripping tower is a high-efficiency countercurrent stripping technology. A single-stage unit typically reduces VOCs in water by up to 99.99%. The soil vacuum extraction system uses vacuum to treat a VOC-contaminated soil mass, with a flow of air through the soil that removes vapor-phase VOCs with the extracted soil gas. The soil gas is then treated by carbon beds to remove the VOCs. The two systems share a single GAC unit. Noncondensable vapor from the stripping system is combined with the vapor from the soil vacuum system and decontaminated by the GAC unit. Byproducts of the system are a free-phase recyclable product and the treated water. The granulated carbon will have to be replaced and the used carbon disposed of every three years.

8.8.3.5   *Ex Situ* Biological Treatment for Groundwater Decontamination

The processes of *ex situ* biological treatment for pumped contaminated groundwater is similar to the processes used in biological wastewater treatment plants. These include activated sludge, waste stabilization ponds and lagoons, trickling filters, rotating biological contactors, and land application.[44,45]

The immobilized cell bioreactor system developed by Allied Signal is an aerobic fixed-film bioreactor system (Figure 8.13). The system offers improved treatment efficiency through the use of a unique proprietary reactor that maximizes the biological activity, and a proprietary design that maximizes contact between the biofilm and the contaminants. The advantages include a fast and complete degradation of target contaminants to carbon dioxide, water, and biomass; high treatment capacity; compact system design; and reduced operation and maintenance cost resulting from simplified operation and slow sludge production.[65]

After further polishing, such as clarifying and filtering, if necessary, the biologically treated groundwater may be reinjected into the aquifer in an operation similar to deep well injection.[66]

The advantage of *ex situ* biological treatment is the ability to control the effluent quality. The use of air for aerobic treatment is easier to control and costs less. Nutrient can be added more effectively and the temperature can be controlled.

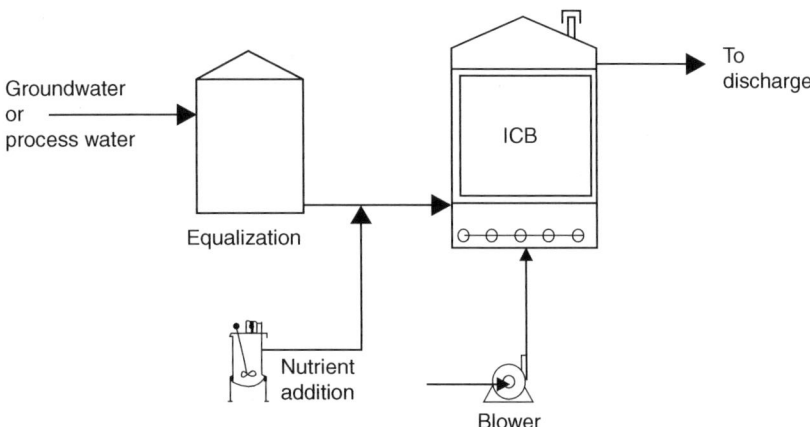

FIGURE 8.13   Allied Signal Immobilized Cell Bioreactor (ICB).

The disadvantages of *ex situ* biological treatment in comparison to *in situ* biological treatment are as follows:

1. After shutting off the system, biological treatment cannot continue in the contaminated site.
2. The contaminants in places where they are strongly adsorbed or where permeability is locally low, or where microcracks are developed in rocks, cannot be efficiently drawn out with water using the pumping method.
3. The emitted gas containing VOCs may cause air pollution problems.

A biological process developed by Wang and colleagues[57] does not cause air pollution problems and is highly efficient for the biodegradation of organics present in water.

8.8.3.6   Oxidation

Oxidation is a means of decontamination. There are several methods that can facilitate oxidation to treat contaminated groundwater. In the following we describe two examples of such technologies.

The perox-pure system developed by Peroxidation Systems is designed to destroy dissolved organic contaminants in groundwater or wastewater through an advanced chemical oxidation process using ultraviolet (UV) radiation and hydrogen peroxide. Hydrogen peroxide is added to the contaminated water, and the mixture is then fed into the treatment system. The treatment system contains four or more compartments in the oxidation chamber. Each compartment contains one high-intensity lamp mounted in a quartz sleeve. The contaminated water flows in the space between the chamber wall and the quartz tube in which each UV lamp is mounted. UV light catalyzes the chemical oxidation of the organic contaminants in water by its combined effect upon the organics and its reaction with hydrogen peroxide. This technology can treat water contaminated with chlorinated solvents, pesticides, polychlorinated biphenyls, phenolics, fuel hydrocarbons, and other toxic compounds at concentrations ranging from a few thousand mg/L to 1 µg/L. For higher organic concentrations, UV light combined with other processes such as air stripping, steam stripping, biological treatment, or air flotation may be more cost effective.[29,58,59]

Chemical Waste Management have developed a technique using evaporation and catalytic oxidation to treat contaminated water.[65] Contaminated water is concentrated in an evaporator by boiling off most of the water and the volatile contaminants, both organic and inorganic. Air or oxygen is added to the vapor, and the mixture is forced through a catalyst bed, where the organic and inorganic compounds are oxidized. This stream, composed mainly of steam, passes through a scrubber, if necessary, to remove any acid gases formed during oxidation. The stream is then condensed or vented to the atmosphere. Suitable wastes include leachates, contaminated groundwater, and process waters. This technique can also be used to treat complex wastewaters that contain volatile and nonvolatile organic compounds, salts, metals, and volatile inorganic compounds.

8.8.3.7 Solvent Extraction

Solvent extraction uses an organic solvent to extract toxic substances from contaminated liquid or solid.[67] Examples can be found in the section dealing with the treatment of contaminated soil.

8.8.3.8 Dissolved Air Flotation (DAF)

Perhaps the most efficient but least recognized process for groundwater decontamination is dissolved gas flotation, also known as dissolved air flotation (DAF), in which air is used for the generation of extremely fine air bubbles having diameters less than 80 μm.

DAF is used to remove suspended solids by decreasing their apparent density; they then rise and float on the water surface. DAF is also used to remove soluble iron, VOCs, oils, and surface active agents by oxidation, air stripping, and surface adsorption. The flotation technology is becoming one of the most important technologies for groundwater decontamination, industrial effluent treatment, and water purification.[58–61,70]

A typical DAF process consists of saturating a portion or all of the influent feed, or a portion of recycled effluent with air at a pressure of 1.76 to 6.33 kg/cm^2 (25 to 90 psi). The pressurized influent is held at this pressure for 0.2 to 3 min in a pressure vessel and then released to atmospheric pressure in a flotation chamber. A controlled reduction in pressure results in the release of microscopic air bubbles, which oxidize the soluble ferrous iron (Fe^{2+}) to form insoluble ferric iron (Fe^{3+}) and attach themselves to VOCs, surfactants, oil, and suspended particles in the influent water in the flotation chamber. This results in agglomeration, air stripping, and surface adsorption due to the generated air bubbles. The VOCs are removed by air stripping and discharged to a gas-phase GAC adsorber for purification. The floated material (oil, surfactants, TSS) rises to the surface with vertical rise rates ranging between 0.15 and 0.6 m/min (0.5 to 2.0 ft/min) and forms a floating scum layer. Specially designed sludge scoops, flight scrapers, and other skimming devices continuously remove the floating scum. The clarified effluent water that is almost free of suspended solids and oil is discharged near the bottom of the flotation chamber. The retention time in the flotation chamber used to be about 20 to 60 min but has been reduced to 3 to 15 min by innovative design.

The effectiveness of DAF depends upon efficient air oxidation and the attachment of bubbles to the oil, VOCs, surfactants, and other particles that are to be removed from the influent water stream. Flotation can be induced in at least three ways:

1. Air bubbles adhering to the insoluble solids by electrical attraction
2. Air bubbles becoming physically trapped in the insoluble solids original or flocculated structure
3. Air bubbles being chemically adsorbed to the insoluble solids in their original form or their flocculated structure

The attraction between the air bubble and contaminants is believed to be primarily a result of particle surface charges and bubble size distribution. The more uniform the distribution of water and microbubbles, the shallower the flotation chamber can be. Generally, the depth of effective flotation chambers is between 0.9 and 2.7 m (3 and 9 ft). Flotation units can be round, square, or rectangular. Gases other than air can be used. The petroleum industry has used nitrogen, with closed vessels, to reduce the possibilities of fire. Ozone can be fed through with air for more efficient reduction of soluble iron, VOCs, and so on.[57] Ozone-UV flotation is another alternative for groundwater decontamination.

Several high-rate air flotation clarifiers (both DAF and dispersed air flotation) with less than 15 min of detention times have been developed for groundwater decontamination, industrial effluent treatment, resources recovery, and water reclamation. Both insoluble and soluble impurities such as

VOCs, activated sludge, fibers, free oil and grease, emulsified oil, lignin, protein, humic acid, tannin, algae, BOD, TOC, iron ions, manganese ions, hardness, titanium dioxide, phosphate, and heavy metals can be separated from a target water stream. Addition of flotation aids to a flotation clarifier is required. Flotation aids include, but are not limited to, aluminum sulfate, ferric chloride, organic polymer, poly aluminum chloride, calcium chloride, ferrous sulfate, calcium hydroxide, ferric sulfate, powdered activated carbon, sodium aluminate, surfactants, and pH adjustment chemicals. Design equations and examples of high-rate DAF clarifiers can be found in the literature.[58,59,69,71]

Toxic organic compounds commonly found in groundwater are presented in Table 8.4. Other toxic organic compounds (representing 1% of cases) include PCBs (polychlorinated biphenyls), 2,4-D, 2,4,5-TP (silvex), toxaphene, methoxychlor, lindane, and endrin, of which 2,4-D and silvex are commonly used for killing aquatic and land weeds. Inorganic toxic substances commonly found in

TABLE 8.4
Toxic Organic Compounds Commonly Found in U.S. Groundwater

Organic Compounds in Groundwater	Percent of Occurrences	Concentration Range
Carbon tetrachloride	5	130 µg/L–10 mg/L
Chloroform	7	20 µg/L–3.4 mg/L
Dibromochloropropane	1	2.5 mg/L
DDD	1	1 µg/L
DDE	1	1 µg/L
DDT	1	4 µg/L
cis-1,2-Dichloroethylene	11	5 µg/L–4 mg/L
Dichloropentadiene	1	450 µg/L
Diisopropyl ether	3	20–34 µg/L
Tertiary methyl butylether	1	33 µg/L
Diisopropyl methyl phosphonate	1	1250 µg/L
1,3-Dichloropropene	1	10 µg/L
Dichloroethyl ether	1	1.1 mg/L
Dichlorosopropyl ether	1	0.8 mg/L
Benzene	3	0.4 µg/L–4.11 mg/L
Acetone	1	10–100 µg/L
Ethyl acrylate	1	200 mg/L
Trichlorotrifluoroethane	1	6 mg/L
Methylene chloride	3	1.21 mg/L
Phenol	3	63 mg/L
Orthochlorophenol	1	100 mg/L
Tetrachloroethylene	13	5 µg/L–70 mg/L
Trichloroethylene	20	5 µg/L–16 mg/L
1,1,1-Trichloroethane	8	60 µg/L–25 mg/L
Vinylidiene chloride	3	5 µg/L–4 mg/L
Toluene	1	5.7 mg/L
Xylenes	4	0.2–10 mg/L
EDB	1	10 µg/L
Others	1	Not available

DDD, Dichlorodiphenyl dichloroethane; DDE, dichlorodiphenyl dichloroethylene; DDT, dichlorodiphenyl trichloroethane; EDB, ethylene dibromide.

Source: Wang, L.K. and Wang, M.H.S., Decontamination of groundwater and hazardous industrial effluents by high-rate air flotation process, *Proc. Great Lakes Conf.*, Hazardous Materials Control Research Institute, Silver Springs, MD, September 1990. With permission.

groundwater include lead, arsenic, copper, cadmium, barium, chromium, mercury, selenium, silver, and nitrate. In a typical groundwater decontamination project, additional industries that are nontoxic but require pretreatment for their removal include iron, manganese, total dissolved solids, and color.

Innovative air flotation technologies have been developed for more cost-effective groundwater decontamination in comparison with the state-of-the-art technologies.[68,69] DAF is very efficient and cost-effective for decontamination of groundwater in which heavy metals, color, TDS, iron, manganese, coliforms, and hardness can all be significantly removed, aiming at not only the decontamination of groundwater but also elimination of biological and chemical fouling for subsequent processes. Furthermore, many VOCs can also be removed by DAF. Table 8.5 represents the U.S. EPA's removal data for DAF processes. The capability of DAF for the treatment of various liquid streams has been well established.[58,59,69] However, its application for the decontamination of groundwater is comparatively new.

Special chemicals may be required for the groundwater decontamination process. For instance, PAC may be dosed into a DAF system for enhancement of contaminant removal efficiency. In such a case, the process is called adsorption flotation (PAC-DAF process). In a pilot plant study, a system consisting of adsorption flotation and sand filtration has proved to be feasible for groundwater decontamination.[70] PAC was added as an adsorbent for the removal of color, odor, EDB (ethylenedibromide), TTHM (total trihalomethane), and other toxic substances from groundwater. Next, the spent PAC was flocculated by coagulants and floated to the water surface by DAF. Finally, the flotation clarified water was polished using the automatic backwash filtration (ABF) process. The results of both bench-scale and pilot plant studies have indicated that using 250 mg/L of PAC at 15 min of detention time can remove color by 100% (from 25 CU [color units]), iron by 100% (from 25 μg/L), humic acid by 98% (from 3200 μg/L), EDB by 100% (from 1.2 μg/L), TTHM by 98% (from 1265 μg/L), odor by 99.6% (from 500 TON [threshold odor numbers]), mercaptans by 100% (from 730 μg/L S), lead by 100% (from 6 μg/L), and arsenic by 100% (from 1000 μg/L). The plant was operated at 40 L/min (10.6 gal/min) for the separation of 250 mg/L of spent PAC. Nearly 100% of spent PAC (from 250 mg/L) and total coliform (from 3/100 mL) and over 95% of turbidity (from 4.5 NTU [nephelometric turbidity units]) were removed by the addition of 1.5 mg/L of anionic polymer and 2.5 mg/L of coagulant. The process was operated at 30% recycle flow rate and 0.014 m^3/h (0.5 ft^3/h) air flow. The sand filter consisted of 28 cm (11 in.) of quartz sand (E = 0.36 mm, U = 1.65) and operated at 102 L/min/m^2 (2.5 gal/min/ft^2).

A DAF-GAC system involving the use of DAF and GAC has also proved to be equally effective for complete groundwater decontamination for the same influent water mentioned above.

For the treatment of a contaminated groundwater source containing a high concentration of hardness, DAF filtration is also an excellent pretreatment process system for the reduction of scale formation in subsequent processes. In a study, groundwater having 12 units of color, 13 NTU of turbidity, and 417 mg/L of carbon hardness as $CaCO_3$ was successfully treated by a continuous DAF filtration plant consisting of hydraulic flocculation, a DAF clarifier, a recarbonation facility, and three sand filters. The added chemicals were 42 mg/L of magnesium carbonate as a coagulant and a small amount of lime for pH adjustment (to pH 11.3). The plant's treatment efficiency in terms of removal had the following values: color, 100%; turbidity, 98%; total hardness, 62%. Recarbonation with CO_2 maintained the effluent pH at 7.2. This plant's operational conditions included a flocculation detention time of 5.6 min, DAF detention time of 3.0 min, flotation clarification rate of 102 L/min/m^2 (2.5 gal/min/ft^2), sand depth of 28 cm (11 in.), influent water flow rate of 45.5 L/min (12 gal/min), recycle water flow rate of 11.4 L/min (3 gal/min), air flow rate of 0.028 m^3/h (1 ft^3/h) at 6.33 kg/cm^2 (90 psig) pressure. Soda ash (Na_2CO_3) may be needed only if permanent hardness ($CaSO_4$) is present. The chemical reactions are as follows:

$$Ca(HCO_3)_2 + Ca(OH)_2 = 2\ CaCO_3 + H_2O$$

$$Mg(HCO_3)_2 + Ca(OH)_2 = CaCO_3 + H_2O + MgCO_3$$

TABLE 8.5
Control Technology Summary for Dissolved Air Flotation

| Pollutant | Effluent Concentration | | % Removal |
	Range	Median	
Classical pollutants (mg/L)			
BOD (5-day)	140–1,000	250	68
COD	18–3,200	1,200	66
TSS	18–740	82	88
Total phosphorus	<0.05–12	0.66	98
Total phenols	>0.001–23	0.66	12
Oil & grease	16–220	84	79
Toxic pollutants (μg/L)			
Antimony	ND–2,300	20	76
Arsenic	ND–18	<10	45
Xylene	ND–1,000	200	97
Cadmium	BDL–<72	BDL	98
Chromium	2–620	200	52
Copper	5–960	180	75
Cyanide	<10–2,300	54	10
Lead	ND–1,000	70	98
Mercury	BDL–2	BDL	75
Nickel	ND–270	41	73
Selenium	BDL–8.5	2	NM
Silver	BDL–66	19	45
Zinc	ND–53,000	200	89
Bis (2-ethylhexyl)phthalate	30–1,100	100	72
Butyl benzyl phthalate	ND–42	ND	>99
Carbon tetrachloride	BDL–210	36	75
Chloroform	ND–24	9	58
Dichlorobromomethane		ND	>99
Di-*n*-butyl phthalate	ND–300	20	97
Diethyl phthalate		ND	>99
Di-*n*-octyl phthalate	ND–33	11	78
N-nitrosodiphenylamine		620	66
2,4-Dimethylphenol	ND–28	14	>99
Pentachlorophenol	5–30	13	19
Phenol	9–2,400	71	57
Dichlorobenzene	18–260	140	76
Ethylbenzene	ND–970	44	65
Toluene	ND–2,100	580	39
Naphthalene	ND–840	96	77
Anthracene/phenanthrene	0.2–600	10	81

ND, non-detectable; BDL, below detection limit; NM, not measured.

Source: Wang, L.K. and Wang, M.H.S., Decontamination of groundwater and hazardous industrial effluents by high-rate air flotation process, *Proc. Great Lakes Conf.*, Hazardous Materials Control Research Institute, Silver Springs, MD, September 1990. With permission.

$$MgCO_3 + Ca(OH)_2 = Mg(OH)_2 + CaCO_3$$

$$CaSO_4 + MgCO_3 = CaCO_3 + MgSO_4$$

$$MgSO_4 + Ca(OH)_2 = Mg(OH)_2 + CaSO_4$$

$$CaSO_4 + Na_2CO_3 = CaCO_3 + Na_2SO_4$$

$$CO_2 + Ca(OH)_2 = CaCO_3 + H_2O$$

$$CO_2 + Mg(OH)_2 = MgCO_3 + H_2O \text{ (coagulant regeneration)}$$

DAF is controlled under laminar hydraulic flow conditions using a very small volume of air flow amounting to about 1 to 3% of the influent groundwater flow. DAF only requires 3 to 5 min of detention time; therefore it is a low-cost process for the decontamination of groundwater.

8.8.3.9 Dispersed or Induced Air Flotation (IAF)

Another innovative process, induced air flotation (IAF), operates under turbulent hydraulic flow conditions by using a large volume of air flow amounting to 400% of the influent groundwater flow. The air bubbles are coarse and large, similar to the air bubbles used in an activated sludge aeration basin. IAF requires only 4 to 10 min of detention time, so it is also a very cost-effective process.[58,59] Unlike DAF, IAF is not an effective pretreatment process for the removal of heavy metals, color, turbidity, TDS, hardness, and coliforms, but it is as efficient as conventional air-sparging and air-stripping processes for the removal of iron, manganese, surfactants, and VOCs.

IAF itself is an aeration process, so soluble iron and manganese ions may be oxidized to form insoluble suspended particles that can be separated easily from the liquid phase. The aeration efficiency of IAF is higher than that of DAF. If groundwater's soluble ferrous iron content is 8 mg/L or below, DAF alone using conventional coagulants will be able to remove the soluble iron.[69] When groundwater's soluble ferrous iron is higher than 8 mg/L, either IAF or an oxidizing agent (ozone, hydrogen peroxide, oxygen, potassium permanganate, and so on) will be required for iron removal.

In the conventional air-stripping process, groundwater is introduced into a gas phase for stripping VOCs; in IAF, air bubbles are injected into the groundwater. An air-stripping tower is over 3 m (10 ft) tall, and an IAF cell can be as shallow as 1 m (3 ft). An important feature of an enclosed IAF cell for VOC reduction is its capability of recycling and reusing its purified air streaming, thus eliminating any possibility of air pollution.[29,30]

In summation, both DAF and IAF are good innovative processes for more efficient and more cost-effective groundwater decontamination.

8.8.4 REMOVAL OF GASOLINE FROM CONTAMINATED SOIL

8.8.4.1 *In Situ* Soil Vapor Extraction

The technologies for *in situ* treatment for groundwater can usually be applied to *in situ* soil remediation, although some of the technologies may have varying suitability for soil. As soil contamination involves a more contaminated phase, the vapor phase, thus vapor extraction is uniquely developed for soil vapor remediation. The decreasing of soil vapor pressure by extraction would cause the free gasoline product to vaporize, so the vapor extraction method also plays a role in the remediation of the liquid phase of VOCs. Based on these observations, the technologies presented in the following discussion will focus mainly on the SVE systems, although other technologies, such as *in situ* soil flushing and *in situ* biological treatment, will also be addressed.

SVE has been an effective technique for removing VOCs such as TCE and some petroleum compounds from the vadose zone of contaminated soil.[72] The following presents some of the newly developed technologies.

Vacuum extraction

The vacuum extraction process involves using vapor extraction wells alone or in combination with air injection wells. Vacuum blowers are used to create the movement of air through the soil. The air flow strips the VOCs from the soil and carries them to the surface. Figure 8.14 shows the flow diagram for such a process. During extraction, water may also be extracted along with vapor. The mixture should be sent to a liquid–vapor separator. The separation process results in both liquid and vapor residuals that require further treatment. Carbon adsorption is used to treat the vapor and water streams, leaving clean water and air for release, and spent GAC for reuse or disposal. Air emissions from the system are typically controlled by adsorption of the volatiles onto activated carbon, by thermal destruction, or by condensation.

The vacuum extraction method has been effectively applied to removing VOCs with low organic carbon content from well-drained soil, although it may also be effective for finer and wetter soils, but with comparatively slower removal rates. There are generally significant differences in the air permeability of various strata, which can influence process performance. Contaminants with low vapor pressure or high water solubilities are difficult to remove.

Soil vacuum extraction is cost-effective if the volume of contaminated soil exceeds 382 m^3 (500 yd^3), and if the contaminated area is more than 6 m (20 ft) deep; otherwise, soil excavation and

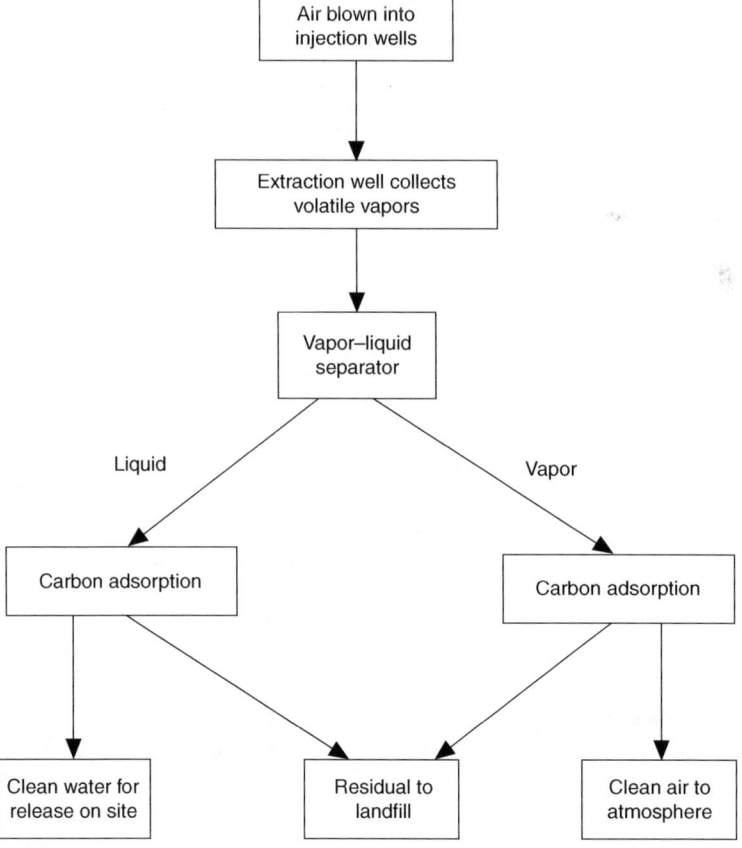

FIGURE 8.14   *In situ* soil vapor extraction.

treatment may be more cost-effective. The level of the groundwater is also important. Rising of the water table that occurs as a result of vacuum extraction wells has to be controlled to avoid water entering the contaminated vadose zone. The water infiltration rate can be controlled by placing an impermeable cap over the site, and a pump may be required to draw the water table down and allow efficient vapor venting. Usually, soil washing follows vapor extraction of volatile contaminants.

In an example, as a result of an impending property transfer that necessitated rapid remediation of diesel-affected soil at a former service station, thermal-enhanced SVE (TESVE) was used to accelerate remediation. The recovery system included a network of TESVE units and injection wells that were separately connected to two regenerative blowers. The recovered vapors were treated in a thermal incinerator, an oxidizing unit to destroy the recovered hydrocarbons. Treated air at a temperature of 1800°F was passed through a heat exchanger and ambient air was simultaneously pumped through the heat exchanger, which increased the temperature of the ambient air to 350°F. The heated ambient air was then injected into the affected soil through a network of four carbon steel injection wells using a regenerative blower.[73]

Soil venting

Soil venting is a technique that removes contaminant vapors from unsaturated soil without excavation. A vacuum extraction system usually consists of gravel packs extending to the soil surface, and a slotted or unslotted well casing that allows gases to move out of the soil. Passive systems consist of vents that are open to the atmosphere and do not require energy for extraction of the gases. Active systems use pressure or vacuum pumps to accelerate the removal of gasoline vapors from the soil. With venting, the vapors are either discharged to the atmosphere or treated before discharge depending on vapor concentrations and regulatory requirements.

Enhanced volatilization

The enhanced volatilization process is operated by putting contaminated soil in contact with clean air in order to transfer the contaminants from the soil into an air stream. The air stream is further treated through the use of carbon canisters, water scrubbers or afterburners to reduce air emission impacts. Four methods are available that can achieve this effect[19]:

1. Mechanical rototilling
2. An enclosed mechanical aeration system
3. A low-temperature thermal stripping system
4. A pneumatic conveyer system

The mechanical rototilling method involves turning over soils to a depth of about 0.30 m (1 ft) below the surface to increase the rate of volatilization. Following treatment, the topsoil is moved to a nearby pile and rototilling is performed on the next 0.30 m (1 ft) of soil. The effectiveness of this mechanical rototilling method is highly dependent on weather conditions. High-speed rototillers and soil shredders can enhance the rate of volatilization.

For effective volatilization using an enclosed mechanical aeration system, contaminated soil is mixed in a pug mill or rotary drum. The gasoline components are released from the soil matrix by the churning action of the air/soil contact. The induced airflow within the chamber captures the gasoline emissions and passes them through an air pollution control device (e.g., a water scrubber or vapor-phase carbon adsorption system) before they are discharged through a properly sized stack.

The configuration of a low-temperature thermal stripping system is similar to the enclosed mechanical aeration system except that additional heat transfer surfaces allow the soil to heat by coming into contact with a screw auger device or rotary drum system. The induced airflow conveys the desorbed volatile organics/air mixture through an afterburner where organic contaminants are destroyed. The over air stream is then discharged through a properly sized stack.

A pneumatic conveyer system consists of a long tube or duct to carry air at high velocities, an induced draft fan to propel the air, a suitable feeder for addition and dispersion of particulate solids into the air stream, and a cyclone collector or other separation equipment for final recovery of the solids from the gas stream. Several such units heat the inlet air to 300°F to induce volatilization of organic contaminants. Pneumatic conveyers are primarily used in the manufacturing industry for drying solids with up to 90% initial moisture content.

Of the four enhanced volatilization methods described above, documentation exists to support the contention that the low-temperature thermal stripping system has the greatest ability to successfully remove contaminants that are similar to gasoline constituents (i.e., compounds with high vapor pressures) from soil. The limitations of some enhanced volatization techniques can be attributed to the following:

1. Associated soil characteristics that inhibit the mobility of gasoline vapors from the soil to the air
2. Contaminant concentrations that may cause an explosion or fire
3. The need to control dust and organic vapor emissions

Some integrated techniques may be more economical if they can be used simultaneously for soil and groundwater treatment, such as integrated vapor extraction and steam vacuum stripping.

8.8.4.2   *In Situ* Soil Flushing

Soil flushing treatment is a technique that removes gasoline constituents from the soil matrix by actively leaching the contaminants from the soil into a leaching medium. The most common washing medium is water, which may contain additives such as acids, alkalis, and detergents. The washing fluid can also be composed of pure organic solvents such as hexane and triethylamine.

The washing media are recharged into soil using a spray recharge system or injection wells. Withdraw wells convey the after-washing liquid to an aboveground treatment facility. The after-washing liquid is treated using biological treatment or physical-chemical methods such as air stripping.

Surfactants have been widely used to reduce the interfacial tension between oil and soil, thus enhancing the efficiency of rinsing oil from soil. Numerous environmentally safe and relatively inexpensive surfactants are commercially available. Table 8.6 lists some surfactants and their chemical properties.[74] The data in Table 8.6 are based on laboratory experimentation; therefore, before selection, further field testing on their performance is recommended. The Texas Research Institute[75] demonstrated that a mixture of anionic and nonionic surfactants resulted in contaminant recovery of up to 40%. A laboratory study showed that crude oil recovery was increased from less than 1% to 86%, and PCB recovery was increased from less than 1% to 68% when soil columns were flushed with an aqueous surfactant solution.[74,76]

Contained recovery of an oily waste process has been developed by the Western Research Institute.[65] It uses steam and hot water (through injection wells) to displace oily waste from the soil, which is then conveyed (by production wells) aboveground for treatment (Figure 8.15). Low-quality steam is injected below the deepest penetration of organic fluids. The steam condenses, causing rising hot water to dislodge and sweep the buoyant organic fluid upward into more permeable soil regions. Hot water is injected above the impermeable soil regions to heat and mobilize the oily waste accumulations, which are recovered by hot water displacement. When oily wastes are displaced, the organic fluid saturation in the subsurface pore space increases, forming an oil bank. The oil saturation is reduced to an immobile residual saturation in the subsurface pore space. The produced oil and water are treated for reuse or discharge. In the process, contaminants are contained laterally by groundwater isolation, and vertically by organic fluid flotation.

The contained recovery method is claimed to have the following advantages:

1. It removes large portions of oily waste accumulations.
2. It stops the downward migration of organic contaminants.
3. It immobilizes any residual saturation of oily wastes.

TABLE 8.6
Surfactants Characteristics

Surfactant Type		Selected Properties and Uses	Solubility	Reactivity
Anionic	Carboxylic acid salts	Good detergency	Generally water-soluble	Electrolyte-tolerant
	Sulfuric acid ester salts	Good wetting agents		Electrolyte-sensitive
	Phosphoric and polyphosphoric acid esters	Strong surface tension reducers	Soluble in polar organics	Resistant to biodegradation
	Perfluorinated anionics			High chemical stability
	Sulfonic acid salts	Good oil in water emulsifiers		Resistant to acid and alkaline hydrolysis
Cationic	Long-chain amines	Emulsifying agents	Low or varying water solubility	Acid stable
	Diamines and polyamines	Corrosion inhibitor	Water-soluble	Surface adsorption to silicaeous materials
	Quaternary ammonium salts			
	Polyoxyethylenated long-chain amines			
Nonionic	Polyoxyethylenated alkyl-phenols, alkylphenol ethoxylates	Emulsifying agents	Generally water-soluble	Good chemical stability
	Polyoxyethylenated straight-chain alcohols and alcohol ethoxylates	Detergents	Water insoluble formulations	Resistant to biodegradation
	Polyoxyethylenated poly-oxypropylene glycols	Wetting agents		Relatively nontoxic
	Polyoxyethylenated mercaptans			
	Long-chain carboxylic acid esters	Dispersents		
	Alkylolamine condensates, alkanolamides	Foam control		Subject to acid and alkaline hydrolysis
	Tertiary acetylenic glycols			
Amphoterics	pH-sensitive	Solublizing agents	Varied	Nontoxic
	pH-insensitive	Wetting agents	(pH-dependent)	Electrolyte-tolerant
				Adsorption to negatively charged surfaces

Source: U.S. EPA, Remedial Action at Waste Disposal Sites, EPA/625/6-85/006, U.S. EPA, Washington, 1985.

4. It reduces the volume, mobility, and toxicity of oily wastes.
5. It can be used for both shallow and deep contaminated areas.
6. It uses the same mobile equipment required for conventional petroleum production technology.

8.8.4.3   *In Situ* Biological Soil Treatment

The technology for *in situ* biological treatment for soil is similar to that for *in situ* biological groundwater treatment. The following sections present three newly developed techniques.

Deep in situ *bioremediation process*
This technique was developed by In-Situ Fixation Company for increasing the efficiency and rate of biodegradation in deep contaminated soils using a dual-auger system. Mixtures of microorganism and required nutrients are injected into the contaminated soils without any excavation. The injection and mixing effectively break down fluid and soil strata barriers and eliminate pockets of

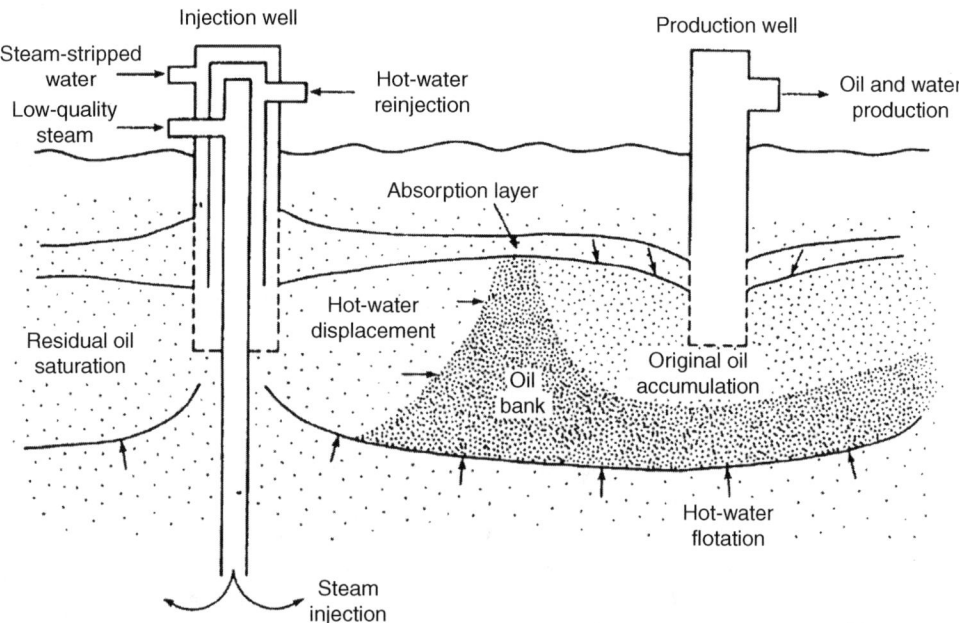

FIGURE 8.15 CROWTM subsurface development.

contaminated soil that would otherwise remain untreated. The drilling is carried out in an overlapping manner to ensure complete treatment of all contaminated soil. The mixing action is continued as the augers are withdrawn. The treatment depth may exceed 30 m (100 ft).[65]

In situ *geolock and biodrain treatment platform*

This system consists of an *in situ* polyethylene tank, an application system, and a bottom water recovery system.[65] An underlying, permeable, water-bearing zone facilitates the creation of ingradient water flow conditions. The tank defines the treatment area, minimizes the potential for release of bacterial cultures to the aquifer, and maintains contaminant concentration levels that facilitate treatment. The ingradient conditions facilitate reverse leaching or soil washing and minimize the potential for outmigration of contaminants.

The application system, called the biodrain, is installed within the treatment area. The biodrain aerates the soil column and any standing water. This cerates an aerobic environment in the pore spaces of the soil. Other gas mixtures can also be introduced to the soil column, such as the air/methane mixtures used in the biodegradation of chlorinated organics. The treatment platforms can be placed in very dense configurations. International Environmental Technology claims that the cost of installation is low.

The bottom water recovery system uses existing wells or new wells to create the water recovery system for removal of the water used to wash the contaminated soil. Reverse leaching or soil washing can be conducted by controlling the water levels within the tank. This design minimizes the volume of clean *ex situ* water entering the system for treatment. Extremely dense clays may be difficult to treat with this technology.

In situ *bioventing technology*

Bioventing technology was developed by the U.S. EPA Risk Reduction Engineering Laboratory to treat soil contaminated by numerous industrial wastes, which is subjected to aerobic microbial degradation, especially to promote the degradation of polycyclic aromatic hydrocarbons.[65] It uses a series of air injection probes, each of which is attached to a low-pressure air pump. The air pump operates at extremely low pressures to allow the inflow of oxygen without volatilization of contaminants. Additional additives such as ozone or nutrients may also be supplied to stimulate microbial growth.[77]

8.8.4.4 *Ex Situ* Soil Treatment

All *ex situ* soil treatment methods involve a two-step approach: soil excavation and aboveground treatment of the excavated soil. The differences in the various *ex situ* excavation/treatment methods for soil remediation lie only in the methods of soil treatment aboveground, such as soil washing plus extraction, and slurry biodegradation.

Soil washing technology
The excavated soil is removed from the site and screened to remove large solid objects. The screened soil is washed and the washing water is treated.[78] Clearly, the washing media used in *in situ* soil-flushing treatment can be used here. The most common washing medium is water. Surfactants are used to reduce the affinity of contaminants to the soil.

Several unit processes can be used in the washing process. The soil is mixed with washing agents and extraction agents that remove the contaminants from the soil and transfer them to the extraction fluid. The soil and washwater are then separated. The soil can be further rinsed with clean water. The soil is removed as clean product, ready to put back into the original excavation, and the washwater is ready to be treated by conventional wastewater treatment processes as addressed in the next subsection.

The big difference in application from the *in situ* flushing method is that this *ex situ* method can apply to soils with lower permeability, because soil is excavated and can be sufficiently washed. The following presents two *ex situ* soil washing processes for organic contaminants: the BioGenesis Soil Cleaning process and the BioTrol Soil Washing System.

The BioTrol Soil Washing System developed by BioTrol, Inc., is shown in Figure 8.16. After debris is removed, the excavated soil mixed with water and is subjected to various unit operations common to the mineral processing industry. Process steps include mixing units, pug mills, vibrating screens, froth flotation or induced air flotation (IAF) cells, scrubbing machines, hydrocyclones, screw classifiers, and various dewatering operations. The core of the system is a multistage, countercurrent, intensive scrubbing circuit with interstage classification. The scrubbing action disintegrates soil aggregates, freeing contaminated fine particles from the coarser sand and gravel. In addition, superficial contamination is removed from the coarse fraction by the abrasive scouring action of the particles themselves. Contaminants may also be solubilized, as dictated by solubility characteristics or partition coefficients. This technology is a water-based volume reduction process for treating excavated soil. Soil washing may be applied to contaminants concentrated in the fine-size fraction of soil (silt, clay, and soil organic matter) and the superficial contamination associated with the coarse soil fraction. This technology can be applied to soils contaminated with PAH (polycyclic aromatic hydrocarbons) and PCP (pentachlorophenol), PCB (polychlorinated biphenyl), petroleum hydrocarbons, and pesticides.[65,78]

The BioGenesis Soil Cleaning process developed by BioVersal USA, Inc., uses a specialized truck, water, and a complex surfactant (a light alkaline mixture of natural and organic materials containing no hazardous ingredients) to clean contaminated soil. Ancillary equipment includes gravity oil/water separators, coalescing filters, and a bioreactor. Figure 8.17 shows the soil washing procedure. After washing, the extracted oil is reclaimed, the wash water is recycled or treated, and the soil is dumped for refill. Hazardous organics are extracted in the same manner and then further treated. It was shown that the clean rate is ca. 25 t/h for 5000 mg/L oil contamination and lower rates for more contaminated soils. One single wash removes 95 to 99% of hydrocarbon contamination levels up to 15,000 mg/L. The main advantages of the process are as follows[65]:

1. Treatment is applicable to soils containing both volatile and nonvolatile oils.
2. Soil containing clay may be treated.
3. The process rate is high.
4. Contaminants are transformed into reusable oil, treatable water, and soil suitable for on-site treatment.
5. There is no air pollution, except during excavation.

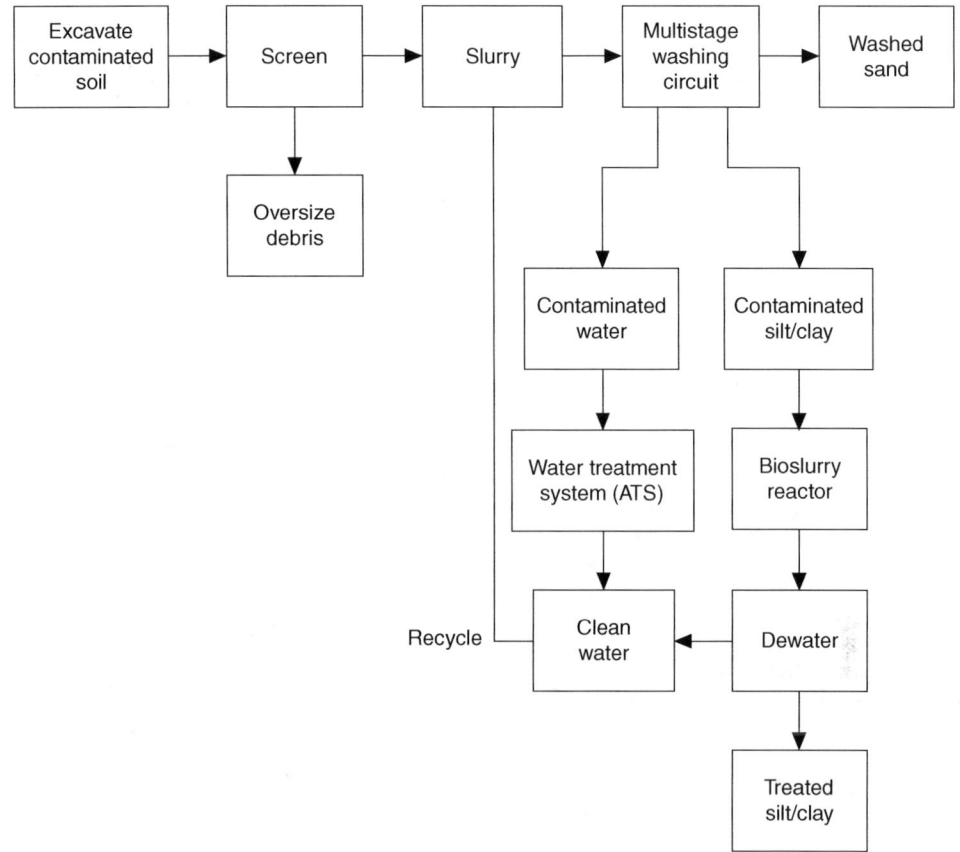

FIGURE 8.16 BioTrol soil washing system process diagram.

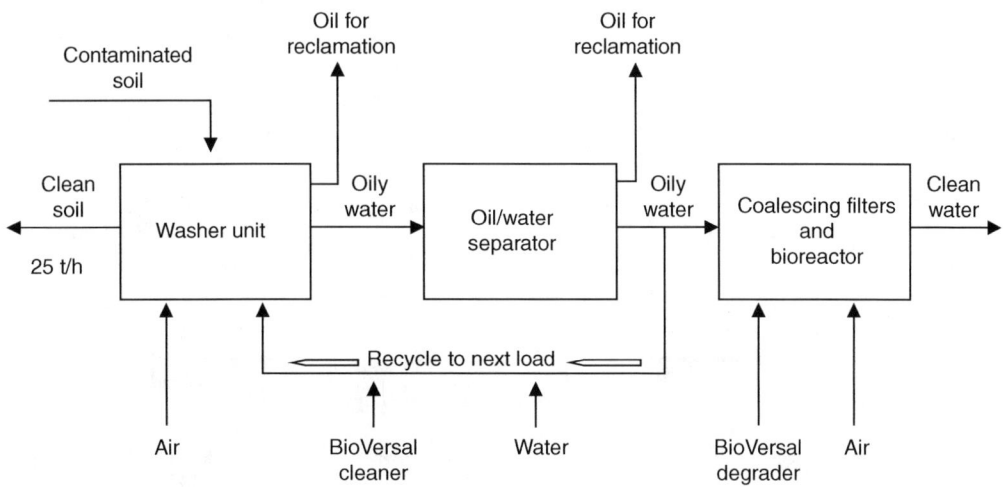

FIGURE 8.17 Soil washing procedure.

A solvent extraction technology developed by CF System Corporation uses liquefied gases as solvent to extract organics (such as PCB, dioxins, PCP, petroleum wastes) from sludges, contaminated soils, and wastewater.[65,79] Propane is the solvent most typically used for contaminated soils, and carbon dioxide is used for wastewater streams. The system is available as either a continuous flow unit for liquid wastes or a batch system for soils. Contaminated soils, slurries, or wastewaters are fed into an extractor along with the solvent. Typically, more than 99% of organics are extracted from the feed. Following phase separation of the solvent and organics, treated water is removed from the extractor while the mixture of solvent and organics passes to the solvent recovery system. In the solvent recovery system, the solvent is vaporized and recycled as a fresh solvent. The organics are drawn out and either reused or disposed of.

Treatment technologies for washing water
Washing fluid can be separated from soil by conventional techniques such as sedimentation, flotation, and filtration.[69] Slurry of soil can be dewatered. The treated soils can then be returned into the original excavation or sent to a sanitary landfill. Treatment of washing water is similar to the treatment of pumped contaminated groundwater, including air stripping of the volatile organics or biological treatment.

8.8.4.5   *Ex Situ* Biological Treatment on Excavated Soil by Slurry Biodegradation

The procedure for slurry biodegradation is not different from conventional biological treatment. The first step is cleaning the soil and separating it from the washing liquid, which is followed by separate biological treatment for the liquid and the soil slurry. The treated soil is then separated from the slurry. Figure 8.18 shows the slurry biodegradation steps in processing the soil.[80]

Waste preparation for slurry biodegradation
Several preparation steps after soil excavation are required to achieve the optimum inlet feed characteristics for maximum contaminant removal:

1. Screening of the soil to remove large objects
2. Size reduction for large particles
3. Water addition
4. pH and temperature adjustments

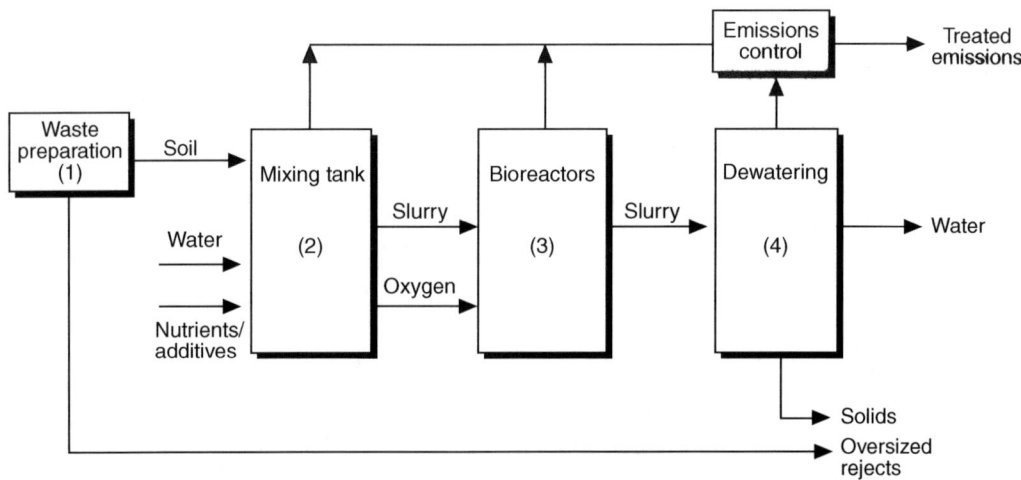

FIGURE 8.18   Slurry biodegradation process.

Treatment by slurry biodegradation

The pretreated soil is mixed with water in a tank to form a slurry. Sufficient mixing is necessary to ensure contact between the contaminants and the microorganisms to facilitate mass transfer from the contaminants to the microorganisms. The well-mixed slurry is conveyed to a bioreactor or a lined lagoon where the aerobic process takes place. Aeration is provided by either floating or submerged aerators. Once the biodegradation of the contaminants is completed, the treated slurry is sent to a dewatering system to separate the soil phase from the aqueous phase of the slurry. Figure 8.19 shows the process for a slurry bioreactor developed by ECOVA Corporation.[65]

U.S. EPA has shown that 90% of process water can be recycled to the front end of the system for slurry preparation, and the rest must be treated on site or transported to an off-site facility.[80] During the aerobic process, some contaminated air may be formed and emitted from the reactor. Depending on the air characteristics, a compatible air pollution control device may be used, such as activated carbon. Slurry biodegradation has been shown to be successful in treating soils contaminated with soluble organics, PAHs, and petroleum waste. The process has been most effective with contaminant concentrations ranging from 2500 mg/kg to 250,000 mg/kg.

The slurry bioreactor developed by ECOVA Corporation[65] showed a 93.4% reduction in PAHs over a 12-week treatment period with an initial 89.3% reduction in the first two weeks.

8.8.4.6 *Ex Situ* Soil Desorption

In situ SVE methods can be used for desorption of VOC from excavated soils. The excavated soil has the advantage that assist technologies may be applied to enhance vaporization, for example, through venting and heating.

One of the desorption technologies, the anaerobic thermal process, is a thermal desorption process. In this process, heating and mixing of the contaminated soils, sludges, and liquids take place in a special rotary kiln that uses indirect heat for processing.[65] The SoilTech anaerobic thermal process is designed to both desorb and treat organic contaminants in soil. The kiln portion of the system contains four separate internal thermal zones: preheat, retort, combustion, and cooling. From the preheat zone, the hot granular solids and unvaporized hydrocarbons pass through a sand seal to the retort zone. Heavy soils vaporize in the retort zone and thermal cracking of the

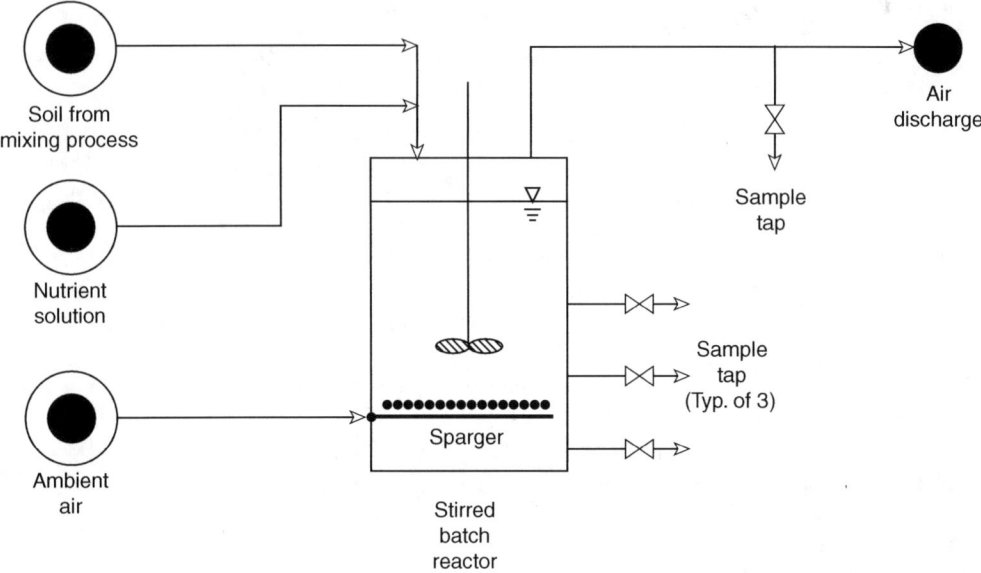

FIGURE 8.19 Process flow diagram.

hydrocarbons forms coke and low-molecular-weight gases. The vaporized contaminants are removed by vacuum to a retort gas-handling system. After cyclones remove the dust from the gases, the gases are cooled, and condensed, and the oil is separated into its various fractions. The coke (with the coked soil) is burned, and the hot soil is either recycled back to the retort zone or sent to the cooling zone. Flue gases from the combustion zone are treated prior to discharge in a cyclone and a baghouse for particle removal, wet scrubber for removal of acid gases, and carbon adsorption bed for removal of trace organic compounds.

The unit desorbs, collects, and recondenses hydrocarbons from the solids. The unit can also be used in conjunction with a dehalogenation process to destroy halogenated hydrocarbons through a thermal chemical process. The technology can be used for oil recovery from tar sands and shales, dechlorination of PCBs and chlorinated pesticides in soil and sludges, separation of oils and water from refinery wastes and spills, and general removal of hazardous organic compound from soils and sludges.

8.8.4.7   *Ex Situ* Thermal Destruction

Incineration
Incineration can effectively eliminate gasoline from soils by complete oxidation. Rotary kilns, fluidized beds, and other systems, either fixed or mobile types, may achieve 99.99% removal. One of the limitations of *ex situ* thermal destruction is associated with soil excavation. Mobile units may be further limited by the permitting process. Costs for incineration vary significantly depending on the particular characteristics of the soil and waste material. Soil containing higher gasoline waste is more economical to treat than a soil with lower gasoline waste, especially when compared with other treatment methods.

Infrared thermal destruction
Infrared thermal destruction technology is a thermal processing system that uses electrically powered silicon carbide rods to heat organic wastes to combustible temperatures. Any remaining combustibles are incinerated in an afterburner. One configuration made by ECOVA Corporation consists of four components[65]:

1. An electric-powered infrared primary chamber
2. A gas-fired secondary combustion chamber
3. An emissions control system
4. A control center

Waste is fed into the primary chamber and exposed to infrared radiant heat up to 1010°C (1850°F) provided by silicon carbide rods above the belt. A blower delivers air to selected locations along the belt to control the oxidation rate of the waste feed. The ash material in the primary chamber is quenched by using the scrubber water effluent. The ash is then conveyed to an ash hopper, where it is removed to a holding area and analyzed for organic contaminants, such as PCB content. Volatile gases from the primary chamber flow into a secondary chamber, which uses higher temperatures, greater residence time, turbulence, and supplemental energy (if required) to destroy these gases. Gases from the secondary chamber are passed through the emissions control system.

This technology is suitable for soils or sediments with organic contaminants. The optimal waste characteristics are as follows:

1. Particle size, 5 μm to 50 mm
2. Moisture content, up to 50% by weight
3. Density, 481 to 2083 kg/m^3 (30 to 130 lb/ft^3)
4. Heating value, up to 5556 kg-cal/kg, or 5556 cal-g/g (10,000 Btu/lb)
5. Chlorine content, up to 5% by weight

6. Sulfur content, up to 5% by weight
7. Phosphorus, 0 to 300 mg/L
8. pH, 5 to 9
9. Alkali metals, up to 1% by weight

Plasma arc vitrification

Plasma arc vitrification, developed by Retech, uses a plasma centrifugal furnace, where heat from transferred arc plasma creates a molten bath that detoxifies the feed material. Organic contaminants vaporize and react at temperatures between 2000 and 2500°F to form innocuous products. Solids are melted and vitrified in the molten bath at 2800 to 3000°F. When metals are cooled, they are rendered to a nonleachable, glassy residue that meets the toxicity characteristic leachate procedure (TCLP) criteria.

This technique can treat soils contaminated with organic compounds and is also suitable for treating liquids and solids containing organic compounds and metals.

8.9 PHENOMENA RELATED TO THE RELEASE OF DNAPLs AND OTHER HAZARDOUS SUBSTANCES

Besides petroleum products, other hazardous substances (see Tables 8.7–8.9) are also stored in USTs. Among them, a common and important group is the dense nonaqueous phase liquids (DNAPLs). This group has some different physical properties from petroleum (especially gasoline) that make them behave differently in the way they move underground. This section presents the important factors associated with the cleanup of DNAPLs.

The relative vapor density (RVD) values in Table 8.9 have been calculated as the density of dry air saturated with the compound of interest at 20°C. This represents the weighted mean molecular weight of the compound-saturated air relative to the mean molecular weight of dry air, which is 29 g/mol. The RVD value may be calculated from Equation 8.23:

$$\text{RVD} = \frac{P_0(\text{MW}) + (760 - P_0)(29)}{(760)(29)} \tag{8.23}$$

where RVD = relative vapor density (dimensionless), MW = molecular weight of the compound of interest, and P_0 = vapor pressure (torr or mmHg).

8.9.1 CHEMICAL AND PHYSICAL PROPERTIES OF DNAPLs

DNAPLs are mainly liquid hydrocarbons such as chlorinated solvents, wood preservatives, coal tar wastes, and pesticides. Table 8.7 lists some common such chemicals.[81]

DNAPLs have higher densities than water, most between 1 and 2 g/mL, some are near 3 g/mL, for example, bromoform, which has a density of 2.89 g/mL. They have limited water solubilities, and are usually found as the free-phase immiscible with water or as residuals trapped by soil. Most DNAPLs are volatile or semivolatile; Pankow[82] has listed information on their physical and chemical properties, such as molecular weight, density, boiling points, solubility in water, vapor pressure, sediment/water partition coefficient, viscosity, Henry's law constant, and so on (see Tables 8.8 and 8.9).

8.9.2 FATE OF DNAPL RELEASE UNDERGROUND

Similar to gasoline, the properties of DNAPLs such as immiscibility with water, volatility, and solubility of some of its components cause the presence of multiphase (pure product, solute, gas, and adsorbate) products and movement that is typical of the phenomena associated with DNAPL release. The theory associated with the interaction of gasoline with soil is applicable to DNAPLs. However,

TABLE 8.7
DNAPL-Related Chemicals

Halogenated Volatiles	Nonhalogenated Semivolatiles
Chlorobenzene	2-Methyl naphthalene
1,2-Dichloropropane	o-Cresol
1,1-Dichloroethane	p-Cresol
1,1 Dichloroethylene	2,4-Dimethylphenol
1,2-Dichloroethane	m-Cresol
Trans-1,2-dichloroethylene	Phenol
Cis-1,2-dichloroethylene	Naphthalene
1,1,1-Trichloroethane	Benzo(a) anthracene
Methylene chloride	Fluorene
1,1,2-Trichloroethane	Acenaphthene
Trichloroethylene	Anthracene
Chloroform	Dibenzo (a,h) anthracene
Carbon tetrachloride	Fluoranthene
1,1,2,2-Tetrachloroethane	Pyrene
Tetrachloroethylene	Chrysene
Ethylene dibromide	2,4-Dinitrophenol
Halogenated Semivolatiles	**Miscellaneous**
1,4-Dichlorobenzene	Coal tar
1,2-Dichlorobenzene	Creosote
Aroclor 1242, 1254, 1260	
Dieldrin	
2,3,4,6-Tetrachlorophenol	
Pentachlorophenol	

Many of these chemicals are found mixed with other chemicals or carrier oils.
Source: U.S. EPA, Estimating Potential for Occurrence of DNAPL at Superfund
         Sites, EPA Publication: 9355.4-O7FS, U.S. EPA, Washington, January 1992.

the gas phase may not be detected as significantly as in the case of gasoline, because the main part of the DNAPL plume sinks below the water table. Therefore the vapor phase does not exist in equilibrium with the free DNAPL phase. Figure 8.20 illustrates such a phenomenon.

The property of DNAPLs that most contrasts to that of gasoline is density, which is higher than water. Thus, the DNAPL plume tends to sink to the bottom of the groundwater body and penetrate down through rock openings instead of floating above the water table. Therefore, DNAPLs are more difficult than gasoline to access and clean up. However, most DNAPLs have higher viscosities than gasoline, and thus may be less transportable with groundwater flow. Moreover, because DNAPLs do not float on the surface of the water table, their capillary movement (though they are important residuals in the vadose zone) has less impact than gasoline release.

8.9.3 SITE REMEDIATION

The remedial technologies[83–85,90–93] described in previous sections for gasoline release are applicable, for the most part, for remediation of DNAPLs. For example, the pumping or trench method for free products, vacuum extraction, biodegradation, pumping and treatment, soil flushing, and soil excavation and treatment are suitable for cleanup of various phases of DNAPLs. Again, because of

TABLE 8.8

Physical and Chemical Properties of Dense Solvent Compounds

Compound	MW (g)	S (mg/L)	P_o (torr)	K_{oc} (mL/g)	d (g/cm³)	BP (°C)
Nonaromatics						
Dichloromethane (DCM)	84.9	20,000	349	8.8	1.33	40
Chloroform	119.4	8,200	151	44	1.49	62
Bromodichloromethane	163.8	4,500	50	61	1.97	90
Dibromochloromethane	208.3	4,000	76	84	2.38	119
Bromoform	252.8	3,010	5	116	2.89	150
Trichlorofluoromethane	137.4	1,100	667	159	1.49	24
Carbon tetrachloride	153.8	785	90	439	1.59	77
1,1-Dichloroethane	99.0	5,500	180	30	1.17	57
1,2-Dichloroethane	99.0	8,690	61	14	1.26	83
1,1,1-Trichloroethane (1,1,1-TCA)	133.4	720[a]	100	152	1.35	74
1,1,2-Trichloroethane	133.4	4,500	19	56	1.44	114
1,1,2,2-Tetrachloroethane	167.9	2,900	5	118	1.60	146
1,1-Dichloroethylene	97.0	400	590	65	1.22	32
1,2-Dibromoethane (EDB)	187.9	4	11[a]	—	2.18	132
1,2-*Cis*-dichloroethylene	97.0	800[a]	200[a]	—	1.28	60
1,2-*Trans*-dichloroethylene	97.0	600	326	59	1.26	48
Trichloroethylene (TCE)	131.5	1,100	58	126	1.46	87
Tetrachloroethylene (PER)	165.8	200	14	364	1.63	121
1,2-Dichloropropane	113.0	2,700	42	51	1.16	97
Trans-1,3-dichloropropylene	110.0	1,000	25	48	1.22	112
Ethers						
Bis(chloromethyl) ether	115.0	22,000	30[a]	1.2	1.32	104
Bis(2-chloroethyl) ether	143.0	10,200	0.7	14	1.22	178
Bis(2-chloroisopropyl) ether	171.1	1,700	0.9	61	1.11	187
2-Chloroethyl vinyl ether	106.6	15,000	27	6.6	1.05	108
Monocyclic aromatics						
Chlorobenzene	112.6	488[a]	12	330	1.11	132
o-Dichlorobenzene	147.0	100	1.0	1700	1.31	180
m-Dichlorobenzene	147.0	123[a]	2.3[a]	1700	1.29	172

Temperature of measurement is 20°C unless otherwise noted.

MW = molecular weight (g); S = solubility in water (mg/L or ppm); P_o = Vapor pressure (torr or mmHg);

K_{oc} = sediment/water partition coefficient (mL/g); d = density (g/cm³); BP = boiling point at 760 torr pressure (°C).

[a]Value measured at 25°C.

Source: Pankow, J.F., *Dense Chlorinated Solvents in Porous and Fractured Media*, Lewis Publisher, Freidrich Schwille, 1988, p. 146. With permission.

the higher density of DNAPLs, some remedial processes may have to be arranged differently, such as in the pumping systems for free product and bioremediation.

8.9.3.1 Pumping Systems for Free Product Recovery

Similar to pumping systems for gasoline remediation, single or dual arrangements can be used. The difference is that the product screen is located below the groundwater in the aquifer. Furthermore, in the dual-pumping systems, an additional screen interval is used in the groundwater zone, located vertically upward from the DNAPL screen intake. Groundwater is withdrawn from the upper screen, resulting in an upwelling of DNAPLs (see Figure 8.21), improving the rate of recovery and resulting

TABLE 8.9

More Physical and Chemical Properties of Dense Solvent Compounds

Compound	μ Absolute Viscosity (cp)	ν Kinematic Viscosity (cs)	H (atm-m³/mol)	Relative Vapor Density[a]
Nonaromatics				
Dichloromethane (DCM)	0.44	0.32	0.0017	1.89
Chloroform	0.56	0.38	0.0028	1.62
Bromodichloromethane	1.71	0.87	0.0024	1.31
Dibromochloromethane	—	—	0.00099	1.62
Bromoform	2.07	0.72	0.00056	1.05
Trichlorofluoromethane	—	—	0.11	4.28
Carbon tetrachloride	0.97	0.61	0.023	1.51
1,1-Dichloroethane	0.50	0.43	0.0043	1.57
1,2-Dichloroethane	0.84	0.67	0.00091	1.19
1,1,1-Trichloroethane (1,1,1-TCA)	0.84	0.62	0.013	1.47
1,1,2-Trichloroethane	—	—	0.00074	1.09
1,1,2,2-Tetrachloroethane	1.76	1.12	0.00038	1.03
1,1-Dichloroethylene	0.36	0.30	0.021	2.54
1,2-*Cis*-dichloroethylene	0.48	0.38	0.0029	1.62
1,2-*Trans*-dichloroethylene	0.40	0.32	0.072	2.01
Trichloroethylene (TCE)	0.57	0.39	0.0071	1.27
Tetrachloroethylene (PER)	0.90	0.54	0.0131	1.09
1,2-Dichloropropane	—	—	0.0023	1.16
Trans-1,3-dichlororpropylene	—	—	0.0013	1.09
Ethers				
Bis(chloromethyl) ether	—	—	0.00021	1.12
Bis(2-chloroethyl) ether	2.41	1.98	0.000013	1.004
Bis(2-chloroisopropyl) ether	—	—	0.00011	1.005
2-Chloroethyl vinyl ether	—	—	0.00025	1.10
Monocyclic aromatics				
Chlorobenzene	0.80	0.72	0.0036	1.05
o-Dichlorobenzene	1.41	1.28	0.0019	1.005
m-Dichlorobenzene	1.08	0.84	0.0036	1.01

μ = absolute viscosity (cP); ν = kinematic viscosity (cSt); H = Henry's Law constant for partitioning between air and water (atm-m³/mol); and RVD = vapor density relative to dry air (dimensionless).

Source: Pankow, J.F., *Dense Chlorinated Solvents in Porous and Fractured Media*, Lewis Publisher, Freidrich Schwille, 1988, p. 146. With permission.

in a more efficient operation. The groundwater withdrawal rate must be carefully determined; too much will result in DNAPLs rising excessively and mixing with water or being suppressed by the higher water velocity; too low will not cause the required upwelling.

Other enhanced DNAPL recovery techniques have been implemented utilizing both water flooding and well bore vacuum. Essentially, this minimizes drawdown, allowing a maximum pumping rate of the DNAPL/water mixture.

8.9.3.2 Biodegradation

As stated previously in Section 8.8, one of the advantages of biodegradation is that it imposes a permanent solution, especially if the release is trapped in cracks or is highly adsorbed. Because

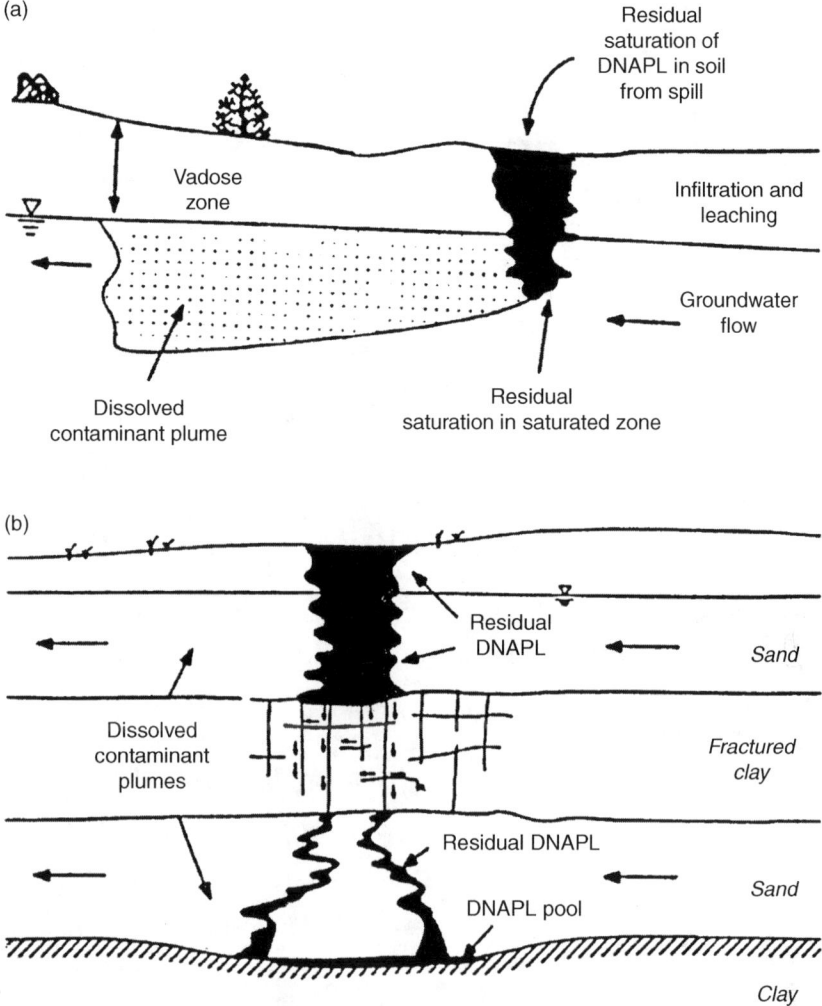

FIGURE 8.20 Groundwater contamination by DNAPL.

DNAPL release has a higher tendency to sink deep down, entering cracks, biodegradation may be more significant for DNAPL release than for gasoline.

The reader may have noticed that Section 8.8 relates biodegradation only to the contaminated groundwater and soil, but not to the free products. In fact in most cases, biodegradation is not suitable for free products. The pure phase, either DNAPLs or gasoline, creates a highly hostile environment for the survival of most microorganisms. Thus, the bioremediation technique is more applicable to groundwater or residuals than to pure free product, and biodegradation is used after the main DNAPL product is recovered. When toxicity has been reduced by product recovery, biodegradation or bioremediation can then be used to further reduce the contaminants at the site.[90-93]

8.9.3.3 U.S. EPA Corrective Action Measures through 2006

Figure 8.22 provides an illustration of historical cleanup backlog trends in the U.S. from 1989 to 2006.[86] Since the beginning of the program, U.S. EPA has cleaned up almost 75% of all releases, and reduced the cleanup backlog to 113,914 cases, a 33% decrease from a peak backlog of 170,000 cases for the 5-yr period 1995 to 2000 (see Figure 8.22).

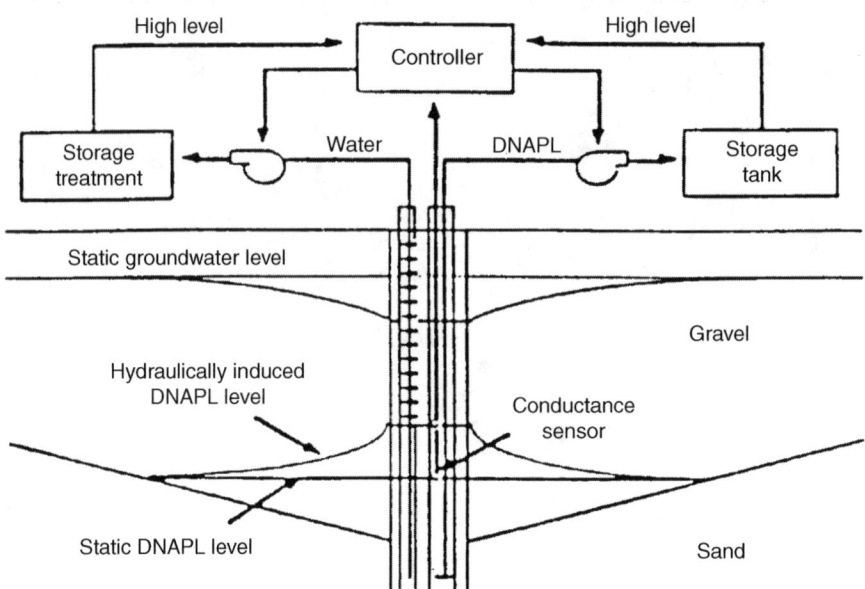

FIGURE 8.21 A DNAPL recovery system where deliberate upwelling of the static coal-tar surface is used to increase the flow of product into the recovery wells.

8.9.4 PRACTICAL EXAMPLES

8.9.4.1 Conversion between Kinetic Viscosity and Absolute Viscosity for Air

Kinematic viscosity of air at 1 bar and 40°C is 16.97 cSt (16.97×10^{-6} m²/sec) (cSt = centistokes). Determine the air's dynamic viscosity or absolute viscosity.

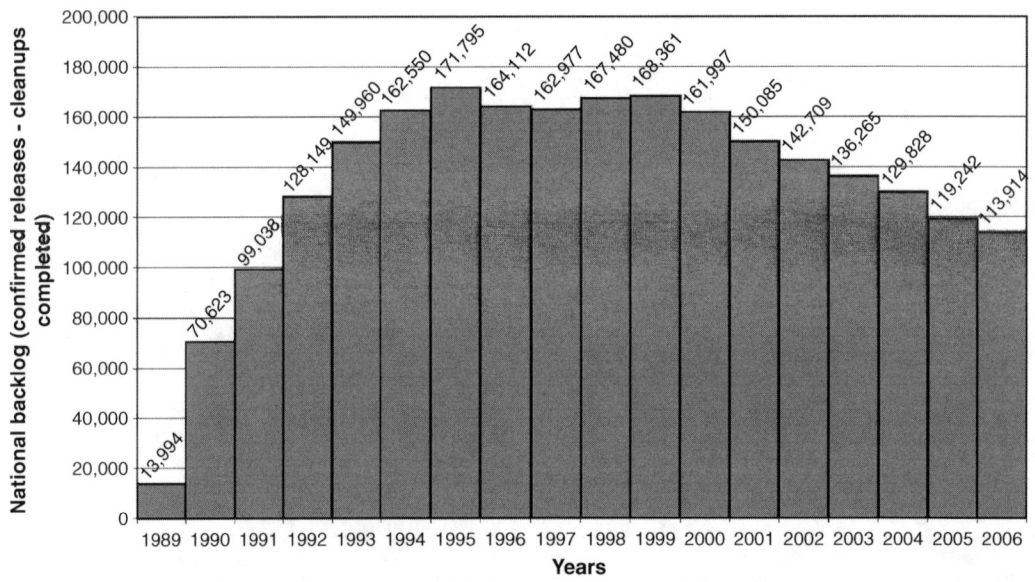

FIGURE 8.22 UST national backlog: 1989 through 2006. (Taken from U.S. EPA, Underground Storage Tanks Corrective Action Measures Archives, FY 2006 End-of-Year Activity Report, U.S. EPA, Washington, 2007. Available at http://www.epa.gov/OUST/cat/camarchv.htm.)

Solution
The density of air may be estimated by the Ideal Gas Law:

$$P = \frac{10^5}{(287)(313)} = 1.11\,kg/m^3$$

The absolute viscosity can then be calculated using Equation 8.2:

$$\mu = (v)(p)$$
$$= (1.11\ kg/m^3)\,(16.97 \times 10^{-6}\ m^2/sec)$$
$$= 1.88 \times 10^{-5}\ kg/m\text{-}sec$$
$$= 1.88 \times 10^{-5}\ N\text{-}sec/m^2$$
$$= 1.88 \times 10^{-5}\ Pa\text{-}sec$$
$$= 1.88 \times 10^{-4}\ poise$$
$$= 1.88 \times 10^{-4}\ dyne\text{-}sec/cm^2$$
$$= 1.88 \times 10^{-4}\ g/cm\text{-}sec$$
$$= 1.88 \times 10^{-2}\ cP$$

8.9.4.2 Conversion between Kinetic Viscosity and Absolute Viscosity for Water

Viscosity is a measure of a fluid's resistance to flow. The knowledge of viscosity is needed for proper design of required temperatures for storage, pumping, or injection of hazardous fluids. Define the viscosity terminologies, and provide technical data of typical liquid pollutants for illustration.

Solution
The viscosity of a fluid is an important property in the analysis of liquid behavior and fluid motion near solid boundaries. Viscosity is the fluid resistance to shear or flow and is a measure of the adhesive/cohesive or frictional fluid property. The resistance is caused by intermolecular friction exerted when layers of fluids attempt to slide by one another.

The dynamic (absolute) viscosity is the tangential force per unit area required to move one horizontal plane with respect to the other at a unit velocity when maintained at a unit distance apart by the fluid. The readers are referred to Equations 8.1 and 8.2 for the dynamic viscosity.

Kinematic viscosity is the ratio of absolute or dynamic viscosity to density—a quantity in which no force is involved. Kinematic viscosity can be obtained by dividing the absolute viscosity of a fluid by its mass density, as shown in Equation 8.3.

Commonly used units for viscosity include the following:

1. CentiPoises (cP) = centistokes (cSt) × density
2. Saybolt Universal Seconds (SUS) = centistokes (cSt) × 4.55
3. Degree Engler × 7.45 = centistokes (cSt)
4. Seconds Redwood × 0.2469 = centistokes (cSt)

Saybolt Universal Seconds (SUS) are used to measure viscosity. The efflux time is the SUS required for 60 mL of a petroleum product to flow through the calibrated orifice of a Saybolt Universal viscometer, under carefully controlled temperature and as prescribed by test method ASTM D 88. This method has largely been replaced by the kinematic viscosity method. SUS is also called the SSU number (Seconds Saybolt Universal) or SSF number (Saybolt Seconds Furol).

TABLE 8.10
Viscosity and Specific Gravity of Liquid Pollutants

CentiPoise (cP)	CentiStoke (cSt)	Saybolt Universal Seconds (SUS)	Liquid Pollutant	Specific Gravity
1	1	31	Water	1.0
3.2	4	40	Milk	—
12.6	15.7	80	No. 4 fuel oil	0.82–0.95
16.5	20.6	100	Cream	—
34.6	43.2	200	Vegetable oil	0.91–0.95
88	110	500	SAE 10 oil	0.88–0.94
176	220	1,000	Tomato juice	—
352	440	2,000	SAE 30 oil	0.88–0.94
820	650	5,000	Glycerine	1.26
1561	1,735	8,000	SAE 50 oil	0.88–0.94
1760	2,200	10,000	Honey	—
5000	6,250	28,000	Mayonnaise	—
15,200	19,000	86,000	Sour cream	—
17,640	19,600	90,000	SAE 70 oil	0.88–0.94

Source: The Engineering Toolbox, Dynamic, Absolute and Kinematic Viscosity, 2007. Available at http://www.engineeringtoolbox.com/dynamic-absolute-kinematic-viscosity-d_412.html. With permission.

Degree Engler is used in Great Britain as a scale to measure kinematic viscosity. Unlike the Saybolt and Redwood scales, the Engler scale is based on comparing a flow of the substance being tested to the flow of another substance—water. Viscosity in Engler degrees is the ratio of the time of a flow of 200 cm³ of the fluid whose viscosity is being measured to the time of flow of 200 cm³ of water at the same temperature (usually 20°C but sometimes 50°C or 100°C) in a standardized Engler viscosity meter.[89]

Saybolt Universal Seconds (SUS), Degree Engler, and Seconds Redwood are applicable to fluids with centistokes greater than 50.

The viscosity of a fluid is highly temperature dependent and, for either dynamic or kinematic viscosity to be meaningful, the reference temperature must be quoted. In ISO 8217 the reference temperature for a residual fluid is 100°C. For a distillate fluid the reference temperature is 40°C.

The physical and chemical properties of hazardous dense solvent compounds are given in Tables 8.8 and 8.9, in which the absolute viscosity and kinematic viscosity are expressed in centipoises and centistokes, respectively.

For the purpose of illustration, the viscosity and specific gravity of some typical liquids (including hazardous No. 4 fuel oil, vegetable oil, SAE-10 oil, glycerine, SAE-50 oil, SAE-70 oil) are listed in Table 8.10 for reference.[89]

NOMENCLATURE

B	Carbon bed (M), (lb)
B_h	Carbon bed depth (L), (ft)
B_{hm}	Carbon bed depth (L), (m)
B_m	Carbon bed (M), (kg)
BP	Boiling point at 760 torr pressure (°C)
C	Unabsorbed concentration of contaminant left in solution (M/L^3)
C_c	Actual carbon dosage (M/L^3), (lb/1000 gal)

C_{cm}	Actual carbon dosage (M/L^3), (kg/m^3)
d	Density (M/L^3), (lb/ft^3), (g/cm^3)
d_m	Carbon density (kg/m^3)
dV_s/dy	Velocity gradient across the surfaces
D_c	Carbon column diameter (in.)
D_{cm}	Carbon column diameter (cm)
D_p	Mean carbon particle diameter (mm)
f	Removal efficiency of a stripper
g	Gravitational acceleration (L/T^2)
G	Volumetric rate of air (L^3/T)
h_A	Head added to the fluid with a mechanical device such as a pump (L)
h_L	Head losses from the system due to friction (L)
h_R	Head removed from the fluid with a mechanical device such as fluid motor (L)
h_1	Groundwater elevation at beginning of the aerated zone, L
h_2	Groundwater elevation at end of aerated zone, L
H	Henry's law constant for partitioning between air and water $(atm\text{-}m^3/mol)$
k	Constant
K	Hydraulic conductivity, L/T
K_c	Carbon adsorption coefficient
K_{oc}	Sediment/water partition coefficient (L^3/M), (mL/g)
L	Volumetric rate of contaminated groundwater (L^3/T)
L_a	Length of aerated zone, L
M_c	Weight of contaminant (mol)
M_w	Weight of water (mol)
MW	Molecular weight of the compound of interest (g)
n	Constant
$1/n$	Slope of the straight-line isotherm
ρ	Density of liquid (M/L^3)
P_1	Pressure at point 1 (M/L^2)
P_2	Pressure at point 2 (M/L^2)
P_a	Particle vapor pressure of VOC (atm)
P_{atm}	Pressure (atm)
P_d	Pressure drop (in.Hg)
P_{dm}	Pressure drop (mmHg)
P_o	Vapor pressure (torr or mmHg)
P_1/γ	Pressure head at point 1 (L)
P_2/γ	Pressure head at point 2 (L)
Q	Flow rate (gal/min)
Q_m	Influent rate (m^3/min)
R	Stripping factor
RVD	Relative vapor density (dimensionless)
S	Solubility in water (M/L^3) (mg/L)
Sg	Specific gravity (dimensionless)
S_s	Shearing stress (M/LT^2)
t	Contact time (min)
t_r	Residence time, T
T	On-stream cycle time (d)
$v_1^2/2g$	Velocity head at point 1 (L)
$v_2^2/2g$	Velocity head at point 2 (L)
V	Volume (m^3)
V_s	Velocity along the surfaces (L/T)

W	Weight of activated carbon (M)
X	Amount of contaminant adsorbed (M)
X_a	Mole fraction of VOC in water (mol/mol)
X_{in}	Influent contaminant concentration in water (M/L^3)
X_{out}	Effluent contaminant concentration in water (M/L^3)
y	Distance (L) between the moving and fixed surfaces
Y_{in}	Influent contaminant concentration in air (M/L^3)
Y_{out}	Effluent contaminant concentration in air (M/L^3)
Z_1	Elevation of point 1 (L)
Z_2	Elevation of point 2 (L)
γ	Density of liquid (M/L^3)
μ	Proportionality constant = dynamic viscosity or absolute viscosity of the fluid (M/LT)
ν	Kinematic viscosity (centiStoke)

ACRONYMS

ABF	Automatic backwash filtration
API	American Petroleum Institute
bgs	Below ground surface
BOD	Biochemical oxygen demand
BTEX	Benzene, toluene, ethylbenzene, and xylenes (collectively)
CERCLA	Comprehensive Environmental Response, Compensation, and Liability Act—Superfund
CWA	Clean Water Act
DAF	Dissolved air flotation
DCE	1,2-Dichloroethane
DIPE	Disopropyl ether
DNAPL	Dense nonaqueous phase liquid
DO	Dissolved oxygen
EDB	Ethylenedibromide
F/M	Food/microorganisms ratio
GAC	Granular activated carbon
IAF	Induced (dispersed or froth) air flotation
MTBE	Methyl tertiary-butyl ether
NAPL	Nonaqueous phase liquids
OSHA	Occupational Safety and Health Administration
PAC	Powdered activated carbon
PAHs	Polycyclic aromatic hydrocarbons
PCBs	Polychlorinated biphenyls
PCP	Pentachlorophenol
RCRA	Resource Conservation and Recovery Act
SVE	Soil vapor extraction
TCLP	Toxicity characteristic leachate procedure
TDS	Total dissolved solids
TESVE	Thermal-enhanced soil vapor extraction
TOC	Total organic carbon
TSS	Total suspended solids
TTHM	Total trihalomethane
U.S. EPA	United States Environmental Protection Agency
USTs	Underground storage tanks
UV	Ultraviolet
VOCs	Volatile organic compounds

REFERENCES

1. Federal Register, Resource Conservation and Recovery Act (RCRA), 42 US Code s/s 6901 et seq., 1976, U.S. Government Printing Office, January 2004. Available at www.access.gpo.gov/uscode/title42/chapter82_.html.
2. U.S. EPA, Resource Conservation and Recovery Act (RCRA)—Orientation Manual, Report EPA 530-R-02-016, U.S. EPA, Washington, January 2003.
3. U.S. EPA, Superfund, CERCLA Overview, U.S. EPA, Washington, March 2006. Available at www.epa.gov/superfund/action/law/cercla.htm.
4. Federal Register, Comprehensive Environmental Response, Compensation, and Liability Act (CERCLA or Superfund) 42 U.S.C. s/s 9601 et seq. (1980), U.S. Government, Public Laws, January 2004. Full text available at www.access.gpo.gov/uscode/title42/chapter103_.html.
5. U.S. EPA, More about Leaking Underground Storage Tanks, A Background Booklet for Chemical Advisory, U.S. EPA, Washington, 1984.
6. U.S. EPA, RCRA Orientation Manual, EPA/530-SW-90-036, U.S. EPA, Washington, 1990.
7. U.S. EPA, Underground Storage Tanks, U.S. EPA, Washington, November 2006. Available at http://www.epa.gov/OUST.
8. U.S. EPA, Underground Storage Tanks, Overview of the Federal Underground Storage Tank Program, U.S. EPA, Washington, March 2006. Available at http://www.epa.gov/OUST/overview.htm.
9. U.S. EPA, Rules and Regulations, 40 CFR part 280, Federal Register, U.S. EPA, Washington, 53, 218, November 1988.
10. Rutgers University, Underground Storage Tank Checklist for Compliance with UST Requirements, Rutgers University, New Brunswick, NJ, 2007. Available at http://rehs.rutgers.edu/pdf_files/ustcheck-list.html.
11. Federal Register, Clean Water Act (CWA), 33 U.S.C. ss/1251 et seq. (1977), U.S. Government Printing Office, May 2002. Available at www.access.gpo.gov/uscode/title33/chapter26_.html.
12. Federal Register, Occupational Safety and Health Act (OSHA), 29 U.S.C. 651 et seq. (1970), U.S. Government Printing Office, May 2002. Available at www.access.gpo.gov/uscode/title29/chapter15_.html.
13. Federal Register, Regulations for USTs, U.S. Government, Public Laws, Vol. 53, No. 218, November 23, 1988.
14. ChemAlliance, Background: Details on USTs, Regulatory Information for the Chemical Process Industries, May 2005. Available at http://www.chemalliance.org/Handbook/background/details-usts.asp.
15. U.S. EPA, Underground Storage Tanks: Requirements and Options, EPA 510-F-97-005, U.S. EPA, Washington, 1997.
16. U.S. EPA, Underground Storage Tanks, Preventing Underground Storage Tank Systems from Leaking, U.S. EPA, Washington, DC, March 2006. Available at http://www.epa.gov/OUST/overview.htm.
17. U.S. EPA, Expedited Site Assessment Tools for Underground Storage Tank Sites: A Guide for Regulators, EPA 510-B-97-001, U.S. EPA, Washington, 1997.
18. U.S. EPA, Straight Talk on Tanks: Leak Detection Methods for Petroleum Underground Storage Tanks and Piping, EPA-510-B-05-001, U.S. EPA, Washington, 2005.
19. U.S. EPA, Cleanups of Releases from Petroleum USTs: Selected Technologies, EPA/530/UST-88/001, U.S. EPA, Washington, 1988.
20. API, Metering assemblies, in *Manual of Petroleum Measurement Standards*, Chap. 6, American Petroleum Institute, 1983.
21. U.S. EPA, *Automatic Tank Gauging Systems for Release Detection: Reference Manual for Underground Storage Tank Inspectors*, EPA 510-B-00-009, U.S. EPA, Washington, 2000.
22. Wise, R.F., Starr, J.W., and Maresca, J.W. Jr, Underground storage tanks containing hazardous chemicals, in *Remedial Action, Treatment, and Disposal of Hazardous Water*, EPA/600/9-91/002, U.S. EPA, Washington, 1991, pp. 43–56.
23. Todd, D.K., *Groundwater Hydrogeology*, Wiley, New York, 1960.
24. Fetter, C.W., *Applied Hydrology*, 2nd ed., Merrill Publishing, Columbus, OH, 1988, p. 592.
25. Burdon, R.S., *Surface Tension and the Spreading of Liquids*, Cambridge, UK, 1940, p. 85.
26. Anderson, W.G., *J. Petrol. Technol.*, 58, 1246–1262, Nov. 1986.
27. Cuiec, L., Wettability and oil reservoirs, in *North Sea Oil and Gas Reservoirs*, J. Kiepper, Ed., Graham & Trotman, London, UK, 1987, pp. 193–207.
28. Toulhoat, H., Wetting of pore walls by oil in reservoir rocks, in *Capillarity Today*, Petre, G. and Sanfeld, A., Eds., Proc. Advanced Workshop on Capillarity, Springer-Verlag, New York, NY, 1991, pp. 318–347.

29. Wang, L.K., Hrycyk, O., and Hurylko, L., Removal of Volatile Compounds and Surfactants from Liquid, U.S. Patent 5,122,165, 1992.
30. Hrycyk, O., Kurylko, L., and Wang, L.K., Removal of Volatile Compounds and Surfactants from Liquid, U.S. Patent 5,122,166, 1992.
31. Goring, C.A. and Hamaker, J.W., *Organic Chemicals in the Soil Environment*, Marcel-Dekker, New York, 1972.
32. Mott, B.L., *Applied Fluid Mechanics*, 3rd ed., Merrill Publishing, Columbus, OH, 1990, p. 645.
33. Davies, C.W. and James, A.M., *A Dictionary of Electrochemistry*, John Wiley & Sons, New York, 1976, p. 246.
34. Hamed, J., Acar, Y.B., and Gale, R.J., Pb(II) removal from kaolinite by electrokinetics, *J. Geotech. Eng.*, 117, 241–370, 1991.
35. CONCAWE, *Protection of Groundwater from Oil Pollution*, The Hague, Netherlands, 1979.
36. STEP Inc, ASTs/USTs, Solutions to Environmental Problems (STEP), 2007. Available at www.stepenv.com/ast_ust.html.
37. Gutierrez, C.J. and Caravanos, J., *Managing Underground Storage Tanks in the United States*, 2007. Available at http://www.hunter.cuny.edu/rome-nyc/Gutierrez.do.
38. O'Brien & Gere Engineerings, Inc., *Hazardous Waste Site Remediation, the Engineering Perspective*, Bellandi, R., Ed., Van Nostrand Reinhold, New York, 1988, p. 422.
39. U.S. ACE, Removal of Underground Storage Tanks (USTs), Engineering Manual on CD-ROM by U.S. Army Corps of Engineers (U.S. ACE), EM 1110-1-4006, September 1998, p. 260.
40. API, Installation of Underground Petroleum Storage Systems, American Petroleum Institute, Publication 1615, 1979.
41. U.S. EPA, How to Effectively Recover Free Product at Leaking Underground Storage Tank Sites: A Guide for State Regulators, EPA 510-R-96-001, U.S. EPA, Washington, 1996.
42. Weber, R.E., Wang, L.K., and Pavlovich, J., Separation of Liquids with Different Boiling Points with Nebulizing Chamber, U.S. Patent 5,156,747, 1992.
43. Wang, L.K., Wang, M.H.S., and Pavlovich, J., Method and Apparatus for Separation of Toxic Contaminants by Nebulization, U.S. Patent 5,171,455, 1992.
44. Wang, L.K., Pereira, N., Hung, Y.T., and Shammas, N.K., Eds., *Biological Treatment Processes*, The Humana Press, Totowa, NJ, 2008.
45. Wang, L.K., Shammas, N.K., and Hung, Y.T., Eds., *Advanced Biological Treatment Processes*, The Humana Press, Totowa, NJ, 2008.
46. Amdurer, M., Fellnam, R.T., Roetzer, J., and Russ, C., Systems to Accelerate *In Situ* Stabilization of Waste, EPA/590/2-86/002, U.S. EPA, Washington, 1986.
47. NJ DEP, Underground Storage Tanks (USTs), Site Remediation and Waste Management, NJ Department of Environmental Protection, Trenton, NJ, November 2006.
48. U.S. EPA, Underground Storage Tanks, Cleaning Up Underground Storage Tank System Releases, U.S. EPA, Washington, March 2006. Available at http://www.epa.gov/OUST/overview.htm.
49. U.S. EPA, How to Evaluate Alternative Cleanup Technologies for Underground Storage Tank Sites: A Guide for Corrective Action Plan Reviewers, EPA 510-R-04-002, U.S. EPA, Washington, 2004.
50. Noonan, D.C. and Curtis, J.T., *Groundwater Remediation and Petroleum, A guide for Underground Storage Tanks*, Lewis Publishers, Chelsea, MI, 1990, p.142.
51. Shammas, N.K., Fine pore aeration of water and wastewater, in *Advanced Physicochemical Treatment Technologies*, Wang, L.K., Hung, Y.T., and Shammas, N.K., Eds., The Humana Press, Totowa, NJ, 2007, pp.391–448.
52. Kavanaugh, M.C. and Trussel, R.R., Air stripping as a pre-treatment process, in *Organic Chemical Contaminants in Groundwater: Transport and Removal*, American Water Works Association, Denver, CO, 1981, pp. 83–106.
53. Hess, A.F., Dyksen, J.E., and Dunn, H.J., Control strategy—aeration treatment techniques, in *Occurrence and Removal of Volatile Organic Chemicals from Drinking Water*, American Water Works Association Research Foundation, Denver, CO, 1983, pp. 87–155.
54. Perry, R.H. and Chilton, C.H., *Chemical Engineer's Handbook*, 5th ed., McGraw-Hill, New York, 1973.
55. Munz, C. and Roberts, P.V., Air–water phase equilibria of volatile organic solutes, *J. Am. Water Work Assoc.*, 79, 62–69, 1987.
56. Ruddy, E.N. and Carroll, L.A., Select the best VOC control strategy, *Chem. Eng. Progr.*, July, 28–35, 1993.
57. Wang, L.K., Kurylko, L., and Wang, M.H.S., Gas Dissolving and Releasing Liquid Treatment System, U.S. Patent 5,167,806, 1992.

58. Wang, L.K., Shammas, N.K., Selke, W.A., and Aulenbach, D.B., *Flotation Technology*, The Humana Press, Totowa, NJ, 2009.

59. Wang, L.K., Shammas, N.K., Selke, W.A., and Aulenbach, D.B., *Flotation Engineering*, The Humana Press, Press, Totowa, NJ, 2009.

60. Wang, L.K. and Mahoney, W.J., Treatment of storm runoff by oil–water separation, flotation, filtration and adsorption, Part A, *The 44th Purdue Industrial Waste Conf. Proc.*, Lewis Publishers, Chelsea, MI, 1990, pp. 655–666.

61. Wang, K., Wang, M.H.S., and Mahoney, W.J., Treatment of storm runoff by oil-water separation, flotation, filtration and adsorption, Part B, *The 44th Purdue Industrial Waste Conf. Proc.*, Lewis Publishers, Chelsea, MI, 1990, pp. 667–674.

62. Wang, L.K., Hung, Y.T., and Shammas, N.K., Eds., *Physicochemical Treatment Processes*, The Humana Press, Totowa, NJ, 2005, 723 p.

63. Wang, L.K., Hung, Y.T., and Shammas, N.K., Eds., *Advanced Physicochemical Treatment Processes*, The Humana Press, Totowa, NJ, 2006, 690 p.

64. O'Brien, R.P. and Stenzel, M.H., Combining granular activated carbon and air stripping, *Public Works J.*, 115, 54–62, 1984.

65. U.S. EPA, Superfund Innovative Technology Evaluation Program, 4th ed., EPA/540/5-91/008, U.S. EPA, Cincinnati, OH, November 1991.

66. Wang, L.K. and Pereira, N.C., Eds., *Water Resources and Natural Control Processes*, The Humana Press, Totowa, NJ, 1986, 496 p.

67. Scovazzo, P., Chen, W.Y., Wang, L.K., and Shammas, N.K., Solvent extraction, leaching and supercritical extraction, in *Advanced Physicochemical Treatment Processes*, Wang, L.K., Hung, Y.T., and Shammas, N.K., Eds., The Humana Press, Totowa, NJ, 2006, pp. 581–614.

68. Wang, L.K., *Theory and Applications of Flotation Processes*, Report PB86-194198/AS, U.S. Department of Commerce, National Technical Information Service, Springfield, VA, November 1985, p. 15.

69. Wang, L.K. and Wang, M.H.S., Decontamination of groundwater and hazardous industrial effluents by high-rate air flotation process, *Proc. Great Lakes Conf.*, Hazardous Materials Control Research Institute, Silver Springs, MD, September 1990.

70. Wang, L.K., Weber, R.E., and Hoagland, F.M., Reduction of color, odor, humic acid and toxic substances by adsorption, flotation and filtration, *Water Treatment*, 7, 1–16, 1992.

71. Krofta, M. and Wang, L.K., Bubble dynamics and air dispersion mechanisms of air flotation process system, Part B, *The 44th Purdue Industrial Waste Conference Proceedings*, Lewis Publishers, Chelsea, MI, 1990, pp. 505–515.

72. U.S. EPA, Engineering Bulletin: *In Situ* Soil Vapor Extraction Treatment, EPA/540/2-91/006, Hazardous Waste Engineering Laboratory, U.S. EPA, Washington, 1991.

73. Sittler, S.P. and Swinford, G.L., Thermal-enhanced soil vapor extraction accelerates cleanup of diesel-affected soils, *Natl. Environ. J.*, January, 40–43, 1993.

74. U.S. EPA, Remedial Action at Waste Disposal Sites, EPA/625/6-85/006, U.S. EPA, Washington, 1985.

75. Texas Research Institute, *Underground Movement of Gasoline on Groundwater and Enhanced Recovery by Surfactants*, American Petroleum Institute, Washington, 1979.

76. Ellis, W.D. and Payne, J.R., Chemical Counter Measures for *In Situ* Treatment of Hazardous Material Releases, U.S. EPA, Edison, NJ, 1983.

77. Vance, D.B., Remediation by *In Situ* Aeration, *Natl. Environ. J.*, 59–62, July–August 1993.

78. U.S. EPA, Engineering Bulletin: Soil Washing Treatment, EPA/540/2-90/017, Hazardous Waste Engineering Laboratory, U.S. EPA, Washington, 1990.

79. U.S. EPA, Engineering Bulletin: Solvent Extraction Treatment, EPA/540/2-90/013, Hazardous Waste Engineering Laboratory, U.S. EPA, Washington, 1990.

80. U.S. EPA, Engineering Bulletin: Slurry Biodegradation, EPA/540/2-90/016, Hazardous Waste Engineering Laboratory, U.S. EPA, Washington, 1990.

81. U.S. EPA, Estimating Potential for Occurrence of DNAPL at Superfund Sites, EPA Publication: 9355. 4-O7FS, U.S. EPA, Washington, January 1992.

82. Pankow, J.F., *Dense Chlorinated Solvents in Porous and Fractured Media*, Lewis Publisher, Freidrich Schwille, 1988, p. 146.

83. U.S. EPA, Operating and Maintaining Underground Storage Tank Systems: Practical Help and Checklists, EPA-510-B-05-002, U.S. EPA, Washington, September 2005.

84. Subsurface Geotechnical Inc., Finding Underground Storage Tanks (USTs), 2007. Available at http://www.geophysical.biz/tank1.htm.

85. GeoModel Inc., Ground Penetrating Radar Survey for Locating Underground Storage Tanks, 2007. Available at http://www.geomodel.com/ust.

86. U.S. EPA, Underground Storage Tanks Corrective action Measures Archives, FY 2006 End-of-Year Activity Report, U.S. EPA, Washington, 2007. Available at http://www.epa.gov/OUST/cat/camarchv. htm.

87. Hillel, D., *Soil and Water: Physical Princples and Processes*, Academic Press, New York, 1971, 288.

88. Villaume, J.F., Investigations at sites contaminated with dense non-aqueous phase liquids (DNAPLs), *Groundwater Monitoring Rev.*, 5, 2, 1985.

89. The Engineering Toolbox, Dynamic, Absolute and Kinematic Viscosity, 2007. Available at http://www. engineeringtoolbox.com/dynamic-absolute-kinematic-viscosity-d_412.html.

90. Wang, L.K., Hung, Y.T., and Shammas, N.K., Eds., *Advanced Physicochemical Treatment Technologies*, Humana Press, Totowa, NJ, 2007, 710 p.

91. Wang, L.K., Hung, Y.T., Lo, H.H., and Yapijakis, C., Eds., *Hazardous Industrial Waste Treatment*, CRC Press, Boca Raton, FL, 2007, 516 p.

92. Wang, L.K., Hung, Y.T., Lo, H.H., and Yapijakis, C., Eds., *Handbook of Industrial and Hazardous Wastes Treatment*, Marcel-Dekker and CRC Press, New York, 2004, 1345 p.

93. Wang, L.K., Kurylko, L., and Hrycyk, O., Site Remediation Technology, U.S. Patent 5,552,051, Sept. 3, 1996.

9 Biological Treatment Processes for Urea and Formaldehyde Containing Wastewater

José Luis Campos Gómez, Anuska Mosquera Corral, Ramón Méndez Pampín, and Yung-Tse Hung

CONTENTS

9.1 GENERATION OF WASTEWATER

9.1.1 PRODUCTION PROCESS

Approximately one million metric tons of urea-formaldehyde resin are produced annually all over the world. More than 70% of this urea-formaldehyde resin is consumed by the forest products industry. The resin is used in the production of an adhesive for bonding particleboard (61% of the urea-formaldehyde used in the industry), medium-density fiberboard (27%), hardwood plywood (5%), and as a laminating adhesive (7%) for bonding furniture case goods, overlays to panels, and interior flush doors, for example.

Urea-formaldehyde resins are the most prominent examples of the thermosetting resins usually referred to as amino resins, comprising ca. 80% of the amino resins produced worldwide. Melamine-formaldehyde resins constitute most of the remainder of this class of resins, with other minor amounts of resins being produced from the other aldehydes or amino compounds (especially aniline), or both.

Amino resins are often used to modify the properties of others materials. These resins are added during the processing of diverse products such as textiles (to impart permanent press characteristics), automobile tires (to improve the bonding of the rubber to the tire cord), paper (to improve its strength, especially when wet), and alkyds and acrylics (to improve their cure). Amino resins are also used for molding products, such as electrical devices, jar caps, buttons, dinnerware, and in the production of countertops.

Urea-formaldehyde resins are used as the main adhesive in the forest product industry because they have a number of advantages, including low cost, ease of use under a wide variety of curing conditions, low cure temperatures, water solubility, resistance to microorganisms and to abrasion, hardness, excellent thermal properties, and a lack of color, especially in the cured resin.

The major disadvantage associated with urea-formaldehyde adhesives as compared with the other thermosetting wood adhesives, such as phenol-formaldehyde and polymeric diisocyanates, is their lack of resistance to moist conditions, especially in combination with heat. These conditions lead to a reversal of the bond-forming reactions and the release of formaldehyde, so these resins are usually used for the manufacture of products intended for interior use only. However, even when used for interior purposes, the slow release of formaldehyde (a suspected carcinogen) from products bonded with urea-formaldehyde adhesives is observed.

9.1.1.1 Chemistry of Urea-Formaldehyde Resin Formation

The synthesis of urea-formaldehyde resin takes place in two stages. In the first stage, urea is hydroxy-methylolated by the addition of formaldehyde to the amino groups of urea (Figure 9.1). This reaction is in reality a series of reactions that lead to the formation of mono-, di-, and trimethylolureas. Tetramethylolurea does not appear to be produced, at least not in a detectable quantity. The addition of formaldehyde to urea takes place over the entire pH range, but the reaction rate is dependent on the pH.

The second stage of urea-formaldehyde synthesis consists of the condensation of methylolureas to low-molecular-weight polymers. The rate at which these condensation reactions occur is very dependent on pH (Figure 9.2) and, for all practical purposes, occurs only at acidic pHs. The increase in the molecular weight of the urea-formaldehyde resin under acidic conditions is thought to be a combination of reactions leading to the formation of the following:

1. Methylene bridges between amido nitrogens by the reaction of methylol and amino groups on reacting molecules (Figure 9.3a)
2. Methylene ether linkages by the reaction of two methylol groups (Figure 9.3b)
3. Methylene linkages from methylene ether linkages by the splitting out of formaldehyde (Figure 9.3c)

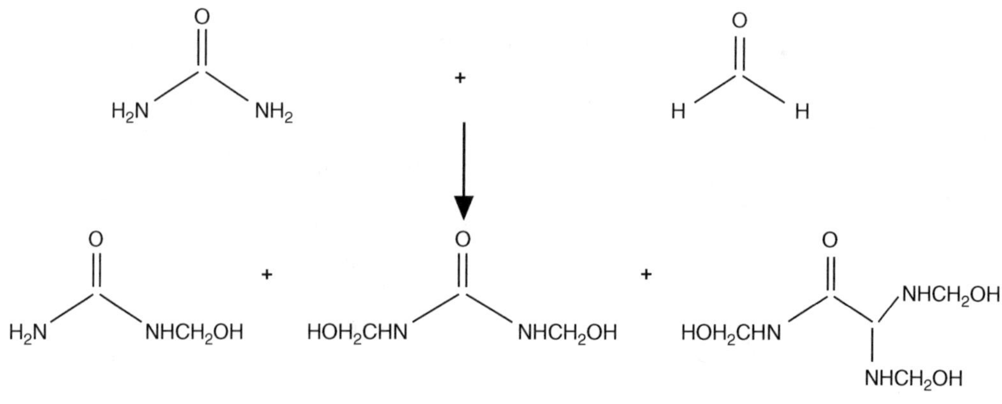

FIGURE 9.1   Formation of mono-, di-, and trimethylolurea by addition of formaldehyde to urea.

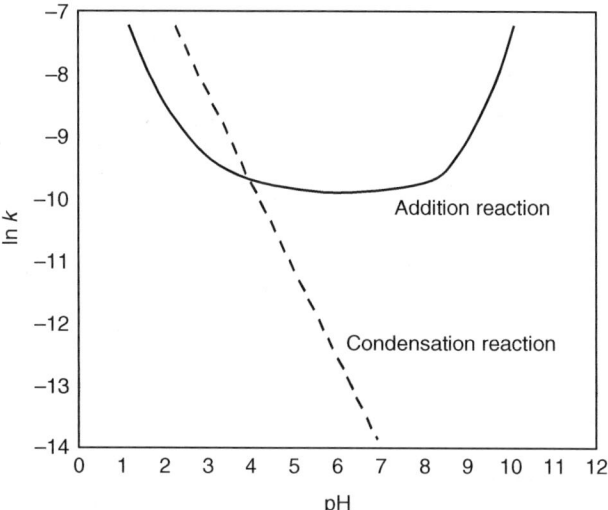

FIGURE 9.2 Influence of pH on the rate constant k of addition (solid line) and condensation (dashed line) reactions of urea and formaldehyde.

4. Methylene linkages by the reaction of methylol groups splitting out water and formaldehyde (Figure 9.3d)

The difference between the pH profiles of the two stages of urea-formaldehyde resin synthesis is taken advantage of in the production of these resins (Figure 9.2). In general, the commercial production of urea-formaldehyde adhesive resins is carried out in two major steps. The first step consists of the formation of methylolureas under basic conditions (pH 8 to 9), to allow the methylolation reactions to proceed in the absence of reactions involving the condensation of the methylolureas.

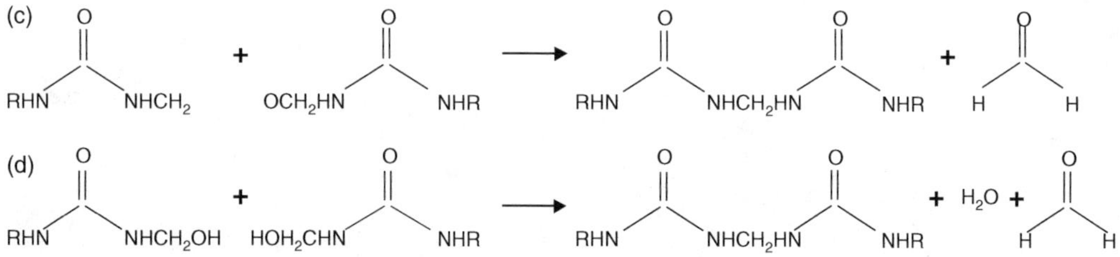

FIGURE 9.3 Condensation reactions of methylolureas to form (a) methylene bridges between amido nitrogens, (b) methylene ether linkages, and (c) and (d) methylene linkages. Reactions of these types produce higher molecular weight oligomers and polymers.

In the second step, the reaction mixture is brought to acid conditions, at ca. pH 5, and the condensation reactions are carried out until the desired viscosity is reached. The reaction mixture is then cooled and neutralized.

An acidic-cure catalyst is added to the urea-formaldehyde resin before it is used as an adhesive. Ammonium chloride and ammonium sulfate are the most widely used catalysts for resins in the forest products industry. A variety of other chemicals can be used as a catalyst, including formic acid, boric acid, phosphoric acid, oxalic acid, and acid salts of hexamethylenetetramine.

Resin cure is normally conducted at a temperature of 120°C and pH < 5. The reactions that occur during the final cure of the resin are thought to be similar to those that occur during the acid condensation of the methylolureas. These reactions lead to the formation of the crosslinked polymeric network characteristic of the hardened, cured resin.

9.1.2 CHARACTERIZATION OF THE EFFLUENT

The effluent generated during the production of the resins arises from different operations within the factory. The effluent of the production processes comes mainly from cleaning operations of reactors, storage tanks, filters from the towers of formaldehyde production, and the filters from the reactors. Another source for disposal comprises the spills occurring during the transfer of the resins from the reactors to the storage tanks and from these to the truck used to distribute them to other factories.

Because of the processes carried out in the plant, the expected compounds in wastewater are formaldehyde, urea, and polymers of these compounds. The global effluent of this kind of factory is characterized by a high chemical oxygen demand (COD) (due mainly to formaldehyde), relatively high values of nitrogen (arising from urea and copolymers) and a low content of phosphorus and inorganic carbon. The main characteristics of the effluent of a resin factory are showed in Table 9.1.

9.2 BIOLOGICAL TREATMENT

9.2.1 BIOLOGICAL PROCESSES AND STRATEGIES

The wastewaters generated by the adhesive industries contain high concentrations of both carbon and nitrogen compounds. The process chosen to treat these wastewaters will depend on their COD/N ratio. When the COD/N ratio is high, an anaerobic treatment is the best option as it will save costs

TABLE 9.1
Characteristics of the Effluent from a Resin Factory

	Vidal et al.[1]	Garrido et al.[2]	Garrido et al.[3]	Eiroa et al.[4]
COD (g/L)	0.46–3.9	1.1–4.1	0.46–4.0	0.12–6.85
Formaldehyde (g/L)	0.22–4.0	0.20–2.8	0.22–4.0	0.007–2.7
TKN (g/L)	0.12–0.81	0.13–0.70	0.11–0.80	0.056–1.46
N-NH$_4^+$ (g/L)	0.003–0.018	—	—	0.006–0.36
TSS (mg/L)	19–150	—	—	12–664
VSS (mg/L)	16–140	—	—	—
pH	—	—	7.1–11.2	6.5–9.6
TOC (g/L)	—	0.30–2.08	—	—
Alkalinity (mg CaCO$_3$/L)	—	—	—	167–2000
P-PO$_4^{3-}$ (mg/L)	—	—	—	0.1–31

COD, chemical oxygen demand; TKN, total Kjeldahl nitrogen; TSS, total suspended solids; VSS, volatile suspended solids; TOC, total organic carbon.

(less energy, less sludge production). In this process, formaldehyde is degraded to methane and carbon dioxide and urea is hydrolyzed to ammonium:

Anaerobic degradation of formaldehyde

$$2\ CH_2O \longrightarrow CH_4 + CO_2 \tag{9.1}$$

Urea hydrolysis

$$H_2N\text{-}CO\text{-}NH_2 + 2\ H_2O \longrightarrow 2\ NH_4^+ + CO_2 + 2\ OH^- \tag{9.2}$$

Generally, the sole use of an anaerobic stage is not enough to reduce COD sufficiently to reach the required concentration for disposal, and the concentration of nitrogen compounds remains practically constant. Therefore, to remove the nitrogen compounds and the remaining COD, a posttreatment based on the nitrification–denitrification process is necessary. This process can be used in a post-denitrifying or predenitrifying configuration (Figure 9.4).

9.2.1.1 Postdenitrifying Configuration

In this case, the wastewater is fed to the aerobic reactor where the remaining formaldehyde is oxidized to CO_2 (Equation 9.3) and urea is hydrolyzed to ammonia. This ammonia is then oxidized to nitrate (Equation 9.4). Nitrate goes to the denitrifying unit where it is reduced to dinitrogen gas in the presence of an electron donor, which is generally provided by organic matter (Equation 9.5). Because formaldehyde is oxidized in the first unit, methanol is commonly added to carry out this process, which produces an increase in operational costs.

Aerobic degradation of formaldehyde

$$CH_2O + O_2 \longrightarrow CO_2 + H_2O \tag{9.3}$$

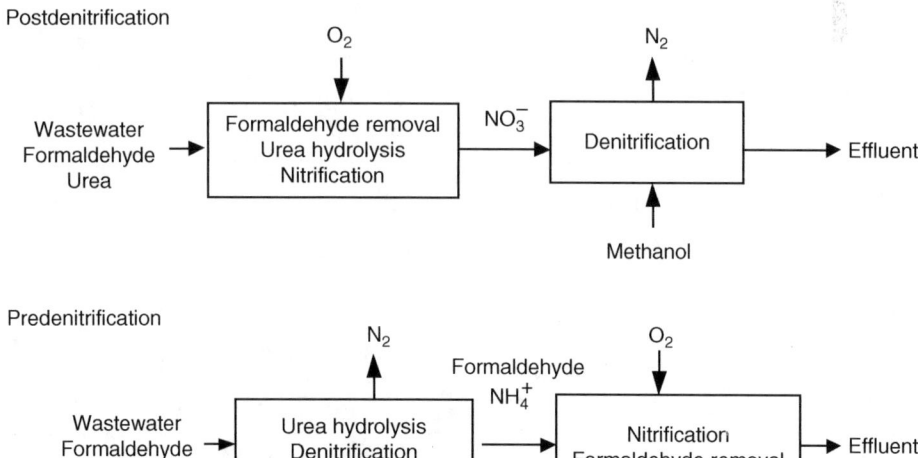

FIGURE 9.4 Postdenitrification and predenitrification configurations for the treatment of wastewaters containing formaldehyde and urea.

Nitrification

$$NH_4^+ + 2\,O_2 \longrightarrow NO_3^- + H_2O + 2\,H^+ \tag{9.4}$$

Denitrification

$$4\,NO_3^- + 5\,CH_2O \longrightarrow 2\,N_2 + 5\,CO_2 + 3\,H_2O + 4\,OH^- \tag{9.5}$$

9.2.1.2 Predenitrifying Configuration

Wastewater is supplied to the anoxic unit, where the nitrate recycled from the nitrifying unit is denitrified using the formaldehyde as the electron donor. When the COD/N ratio of the wastewater is high, the anaerobic degradation of formaldehyde and denitrification can occur in the same unit, this last process having preference for thermodynamic reasons.[3] The hydrolysis of urea is also carried out in the anoxic reactor. The wastewater containing ammonia and a low concentration of formaldehyde is fed to the aerobic tank, where ammonia is nitrified to nitrate and the remaining formaldehyde is oxidized. The disadvantage of this configuration is the dependence of the percentage of nitrogen removal on the recycling ratio between the aerobic and anoxic units:

$$\eta = \frac{R}{R+1} \times 100 \tag{9.6}$$

where η is the percentage of nitrogen removal and R is the recycling ratio between the aerobic and anoxic units.

If the COD/N ratio of the wastewater is low, a better option is the use of a nitrification–denitrification stage without a previous anaerobic digestion in order to preserve organic matter for denitrification.

9.2.2 Interactions between Biological Processes and Compounds

Maintaining the stability of a biological treatment of wastewaters containing formaldehyde and urea is complicated because some compounds exert a toxic effect on the processes involved. Figure 9.5 shows the possible toxic interactions between the different compounds and processes.

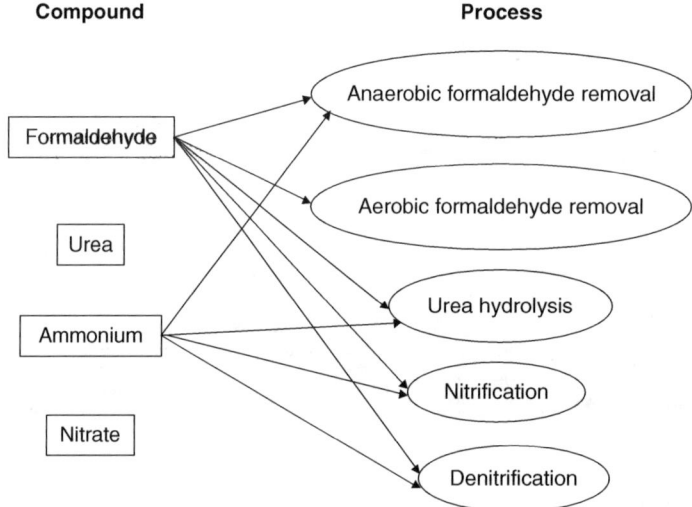

FIGURE 9.5 Compounds and intermediates of wastewater treatment, with arrows indicating the inhibitory effects of them on the different processes.

9.2.2.1   Anaerobic Formaldehyde Removal

Biodegradation pathway

Different pathways have been proposed to explain the anaerobic biodegradation of formaldehyde according to the intermediate products observed.[5–7]

González-Gil and colleagues[5] carried out anaerobic activity tests using formaldehyde as the only carbon source and found that part of this compound was readily transformed into methanol. These authors could recover all substrate COD as methane when assayed for initial formaldehyde doses of 200 and 600 mg/L COD, but no methane production was observed for an initial dose of 1400 mg/L COD due to the toxic effect of the formaldehyde on the reaction. Nevertheless, the conversion of formaldehyde into methanol was not inhibited. During formaldehyde conversion, a peak of hydrogen was observed, this peak being related to the initial amount of formaldehyde dosed. It is likely that formaldehyde was first oxidized to formate and then reduced to methanol. Considering that all formaldehyde is converted into methanol and formate, the following reactions are proposed:

Oxidation

$$HCHO + H_2O \longrightarrow HCOOH + H_2 \tag{9.7}$$

Reduction

$$HCHO + H_2 \longrightarrow CH_3OH \tag{9.8}$$

Total

$$2\,HCHO + H_2O \longrightarrow CH_3OH + HCOOH \tag{9.9}$$

Omil and colleagues[6] also carried out anaerobic activity tests to study the biodegradation of formaldehyde in the presence and absence of cosubstrate. In the absence of cosubstrate, these authors suggest that the hydrogen generated during formaldehyde removal was consumed both for direct methane conversion and for methanol generation. In Table 9.2, two possible formaldehyde degradation reactions are shown (reactions *i* and *ii*). Both are dependent on hydrogen concentration. Although degradation via methanol is thermodynamically favored in standard conditions, this pathway would imply, for the complete mineralization of formaldehyde, the synergistic action of methylotroph methanogenic bacteria (reaction *iii*) which suggests a situation of certain competition between reaction *iii* and the consecutive reactions *iv* and *vi*. In this way, when autotrophic methanogenic bacteria become inhibited and hydrogen concentration begins to accumulate in the medium, methanol generation becomes more favorable.

In the presence of a cosubstrate, formaldehyde removal was highly enhanced especially by acetate (*vii*) but not by propionate (*viii*) and butyrate (*ix*). The effect that acetate degradation exerts on formaldehyde removal may be related to the inorganic carbon generated, which allows autotrophic bacteria to convert hydrogen into methane (reaction *vi*) under more favorable conditions. When low concentrations of volatile fatty acids (VFA) and methanol are present, the direct conversion into methane by methylotrophic methanogenic bacteria should be the most favorable pathway for methanol degradation (in competition with the acetogenic bacteria (*v*)), and autotrophic methane generation would not be favorable; therefore, both direct conversion into methane and the acetogenic pathway should be considered for methanol consumption. Neither are hydrogen dependent.

Oliveira and colleagues[7] found intermediate compounds with 2 to 5 carbons during the degradation of formaldehyde (with 12% methanol) as the sole carbon source and attribute this to a chemical reaction as formaldehyde can form polymers in aqueous solution. The reactions are rapid in the absence of methanol, which is added to formaldehyde solutions to prevent such polymerization. In aqueous solution, when methanol is consumed, formaldehyde is almost completely hydrated to

TABLE 9.2

Estimated Free Energy Changes of Selected Biological Reactions Involved in the Anaerobic Degradation of Formaldehyde and Methanol

Reactions	ΔG^0 (kJ)
Formaldehyde	
(i) $HCOH + H_2O \rightarrow 2\,H_2 + CO_2$	−21.8
(ii) $HCOH + H_2 \rightarrow CH_3OH$	−44.0
Methanol	
Methanogenesis	
(iii) $4\,CH_3OH \rightarrow 3\,CH_4 + HCO_3^- + H^+ + H_2O$	−314.6
Hydrogen generation	
(iv) $CH_3OH + 2\,H_2O \rightarrow 3\,H_2 + HCO_3^- + H^+$	23.0
Acetogenesis	
(v) $4\,CH_3OH + 2\,HCO_3^- \rightarrow 3\,CH_3COO^- + H^+ + 4\,H_2O$	−221.6
Hydrogen and VFA	
(vi) $4\,H_2 + HCO_3^- + H^+ \rightarrow CH_4 + 3\,H_2O$	−135.6
(vii) $CH_3COO^- + 2\,H_2O \rightarrow CH_4 + HCO_3^-$	−31.0
(viii) $CH_3CH_2COO^- + 3\,H_2O \rightarrow CH_3COO^- + HCO_3^- + H_2 + H^+$	+76.0
(ix) $CH_3CH_2CH_2COO^- + 2\,H_2O \rightarrow 2\,CH_3COO^- + 2\,H_2 + H^+$	+48.1

VFA, volatile fatty acids.

methylene glycol, which may polymerize to form a series of polyoxymethylene glycols. These authors suggested that the intermediate compounds come from the anaerobic degradation of the formed polymers. Another possibility is an aldo condensation, which occurs in the presence of weak bases, forming glycolic aldehyde and carbohydrates.

Toxic effects

The toxicity of formaldehyde during anaerobic treatment has been reported by several authors.[5,8] Its toxicity depends of several different parameters:

1. Nature of the cosubstrates
2. Operational mode (batch or continuous)
3. Type of reactor (suspended or attached growth systems)
4. Formaldehyde/microorganisms ratio
5. COD/formaldehyde ratio

The importance of the nature of the cosubstrate was shown by Todini and Hulshoff Pol,[9] who determined the specific activity of formaldehyde-degrading microorganisms with different cosubstrates such as hydrogen, sodium butyrate, and sucrose. They obtained the highest degradation rates with sucrose. Vidal and colleagues[1] treated wastewater containing formaldehyde in an upflow anaerobic sludge blanket (UASB) reactor using glucose as the cosubstrate, as this compound enhances the reduction of aldehyde to methanol, which is less toxic for the bacteria. Values of 80 mg/L of formaldehyde were reported when acetate was used,[10] whereas Todini and Hulshoff Pol[9] reported a 50% inhibition at 238 mg/L of formaldehyde when sucrose was fed as the main cosubstrate.

Bhattacharya and Parkin[11] studied the influence of the operational mode on formaldehyde degradation. They showed that higher formaldehyde concentrations were tolerated when they were added continuously to acetate and propionate enriched systems rather than when slug doses were used, thus indicating that continuous operation is more favorable for bacterial acclimation.

Sharma and colleagues[12] evaluated the anaerobic biodegradation of petrochemical wastewater containing 4.5 g/L of formaldehyde, and found a higher resistance to microorganisms when biomass concentration was increased by immobilization using biomass-supporting particles. Formaldehyde was observed to exert toxicity at 375 mg/L in the reactor working with these supporting particles, whereas only 125 mg/L was tolerated in the control reactor with suspended biomass.

The formaldehyde/biomass ratio is also reported to be a key factor, as demonstrated by de Bekker and colleagues,[13] who determined that 0.89 g formaldehyde/g VSS in batch operation exerted a complete inhibition of anaerobic bacteria. They also reported that anaerobic treatment of wastewaters containing formaldehyde is possible only when the COD/formaldehyde ratio is higher than 1000.[13] However, other authors such as Parkin and colleagues[14] achieved a stable operation with lower COD/formaldehyde ratios (about 6), using acetate as the main substrate and a continuous addition of 400 mg/L formaldehyde to an anaerobic filter. They found that formaldehyde exhibits reversible inhibition, recovery being accelerated by removing the toxicant from the liquid phase. These results agree with those of Gonzalez-Gil and colleagues,[15] who found that the inhibition exerted by formaldehyde on the methane production rate is partially reversible when formaldehyde concentrations in the reactor are lower than 22 mg/L C-formaldehyde.

There is no information about the mechanisms of formaldehyde toxicity, and the information available in the literature about formaldehyde toxicity in batch and continuous systems is difficult to extrapolate for design purposes (Tables 9.3 and 9.4).

The hydrolysis of urea releases ammonia to the liquid bulk, which can cause inhibition of the methanogenic sludge.[21] Nevertheless, these authors found a toxic effect at 1300 mg/L N-NH$_3$, this value being higher than the nitrogen concentration of these wastewaters and no inhibitory effects could be expected.

9.2.2.2 Aerobic Degradation of Formaldehyde

Biodegradation pathway
The aerobic degradation of formaldehyde in wastewater has been studied by different authors in both continuous[22] and batch experiments.[23–25] The degradation can occur by two possible paths (see Equations 9.10 and 9.11):

1. Initiated by a dismutation reaction, yielding formic acid and methanol as products, if the microorganism has a formaldehyde dismutase enzyme
2. Via formic acid if the microorganism has the enzymes formaldehyde and formate dehydrogenase[24]

The biodegradation of the metabolites starts after exhaustion of formaldehyde in the medium.

$$2\,HCHO \xrightarrow[\text{dismutase}]{\text{Formaldehyde}} CH_3OH + HCOOH \tag{9.10}$$

$$HCHO \xrightarrow[\text{dehydrogenase}]{\text{Formaldehyde}} HCOOH \xrightarrow[\text{dehydrogenase}]{\text{Formate}} CO_2 \tag{9.11}$$

Toxic effects
Zagornaya and colleagues[22] reported the complete biodegradation of 2300 mg/L of formaldehyde in wastewater treated in an activated sludge plant, whereas Gerike and Gode[26] observed that 30 mg/L

TABLE 9.3

Formaldehyde Studies in Batch Systems at 35°C

Biomass	Main Substrate	Tested HCHO (mg/L)	IC_{50} (mg/L)	Reference
Anaerobic digested sludge	—	10–100	—	(16)
Domestic wastewater	—	1–10,000	200	(17)
Sludge treating water from seafood processes	VFA	50–200	125	(1)
Granular sludge from a UASB	Sucrose	—	254	(9)
Activated sludge from a plant treating wood-processing-industry wastewater	Glucose	2–400	300	(18)

IC_{50}, 50% inhibition concentration; VFA, volatile fatty acids.

formaldehyde inhibited oxygen consumption in activated sludge. Eiroa and colleagues[25] studied the inhibitory effect of formaldehyde in batch tests; they found no inhibition and also that high concentrations of formaldehyde up to 3890 mg/L could be removed using it as the single carbon source. When the same formaldehyde concentrations in the presence of methanol as cosubstrate were tested, higher formaldehyde biodegradation rates were obtained. This possibility of formaldehyde biodegradation despite the presence of an alternative readily metabolizable carbon source is a characteristic of significant practical interest when formaldehyde needs to be removed in environments containing other carbon sources, as in the case of wastewaters from synthetic resin-producing factories. Glancer-Soljan and colleagues[24] also found no inhibitory effects of formaldehyde biodegradation in batch assays with an initial concentration of 1000 mg/L using a mixed culture containing two bacterial strains.

9.2.2.3   Urea Hydrolysis

A wide variety of aerobic and anaerobic microorganisms are able to express the enzyme urease (urea amidohydrolase), which catalyses the hydrolysis of urea to ammonia and carbon dioxide.[27] So far,

TABLE 9.4

Some Results from Literature Obtained in Continuous Systems Treating Formaldehyde-Containing Wastewater

Reactor	Tested HCHO (mg/L)	Limiting Dose (mg HCHO/L)	Formaldehyde Removal Efficiency (%)	Reference
Anaerobic filter	100–400	400	—	(14)
CST	—	125	85–88	(12)
CST immobilized biomass	—	375	88–95	(12)
Chemostat	100–1110	1110	99.9	(19)
EGSB	333	—	>93	(20)
EGSB	200/400/600	—	High	(5)
UASB	50–2000	100	98	(1)
UASB	95–950	380	95	(1)
HAIB	26–1158	Not observed	>95	(7)

CST, continuous stirred tank; EGSB, expanded granular sludge blanket; HAIB, horizontal-flow anaerobic immobilized biomass.

most authors have preferred anaerobic conditions for the biological treatment of high-strength urea wastewaters with urea concentrations of up to 2g/L.[1,28] Also, wastewaters containing high loads of urea together with ammonia and formaldehyde have been treated under anoxic conditions,[3] and an aerobic urea hydrolysis has been described by Gupta and Sharma[29] and Hamoda.[30] Rittstieg and colleagues,[31] treating an industrial wastewater containing high concentrations of urea and sulfate, proposed the use of the aerobic process to avoid the production of sulfide if an anaerobic stage were used.

There are no clear results as to which microorganism causes hydrolysis in aerobic conditions. Prosser[32] reported that *Nitrosomonas* or *Nitrospira* were not ureolytic, which agrees with the conclusion of Campos and colleagues,[33] who observed no degradation of urea when this compound was fed to a nitrification reactor. However, Koops and Chritian[34] pointed out that the five genera of ammonia-oxidizing bacteria might use urea as an ammonia source. Gupta and Sharma[29] and Hamoda[30] observed hydrolysis of urea and high nitrification percentages when they treated effluents from fertilizer industries aerobically. Recently, Sliekers and colleagues[35] have observed that anaerobic ammonium oxidation (anammox) bacteria did not hydrolyze urea by themselves.

Toxic effects
Different effects of formaldehyde on the hydrolysis of urea are reported. On the one hand, Garrido and colleagues,[3] applying anoxic conditions, observed that an inhibitory effect started at 50 mg/L formaldehyde and the levels of inhibition were 50% and 90% for concentrations of formaldehyde of 100 mg/L and 300 mg/L, respectively. Similar effects were found by Campos and colleagues,[33] working with an anoxic USB, who observed that formaldehyde concentrations in the reactor of 250 to 300 mg/L caused an inhibition of around 53%. This inhibition on the ureolytic activity was also reported by Walker.[36] On the other hand, Eiroa and colleagues[37] carried out batch assays at different initial urea concentrations from 90 to 370 mg/L N-urea in the presence of 430 mg/L formaldehyde. They observed that a complete hydrolysis was achieved and initial urea hydrolysis rates remained constant.

Eiroa and colleagues[37] operated a denitrifying granular sludge blanket with inlet urea concentrations between 100 and 800 mg/L N-urea, and always maintained the efficiency of the hydrolysis in spite of the presence of concentrations of ammonia up to 730 mg/L N (110 mg/L N-NH$_3$). The ammonia levels in the effluent corresponded to ca. 77.5% of the amount of urea fed, the unaccounted portion being attributed to microbial assimilation. However, Garrido and colleagues,[3] when increasing the urea loading rate in a multifed upflow filter (MUF) by increasing the inlet concentration, observed that fully hydrolytic efficiency was maintained for a short period of time but later decreased to 55%. These authors attribute the loss of ureolytic activity of the sludge to the higher ammonia concentrations.

9.2.2.4 Nitrification

Nitrification is a two-step process where ammonia is first oxidized to nitrite by ammonia-oxidizing bacteria (*Nitrosomonas*, *Nitrosococcus*, *Nitrosospira*, and so on) and the produced nitrite is finally oxidized to nitrate by nitrite-oxidizing bacteria (*Nitrobacter*, *Nitrospina*, *Nitrospira*, and so on) (Equation 9.12 and Equation 9.13). Both ammonia- and nitrite-oxidizing bacteria are autotrophic microorganisms, which supposes a low growth rate, nitrification being the limiting process during nitrogen removal.

$$NH_4^+ + 3/2\ O_2 \longrightarrow NO_2^- + H_2O + 2\ H^+ \tag{9.12}$$

$$NO_2^- + \frac{1}{2}O_2 \longrightarrow NO_3^- \tag{9.13}$$

Generally, ammonia oxidation is slower than nitrite oxidation and, therefore, no nitrite production is observed. However, when the amount of carbon source available in the effluent is not high enough to complete the denitrification process (low COD/N ratio), the addition of external organic matter is necessary, which produces an increase in treatment costs. In this case the partial nitrification of ammonia to nitrite reduces not only the oxygen requirements for the oxidation, but also the amount of added organic matter required for denitrification.[3]

Toxic effects

Osislo and Lewandowski[38] studied the effects of several organic compounds on nitrification (acetone, methanol, formaldehyde, and glucose) and found that formaldehyde was the most inhibitory. This inhibition was not due to heterotrophic growth, but to a toxic effect. Campos and colleagues[33] shocked a nitrifying system with different concentrations of formaldehyde (100, 200, and 300 mg/L formaldehyde) over 3h. These shocks caused ammonia to appear in the effluent for a short time, but nitrite was never detected. These authors observed a linear tendency between formaldehyde concentration in the reactor and the decrease in the nitrification rate. They also found that most of this compound was consumed in the reactor. Eiroa and colleagues[25] studied the effect of formaldehyde on nitrification in batch assays. These authors found that initial concentrations of formaldehyde above 350 mg/L start to decrease the nitrification rate, with complete inhibition at an initial concentration of 1500 mg/L. An increase in the lag phase before nitrification started was also observed. When the authors repeated the experiments in presence of methanol, they found that the inhibitory effect was greater at lower formaldehyde concentrations. In the presence of methanol, at initial formaldehyde concentrations of 175 mg/L, nitrification started to decrease and was completely inhibited at 500 mg/L. The authors explained the differences by the fact that the COD/total Kjeldahl nitrogen (TKN) ratio was higher in the assays with formaldehyde and methanol as carbon sources than in assays without methanol. Therefore, the competition between heterotrophic bacteria and nitrifiers for oxygen and ammonium was higher. However, Eiroa and colleagues[39] observed that the simultaneous removal of formaldehyde and ammonium may be carried out in an activated sludge unit, maintaining a nitrification efficiency of 99.9%.

Anthonisen and colleagues[40] found that free ammonia (NH_3) is an inhibitory compound for both steps of nitrification, nitrite oxidation being more sensitive. Concentrations of this compound depend on dissolved NH_4^+ and pH; therefore, for a certain concentration of NH_4^+, pH can be a suitable parameter to control inhibition by the substrate. Gupta and colleagues,[41] treating wastewaters containing both ammonia and urea, found nitrite in the effluent due to the inhibition of nitrite oxidation. Eiroa and colleagues,[25] during batch assays with wastewaters containing ammonia and formaldehyde, observed the transitory accumulation of nitrite, probably as a result of the high initial free ammonium (3.9 mg/L N-NH_3).

9.2.2.5   Denitrification

The denitrification process is carried out by heterotrophic bacteria such as *Pseudomonas, Acinetobacter, Paracoccus, Alcaligenes,* and *Thiobacillus.* The route of nitrogen reduction is showed in Equation 9.14. Generally, dinitrogen gas is the final product, but nitrous oxide may be the final product of denitrification if the denitrifying microorganisms lack N_2O reductase,[42] at low pH values,[42] or in the presence of toxic compounds.[43] The presence of low dissolved oxygen concentrations during denitrification also causes the accumulation of N_2O[44]:

$$NO_3^- \xrightarrow{r_1} NO_2^- \xrightarrow{r_2} NO \xrightarrow{r_3} N_2O \xrightarrow{r_4} N_2 \qquad (9.14)$$

Garrido and colleagues,[43] treating wastewaters containing formaldehyde and urea, observed a relation between the formaldehyde concentration in the reactor and the percentage of nitrous oxide produced in the gas phase, which indicates that, probably, the reduction of nitrous oxide to nitrogen is inhibited by the presence of formaldehyde. Therefore, nitrous oxide measurement might serve to check for the presence of formaldehyde or other toxic or inhibitory compounds in denitrifying reactors and consequently to advise the plant supervisor about a possible failure in the system.

As trace gases concentration in biological processes changes rapidly with operating conditions, nitrous oxide could serve to monitor denitrifying systems as well as it was proposed for hydrogen or carbon monoxide for monitoring methanogenic systems.

Because wastewater may contains a low COD/N ratio, the oxidation of ammonia to nitrite during nitrification contributes to decrease the amount of organic matter needed during denitrification:

$$4NO_2^- + 3\,CH_2O \longrightarrow 2\,N_2 + 3\,CO_2 + H_2O + 4\,OH^- \tag{9.15}$$

The theoretical formaldehyde requirements for denitrifying nitrite or nitrate, if biomass production is not considered, are 0.64 and 1.07 kg C/kg N-NO_x^-, respectively. Garrido and colleagues[3] found C/N ratios of 0.8 and 1.3 kg C/kg N-NO_x^- for denitrification of nitrite and nitrate, respectively, these values being 20% higher than the theoretical ones.

Toxic effects

A negative effect of formaldehyde on the denitrification process has been observed by several authors.[3,33] Campos and colleagues[33] found a decrease of 85% in nitrate consumption when formaldehyde accumulated in a denitrifying USB reactor (up to 300 mg/L formaldehyde) with an increase of the formaldehyde loading rate. The efficiency of denitrification was totally restored after the formaldehyde accumulation was eliminated by decreasing the loading rate, showing a reversible inhibitory effect. However, Garrido and colleagues[3] found only a slight decrease in the denitrification efficiency, from 90 to 80% at concentrations of 700 mg/L of formaldehyde, during the operation of a MUF. Nevertheless, these authors detected nitrous oxide in the off-gas at concentrations higher than 100 mg/L of formaldehyde, this probably being related to a partial inhibition by this compound in the last step of denitrification.

Eiroa and colleagues[37] carried out batch denitrifying assays with an initial concentration of 430 mg/L of formaldehyde. They found that formaldehyde was completely biodegraded in less than 30 h, but the denitrification process lasted several days. Therefore, formaldehyde was transformed into other organic compounds (methanol and formic acid), which were then used as carbon sources for denitrification. These authors operated a denitrifying granular sludge blanket reactor at different COD/N-NO_3^- ratios and at formaldehyde inlet concentrations up to 5000 mg/L, and obtained a mean denitrification efficiency of 98.4%. This high efficiency can be related to the low formaldehyde concentration in the reactor (below 10.3 mg/L), even when the formaldehyde inlet concentrations were increased. Meanwhile, Zoh and Stenstrom[45] carried out batch tests to determine the denitrifying kinetics of nitrite using different carbon sources. These authors found that acetate and formaldehyde showed similar rates.

Denitrification can be affected by free ammonia, but this inhibition does not appear up to 300 to 400 mg/L NH_3.[46] This high concentration can justify that no inhibition of the denitrification process has been reported for this kind of wastewater.[3,4] Eiroa and colleagues[37] observed that nitrate was eliminated much faster at higher initial urea concentrations. However, they also found an increase of nitrite accumulation, which was later removed, due to high urea concentrations.

9.3 TECHNOLOGIES FOR WASTEWATER TREATMENT

Different kinds of bioreactors and configurations have been used to treat wastewater containing formaldehyde and urea, and three different kinds of treatments can be applied: anaerobic treatment, aerobic treatment, and combined nitrification and denitrification treatments.

9.3.1 ANAEROBIC TREATMENT

Anaerobic treatment is recommended for highly concentrated COD wastewater, as the amount of methane generated can compensate for the energy cost in maintaining the temperature of the reactor.

Moreover, this process produces less sludge compared to aerobic treatment. During the anaerobic process, formaldehyde is converted to CO_2 and CH_4 and urea is hydrolyzed to ammonia; therefore, this process only removes organic matter and a small amount of nitrogen due to ammonia assimilation by anaerobic microorganisms. Most of the time, in order to fulfill disposal targets, a posttreatment to remove nitrogen and the remaining organic matter is necessary.

Different kinds of reactors have been used at the laboratory scale to anaerobically treat wastewater containing formaldehyde. Qu and Bhattacharya,[19] using a chemostate, treated a synthetic influent with formaldehyde concentrations up to 1100 mg/L. These authors obtained efficiencies for formaldehyde removal of 99% at volumetric loading rates up to $0.38 kg/m^3 \cdot d$ CH_2O. Vidal and colleagues[1] and Garrido and colleagues[3] used a UASB reactor and a MUF to treat synthetic influents with formaldehyde and urea. Vidal and colleagues,[1] using glucose as cosubstrate, managed to treat up to $3 kg/m^3 \cdot d$ of formaldehyde, while Garrido and colleagues[3] removed $0.5 kg/m^3 \cdot d$ of formaldehyde. The discrepancies between the values might be due to the presence of the cosubstrate, which favors the reduction of the aldehyde to methanol, which is less toxic to the biomass. Nevertheless, the volumetric hydrolytic rates of urea achieved in both systems were similar ($0.46 kg/m^3 \cdot d$ N-urea[3] and $0.58 kg/m^3 \cdot d$ N-urea[1]), being lower than the value of $1.5 kg/m^3 \cdot d$ obtained by Latkar and Chakrabarti[47] in a UASB.

At an industrial scale, Zoutberg and de Been[20] treated wastewaters from a chemical factory containing up to 10 g/L of formaldehyde and 40 g/L of COD. These authors used a Biobed® EGSB (expanded granular sludge blanket) of 275 m³ with a hydraulic retention time (HRT) of 1.25 d, achieving efficiencies up to 98% (Figure 9.6). To avoid the inhibitory effect of high concentrations of formaldehyde, they operated at a recycle ratio of 30, that is, a superficial upflow liquid velocity of 9.4 m/h, which is rather higher than the 1 m/h used in conventional UASBs. The effluent of the Biobed EGSB was posttreated in a low loaded carrousel to meet the strict demands (overall COD efficiency higher than 99.8%).

9.3.2 Aerobic Treatment

During aerobic treatment formaldehyde is oxidized to CO_2 and urea is hydrolyzed, the generated ammonia being oxidized to nitrate if the operational conditions are suitable for nitrification. During this treatment organic matter can be removed, but only a small amount of nitrogen is removed by assimilation; therefore, this treatment is not good enough to fulfill disposal requirements with regard to nitrogen compounds.

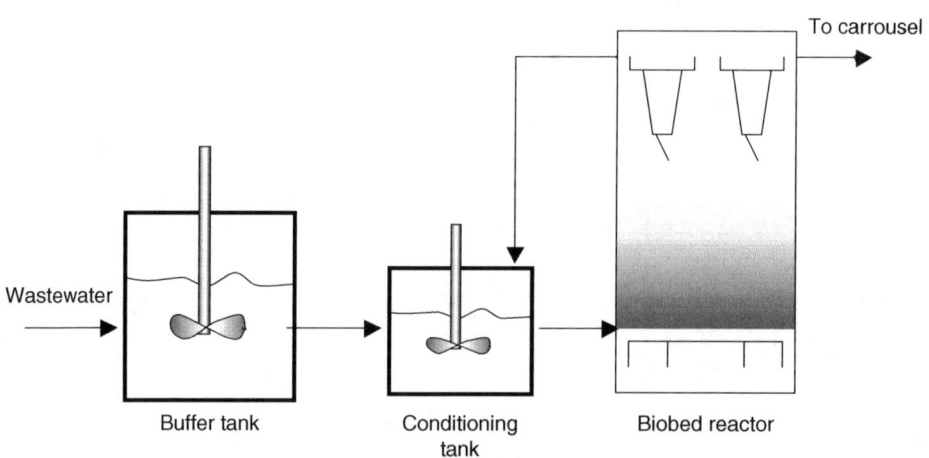

FIGURE 9.6 Schematic of a plant to treat wastewater containing formaldehyde.

Canals[48] managed to treat wastewater from a petrochemical factory at concentrations of up to 2000 mg/L of formaldehyde using an activated sludge reactor, and Zagornaya and colleagues[22] obtained a good removal of this compound when treating resin wastewater in an aerobic reactor.

Garrido and colleagues[2] treated wastewaters from a formaldehyde-urea factory using three activated sludge units operating with solids retention times of 10, 17 and 25 d. These authors applied an organic loading rate (OLR) between 0.2 and 1.2 kg/m³·d COD and obtained removal efficiencies of 80 to 95% and 99.4% for COD and formaldehyde, respectively. Their system achieved a nitrification rate ($0.1\ kg\ N\text{-}NO_x^-/m^3 \cdot d$), the percentage of TKN removal being 45 to 65% due to the biomass growth.

9.3.3 Treatment Combining Nitrification and Denitrification Units

In order to fulfill disposal requirements the best option to treat wastewaters containing formaldehyde and urea is the combination of nitrification and denitrification units in a predenitrifying configuration. In the denitrifying tank, nitrate recycled from the nitrifying unit is denitrified using formaldehyde as the electron donor and urea is hydrolyzed to ammonia. In the nitrification unit, ammonia and the remaining formaldehyde are oxidized to nitrate and CO_2, respectively. The nitrogen removal percentage will depend on the recycling ratio between both units.

Garrido and colleagues[3] operated a MUF under anaerobic and anoxic conditions and achieved, under anoxic conditions, the treatment of up to 2 kg/m³·d of formaldehyde and a hydrolysis rate of up to 0.37 kg/m³·d N-urea. These authors observed that formaldehyde biodegradation is more stable under anoxic conditions than under anaerobic conditions, but only 80% of urea was hydrolyzed in an anoxic environment while a complete conversion occurred under anaerobic conditions. Eiroa and colleagues[49] obtained similar values operating a denitrifying granular sludge blanket reactor with synthetic wastewaters containing formaldehyde and urea. They applied up to 2.8 kg/m³·d of formaldehyde and 0.44 kg/m³·d N-urea, obtaining efficiencies of 99.5 and 77.5% for formaldehyde removal and urea hydrolysis, respectively. Campos and colleagues,[33] using an anoxic USB, achieved a loading rate of hydrolyzed urea of 0.94 kg/·m³·d N-urea and a loading rate of 2.35 kg/m³·d for formaldehyde.

In systems treating formaldehyde, the loading rates of removed nitrate ranged from 0.44 kg/m³·d to 0.94 kg/m³·d $N\text{-}NO_3^-$.[33,49] These values are in the range of denitrifying loading rates obtained for other kinds of wastewaters (1.1 kg/m³·d or 1.5 kg/m³·d $N\text{-}NO_3^-$),[50,51] which means formaldehyde can be used efficiently as an electron donor for denitrification.

Garrido and colleagues[2] used an activated sludge nitrification–denitrification system to treat wastewater from a formaldehyde-urea adhesive factory (Figure 9.7). The treated wastewater contained 590 to 1545 mg/L COD, 197 to 953 mg/L formaldehyde and 129 to 491 mg/L TKN and was also characterized by the presence of polymers with a molecular weight higher than 8000 g/mol, which are not biodegradable. The system was capable of achieving removal efficiencies of 99, 70 to 85, and 30 to 50% for formaldehyde, COD, and TKN, respectively. The COD removal percentage was

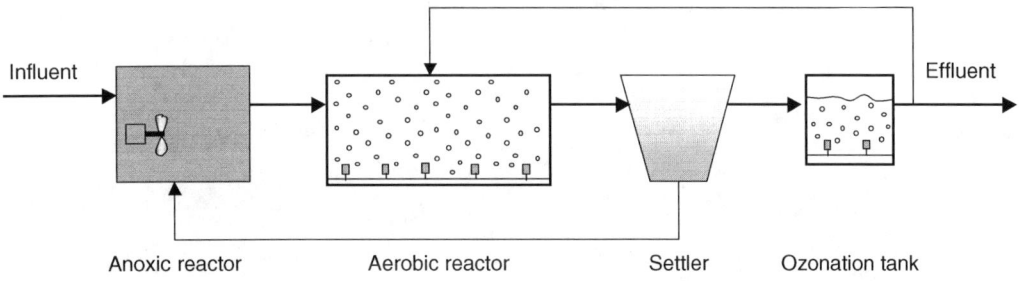

Influent   Effluent

Anoxic reactor   Aerobic reactor   Settler   Ozonation tank

FIGURE 9.7 Schematic representation of a nitrification–denitrification activated sludge plant.

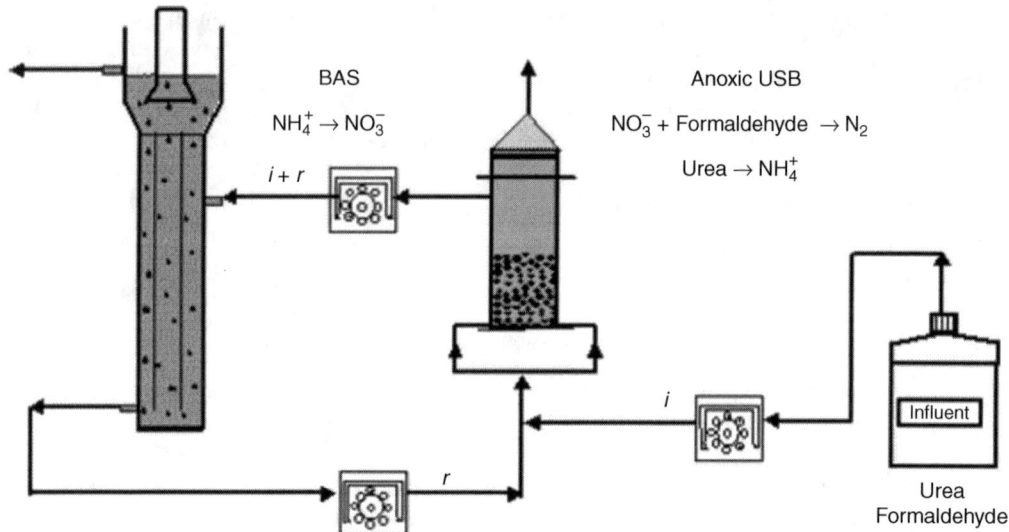

FIGURE 9.8   Plant for the integral treatment of wastewaters containing formaldehyde and urea.

not related to the operational conditions but to the percentage of COD from the formaldehyde. COD removal essentially took place in the anoxic stage, as was the case for formaldehyde, and only nitrification was carried out in an aerobic reactor.

To remove urea and formaldehyde from synthetic wastewater, Campos and colleagues[33] operated a coupled system consisting of a biofilm airlift suspension (BAS) reactor to carry out nitrification and an anoxic USB reactor to carry out the denitrification and urea hydrolysis (Figure 9.8).

These authors studied the effect of the recycling ratio (calculated as the ratio r/i of the flows) for different fed C/N ratios (0.58, 1.0, and 1.5 g C-formaldehyde/g $N-NH_4^+$), always using a constant urea inlet concentration of 400 g/L N-urea. The nitrogen removal percentages achieved are shown in Table 9.5. The maximum nitrogen removal percentages were achieved at a C/N ratio of 1.0 g C-formaldehyde/g $N-NH_4^+$ for both recycling ratios. When this ratio is lower (0.58) not enough organic matter is present to remove nitrate in the anoxic stage, whereas a fed C/N ratio of 1.5 caused a decrease in the efficiency of the system with respect to nitrogen removal, due to the presence of formaldehyde in the BAS reactor, which decreased the nitrification.

When the system was operated at a high inlet C/N ratio, part of the formaldehyde was not removed in the anoxic reactor and entered the nitrification reactor. This led to a heterotrophic layer

TABLE 9.5
Percentages of Nitrogen Removal

r/i	C/N	Nitrogen Removal (%)
3	0.58	43.5 ± 10.2
3	1.00	66.2 ± 7.3
3	1.50	8.4 ± 1.8
9	0.58	51.2 ± 3.1
9	1.00	82.4 ± 3.8
9	1.50	68.6 ± 7.4

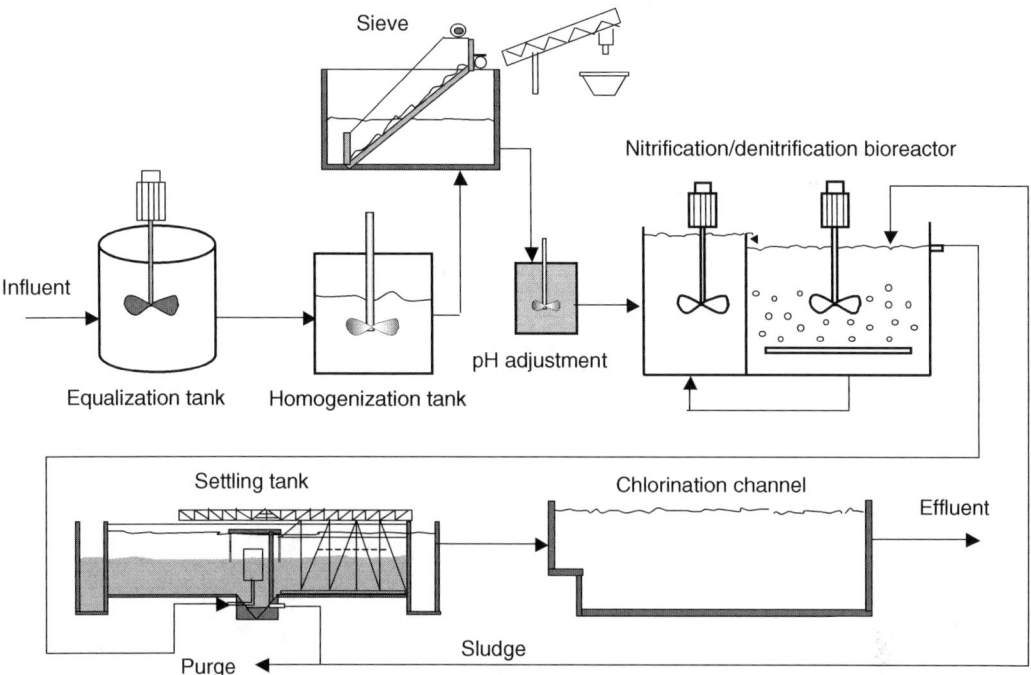

FIGURE 9.9   Industrial plant for the integral treatment of wastewaters from an adhesive factory.

being formed around the nitrifying biofilm, which consumed formaldehyde, and depleted the oxygen for the nitrifiers. The loss of nitrification capacity caused a snowball effect, as no nitrate was available for denitrification, which caused the presence of higher concentrations of formaldehyde in the anoxic system and, then, instability of the denitrification and urea hydrolysis processes. These negative effects of formaldehyde can be reduced by operating at higher recycling ratios, because the increase of the recycling ratio causes a dilution effect in the streams, the formaldehyde concentration in the reactors being lower.

Cantó and colleagues[52] operated an integrated anoxic–aerobic treatment of wastewaters from a synthetic resin producing factory (Figure 9.9). These authors managed to treat up to 2.01 kg/m^3·d COD and up to 0.93 kg/m^3·d TKN with removal efficiencies of 80 to 95% and 58 to 93% for COD and TKN, respectively.

As wastewater from resin-producing factories contains recalcitrant compounds, the removal efficiencies achieved by means of the nitrification–denitrification systems could not reach the required disposal values and a posttreatment, such as ozonation, would be necessary to enhance the biodegradability of those compounds.[2,53]

9.4   GUIDELINES FOR THE DESIGN OF A WASTEWATER TREATMENT PLANT FOR WASTEWATER CONTAINING FORMALDEHYDE AND UREA

9.4.1   DECISION TREE STRUCTURE

The technology chosen to treat wastewater containing formaldehyde and urea will basically depend on the COD concentration and COD/N ratio. The following decision tree structure can be used in the choice of an approach for wastewater treatment (Figure 9.10).

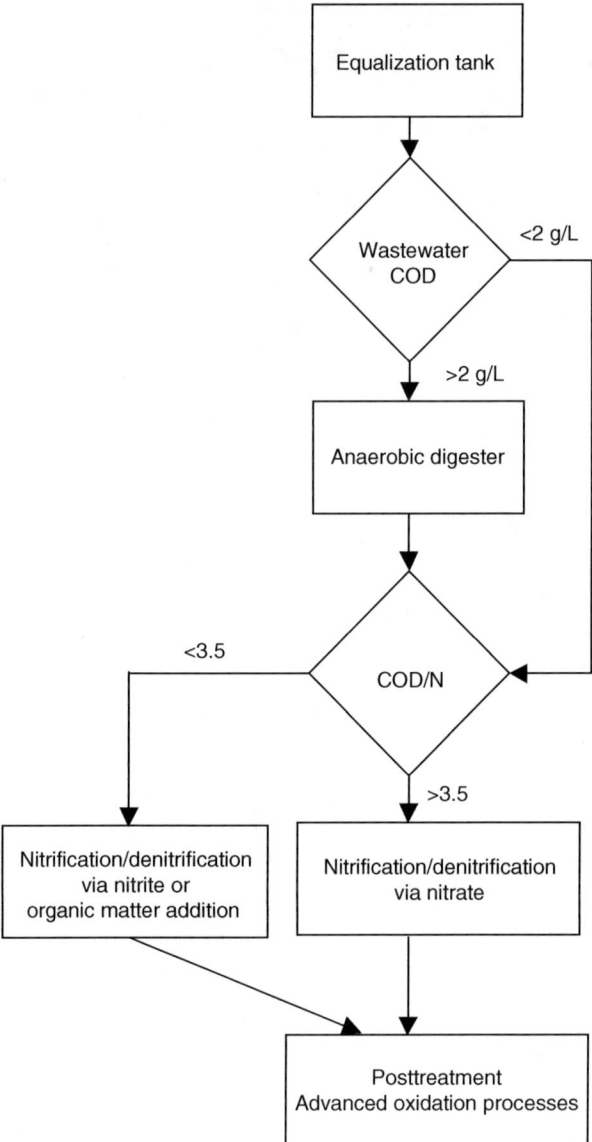

FIGURE 9.10 Decision tree structure.

9.4.2 RECOMMENDATIONS

Because formaldehyde is the most toxic compound present in this kind of wastewater, to control its concentration in reactors is important in order to maintain the stability of the wastewater treatment plant. For this reason the following are recommended:

1. To use an equalization tank to minimize the possible inlet of a peak of formaldehyde.
2. To use anaerobic digesters with high internal recycling ratios to maintain a low concentration of formaldehyde inside the system.
3. To maintain high recycling ratios between the nitrification and denitrification units. This recommendation is also useful to increase the efficiency of nitrogen removal.

When denitrification via nitrate is not possible (COD/N ratios lower than 3.5) there are two possible options to remove nitrogen:

1. To control the dissolved oxygen in the nitrification unit to obtain a partial oxidation of ammonia to nitrite
2. To add an external carbon source

As the adhesive factory will consume a large amount of methanol in its processes, the addition of this compound to carry out nitrogen removal would have a low cost, and is one of the most feasible options.

REFERENCES

1. Vidal, G., Jiang, Z.P., Omil, F., Thalasso, F., Mendez, R., and Lema, J.M., Continuous anaerobic treatment of waste-waters containing formaldehyde and urea, *Biores. Technol.*, 70, 283–291, 1999.
2. Garrido, J.M., Méndez, R., and Lema, J.M., Treatment of wastewaters from a formaldehyde-urea adhesives, *Water Sci. Technol.*, 42, 293–300, 2000.
3. Garrido, J.M., Méndez, R., and Lema, J.M., Simultaneous urea hydrolysis, formaldehyde removal and denitrification in a multifed upflow filter under anoxic and anaerobic conditions, *Water Res.*, 35, 691–698, 2001.
4. Eiroa, M., Biological Removal of Organic and Nitrogen Compounds Present in the Effluents of a Resin Factory (in Spanish), PhD thesis, University of A Coruña, 2004.
5. Gonzalez-Gil, G., Kleerebezem, R., van Aelst, A., Zoutberg, G.R., Versprille, A.I., and Lettinga, G., Toxicity effects of formaldehyde on methanol degrading sludge and its anaerobic conversion in biobed expanded granular sludge bed (EGSB) reactors, *Water Sci. Technol.*, 40, 195–202, 1999.
6. Omil, F., Méndez, D., Vidal, G., Méndez, R., and Lema, J.M., Biodegradation of formaldehyde under anaerobic conditions, *Enzyme Microb. Technol.*, 24, 255–262, 1999.
7. Oliveira, S.V.W.B., Moraes, E.M., Adorno, M.A.T., Varesche, M.B.A., Foresti, E., and Zaiat, M., Formaldehyde degradation in an anaerobic packed-bed bioreactor, *Water Res.*, 38, 1685–1694, 2004.
8. Zoutberg, G.R. and Frankin, R., Anaerobic treatment of chemical and brewery waste water with a new type of anaerobic reactor: the Biobed® EGSB reactor, *Water Sci. Technol.*, 34, 375–381, 1996.
9. Todini, O. and Hulshoff Pol, L., Anaerobic degradation of benzaldehyde in methanogenic granular sludge: the influence of additional substrates, *Appl. Microbiol. Biotechnol.*, 38, 417–420, 1992.
10. Chou, W., Speece, R., Siddiqi, R., and McKeon, R., The effect of petrochemical structure on methane fermentation toxicity, *Program Water Technol.*, 10, 545–550, 1987.
11. Bhattacharya, S.K. and Parkin, G.F., Fate and effect of methylene chloride and formaldehyde in methane fermentation systems, *J. Water Pollut. Control Fed.*, 60, 531–536, 1988.
12. Sharma, S., Ramakrishna, C., Desai, J.D., and Batt, N.M., Anaerobic biodegradation of petrochemical wastewater using biomass support particles, *Appl. Microbiol. Biotechnol.*, 40, 768–771, 1994.
13. Bekker de, P., Jans, T., and Piscaer, P., Anaerobic treatment of formaldehyde containing waste water, in *Proc. European Symposium AW*, van den Brink, W.J., Ed., November 23–25, 1983, Noordwijkerhout, TNO Corporate Communication Department, The Hague, The Netherlands, pp. 449–463, 1983.
14. Parkin, G.F., Speece, R.E., Yang, C.H.J., and Kocher, W.M., Response of methane fermentation systems to industrial toxicants, *J. Water Pollut. Control Fed.*, 55, 44–53, 1983.
15. Gonzalez-Gil, G., Kleerebezem, R., and Lettinga, G., Formaldehyde toxicity in anaerobic systems, *Water Sci. Technol.*, 42, 223–229, 2000.
16. Hickey, R.F., Vanderwilen, J., and Switzenbaum, M.S., The effects of organic toxicants on methane production and hydrogen gas levels during the anaerobic digestion of the waste activated sludge, *Water Res.*, 21, 1417–1427, 1987.
17. Pearson, F., Shiun-Chung, C., and Gautier, M., Toxic inhibition of anaerobic biodegradation, *J. Water Pollut. Control Fed.*, 52, 472–482, 1980.
18. Lu, Z. and Hegemann, W., Anaerobic toxicity and biodegradation of formaldehyde in batch cultures, *Water Res.*, 32, 209–215, 1988.
19. Qu, M. and Bhattacharya, S.K., Toxicity and biodegradation of formaldehyde in anaerobic methanogenic culture, *Biotechnol. Bioeng.*, 55, 727–736, 1997.

20. Zoutberg, G.R. and de Been, P., The Biobed® EGSB (expanded granular sludge bed) system covers shortcomings of the up flow anaerobic sludge blanket reactor in the chemical industry, *Water Sci. Technol.*, 35, 183–188, 1997.

21. Guerrero, L., Omil, F., Méndez, R., and Lema, J.M., Treatment of saline wastewaters from fish meal factories in an anaerobic filter under extreme ammonia concentrations, *Biores. Technol.*, 61, 69–78, 1997.

22. Zagornaya, P.B., Denis, A.D., Gvozdyak, P.I., Nikonendo, V.U., and Chehovskaya, T.P., Microbiological purification of above resin waste-waters, *Biotekhnologya*, 2, 51–53, 1990.

23. Adroer, N., Casa, C., de Mas, C., and Solá, C., Mechanism of formaldehyde biodegradation by *Pseudomonas putida*, *Appl. Microbiol. Biotechnol.*, 33, 217–220, 1990.

24. Glancer-Soljan, M., Soljan, V., Landeka, T., and Cacic, L., Aerobic degradation of formaldehyde in wastewater, *Food Technol. Biotechnol.*, 39, 197–202, 2001.

25. Eiroa, M., Kennes, C., and Veiga, M.C., Formaldehyde biodegradation and its inhibitory effect on nitrification, *J. Chem. Technol. Biotechnol.*, 79, 499–504, 2004.

26. Gerike, P. and Gode, P., The biodegradability and inhibitory threshold concentration of some disinfectants, *Chemosphere*, 21, 799–812, 1990.

27. Mobley, H.L.T. and Hausinger, R.P., Microbial ureases: Significant, regulation, and molecular characterization, *Microbiol. Rev.*, 53, 85–108, 1989.

28. Chakrabarti, T. and Subrahmanyam, P.V.R., Biological hydrolysis of urea in a continuous flow stirred tank reactor under laboratory conditions—a bench scale study, *Proc. 36th Industrial Waste Conference*, Purdue University, pp. 477, 1981.

29. Gupta, S.K. and Sharma, R., Biological oxidation of high strength nitrogenous waste-water, *Water Res.*, 30, 593–600, 1996.

30. Hamoda, M.F., Aerobic treatment of ammonium fertilizer effluent in a fixed-film biological system, *Water Sci. Technol.*, 22, 75–84, 1990.

31. Rittstieg, K., Robra, K.H., and Somitsch, W., Aerobic treatment of a concentrated urea wastewater with simultaneous stripping of ammonia, *Appl. Microbiol. Biotechnol.*, 56, 820–825, 2001.

32. Prosser, J.I., Autotrophic nitrification in bacteria, in *Advances in Microbial Physiology*, Vol. 30, Rose, A.H. and Tempest, D.W., Eds., Academic Press, London, 125–181, 1989.

33. Campos, J.L., Sánchez, M., Mosquera-Corral, A., Méndez, R., and Lema, J.M., Coupled BAS and anoxic USB system to remove urea and formaldehyde from wastewater, *Water Res.*, 37, 3445–3451, 2003.

34. Koops, H.P. and Chritian, U., The lithotrophic ammonia-oxidizing bacteria, in *Variations in Autotrophic Life*, Shively, J.M. and Burton, L.L., Eds., Harcourt Brace Jovanovich Pub., New York, 1991.

35. Sliekers, A.O., Haaijer, S., Schmid, M., Harhangi, H., Verwegen, K., Kuenen, J.G., and Jetten, M.S.M., Nitrification and anammox with urea as the energy source, *System. Appl. Microbiol.*, 27, 271–278, 2004.

36. Walker, J.D., Effects of chemicals on microorganisms, *J. Water Pollut. Control Fed.*, 61, 1077–1097, 1989.

37. Eiroa, M., Kennes, C., and Veiga, M.C., Formaldehyde and urea removal in a denitrifying granular sludge blanket reactor, *Water Res.*, 38, 3495–3502, 2004.

38. Osislo, A. and Lewandowski, Z., Inhibition of nitrification in the packed bed reactors by selected organic compounds, *Water Res.*, 19, 423–426, 1985.

39. Eiroa, M., Kennes, C., and Veiga, M.C., Simultaneous nitrification and formaldehyde biodegradation in an activated sludge unit, *Biores. Technol.*, 96, 1914–1918, 2005.

40. Anthonisen, A.C., Loehr, R.C., Prakasam, T.B.S., and Srinath, E.G., Inhibition of nitrification by ammonia and nitrous acid, *J. Water Pollut. Control Fed.*, 48, S35–S52, 1976.

41. Gupta, S.K., Raja, S.M., and Gupta, A.B., Simultaneous nitrification-denitrification in a rotating biological contactor, *Environ. Technol.*, 15, 145–153, 1994.

42. Knowles, R., Denitrification, *Microbiol. Rev.*, 46, 43–70, 1982.

43. Garrido, J.M., Moreno, J., Méndez-Pampín, R., and Lema, J.M., Nitrous oxide production under toxic conditions in a denitrifying anoxic filter, *Water Res.*, 32, 2550–2552, 1998.

44. Schulthess, R., Wild, D., and Gujer, W., Nitric and nitrous oxide from denitrifying activated sludge at low oxygen concentration, *Water Sci. Technol.*, 30, 123–132, 1994.

45. Zoh, K.D. and Stenstrom, M.K., Biological denitrification of high explosives processing wastes, *Water Sci. Technol.*, 36, 47–54, 1997.

46. Sánchez, M., Mosquera-Corral, A., Méndez, R., and Lema, J.M., Effect of ammonia on the performance of denitrifying reactors, European Conference on New Advances in Biological Nitrogen and Phosphorus Removal for Municipal or Industrial Wastewaters, Narbonne, France, pp. 119–126, 1998.

47. Latkar, M. and Chakrabarti, T., Performance of upflow anaerobic sludge blanket reactor carrying out biological hydrolysis of urea, *Water Environ. Res.*, 66, 12–15, 1994.

48. Canals, J. Biological degradation of formaldehyde. Studies at pilot plant and industrial application (in Spanish), *Ingeniería Química*, 85–88, January 1983.
49. Eiroa, M., Vilar, A., Amor, L., Kennes, C., and Veiga, M.C., Biodegradation and effect of formaldehyde and phenol on the denitrification process, *Water Res.*, 39, 449–455, 2005.
50. Rahmani, H., Rols, J.L., Capdeville, B., Cornier, J.C., and Deguin, A., Nitrite removal by a fixed culture in a submerged granular biofilter, *Water Res.*, 29, 1745–1753, 1995.
51. Hanaki, K. and Polpraset, C., Contribution of methanogenesis to denitrification with an upflow filter, *J. Water Pollut. Control Fed.*, 61, 1604–1611, 1989.
52. Cantó, M., Gómez, J., Kennes, C., and Veiga, M.C., Integrated anoxic-aerobic treatment of wastewaters from a synthetic resin producing factory, European Conference on New Advances in Biological Nitrogen and Phosphorus Removal for Municipal or Industrial Wastewaters, Narbonne, France, October 12–14, 1998.
53. Aparicio, M.A., Eiroa, M., Kennes, C., and Veiga, M.C., Combined post-ozonation and biological treatment of recalcitrant wastewater from a resin-producing factory, *J. Hazard. Mater.*, 143, 285–290, 2007.

10 Hazardous Waste Deep-Well Injection

Nazih K. Shammas and Lawrence K. Wang

CONTENTS

10.1 INTRODUCTION

The technology of deep-well injection has been around for more than 70 years. "Most Americans would be surprised to know that there is a waste management system already in operation in the U.S. that has no emissions into the air, no discharges to surface water, and no off-site transfers, and exposes people and the environment to virtually no hazards."[1] The U.S. Environmental Protection Agency (U.S. EPA) has stated that Class 1 wells are safer than virtually all other waste disposal practices for many chemical industry wastes.

A typical injection well consists of concentric pipes that extend several thousand feet down from the surface level into highly saline, permeable injection zones that are confined vertically by impermeable strata. The outermost pipe or surface casing extends below the base of any underground sources of drinking water (USDW) and is cemented back to the surface to prevent contamination of the USDW. Directly inside the surface casing is a long string casing that extends to and sometimes into the injection zone. This casing is filled with cement all the way to the surface in order to seal off the injected waste from the formations above the injection zone back to the surface. The casing provides a seal between the wastes in the injection zone and the upper formations. The waste is injected through the injection tubing inside the long string casing either through perforations in the long string or in the open hole below the bottom of the long string. The space between the string casing and the injection tube, called the annulus, is filled with an inert, pressurized fluid, and is sealed at the bottom by a removable packer preventing injected wastewater from backing up into the annulus.[2]

The geochemical fate of deep-well-injected wastes must be thoroughly understood to help avoid problems when incompatibility between the injected wastes and the injection-zone formation is a possibility. An understanding of geochemical fate will also be useful when a geochemical no-migration demonstration must be made. This chapter was written to address both of these needs by presenting state-of-the-art information on the geochemical fate of hazardous deep-well-injected wastes. Furthermore, operators of any new industrial-waste injection well who must consider the possibility of incompatibility will find this chapter helpful in identifying geochemical reactions of potential concern and methods for testing incompatibility.

U.S. EPA regulations (53 *Federal Register* 28118–28157, July 26, 1988) stipulate that deep-well injection of hazardous wastes is allowed only if either of the following two no-migration standards is met[3]:

1. Fluid movement conditions are such that the injected fluids will not migrate within 10,000 years vertically upward out of the injection zone; or laterally within the injection zone to a point of discharge or interface with an Underground Source of Drinking Water (USDW).
2. Before the injected fluids migrate out of the injection zone or to a point of discharge or interface with USDW, the fluid will no longer be hazardous because of attenuation, transformation, or immobilization of hazardous constituents within the injection zone by hydrolysis, chemical interactions, or other means.

According to the Federal Remediation Technology Roundtable (FRTR) the factors that may limit the applicability and effectiveness of this technology include the following[2]:

1. Injection will not be used for hazardous waste disposal in any areas where seismic activity could potentially occur.
2. Injected wastes must be compatible with the mechanical components of the injection well system and the natural formation water. The waste generator may be required to perform physical, chemical, biological, or thermal treatment for removal of various contaminants or constituents from the waste to modify the physical and chemical character of the waste to assure compatibility.

3. High concentrations of suspended solids (typically >2 mg/L) can lead to plugging of the injection interval.
4. Corrosive media may react with the injection well components, with injection zone formation, or with confining strata with very undesirable results. Wastes should be neutralized.
5. High iron concentrations may result in fouling when conditions alter the valence state and convert soluble species to insoluble species.
6. Organic carbon may serve as an energy source for indigenous or injected bacteria, resulting in rapid population growth and subsequent fouling.
7. Wastestreams containing organic contaminants above their solubility limits may require pretreatment before injection into a well.
8. Site assessment and aquifer characterization are required to determine the suitability of a site for wastewater injection.
9. Extensive assessments must be completed prior to receiving approval from regulatory authority.

State-of-the-art fluid-transport modeling is considerably more advanced than that of geochemical-fate and transport modeling. Consequently, geochemical-fate modeling is most likely to be used if a fluid-flow no-migration standard cannot be met. Geochemical-fate transport modeling of deep-well-injected hazardous wastes is in the early stages of development, and its use in meeting current U.S. EPA Underground Injection Control regulations is unbroken ground. However, where the no-migration standard must be considered, there is a U.S. EPA guide[3] that can help determine whether geochemical-fate/transport modeling of a specific waste is even feasible, and what approaches might be taken.

10.2 CHARACTERISTICS OF INJECTED HAZARDOUS WASTES

This section discusses the characteristics of hazardous wastes typically injected into Class I injection wells. It includes the following:

1. The properties that define a waste as hazardous
2. The sources, amounts, and composition of existing deep-well-injected hazardous wastes
3. Trends and distribution of industrial and hazardous waste injection
4. The design and construction of deep-injection wells

10.2.1 IDENTIFYING HAZARDOUS WASTES

Wastes are defined as hazardous for the purposes of regulatory control in 40 CFR Part 261.[4] In this regulation, wastes are classified as hazardous either by being listed in tables within the regulation or by meeting certain specified characteristics. Thus, under 40 CFR Part 261 hazardous wastes are known either as "listed" or "characteristic" wastes. Some listed wastestreams, such as spent halogenated solvents, come from many industries and processes. Other listed wastestreams, such as American Petroleum Institute (API) separator sludges from the petroleum-refining industry, come from one particular industry and one process. A characteristic waste is not listed, but is classified as hazardous because it exhibits one or more of the following characteristics[4,5]:

1. Toxicity to living organisms
2. Reactivity
3. Corrosivity
4. Ignitability

Listed wastes also exhibit one or more of these characteristics. The significance of each of the characteristics listed above is discussed below and is summarized in Table 10.1.[3] Deep-well-injected

TABLE 10.1

Hazardous and Physicochemical Properties of Injected Wastes

Characteristic	Comment
Hazardous characteristics	
Toxicity	Has toxic properties that result in classification as a hazardous waste, but specific properties may vary greatly
Reactivity	Reactivity usually reduced by dilution; actual concentration may affect toxicity and mobility
Corrosivity	May be a significant consideration in well design and geochemical fate
Ignitability	Not a significant consideration under injection conditions
Physical/chemical properties	
Normal physical state	Liquids or dissolved solids
Molecular weight	May affect structure–activity relationships
Density/specific gravity	Must be miscible in water
Solubility	Must be soluble or miscible in water
Boiling point	Greater than ambient temperatures
Melting point	Less than ambient temperatures
Vapor pressure/density	Water-soluble volatile compounds may be involved, but vapor pressure and vapor density are not significant considerations in deep-well injection
Flash point/autoignition point	Greater than ambient temperatures.

Source: U.S. EPA, Assessing the Geochemical Fate of Deep-Well-Injected Hazardous Waste: A Reference Guide, EPA/625/6-89/025a, U.S. EPA, Cincinnati, OH, June 1990.

wastes commonly contain several components that classify the waste as hazardous, along with other nonhazardous components.

10.2.1.1   Toxicity

A waste is toxic under 40 CFR Part 261 if the extract from a sample of the waste exceeds specified limits for any one of eight elements and five pesticides (arsenic, barium, cadmium, chromium, lead, mercury, selenium, silver, endrin, methoxychlor, toxaphene, 2,4-D and 2,4,5-TP Silvex using extraction procedure (EP) toxicity test methods. Note that this narrow definition of toxicity relates to whether a waste is defined as hazardous for regulatory purposes; in the context of this chapter, toxicity has a broader meaning because most deep-well-injected wastes have properties that can be toxic to living organisms.

10.2.1.2   Reactivity

Reactivity describes a waste's tendency to interact chemically with other substances. Many wastes are reactive, but it is the degree of reactivity that defines a waste as hazardous. Hazardous reactive wastes are those that are normally unstable and readily undergo violent change without detonating, react violently with water, form potentially explosive mixtures with water, generate toxic gases or fumes when combined with water, contain sulfide or cyanide and are exposed to extreme pH conditions, or are explosive. Because deep-well-injected wastestreams are usually dilute (typically less than 1% waste in water), hazardous reactivity is not a significant consideration in deep-well injection, although individual compounds may exhibit this property at higher concentrations than those that exist in the wastestream. Nonhazardous reactivity is, however, an important property in deep-well injection, because when a reactive waste is injected, precipitation reactions that can lead to well plugging may occur.

10.2.1.3   Corrosivity

Corrosive wastes are defined as those wastes with a pH ≤ 2 or pH ≥ 12.5 (i.e., the waste is very acidic or very basic). Beyond its importance in defining a waste as hazardous, the corrosivity of wastes is also a property of concern to deep-well injection systems and operations. Corrosive wastes may damage the injection system, typically by electrochemical or microbiological means. Corrosion of injection-well pumps, tubing, and other equipment can lead to hazardous waste leaking into strata not intended for injection. For information on various types of electrochemical corrosion relevant to the injection-well system, the reader is referred to Warner and Lehr.[6] Other recommended sources include references 7 to 10. These sources discuss saturation and stability indexes for predicting the potential for corrosion or scaling (accumulation of carbonate and sulfate precipitates) in injection wells. The Stiff and Davis index[10] is recommended by Warner and Lehr[6] as most applicable to deep-well injection of hazardous wastes, because it is intended for use with highly saline groundwaters. Additionally, Ostroff[11] provides examples of how to use the index, Watkins[12] describes procedures that test for corrosion, and Davis[13] thoroughly discusses microbiological corrosion of metals.

10.2.1.4   Ignitability

As noted, deep-well-injected wastes are relatively dilute. Therefore, ignitability is not a significant consideration in deep-well injection, although in a concentrated form, individual compounds may exhibit this property. Ignitability has no further implications for the fate of deep-well-injected waste.

10.2.2   Sources, Amounts, and Composition of Deep-Well-Injected Wastes

The sources, amounts, and composition of injected hazardous wastes are a matter of record, because the Resource Conservation and Recovery Act (RCRA)[5,14] requires hazardous waste to be manifested (i.e., a record noting the generator of the waste, its composition or characteristics, and its volume must follow the waste load from its source to its ultimate disposal site). The sources and amounts of injected hazardous waste can be determined, therefore, based on these records. Table 10.2 shows the estimated volume of deep-well-injected wastes by industrial category.[3] More than 11 billion gallons of hazardous waste were injected in 1983. Organic chemicals (51%) and petroleum-refining and petrochemical products (25%) accounted for three-quarters of the volume of injected wastes that

---

TABLE 10.2
Estimated Volume of Deep-Well Injected Wastes by Industrial Category

Industrial Category	Volume (MG/yr)	Percent of Total
Organic chemical	5,868	51
Petroleum refining and petrochemical products	2,888	25
Miscellaneous chemical products	687	6
Agricultural chemical products	525	4.5
Inorganic chemical products	254	2.2
Commercial disposal	475	4
Metals and minerals	672	5.8
Aerospace and related industry	169	1.5
Total	11,538	100.0

Source: U.S. EPA, Report to Congress on Injection of Hazardous Wastes, EPA 570/9-85-003, NTIS PB86-203056, U.S. EPA, Washington, 1985.

year. The remaining 24% was divided among six other industrial categories: miscellaneous chemical products, agricultural chemical products, inorganic chemical products, commercial disposal, metals and minerals, and aerospace and related industry.

Although the general composition of each shipment of wastes to an injection well may be known, a number of factors makes it difficult to characterize fully the overall composition of industrial wastewaters at any one well. These factors include the following[15]:

1. Variations in flow, in concentrations, and in the nature of organic constituents over time
2. Biological activity that may transform constituents over time
3. Physical inhomogeneity (soluble and insoluble compounds)
4. Chemical complexity; an example of the complexity of organic wastes is illustrated in the work of Roy and colleagues,[16] which presents an analysis of an alkaline pesticide-manufacturing waste—this waste contained more than 50 organic compounds, two-fifths of which could not be precisely identified

Although no systematic database exists on the exact composition of deep-well-injected wastes, in a survey of 209 operating waste-injection wells, Reeder[17] found that 53% injected one or more chemicals identified in that study as hazardous. The U.S. EPA gathered data for 108 wells (55% of total active wells) that were under operation.[3] A little more than half of the undiluted waste volume was composed of nonhazardous inorganics (52%). Acids were the most important constituent by volume (20%), followed by organics (17%). Heavy metals and other hazardous inorganics made up less than 1% of the total volume in the 108 wells. About a third of the wells injected acidic wastes and about two-thirds injected organic wastes. Although the percentage of heavy metals by volume was low, almost one-fifth of the wells injected wastes containing heavy metals. An injected wastestream is composed of the waste material and a large volume of water. It is reported that typical ratios in the total volume of injected fluids are 96% water and 4% waste.

The U.S. EPA gathered data also showed that the average concentration of all the acidic wastes exceeded 40,000 mg/L. Concentrations of metals ranged from 1.4 mg/L (chromium) to 5500 mg/L (unspecified metals, probably containing multiple species). Five of the 18 organic constituents exceeded 10,000 mg/L (total organic carbon, organic acids, formaldehyde, chlorinated organics, and formic acid); four others exceeded 1000 mg/L (oil, isopropyl alcohol, urea nitrogen, and organic peroxides).

10.2.3 Geographic Distribution of Hazardous Waste Injection Wells

The use of wells for disposal of industrial wastes dates back to the 1930s, but this method was not used extensively until the 1960s, when it was implemented primarily in response to more stringent water pollution control regulations.

The number of industrial-waste injection wells more than doubled between 1967 and 1986.[3] In 1986, Class I injection wells were concentrated in two states, Texas (112 wells) and Louisiana (70 wells), which between them had a total of 69% of all wells (263 wells). Growth from 1984 to 1986 was concentrated in Texas, with a 38% increase from 81 to 112 wells. The only other states to show a significant increase from 1984 to 1986 were Indiana (13 proposed wells) and California (7 proposed wells). Nine states had had industrial-waste injection wells in the past but did not have any permitted Class I wells in 1986 (Alabama, Colorado, Iowa, Mississippi, Nevada, North Carolina, Pennsylvania, Tennessee, and Wyoming). One state (Washington) had a Class I well in 1986, but no record of industrial wastewater injection before that year. The total number of industrial-waste injection wells increased to 300 at the end of the 1990s and beginning of this century, approximately 100 Class I hazardous waste injection wells and about 200 Class I wells that hold nonhazardous waste.[1,18]

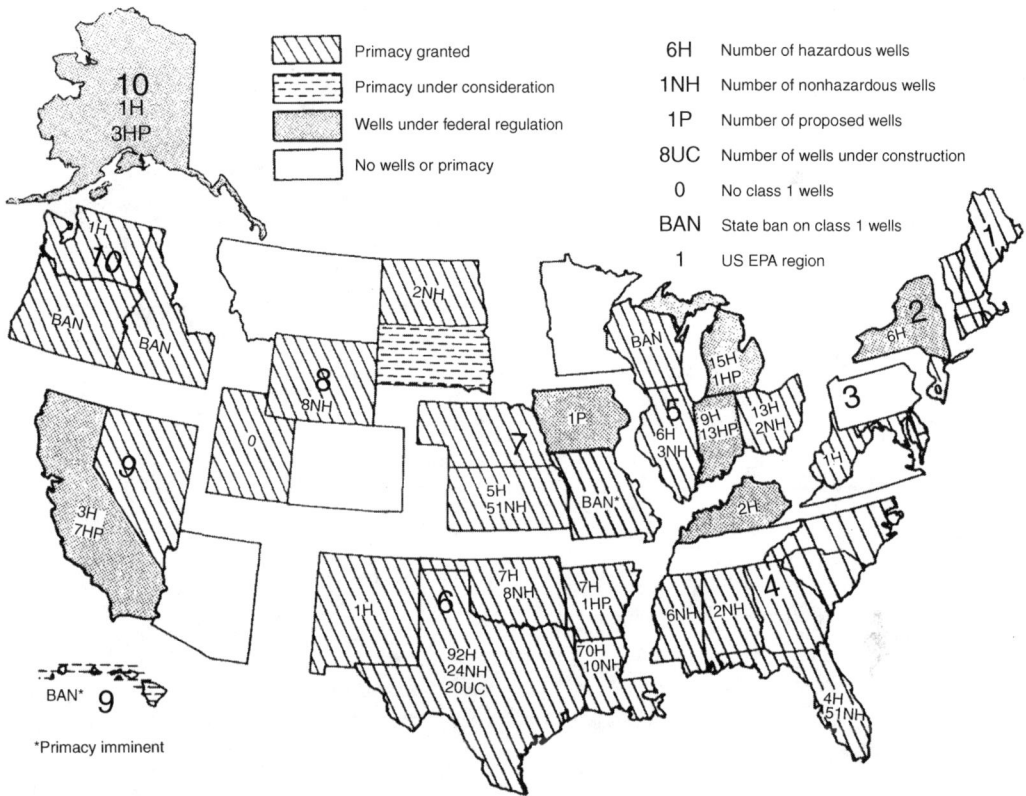

Primacy granted	6H Number of hazardous wells
Primacy under consideration	1NH Number of nonhazardous wells
Wells under federal regulation	1P Number of proposed wells
No wells or primacy	8UC Number of wells under construction
	0 No class 1 wells
	BAN State ban on class 1 wells
	1 US EPA region

*Primacy imminent

FIGURE 10.1 Regulatory status and number of Class I wells in the U.S. (From U.S. EPA, Assessing the Geochemical Fate of Deep-Well-Injected Hazardous Waste: A Reference Guide, EPA/625/6-89/025a, U.S. EPA, Cincinnati, OH, June 1990.)

Figure 10.1 shows the number of Class I wells in the 1986 survey by state, divided into U.S. EPA regions, and also indicates the regulatory status of such wells in each state as of 1989. The map shows the heavy concentration of hazardous waste injection wells in three geologic basins: Gulf Coast, Illinois Basin, and the Michigan Basin.[3]

10.2.4 Design and Construction of Deep-Injection Wells

The following subsections give a description of the design and construction of deep-injection wells.[19–23]

10.2.4.1 Surface Equipment Used in Waste Disposal

Figure 10.2 shows the surface equipment used in a typical subsurface waste-disposal system. Detailed discussion of surface treatment methods can be found in Warner and Lehr.[6] The individual elements are listed in the following:

1. *Sump tank.* A sump tank or an open 113,550 to 189,250 L (30,000 to 50,000 gal) steel tank is commonly used to collect and mix wastestreams. An oil layer or, in a closed tank, an inert gas blanket is often used to prevent air contact with the waste. Alternatively, large, shallow, open ponds may provide sufficient detention time to permit sedimentation of particulate matter. Such ponds are often equipped with cascade, spray, or forced-draft

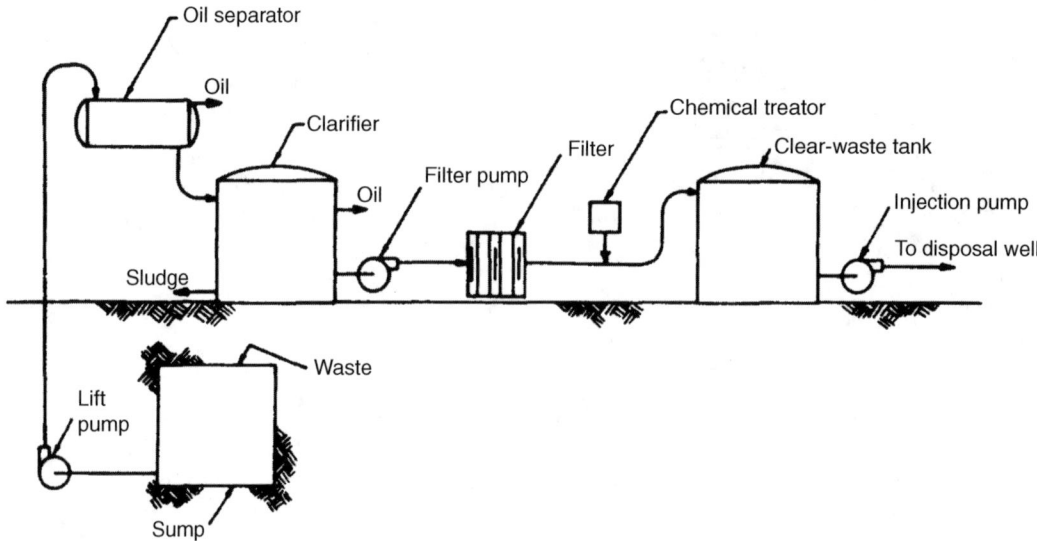

FIGURE 10.2   Above-ground components of a subsurface waste disposal system. (From U.S. EPA, Assessing the Geochemical Fate of Deep-Well-Injected Hazardous Waste: A Reference Guide, EPA/625/6-89/025a, U.S. EPA, Cincinnati, OH, June 1990.)

aerators to oxidize iron and manganese salts to insoluble forms that precipitate in the aeration ponds.

2. *Oil separator.* An oil separator is used when the waste contains oil, because oil tends to plug the disposal formation. The waste is passed through a settling tank equipped with internal baffles to separate the oil from the waste.[24]

3. *Clarifier.* A clarifier removes such particulate matter as polymeric flocs, dirt, oil, and grease. It is often a tank or a pond in which the detention time is long enough to allow suspended particles to settle gradually.[25] The process may also be accelerated by adding a flocculating agent such as aluminum sulfate, ferric sulfate, or sodium aluminate.[26] Tank clarifiers are often equipped with a mechanical stirrer, sludge rake, and surface skimmer that continuously remove sludge and oil.

4. *Filter.* A filter is used in some cases when coagulation and sedimentation do not completely separate the solids from the liquid waste in areas where sand and sandstone formations are susceptible to plugging. Filters with a series of metal screens coated with diatomaceous earth or cartridge filters are typically used.[27] Where limestone formations with high solution porosity are used for injection, filtration is usually not required.

5. *Chemical treater.* A chemical treater is used to inject a bactericide if microorganisms could cause fouling of injection equipment and plugging of the injection reservoir.

6. *Clear-waste tank.* An unlined steel clear-waste tank is typically used to hold clarified waste before injection. The tank is equipped with a float switch designed to start and stop the injection pump at predetermined levels.

7. *Injection pump.* An injection pump is used to force the waste into the injection zone, although in very porous formations, such as cavernous limestone, the hydrostatic pressure of the waste column in the well is sufficient. The type of pump is determined primarily by the well-head pressures required, the volume of liquid to be injected, and the corrosiveness of the waste. Single-stage centrifugal pumps are used in systems that require well-head pressures up to about 10.5 kg/cm^2 (150 psi), and multiplex piston pumps are used to achieve higher injection pressures.

10.2.4.2 Injection-Well Construction

Most injection wells are drilled using the rotary method, although, depending on the availability of equipment and other site-specific factors, reverse-rotary or cable-tool drilling may be used. The construction of an injection well incorporates several important elements[28]:

1. Bottom-hole and injection-interval completion
2. Casing and tubing
3. Packing and cementing
4. Corrosion control
5. Mechanical integrity testing.

A detailed discussion of the technical aspects of industrial-waste injection-well construction can be found in Warner and Lehr.[6] U.S. EPA[20] also presents a survey of well construction methods and materials used for 229 hazardous waste injection wells. Two types of injection well completions are used with hazardous waste injection wells:

1. *Open hole.* Open hole completion is typically used in competent formations such as limestone, dolomite, and consolidated sandstone that will stand unsupported in a borehole. In 1985, 27% of Class I wells were of this type, with most located in the Illinois Basin.
2. *Gravel pack.* Gravel pack and perforated completions are used where unconsolidated sands in the injection zone must be supported. In gravel-pack completions the cavity in the injection zone is filled with gravel or, more typically, a screen or liner is placed in the injection-zone cavity before the cavity is filled with gravel. In perforated completions, the casing and cement extend into the injection zone and are then perforated in the most permeable sections. In 1985, 53% of Class I wells were perforated and 17% were screened.[20]

Casing and tubing are used to prevent the hole from caving in and to prevent aquifer contamination by confining wastes within the well until they reach the injection zone. Lengths of casing of the same diameter are connected together to form casing strings. Usually, two- or three-casing strings are used. The outer casing seals the near-surface portion of the well (preferably to below the point where aquifers containing less than 10,000 mg/L total dissolved solids, potential underground sources of drinking water, are located). The inner casing extends to the injection zone. Tubing is placed inside the inner casing to serve as the conduit for injected wastes, and the space between the tubing and casing is usually filled with kerosene or diesel oil after packing and cementing are completed.

Packers are used at or near the end of the injection tubing to plug the space, called the annulus, between the injection tubing and the inner casing. Cement is applied to the space between the outer walls of the casing and the borehole or other casing. Portland cement is used most commonly for this purpose, although when acidic wastes are injected, special acid-resistant cements are sometimes used in the portion of the well that passes through the confining layers.

Corrosion control can be handled several ways:

1. By using corrosion-resistant material in constructing the well
2. By treating the wastestream through neutralization or other measures
3. By cathodic protection

Mechanical integrity testing is required by U.S. EPA regulations (40 CFR 146.08) to ensure that an injection well has been constructed or is operating without significant leakage from the casing, tubing, or packer or upward movement of fluid through vertical channels adjacent to the well bore. A detailed discussion of mechanical integrity can be found in reference 28.

10.3   PROCESSES AFFECTING THE GEOCHEMICAL FATE OF DEEP-WELL-INJECTED WASTES

This section examines the major processes that affect the fate of deep-well-injected hazardous wastes. The focus is on processes that (1) are known to occur in the deep-well environment or (2) have not been directly observed but are theoretically possible.

10.3.1   OVERVIEW OF FATE-INFLUENCING PROCESSES IN CHEMICAL SYSTEMS

10.3.1.1   Key Characteristics of Chemical Systems

A chemical system is a mixture of individual components. Chemical systems can be described by interactions that occur within the system and by the effect these processes have on the chemical composition and phases of the system. Interactions that change the chemical structure of system components are called chemical reactions. (Other interactions, such as processes that alter the solubility of system components, change the system without altering chemical structures.) Whether one reaction or a set of reactions occurs and how quickly the reaction proceeds are determined by the thermodynamics and kinetics of the system.

A substance may exist in one of three phases—solid, liquid, or gas. The mobility of a substance in the subsurface is influenced by which of several forms or species it may take. Species in deep-well-injection formations fall into six main categories[3]:

1. Free ions are surrounded only by water molecules and are very mobile in groundwater. Acid–base and dissolution reactions create free ions.
2. Species with low solubility in water may exist in solid form (e.g., Ag_2S, $BaSO_4$) or liquid form (e.g., chlorinated solvents). Precipitation reactions and immiscible-phase separation are important processes affecting this type of speciation.
3. Metal/ligand complexes (such as $Al[OH]^{2+}$, Cu–humate) and organic/ligand complexes tend to be mobile in groundwater.
4. Physically adsorbed species are immobile in groundwater but may be remobilized if replaced by other species with a stronger affinity to the solid surface.
5. Species held on a surface by ion exchange (such as calcium ions on clay) are also immobile in groundwater. As with physically adsorbed species, they may be replaced by ions with a greater affinity to the solid surface.
6. Species may differ by oxidation state: for example, manganese(II) and (IV); iron(II) and (III); and chromium(III) and (VI). Oxidation state is influenced by the redox potential. Mobility is affected because oxidation state influences precipitation–dissolution reactions and also toxicity in the case of heavy metals.

Dissolved species may be ionic or nonionic. In ionic species, an excess or shortage of electrons in the chemical structure creates a net positive or negative charge. In nonionic species, all negative and positive charges cancel each other out to form a neutral molecule. Cations are positively charged ions (Na^+, Ca^{2+}) and anions are negatively charged (SO_4^{2-}). The ability of a neutral substance to dissociate into ionic species is more common with inorganic substances than with organic substances. Acid–base reactions determine the distribution between ionic and nonionic species.

Neutral species may be nonpolar or polar. In nonpolar species, positively charged protons and negatively charged electrons are arranged in the molecular structure so as to create a uniform neutral charge on the molecule's surface. In polar species, the molecular structure creates chained poles on the molecule, even though the net charge is zero. Water (H_2O) is a polar molecule, with the positive pole on the side of the hydrogen atom and the negative pole on the side of the oxygen atoms. Nonpolar molecules tend to be hydrophobic (water-avoiding).

The thermodynamics of a system relate to the stability of substances within the system, that is, whether a reaction can occur. Kinetics relates to reaction mechanisms and rate, that is, how fast reactions can occur. Bedient and colleagues[29] reviewed the basic empirical equations defining zero-order, first-order, and second-order kinetic rate laws. An equilibrated state implies that as long as there are no significant changes in environmental factors affecting the system, the chemical composition and phases of the system will not change. Equilibrium does not necessarily imply that chemical processes cease. However, it does mean that for every reaction in one direction, a compensating reaction occurs in the opposite direction.

In nonequilibrium systems, chemical processes spontaneously alter the composition or phase of the system until equilibrium is attained. Simple systems, such as a mixture of sodium chloride and water, attain equilibrium quickly, whereas complex systems may reach equilibrium only after decades or eons.

The term "steady state" is sometimes used to describe chemical systems where thermodynamically unstable species exist but the rate of conversion to stable species is so slow that a quasi-equilibrated state exists. Because deep-well-injected wastes may be very complex chemical systems, the attainment of true equilibrium is uncertain.

Chemical reactions may result from interactions among and between the three phases of matter: solid, liquid, and gas. The major interactions that occur in the deep-well environment are those between different liquids (injected waste with reservoir fluids) and those between liquids and solids (injected wastes and reservoir fluids with reservoir rock). Although gases may exist, they are usually dissolved in liquid at normal deep-well pressures.

Two chemical properties important in predicting fate in the deep-well environment are homogeneity and reversibility. Chemical processes can be broadly classified as either homogeneous or heterogeneous and either reversible or irreversible.

Homogeneous reactions in the deep-well environment take place in only one phase (aqueous). These reactions generally occur uniformly throughout the phase and are easier to study and predict than heterogeneous reactions. Heterogeneous reactions (for example, adsorption) tend to occur at the interface between phases. Some reactions (such as precipitation) may result in phase changes. Heterogeneous reactions also tend to occur more actively at some locations in the chemical system than at others. Bacterial decomposition of wastes is a heterogeneous process that will be more active in locations with conditions favorable to organisms and less active elsewhere.

The reversibility of reactions is another important characteristic in assessing the fate of deep-well-injected wastes. Depending on environmental conditions, reversible reactions readily proceed in either or both directions. Most acid–base reactions exemplify reversible processes. In aqueous solutions, relatively minor changes in such factors as pH or concentration can change the direction of these reactions. Irreversible reactions, typified by hydrolysis, have a strong tendency to go in one direction only.

Table 10.3 lists the reversible and irreversible processes that may be significant in the deep-well environment.[3] The characteristics of the specific wastes and the environmental factors present in a well strongly influence which processes will occur and whether they will be irreversible. Irreversible reactions are particularly important. Waste rendered nontoxic through irreversible reactions may be considered permanently transformed into a nonhazardous state. A systematic discussion of mathematical modeling of groundwater chemical transport by reaction type is provided by Rubin.[30]

10.3.1.2 Fate-Influencing Processes in the Deep-Well Environment

At the simplest level, the processes that most influence geochemical fate can be divided into three groups: partition, transformation, and transport:

1. *Partition processes* affect the form or state of a specific chemical substance at a given time or under specific environmental conditions, but not its chemical structure or toxicity.

TABLE 10.3

Characteristics of Chemical Processes That May Be Significant in the Deep-Well Environment

Characteristic	Types of Reactions
Homogeneous	Acid–base, hydrolysis, hydration, neutralization, oxidation–reduction, polymerization, thermal degradation
Heterogeneous	Adsorption–desorption, precipitation-dissolution, immiscible-phase separation, biodegradation, complexation
Reversible	Acid–base, neutralization, oxidation–reduction (most inorganic and some biologically mediated), adsorption–desorption, precipitation-dissolution, complexation
Irreversible	Hydrolysis, oxidation–reduction (biodegradation of anthropogenic inorganics), immiscible-phase separation

Source: U.S. EPA, Assessing the Geochemical Fate of Deep-Well-Injected Hazardous Waste: A Reference Guide, EPA/625/6-89/025a, U.S. EPA, Cincinnati, OH, June 1990.

Thus, a substance may be in a solid form or in solution (described by the precipitation–dissolution process), but its toxicity remains unaltered regardless of form. The form or state of a substance, however, influences the transformation and transport processes that can occur. For this reason, partition processes are important to define in a fate assessment.

2. *Transformation processes* alter the chemical structure of a substance. In the deep-well environment, the transformation processes that may occur are largely determined by the conditions created by partition processes and the prevalent environmental factors. Transport processes do not need to be considered if transformation processes irreversibly change a hazardous waste to a nontoxic form.

3. *Transport processes* carry wastes through the subsurface environment and must be considered in a fate assessment if the interaction of partition and transformation processes does not immobilize or alter the hazardous waste. Waste migration can take place either in solution or in solid form (particle migration).

Table 10.4 presents the partition and transformation processes known to occur in the near-surface environment along with the special factors that should be considered when evaluating data in the context of the deep-well environment. Geochemical processes affecting hazardous wastes in deep-well environments have been studied much less than those occurring in near-surface environments (such as soils and shallow aquifers). Consequently, laboratory data and field studies for a particular substance may be available for near-surface conditions, but not for deep-well conditions.

As Table 10.4 shows, several processes can occur in both the near-surface and deep-well environments. For example, neutralization of acidic or alkaline wastes is a straightforward process, and although temperature differences between the two environments may need to be considered, no other factors make the deep-well setting distinctly different. The same holds true for oxidation–reduction (redox) processes.

The remaining processes, although they occur under near-surface and deep-well conditions, are less applicable to the latter. Distinct differences between the two environments, however, can lead to significant differences in how the processes affect a specific hazardous substance. Compared with the near-surface environment, the deep-well environment is characterized by higher temperatures, pressures, and salinity, and lower organic matter content and Eh (oxidation–reduction potential).

Table 10.5 lists the partition and transformation processes applicable in the deep-well environment and indicates whether they significantly affect the toxicity or mobility of hazardous wastes. None of the partition processes results in detoxification (decomposition to harmless inorganic constituents), but all affect mobility in some way. All transformation processes except complexation can result in detoxification; however, because transformation processes can create new toxic substances, the mobility of the waste can be critical in all processes except neutralization.

TABLE 10.4

Near-Surface Geochemical Processes and Their Relevance to the Deep-Well Environment

Process	Surface Data Applicability to Deep-Well Environment	Comments
Partition processes		
Acid–base equilibria	Partly	Near-surface studies tend to investigate fresh or moderately saline water, which creates quite different conditions for acid–base equilibria. Studies of ocean geochemistry come closest to approximating deep-well conditions.
Adsorption–desorption	Partly	Mechanisms for adsorption on similar materials will be similar. Soil adsorption data generally do not reflect the saturated conditions of the deep-well environment. Organic-matter content is a major factor affecting adsorption in the near-surface; its significance in the deep-well environment is less clear. Fate studies involving artificial recharge are probably useful, but differences between fresh waters and deep brines may reduce relevance.
Precipitation–dissolution	Partly	Higher temperatures, pressures, and salinity of the deep-well environment may result in significant differences between reactions in the two environments.
Immiscible-phase separation	No	Fluids (such as gasoline) that are immiscible in water are a significant consideration in near-surface contamination. Deep-well injection is limited to wastestreams that are soluble in water. Well blowout from gaseous carbon dioxide formation is an example of this process that is distinct to the deep-well environment.
Transformation processes		
Volatilization	No	No atmosphere.
Photolysis	No	No sunlight.
Biodegradation	Partly	Some near-surface bacteria appear capable of entering and surviving in the deep-well environment. However, in general, temperature and pressure conditions in the deep-well environment are unfavorable for microbiota that are adapted to near-surface conditions. Biological transformations are primarily anaerobic.
Complexation	Partly	Humic substances are very significant factors in near-surface complexation processes, probably less so in the deep-well environment. Data on complexation in saline waters are probably most relevant.
Hydrolysis	Partly	Basic processes will be the same. Higher salinity of deep-well environment may affect rate constants.
Neutralization	Partly	Basic process is the same, but some adjustments may be required for pressure/temperature effects.
Oxidation–reduction	Partly	The deep-well environment tends to be more reducing than the near-reduction surface environment, but equally reducing conditions occur in the near-surface. Some adjustments may be required for pressure/temperature effects.

Source: U.S. EPA, Assessing the Geochemical Fate of Deep-Well-Injected Hazardous Waste: A Reference Guide, EPA/625/6-89/025a, U.S. EPA, Cincinnati, OH, June 1990.

TABLE 10.5
Significance of Chemical Processes in the Deep-Well Environment

Process	Detoxification	Mobility	Biotic/Abiotic
Partitioning			
Acid–base equilibrium	No	Yes	Both
Adsorption–desorption	No	Yes	Abiotic
Precipitation–dissolution	No	Yes	Abiotic
Immiscible-phase separation	No	Yes	Both
Transformation			
Biodegradation	Yes	Yes	Biotic
Complexation	No	Yes	Abiotic
Hydrolysis	Yes	Yes	Both
Neutralization	Yes	No	Abiotic
Oxidation–reduction	Yes	Yes	Both

Source: U.S. EPA, Assessing the Geochemical Fate of Deep-Well-Injected Hazardous Waste: A Reference Guide, EPA/625/
6-89/025a, U.S. EPA, Cincinnati, OH, June 1990.

Table 10.5 also indicates whether a process is biotic (mediated or initiated by organisms in the environment), abiotic (not involving biological mediation), or both. Biotic processes are limited to environmental conditions that favor growth of mediating organisms. Abiotic processes occur under a wide range of conditions. Adsorption, precipitation, complexation, and neutralization are abiotic; all other processes in Table 10.5 may be either.

10.3.2 PARTITION PROCESSES

Partition processes determine how a substance is distributed among the liquid, solid, and gas phases and determine the chemical form or species of a substance. Partitioning usually does not affect the toxic properties of the substance. Partitioning can, however, affect the mobility of the waste, its compatibility with the injection zone, or other factors that influence fate in the deep-well environment. The major partition processes are as follows:

1. Acid–base reactions
2. Adsorption–desorption
3. Precipitation–dissolution
4. Immiscible-phase separation

10.3.2.1 Acid–Base Reactions

Acid–base reactions affect pH (the concentration of hydrogen ions in solution), which is a controlling factor in the type and rate of many other chemical reactions.

Acids dissociate in solution yielding hydrogen ions and anions according to the general reaction

$$\text{HA (neutral)} \longleftrightarrow \text{H}^+ \text{ (cation)} + \text{A}^- \text{ (anion)} \qquad (10.1)$$

The ionization is reversible. The anion (acting as a weak base) can recombine with the hydrogen ion to reform neutral HA. Both reactions occur continuously in solution, with the extent of ionization dependent on the strength of the acid. Strong acids, such as HCl, ionize completely in dilute aqueous solution. Thus a 0.01 molar (10^{-2} molar) solution has a pH of 2. Weak acids, such as acetic and other organic acids, ionize only slightly in solution and form solutions with pH from 4 to 6.

In the above example, the anion (A⁻) functions as a base when it combines with a hydrogen ion. (By definition, any substance that combines with hydrogen ions is a base. Like strong acids, strong bases ionize completely in a dilute aqueous solution.) Thus NaOH dissolves in water to form hydroxide ions, which in turn function as a base when they combine with hydrogen ions to form water, as shown by the general equations

$$MOH \longleftrightarrow M^+ + OH^- \qquad (10.2a)$$

$$OH^- + H^+ \longleftrightarrow H_2O \qquad (10.2b)$$

Strong acids (those that ionize completely in solution) are more likely to dissolve solids because charged particles such as hydrogen ions will interact more strongly with solids than will neutral particles. Weak acids do not readily donate hydrogen ions and consequently remain mostly in the neutral form. As a result, weak acids do not dissolve solids as readily as strong acids.

Strong bases (those that most readily extract hydrogen ions from solution) are also found predominantly in ionic forms and are similarly more reactive with solids than weak bases, which remain mostly in neutral form. The extent to which any base will extract hydrogen ions from solution depends on pH and the strength of the base. Acid–base reactions occur quickly. When the pH of a solution changes, acids and bases readily attain a new equilibrium between neutral and ionic forms. Because toxic organics almost always exist in very low concentrations and tend to be weak acids or weak bases, they have little, if any, influence on the pH of water. Acid–base equilibrium reactions involving hazardous organic compounds do not affect the toxicity of the waste and, as noted above, do not strongly influence pH.

When weak acids and bases ionize in wastestreams, pH is affected very little, but when strong acids and bases ionize in wastestreams, pH is affected dramatically. By definition, wastestreams having a pH ≤ 2 (highly acidic) or a pH ≥ 12.5 (strongly basic) are highly corrosive and are regulated as hazardous. Acid–base reactions can neutralize acidic or basic hazardous waste by raising or lowering its pH.

10.3.2.2 Adsorption and Desorption

Adsorption is a physicochemical process whereby ionic and nonionic solutes become concentrated from solution at solid–liquid interfaces.[31,32] Adsorption and desorption are caused by interactions between and among molecules in solution and those in the structure of solid surfaces. Adsorption is a major mechanism affecting the mobility of heavy metals and toxic organic substances and is thus a major consideration when assessing transport. Because adsorption is usually fully or partly reversible (desorption), only rarely can it be considered a detoxification process for fate-assessment purposes. Although adsorption does not directly affect the toxicity of a substance, the substance may be rendered nontoxic by concurrent transformation processes such as hydrolysis and biodegradation. Many chemical and physical properties of both aqueous and solid phases affect adsorption, and the physical chemistry of the process itself is complex. For example, adsorption of one ion may result in desorption of another ion (known as ion exchange).

Adsorption is typically exothermic (i.e., releases energy in the process of bonding), but can be endothermic, and can be classified into two groups, based on the energies involved: chemical adsorption and physical adsorption. Chemical adsorption is more significant for heavy metals, either in the form of ion exchange or interactions involving metal complexes.

In chemical adsorption (also called chemisorption), chemical bonds are formed between the adsorbate molecule and the adsorbent. These bonds typically involve energies on the order of 7 kcal/mol or greater.[33] These energies distinguish them from physical bonds, which typically involve energies less than 7 kcal/mol. Ion exchange, ligand exchange, protonation, and hydrogen bonds typically fall in the category of chemical bonds. Depending on the classification scheme used, numerous distinct types of chemical bonds have been identified in the laboratory under controlled conditions. Determining bonding mechanisms in the natural environment is much more difficult because of the diversity and complexity of adsorption surfaces.

10.3.2.3   Precipitation and Dissolution

Precipitation is a phase-partitioning process whereby solids separate from a solution.[34] Dissolution involves movement from the solid or gaseous phase to the aqueous phase. Solids dissolve into ions, whereas gases retain their original chemical structure when dissolved. The solubility of a compound (its tendency to dissolve in water or other solutions) is the main property affecting the precipitation–dissolution process.

The concentration of a compound in water is controlled by its equilibrium solubility or solubility constant (the maximum amount of a compound that will dissolve in a solution at a specified temperature and pressure). Equilibrium solubility will change with environmental parameters such as temperature, pressure, and pH; for example, the solubility of most organic compounds triples when temperature rises from $0°C$ to $30°C$. Each type of waste has a specific equilibrium solubility at a given temperature and pressure. The solubility of toxic organic compounds is generally much lower than that of inorganic salts. This characteristic is particularly true of nonpolar compounds because of their hydrophobic character.

Precipitation usually occurs when the concentration of a compound in solution exceeds the equilibrium solubility, although slow reaction kinetics may result in "supersaturated" solutions. For organic wastes in the deep-well environment, precipitation is not generally a significant partitioning process; in certain circumstances, however, it may need to be considered. For example, pentachlorophenol precipitates out of solution when the solution has a pH of <5,[35,36] and polychlorophenols form insoluble precipitates in water high in Mg^{2+} and Ca^{2+} ions.[37] Also, organic anions react with such elements as Ca^{2+}, Fe^{2+}, and Al^{3+} to form slowly soluble to nearly insoluble compounds.

Precipitation may be significant for heavy metals and other inorganic constituents in injected wastes. For example, sulfide ions have a strong affinity for metal ions, precipitating as metal sulfides. The dissolved constituents in injected wastes and reservoir fluids would not be in equilibrium with the *in situ* brines because of the fluids' different temperature, pH, and Eh. When the fluids are mixed, precipitation reactions can lead to injection-well plugging.

Coprecipitation is a partitioning process whereby toxic heavy metals precipitate from the aqueous phase even if the equilibrium solubility has not been exceeded. This process occurs when heavy metals are incorporated into the structure of silicon, aluminum, and iron oxides when these latter compounds precipitate out of solution. Iron hydroxide collects more toxic heavy metals (chromium, nickel, arsenic, selenium, cadmium, and thorium) during precipitation than aluminum hydroxide.[38] Coprecipitation is considered to effectively remove trace amounts of lead and chromium from solution in injected wastes at New Johnsonville, Tennessee.[39] Coprecipitation with carbonate minerals may be an important mechanism for dealing with cobalt, lead, zinc, and cadmium.

Dissolution of carbonates (acidic wastes), sand (alkaline wastes), and clays (both acidic and alkaline wastes) can neutralize deep-well-injected wastes.[39] Because precipitation–dissolution reactions are highly dependent on environmental factors such as pH and Eh, changes in one or more factors as a result of changes in injected-waste characteristics, or varying percentages of injected waste and reservoir fluids concentrations, may result in re-solution or reprecipitation of earlier reaction products. This sensitivity to environmental factors increases the complexity of predicting precipitation–dissolution reactions, because different equilibrium solubilities of a compound may exist in different parts of the injection zone depending on the proportions of waste and reservoir fluid. Similarly, a sequence of precipitation and dissolution reactions may take place at a given location of the injection zone as the concentration of injected wastes increases.

10.3.2.4   Immiscible-Phase Separation

An insoluble liquid or gas will separate from water, resulting in immiscible-phase separation. The behavior of nonaqueous-phase liquids (NAPLs) that may be lighter (LNAPLs) or denser (DNAPLs) than water is important in near-surface groundwater contamination studies.[40] However, aqueous-phase separation is generally not an issue in the deep-well environment because injected hazardous wastes are usually dilute. Failure to remove immiscible oily fluids from injected wastes potentially may cause plugging in the injection zone. Density and viscosity differences between injected and reservoir fluids, however, may need to be considered in transport modeling. Generally, pressures are high enough in the deep-well environment to keep gases such as carbon dioxide, generated as products of waste–reservoir interactions, in solution. Under certain conditions of high temperature and high waste concentrations, however, injected hydrochloric acid can cause carbon dioxide to separate from the liquid and produce a well blowout.

10.3.3 Transformation Processes

Transformation processes change the chemical structure of a compound. Because not all transformation processes convert hazardous wastes to nonhazardous compounds, geochemical fate assessment must consider both the full range of transformation processes that may occur and the toxicity and mobility of the resulting products. For deep-well-injected wastes, transformation processes and subsequent reactions may lead to one or more of the following:

1. Detoxification
2. Transtoxification
3. Toxification

Detoxification is an irreversible change in a substance from toxic to nontoxic form. For example, when an organic substance breaks down into its inorganic constituents, detoxification has taken place. Transtoxification occurs when one toxic compound is converted into another toxic compound. Toxification is the conversion of a nontoxic compound to a toxic substance. Table 10.6 lists some examples of each.

Transformation processes that may be significant in deep-well-injection fate assessments are as follows:

1. Neutralization
2. Complexation
3. Hydrolysis
4. Oxidation–reduction
5. Catalysis
6. Polymerization
7. Thermal degradation
8. Biodegradation

Two other processes that may transform hazardous wastes are photolysis and volatilization, but they are not considered here because they do not occur in the deep-well environment.

10.3.3.1 Neutralization

Acidic wastes with a pH of ≤ 2.0 and alkaline wastes with a pH of ≥ 12.5 are defined as hazardous (40 CFR Part 261). To meet the regulatory definition of nonhazardous, acidic wastes must be neutralized to a pH of >2.0 by reducing the hydrogen ion concentration, and alkaline wastes must be neutralized to a pH of ≤ 12.5 by increasing the hydrogen ion concentration.

Carbonates (limestone and dolomite) will dissolve in and neutralize acidic wastes with the following process:

$$CaCO_3 \longrightarrow Ca^{2+} + CO_3^{2-} \text{ (dissolution)} \tag{10.3}$$

$$CO_3^{2-} + 2H^+ \longrightarrow CO_2 + H_2O \text{ (neutralization)} \tag{10.4}$$

TABLE 10.6

Examples of the Effects of Transformation Processes on the Toxicity of Substances

Examples

Type of Transformation	Process
Detoxification	
Cyanide $\rightarrow$ amide $\rightarrow$ acids + ammonia	Hydrolysis
Cyanide $\rightarrow$ sulfate + carbon + nitrogen	Biooxidation
Nitrile $\rightarrow$ amide $\rightarrow$ acids + ammonia	Hydrolysis
Alkyl halide $\rightarrow$ alcohol + halide ion	Hydrolysis
Chlorobenzene $\rightarrow CO_2$ + Cl^- + H_2O	Biooxidation
1,3-Dichlorobenzene $\rightarrow CO_2$ + Cl^- + H_2O	Biooxidation
1,4-Dichlorobenzene $\rightarrow CO_2$ + Cl^- + H_2O	Biooxidation
Vinyl chloride $\rightarrow CO_2$ + Cl^- + H_2O	Bioreduction
Transtoxification	
2,4-D ester $\rightarrow$ 2,4-D acid (increased)	Hydrolysis
Phenol + formaldehyde $\rightarrow$ phenolic resins	Polymerization
Aldrin $\rightarrow$ dieldrin	Oxidation
DDT $\rightarrow$ DDD	Reduction
o-Xylene $\rightarrow$ o-toluic acid	Cometabolism
Benzene $\rightarrow$ phenol	Biooxidation
Carbon tetrachloride $\rightarrow$ chloroform $\rightarrow$ methylene chloride	Bioreduction
Ethylbenzene $\rightarrow$ phenylacetic acid	Cometabolism
1,1,1 -Trichloroethane $\rightarrow$ 1,1-dichloroethane $\rightarrow$ chloroethane	Bioreduction
Tetrachloroethylene $\rightarrow$ trichloroethylene $\rightarrow$ various dichloroethenes $\rightarrow$ vinyl chloride	Bioreduction
1,2-Dichloroethane $\rightarrow$ vinyl chloride	Hydrolysis
Inorganic mercury $\rightarrow$ methyl mercury	Bioreduction
Nitrilotriacetate $\rightarrow$ nitrosamines	Bioreduction
Toxification	
Amines $\rightarrow$ nitrosamines	Biooxidation

Source: U.S. EPA, Assessing the Geochemical Fate of Deep-Well-Injected Hazardous Waste: A Reference Guide, EPA/625/6-89/025a, U.S. EPA, Cincinnati, OH, June 1990.

When calcium carbonate goes into solution, it releases basic carbonate ions (CO_3^{2-}), which react with hydrogen ions to form carbon dioxide (which will normally remain in solution at deep-well-injection pressures) and water. Removal of hydrogen ions raises the pH of the solution. However, aqueous carbon dioxide serves to buffer the solution (i.e., re-forms carbonic acid in reaction with water to add H^+ ions to solution). Consequently, the buffering capacity of the solution must be exceeded before complete neutralization will take place. Nitric acid can react with certain alcohols and ketones under increased pressure to increase the pH of the solution, and this reaction was proposed by Goolsby[41] to explain the lower-than-expected level of calcium ions in backflowed waste at the Monsanto waste injection facility in Florida.

Quartz (SiO_2) and other silicates are generally stable in acidic solutions but will dissolve in highly alkaline waste solutions, decreasing the pH of the waste. The process by which this reaction occurs is complicated because it creates complex mixtures of nonionic and ionic species of silica. Scrivner and colleagues[39] discuss these reactions in some detail. They observe that the silicates in solution buffer the liquid. Also, laboratory experiments in which alkaline wastes have been mixed with sandstone have shown relatively small reductions in pH. At near-surface temperature and pressure conditions, an alkaline waste remains hazardous, but at simulated subsurface temperatures and pressures, the waste is rendered nonhazardous, ranging in pH from 11.5 to 12.4 in the experiments performed by Roy and colleagues.[33] However, the pH of the sandstone–waste mixture remained above 12.5 in other investigations, possibly because a higher solid/liquid ratio (less sandstone per volume of liquid) was used.

Reactions with clay minerals can neutralize both low-pH and high-pH solutions. Neutralization of acids occurs when hydrogen ions replace Al, Mg, and Fe. In alkaline solutions, neutralization is more complex and may involve cation exchange, clay dissolution, and reaction of cations with hydroxide ions to form new minerals called zeolites.[39]

10.3.3.2 Complexation

A complex ion is one that contains more than one ion. Because of its effect on mobility, complexation, the process by which complex ions form in solution, is very important for heavy metals and may be significant for organic wastes. Heavy metals are particularly prone to complexation because their atomic structure (specifically the presence of unfilled d-orbitals) favors the formation of strong bonds with polar molecules, such as water and ammonia (NH_3), and anions, such as chloride (Cl^-) and cyanide (CN^-). Depending on the chemistry of an injected waste and existing conditions, complexation can increase or decrease the waste's mobility.

Complexation is more likely in solutions with high ionic strength (which is typical of fluids found in the deep-well-injection environment). This is true because the large number of ions present in solution increases the number of chemical species that can form.[42] Many variables affect the stability of a complex ion relative to ions and metals that can serve as potential ligands to the central metal, the most important of which is the valence (charge) of the central cation and its radius. As a rule, the stability of complexes formed with a given ligand increases with cation charge and decreases with cation radius.[43]

The solubility of most metals is much higher when they exist as organometallic complexes.[44,45] Naturally occurring chemicals that can partially complex with metal compounds and increase the solubility of the metal include aliphatic acids, aromatic acids, alcohols, aldehydes, ketones, amines, aromatic hydrocarbons, esters, ethers, and phenols. Several complexation processes, including chelation and hydration, can occur in the deep-well environment.

10.3.3.3 Hydrolysis

Hydrolysis occurs when a compound reacts chemically with water (i.e., new chemical species are formed by the reaction), and can be a significant transformation process for certain hazardous wastes in the deep-well environment (see Table 10.7). Hydrolysis reactions fall into two major categories: replacement and addition. The rates at which these reactions occur are also significant in a fate assessment because some take so long to occur that they will not take place during the analytical time frame (10,000 years).

10.3.3.4 Oxidation–Reduction

Oxidation–reduction (redox) reactions involve the loss of electrons and increase in oxidation number (oxidation) by one substance or system, with an associated gain of electrons and decrease in oxidation number (reduction) by another substance or system. Thus for every oxidation reaction there must be a reduction reaction. The oxidation number of an atom represents the hypothetical charge an atom would have if the ion or molecule were to dissociate.[46,47]

Because redox reactions involve the transfer of electrons, the intensity of redox reactions is measured by electrical potential differences, termed Eh. Highly oxidizing conditions will have an Eh of about 0.8 V; highly reducing conditions will have an Eh of about −0.4 V. Eh is difficult to

TABLE 10.7

Listed Hazardous Organic Wastes for Which Hydrolysis May Be a Significant Transformation Process in the Deep-Well Environment

Group/Compound	Half-Life (d)
Pesticides	
DDT	81–4400
Dieldrin	3800
Endosulfan/endosulfan sulfate	21
Heptachlor	1
Halogenated aliphatic hydrocarbons	
Chloroethane (ethyl chloride)	38
1,2-Dichloropropane	180–700
1,3-Dichloropropene	60
Hexachlorocyclopentadiene	14
Bromomethane (methyl bromide)	20
Bromodichloromethane	5000
Halogenated ethers	
bis(Chloromethyl) ether	<1
2-Chloroethyl vinyl ether	1800
Monocyclic aromatics	
Pentachlorophenol	200
Phthalate esters	
Dimethyl phthalate	1200
Diethyl phthalate	3700
Di-*n*-butyl phthalate	7600
Di-*n*-octyl phthalate	4900

Source: U.S. EPA, Assessing the Geochemical Fate of Deep-Well-Injected Hazardous Waste: A Reference Guide, EPA/625/6-89/025a, U.S. EPA, Cincinnati, OH, June 1990.

measure accurately, and groundwater systems are often not in equilibrium with respect to redox reactions. Consequently, the Eh of a chemical system indicates the types of redox reactions that may occur rather than predicting the specific reactions that are occurring. In inorganic chemical systems, redox reactions tend to be reversible, whereas microbiologically mediated redox reactions involving hydrocarbons tend to be irreversible. Therefore, inorganic oxidation–reduction equilibria are somewhat analogous to acid–base equilibria. Examples of redox reactions are given in Table 10.8 and the relative oxidation states of organic groups are shown in Table 10.9.

10.3.3.5 Catalysis

The rate of many reactions increase in the presence of a catalyst, which itself remains unchanged in quantity and composition afterward. Although the catalyst itself is not transformed, the catalyst speeds up reactions that would occur naturally or promotes reactions that would not occur otherwise. For example, metal ions catalyze the hydrolysis and oxidation reactions in biochemical systems.[47] Phenol and phenol derivatives are normally resistant to oxidation in wastewaters, but the reaction can be accomplished by metal-ion catalysis when Fe^{2+}, Mn^{2+}, Cu^{2+}, and Co^{2+} are combined with chelating agents.[48,49] The reactions involved in destroying the aromatic ring in these compounds are complex and more likely to occur during waste pretreatment than as a result of processes in the deep-well environment. Certain metals in the presence of clays can also catalyze the polymerization of phenols and benzenes. Organic reactions that are catalyzed by clay minerals have been reviewed by Laszlo.[50]

TABLE 10.8

Redox Reactions in a Closed Groundwater System

Reaction	Equation
Aerobic respiration	$CH_2O + O_2 \rightarrow CO_2 + H_2O$
Denitrification	$5\ CH_2O + \text{nitrate}\ (4\ NO_3^-)\ +\ 4\ H^+ \rightarrow \text{nitrogen}\ (2\ N_2) + 5\ CO_2\ +\ 7\ H_2O$
Mn(IV) reduction	$CH_2O + 2\ MnO_2 + 4\ H^+ \rightarrow 2\ Mn^{2+} + CO_2 + 3\ H_2O$
Fe(III) reduction	$CH_2O + 8\ H^+ + 4\ Fe(OH)_3 \rightarrow 4\ Fe^{2+} + CO_2 + 11\ H_2O$
Sulfate reduction	$2\ CH_2O + \text{sulfate}\ (SO_4^{2-}) + H^+ \rightarrow HS^- + 2\ CO_2 + 2\ H_2O$
Methane fermentation	$2\ CH_2O + CO_2 \rightarrow \text{methane}\ (CH_4) + 2\ CO_2$
Nitrogen fixation	$3\ CH_2O + 3\ H_2O + 2\ N_2 + 4\ H^+ \rightarrow \text{ammonia}\ (4\ NH_4^+) + 3\ CO_2$

Source: U.S. EPA, Assessing the Geochemical Fate of Deep-Well-Injected Hazardous Waste: A Reference Guide, EPA/625/6-89/025a, U.S. EPA, Cincinnati, OH, June 1990.

10.3.3.6 Polymerization

Polymerization is the formation of large molecules (polymers) by the bonding together of many smaller molecules. For example, styrene polymerizes to form polystyrene. Polymerization can enhance the tendency of a substance to be adsorbed on mineral surfaces by increasing the molecular weight, but is not likely to result in detoxification of hazardous wastes.

Polar organic compounds such as amino acids normally do not polymerize in water because of dipole–dipole interactions. However, polymerization of amino acids to peptides may occur on clay surfaces. For example, Degens and Metheja[51] found kaolinite to serve as a catalyst for the polymerization of amino acids to peptides. In natural systems, Cu^{2+} is not very likely to exist in significant concentrations. However, Fe^{3+} may be present in the deep-well environment in sufficient amounts to enhance the adsorption of phenol, benzene, and related aromatics. Wastes from resin-manufacturing facilities, food-processing plants, pharmaceutical plants, and other types of chemical plants occasionally contain resin-like materials that may polymerize to form solids at deep-well-injection pressures and temperatures.

10.3.3.7 Thermal Degradation

Thermal degradation occurs when heat causes compounds to undergo structural changes, leading to the formation of simpler species. For example, many organophosphorus esters isomerize when heated

TABLE 10.9

Relative Oxidation States of Organic Functional Groups

		Functional Group Oxidation State		
-4	**-2**	**0**	**+2**	**+4**
Least Oxidized			**Most Oxidized**	
RH	ROH	RC(O)R	RCOOH	CO_2
	RCl	$(R)_2CCl_2$	$RC(O)NH_2$	CCl_4
RNH_2			$RCCl_3$	
	C=C	$-C\equiv C-$		

Source: U.S. EPA, Assessing the Geochemical Fate of Deep-Well-Injected Hazardous Waste: A Reference Guide, EPA/625/6-89/025a, U.S. EPA, Cincinnati, OH, June 1990.

and break down into component molecules. Temperatures and pressures common in the deep-well environment are normally too low to initiate high-temperature reactions, but if the right chemicals (not necessarily hazardous) are present, thermal degradation might be initiated. For example, thermal decarboxylation is probably the mechanism of acetate degradation in oilfield waters[52] where temperatures exceed 200°C; however, injection zones usually do not reach this temperature. At depths of 900 m (approximately 3000 ft), temperatures range from 50°C to 100°C.[33]

Smith and Raptis[53] have suggested using the deep-well environment as a wet-oxidation reactor for liquid organic wastes. This process, however, does not involve deep-well injection of wastes but rather uses temperatures and pressures in the subsurface to increase the oxidation rate of organic wastes, which are then returned to the surface.

10.3.3.8 Biodegradation

Biotransformation is the alteration of a compound as a result of the influence of organisms. It is one of the most prevalent processes causing the breakdown of organic compounds in the near-surface environment. Biodegradation is a more specific term used to describe the biologically mediated change of a chemical into simpler products. The term includes, and sometimes obscures, a series of distinctive processes of toxicological significance in natural ecosystems. Biodegradation is probably more significant in the decomposition of the nonhazardous components of deep-well-injected organic wastes, although a few hazardous compounds, such as acrylonitrile and some monocyclic aromatic hydrocarbons and halogenated aliphatics, may be subject to biodegradation in the deep-well environment.

Microorganisms are by far the most significant group of organisms involved in biodegradation. They can mineralize (convert to CO_2 and H_2O) many complex organic molecules that higher organisms, such as vertebrates, cannot metabolize. They are often the first agents in biodegradation, converting compounds into the simpler forms required by higher organisms. Most biodegradation in near-surface environments is carried out by heterotrophic bacteria (microorganisms that require organic matter for energy and oxygen).[54]

Biodegradation in deep-well environments is performed predominantly by anaerobic microorganisms, which do not consume oxygen and are either obligate (oxygen is toxic to the organism) or facultative (the organism can live with or without oxygen or prefers a reducing environment). The two main types of anaerobic bacteria, methanogenic (methane-producing) and sulfate-reducing do not degrade the same compounds. The byproducts of sulfate reduction are hydrogen sulfide, carbon dioxide, and water. Methanogenic bacteria produce methane and carbon dioxide (see Table 10.9). The extent to which either type proliferates is strongly influenced by pH. As a group, anaerobic organisms are more sensitive and susceptible to inhibition than aerobic bacteria. Typically, aerobic degradation is also more efficient than anaerobic degradation, and high temperatures are not as limiting for aerobes as for anaerobes.[54]

Alexander[55] identifies six major kinds of biodegradation: mineralization, cometabolism, detoxification, transtoxification, activation, and defusing. Table 10.10 describes each of these processes and gives examples.

For several reasons, mineralization (decomposition to inorganic constituents) is generally a more effective form of biodegradation than cometabolism (conversion to another compound without using the original compound for energy or growth). First, detoxification is more likely to occur during mineralization. Second, mineralizing populations will increase until the compound is completely degraded, because they use the compound as a source of energy. In contrast, cometabolized compounds tend to change slowly, and the original compound and its reaction products tend to remain in the environment because the cometabolized compounds are not used for energy.

Almost all the specific chemical reactions in biodegradation can be classified as oxidation–reduction, hydrolysis, or conjugation. Hydrolysis and oxidation–reduction have been discussed before. Conjugation involves the addition of functional groups or a hydrocarbon moiety to an organic

TABLE 10.10
Descriptions of the Major Types of Biological Transformation Processes

Process	Description
Mineralization	The complete conversion of an organic compound to inorganic constituents (water, carbon dioxide). Generally results in complete detoxification unless one of the products is of environmental concern, such as nitrates and sulfides under certain conditions.
Cometabolism	Conversion of an organic compound to another organic compound without the microorganism using the compound as a nutrient. Resulting compounds may be as toxic (DDT to DDE or DDD) or less toxic (xylenes to toluic acid).
Detoxification	Conversion of a toxic organic compound to a nontoxic organic compound. The pesticide 2,4-D can be detoxified microbially to 2,4-dichlorophenol.
Transtoxification	Conversion of a toxic compound to another toxic compound with similar, increased, or reduced toxicity.
Activation	Conversion of a nontoxic molecule to one that is toxic, or a molecule with low potency to one that is more potent. Examples include the formation of the phenoxy herbicide 2,4-D from the corresponding butyrate, formation of nitrosamines, and methylation of arsenicals to trimethylarsine.
Defusing	Conversion of a compound capable of becoming hazardous to another nonhazardous compound by circumventing the hazardous intermediate. This has been observed in the laboratory, but not identified in the environment. An example is the direct formation of 2,4-dichlorophenol from the corresponding butyrate of 2,4-D.

Source: U.S. EPA, Assessing the Geochemical Fate of Deep-Well-Injected Hazardous Waste: A Reference Guide, EPA/625/6-89/025a, U.S. EPA, Cincinnati, OH, June 1990.

molecule or inorganic species. For example, conjugation occurs when microbial processes transform inorganic mercury into dimethyl mercury.

At least 26 oxidative, 7 reductive, and 14 hydrolytic transformations of pesticides had been identified. Detailed identification and discussion of specific reactions can be found in the works of Alexander[56] and Scow.[57]

10.3.4 Transport Processes

Many factors and processes must be considered when evaluating the movement of deep-well-injected hazardous wastes. Four factors are relevant to geochemical characteristics:

1. Hydrodynamic dispersion
2. Osmotic potential
3. Particle migration
4. Density and viscosity

10.3.4.1 Hydrodynamic Dispersion

Hydrodynamic dispersion refers to the net effect of a variety of microscopic, macroscopic, and regional conditions that affect the spread of a solute front through an aquifer.[58] Quantifying the dispersion is important to fate assessment because contaminants can move more rapidly through an aquifer by this process than would be predicted by simple plugflow (i.e., uniform movement of water through an aquifer with a vertical front). In other words, physical conditions (such as more-permeable zones, where water can move more quickly) and chemical processes (e.g., movement of dissolved species at greater velocities than the water moves by molecular diffusion) result in more rapid movement of contaminants than would be predicted by groundwater equations for physical flow, which must assume average values for permeability. Dispersion on the microscopic scale is caused by the following:

1. Velocity variations resulting from variations in pore geometry and the fact that water velocity is higher in the center of a pore space than that for water moving near the pore wall
2. Molecular diffusion along concentration gradients
3. Variations in fluid properties such as density and viscosity

Dispersion on the macroscopic scale is caused by variations in hydraulic conductivity and porosity, which create irregularities in the seepage velocity with consequent mixing of the solute. Finally, over large distances, regional variations in hydrogeologic units can affect the amount of dispersion. In hydrogeologic modeling, the hydrodynamic dispersion coefficient D is often expressed as the sum of a mechanical dispersion coefficient D_m and molecular (Fickian) diffusion D^*.

In most instances, hydrodynamic dispersion is not great enough to require detailed consideration in hydrogeologic modeling for fate assessment of deep-well-injected wastes. However, regional variations (such as the presence of an USDW in the same aquifer as the injection zone, as is the case in parts of Florida) should be evaluated before a decision is made to exclude it.

10.3.4.2 Osmotic Potential

Osmotic potential refers to the energy required to pull water away from ions in solution that are attracted to the polar water molecules. In the presence of a semipermeable membrane between two solutions, water molecules will move through the membrane to the side with the higher concentration. This property may be important to fate assessment because in the deep-well environment, shales that serve as confining layers can act as semipermeable membranes if the injected waste significantly changes the solute concentrations.[59] In laboratory experiments, Kharaka[60] found that retardation sequences across geologic membranes varied with the material, but that monovalent and divalent cations generally followed identical sequences: $Li^+ < Na^+ < NH_4^+ < K^+ < Rb^+ < Cs^+$ and $Mg^{2+} < Ca^{2+} < Sr^{2+} < Ba^{2+}$.

If osmotic effects are possible, several other effects would need to be considered in a geochemical-fate assessment, depending on whether the solute concentration is increased or decreased. If solute concentrations are increased, pressures associated with injection would increase beyond those predicted without osmotic effects. Also, the movement of ions to the injection zone from the aquifer with lower salinity (above the clay confining layer) would increase the salinity above those levels predicted by simple mixing of the reservoir fluid and the injected wastes. This action could affect the results of any geochemical modeling.

If solute concentrations are decreased, the remote possibility exists that wastes would migrate through the confining layer. For this to occur, solute concentrations above the confining layer would have to be higher than those in the injection zone, and movement, in any event, would be very slow. As USDWs have salinities less than 10,000 mg/L, compared with typical salinities in injection zones of 20,000 to 70,000 mg/L, even if this process were to occur it would cause migration only to overlying aquifers that are not USDWs.

10.3.4.3 Particle Migration

Particle migration can occur when the mixing of incompatible fluids mobilizes clays or very fine particles precipitate out of solution. This process is most likely to occur when solutions with low concentrations of salts are mixed with reservoir fluids containing high concentrations, or when highly alkaline solutions dissolve silica and release fines. This type of reaction is of concern primarily when it occurs near the injection zone, because particle migration can clog pores and drastically reduce permeability. McDowell-Boyer and colleagues[61] provide a good review of the literature on subsurface particle migration.

TABLE 10.11
Physical Parameters Affecting Particle Migration in Porous-Media Flow

Parameter	Significance
Matrix	
Porosity	Indicates voids; space available for retention of clogging material.
Particle size for which 10% of the matrix is smaller than that size	Termed the effective size for filter sands.
Particle size for which 60% of the matrix is smaller than that size	The ratio of the 60% size to the 10% size is an indicator of the uniformity.
Bulk density	For a given material, indicates the closeness of packing and propensity for material movement under stress.
Specific surface area	Relates to surface-active phenomena and adsorption rate.
Grain shapes	Affects shape of pores and thus fluid-flow patterns.
Surface roughness of grains	Affects retention of suspension on the particle surface.
Pore-diameter size and size distribution	Propensity for entrapment or filtration of suspension.
Surface charge of grains	Negatively charged surface grains will attract a suspended particle with a positive charge.
Fluid	
Viscosity	Shear forces and fluid resistance to flow.
Density	Mixing effects when different densities are involved; may affect direction and rate of flow.
Velocity of flow	Hydrodynamic forces on the medium and suspension.
Pressure	Driving force moving the liquid and suspension into and through the medium.
Suspended particles	
Concentration (inflow, within medium, outflow)	Material available for inflow, retention, and through-flow.
Size	Ability to pass through pore openings.
Shape	Effect on retention or through-flow due to orientation.
Electric charge	Attraction or repulsion to medium or intermediate materials.

Source: U.S. EPA, Assessing the Geochemical Fate of Deep-Well-Injected Hazardous Waste: A Reference Guide, EPA/625/ 6-89/025a, U.S. EPA, Cincinnati, OH, June 1990.

It is possible for complex metals ions that are adsorbed onto very small particles of clay to migrate as metal-clay particles. Laboratory experiments found that radioisotope-clay particles at a low salinity were retained in a sand core, but passed through it at a high salinity.[44] Clay-metal particles would not be expected to travel long distances in deep-well reservoir rocks because the pores would be too small.

Injection of highly acid or alkaline wastes has the potential to dissolve some reservoir rock to create channels that would allow more distant transport of small particles. Table 10.11 summarizes the various physical parameters that affect particle migration in porous-media flow.

10.3.4.4 Density/Viscosity Differences

Wastes having different densities or viscosities (tendency to resist internal flow) than the injection zone fluids will tend to concentrate in the upper (lower density/viscosity) or lower (higher density/ viscosity) portions of the injection zone. Frind[62] and Larkin and Clark[63] examined the basic requirements for the mathematical simulation of density-dependent transport in groundwater. Miller and colleagues[64] described a density-driven flow model designed specifically for evaluating the potential for upward migration of deep-well-injected wastes.

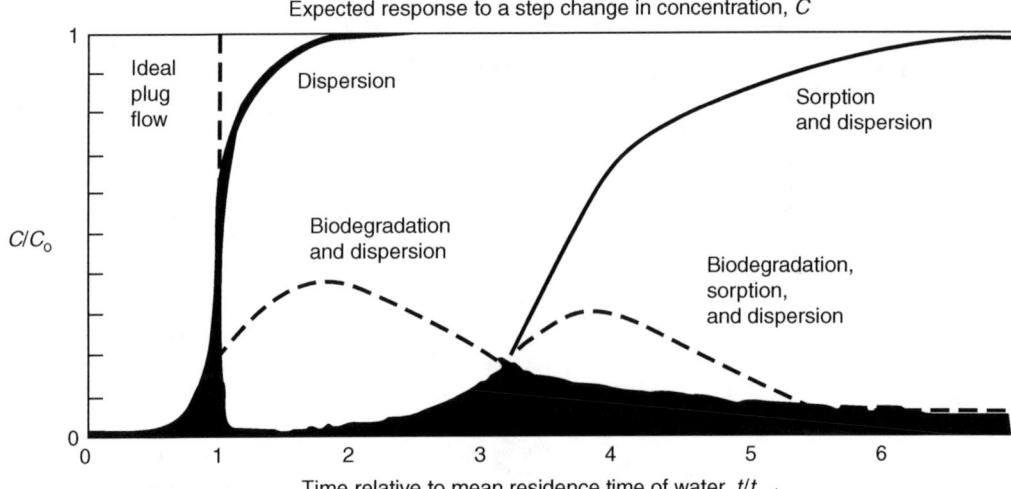

FIGURE 10.3 Effects of dispersion, adsorption, and biodegradation on the time change in concentration of an organic compound in an aquifer observation well. (From U.S. EPA, Assessing the Geochemical Fate of Deep-Well-Injected Hazardous Waste: A Reference Guide, EPA/625/6-89/025a, U.S. EPA, Cincinnati, OH, June 1990.)

10.3.5 INTERACTION OF PARTITION, TRANSFORMATION, AND TRANSPORT PROCESSES

The actual movement of a specific deep-well-injected hazardous substance depends on the types of processes that act on the waste and on the ways in which different processes interact. Figure 10.3 shows the expected change in concentration over time of a deep-well-injected organic compound in an observation well at an unspecified distance from the original point of injection.

With only dispersion operating, low concentrations are observed before the arrival of a fluid exhibiting ideal plug flow, but dispersion also serves to delay the time it takes for 100% of the initial concentration to be observed. Adsorption combined with dispersion delays the arrival of the compound, and eventually the contaminant will reach full concentration when adsorption capacity is reached. When biodegradation occurs, initial concentrations might well be governed by dispersion alone, until sufficient time has passed for an acclimated bacterial population to establish itself and become large enough to change the organic concentration significantly. If this occurs, the concentration would decrease and level out at some minimum value. When adsorption acts with biodegradation, the arrival of the contaminant is delayed, as with adsorption alone; then the concentration of the contaminant rises to a maximum level below that of the original concentration and declines as biodegradation becomes active.

10.4 ENVIRONMENTAL FACTORS AFFECTING DEEP-WELL-INJECTION GEOCHEMICAL PROCESSES

Environmental conditions determine in large part the chemical reactions that will occur when waste is injected. For example, precipitation–dissolution reactions are strongly controlled by pH. Thus, iron oxides, which may be dissolved in acidic wastes, may precipitate when injection-zone mixing increases the pH of the waste. Similarly, redox potential (Eh) exerts a strong control on the type of microbiological degradation of wastes.

The most variable and site-specific factor is the reservoir rock matrix. Geologic formations vary greatly in chemical and physical properties depending on the conditions under which they formed and the geologic processes to which they have been subjected.

10.4.1 Major Environmental Factors Influencing Geochemical-Fate Processes

The previous chapter examined the geochemical processes that can occur in the deep-well environment. The type and outcome of reactions that will actually occur when a waste is injected, however, depend on its chemical characteristics and on injection-zone conditions. This chapter examines six major environmental factors that must be taken into consideration.

10.4.1.1 pH

The pH of a system greatly influences what chemical processes will occur in the deep-well environment. Directly or indirectly, pH also affects most of the other environmental factors. Table 10.12 summarizes the significance and some major effects of changes in pH on chemical processes and environmental factors in the deep-well environment.

TABLE 10.12

Effects of pH on Deep-Well Geochemical Processes and Other Environmental Factors

Process/Factor	Significance of pH
Partition processes	
Acid–base	Measures acid–base reactions. Strong acids (bases) will tend to change pH; weak acids (bases) will buffer solutions to minimize pH changes.
Adsorption–desorption	Strongly influences adsorption, because hydrogen ions play an active role in both chemical and physical bonding processes. Mobility of heavy metals is strongly influenced by pH. Adsorption of some organics is also pH-dependent.
Precipitation–dissolution	Strongly influences precipitation–dissolution reactions. Mixing of solutions with different pH often results in precipitation reactions. See also reservoir matrix below.
Transformation processes	
Complexation	Strongly influences positions of equilibria involving complex ions and metal-chelate formation.
Hydrolysis	Strongly influences rates of hydrolysis. Hydrolysis of aliphatic and alkylic halides optimum at neutral to basic conditions.[43] Other hydrolysis reactions tend to be faster at either high or low pH.[186]
Oxidation–reduction	Redox systems generally become more reducing with increasing pH.[74]
Environmental factors	
Biodegradation	In combination with Eh, pH strongly influences the types of bacteria that will be present. High- to medium-pH, low-Eh environments will generally restrict bacterial populations to sulfate reducers and heterotrophic anaerobes.[187] In reducing conditions, pH strongly affects whether methanogenic or sulfate-reducing bacteria predominate.[43]
Eh	Increasing pH generally lowers Eh.
Salinity	pH-induced dissolution increases salinity; pH-induced precipitation decreases salinity.
Reservoir matrix	Acidic solutions tend to dissolve carbonates and clays; highly alkaline solutions tend to dissolve silica and clays. Greater pH generally increases cation-exchange capacity of clays.
Temperature	pH-driven exothermic (heat-releasing) reactions will increase fluid temperature; pH-driven endothermic (heat-consuming) reactions will decrease fluid temperature.
Pressure	Will not influence pressure unless pH-induced reactions result in a significant change in the volume of reaction products.

Source: U.S. EPA, Assessing the Geochemical Fate of Deep-Well-Injected Hazardous Waste: A Reference Guide, EPA/625/6-89/025a, U.S. EPA, Cincinnati, OH, June 1990.

Very small changes in acidity greatly affect chemical reactions and the form of chemical species in solution. For example, the hydrolysis half-life of hydrogen cyanide is greater than 100,000 years at pH 4 but drops to about 10 years at pH 9.[39]

Buffer capacity is a measure of how much the pH changes when a strong acid or base is added to a solution. A highly buffered solution will show little change; conversely, the pH of a solution with low buffering capacity will change rapidly. Weak acids or bases buffer a solution, and the higher their concentration in solution, the greater the buffering capacity. Alkalinity (usually expressed in calcium carbonate equivalents required to neutralize acid to a specified pH) is a measure of the buffering capacity of a solution.[65]

Acid–base reactions of buffers act either to add or to remove hydrogen ions to or from the solution so as to maintain a nearly constant equilibrium concentration of H^+. For example, carbon dioxide acts as a buffer when it dissolves in water to form carbonic acid, which dissociates to carbonate and bicarbonate ions:

$$CO_2 + H_2O \longrightarrow H_2CO_3 \longleftrightarrow HCO_3^- + H^+ \longleftrightarrow CO_3^{2-} + 2H^+ \tag{10.5}$$

At equilibrium, the concentration of H^+ will remain constant. When a strong acid (represented by H^+ or HA) is introduced into solution, the concentration of H^+ is increased. The buffer compensates by reacting with the excess H ions, moving the direction of the above reaction to the left. By combining with bicarbonate and carbonate ions to form the nonionic carbonic acid, equilibrium is reestablished at a pH nearly the same as that existing before. The buffer capacity in this case is determined by the total concentration of carbonate and bicarbonate ions. When no more carbonate or bicarbonate ions are available to combine with excess H^+ ions, the buffer capacity has been exceeded and pH will change dramatically upon addition of further acid.

10.4.1.2 Eh and Other Redox Indicators

The term Eh, which is the oxidation–reduction potential (often referred to as redox potential), is an expression of the tendency of a reversible redox system to be oxidized or reduced. It is especially significant in its influence on biodegradation processes. The energy of oxidation (electron-escaping tendency) present in a reversible oxidation–reduction system (in volts [V] or millivolts [mV]) is measured as the potential difference between a standard hydrogen electrode and the system being measured. Large positive values (up to ca. +800 mV) indicate an oxidizing tendency, and large negative values (down to ca. –500 mV) indicate a strong reducing tendency. Eh values of +200 mV and lower indicate reducing conditions in near-surface soils and sediments.[16]

The Eh of connate waters (water entrapped in the interstices of sediment at the time of deposition) ranges from 0 to –200 mV. For example, formation water from two monitoring wells in the lower limestone of the Florida aquifer near Pensacola ranged from +23 to –32 mV,[67] and formation fluids from a Devonian limestone in Illinois used for injection at a depth of about 3200 ft had an Eh of –154 mV.[16]

Several measures of organic pollutant loading to waters have been developed to indicate the redox status of a system:

1. Biochemical oxygen demand (BOD)
2. Chemical oxygen demand (COD)
3. Total organic carbon (TOC)
4. Dissolved organic carbon (DOC)
5. Suspended organic carbon (SOC)

When values for any of these parameters are high, oxygen is rapidly depleted in groundwaters and reducing conditions will develop. BOD and COD were designed to measure oxygen consumption during the microbial degradation of municipal sewage. They are only semiquantitative indicators of organic loading because measurement procedures for these parameters have no direct geochemical significance.[65] Malcolm and Leenheer[68] recommend the use of DOC and SOC, which are independent of microbial effects, toxic substance, and variability with diverse organic constituents. TOC, when measured as a single parameter (rather than as the sum of DOC and SOC), provides less information for geochemical interpretation.

Reducing conditions predominate in the deep-well environment for several reasons:

1. No source of oxygen replenishment exists.
2. Higher temperatures in the deep-well environment are associated with decreases in Eh.
3. Neutral to slightly alkaline water in the deep-well environment favors lower Eh values.

Deep-well injection of wastes can change, at least temporarily, the Eh of the injection zone. For example, Ragone and coleagues[69] observed a change from reducing to oxidizing conditions when tertiary-treated sewage (reclaimed water) was injected into the Magothy aquifer, Long Island, NY, at a depth of 400 ft. The reclaimed water had 6.6 mg/L dissolved oxygen compared with no dissolved oxygen in the formation water. On the other hand, the Eh of an acidic waste dropped dramatically, from +800 mV to ca. +100 mV, when mixed with siltstone under conditions of low oxygen and simulated deep-well temperature and pressure.[67] Similarly, the Eh of an alkaline waste dropped from +600 mV to ca. +200 mV.[67]

10.4.1.3 Salinity and Specific Conductance

Salinity is defined as the concentration of total dissolved solids (TDS) in a solution, usually expressed in mg/L. The TDS concentration in water is usually determined from the weight of the dry residue remaining after evaporation of the volatile portion of the original solution. Groundwater may be classified into four salinity classes[64]:

1. Slightly saline (1000 to 3000 mg/L)
2. Moderately saline (3000 to 10,000 mg/L)
3. Very saline (10,000 to 35,000 mg/L)
4. Brine (more than 35,000 mg/L) (seawater is about 35,000 mg/L)

Water with a salinity of less than 10,000 mg/L is considered to be a potential underground source of drinking water. By regulatory definition, deep-well injection of hazardous waste can occur only in very saline waters or brines. Actual salinities of waters in currently used deep-well injection zones vary greatly.[70] Normally, the term brine is used to refer to the natural waters in deep-well injection zones. As noted above, however, this term is not technically correct if TDS levels are less than 35,000 mg/L.

Solutions of substances that are good conductors of electricity are called electrolytes. Sodium chloride, the major constituent of seawater, is a strong electrolyte. Most salts, as well as strong acids and bases, are strong electrolytes because they remain in solution primarily in ionic (charged) forms. Weak acids and bases are weak electrolytes because they tend to remain in nonionic forms. Pure water is a nonconductor of electricity.

The conductivity of solutions is measured as specific conductance, which may be expressed as μmhos/cm or mmhos/cm at 25°C. Seawater has a specific conductance of about 50 mmhos/cm. Salinity shows a high correlation with specific conductance at low to moderate TDS levels, but the concentrations of ions in brines are so high that the relationship between concentration and conductance becomes ill-defined.[64]

10.4.1.4 Reservoir Matrix

With few, if any, exceptions, deep-well injection zones will be sedimentary rock, and the reactions that take place when hazardous wastes are injected are determined largely by the physical and chemical properties of that rock. The most important physical properties of sedimentary rocks in relation to deep-well geochemical interactions are texture (the proportions of different sized particles in sediment) and specific surface area. The most important chemical property is mineralogy, defined by the types and proportions of minerals present.

10.4.1.5   Temperature and Pressure

Temperature and pressure are the primary influences on the rate of chemical reactions. Both temperature and pressure increase with depth below the Earth's surface. Consequently, temperatures and pressures in the deep-well environment are significantly higher than those in the near-surface environment.

Geothermal gradients in the subsurface typically range from 1°C per 15 m (50 ft) to 1°C per 45 m (150 ft), with most regions having a gradient of around 1°C per 30 m (100 ft). Tables giving data on temperature gradients for 679 wells located in 23 states can be found in reference 71. Temperature can vary greatly at the same depth in different locations. For example, temperatures at approximately the same depth in Florida differ by almost 26°C.

The velocity of most acid–base and dissolution reactions increases as temperature increases. Higher temperatures generally also increase the rate of redox reactions; however, the effect is difficult to predict exactly because the interactions among competing reactions may offset the effect of the increase. In contrast, higher temperatures usually decrease the amount and rate of adsorption, because these reactions are generally exothermic (heat-producing). An exception has been noted by Choi and Aomine,[72] who found that adsorption rates of pentachlorophenol on soil increase 6% to 12% when samples of three different soils are subjected to an increase in temperature from 4°C to 33°C. Adsorption decreased by 9% in a fourth sample. Laboratory adsorption experiments at constant, simulated deep-well pressure with phenol and 1,2-dichloroethane result in decreased adsorption with increased temperature.[73]

Greater pressures tend to decrease the growth and survival of bacteria, but for certain species increased temperature counters this effect. For example, the growth and reproduction of *E. coli* essentially stops in nutrient cultures at 20°C and 400 atm (40.5 MPa). When the temperature is increased to 40°C, however, growth and reproduction are about the same as at near-surface conditions.[74]

10.4.2   Geochemical Characteristics of Deep-Well-Injection Zones

This section provides information on the range of environmental conditions that occur in deep-well-injection zones in different geologic regions of the U.S. The section on lithology discusses the types of sedimentary formations that are suitable for deep-well injection and confining layers and provides some information on geologic formations that are used for deep-well injection of wastes. The section on brine chemistry discusses the typical range of chemical characteristics of formation waters found in injection zones.

10.4.2.1   Lithology

Rock that can be mapped over a large area based on mineralogy, fossil content, or other recognizable characteristic is called a formation. The lithology (texture and mineralogy) of a geologic formation influences its suitability for deep-well injection. Sedimentary carbonates and sandstones usually have suitable geologic and engineering characteristics for disposal of hazardous wastes by deep-well injection. These characteristics include sufficient porosity, permeability, thickness, and extent to permit use as a liquid-storage reservoir at safe injection pressures.[75] In 1981, 62% of the injection wells in the U.S. were drilled into two types of reservoir rocks, either consolidated sandstone or unconsolidated sands that had not yet been altered by cementation to form strongly cohesive sandstone. The latter were usually of Tertiary age. At that time (1981), 34% of all wells used limestones and dolomites as reservoir rock and 4% used miscellaneous formations.

   Sedimentary-rock formations that overlie the injection formation are called confining layers. To prevent injected wastes from migrating to higher strata or to potential underground sources of drinking water, a confining layer must have certain geologic and engineering characteristics[3,76,77]:

1. Sufficient thickness and area to prevent upward migration of wastes
2. Low porosity and permeability and the ability to maintain low porosities and permeabilities when interacting with wastes that may dissolve minerals through neutralization
3. Lack of natural continuous fracturing or faulting, and resistance to artificial fracturing in response to injection pressures
4. No abandoned unplugged or improperly plugged wells

   Sedimentary rocks that are most likely to meet the first three criteria are unfractured shale, clay, siltstone, anhydrite, gypsum, and salt formations. Massive limestones and dolomites (i.e., carbonates with no continuous fracturing and solution channels) can also serve as confining layers. Their suitability must be determined on a case by case basis. The fourth criterion has no relationship to lithology.

   Formations from all geologic periods have been used for deep-well injection, but Paleozoic rocks are used for most injection zones (53%), followed by Tertiary-age formations (39%). Older Paleozoic rocks have been more frequently used for injection primarily because they tend to be more deeply buried. However, the more recent Tertiary-age Gulf Coast sediments are also very thick, and most injection in rocks of this age takes place there.

   Figure 10.4 provides a general indication of site suitability based on geologic factors.

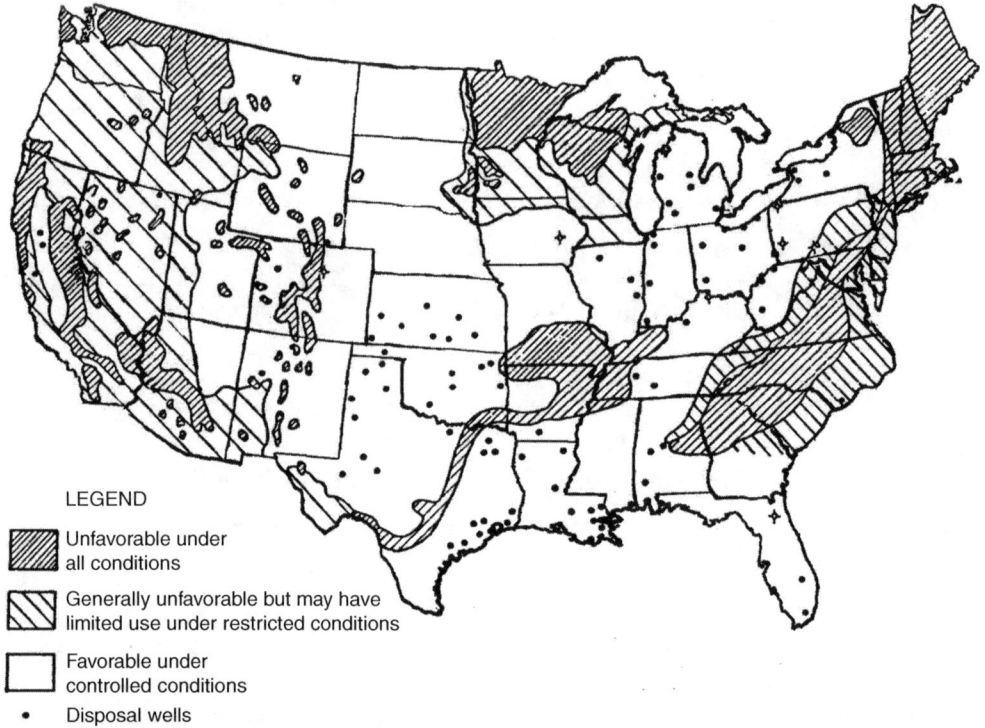

LEGEND

Unfavorable under all conditions

Generally unfavorable but may have limited use under restricted conditions

Favorable under controlled conditions

• Disposal wells
✛ Abandoned or plugged disposal wells

FIGURE 10.4   Site suitability for deep-well injection and locations of industrial waste disposal wells. (From U.S. EPA, Assessing the Geochemical Fate of Deep-Well-Injected Hazardous Waste: A Reference Guide, EPA/625/6-89/025a, U.S. EPA, Cincinnati, OH, June 1990.)

10.4.2.2 Brine Chemistry

Brines are classified according to their chemical constituents. At least nine distinct types are recognized by petroleum geologists, but most brines encountered in injection operations are either Na-Cl or Na-Ca-Cl brines.[78] None is similar to seawater, and the geochemical mechanisms by which such brines develop are not well-understood. Three mechanisms have been proposed to explain the high concentrations of dissolved solids and the chemical composition of brines, but at present there is no consensus on their relative importance in explaining brine chemistry.[78] The dominant mechanism at work in a deep-well environment has important implications for the hydrodynamic conditions affecting the movement of injected wastes. The mechanisms and their implications are summarized in Table 10.13. The salinity, pH, and chemical composition of the very saline and briny waters into which hazardous wastes are injected can vary greatly, both among geologic basins and within a single formation.

The maximum salinities in the Tertiary section of the Gulf of Mexico basin (the most extensively used strata for deep-well injection) reach almost four times that of seawater. The Michigan basin has the highest salinity, reaching 400,000 mg/L TDS, more than 11 times that of seawater. In Florida, however, where seawater circulates through the Floridan aquifer, maximum salinities tend to be controlled by the salinity of the seawater.[79]

The Frio formation, in Texas, receives more hazardous waste by volume through deep-well injection than any other geologic formation in the U.S. The average salinity of this formation is about twice that of seawater (72,185 mg/L TDS), but individual samples range from a low of 10,528 mg/L TDS (barely above the salinity cutoff for potential USDWs) to a high of more than 118,000 mg/L TDS. Data from sites in Illinois and North Carolina indicate the presence of very saline water (around 20,000 mg/L TDS, but still less saline than seawater).

The pH of formation waters in the Frio formation varies widely from moderately acidic (5.7) to moderately alkaline (8.2), with nearly neutral averages (6.8). The pH of formation waters from other injection sites tends to be more alkaline, ranging from slightly alkaline (Belle Glade, Florida, pH 7.5) and moderately alkaline (Wilmington, North Carolina, pH 8.6), to very alkaline (Marshall, Illinois, pH 7.1 to 10.7).

10.4.3 Influence of Environmental Factors on Waste/Reservoir Compatibility

This section focuses on environmental conditions that may result in physical or chemical incompatibilities between wastes and reservoirs. Determining the potential for incompatibility is a part of the geochemical fate assessment that must be undertaken for any injection project because of possible

---

TABLE 10.13

Implications of Brine-Formation Mechanisms on Movement of Injected Wastes

Mechanism	Brine Type	Implications
Residual left after precipitation of evaporites (salt deposits).	Na-Ca-Cl	Brines are as old as the formation in which they occur; stagnant conditions exist.
Solution of halite present as bedded or domal salt-evaporite deposits.	Na-Cl Na-Ca-Cl	Active hydrologic conditions exist, although neither the mechanism nor the rate of fluid movement is indicated.
Reverse osmosis. Basinal waters forced through low-permeability shales, leaving the high-pressure side.	Na-Cl Na-Ca-Cl	Active hydrologic conditions exist because large volumes of water would have to pass brine through a basin to reach observed brine concentrations.

Source: U.S. EPA, Assessing the Geochemical Fate of Deep-Well-Injected Hazardous Waste: A Reference Guide, EPA/625/6-89/025a, U.S. EPA, Cincinnati, OH, June 1990.

TABLE 10.14

Processes Significant in Different Types of Waste–Reservoir Interactions

Interaction	Process
Waste with *in situ* fluids	Precipitation may result from incompatible brine. Hydrolysis may detoxify wastes. Complexation may increase or decrease mobility depending on condition. Oxidation or reduction of wastes may occur.
Waste with rock	Dissolution by highly acidic or alkaline wastes may threaten well and rock integrity. Gases generated by dissolution of carbonates may cause immiscible phase separation and well blowout. Adsorption on mineral surfaces may immobilize wastes. Clays may be mobilized and clog pores.
Waste/brine with rock	Waste/brine precipitates may clog pores. Successive adsorption/desorption reactions mayoccur at a particular location as waste/brine mixtures of varying proportions come in contact with the rock.
Microbiota	May form mats that clog pores near the injection well. May transform waste to nontoxic or other toxic forms.

Source: U.S. EPA, Assessing the Geochemical Fate of Deep-Well-Injected Hazardous Waste: A Reference Guide, EPA/625/6-89/025a, U.S. EPA, Cincinnati, OH, June 1990.

operational problems that may result from waste/reservoir incompatibility. The following are the major operational problems that can occur[3]:

1. Well plugging
2. Casing/confining layer failure
3. Well blowout

In extreme situations, incompatibility between injection fluids and reservoir components can be so great that deep-well disposal will not be the most cost-effective approach to waste disposal. In other situations, such remedial measures as pretreatment or controlling fluid concentrations or temperatures can permit injection even when incompatibilities exist. In addition to operational problems, waste–reservoir incompatibility can cause wastes to migrate out of the injection zone (casing/confining-layer failure) and even cause surface-water contamination (well blowout).

Four major types of chemical interactions are important when evaluating compatibility:

1. Waste interactions with brine
2. Waste interactions with rock
3. Waste–brine mixture interactions with rock
4. Microbiological interactions with the waste/brine/rock system

Each interaction involves numerous chemical processes. The dominance of a specific interaction depends on the type of waste, the characteristics of the brine and rock in the reservoir, and environmental conditions. Table 10.14 describes some of the more common processes that may result in incompatibility.

10.4.3.1 Well Plugging

The term well plugging refers to any of a variety of processes that reduce the permeability of the injection formation or the screens that are placed in the well's injection interval. When permeability is reduced, injection rates must be reduced or injection pressures increased. Table 10.15 lists a number of ways in which plugging may occur. One or more of these situations will probably take place in most injection wells; the number and severity of reactions will determine whether serious operational problems arise. If plugging is confined to the immediate vicinity of the injection well,

TABLE 10.15
Causes of Well Plugging and Possible Remedial Actions

Cause	Possible Action
Particulate solids and/or colloids.	Filter before injection.
Bacterial growth on well screen and formation.	Treat with bactericides.
Emulsification of two fluid phases.	Do not exceed solubility limits of organic wastes in water.
Precipitates resulting from mixing of injection and reservoir fluids.	Use pretreatment or buffer of non reactive water.
Expansion and dispersion of water-sensitive clays	Avoid injection of low-salinity solutions in water-sensitive (particularly montmorillonite) formations. Use clay stabilizers.
Migration of fines (very small particles) released by dissolution.	Neutralize before injection.
Reprecipitation of dissolved material (iron or calcium sulfate).	Use pretreatment.
Change in wettability or reduction in pore dimensions by adsorption (organics with large molecular weight).	Difficult to remedy.
Flow of unconsolidated sands into bore.	Use gravel-pack well screen. Inject a slug of brine after every period of interrupted flow.
Scaling on injection equipment by precipitation from injection fluid.	Use pretreatment; flush with solutions to remove accumulated scale.
Entrapped gases.	Remove gases from waste before injection or treat to prevent gas formation in the injection zone.

Source: U.S. EPA, Assessing the Geochemical Fate of Deep-Well-Injected Hazardous Waste: A Reference Guide, EPA/625/6-89/025a, U.S. EPA, Cincinnati, OH, June 1990.

wastes will not migrate into the injection zone until permeability is reestablished by physical or chemical means (see Table 10.15). Partial reductions in permeability may allow wastes to move into the injection zone but at increased pressures. This latter situation may contribute to well-casing or confining-layer failure. Clay swelling, mobilization of fine particles by dissolution, and precipitation are the common causes of well plugging.

10.4.3.2   Well-Casing and Confining-Formation Failure

Interactions between corrosive wastes and casing and packing can threaten the integrity of a well if proper materials have not been used in construction. Of equal concern is the potential for failure of the confining zone due to physical or chemical effects. For example, dissolution of an overlying carbonate confining layer may allow upward migration of wastes. This process was observed when hot acidic wastes were injected in a Florida well.

Chemically active injected fluids can also have negative impacts on the mechanical properties of the reservoir rock. For example, adsorption of aluminum and iron hydroxides and ferric chloride on quartz and other silicates can weaken the surface silicon–oxygen bonds by hydrolysis, reducing the surface energy, surface cohesion, and breaking strength of the formation. In addition, stress changes caused by increased injection pressures can fracture rock, forming permeability channels in a confining formation through which injected fluids could escape.[80]

10.4.3.3   Well Blowout

Gases entrapped in pore spaces resulting from phase separation of gases from liquids can reduce the permeability of a formation. This process was the major cause of clogging at groundwater recharge wells in the Grand Prairie Region in Arkansas.[81] Normally, pressures in deep-well-injection zones are high enough to keep gases in solution, so phase separation is not a problem. However, it is possible for permeability to be reduced by air entrainment at the same time gases are generated by reactions between the injected waste and reservoir formation. The resulting pressure then forces waste and reservoir fluid up the injection well to the surface, causing a well blowout.

The hazard of well blowout is greatest if hydrochloric acid wastes exceeding certain temperature and concentration limits are injected into a carbonate formation. When carbonate dissolves in acid, carbon dioxide is formed. Normally, this gas remains dissolved in the formation waters at deep-well temperatures and pressures, but if the temperature exceeds 88°F or acid concentration exceeds 6% HCl, carbon dioxide will separate from the formation waters as a gas. The resulting gas accumulation can increase pressures to a point where, if injection stops or drops below the subsurface carbon dioxide pressure, a blowout can occur.

10.4.4 Influence of the Deep-Well Environment on Biodegradation

Biodegradation of hazardous organic compounds in groundwaters has been the subject of much research in recent years. Ghiorse and Wilson[82] provide good general reviews of the topic. Unpolluted near-surface aquifers typically contain enough oxygen for aerobic processes to prevail. For example, Ghiorse and Wilson[82] summarize biodegradation data on 38 trace organic contaminants in subsurface materials from pristine sites. At most sites aerobic degradation is observed. In contrast, the deep-well-injection environment is typically anaerobic.

10.4.4.1 Occurrence of Microbes

Messineva[83] classifies subsurface sediments and rocks into geochemically active and geochemically inactive categories, based on microbial activity. Geochemically active sediments and rocks tend to be heterogeneous, containing organic material, nitrogen, and phosphorus, and support indigenous bacteria populations. Geochemically inactive formations do not maintain *in situ* microbial populations and lack fermentive properties when microorganisms are added. Such rocks are typically homogeneous, well-sorted clays.[83] Sinclair and Ghiorse[84] describe similar relationships between microbiological activity and the saturated zone in near-surface aquifers: gravelly sand was the most biologically active and clayey layers the least.

It is now generally accepted that microorganisms are ubiquitous in the deep subsurface, although, as noted, not all strata are biologically active.[85] Microorganisms have adapted to the complete range of environmental conditions that exist on and below the Earth's surface. They have been observed at pressures up to 1760 kg/cm^2 (25,000 psi), temperatures up to 100°C, and salt concentrations up to 300,000 mg/L.[86]

Most pre-1970 research on microorganisms in the deep-surface was done by petroleum microbiologists. Dunlap and McNabb[87] summarize data from 30 studies reporting isolation of microorganisms from deep-subsurface sediments. Because deep-well injection zones in the Gulf Coast region (where most deep-well injection of hazardous wastes occurs) are commonly associated with petroleum-producing strata, this research probably has some relevance. Kuznetsov and colleagues,[86] in an analysis of 50 samples of oilfield waters in Russia, found methanogenic organisms in 23 samples.

Ghiorse and Wilson[82] reviewed 14 studies characterizing subsurface microorganisms in pristine aquifers; only three studies involve samples deeper than 300 m below the surface. Olson and colleagues[88] found sulfate-reducing and methanogenic bacteria in waters from wells 1800 m deep in the Madison Limestone in Montana. In a comparison of microbial activity in the Bucatanna clay at 410 m near Pensacola, Florida, with that in the shallow Fort Polk aquifer, Louisiana, it was found that the biomass had to be about half that in the shallow aquifer and that there was a greater evidence of the byproducts of anaerobic bacterial activity.

Ehrlich and colleagues[89] examined microbial populations in samples of industrial wastes containing acrylonitrile and inorganic sodium salts (nitrate, sulfate, and thiocyanate) that had been injected to a depth of 375 to 425 m at a second waste-injection facility at Pensacola, Florida. Samples were obtained by allowing the injected waste to backflow, with a maximum estimated aquifer residence time of 107 h. Denitrifying bacteria dominated in the waste/formation-water mixture (10^5 to $>10^6$ organisms/mL), although substantial populations of both aerobes and anaerobes were also present (10^3 to 10^6 organisms/mL).

10.4.4.2 Degradation of Organic Compounds in Anaerobic Conditions

The three most significant groups of bacteria that may mineralize hazardous organic compounds are as follows:

1. Denitrifiers, which reduce nitrate to nitrogen
2. Sulfate reducers, which reduce sulfate to hydrogen sulfide
3. Methanogens, which reduce carbon dioxide to methane

Biodegradation of organic compounds under denitrifying conditions has been the least-studied of the three groups. Ehrlich and colleagues[89] inferred that acrylonitrile injected into a carbonate aquifer was completely degraded because the waste was not found in samples taken from a monitoring well where the waste arrived about 260 d after injection began, or in any subsequent samples. Bouwer and McCarty[90] observed partial to almost complete degradation of carbon tetrachloride ($>95\%$), bromodichloromethane ($>55\%$), dibromochloromethane ($>85\%$), and bromoform ($>90\%$) in laboratory batch experiments simulating denitrifying conditions. Compounds studied that did not show significant degradation under these conditions include chlorinated benzenes, ethylbenzene, naphthalene, chloroform, 1,1,1-trichloroethane, and 1,2-dibromomethane. Phthalic acids, phenol, tri-sodium nitrilotriacetate, and o- and m-xylene[3] are other compounds for which degradation has been observed under denitrifying conditions.

Degradation of organic compounds by sulfate-reducing bacteria has been studied mostly in the context of petroleum deposits.[91,92] These microbes are good scavengers of organic waste products regardless of the source of the waste. Novelli and ZoBell[91] reported finding some strains of sulfate-reducing bacteria that use hydrocarbons, beginning with decane and higher forms, paraffin oil and paraffin wax. In this study, the aromatic hydrocarbons—benzene, xylene, anthracene, and naphthalene—are not degraded, nor are aliphatic hydrocarbons, hydrocarbons with molecular weight lower than that of decane, or hydrocarbons of the naphthene series (cyclohexane). Rosenfeld[92] reported that high-molecular-weight aliphatic hydrocarbons are quickly decomposed by sulfate-reducing bacteria. However, the thinking is that molecular oxygen is required to degrade saturated hydrocarbons and that the experiments in the above-cited papers did not fully simulate anoxic conditions.

Degradation of organic compounds by methanogens has been the most extensively studied of the three groups. Methanogenic bacteria can readily degrade a number of monocyclic aromatics, phenol and some chlorophenols, benzene, ethyl benzene and a number of C_1 and C_2 halogenated aliphatic compounds.[3] However, the amount of degradation depends on the specific compound and conditions favorable for bacteria that can adapt to degrade the compound.

Biodegradation in groundwater systems may involve complex interactions among many types of bacteria, including denitrifying, sulfate-reducing, methanogenic, and others. Whether complete mineralization occurs depends on the compound, environmental conditions at the site, and the microorganisms that are best adapted to those conditions.

Iron- and manganese-reducing and ammonia-producing bacteria may also be significant in biochemical reactions that occur in the subsurface environment. Iron and manganese oxides are usually broken down through microbial reduction. Consequently, the possibility of this process should be considered when evaluating chemical reactions of iron and manganese species in the deep-well environment. Lovley[93] reviews the literature on biomineralization of organic matter with the reduction of ferric iron, and Ehrlich[94] reviews the literature on manganese oxide reduction through anaerobic respiration.

10.4.4.3 Microbial Ecology

The dissolved organic carbon content of subsurface waters is sufficient to maintain a small but diverse population of microorganisms. Denitrifiers, sulfate-reducers, and methanogens are likely to be present in low numbers in most groundwater unless conditions strongly favoring one group exist. Consequently, when a potential energy source in the form of an organic contaminant enters the water, the group most capable of utilizing the substrate at the environmental conditions existing in the aquifer will adapt and increase in population, while the population of other indigenous microbes will remain small or possibly be eliminated.

Effects of salinity

Typical salinities in deep-well injection zones range from about 20,000 to 70,000 mg/L, which is within the optimum range (50,000 to 60,000 mg/L) for halophilic organisms.[86] Many nonhalophilic bacteria can also live within this range. For example, a test of 14 microbe genera representing widely varying groups showed that most grew in salt concentrations of up to 60,000 mg/L.[95] Nitrification readily occurs at high salinities. Rubentschik[96] observed the conversion of ammonia to nitrate at concentrations of 150,000 mg/L NaCl, and isolated a culture of *Nitrosomonas* showing optimal growth at 40,000 mg/L. However, very high concentrations may slow denitrification. Hof[95] found that it took more than three times as long for the same amount of gas to be generated from denitrification at 300,000 mg/L NaCl as at 30,000 mg/L NaCl (10 vs. 3 d).

Effects of pressure

In general, growth and reproduction of both aerobic and anaerobic bacteria occurring at near-surface conditions decrease with increasing pressures.[74] However, certain barophilic (pressure-loving) bacteria have adapted to the temperature and pressure conditions in the deep-well environment. For example, aliphatic acids (acetate ions) are degraded by methanogenic bacteria in oilfield waters as long as temperatures are lower than 80°C.[97] Additionally, ZoBell and Johnson[74] found that certain sulfate-reducing bacteria isolated from oil-well brines located several thousand feet below the surface are metabolically more active when compressed to 400 to 600 atm (40.5 to 60.8 MPa) than at 1 atm. On the other hand, the pressures in deep-well waste injection formations may be sufficiently high to kill or otherwise severely affect the metabolic activity of microbes from surface habitats that may be indigenous to the injected wastes.[98]

Interactions among microbial groups

Decomposition of organic matter in anaerobic environments often depends on the interaction of metabolically different bacteria. Degradation in this situation is a multistep process in which complex organic compounds are degraded to short-chain acids by facultative bacteria and then to methane and carbon dioxide by methanogenic bacteria. In these interactions, methanogens may function as electron sinks during organic decomposition by altering electron flow in the direction of hydrogen production.[99] The altered flow of interspecies hydrogen transfer that occurs during coupled growth of methanogens and nonmethanogens may result in increased substrate utilization; different proportions of reduced end products; increased growth of both organisms; and displacement of unfavorable reaction equilibria.[99]

Redox conditions favoring denitrification lie somewhere between those for aerobic and methanogenic decomposition. However, denitrification and methanogenesis are not entirely mutually exclusive. Ehrlich and colleagues[100] observed evidence of both denitrifying and methanogenic bacteria in phenol-depleted zones of a creosote-contaminated aquifer and concluded that the denitrifying bacteria contributed to degradation. In this study, denitrifiers and iron reducers were the dominant anaerobes in contaminated wells. Methane production was highest in the closest wells downgradient from the contaminated site, indicatin g the development of redox zones with methanogenic conditions strongest where contaminant concentrations were highest, changing to stronger denitrifying conditions where contaminant concentrations were lower.

Studies by the U.S. Geological Survey at the Wilmington, NC, deep-well waste-injection facility also provide evidence of simultaneous degradation of organics by denitrifying and methanogenic organisms.[101,102] When the dilute waste front, containing organic acids, formaldehyde, and methanol, reached the first observation well, production of gases increased dramatically. For a period of about 6 weeks, about half the gas volume was methane and about a quarter, nitrogen. Two weeks later, nitrogen had increased to 62% and methane dropped to 33%, and after another three weeks nitrogen had increased to 68%, and methane had dropped to 12%. These relationships indicate that the methanogens were more sensitive to the increases in waste concentration as the dilute front passed the observation well and more concentrated waste reached the site. The inhibiting effects of sulfates on methane production would seem to indicate that sulfate-reduction will take place in preference to methanogenesis as long as sulfates are present.

10.5   GEOCHEMICAL CHARACTERISTICS OF HAZARDOUS WASTES

This section relates the chemical characteristics of inorganic and organic hazardous wastes to the important fate-influencing geochemical processes occurring in the deep-well environment.

10.5.1   INORGANIC VERSUS ORGANIC HAZARDOUS WASTES

Hazardous wastes are broadly classified as either organic or inorganic. Carbon is the central building block of organic wastes, whereas inorganic wastes are compounds formed by elements other than carbon (except for a few carbon-containing compounds such as metal carbonates, metal cyanides, carbon oxides, and metal carbides). Heavy metals may straddle the definition; although usually associated with inorganics, they can also be incorporated into organic compounds. In fact, organic forms of heavy metals, such as dimethyl mercury, are often more toxic than inorganic compounds formed by the same metal.

A major difference between organic and inorganic hazardous wastes is that, with the exception of cyanide, inorganics cannot be destroyed by being broken down into nonhazardous component parts, because at least one element in the compound is toxic. Inorganic hazardous wastes containing toxic elements can be transformed from a more to a less toxic form, but can never be transformed to a nontoxic form.

Toxic organic compounds (with the exception of organometallic compounds containing toxic metals), however, may be rendered harmless in some cases by being broken down into their inorganic components: carbon, hydrogen, oxygen, and other nontoxic elements. Most hazardous organic substances must be manufactured under carefully controlled conditions and are highly unlikely to form from the basic elements of hydrogen, oxygen, and others under uncontrolled deep-well environmental conditions. Therefore, once these wastes have completely broken down, their detoxification can be considered permanent.

Another major difference between inorganic and organic compounds is the number of compounds. Inorganic elements that exhibit toxic properties at levels of environmental concern number in the dozens, and only ten are regulated as hazardous wastes under the UIC program (arsenic, barium, cadmium, chromium, lead, mercury, nickel, selenium, thallium, and cyanide). Additionally, the number of inorganic compounds that any individual toxic element may form is limited (fewer than 50). On the other hand, the extreme versatility of carbon as a building block for organic compounds means that literally millions are possible, and the number that exhibit toxic properties is probably on the order of thousands or tens of thousands.

Regardless of whether a waste is classified as organic or inorganic, it must have certain physical and chemical properties to be suited for deep-well injection. Because water is the medium for injection, injected wastes, whether organic or inorganic, will typically be liquid or water-soluble or miscible, and relatively nonvolatile.

10.5.2 Chemical Properties of Inorganic Hazardous Wastes

The only means by which inorganic wastes can be rendered nonhazardous are dilution, isolation (as in deep-well injection), in some cases changes in oxidation state, and neutralization. Acidic wastes made up one-fifth of the injected waste volume and involved one-third of the injection wells in 1983. Most of the volume was from inorganic acids (hydrochloric, sulfuric, and nitric). Acid–base characteristics and neutralization were discussed in detail earlier, so the remainder of this section will focus on heavy metals and other hazardous inorganics (selenium and cyanide).

Inorganic elements can be broadly classified as metals and nonmetals. Most metallic elements become toxic at some concentration. Nine elements (arsenic, barium, cadmium, chromium, lead, mercury, nickel, selenium, and thallium) and cyanide are defined as hazardous inorganics for the purposes of deep-well injection.

In aqueous geochemistry, the important distinguishing property of metals is that, in general, they have a positive oxidation state (donate electrons to form cations in solution), but nonmetals have a negative oxidation state (receive electrons to form anions in solution). In reality, there is no clear dividing line between metals and nonmetals. For example, arsenic, which is classified as a nonmetal, behaves like a metal in its commonest valence states and is commonly listed as such. Other nonmetals, such as selenium, behave more like nonmetals.

Metals are divided into light (also called alkali-earth metals) and heavy. All toxic metals are heavy metals except for beryllium and barium. Additionally, other categories of elements that are or may be significant chemically as dissolved species in deep-well-injection zones include the following:

1. *Alkali-earth metals*: sodium, magnesium, potassium, calcium, and strontium
2. *Heavy metals*: manganese, iron, and aluminum, which may be significant in precipitation reactions
3. *Nonmetals*: carbon, nitrogen, oxygen, silicon, phosphorus, sulfur, chlorine, bromine, and iodine

10.5.2.1 Major Processes and Environmental Factors Affecting Geochemical Fate of Hazardous Inorganics

The major processes affecting the geochemical fate of hazardous inorganics are acid–base adsorption–desorption, precipitation–dissolution, complexation, hydrolysis, oxidation–reduction, and catalytic reactions. The significance of these processes to inorganic wastes is discussed only briefly here; additional information on individual elements is given in Table 10.16.

Acid–base equilibrium is very important to inorganic chemical reactions. Adsorption–desorption and precipitation–dissolution reactions are also of major importance in assessing the geochemical fate of deep-well-injected inorganics. Interactions between and among metals in solution and solids in the deep-well environment can be grouped into four types[3]:

1. Adsorption (including both physical adsorption and ion exchange) by clay minerals and silicates
2. Adsorption and coprecipitation by hydrous iron and manganese oxides
3. Complexation by organic substances such as fulvic and humic acids
4. Precipitation or coprecipitation by incorporation in crystalline minerals

TABLE 10.16

Geochemical Properties of Listed Metals and Nonmetals

Property	Forms/Conditions
Mobility	Cr is very mobile in neutral to alkaline conditions.
	As is more mobile under anaerobic than aerobic conditions and in alkaline conditions.
	Pb^{2+2} is relatively immobile except in highly acidic environments.
Strong adsorption on Fe and Mn oxides and hydrous oxides	Cd, Cr(IV), Hg, Ni, Se.
Precipitation	$Cd + H_2S \rightarrow CdS$.
	Cr + organic material $\rightarrow$ insoluble (aerobic conditions) precipitates.
	Cr(III) hydroxide, carbonate, and sulfide precipitate (pH > 6); Cr(VI) does not precipitate in these conditions.
	Pb typically precipitates as $Pb(OH)_2$, $PbCO_3$, $Pb_5(PO_4)_3OH$. NaCl increases solubility.
	Ni carbonates, hydroxides, and sulfides are relatively insoluble; Ni oxides in acidic solution may precipitate with neutralization.
Oxidation–reduction	Many selenium compounds can be reduced to produce elemental selenium when exposed to organic matter in subsurface environment.
Bioconversion	$As(OH)_3$ to $As(CH_3)_3$ (anaerobic); Hg (inorganic) to methyl mercury (anaerobic).

Source: U.S. EPA, Assessing the Geochemical Fate of Deep-Well-Injected Hazardous Waste: A Reference Guide, EPA/625/6-89/025a, U.S. EPA, Cincinnati, OH, June 1990.

Solution complexation is of major importance for the fate of metals in the deep-well environment. Soluble metal ions in solution can be divided into three major groups: simple hydrated metal ions,[103] metals complexed by inorganic anions, and organometallic complexes.[104] The stability of complexes between metals and organic matter is largely independent of ligand, and follows the following general relationships[105]:

1. Monovalent ions: Ag > Tl > Na > K > Pb > Cs
2. Divalent ions: Pt > Pd > Hg > UO_2 > Cu > Ni > Co > Pb > Zn > Cd > Fe > Mn > Sr > Ba
3. Trivalent ions: Fe > Ge> Sc > In > Y > Pl > Ce > La

Hydration reactions between metal ions and water affect mobility and adsorption but not toxicity. Hydrolysis is particularly important in the chemistry of cyanide.

Oxidation–reduction reactions may affect the mobility of metal ions by changing the oxidation state. The environmental factors of pH and Eh (oxidation–reduction potential) strongly affect all the processes discussed above. For example, the type and number of molecular and ionic species of metals change with a change in pH (see Figures 10.5–10.7). A number of metals and nonmetals (As, Be, Cr, Cu, Fe, Ni, Se, V, Zn) are more mobile under anaerobic conditions than aerobic conditions, all other factors being equal.[104] Additionally, the high salinity of deep-well injection zones increases the complexity of the equilibrium chemistry of heavy metals.[106]

Förstner and Wittmann[107] reported the following observations about the general mobility of heavy metals in groundwater:

1. Mobility tends to increase with increasing salinity because alkali- and alkaline-earth cations compete for adsorption sites on solids.
2. A change in redox conditions (lower Eh) can partly or completely dissolve Fe and Mn oxides and liberate other coprecipitated metals.
3. When natural or synthetic complexing agents are added, soluble metal complexes may form.

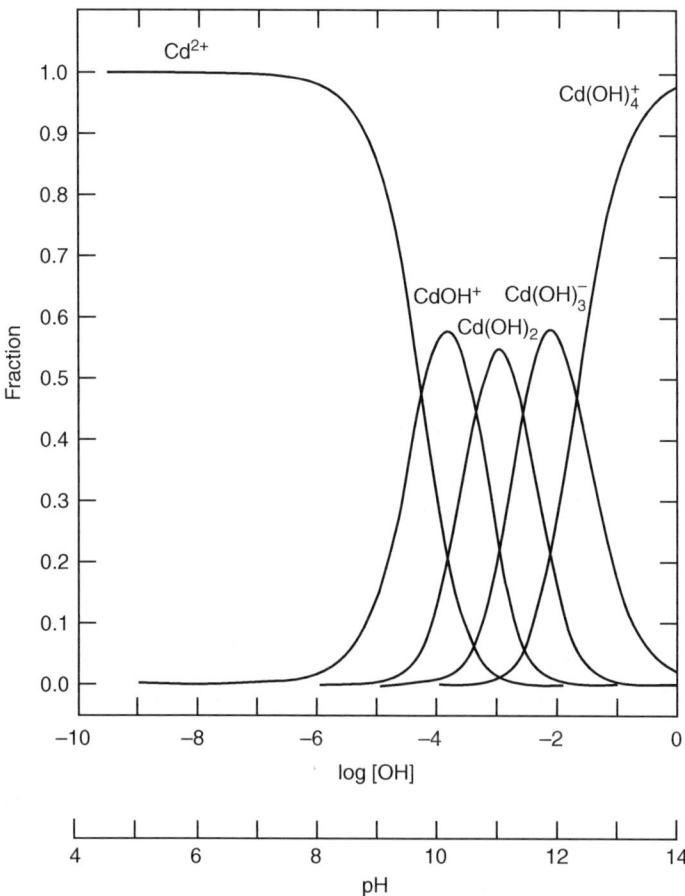

FIGURE 10.5 Distribution of molecular and ionic species of divalent cadmium at different pH values. (From U.S. EPA, Assessing the Geochemical Fate of Deep-Well-Injected Hazardous Waste: A Reference Guide, EPA/625/6-89/025a, U.S. EPA, Cincinnati, OH, June 1990.)

10.5.2.2 Known Properties of Listed Hazardous Inorganics

An extensive body of literature is available on the chemistry of listed inorganic wastes, although most of it is oriented toward near-surface environments. For example, Förstner and Wittmann[107] present a good overview of the aqueous geochemistry of metal contaminants, and the various reports of the National Research Council of Canada provide summaries of the geochemistry of individual metals. Fuller[105] contains over 200 citations on the movement of metals in soil, and Moore and Ramamoorthy[108] devote individual chapters to the chemistry of As, Cd, Cr, Cu, Pb, Hg, Ni, and Zn in natural waters. One source that does discuss the chemistry of listed wastes in the deep-well environment is Strycker and Collins.[109] The information on listed inorganic wastes is summarized in Table 10.16.

10.5.3 CHEMICAL PROPERTIES OF ORGANIC HAZARDOUS WASTES

Because carbon atoms can form strong bonds with one another while combining with other elements, the number of organic compounds is enormous. More than two million such compounds have been described and characterized,[3] which is more than ten times the total number of known compounds of all other elements except hydrogen.

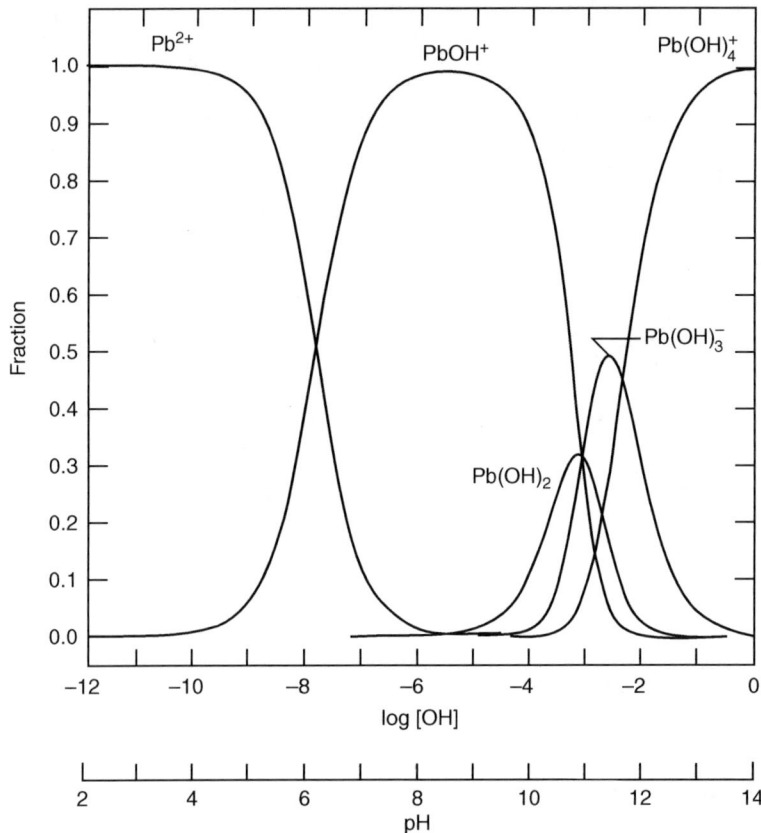

FIGURE 10.6   Distribution of molecular and ionic species of divalent lead at different pH values. (From U.S. EPA, Assessing the Geochemical Fate of Deep-Well-Injected Hazardous Waste: A Reference Guide, EPA/625/6-89/025a, U.S. EPA, Cincinnati, OH, June 1990.)

Organic compounds can be broadly grouped into hydrocarbons (compounds formed from only carbon and hydrogen atoms) and their derivatives, in which a hydrogen atom is replaced with another atom or group of atoms, such as a functional group (e.g., an atom or atom group that imparts characteristic chemical properties to the organic molecules containing it). Structurally, organic compounds can also be classified as straight-chain compounds, branched-chain compounds, and cyclic compounds. Another classification of organic compounds divides these compounds between aromatics (those with a six-member ring structure in which single and double carbon bonds alternate) and aliphatics (those containing chains or nonaromatic rings of carbon atoms). There are seven major groups of hazardous organics:

1. Halogenated aliphatic hydrocarbons
2. Halogenated ethers
3. Monocyclic aromatics
4. Phthalate esters
5. Polycyclic aromatic hydrocarbons
6. Nitrogenous compounds
7. Pesticides

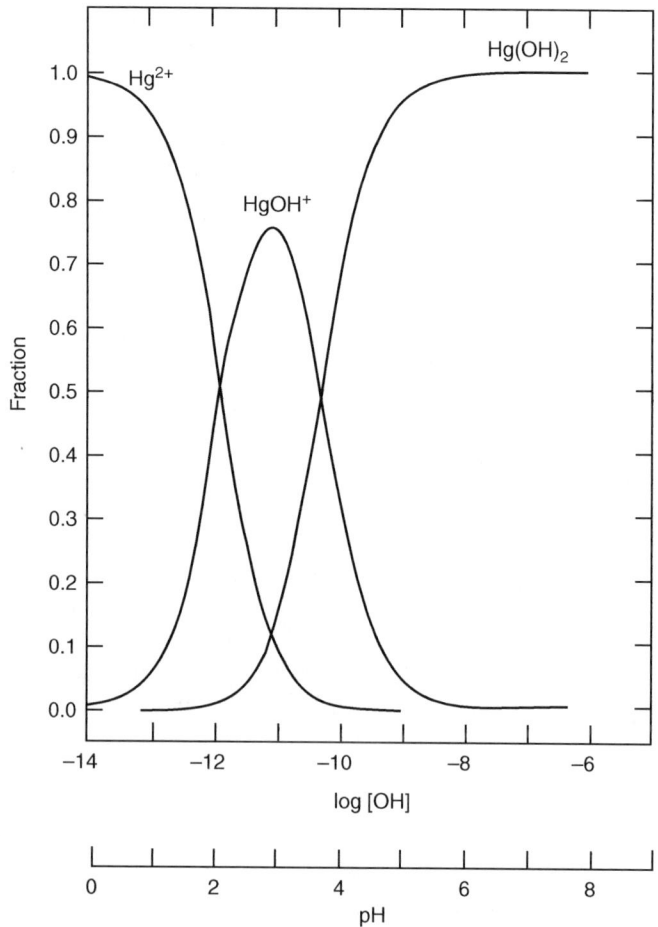

FIGURE 10.7 Distribution of molecular and ionic species of divalent mercury at different pH values. (From U.S. EPA, Assessing the Geochemical Fate of Deep-Well-Injected Hazardous Waste: A Reference Guide, EPA/625/6-89/025a, U.S. EPA, Cincinnati, OH, June 1990.)

10.5.3.1   Halogenated Aliphatic Hydrocarbons

Hazardous halogenated aliphatic hydrocarbons include mostly straight-chain hydrocarbons (alkanes containing single bonds, such as methane and ethane, and alkenes containing one double bond between carbon atoms, such as ethene and propene) in which one or more hydrogen atoms are replaced by atoms of the halogen group of elements (fluorine, chlorine, or bromine). Moore and Ramamoorthy[110] reviewed the behavior of aliphatic hydrocarbons in natural waters.

Tabak and colleagues[111] found most compounds in the group to be subject to significant degradation under experimental aerobic conditions. At least ten of the compounds are subject to biodegradation under anaerobic conditions. Britton[112] discusses microbial degradation of aliphatic hydrocarbons in more detail.

10.5.3.2   Halogenated Ethers

Ethers are either aliphatic (chain-structure) or aromatic (ring-structure) hydrocarbons containing an oxygen atom connected to two carbon atoms by single bonds. In halogenated ethers, one or more halogens (chlorine or bromine) replace hydrogen in the aliphatic or aromatic portion of the molecule. This group contains mostly aliphatic ethers except for 4-chlorophenyl phenyl ether and 4-bromophenyl phenyl ether, which are aromatic hydrocarbons.

Adsorption is very likely to be a more significant process for the aromatic halogenated ethers than for the aliphatic halogenated ethers. Hydrolysis is important for two of the aliphatic ethers: *bis*(chloromethyl) ether and 2-chloroethyl vinyl ether. The group appears generally resistant to biodegradation, although under certain conditions several may be degraded.

10.5.3.3   Monocyclic Aromatic Hydrocarbons and Halides

As mentioned, aromatic hydrocarbons have a six-member ring structure in which single and double carbon bonds alternate. This ring structure tends to be stable, so chemical reactions tend to result in the substitution of hydrogen atoms for another atom or functional group. Five of these compounds are hydrocarbons (benzene, ethylbenzene, toluene, phenol, and 2,4-dimethyl phenol) and the rest are halogenated or nitrogenated derivatives of benzene, toluene, and phenol. Moore and Ramamoorthy[110] reviewed the behavior of monocyclic aromatics and phenols in natural waters.

Adsorption may be important for most of the compounds in this group, whereas hydrolysis may not be a significant process except for pentachlorophenol. Tabak and colleagues[111] found that significant degradation with rapid or gradual adaptation occurred for 15 of a listed 23 compounds. Anaerobic degradation has been reported for five compounds in this group (benzene, ethylbenzene, phenol, 2-chlorophenol, and 2,4-dichlorophenol). Chapman[113] discusses in some detail the reaction sequence used for the bacterial degradation of phenolic compounds; Gibson and Subramanian[114] provide a general review of microbial degradation of aromatic hydrocarbons; and Reinke[115] reviews microbial degradation of halogenated aromatics.

10.5.3.4   Phthalate Esters

Esters contain a single oxygen atom attached to a single carbon atom by a single bond, and a second oxygen atom attached to the same carbon atom by a double bond. Phthalate esters form when aliphatic hydrocarbon groups replace the acidic hydrogen atoms in phthalic acid (benzenedicarboxylic acid). All phthalate esters are subject to adsorption and are readily biodegraded under aerobic conditions, but apparently not under anaerobic conditions. Ribbons and colleagues[116] review mechanisms for microbial degradation of phthalates. Hydrolysis half-lives of four phthalate esters (dimethyl phthalate, diethyl phthalate, di-*n*-butyl phthalate, and di-*n*-octyl phthalate) are on the order of thousands of days, which may be significant in the timeframe of deep-well injection.

10.5.3.5   Polycyclic Aromatic Hydrocarbons

Polycyclic (also called polynuclear) aromatic hydrocarbons (PAHs) are composed of multiple rings connected by shared carbon atoms (i.e., separate rings are combined by sharing two carbon atoms). All these compounds are pure hydrocarbons except for the two benzo-fluoranthenes, polychlorinated biphenyls (PCBs), and 2-chloronaphthalene. Moore and Ramamoorthy[110] review the behavior of PAHs in natural waters.

Adsorption and biodegradation under aerobic conditions are significant for the entire group, but PAHs are generally resistant to anaerobic degradation. Safe[117] reviews the literature on microbial degradation of PCBs. Hydrolysis is not significant for any compounds in the group.

10.5.3.6   Nitrogenous Compounds

The diverse nitrogenous compounds group is composed of substances that have in common the substitution of one or more nitrogen-containing functional groups for hydrogen in the structure. Amines are derivatives of ammonia and contain a nitrogen atom bonded to at least one carbon atom. Nitrosamines are amines with a nitro ($-NO_2$) functional group; two are aliphatic (dimethylnitrosamine and di-*n*-propyl nitrosamine) and one is aromatic (diphenylnitrosamine). The two benzidines and 1,2-diphenyl hydrazine are aromatic amines. Acrylonitrile contains the nitrile (–CN) functional group. Adsorption is a significant process for all four of the aromatic amines; hydrolysis is not. Compounds in the group are generally not amenable to biodegradation. Acrylonitrile, however, is readily mineralized by anaerobic denitrifying bacteria.

10.5.3.7   Pesticides

By definition, any pesticide has toxic effects on organisms. Listed pesticides are those that combine high toxicity with resistance to degradation in the environment. Moore and Ramamoorthy[109] review the behavior of chlorinated pesticides in natural waters.

Most of the common 15 hazardous pesticides are chlorinated hydrocarbons. Adsorption can be an important process for most. All except DDT, endosulfan, and heptachlor resist hydrolysis, and most are also resistant to biodegradation. Kearney and Kaufman[118] review conditions under which chlorinated pesticides are biodegraded.

10.6   METHODS AND MODELS FOR PREDICTING THE GEOCHEMICAL FATE OF DEEP-WELL-INJECTED WASTES

10.6.1   BASIC APPROACHES TO GEOCHEMICAL MODELING

The geochemical interactions possible between an injected waste and the reservoir rock and its associated fluids can be quite complex. Thus a combination of computer modeling, laboratory experimentation, and field observation will inevitably be necessary to satisfy current regulatory requirements for a geochemical no-migration deep-well injection. This section covers the computer methods and models available for predicting geochemical fate.

The American Society for Testing and Materials (ASTM)[119] has developed a standard protocol for evaluating environmental chemical-fate models, along with the definition of basic modeling terms, shown in Table 10.17. Predicting fate requires natural phenomena to be described mathematically.

TABLE 10.17
Definitions of Terms Used in Chemical Fate Modeling

Term	Definition
Algorithm	The numerical technique embodied in the computer code.
Calibration	A test of a model with known input and output information that is used to adjust or estimate factors for which data are not available.
Computer code	The assembly of numerical techniques, bookkeeping, and control languages that represents the model from acceptance of input data and instruction to delivery of output.
Model	An assembly of concepts in the form of a mathematical equation that portrays understanding of a natural phenomenon.
Sensitivity	The degree to which the model result is affected by changes in a selected input parameter.
Validation	Comparison of model results with numerical data independently derived from experiment or observation of the environment.
Verification	Examination of the numerical technique in the computer code to ascertain that it truly represents the conceptual model and that there are no inherent numerical problems associated with obtaining a solution.

Source: U.S. EPA, Assessing the Geochemical Fate of Deep-Well-Injected Hazardous Waste: A Reference Guide, EPA/625/6-89/025a, U.S. EPA, Cincinnati, OH, June 1990.

The expression of chemical fate can be computerized using a code to perform the computations and predict the results when inputs simulating conditions of interest are provided. Two critical aspects of the use of computer codes for predicting geochemical fate are the verification and validation of the models on which the codes are based.

In addition to the limited availability of validation, the following are some of the problems found in computer and mathematical modeling[120]:

1. The data on thermodynamic properties of many relevant water-miscible organic species are either incomplete or unavailable.
2. Many minerals are solid solutions (e.g., clays, amphiboles, and plagioclase feldspars). Solid-solution models are either not available or appropriate algorithms have not been incorporated into computer codes.
3. Models describing the adsorption of water-miscible organic compounds on natural materials have not been correlated with field observations under typical injection-zone conditions. Few computer codes contain algorithms for calculating the distribution of species between the adsorbed and aqueous states.
4. Calcium-sodium-chloride-type brines (which typically occur in deep-well-injection zones) require sophisticated electrolyte models to calculate their thermodynamic properties. Many parameters for characterizing the partial molal properties of the dissolved constituents in such brines have not been determined. (Molality is a measure of the relative number of solute and solvent particles in a solution and is expressed as the number of gram-molecular weights of solute in 1000 g of solvent.) Precise modeling is limited to relatively low salinities (where many parameters are unnecessary) or to chemically simple systems operating near 25°C.
5. Computer codes usually calculate only the thermodynamically most stable configuration of a system. Modifications can simulate nonequilibrium, but there are limitations on the extent to which codes can be manipulated to simulate processes that are kinetically (rate) controlled; the slow reaction rates in the deep-well environment compared with groundwater movement (i.e., failure to attain local homogeneous or heterogeneous reversibility within a meter or so of the injection site) create particular problems.
6. Little is known about the kinetics of dissolution, precipitation, and oxidation–reduction reactions in the natural environment. Consequently, simulating the kinetics of even more complicated injection- zone chemistry is very difficult.

Bergman and Meyer[121] point out a particularly relevant problem with mathematical models. The relative reliability of mathematical models (compared with physical models based on empirical field or laboratory studies) decreases rapidly as the number of environmental pollutants being modeled increases (see Figure 10.8). Consequently, mathematical models tend to be less cost-effective for complex wastestreams than physical (empirical) models.

10.6.2 SPECIFIC METHODS AND MODELS

Most of the chemical processes discussed before (acid–base equilibria, precipitation–dissolution, neutralization, complexation, and oxidation–reduction) are interrelated; that is, reactions of one type may influence other types of reactions, and consequently must be integrated into aqueous- and solution-geochemistry computer codes.

10.6.2.1 Aqueous- and Solution-Geochemistry Computer Codes

More than 50 computer codes that calculate chemical equilibrium in natural waters or similar aqueous systems are described in the literature.[122] Most are not suitable for modeling the deep-well

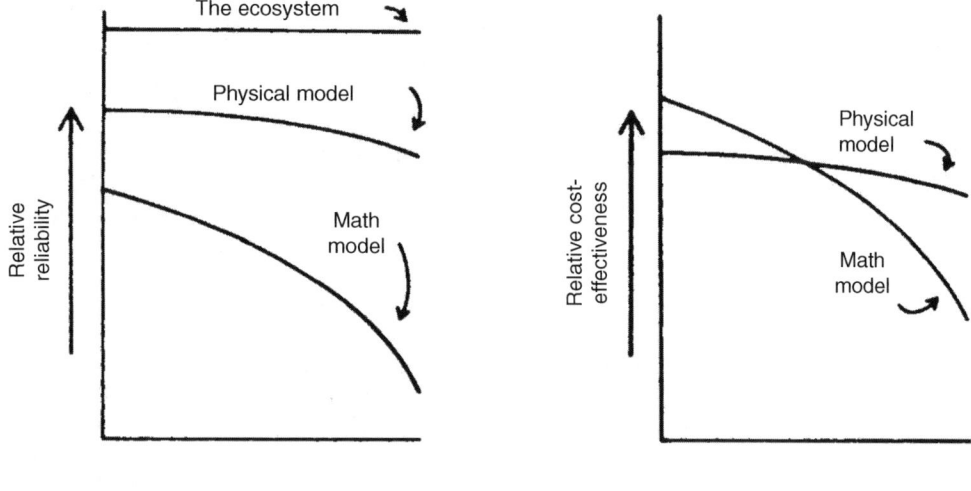

FIGURE 10.8 Relative tradeoffs between physical (microcosm) and mathematical models as affected by effluent complexity. (From U.S. EPA, Assessing the Geochemical Fate of Deep-Well-Injected Hazardous Waste: A Reference Guide, EPA/625/6-89/025a, U.S. EPA, Cincinnati, OH, June 1990.)

injection of hazardous wastes, because they are limited to simulating reactions under one or more of the following conditions:

1. Ambient temperatures (25°C)
2. Low pressures (1 atm)
3. Relatively low salinities

When the simulation of deep-well temperatures, pressures, and salinities is imposed as a condition, the number of codes that may be of value is reduced to a much smaller number. Nordstrom and Ball[121] recommend six references as covering virtually all the mathematical, thermodynamic, and computational aspects of chemical-equilibrium formulations (see references 123–128). Recent references on modeling include references 45, 63, 70, 129 and 130.

10.6.2.2 Adsorption

Mineral surfaces on which adsorption may occur are diverse and complex, and the mechanisms by which a hazardous constituent may attach to the solid surface vary substantially. Therefore, theoretical models that can be used readily to predict adsorption for a variety of compounds over a range of conditions are difficult to develop. Table 10.18 summarizes the applicability of three major methods for predicting adsorption in the deep-well environment. These methods include the following:

1. Adsorption isotherms
2. The clay ion-exchange model
3. The triple-layer model

Adsorption isotherms
The simplest and most widely used method for predicting adsorption is to measure adsorption isotherms (the variations in the amount of a substance adsorbed at different concentrations measured at a constant temperature). Empirical constants can be calculated from such measurements.

TABLE 10.18

Applicability of Methods and Models for Predicting Adsorption in the Deep-Well Environment

Method/Model	Applicability
Methods	
Adsorption isotherms	Relatively easy to measure. The main disadvantage is that the empirical coefficients may change with changing environmental conditions, requiring measurement.
Linear distribution coefficient	Applicable only at very dilute concentrations of organic compounds and where >0.1% organic matter is present. Usefulness is uncertain.
Langmuir	Underlying assumptions for the derivation of the equation typically will not apply.
Freundlich	Limited available data on adsorption under simulated deep-well conditions are best described by the formula; however, the disadvantage of all adsorption isotherms applies.
Models	
Clay ion-exchange model	May be useful for predicting adsorption of heavy metals. Aqueous-phase-activity solid-solution model coefficients can be obtained from distribution-of-species models. Estimating clay-phase activity coefficients is more problematic.
Triple-layer model	Of limited value because of the complexity of adsorption sites, unpredictable interactions among adsorbents, and complications introduced by high salinities.

Source: U.S. EPA, Assessing the Geochemical Fate of Deep-Well-Injected Hazardous Waste: A Reference Guide, EPA/625/6-89/025a, U.S. EPA, Cincinnati, OH, June 1990.

The amount of adsorption at concentrations other than those that were measured can then be predicted using the empirical constants in an appropriate formula. The correct application of this method requires acknowledging such effects as matrix and temperature.

Three types of adsorption isotherms are discussed in this section:

1. The linear distribution coefficient
2. The Langmuir adsorption isotherm
3. The Freundlich adsorption isotherm

The distribution coefficient assumes that adsorption is linear (i.e., the amount of adsorption is directly proportional to the concentration of the compound in solution) and is actually a special case of the Langmuir and Freundlich isotherms, which are nonlinear.[31,32]

The simplest type of isotherm is the linear-distribution coefficient, K_d.[121] It is also called the partition coefficient, K_p.[58] The equation for calculating adsorption at different concentrations is

$$S = K_d C \tag{10.6}$$

where S = amount adsorbed (μg/g solid), C = concentration of adsorbed substance in solution (μg/mL), and K_d = linear distribution coefficient = partition coefficient = K_p.

This equation is widely used to describe adsorption in soil and near-surface aquatic environments. Another widely used linear coefficient is the organic-carbon partition coefficient K_{oc}, which is equal to the distribution coefficient divided by the percentage of organic carbon present in the system as proposed by Hamaker and Thompson.[131]

$$K_{oc} = \frac{K_d}{\% \text{ organic carbon}} \tag{10.7}$$

Sabljiã[132] presents very accurate equations for predicting the K_{oc} of both polar and nonpolar organic molecules based on molecular topology, provided the organic matter percentage exceeds 0.1%. Karickhoff[133] discusses in detail adsorption processes of organic pollutants in relation to K_{oc}.

Winters and Lee[134] describe a physically based model for adsorption kinetics for hydrophobic organic chemicals to and from suspended sediment and soil particles. The model requires determination of a single effective diffusivity parameter, which is predictable from compound solution diffusivity, the octanol–water partition coefficient, and the adsorbent organic content, density, and porosity.

Major problems are associated with using the linear distribution coefficient for describing adsorption–desorption reactions in groundwater systems. Some of these problems include the following[135,136]:

1. The coefficient actually measures multiple processes (reversible and irreversible adsorption, precipitation, and coprecipitation). Consequently, it is a purely empirical number with no theoretical basis on which to predict adsorption under differing environmental conditions or to give information on the types of bonding mechanisms involved.
2. The waste-reservoir system undergoes a dynamic chemical evolution in which changing environmental parameters may result in variations of K_d values by several orders of magnitude at different locations and at the same location at different times.
3. All methods used to measure the K_d value involve some disturbance of the solid material and consequently do not accurately reflect *in situ* conditions.

The Langmuir equation was originally developed to describe adsorption of gases on homogeneous surfaces and is commonly expressed as follows:

$$\frac{C}{S} = \frac{1}{kS_{max}} + \frac{1}{CS_{max}} \tag{10.8}$$

where S_{max} = maximum adsorption capacity (µg/g soil), k = Langmuir coefficient related to adsorption bonding energy (mL/g), S = amount adsorbed (µg/g solid), and C = concentration of adsorbed substance in solution (µg/mL).

A plot of C/S versus $1/C$ allows the coefficients k and S_{max} to be calculated. When $kC \ll 1$, adsorption will be linear, as represented by Equation 10.6.

The Langmuir model has been used to describe adsorption behavior of some organic compounds at near-surface conditions.[137] However, three important assumptions must be made:

1. The energy of adsorption is the same for all sites and is independent of degree of surface coverage.
2. Adsorption occurs only on localized sites with no interactions among adjoining adsorbed molecules.
3. The maximum adsorption capacity (S_{max}) represents coverage of only a single layer of molecules.

In a study of adsorption of organic herbicides by montmorillonite, Bailey and colleagues[138] found that none of the compounds conformed to the Langmuir adsorption equation. Of the 23 compounds tested, only a few did not conform well to the Freundlich equation.

The assumptions mentioned above for the Langmuir isotherm generally do not hold true in a complex heterogeneous medium such as soil. The deep-well environment is similarly complex and consequently the studies of adsorption in simulated deep-well conditions[139,140] have followed the form of the Freundlich equation:

$$S = KC^N \qquad (10.9)$$

where S and C are as defined in Equation 10.6, and K and N are empirical coefficients.

Taking the logarithms of both sides of Equation 10.9:

$$\log S = \log K + N \log C \qquad (10.10)$$

Thus, log–log plots of S versus C provide an easy way to obtain the values for K (the intercept) and N (the slope of the line). The log–log plot can be used for graphic interpolation of adsorption at other concentrations, or, when values for K and N have been obtained, the amount of adsorption can be calculated from Equation 10.9. Figure 10.9 shows an example of adsorption isotherms for phenol adsorbed on Frio sandstone at two different temperatures. Note that when $N = 1$, Equation 10.9 simplifies to Equation 10.6 (i.e., adsorption is linear).

The Langmuir equation has a strong theoretical basis, whereas the Freundlich equation is an almost purely empirical formulation because the coefficient N has embedded in it a number of thermodynamic parameters that cannot easily be measured independently.[120] These two nonlinear isotherm equations have most of the same problems discussed earlier in relation to the distribution-coefficient equation. All parameters except adsorbent concentration C must be held constant when measuring Freundlich isotherms, and significant changes in environmental parameters, which would be expected at different times and locations in the deep-well environment, are very likely to result in large changes in the empirical constants.

An assumption implicit in most adsorption studies is that adsorption is fully reversible. In other words, once the empirical coefficients are measured for a particular substance, Equations 10.6 to 10.10 describe both adsorption and desorption isotherms. This assumption is not always true. Collins and Crocker[140] observed apparently irreversible adsorption of phenol in flowthrough adsorption experiments involving phenol interacting on a Frio sandstone core under simulated deep-well

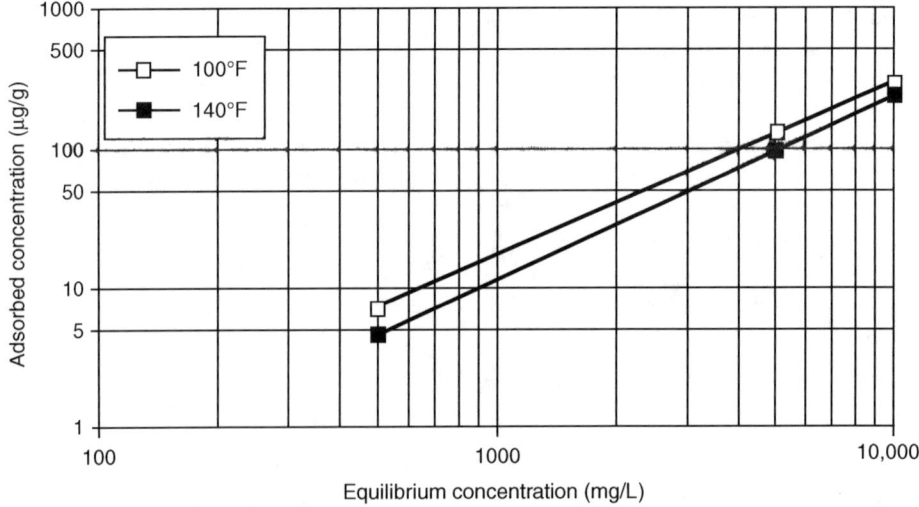

FIGURE 10.9   Freundlich isotherm for phenol adsorbed on Frio Core. (From U.S. EPA, Assessing the Geochemical Fate of Deep-Well-Injected Hazardous Waste: A Reference Guide, EPA/625/6-89/025a, U.S. EPA, Cincinnati, OH, June 1990.)

temperatures and pressures. If adsorption–desorption is not fully reversible, it may be necessary to use separate Freundlich adsorption- and desorption-isotherm equations to model these processes in the deep-well environment.[120]

Clay ion-exchange model

As noted above, adsorption isotherms are largely derived empirically and give no information on the types of adsorption that may be involved. Scrivner and colleagues[39] have developed an adsorption model for montmorillonite clay that can predict the exchange of binary and ternary ions in solution (two and three ions in the chemical system). This model would be more relevant for modeling the behavior of heavy metals that actively participate in ion-exchange reactions than for organics, in which physical adsorption is more important.

The clay ion-exchange model assumes that the interactions of the various cations in any one clay type can be generalized and that the amount of exchange will be determined by the empirically determined cation-exchange capacity (CEC) of the clays in the injection zone. The aqueous-phase activity coefficients of the cations can be determined from a distribution-of-species code. The clay-phase activity coefficients are derived by assuming that the clay phase behaves as a regular solution and by applying conventional solution theory to the experimental equilibrium data in the literature.[3]

Scrivner and colleagues[39] compared the ion-exchange model predictions with several sets of empirical data. The model predictions are very accurate for binary-exchange reactions involving the exchange of nickel ions for sodium and potassium ions on illite and less accurate for ternary reactions involving hydrogen, sodium, and ammonia ions. The deep-well environment, however, is very likely to have multiple exchangeable species (such as Na^+, K^+, Ca^{2+}, and Mg^{2+}), and injected wastes commonly have elevated concentrations of more than one heavy metal. These concentrations result in complex ion-exchange interactions that probably exceed the capabilities of the model.

Triple-layer model

One of the more sophisticated models for describing adsorption phenomena in aqueous solutions is the triple-layer model (TLM), also called the Stanford General Model for Adsorption (SGMA) because it has been developed, refined, and tested over a number of years by faculty and researchers at Stanford University.[141–143] The TLM separates the interface between the aqueous phase and the adsorbent surface into three layers: surface layer, inner diffuse layer, and outer diffuse layer. Each has an electrical potential, charge density, capacitance, and dielectric constant. Hydrogen ions are assumed to bind at the surface plane; electrolyte ions (such as Na^+) bind at the inner diffuse plane. The surface is assumed to be coated with hydroxyl groups (OH^-), with each surface site associated with a single hydroxyl group. The hydroxyl-occupied surface sites may either react with other ions in solution or dissociate according to a series of reactions, with each having an associated equilibrium constant. Experimental terms relate the concentrations of the ions at their respective surface planes to those in the bulk solution. The sum of the charges of the three layers is assumed to be zero (i.e., the triple layer is electrically neutral). For all its sophistication, TLM is of limited value for predicting adsorption in deep-well environments[120]:

1. Site-binding constants have been determined for only a limited range of simple oxides with only one type of surface site. Multiple-surface site minerals occurring in the deep-well environment such as silicates, aluminosilicates, and complex oxides (such as manganese oxide) will require much more complex TLMs.
2. Fixed-charge minerals such as clay are even more complex than the multiple-surface site minerals, and both ion exchange and other types of adsorption must be measured to characterize absorption reactions fully.
3. Minerals with different adsorptive properties in the injection zone may interact to produce results different from those that would be obtained if each mineral were tested separately. No satisfactory model has been developed that predicts adsorption properties of mixtures based on the properties of individual adsorbents.

4. The TLM is based on laboratory measurements of adsorption on materials that are suspended in solution. No satisfactory methods for measuring and interpreting the adsorptive properties of intact host rock have been developed for TLM application.
5. The TLM has been developed using studies based on solutions of relatively low concentrations of dissolved compounds. The very saline and briny conditions found in the deep-well environment may require an entirely different model.

10.6.2.3   Biodegradation

This section examines two quantitative models for predicting biodegradation: the kinetic rate expressions and the biofilm model. It also examines several qualitative models for describing biodegradation in the deep-well environment.

Kinetic rate expressions
When microorganisms use an organic compound as a sole carbon source, their specific growth rate is a function of chemical concentration and can be described by the Monod kinetic equation. This equation includes a number of empirical constants that depend on the characteristics of the microbes, pH, temperature, and nutrients.[54] Depending on the relationship between substrate concentration and rate of bacterial growth, the Monod equation can be reduced to forms in which the rate of degradation is zero order with substrate concentration and first order with cell concentration, or second order with concentration and cell concentration.[144]

The Monod equation assumes a single carbon source. The difficulty in handling multiple carbon sources, which are typical in nature, has led to the use of an empirical biodegradation rate constant k_1:

$$S = k_1 BC \qquad (10.11)$$

where B = bacterial concentration, k_1 = an empirical biodegradation rate constant. This equation is of the same form as Equation 10.6 for linear adsorption. Predicting biodegradation using such a rate constant is complicated when multiple biodegradable compounds are present. For example, phenol and naphthalene are both rapidly biodegraded in single-compound laboratory shake-flask experiments when seeded with bacteria from an oil-refinery settling pond, but when the two compounds are combined, naphthalene is not degraded until the phenol is gone.[3]

When a compound is cometabolized (degraded but not used as a nutrient), a second-order biodegradation coefficient can be used to estimate, k_B:

$$k_B = k_{B2} B \qquad (10.12)$$

where k_B = first-order biodegradation coefficient, k_{B2} = second-order biodegradation coefficient, and B = bacterial concentration.

Mills and colleagues[58] describe the use of these formulations to predict aerobic biodegradation in surface waters and present methods of adjusting for temperature and nutrient limitations. This approach to predicting biodegradation is problematic because it is difficult to obtain empirical coefficients in the deep-well setting.

Baughman and colleagues[145] derive a second-order kinetic rate expression as a special case of the Monod kinetic equation. It appears to describe biodegradation of organics in natural surface waters reasonably well:

$$-dC/dt = k[B][C] \qquad (10.13)$$

Paris and colleagues[144] found that degradation of several pesticides in samples from over 40 lakes and rivers fits this second-order model of microbial degradation.

General degradation rate models of organics in soils have been described by Hamaker,[146] Larson,[147] and Rao and Jessup.[148] In most instances, biodegradation is the major, but not necessarily the only, process affecting the rate of degradation.

Biofilm model

The most sophisticated model available for predicting biodegradation of organic contaminants in subsurface systems is the biofilm model, presented by Williamson and McCarty[149,150] which has been refined over several years by researchers at Stanford University and the University of Illinois/Urbana.[151–157]

The biofilm model is based on two important features of the groundwater environment:

1. The nutrient concentrations tend to be low.
2. The solid matrix has a high specific surface area.

These characteristics favor the attachment of bacteria to solid surfaces in the form of biofilm so that nutrients flowing in the groundwater can be used. The presence of low nutrient levels in the groundwater also implies that bacteria must regularly use many different compounds as energy sources and, consequently, may select organic contaminants more readily as nutrients.

The basic biofilm model[149,150] idealizes a biofilm as a homogeneous matrix of bacteria and the extracellular polymers that bind the bacteria together and to the surface. A Monod equation describes substrate use; molecular diffusion within the biofilm is described by Fick's second law; and mass transfer from the solution to the biofilm surface is modeled with a solute-diffusion layer. Six kinetic parameters (several of which can be estimated from theoretical considerations and others of which must be derived empirically) and the biofilm thickness must be known to calculate the movement of substrate into the biofilm.

Rittmann and McCarty[152,153] have developed equations for incorporating bacterial growth into the model, allowing the steady-state utilization of substrate materials to be predicted. They also show theoretically and verify experimentally that there is a substrate concentration threshold S_{min} below which no significant activity occurs. McCarty and colleagues[154] introduce the idea of secondary substrate utilization by a biofilm, in which microbes can metabolize trace compounds ($S < S_{min}$) in the presence of another substrate that is in sufficient concentrations to support biofilm growth. Bouwer and McCarty[155] incorporate steady-state utilization of secondary substrates into the model by coupling the biofilm mass (controlled by degradation of the primary substrate) with concentration and individually determine rate parameters for each secondary substrate. Laboratory tests of degradation on a variety of chlorinated benzenes, nonchlorinated aromatics, and halogenated aliphatics as secondary substrates agree reasonably well with predicted values.[155] The later refinement of the model incorporates the effects of adsorption of material substrate to the surface on which the biofilm is attached, but is restricted to biofilm on activated carbon.[156,157]

When water containing substrate concentrations greater than S_{min} is injected into the subsurface, the model predicts that biofilm development will occur only in the first meter or so of the injection zone.[151] Low concentrations of hazardous compounds will be significantly degraded as secondary substrates only if they are readily biodegraded in the biofilm zone. Any amount not biodegraded in the biofilm zone will tend to persist once it leaves the zone of concentrated biological activity. When substrate concentrations are not sufficient to sustain biofilm development, Bouwer and McCarty[155] suggest that a simple biodegradation coefficient such as that discussed earlier (Equation 10.11) is probably adequate.

Qualitative models

Several qualitative models for biodegradation in the deep-well environment have been suggested. They do not allow quantitative predictions to be made, but they do provide insight into the types of biodegradation processes that may occur. These models have not been expressed quantitatively to

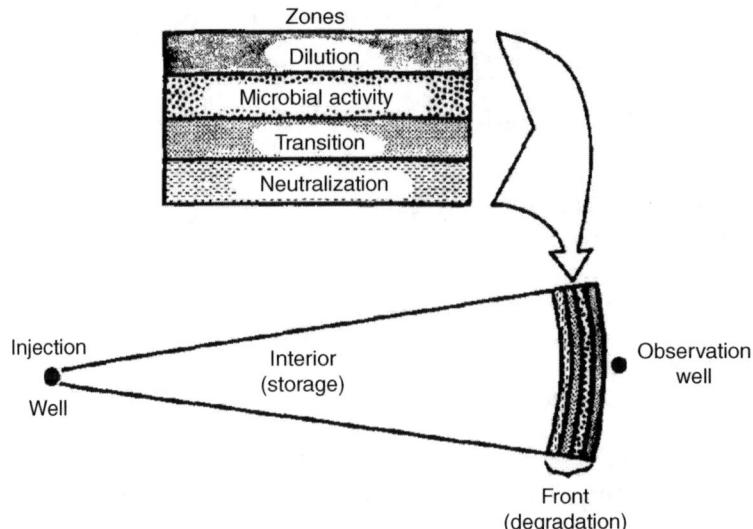

FIGURE 10.10 Proposed geochemical model of waste after injection into the subsurface. (From U.S. EPA, Assessing the Geochemical Fate of Deep-Well-Injected Hazardous Waste: A Reference Guide, EPA/625/ 6-89/025a, U.S. EPA, Cincinnati, OH, June 1990.)

simulate degradation, although relatively simple codes using first-order biodegradation constants k_B could probably be developed without much difficulty. In the absence of quantitative models for predicting biodegradation, laboratory simulations must be used to assess biodegradation potential.

The conceptual geochemical model of acidic waste after injection into the subsurface, proposed by Leenheer and Malcolm,[102] involves a moving front of microbial activity with five zones as shown in Figure 10.10:

1. The dilute zone, controlled by diffusion
2. A zone where substrate concentrations are sufficiently high to allow significant microbial activity
3. The transition zone, where increasing waste concentrations create unfavorable conditions for microbial growth
4. The neutralization zone, where abiotic chemical reactions predominate
5. The waste storage zone where undiluted waste no longer reacts with the host rock.

This model implies that the rate of injection far exceeds the zone's capacity for biodegradation. Bouwer and McCarty[155] suggest a qualitative model that represents nonbiofilm microbial biodegradation over increasing distances from the injection point. This model follows the redox reaction sequence. This model implies that most compounds not degraded in their appropriate zone will move through the groundwater system without significant additional degradation. The model also implies, however, that those compounds that are biodegraded by methanogenesis will continue to move through the groundwater until degradation is complete.

10.6.2.4 Hydrolysis

Hydrolysis is easily predicted, provided that the rate constants for a compound are known. The rate of abiotic hydrolysis is given by

$$R = -k_H C_T \tag{10.14}$$

where R = the rate of hydrolysis (mole/L/sec or μg/L/sec), k_H = specific hydrolysis rate constant (L/sec), C_T = the dissolved plus adsorbed phase concentration of compound C (mol/L or μg/L).

The hydrolysis rate constant k_H is actually the sum of three rate constants:

$$k_H = k_n + k_a[H^+] + k_b[OH^-] \tag{10.15}$$

where k_n = the natural hydrolysis rate constant for the pH independent reactions of a chemical with water (L/sec), k_a = the acid-catalyzed hydrolysis rate constant (L/mol/sec), $[H^+]$ = the concentration of hydrogen ion (mol/L), k_b = the base-catalyzed hydrolysis rate constant (L/mol/sec), $[OH^-]$ = the concentration of hydroxide ion (mol/L).

Note that in an acid solution, $k_b = 0$, and in an alkaline solution, $k_a = 0$. k_H can be adjusted to include the effects of adsorption by multiplying $(k_a[H^+] + k_b[OH^-])$ times the decimal fraction of the total amount of a dissolved compound, C.[58] At any fixed pH, the half-life of a substance is independent of concentration and can be calculated with the equation

$$t_{1/2} = 0.693/k_h \tag{10.16}$$

Hydrolysis is strongly pH-dependent, with k_a dominant at low pH and k_b dominant at high pH; at pH 7, k_n can often be most important. However, the detailed relationship of pH and rate depends on the specific values of k_n, k_a, and k_b. If these rate constants are known, then the hydrolysis rate at any pH can be readily calculated. Mabey and Mill[158] provide these data for a large number of organic compounds, and Ellington[159–161] provides data on about 70 regulated hazardous pollutants.

Mills[58] describes step-by-step procedures for calculating k_H, and Scrivner and colleagues[39] describe in detail the modeling of cyanide and nitrite hydrolysis in the deep-well environment.

10.6.2.5 Chemical Transport

Basic approaches and important models of chemical transport will be addressed briefly. Three major approaches can be used to modeling chemical transport:

1. *Retardation-factor models*, which incorporate a simple retardation factor derived from a linear- or linearized-distribution coefficient
2. *Integrated models*, in which all mass, momentum, and energy transfer equations, including those in which chemical reactions participate, are solved simultaneously for each time step in the evolution of the system
3. *Two-step models*, which first solve mass momentum and energy balances for each time step and then reequilibrate the chemistry using a distribution-of-species code.

Empirically determined retardation factors (either partition coefficients or breakthrough curve measurements, which are the change in solute concentration measured over time in laboratory or field experiments) have been widely used because of their inherent simplicity.[162] Modeling of specific geochemical partition and transformation processes is not necessary if the retardation factor can be determined empirically.

The problems with linear-distribution coefficients apply equally to any retardation factor derived from them. Field measurements can be made but are expensive to obtain and highly site specific. Nevertheless, retardation factors provide some insight into organic chemical transport.

Integrated and two-step chemical-transport models incorporate distribution-of-species or reaction-progress codes into hydrologic transport codes. The few studies in which the two approaches have been tested using the same set of field data have agreed reasonably well; thus one approach does not have an obvious advantage over the other. The two-step approach tends to be computationally less intensive than the integrated approach but may have difficulty maintaining mass balance when rapid precipitation and dissolution occur.[120]

A number of models of both types have been described in the literature. Of the models, DYNAMIX would appear to have the greatest potential for use in simulating chemical transport in the deep-well environment because it incorporates the reaction-progress code PHREEQE, which can handle deep-well temperatures. PHREEQE, however, does not incorporate pressure equilibria.

10.7 CASE STUDIES OF DEEP-WELL INJECTION OF INDUSTRIAL WASTE

This section discusses how field studies can be used in geochemical fate assessment and includes six cases of deep-well-injection facilities, documenting the geochemistry of the injected hazardous and other industrial wastes. Each case study is organized in the same format, with section headings as follows:

1. *Injection Facility Overview* describes the type of facility, its current status, and the characteristics of the injected wastes, and presents a brief history of injection and monitoring activities, including the distance traveled by the waste.
2. *Injection/Confining-Zone Lithology and Chemistry* provides information on the geology and chemistry of the injection zone formation fluids.
3. *Chemical Processes Observed* briefly describes the types of interactions and major physical effects that have been observed at the site and evaluates their significance.

Table 10.19 summarizes information about each study, including the location of the well, the lithology of the injection zone, waste characteristics, and the major geochemical processes observed. Current commercial-hazardous-waste, deep-well-injection facilities can be found on the Environment, Health and Safety Online (EHSO) web site.[163]

Field studies are an important complement to geochemical modeling and to laboratory studies. The following are two ways to investigate the interactions between injected wastes and reservoir material:

1. Direct observation of the injection zone and overlying aquifers using monitoring wells
2. Backflushing of the injected waste

In both instances, samples of the fluids in the zone are collected at intervals to characterize the nature of geochemical reactions and to track changes over time.

Monitoring wells drilled into the injection zone at selected distances and directions from the injection well allow direct observation of formation water characteristics and the interactions that occur when the waste front reaches the monitoring well. When placed near the injection well in the aquifer above the confining layer, monitoring wells can detect the upward migration of wastes caused by casing or confining-layer failure. Foster and Goolsby[164] describe detailed methods for constructing monitoring wells.

Monitoring wells have several advantages, in that time-series sampling of the formation over extended periods is easy and the passage of the waste front can be observed precisely. Disadvantages include cost and the potential for upward migration of wastes if monitoring well casings fail. A monitoring well at the Monsanto plant had to be plugged when unneutralized waste reached it because of fears that the casing would corrode. The three Florida case studies and the North Carolina case study illustrate the usefulness of monitoring wells.

Backflushing of injected wastes can also be a good way to observe waste/reservoir geochemical interactions. Injected wastes are allowed to backflow (if formation pressure is above the elevation of the wellhead) or are pumped to the surface. Backflowed wastes are sampled periodically (and reinjected when the test is completed); the last sample taken will have had the longest residence time in the injection zone. Keely[165] and Keely and Wolf[166] describe this technique for characterizing

TABLE 10.19
Summary of Case Studies

Location	Lithology	Wastes	Processes Observed
Florida			
Pensacola (Monsanto)	Limestone	Nitric acid	Neutralization
		Inorganic salts	Bacterial denitrification
		Organic compounds	
Pensacola (American Cyanamid)	Limestone	Acrylonitrile	Bacterial denitrification
		Sodium salts (nitrate, sulfate thiocyanate)	No retardation of thiocyanate ions
Belle Glade	Carbonate	Hot acid	Neutralization
		Organic plant wastes	Bacterial sulfate reduction
			Methane production
North Carolina			
Wilmington	Sand	Organic acids	Neutralization
	Silty sand	Formaldehyde	Dissolution–precipitation
	Limestone	Methanol	Complexation
			Adsorption
			Bacterial sulfate and iron reduction
			Methane production
Illinois			
Tuscola	Dolomite	Hydrochlorite acid	Neutralization
			Dissolution
			CO_2 gas production
Texas			
Not specified	Miocene sand	1. Organic acids	Precipitation
		Organic compounds	Adsorption (inferred)
		2. Alkaline salts	
		Organic compounds	

Source: U.S. EPA, Assessing the Geochemical Fate of Deep-Well-Injected Hazardous Waste: A Reference Guide, EPA/625/6-89/025a, U.S. EPA, Cincinnati, OH, June 1990.

contamination of near-surface aquifers and suggest using logarithmic time intervals for chemical sampling. The three Florida studies all present results from backflushing experiments.

The advantages of backflushing are reduced cost compared with that of monitoring wells and reduced sampling time (sampling takes place only during the test period). Disadvantages include less precise time- and distance-of-movement determinations and the need to interrupt injection and to have a large enough area for backflushed fluid storage before reinjection.

10.7.1 CASE STUDY NO. 1: PENSACOLA, FL (MONSANTO)

10.7.1.1 Injection-Facility Overview

Monsanto operates one of the world's largest nylon plants on the Escambia River about 13 miles north of Pensacola, Florida. The construction, operations, and effects of the injection-well system at this site have been extensively documented by the U.S. Geological Survey in cooperation with the Florida Bureau of Geology. Pressure and geochemical effects are reported by Goolsby,[67] Faulkner and Pascale,[167] and Pascale and Martin.[168] Additional microbiological data are reported by Willis and colleagues[169] and Elkan and Horvath.[170] Major chemical processes observed at the site include neutralization, dissolution, biological denitrification, and methanogenesis.

The waste is an aqueous solution of organic monobasic and dibasic acids, nitric acid, sodium and ammonium salts, adiponitrile, hexamethylenediamine, alcohols, ketones, and esters.[67] The waste also contain cobalt, chromium, and copper, each in the range of 1 to 5 mg/L. Wastestreams with different characteristics, produced at various locations in the nylon plant, are collected in a large holding tank; this composite waste is acidic. The specific characteristics of the waste varied somewhat as a result of process changes (e.g., after 1968 more organic acids and nitric acid were added). Until mid-1968, wastes were partially neutralized by pretreatment. After that, unneutralized wastes were injected. No reason was reported for suspending treatment. Goolsby[67] reports pH measurements ranging from a high of 5.6 in 1967 (at which time the pH was raised before injection by adding aqueous ammonia) to a low of 2.4 in 1971, and Eh ranging from +300 mV in 1967 to +700 mV in 1971. The chemical oxygen demand in 1971 was 20,000 mg/L.

Monsanto began injecting wastes into the lower limestone of the Floridan aquifer in 1963. In mid-1964, a second well was drilled into the formation about 300 m (1000 ft) southwest of the first. A shallow monitoring well was placed in the aquifer above the confining layer about 30 m (100 ft) from the first injection well, and a deep monitoring well was placed in the injection zone about 400 m (1300 ft) south of both injection wells. The deep monitoring well (henceforth referred to as the near-deep monitoring well) was plugged with cement in 1969. In late 1969 and early 1970, two additional deep monitoring wells were placed in the injection formation, 2.4 km (1.5 miles) south-southeast (downgradient) and 3 km (1.9 miles) north-northwest (upgradient) of the site. From 1963 to 1977, about 50 billion liters (13.3 billion gal) of waste were injected. During the same period, injection pressures ranged from 8.8 to 16.5 kg/cm^2 (125 to 235 psi). Since then, a third injection well has been added.

Ten months after injection of neutralized wastes began, chemical analyses indicated that dilute wastes had migrated 1300 ft to the nearest deep monitoring well. Injection of unneutralized wastes began in April 1968. Approximately 8 months later, unneutralized wastes reached the near-deep monitoring well, indicating that the neutralization capacity of the injection zone between the injection wells and the monitoring well had been exceeded. At this point, the monitoring well was plugged with cement from bottom to top because operators were concerned that the acidic wastes could corrode the steel casing and migrate upward.[67] The rapid movement of the waste through the limestone indicated that most of it migrated through a more permeable section, which was about 20 m (65 ft) thick. By mid-1973, 10 years after injection began, a very dilute waste front arrived at the south monitoring well, 2.4 km (1.5 miles) away. As of early 1977, there was no evidence that wastes had reached the upgradient monitoring well. The shallow monitoring well remained unaffected during the same period.

Increases in permeability caused by limestone dissolution approximately doubled the injection index (the amount of waste that can be injected at a specified pressure). As of 1974, the effects of the pressure created by the injection were calculated to extend more than 40 miles radially from the injection site.[167] An updip movement of the freshwater/saltwater interface in the injection-zone aquifer, which lies less than 32 km (20 miles) from the injection wells, was also observed.

10.7.1.2   Injection/Confining-Zone Lithology and Chemistry

The lower limestone of the Floridan aquifer is used as the injection zone (at 430 to 520 m), and the Bucatunna clay member of the Byram formation (about 67 m thick) serves as the confining layer. Figure 10.11 shows the stratigraphy of the area, and Figure 10.12 shows the local stratigraphy and the monitoring well installations. The formation water in the injection zone is a highly saline (11,900 to 13,700 mg/L TDS) sodium-chloride solution. The Eh of samples collected from two monitoring wells located in the injection formation ranged from +23 to –32 mV, indicating reducing conditions in the injection zone that would favor anaerobic biodegradation.

The injection zone contains about 7900 mg/L chloride, but less than 32 km (20 miles) northeast of the injection site, chloride concentrations are less than 250 mg/L. Under natural conditions, water

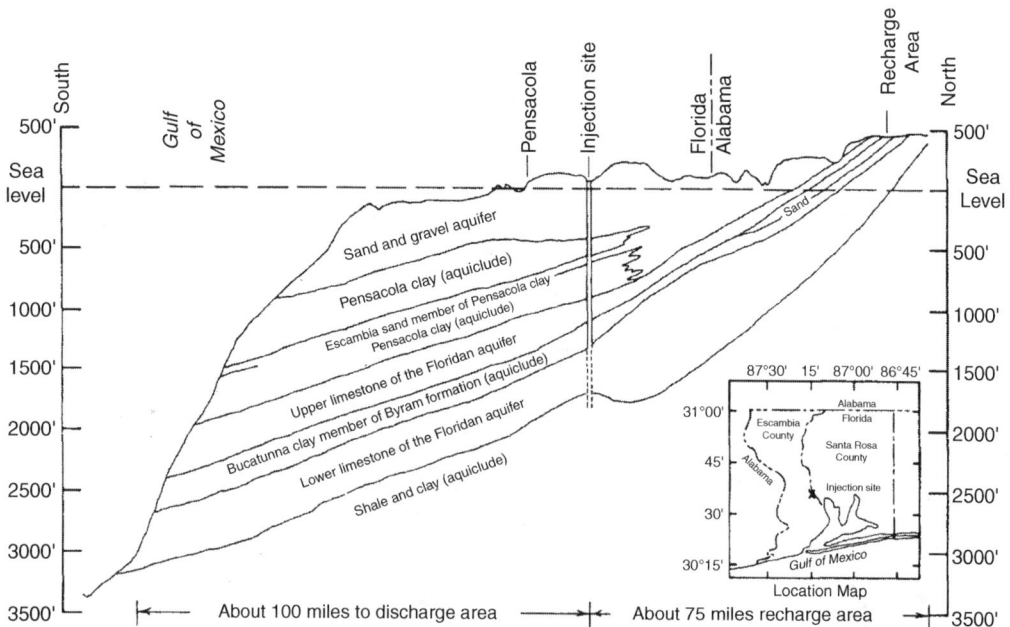

FIGURE 10.11   Generalized north–south geologic section through Southern Alabama and Northwestern Florida. (From U.S. EPA, Assessing the Geochemical Fate of Deep-Well-Injected Hazardous Waste: A Reference Guide, EPA/625/6-89/025a, U.S. EPA, Cincinnati, OH, June 1990.)

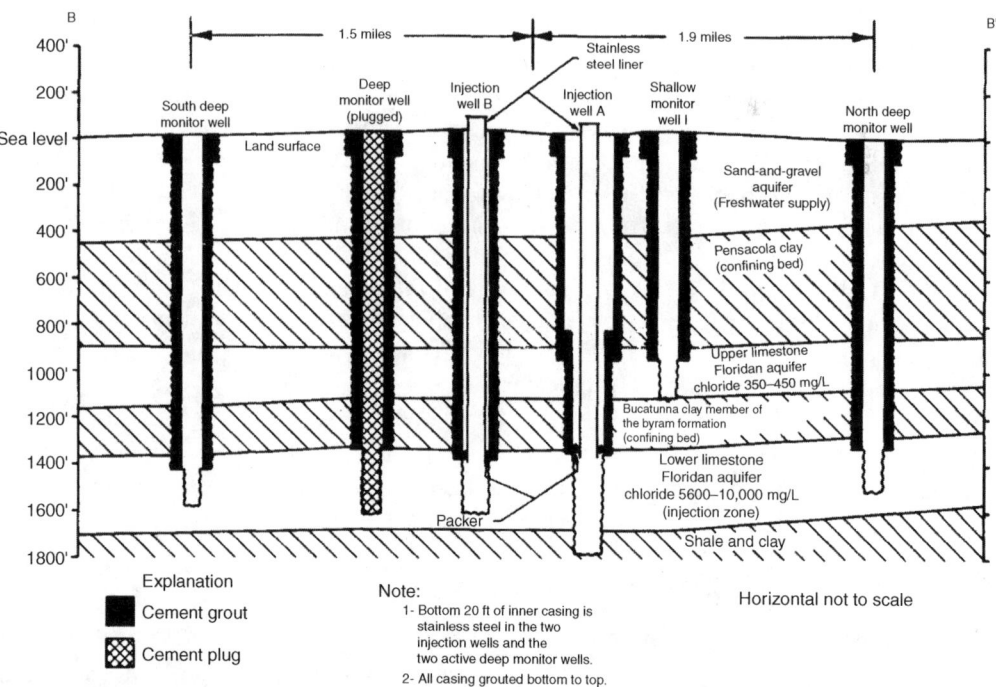

FIGURE 10.12   Monsanto Injection Facility hydrogeologic cross-section. (From U.S. EPA, Assessing the Geochemical Fate of Deep-Well-Injected Hazardous Waste: A Reference Guide, EPA/625/6-89/025a, U.S. EPA, Cincinnati, OH, June 1990.)

in the injection zone moves slowly south-southwestward toward the Gulf of Mexico, where it is assumed to discharge about 160 km (100 miles) offshore. The pre-injection hydraulic gradient was about 0.25 m/km (1.3 ft/mile).

10.7.1.3 Chemical Processes Observed

As a result of dissolution of the limestone by the partly neutralized acid wastes, calcium concentrations more than doubled in the near-deep monitoring well 10 months after injection started in 1963.[67] In early 1966, however, they dropped to background levels (about 200 mg/L), possibly in response to biochemical decomposition of the waste. In September 1968, after about 300 million gallons of the acidic, unneutralized waste had been injected, the calcium concentration began to increase again. An abrupt increase in calcium to 2700 mg/L accompanied by a decrease in pH to 4.75 in January 1969 led to the decision to plug the near-deep monitoring well.

In an attempt to find out how fast the waste was reacting with limestone, a 3-h backflushing experiment, in which waste was allowed to flow back out of the injection well, yielded some unexpected results. The increase in pH of the neutralized waste could not be fully accounted for by the solution of limestone as determined from the calcium content of the backflushed liquid; the additional neutralization apparently resulted from reactions between nitric acid and alcohols and ketones in the original waste induced by increased pressure in the injection zone compared to surface conditions.[41]

The lack of nitrates (which were present at levels of 545 to 1140 mg/L in the waste) in the near-deep monitoring well, combined with the presence of nitrogen gas, indicated that degradation by denitrifying bacteria had taken place.[67] Backflushing shortly before injecting unneutralized wastes confirmed denitrification. Nitrate concentrations decreased rapidly as the backflushed waste was replaced by formation water. Similar backflushing experiments conducted after unneutralized wastes were injected, however, provided no evidence of denitrification, indicating that microbial activity was suppressed in the portion of the zone containing unneutralized wastes.

Elkan and Horvath[170] performed a microbiological analysis of samples taken from the north and south deep monitoring wells in December 1974, about 6 months after the dilute waste front had reached the south well. Both denitrifying and methanogenic bacteria were observed. The lower numbers and species diversity of organisms observed in the south monitoring well compared with those in the north well indicated suppression of microbial activity by the dilute wastes.

Between September 1973 and March 1977 bicarbonate concentrations increased from 282 mg/L to 636 mg/L and dissolved organic carbon increased from 9 mg/L to 47 mg/L. These increases were accompanied by an increase in the dissolved-gas concentration and a distinctive odor like that of the injected wastes. The pH, however, remained unchanged. During the same period, dissolved methane increased from 24 mg/L to 70 mg/L, indicating increased activity by methanogenic bacteria. The observation of denitrification in the near-deep monitoring well and methanogenesis in the more distant south monitoring well fit the redox-zone biodegradation model.

Significant observations made at this site are:

1. Organic contaminants (as measured by dissolved organic carbon) continue to move through the aquifer even when acidity has been neutralized.
2. Even neutralized wastes can suppress microbial populations.

10.7.2 CASE STUDY NO. 2: PENSACOLA, FL (AMERICAN CYANAMID)

10.7.2.1 Injection-Facility Overview

American Cyanamid Company operates a plant near Milton, Florida, which lies about 12 miles northeast of Pensacola and about 8 miles east of the Monsanto plant discussed in the previous sections. Chemical changes caused by the injection of acidic wastes from this plant have been reported by Ehrlich and colleagues[89] and Vecchioli and colleagues,[171] with the former citation providing the most complete information on the site. This case study illustrates the complexity of assessing the geochemical fate of mixed wastes. Acrylonitrile was detoxified by biological reduction, whereas sodium thiocyanate remained unaltered.

The facility combines acidic wastestreams from various plant operations in a holding pond where they are mixed and aerated. The waste is pumped from the pond and neutralized with sodium hydroxide. The neutralized wastes are treated with alum to flocculate suspended solids and then passed through mixed-media filters. A small amount of hydrogen peroxide solution (amount unspecified) is added before filtration to inhibit microbial growth on the filters. The pretreated waste that is injected contains high concentrations of sodium nitrate, sodium sulfate, sodium thiocyanate (an inorganic cyanide compound), and various organic compounds, including acrylonitrile (a listed hazardous waste). The average pH of the waste is 5.8, and the average chemical oxygen demand is 1690 mg/L.

A primary injection well and a standby well are situated about 460 m (1500 ft) apart. A shallow monitoring well is located near the primary injection well in the upper limestone Floridan aquifer that overlies the confining Bucatunna clay. Two deep monitoring wells in the injection zone are located 300 m (1000 ft) southwest and 2492 m (8170 ft) northeast of the primary injection well.

Waste injection began in June 1975, and waste was first detected in the downgradient southwest deep monitoring well about 260 days later. To analyze the waste's physical and chemical properties after injection, the primary injection well was allowed to backflow into a holding pond for 5 days in November 1977. This waste was sampled periodically (and reinjected when the test was completed). About 4 years after injection began; dilute waste arrived at the standby injection well 476 m (1560 ft) south of the primary well.

10.7.2.2 Injection/Confining-Zone Lithology and Chemistry

The injection well is in the same area as the Monsanto well, so the geology and native water chemistry are very similar to that described before. Figure 10.13 shows the stratigraphy of the immediate area

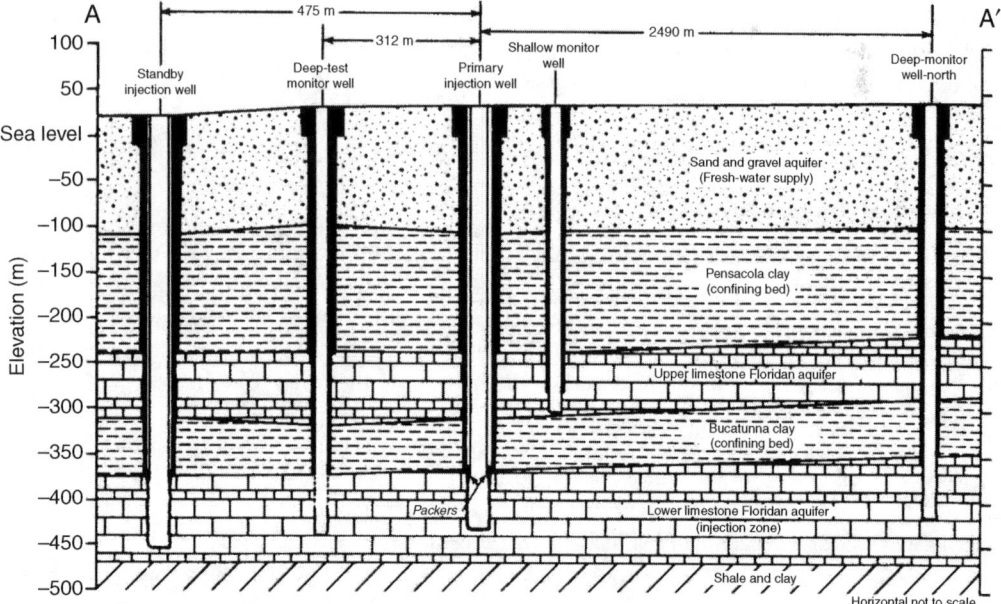

FIGURE 10.13 American Cyanamid Injection Facility hydrogeologic cross-section. (From U.S. EPA, Assessing the Geochemical Fate of Deep-Well-Injected Hazardous Waste: A Reference Guide, EPA/625/6-89/025a, U.S. EPA, Cincinnati, OH, June 1990.)

and distances between the injection and monitoring wells. The lower limestone of the Floridan aquifer is used as the injection zone (1230 to 1440 ft), and the confining Bucatunna clay is about 50 m (165 ft) thick. TDS levels range from 12,000 to 12,700 mg/L, with chloride ion concentrations of 6700 mg/L. The pH ranges from 7.3 to 7.6, and temperature from 30°C to 32°C. Caliper and flowmeter tests made in the injection wells suggest that the waste moves almost exclusively within the top 18 m (55 ft) of the lower limestone. The preinjection groundwater flow direction is south-southwest.

10.7.2.3 Chemical Processes Observed

The Eh of the injected waste dropped rapidly from +40 mV to −80 mV in the first 40 h after injection began and remained at about −80 mV thereafter. Denitrifying bacteria detoxified the acrylonitrile by mineralizing the compound, breaking it down into bicarbonate and ammonia. The nitrates were degraded to nitrogen gas. The backflow test produced data indicating that these transformations were about 90% complete within 25 m (82 ft) of the injection well and virtually 100% complete within 100 m (328 ft). These results are an example of a biodegradation-dispersion curve. Denitrifying-bacteria densities increased from traces (101 organisms/100 mL in the native groundwater) to large populations (10 to 10^8 organisms/100 mL) in injected wastes that had been in the aquifer for several days.

Sodium thiocyanate (NaSCN) was first detected in the closest monitoring well (300 m away) 260 days after injection began. Ammonium ions (a reaction product of biomineralization) did not appear as a contaminant until 580 days after initial injection. This delay was probably the result of ion exchange or other adsorption processes and may be an example of an adsorption–dispersion curve. Because sodium thiocyanate in the waste remained unchanged during its movement through the injection zone, it was used to detect the degree of mixing that took place between the waste liquid and native water in an observation well. Thus the appearance of sodium thiocyanate as well as an increase in chemical oxygen demand in the standby well 4 years after injection began signaled the arrival of wastes at that location.

This case study is interesting in that one hazardous waste (acrylonitrile) was quickly rendered nonhazardous after injection, whereas another (sodium thiocyanate) showed no evidence of decomposition during the duration of the study. The implication for geochemical fate assessment is that research should focus on the compounds likely to be most resistant to decomposition or immobilization, as they will be the ones most critical in demonstrating containment in a no-migration petition.

10.7.3 Case Study No. 3: Belle Glade, FL

10.7.3.1 Injection-Facility Overview

The Belle Glade site, located southeast of Lake Okeechobee in south-central Florida, illustrates some of the problems that can develop with acidic-waste injection when carbonate rock is the confining layer. Contributing factors to the contamination of the aquifer above the confining zone were the dissolution of the carbonate rock and the difference in density between the injected wastes and the formation fluids. The injected waste was less dense than the groundwater because of its lower salinity and higher temperature.[172]

The injected fluids include the effluent from a sugar mill and the waste from the production of furfural, an aldehyde processed from the residues of processed sugar cane. The waste is hot (about 75°C to 93°C), acidic (pH 2.6 to 4.5), and has high concentrations of organics, nitrogen, and phosphorus.[173] The waste is not classified as hazardous under 40 CFR 261, and the well is currently regulated by the State of Florida as a nonhazardous injection well. The organic carbon concentration exceeds 5000 mg/L.

The well was originally cased to a depth of 456 m (1495 ft), and the zone was left as an open hole to a depth of 591 m (1939 ft). The depth of the zone has been increased twice. Seasonal injection (fall, winter, and spring) began in late 1966; the system was inactive during late summer.

Injection rates ranged from 25 to 50 L/h (400 to 800 gal/min), and wellhead injection pressures ranged from 2.1 to 4.2 kg/cm^2 (30 to 60 psi). By 1973, injection had become more or less continuous. From 1966 to 1973, more than 4.16 billion liters (1.1 billion gal) of waste had been injected.[172]

At the time injection began, a shallow monitoring well was placed 23 m (75 ft) south of the injection well in the upper part of the Floridan aquifer above the confining layer. A downgradient, deep monitoring well was placed in the injection zone 300 m (1000 ft) southeast of the injection well. Another shallow well, located 3.2 km (2 miles) southeast of the injection site at the University of Florida's Everglades Experiment Station, has also been monitored for near-surface effects.

Acetate ions from the injected waste were detected in the deep monitoring well 300 m (1000 ft) southeast of the injection well in early 1967, a matter of months after injection began.[174] In 1971, about 27 months after injection began; evidence of waste migration was detected at a shallow monitoring well in the upper part of the Floridan aquifer. Dissolution of the carbonate confining layer by the acidic waste was the main reason for the upward migration. However, the lower density of the injected wastes compared with that of the formation waters (0.98 g/mL vs. 1.003 g/mL) served to accelerate the rate of upward migration.[174] In an attempt to prevent further upward migration, the injection well was deepened to 684 m (2242 ft), and the inner casing was extended and cemented to 591 m (1938 ft). When waste injection was resumed, evidence of upward migration to the shallow aquifer was observed only 15 months later. By late 1973, 7 years after injection began, the waste front was estimated to have migrated 1 to 1.6 km (0.6 to 1 mile) from the injection well.[173]

The injection well was deepened a third time, to a depth of 900 m (3000 ft).[175] A new, thicker confining zone of dense carbonate rock separates the current injection zone from the previous zone. As of early 1989, the wastes were still contained in the deepest injection zone. For details on acid injection into carbonate rock refer to Clark.[176]

10.7.3.2   Injection/Confining-Zone Lithology and Chemistry

The wastes are injected into the lower part of the carbonate Floridan aquifer, which is extremely permeable and cavernous. The natural direction of groundwater flow is to the southeast. The confining layer is 45 m (150 ft) of dense carbonate rocks. The chloride concentration in the upper part of the injection zone is 1650 mg/L, increasing to 15,800 mg/L near the bottom of the formation.[172] The sources used for this case study did not provide any data on the current injection zone. The native fluid was basically a sodium-chloride solution but also included significant quantities of sulfate (1500 mg/L), magnesium (625 mg/L), and calcium (477 mg/L).

10.7.3.3   Chemical Processes Observed

Neutralization of the injected acids by the limestone formation led to concentrations of calcium, magnesium, and silica in the waste solution that were higher than those in the unneutralized wastes. Anaerobic decomposition of the organic matter in the injected waste apparently occurred through the action of both sulfate-reducing and methanogenic bacteria. Sulfate-reducing bacteria were observed in the injected wastes that were allowed to backflow to the surface. Sulfate levels in the native groundwater declined by 45%, and the concentration of hydrogen sulfide increased by 1600%. Methane fermentation (reduction of CO_2 to CH_4) was also inferred from the presence of both gases in the backflow fluid, but the presence of methanogenic bacteria was not confirmed. Increased hydrogen sulfide concentrations produced by the bacteria during biodegradation and the subsequent decrease in sulfate/chloride ratio in the observation wells were taken as indicators of upward and lateral migration. Migration into the shallow monitoring well was also indicated by a decline in pH from around 7.8 to 6.5, caused by mixing with the acidic wastes.

Chemical analyses of the backflowed injected waste that had been in the aquifer for about 2.5 months (for which some dilution had occurred) indicated that COD was about half that of the original waste. Samples that had been in residence for about 5 months had a COD approximately one-quarter that of the original waste (12,200 mg/L in the original waste compared with 4166 mg/L in the samples). The percent reduction in COD resulting from bacterial action rather than dilution was not estimated.

10.7.4 CASE STUDY NO. 4: WILMINGTON, NC

10.7.4.1 Injection-Facility Overview

The Hercules Chemical, Inc. (now Hercufina, Inc.), facility, 4 miles north of Wilmington, North Carolina, attempted deep-well injection of its hazardous wastes from May 1968 to December 1972, but had to discontinue injection because of waste–reservoir incompatibility and unfavorable hydrogeologic conditions. The U.S. Geological Survey conducted extensive geochemical studies of this site until the well was abandoned.[102,177–179] Biodegradation processes were also studied.[170] More geochemical-fate processes affecting injected organic wastes have been documented at this site than at any other.

Hercules Chemical produced an acidic organic waste derived from the manufacture of dimethyl terphthalate, which is used in the production of synthetic fiber. The average dissolved organic carbon concentration was about 7100 mg/L and included acetic acid, formic acid, p-toluic acid, formaldehyde, methanol, terphthalic acid, and benzoic acid. The pH ranged from 3.5 to 4.0. The waste also contained traces (less than 0.5 mg/L) of 11 other organic compounds, including dimethyl phthalate, a listed hazardous waste.

From May 1968 to December 1972, the waste was injected at a rate of about 300,000 gal/d. The first injection well was completed to a depth of 259 to 313 m (850 to 1025 ft) (i.e., cased from the surface to 259 m with screens placed in the most permeable sections of the injection zone to a depth of 313 m). One shallow observation well was placed 15 m (50 ft) east of the injection site at a depth of 210 m (690 ft). Four deep monitoring wells were also placed in the injection zone, one at 15 m (50 ft) and three at 45 m (150 ft) from the injection well.

The injection well became plugged after a few months of operation because of the reactive nature of the wastes and the low permeability of the injection zone. The actual plugging process was caused both by reprecipitation of the initially dissolved minerals and by plugging of pores by such gaseous products as carbon dioxide and methane. When the first well failed, a second injection well was drilled into the same injection zone about 5000 ft north of the first, and injection began in May 1971. Nine additional monitoring wells (three shallow, and six deep) were placed at distances ranging from 450 to 900 m (1500 to 3000 ft) from the second injection well. Injection was discontinued in 1972 after the operators determined that the problems of low permeability and waste–reservoir incompatibility could not be overcome. Monitoring of the waste movement and subsurface environment continued into the mid-1970s in the three monitoring wells located 450 to 600 m (1500 to 2000 ft) from the injection wells.

Within 4 months, the waste front had passed the deep observation wells located within 45 m (150 ft) of the injection well. About 9 months after injection began; leakage into the aquifer above the confining layer was observed. This leakage was apparently caused by the increased pressures created by formation plugging and by the dissolution of the confining beds and the cement grout surrounding the well casing of several of the deep monitoring wells, caused by organic acids.

Eight months after injection began in the second injection well, wastes had leaked upward into the adjacent shallow monitoring well. The leak apparently was caused by the dissolution of the cement grout around the casing. In June 1972, 13 months after injection began in the second well, the waste front reached the deep monitoring well located 450 m (1500 ft) northwest of the injection well. Waste injection ended in December 1972. As of 1977, the wastes were treated in a surface facility.[170]

10.7.4.2 Injection/Confining-Zone Lithology and Chemistry

The injection zone consisted of multiple Upper Cretaceous strata of sand, silty sand, clay, and some thin beds of limestone (see Figure 10.14). The clay confining layer was about 30 m (100 ft) thick.

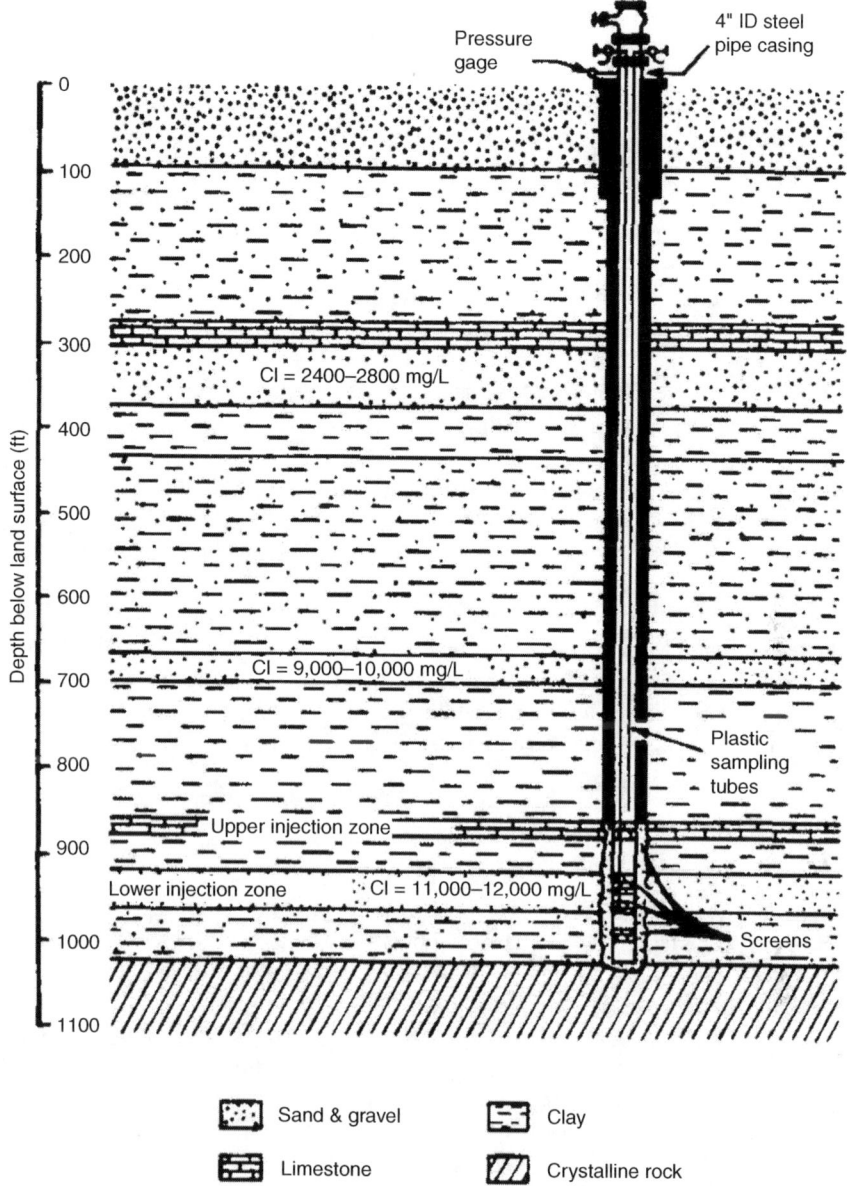

FIGURE 10.14 Diagram showing construction features and lithologic log of North Observation Well, Wilmington, NC. (From U.S. EPA, Assessing the Geochemical Fate of Deep-Well-Injected Hazardous Waste: A Reference Guide, EPA/625/6-89/025a, U.S. EPA, Cincinnati, OH, June 1990.)

The TDS concentration in the injection-zone formation water was 20,800 mg/L, with sodium chloride the most abundant constituent.

10.7.4.3 Chemical Processes Observed

A number of chemical processes were observed at the site[178,179]:

1. The waste organic acids dissolved carbonate minerals, alumino-silicate minerals, and iron/manganese-oxide coatings on the primary minerals in the injection zone.

2. The waste organic acids dissolved and formed complexes with iron and manganese oxides. These dissolved complexes reprecipitated when the pH increased to 5.5 or 6.0 because of neutralization of the waste by the aquifer carbonates and oxides.

3. The aquifer mineral constituents adsorbed most waste organic compounds, with the exception of formaldehyde. Adsorption of all organic acids except phthalic acid increased with a decrease in waste pH.

4. Phthalic acid was complexed with dissolved iron. The concentration of this complex decreased as the pH increased because the complex coprecipitated with the iron oxide.

5. Biochemical waste transformation occurred at low waste concentrations, resulting in the production of methane. Additional microbial degradation of the waste resulted in the reduction of sulfates to sulfides and ferric ions to ferrous ions.

When the dilute waste front reached the North Observation Well in June 1972 microbial populations rapidly increased in this well, with methanogenesis being the major degradative process.[180] Elkan and Horvath[170] found greater numbers and species diversity of microorganisms in the observation well, which contained dilute wastes, than in the observation well, which was uncontaminated. In laboratory experiments, however, DiTommaso and Elkan[180] found that bacterial growth was inhibited as the concentration of waste increased and could not decompose the waste at the rate it was being injected.

This case study illustrates the importance of dissolution/precipitation reactions in determining waste– reservoir compatibility. Adsorption was observed to immobilize most of the organic constituents in the waste except for formaldehyde. As with the Monsanto case study, biodegradation was an important process when wastes were diluted by formation waters, but the process became inhibited when undiluted waste reached a given location in the injection zone.

10.7.5 CASE STUDY NO. 5: ILLINOIS HYDROCHLORIC ACID-INJECTION WELL

10.7.5.1 Injection-Facility Overview

This case study is an example of a well blowout resulting from the neutralization of acid by carbonate rock. Kamath and Salazar[181] and Panagiotopoulos and Reid[182] both discuss the same incident. Although they do not specify the location, Brower and colleagues[183] identify the site as the Cabot Corporation injection well, near Tuscola, Illinois.

The waste hydrochloric acid (HCl) injected at the site was a byproduct of a combustion process at 1633°C (2972°F). When not recovered, the acidic stream was dumped into holding ponds where it was cooled to about 24°C (75°F) before injection. The concentration of injected acid typically varied from 0.5 to 5% HCl, but ranged as high as about 30%. (The pH of injected acid that backflowed during one blowout incident ranged from 0.5 to 1.3.)

The injection well was cased to a depth of about 1495 m (4900 ft) and extended into dolomite to a total depth of 1617 m (5300 ft). Injection began in the early 1960s and averaged around 340 L/min (90 gal/min). The natural fluid level was 60 m (200 ft) below the wellhead, and wastes were injected using gravity flow; that is, the pressure head of the well when filled to the surface with fluid was sufficient to inject fluids without pumping under pressure.[181]

Between 1973 and 1975, several blowouts caused surface water pollution and fish kills. The most serious occurred in 1975 after unusually high concentrations of HCl (ca. 30%) were injected intermittently for several weeks. The well refused to accept additional acid under gravity flow. At first the operators thought the well bore had become plugged, and they pumped a concentrated calcium-chloride solution down the hole to dissolve precipitates that might have formed. Shortly thereafter the well tubing broke, pressure suddenly rose to 37 kg/cm² (450 psi), and a section of the upper tubing was ejected through the wellhead along with acid and annulus fluids. Backflow was stopped for a while by draining cold water from a fire hydrant into the well at 190 L/min

(50 gal/min). The well erupted again the next day, however, with a 3-m (10-ft) gusher discharging at 946 L/min (250 gal/min). The blowout was brought under control 2 d later when a blowout preventer was installed.

10.7.5.2   Injection/Confining-Zone Lithology and Chemistry

The injection zone was a cavernous dolomite, and the native groundwater was very saline, with TDS levels ranging from 21,000 to 26,000 mg/L. No information was provided on the confining layer, but it is discussed in the work by Brower and colleagues[183] in detail.

10.7.5.3   Chemical Processes Observed

The HCl dissolved the dolomite, forming carbon dioxide (CO_2) gas. Under normal circumstances this gas remains in solution, but if the temperature of the acid or the acid concentration exceed certain limits, CO_2 evolves as a gas and accumulates in the upper portion of the cavity. The escape of even small amounts of CO_2 into the injection pipe can serve as a driving force to reverse the flow of the injected liquids, because as the CO_2 rises, pressure decreases and the gas expands.

There is some disagreement as to which parameter is most critical to gas blowout. Based on analysis of CO_2 phase behavior at different temperatures and pressures, Kamath and Salazar[181] concluded that gas blowout becomes hazardous if the temperature of the injected HCl exceeds 88°F. Panagiotopoulos and Reid[182] concluded that HCl concentration is the critical factor and that HCl concentrations exceeding 6% will evolve CO_2 gas and create a blowout hazard. Both sets of investigators explained the circumstances of this case study in terms of their respective models.

10.7.6   CASE STUDY NO. 6: TEXAS PETROCHEMICAL PLANT

10.7.6.1   Injection-Facility Overview

This case study involves an unnamed petrochemical plant located about 15 miles inland from the Texas Gulf Coast, described by Donaldson and Johansen.[184] It illustrates two approaches to injecting incompatible wastestreams to prevent well plugging by precipitation: surface treatment and multiple injection wells.

The plant began full-scale operation in 1962 and produced acetic, adipic, and propionic acids; acetaldehyde; butanol; hexamethyldiamine; vinyl acetate; nylon; and other chemical products from petroleum-base stocks. The effluent was collected at waste treatment facilities as two separate mixtures. Because mixing two wastestreams produced considerable precipitation, the wastestreams were processed and injected separately into two wells.

Organic constituents in the first wastestream totaled about 14,000 mg/L (acetaldehyde, acetaldol, acetic acid, butanol-1, butyraldehyde, chloroacetaldehyde, crotonaldehyde, phenol, and propionic acid) and about 5200 mg/L inorganic constituents. The pH ranged from 4 to 6, and TDS ranged from 3000 to 10,000 mg/L.

The second wastestream contained amines and nitrates generated from the manufacture of nylon, hydrocarbon solvents used in processing, and other minor constituents. Organic constituents (amyl alcohol, cyclohexane, dodecane, hexanol, 1-hexylamine, 1,6-hexylamine, methanol, and valeric acid) totaled about 4700 mg/L. Inorganic constituents in the second wastestream totaled about 21,350 mg/L, including 7500 mg/L nitrate and 4600 mg/L nitrite. The second wastestream was basic, with a pH ranging from 8 to 10. The composition of the wastes changed over time when processes changed or a new unit was installed. Several new process wastes (unspecified) that were incompatible with either wastestream were made compatible by adjusting the pH and diluting them.

Injection began in both wells in mid-1963. The injection zone for Well No. 1 was 13.7 m (45 ft) thick beginning at about 1037 m (3400 ft) below the surface. Well No. 2 was located 824 m (2700 ft) north of Well No. 1, and the injection zone was located between 991 and 1083 m (3520 and 3550 ft).

Donaldson and Johansen[184] mention no monitoring wells at the site. About 6 years after injection began, pressure interference from the two injection wells was observed. During the same period, the fluid front from Well No. 1 was about 223 m (730 ft) from the well bore.

10.7.6.2   Injection/Confining-Zone Lithology and Chemistry

The injection formation was loosely consolidated, fine-grained Miocene sand. The confining strata between the base of the freshwater aquifer and the injection zone included about 366 m (1200 ft) of relatively impermeable shale and clay beds with individual zone thickness ranging from 3 to 75 m (10 to 245 ft).

10.7.6.3   Chemical Processes Observed

Well head pressures increased when injection was stopped at Well No. 1 for more than 24 h, apparently caused by a combination of precipitation reactions and backflow of sand. Injecting a slug of brine after every period of interrupted flow solved this problem. Movement of the main organic constituents (n-hexylamine, butanal, butanol, and phenol) was assumed to be slowed by adsorption. This conclusion was based on laboratory adsorption experiments by involving a different geologic formation (Cottage Grove sandstone); no direct observations were made of the injected waste. For current hazardous waste injection wells in Texas, the reader can refer to Texas Environmental Profiles web site for on-line resources for the State of Texas.[185]

NOMENCLATURE

B	Bacterial concentration
C	Concentration of adsorbed substance in solution, µg/mL
C_T	Dissolved plus adsorbed phase concentration of compound C, mol/L or µg/L
$[H^+]$	Concentration of hydrogen ion, mole/L
k	Langmuir coefficient related to adsorption bonding energy (mL/g)
k_1	Empirical biodegradation rate constant
k_a	Acid-catalyzed hydrolysis rate constant, L/mol/sec
k_b	Base-catalyzed hydrolysis rate constant, L/mol/sec
k_B	First-order biodegradation coefficient
k_{B2}	Second-order biodegradation coefficient
k_H	Specific hydrolysis rate constant, L/sec
k_n	Natural hydrolysis rate constant for the pH-independent reactions of a chemical with water, L/sec
K	Empirical coefficient
K_d	Distribution coefficient
K_{oc}	Organic-carbon partition coefficient
K_p	Partition coefficient
N	Empirical coefficient
$[OH^-]$	Concentration of hydroxide ion, mol/L
R	Rate of hydrolysis, mole/L/sec or µg/L/sec
S	Amount adsorbed (µg/g solid)
S_{max}	Maximum adsorption capacity (µg/g soil)
$t_{1/2}$	Half-life of a substance

ACRONYMS

ASTM	American Society for Testing and Materials
BOD	Biochemical oxygen demand

CEC   Cation-exchange capacity
COD   Chemical oxygen demand
DNAPL   Dense nonaqueous-phase liquid
Eh   Oxidation–reduction potential
EHSO   Environment, Health and Safety Online
EP   Extraction procedure
LNAPL   Light nonaqueous-phase liquid
NAPL   Nonaqueous-phase liquid
PAH   Polycyclic aromatic hydrocarbon
RCRA   Resource Conservation and Recovery Act
SOC   Suspended organic carbon
TDS   Total dissolved solids
TLM   Triple-layer model
TOC   Total organic carbon
USDW   Underground Source of Drinking Water
U.S. EPA   U.S. Environmental Protection Agency

REFERENCES

1. News Staff, Deep Wells a Safe Haven for Hazardous Waste, *Environment News*, The Heartland Institute, January 1, 1998. Available at http://www.heartland.org/Article.cfm?artId=13901.
2. FRTR, Deep Well Injection, Federal Remediation Technology Roundtable, 2007. Available at http://www.frtr.gov/matrix2/section4/4-54.html.
3. U.S. EPA, Assessing the Geochemical Fate of Deep-Well-Injected Hazardous Waste: A Reference Guide, EPA/625/6-89/025a, U.S. EPA, Cincinnati, OH, June 1990.
4. U.S. EPA, Introduction to Hazardous Waste Identification (40 CFR, Part 261), Report U.S. EPA530-K-05-012, U.S. EPA, Washington, September 2005.
5. Federal Register, Resource Conservation and Recovery Act (RCRA), 42 U.S. Code s/s 6901 et seq. (1976), U.S. Government, Public Laws, January 2004. Full text available at www.access.gpo.gov/uscode/title42/chapter82_.html.
6. Warner, D.L and Lehr, J.H., An Introduction to the Technology of Subsurface Waste water Injection, EPA 600/2-77-240, NTIS PB279 207, U.S. EPA, Washington, 1977.
7. Langelier, W.F., The analytical control of anti-corrosion water treatment. *J. Am. Water Works Assoc.*, 28, 1500–1521, 1936.
8. Ryzner, J.W., A new index for determining amount of calcium carbonate scale formed by water, *J. Am. Water Works Assoc.*, 36, 472–486, 1944.
9. Larson, T.E. and Buswell, A.M., Calcium carbonation saturation index and alkalinity interpretations, *J. Am. Water Works Assoc.*, 34, 1667–1684, 1942.
10. Stiff, H.A. and Davis, L.E., A method for predicting the tendency of oil field waters to deposit calcium carbonate, *Am. Inst. Mining Metall. Eng. Trans, Petroleum Div.*, 195, 213–216, 1952.
11. Ostroff, A.G., *Introduction to Oil Field Water Technology*, Prentice-Hall, Englewood Cliffs, NJ, 1965.
12. Watkins, J.W., Analytical Methods of Testing Waters to be Injected into Subsurface Oil-Productive Strata, Report No. 5031, U.S. Bureau of Mines, Washington, 1954.
13. Davis, J.B., *Petroleum Microbiology*, Elsevier, New York, 1967.
14. U.S. EPA, Resource Conservation and Recovery Act (RCRA)—Orientation Manual, Report EPA530-R-02-016, U.S. EPA, Washington, January 2003.
15. Hunter, J.V., Origin of organics from artificial contamination, in *Organic Compounds in Aquatic Environments*, Faust, S.D. and Hunter, J.V., Eds., Marcel Dekker, New York, 1971, pp. 51–94.
16. Roy, W.R., Mravik, S.C., Krapac, I.G., Dickerson, D.R., and Griffin, R.A., *Geochemical Interactions of Hazardous Wastes with Geological Formations in Deep-Well Systems*, Environmental Geology Notes 130, Illinois State Geological Survey, Champaign, Illinois, 1989.
17. Reeder, L.R., Review and Assessment of Deep-Well Injection of Hazardous Wastes, EPA 600/2-77-029a-d, NTIS PB 269 001-004, U.S. EPA, Washington, 1977.
18. Clark, J., Bonura, D.K., and Van Voorhees, R.F., An overview of injection well history in the U.S. of America, in *Underground Injection Science and Technology*, Tsang, C.F. and Apps, J.A., Eds., Elsevier, New York, February 2007.

19. Donaldson, E.C., Thomas, R.D., and Johnston, K.H., Subsurface Waste Injections in the U.S.: Fifteen Case Histories, Circular 8636, U.S. Bureau of Mines, Washington, 1974.

20. U.S. EPA, Report to Congress on Injection of Hazardous Wastes, EPA 570/9-85-003, NTIS PB86-203056, U.S EPA, Washington, 1985.

21. Tsang, C.F. and Apps, J.A., Eds., *Deep Injection Disposal of Hazardous and Industrial Waste: Scientific and Engineering Aspects*, Academic Press, New York, 1996.

22. Kaufman, D.G. and Franz, C.M., *Biosphere 2000: Protecting Our Global Environment*, 3rd ed., Kendall-Hunt Publishing, Dubuque, IA, 2000.

23. Tsang, C.F. and Apps, J.A., Eds., *Underground Injection Science and Technology*, Elsevier, New York, February 2007.

24. Bennett, G.F. and Shammas, N.K., Separation of oil from wastewater by air flotation, in *Flotation Engineering*, Wang, L.K., Shammas, N.K., Selke, W.A., and Aulenbach, D.B., Eds., Humana Press, Totowa, NJ, 2008.

25. Shammas, N.K., Kumar, I.J., Chang, S.Y., and Hung, Y.T., Sedimentation, in *Physicochemical Treatment Processes*, Wang, L.K., Hung Y.T., and Shammas, N.K., Eds., Humana Press, Totowa, NJ, 2005.

26. Shammas, N.K., Coagulation and flocculation, in *Physicochemical Treatment Processes*, Wang, L.K., Hung, Y.T., and Shammas, N.K., Eds., Humana Press, Totowa, NJ, 2005.

27. Wang, L.K., Diatomaceous earth precoat filtration, in *Advanced Physicochemical Treatment Processes*, Wang, L.K., Hung, Y.T., and Shammas, N.K., Eds., Humana Press, Totowa, NJ, 2006.

28. U.S. EPA, Injection Well Mechanical Integrity Testing, EPA 625/9-89/007, U.S. EPA, Washington, 1989.

29. Bedient, R.B., Springer, N.K., Cook, C.J., and Tomson, M.B., Modeling chemical reactions and transport in groundwater systems: A review, in *Modeling the Fate of Chemicals in the Aquatic Environment*, Dickson, K.L.A., Maki, W., and Cairns, J. Jr, Eds., Ann Arbor Science, Ann Arbor, MI, 1982, pp. 215–246.

30. Rubin, J.R., Transport of reacting solutes in porous media: relation between mathematical nature of problem formulation and chemical nature of reactions, *Water Resource Res.*, 19, 1231–1252, 1983.

31. Hung, Y.T., Lo, H.H., Wang, L.K., Taricska, J.R., and Li, K.H., Granular activated carbon adsorption, in *Physicochemical Treatment Processes*, Wang, L.K., Hung, Y.T., and Shammas, N.K., Eds., Humana Press, Totowa, NJ, 2005.

32. Hung, Y.T., Lo, H.H., Wang, L.K., Taricska, J.R., and Li, K.H., Powdered activated carbon adsorption, in *Advanced Physicochemical Treatment Processes*, Wang, L.K., Hung, Y.T., and Shammas, N.K., Eds., Humana Press, Totowa, NJ, 2006.

33. Roy, W.R., Krapac, I.G., Chou, S.F.J., and Griffin, R.A., Batch-Type Procedures for Estimating Soil Adsorption of Chemicals, Technical Resource Document (TRD), EPA/530-SW-90-006-F, U.S. EPA, Washington, 1990.

34. Wang, L.K., Vaccari, D.A., Li, Y., and Shammas, N.K., Chemical precipitation, in *Physicochemical Treatment Processes*, Wang, L.K., Hung, Y.T., and Shammas, N.K., Eds., Humana Press, Totowa, NJ, 2005.

35. Choi, J. and Aomine, S., Adsorption of pentachlorophenol by soils, *Soil Sci. Plant Nutr.*, 20, 135–144, 1974.

36. Choi, J. and Aomine, S., Mechanisms of pentachlorophenol adsorption by soils, *Soil Sci. Plant Nutr.*, 20, 371–379, 1974.

37. Davis, J.B., *Petroleum Microbiology*, Elsevier, New York, 1967.

38. Bunshah, R., *Techniques of Metal Research*, Vol. III, Part 1, Wiley, New York, 1970.

39. Scrivner, N.C., Bennet, K.E., Pease, R.A., Kopatsis, A., Sanders, S.J., Clark, D.M., and Rafal, M., Chemical fate of injected wastes, in *Proc. Int. Symp. Subsurface Injection of Liquid Wastes*, New Orleans, National Water Well Association, Dublin, OH, 1986, pp. 560–609.

40. Palmer, C.D. and Johnson, R.L., Physical processes controlling the transport of nonaqueous phase liquids in the subsurface, in *Transport and Fate of Contaminants in the Subsurface*, EPA 625/4-89/019, U.S. EPA, Washington, 1989.

41. Goolsby, D.A., Hydrogeochemical effects of injecting wastes into a limestone aquifer near pensacola, Florida, *Groundwater*, 19, 13–19, 1971.

42. Langmuir, D., Controls on the amounts of pollutants in subsurface waters, *Earth Miner. Sci.*, 42, 9–13, 1972.

43. Langmuir, D., Techniques of estimating thermodynamic properties for some aqueous complexes of geo-chemical interest, in *Chemical Modeling in Aqueous Systems: Speciation, Sorption, Solubility and Kinetics*, Jenne, E.A., Ed., ACS Symposium, American Chemical Society, Washington, DC, 1979, pp. 353–387.

44. Strycker, A. and Collins, A.G., State-of-the-Art Report: Injection of Hazardous Wastes into Deep Wells, EPAI600/8-87/01 3, NTIS PB87-1 70551, U.S. EPA, Washington, 1987.

45. Dyer, J.A., Predicting trace-metal fate in aqueous systems using a coupled equilibrium-surface-complexation, dynamic-simulation model, in *Underground Injection Science and Technology*, Tsang, C.F. and Apps, J.A., Eds., Elsevier, New York, February 2007.

46. Shammas, N.K., Yang, J.Y., Yuan, P.C., and Hung, Y.T., Chemical oxidation, in *Physicochemical Treatment Processes*, Wang, L.K., Hung, Y.T., and Shammas, N.K., Eds., Humana Press, Totowa, NJ, 2005.

47. Wang, L.K. and Li, Y., Chemical reduction/oxidation, in *Advanced Physicochemical Treatment Processes*, Wang, L.K., Hung, Y.T., and Shammas, N.K., Eds., Humana Press, Totowa, NJ, 2006.

48. Martell, A.E., Principles of complex formation, in *Organic Compounds in Aquatic Environments*, Faust, S.D. and Hunter, J.V., Eds., Marcel Dekker, New York, 1971, pp. 239–263.

49. Spycher, N. and Larkin, R.S., Investigation of chemical interactions between waste, native fluid, and host rock during deep well injection, in *Underground Injection Science and Technology*, Tsang, C.F. and Apps, J.A., Eds., Elsevier, New York, 2007.

50. Laszlo, P., Chemical reactions on clays, *Science*, 235, 1473–1477, 1987.

51. Degens, E.T. and Matheja, J., Formation of organic polymers on minerals and vice versa, in *Organic Compounds in Aquatic Environments*, Faust, S.D. and Hunter, J.V., Eds., Marcel-Dekker, New York, 1971, pp. 29–41.

52. Kharaka, Y.K., Carothers, W.W., and Rosenbauer, R.J., Thermal decarboxylation of acetic acid: implications for origin of natural gas, *Geochimica et Cosmochimica Acta*, 47, 397–402, 1983.

53. Smith J.M. and Raptis, T.J., Supercritical deep well wet oxidation of liquid organic wastes, in *Proc. Int. Symp. Subsurface Injection of Liquid Wastes*, New Orleans, National Water Well Association, Dublin, OH, 1986, pp. 715–732.

54. Shammas, N.K., Liu, Y., and Wang, L.K., Principles and kinetics of biological processes, in *Advanced Biological Treatment Processes*, Wang, L.K. Shammas, N.K., and Hung, Y.T., Eds., Humana Press, Totowa, NJ, 2008.

55. Alexander, M., Biodegradation of toxic chemicals in water and soil, in *Dynamics, Exposure and Hazard Assessment of Toxic Chemicals*, Haque, R., Ed., Ann Arbor Science, Ann Arbor, MI, 1980, pp. 179–190.

56. Alexander, M., Biodegradation of chemicals of environmental concern, *Science*, 211, 132–138, 1981.

57. Scow, K.M., Rate of biodegradation, in *Handbook of Chemical Property Estimation Methods: Environmental Behavior of Organic Compounds*, Lyman, W.J., Reehl, W.F., and Rosenblatt, D.H., Eds., McGraw-Hill, New York, 1982, pp. 9-1–9-85.

58. Mills, W.B., Water Quality Assessment: A Screening Procedure for Toxic and Conventional Pollutants, EPA/600/6-85/002a-b, U.S. EPA, Athens, Georgia, 1985.

59. Hanshaw, B.B., Natural-membrane phenomena and subsurface waste emplacement, in *Symposium on Underground Waste Management and Environmental Implications*, Houston, Texas, Cook, T.D., Ed., Am. Assn. Petr. Geol. Mem., 1972, pp. 308–315.

60. Kharaka, Y.K., Retention of dissolved constituents of waste by geologic membranes, in *Symposium on Underground Waste Management and Artificial Recharge*, Braunstein, J., Ed., publication 110, International Association of Hydrological Sciences, 1973, pp. 420–435.

61. McDowell-Boyer, L.M., Hunt, J.R., and Sitar, N., Particle transport through porous media, *Water Resources Res.*, 22, 1901–1921, 1986.

62. Frind, E.O., Simulation of long-term transient density-dependent transport in groundwater, *Adv. Water Res.*, 5, 73–88, 1982.

63. Larkin, R.G. and Clark, J.E., Modeling density changes in hazardous disposal well plumes, in *Underground Injection Science and Technology*, Tsang, C.F. and Apps, J.A., Eds., Elsevier, New York, 2007.

64. Miller, C.T., Fischer, II, A., Clark, J.E., Porter, W.M., Hales, C.H., and Tilton, J.R., Flow and containment of injected wastes, in *Proc. Int. Symp. Subsurface Injection of Liquid Wastes*, New Orleans, National Water Well Association, Dublin, OH, 1986, pp. 520–559.

65. Hem, J.D., *Study and Interpretation of the Chemical Characteristics of Natural Water*, U.S. Geological Survey, Water supply paper 1473, 1970.

66. Ponnamperuma, F.N., The chemistry of submerged soils, *Adv. Agron.*, 24, 29–98, 1972.

67. Goolsby, D.A., Geochemical effects and movement of injected industrial waste in a limestone aquifer, in *Symposium on Underground Waste Management and Environmental Implications*, Houston, Texas, Cook, T.D., Ed., Am. Assn. Petr. Geol. Mem., 18, pp. 355–368, 1972.

68. Malcolm, R.L. and Leenheer, J.A., The usefulness of organic carbon parameters in water quality investigations, in *Proc. Inst. Env. Sciences 1973 Annual Meeting*, Anaheim, CA, April 1973, 1–6, pp. 336–340.

69. Ragone, S.E., Vecchioli, J., and Ku, H.F.H., Short-term effect of injection of tertiary-treated sewage on iron concentration of water in Magothy Aquifer, Bay Park, New York, in *Symposium on Underground Waste Management and Artificial Recharge*, Braunstein, J., Ed., publication 110, International Association of Hydrological Sciences, 1973, pp. 273–290.
70. Zemke, J., Stower, M., and Borgmeier, M., Injection of brine from cavern leaching into deep saline aquifers: long-term experiences in modeling and reservoir survey, in U*nderground Injection Science and Technology*, Tsang, C.F. and Apps, J.A., Eds., Elsevier, New York, February 2007.
71. Van Orstrand, C.E., Temperature gradients, in *Problems of Petroleum Geology*, Wrather, W.E., and Lahee, F.H., Eds., American Association of Petroleum Geologists, Tulsa, OK, 1934, pp. 989–1021.
72. Choi, J. and Aomine, S., Mechanisms of pentachlorophenol adsorption by soils, *Soil Sci. Plant Nutr.*, 20, 371–379, 1974.
73. Collins, A.G. and Crocker, M.E., Laboratory Protocol for Determining Fate of Waste Disposed in Deep Wells, EPA/600/8-88/008, National Institute for Petroleum and Energy Research, Bartlesville, OK, 1988.
74. ZoBell, C.E. and Johnson, F.H., The influence of hydrostatic pressure on the growth and viability of terrestrial and marine bacteria, *J. Bacteriol.*, 57, 179–189, 1949.
75. Warner, D.L., Davis, S.N., and Syed, T., Evaluation of confining layers for containment of injected wastewater, in *Proc. Int. Symp. Subsurface Injection of Liquid Wastes*, New Orleans, National Water Well Association, Dublin, OH, 1986, pp. 417–446.
76. Knape, B., Applications of deep well injection of industrial and municipal wastewater in Texas, in *Underground Injection Science and Technology*, Tsang, C.F. and Apps, J.A., Eds., Elsevier, New York, February 2007.
77. Kobelski, B., Smith, R.E., and Whitehurst, A.L., An interpretation of the Safe Drinking Water Act's "NonEndangerment" Standard for the Underground Injection Control (UIC) Program, in *Underground Injection Science and Technology*, Tsang, C.F. and Apps, J.A., Eds., Elsevier, New York, February 2007.
78. Kreitler, C.W., Hydrogeology of sedimentary basins as it relates to deep-well injection of chemical wastes, in *Proc. Int. Symp. Subsurface Injection of Liquid Wastes*, New Orleans, National Water Well Association, Dublin, OH, 1986, pp. 398–416.
79. Henry H.R. and Kahout, F.A., Circulation patterns of saline groundwater affected by geothermal heating as related to waste disposal, in *Symposium on Underground Waste Management and Environmental Implications*, Houston, TX, Cook, T.D., Ed., American Association of Petroleum Geologists, 18, 1973, pp. 202–221.
80. Swolf, H.S., Chemical effects of pore fluids on rock properties, in *Symposium on Underground Waste Management and Environmental Implications*, Houston, TX, Cook, T.D., Ed., American Association of Petroleum Geologists 18, 1972, pp. 224–234.
81. Sniegocki, R.T., Problems of artificial recharge through wells in Grand Prairie Region, Arkansas, U.S. Geol. Surv. Water Supply Paper 1615-F, 1963.
82. Ghiorse, W.C. and Wilson, J.T., Microbial ecology of the terrestrial subsurface, *Adv. Appl. Microbiol.*, 33, 107–172, 1988.
83. Messineva, M.A., The geological activity of bacteria and its effect on geochemical processes, in *Geologic Activity of Microorganisms*, Kuznetsov, S.I., Ed., Transactions of the Institute of Microbiology No. IX (trans. from Russian), Consultants Bureau, New York, 1962, pp. 6–24.
84. Sinclair, J.L. and Ghiorse, W.C., Distribution of protozoa in subsurface sediments of a pristine groundwater study site in Oklahoma, *Appl. Environ. Microbiol.*, 53, 1157–1163, 1987.
85. Davis, K. and McDonald, L.K., Potential corrosion and microbiological mechanisms and detection techniques in solution mining and hydrocarbon storage wells, in *Underground Injection Science and Technology*, Tsang, C.F. and Apps, J.A., Eds., Elsevier, New York, 2007.
86. Kuznetsov, S.I., Ivanov, M.V., and Lyalikova, N.N., *Introduction to Geological Microbiology*, McGraw-Hill, New York, 1963.
87. Dunlap, W.J. and McNabb, J.F., Subsurface Biological Activity in Relation to Groundwater Pollution, EPA 660/2-73-014, NTIS PB227 990, U.S. EPA, 1973.
88. Olson, G.J.H., Dockins, W.C., McFeters, G.A., and Iverson, W.P., Sulphate-reducing and methanogenic bacteria from deep aquifers in Montana, *Geomicrobial. J.*, 2, 327–340, 1981.
89. Ehrlich, G.G., Godsy, E.M., Pascale, C.A., and Vecchioli, J., Chemical changes in an industrial waste liquid during post-injection movement in a limestone aquifer, Pensacola, Florida, *Groundwater*, 17, 562–573, 1979.
90. Bouwer, E.J. and McCarty, P.L., Transformations of 1- and 2-carbon halogenated aliphatic organic compounds under methanogenic conditions, *Appl. Environ. Microbiol.*, 45, 1286–1294, 1983.

91. Novelli, G.D. and ZoBell, C.E., Assimilation of petroleum hydrocarbons by sulfate-reducing bacteria, *J. Bacteriol.*, 47, 447–448, 1944.

92. Rosenfeld, W.D., Anaerobic oxidation of hydrocarbons by sulfate-reducing bacteria, *J. Bacteriol.*, 54, 664–668, 1947.

93. Lovley, D.R., Organic matter mineralization with the reduction of ferric iron: a review, *Geomicrobiol. J.*, 5, 375–399, 1987.

94. Ehrlich, H.L., Manganese oxide reduction as a form of anaerobic respiration, *Geomicrobiol. J.*, 5, 423–429, 1987

95. Hof, T., Investigations concerning bacterial life in strong brines, *Rec. Trav. Botan. Neerl.*, 32, 92–173, 1935.

96. Rubentschik, L., Studies on the bios question, *Zentr. Bakteriol, Parasitenk, Abtk. II*, 68,161–179, 1929.

97. Carothers, W.W. and Kharaka, Y.K., Aliphatic acid anions in oil-field waters—implications for origin of natural gas, *Am. Assoc. Petrol. Geol. Bull.*, 62, 2441–2453, 1978.

98. McNabb, J.F. and Dunlap, W.J., Subsurface biological activity in relation to groundwater pollution, *Groundwater*, 13, 33–44, 1975.

99. Zeikus, J.G., The biology of methanogenic bacteria, *Bacteriol. Rev.*, 41, 514–541, 1977.

100. Ehilich, G.G., Godsy, E.M., Goerlitz, D.F., and Hult, M.F., Microbial ecology of a creosote-contaminated aquifer at St. Louis Park, MN, *Dev. Ind. Microbiol.*, 24, 235–245, 1983.

101. Martial, J.S., Join, J.L., and Coudray, J., Injection of organic liquid waste in a basaltic confined coastal aquifer, Reunion Island, in *Underground Injection Science and Technology*, Tsang, C.F. and Apps, J.A., Eds., Elsevier, New York, 2007.

102. Leenheer, J.A. and Malcolm, R.L., Case history of subsurface waste injection of an industrial organic waste, in *Symposium on Underground Waste Management and Artificial Recharge*, Braunstein, J., Ed., publication 110, International Association of Hydrological Sciences, 1973, pp. 565–584.

103. Veley, C.O., How hydrolysable metal ions react with clay to control formation water sensitivity, *J. Petroleum Technol.*, September 1, 1111–1118, 1969.

104. Buffle, J., Tessier, A., and Haerdi, W., Interpretation of trace metal complexation by aquatic organic matter, in *Complexation of Trace Metals in Natural Waters*, Kramer, C.J.M. and Duinker, J.C., Eds., Martinus Nijhoff/Dr. W. Junk Publishers, The Hague, 1984, pp. 301–316.

105. Fuller, W.H., Movement of Selected Metals, Asbestos and Cyanide in Soils: Applications to Waste Disposal Problems, EPA 600/2-77-020, NTIS PB 266 905, U.S. EPA, 1977.

106. Van Luik, A.E. and Jurinak, J.J., Equilibrium chemistry of heavy metals in concentrated electrolyte solution, in *Chemical Modeling in Aqueous Systems: Speciation, Sorption, Solubility and Kinetics*, Jenne, E.A., Ed., ACS Symp. Series 93, American Chemical Society, Washington, 1979, pp. 683–710.

107. Förstner, U. and Wittmann, G.T.W., *Metal Pollution in the Aquatic Environment*, Springer-Verlag, New York, 1979.

108. Moore, J.W. and Ramamoorthy, S., *Heavy Metal in Natural Waters: Applied Monitoring and Impact Assessment*, Springer-Verlag, New York, 1984.

109. Strycker, A. and Collins, A.G., State-of-the-Art Report: Injection of Hazardous Wastes into Deep Wells, EPA/600/8-87/013, NTIS PB87-1 70551, U.S EPA, 1987.

110. Moore, J.W. and Ramamoorthy, S., *Organic Chemicals in Natural Waters: Applied Monitoring and Impact Assessment*, Springer-Verlag, New York, 1984.

111. Tabak, H.H., Quave, S.A., Mashni, C.I., and Barth, E.F., Biodegradability studies with organic priority pollutant compounds, *J. Water Poll. Control Fed.*, 53, 1503–1518, 1981.

112. Britton, L.N., Microbial degradation of aliphatic hydrocarbons, in *Microbial Degradation of Organic Compounds*, Gibson, D.T., Ed., Marcel Dekker, New York, 1984, pp. 89–130.

113. Chapman, P.J., An outline of reaction sequences used for the bacterial degradation of phenolic compounds, in *Degradation of Synthetic Organic Molecules in the Biosphere*, National Academy of Sciences, Washington, 1972, pp. 17–53.

114. Gibson, O.T. and Subramanian, V., Microbial degradation of aromatic hydrocarbons, in *Microbial Degradation of Organic Compounds*, Gibson, D.T., Ed., Marcel Dekker, New York, 1984, pp. 181–252.

115. Reinke, W., Microbial degradation of halogenated aromatic compounds, in *Microbial Degradation of Organic Compounds*, Gibson, D.T., Ed., Marcel Dekker, New York, 1984, pp. 319–360.

116. Ribbons, D.W., Keyser, P., Eaton, R.W., Anderson, B.N., Kunz, D.A., and Taylor, B.F., Microbial degradation of phthalates, in *Microbial Degradation of Organic Compounds*, Gibson, D.T., Ed., Marcel Dekker, New York, 1984, pp. 371–398.

117. Safe, S.H., Microbial degradation of polychlorinated biphenyls, in *Microbial Degradation of Organic Compounds*, Gibson, D.T., Ed., Marcel Dekker, New York, 1984, pp. 261–370.

118. Kearney, P.C. and Kaufman, D.D., Microbial degradation of some chlorinated pesticides, in *Degradation of Synthetic Organic Molecules in the Biosphere*, National Academy of Sciences, Washington, 1972, pp. 166–188.

119. ASTM, Standard practice for evaluating environmental fate models of chemicals, *Annual Book of ASTM Standard*, American Society for Testing and Materials (ASTM), E978-84, Philadelphia, PA, 1984.

120. Apps, J.A., Current Geochemical Models to Predict the Fate of Hazardous Wastes in the Injection Zones of Deep Disposal Wells, Lawrence Berkeley Laboratory, Report LBL-26007, 1988.

121. Bergman, H.L. and Meyer, J.S., Complex effluent fate modeling, in *Modeling the Fate of Chemicals in the Aquatic Environment*, Dickson, K.L., Maki, A.W., and Cairns, J., Jr, Eds., Ann Arbor Science, Ann Arbor, MI, 1982, pp. 247–267.

122. Nordstrom, D.K. and Ball, J.W., Chemical models, computer programs and metal complexation in natural waters, in *Complexation of Trace Metals in Natural Waters*, Kramer, C.J.M. and Duinker, J.C., Eds., Martinus, 1984.

123. Wigley, T.M.L., WATSPEC: A computer program for determining the equilibrium speciation of aqueous solutions, *Brit. Geomorph. Res. Group Tech. Bull.*, 20, 48, 1977.

124. Van Zeggeren, F. and Storey, S.H., *The Computation of Chemical Equilibria*, Cambridge University Press, Cambridge, 1970, p. 176.

125. Smith, W.R. and Missen, R.W., *Chemical Reaction Equilibrium Analysis*, Wiley-Interscience, New York, 1982, p. 364.

126. Zeleznick, F.J. and Gordon, S., Calculation of complex chemical equilibria, *Ind. Eng. Chem.*, 60, 27–57, 1968.

127. Wolery, T.J., Calculation of Chemical Equilibrium between Aqueous Solution and Minerals: The EQK3/6 Software Package, Lawrence Livermore Laboratory Report UCRL-52658, 1979, p. 41.

128. Rubin, J., Transport of reacting solutes in porous media: relation between mathematical nature of problem formulation and chemical nature of reactions, *Water Resources Res.*, 19, 1231–1252, 1983.

129. Pozdniakov, S.P., Modeling of waste injection in heterogeneous sandy-clay formations, in *Underground Injection Science and Technology*, Tsang, C.F. and Apps, J.A., Eds., Elsevier, New York, 2007.

130. Zhu, C. and Anderson, G., *Environmental Applications of Geochemical Modeling*, Cambridge University Press, Cambridge, UK, 2002.

131. Hamaker, J.W. and Thompson, J.M., Adsorption, in *Organic Chemicals in the Soil Environment*, Vol. I, Goring, C.A.I. and Hamaker, J.W., Eds., Marcel-Dekker, New York, 1972, pp. 49–143.

132. Sabljiã, A., On the prediction of soil sorption coefficients of organic pollutants from molecular structure: application of molecular topology model, *Environ. Sci. Technol.*, 21, 358–366, 1987.

133. Karickhoff, S.W., Organic pollutant sorption in aquatic systems, *J. Hydraul. Eng.*, 110, 707–735, 1984.

134. Winters, S.L. and Lee, D.R., *In situ* retardation of trace organics in groundwater discharge to a sandy bed, *Environ. Sci. Technol.*, 21, 1182–1186, 1987.

135. Apps, J.A., Lucas, J., Mathur, A.K., and Tsao, L., Theoretical and Experimental Evaluation of Waste Transport in Selected Rocks, Lawrence Berkeley Laboratory Report LBL-7022, 1977.

136. Reardon, E.J., K_d's—Can they be used to describe reversible ion sorption reactions in contaminant migration? *Groundwater* 19, 279–286, 1981.

137. Alben, K.T., Shpirt, E., and Kaczmarczyk, J.H., Temperature dependence of trihalomethane adsorption on activated carbon: implications for systems with seasonal variations in temperature and concentration, *Environ. Sci. Technol.*, 22, 406–412, 1988.

138. Bailey, G.W., White, J.L., and Rothberg, T., Adsorption of organic herbicides by montmorillonite: role of pH and chemical character of adsorbate, *Soil Sci. Soc. Am. Proc.*, 32, 222–234, 1968.

139. Donaldson, E.C., Crocker, M.E., and Manning, F.S., Adsorption of Organic Compounds on Cottage Grove Sandstone, BERC/RI-75/4, Bartlesville Energy Research Center, OK, 1975.

140. Collins, A.G. and Crocker, M.E., Laboratory Protocol for Determining Fate of Waste Disposed in Deep Wells, EPA 600/8-88/008, National Institute for Petroleum and Energy Research, Bartlesville, OK, 1988.

141. Davis, J.A. and Leckie, J.O., Surface ionization and complexation at the oxide/water interface, II: surface properties of amorphous iron oxyhydroxide and adsorption of metal ions, *J. Colloid Interface Sci.* 67, 90–107, 1978.

142. Davis, J.A and Leckie, J.O., Surface ionization and complexation at the oxide/water interface, II: adsorption of anions, *J. Colloid Interface Sci.*, 74, 32–43, 1980.

143. Kent, D.B., Tripathi, V.S., Ball, N.B., Leckie, J.O., and Siegel, M.D., Surface-Complexation Modeling of Radionuclide Adsorption in Subsurface Environments, U.S. Nuclear Regulatory Commission Report NUREG/CR-4807, 1988, p. 113.

144. Paris, D.F., Steen, W.C., Bauchman, G.L., and Barnett, J.T., Jr, Second-order model to predict microbial degradation of organic compounds in natural waters, *Appl. Env. Microbiol.*, 41, 603–609, 1981.
145. Baughman, G.L., Paris, D.F., and Steen, W.C., Quantitative expression of biotransformation rate, in *Biotransformation and Fate of Chemicals in the Aquatic Environment*, Maki, A.W., Dickson, K.L., and Cairns, J., Jr, Eds., American Society of Microbiology, Washington, 1980, pp. 105–111.
146. Hamaker, J.W., Decomposition: quantitative aspects, in *Organic Chemicals in the Soil Environment*, Vol. I, Goring, C.A.I. and Hamaker, J.W., Eds., Marcel Dekker, New York, 1972, pp. 253–340.
147. Larson, R.J. Role of biodegradation kinetics in predicting environmental fate, in *Biotransformation and Fate of Chemicals in the Aquatic Environment*, Maki, A.W., Dickson, K.L., and Cairns, J., Jr, Eds., American Society of Microbiology, Washington, 1980, pp. 67–86.
148. Rao, P.S.C. and Jessup, R.E., Development and verification of simulation models for describing pesticides dynamics in soils, *Ecol. Modeling*, 16, 67–75, 1982.
149. Williamson, K. and McCarty, P.L., A model of substrate utilization by bacterial films, *J. Water Poll. Control Fed.*, 48, 9–24, 1976.
150. Williamson, K. and McCarty, P.L., Verification studies of the biofilm model for bacterial substrate utilization, *J. Water Poll. Control Fed.*, 48, 281–296, 1976.
151. Rittmann, B.E., McCarty, P.L., and Roberts, P.V., Trace-organics biodegradation in aquifer recharge, *Groundwater*, 18, 236–242, 1980.
152. Rittmann, B.E. and McCarty, P.L., Model of steady-state biofilm kinetics, *Biotech. Bioeng.*, 22, 2343–2357, 1980.
153. Rittmann, B.E. and McCarty, P.L., Evaluation of steady-state biofilm kinetics, *Biotech. Bioeng.*, 22, 2359–2373, 1980.
154. McCarty, P.L., Reinhard, M., and Rittmann, B.E., Trace organics in groundwater, *Environ. Sci. Technol.*, 15, 40–51, 1981.
155. Bouwer, E.J. and McCarty, P.L., Modeling of trace organics biotransformation in the subsurface, *Groundwater*, 22, 433–440, 1984.
156. Changand, H.T. and Rittmann, B.E., Mathematical modeling of biofilm on activated carbon, *Environ. Sci. Technol.*, 21, 273–280, 1987.
157. Chang, H.T. and Rittmann, B.E., Verification of the model of biofilm on activated carbon, *Environ. Sci. Technol.*, 21, 280–288, 1987.
158. Mabey, W. and Mill, T., Critical review of hydrolysis of organic compounds in water under environmental conditions, *J. Phys. Chem. Ref. Data*, 7, 383–415, 1978.
159. Ellington, J.J., Measurement of Hydrolysis Rate Constants for Evaluation of Hazardous Waste Land Disposal, Vol 1: Data on 32 Chemicals, EPA/600/3-86/043, NTIS PB87-140349, U.S. EPA, Washington, 1986.
160. Ellington, J.J., Measurement of Hydrolysis Rate Constants for Evaluation of Hazardous Waste Land Disposal, Vol 2: Data on 54 Chemicals, EPA/600/3-87/019, NTIS PB87-227344, U.S. EPA, Washington, 1987.
161. Ellington, J.J., Measurement of Hydrolysis Rate Constants for Evaluation of Hazardous Waste Land Disposal, Vol 3: Data on 70 Chemicals, EPA/600/3-88/028, NTIS PB88-234042, U.S. EPA, Washington, 1988.
162. Javandal, I., Doughty, C., and Tsang, C.F., *Groundwater Transport: Handbook of Mathematical Models*, Water Resources Monograph Series 10, American Geophysical Union, Washington, 1984, p. 228.
163. EHSO, *Hazardous Wastes Deep Well Injection*, Environment, Health and Safety Online, Atlanta, GA, 2007. Available at http://ehso.com/cssepa/tsdfdeepwells.php.
164. Foster, J.B. and Goolsby, D.A., Construction of Waste-Injection Monitoring Wells near Pensacola, Florida, Florida Bureau of Geology Information Circular 74, 1972.
165. Keely, J.F., Chemical time-series sampling, *Groundwater Monitoring Review*, Fall, 29–38, 1982.
166. Keely, J.F. and Wolf, F., Field applications of chemical time-series sampling, *Groundwater Monitoring Review*, Fall, 26–33, 1983.
167. Faulkner, G.L. and Pascale, C.A., Monitoring regional effects of high pressure injection of industrial waste water in a limestone aquifer, *Groundwater*, 13, 197–208, 1975.
168. Pascale, C.A. and Martin, J.B., Hydrologic Monitoring of a Deep-Well Waste-Injection System near Pensacola, Florida, March 1970–March 1977, U.S. Geological Survey Water Resource Investigation 78-27, 1978.
169. Willis, C.J., Elkan, G.H., Horvath, E., and Dail, K.R., Bacterial flora of saline aquifers, *Groundwater*, 13, 406–409, 1975.
170. Elkan, G. and Horvath, E., The Role of Microorganisms in the Decomposition of Deep Well Injected Liquid Industrial Wastes, NSF/RA-770102, NTIS PB 268 646, 1977.

171. Vecchioli, J., Ehrlich, G.G., Godsy, E.M., and Pascale, C.A., Alterations in the chemistry of an industrial waste liquid injected into limestone near Pensacola, Florida, in *Hydrogeology of Karstic Terrains, Case Histories*, Vol. 1, Castany, G., Groba, E., and Romijn, E., Eds., International Association of Hydrogeologists, 1984, pp. 217–221.

172. Kaufman, M.I., Goolsby, O.A., and Faulkner, G.L., Injection of acidic industrial waste into a saline carbonate aquifer: geochemical aspects, in *Symposium on Underground Waste Management and Artificial Recharge*, Braunstein, J., Ed., publication 110, International Association of Hydrological Sciences, 1973, pp. 526–555.

173. Kaufman, M.I. and McKenzie, D.J., Upward migration of deep-well waste injection fluids in Floridan aquifer, South Florida, *J. Res. U. S. Geol. Surv.*, 3, 261–271, 1975.

174. Garcia-Bengochea, J.I. and Vernon, R.O., Deep-well disposal of waste waters in saline aquifers of South Florida, *Water Resources Res.*, 6, 1464–1470, 1970.

175. Mckenzie, D.J., Injection of Acidic Industrial Waste into the Roridan Aquifer near Belle Glade, Florida: Upward Migration and Geochemical Interactions, U.S. Geological Survey Open File Report 76-626, 1976.

176. Clark, J., Demonstration of presence and size of a CO_2-rich fluid phase after HCL injection in carbonate rock, in *Underground Injection Science and Technology*, Tsang, C.F. and Apps, J.A., Eds., Elsevier, New York, February 2007.

177. Peek, H.M. and Heath, R.C., Feasibility study of liquid-waste injection into aquifers containing salt water, Wilmington, North Carolina, in *Symposium on Underground Waste Management and Artificial Recharge*, Braunstein, J., Ed., publication 110, International Association of Hydrological Sciences, 1973, pp. 851–875.

178. Leenheer, J.A., Malcolm, R.L., and White, W.R., Physical, Chemical and Biological Aspects of Subsurface Organic Waste Injection near Wilmington, North Carolina, U.S. Geological Survey Professional Paper 9871976a, 1976.

179. Leenheer, J.A., Malcolm, R.L., and White, W.R., Investigation of the reactivity and fate of certain organic compounds of an industrial waste after deep-well injection, *Environ. Sci. Technol.* 10, 445–451, 1976.

180. DiTommaso, A. and Elkan, G.H., Role of bacteria in decomposition of injected liquid waste at Wilmington, North Carolina, in *Symposium on Underground Waste Management and Artificial Recharge*, Braunstein, J., Ed., publication 110, International Association of Hydrological Sciences, 1973, pp. 585–599.

181. Kamath, K. and Salazar. M., The role of the critical temperature of carbon dioxide on the behavior of wells injecting hydrochloric acid into carbonate formations, in *Proc. Int. Symp. Subsurface injection of Liquid Wastes*, New Orleans, National Water Well Association, Dublin, OH, 1986, pp. 638–655.

182. Panagiotopoulos, A.Z. and Reid, R.C., Deep-well injection of aqueous hydrochloric acid, in *Proc. Int. Symp. Subsurface Injection of Liquid Wastes*, New Orleans, National Water Well Association, Dublin, OH, 1986, pp. 610–637.

183. Brower, R.D., Visocky, A.P., Krapac, I.G., Hensel, B.R., Peyton, G.R., Nealon, J.S., and Guthrie, M., Evaluation of Underground Injection of Industrial Waste in Illinois, Final Report, Illinois Scientific Surveys Joint Report 2, Illinois State Geological Survey, Champaign, IL, 1989.

184. Donaldson, E.C. and Johansen, R.T., History of a two-well industrial-waste disposal system, in *Symposium on Underground Waste Management and Artificial Recharge*, Braunstein, J., Ed., publication 110, International Association of Hydrological Sciences, 1973, pp. 603–621.

185. Texas Environmental Profiles, *Underground Injection of Hazardous Waste*, On-line resources for the State of Texas, 2007. Available at http://www.texasep.org/html/wst/wst_4imn_injct.html.

186. Kreitler, C.W., Akhter, M.S., and Donnelly, A.C.A., *Hydrologic-Hydrochemical Characterization of Texas Gulf Coast Saline Formations Used for Deep Well Injection of Chemical Wastes*, Prepared for U.S. Environmental Protection Agency by University of Texas at Austin, Bureau of Economic Geology, 1988.

187. Baas-Becking, L.G.M., Kaplan, J.R., and Moore, D., Limits of the natural environment in terms of pH and oxidation-reduction potentials, *J. Geol.*, 68, 3, 243–284, 1960.

11 Waste Management in the Pulp and Paper Industry

Nazih K. Shammas

CONTENTS

11.1   INTRODUCTION

11.1.1   BACKGROUND

The paper and allied products industry comprises three types of facilities: pulp mills that process raw wood fiber or processed fiber to make pulp; paper and board mills that manufacture paper or board; and converting facilities that use these primary materials to manufacture more specialized products such as writing paper, napkins, and other tissue products. The process of converting paper is not a source of water or air pollution, as is the case for the first two facilities. This chapter focuses primarily on the greatest areas of environmental concern within the pulp and paper industry: those from pulping processes.

The specific components in the pulp and paper industry include the following[1,2]:

1. *Pulp mills.* These separate the fibers of wood or other materials, such as rags, linters, waste-paper, and straw, in order to create pulp. Mills may use chemical, semichemical, or mechanical processes, and may create coproducts such as turpentine and tall oil. Most pulp mills bleach the pulp they produce, and, when wastepaper is converted into secondary fiber, it is deinked. The output of some pulp mills is not used to make paper, but to produce cellulose acetate or to be dissolved and regenerated in the form of viscose fibers or cellophane.
2. *Paper mills.* These are primarily engaged in manufacturing paper from wood pulp and other fiber pulp, and may also manufacture converted paper products. Establishments primarily engaged in integrated operations of producing pulp and manufacturing paper are included in this industry if primarily shipping paper or paper products.
3. *Paperboard mills.* These are primarily engaged in manufacturing paperboard, including paperboard coated on a paperboard machine, from wood pulp and other fiber pulp; they may also manufacture converted paperboard products.
4. *Paperboard containers and boxes.* These establishments are engaged in the manufacture of corrugated and solid fiber boxes and containers from purchased paperboard. The principal commodities of this industry are boxes, pads, partitions, display items, pallets, corrugated sheets, food packaging, and nonfood (e.g., soaps, cosmetics, and medicinal products) packaging.

5. *Miscellaneous converted paper products.* These establishments produce a range of paper, paperboard, and plastic products with purchased material. Common products include paper and plastic film packaging, specialty paper, paper and plastic bags, manila folders, tissue products, envelopes, stationery, and other products.

One important characteristic of the pulp and paper industry is the interconnection of operations between pulp mills and downstream processing of pulp into paper, paperboard, and building paper. Another important characteristic of the pulp and paper industry is that the range of processes, chemical inputs, and outputs used are used in pulp manufacture. On the whole, pulp mill processes are chemical intensive and have been the focus of past and ongoing pollution prevention rulemaking. There are also numerous manufacturers of finished paper and paperboard products from paper and paperboard stock. Some companies are involved in both the manufacture of primary products and converting, particularly in the production of tissue products, corrugated shipping containers, folding cartons, flexible packaging, and envelopes.

11.1.2 CHARACTERIZATION OF THE PULP AND PAPER INDUSTRY

The pulp and paper industry produces primary products—commodity grades of wood pulp, printing and writing papers, sanitary tissue, industrial-type papers, containerboard, and boxboard—using cellulose fiber. The two steps involved are pulping and paper or paperboard manufacturing.

11.1.2.1 Pulping

Pulping is the process of separating wood chips into individual fibers by chemical, semichemical, or mechanical methods. The particular pulping process used affects the strength, appearance, and intended use characteristics of the resultant paper product. Pulping is the major source of environmental impacts from the pulp and paper industry. There are more than a dozen different pulping processes in use in the U.S.; each process has its own set of process inputs, outputs, and resultant environmental concerns.[3] Table 11.1 provides an overview of the major pulping processes and the main products that they produce. Kraft pulp, bleached and unbleached, is used to manufacture the majority of paper products. Together, chemical pulping processes account for 84% of the pulp produced in the U.S.[1] Figure 11.1 presents the relative outputs of the major pulping processes.

A bleached kraft pulp mill requires 15,140 to 45,420 L (4000 to 12,000 gal) of water and 8.56 to 12.22 million chu (14 to 20 million Btu) of energy per ton of pulp, of which ca. 4.44 to 5.56 million chu (8 to 10 million Btu) are typically derived from biomass-derived fuel from the pulping process itself.[4] Across all facilities, the pulp, paper, and allied products industry is the largest consumer of process water and the third largest consumer of energy (after the chemicals and metals industries).[5,6] The large amounts of water and energy used, as well as the chemical inputs, lead to a variety of environmental concerns.

11.1.2.2 Paper and Paperboard Manufacturing

The paper or paperboard manufacturing process is similar for all types of pulp. Pulp is spread out as extremely dilute slurry on a moving endless belt of filtering fabric. Water is removed by gravity and vacuum, and the resulting web of fibers is passed through presses to remove more water and consolidate the web. Paper and paperboard manufacturers use nearly identical processes, but paperboard is thicker (more than 0.3 mm).

11.1.3 INDUSTRY SIZE AND GEOGRAPHIC DISTRIBUTION

The pulp and paper industry is characterized by very large facilities; of the 514 pulp and paper mills reported by the Bureau of the Census in 1998, 343 (67%) had 100 or more employees. Across all of

TABLE 11.1
Description of Pulping Processes

Pulping Process	Description/Principal Products
Dissolving kraft	Highly bleached and purified kraft process wood pulp suitable for conversion into products such as rayon, viscose, acetate, and cellophane
Bleached papergrade kraft and soda Unbleached kraft	Bleached or unbleached kraft process wood pulp usually converted into paperboard, coarse papers, tissue papers, and fine papers such as business, writing and printing
Dissolving sulfite	Highly bleached and purified sulfite process wood pulp suitable for conversion into products such as rayon, viscose, acetate, and cellophane
Papergrade sulfite	Sulfite process wood pulp with or without bleaching used for products such as tissue papers, fine papers, and newsprint
Semichemical	Pulp is produced by chemical, pressure, and occasionally mechanical forces with or without bleaching used for corrugating medium (cardboard), paper, and paperboard
Mechanical pulp	Pulp manufacture by stone groundwood, mechanical refiner, thermo-mechanical, chemi-mechanical, or chemi-thermomechanical means for newsprint, coarse papers, tissue, molded fiber products, and fine papers
Secondary fiber deink	Pulps from recovered paper or paperboard using a chemical or solvent process to remove contaminants such as inks, coatings, and pigments used to produce fine, tissue, and newsprint papers
Secondary fiber nondeink	Pulp production from recovered paper or paperboard without deinking processes to produce tissue, paperboard, molded products, and construction papers
Nonwood chemical pulp	Production of pulp from textiles (e.g., rags), cotton linters, flax, hemp, tobacco, and abaca to make cigarette wrap papers and other specialty paper products

Source: U.S. EPA, Profile of the Pulp and Paper Industry, 2nd ed., report EPA/310-R-02-002, U.S. EPA, Washington, November 2002.

these facilities, there are 172,000 employees who produced USD 59 billion in shipments (in 1998 dollars). In 2000, the industry employed 182,000 people and produced USD 79 billion in shipments. In contrast, the downstream facilities (container and specialty product manufacturers) tend to be more numerous but smaller. More than 75% of these facilities have fewer than 100 employees. Table 11.2 presents the employment distribution for both pulp and paper facilities and downstream manufacturers

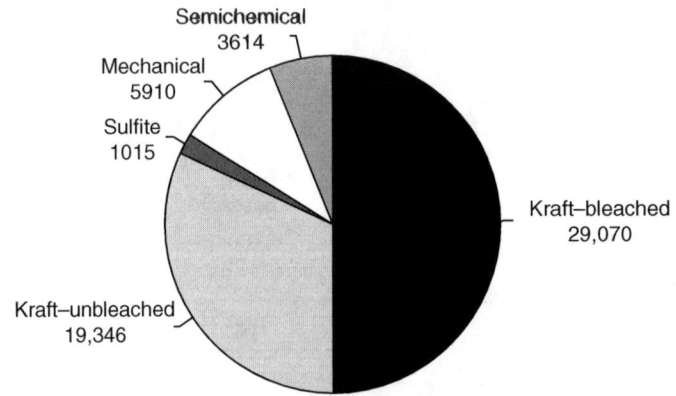

FIGURE 11.1   U.S. pulp production in 1000 t (year 2000). (Taken from U.S. EPA, Profile of the Pulp and Paper Industry, 2nd ed., report EPA/310-R-02-002, U.S. EPA, Washington, November 2002.)

TABLE 11.2

Size of Paper and Allied Products Facilities

Industry	Employees per Facility (% of Total)			
	1–19	20–99	100–499	>499
Pulp mills	3 (7%)	14 (34%)	18 (44%)	6 (15%)
Paper mills	6 (2%)	63 (24%)	107 (41%)	83 (32%)
Paperboard mills	8 (4%)	77 (36%)	96 (45%)	33 (15%)
Paperboard containers and boxes	748 (26%)	1311 (46%)	782 (27%)	14 (<1%)
Misc. converted paper products	1383 (44%)	1116 (36%)	597 (19%)	70 (2%)

Source: U.S. EPA, Profile of the Pulp and Paper Industry, 2nd ed., report EPA/310-R-02-002, U.S. EPA, Washington, November 2002.

in 1997 (the most recent data available) as reported by the U.S. Census Bureau.[7] Because recent years have seen some facility closures, the current number of facilities may be somewhat lower.

The integrated pulp and paper industry is among the top 10 U.S. manufacturing industries in value of shipments. The industry shipments amount to 146 billion USD with an employment of 609,480. Individual pulp and paper mills employ only 28% of the workers in the industry, but produce over 40% of the shipments.[8]

The geographic distribution of pulp and paper mills varies according to the type of mill. As there are tremendous variations in the scale of individual facilities, tallies of the number of facilities may not represent the level of economic activity (nor possible environmental consequences). Pulp mills are located primarily in regions of the country where trees are harvested from natural stands or tree farms, such as the Southeast, Northwest, Northeast, and North Central regions.[9] Pulp mills that process recycled fiber are generally located near sources of waste paper. Paper mills, however, are more widely distributed. They are located near pulping operations or near converting markets. The distribution of paperboard mills reflects the location of manufacturing in general, as such operations are the primary market for paperboard products. Figure 11.2 presents the locations of pulp and paper mills in the U.S.

11.1.4 ECONOMIC TRENDS

The U.S. produces roughly 30% of the world's paper and paperboard. The pulp and paper industry is one of the most important industries for the balance of trade in the U.S. This trade balance increased through most of the 1990s. In 1999, exports were USD 8.5 billion. In recent years, however, exports have been declining and imports have been increasing. Between 1997 and 2000, exports declined 5.5% and imports increased by more than 20%. The declining exports and increasing imports are partly due to a strong dollar in this period and the recent slow down of the U.S. economy.[1]

The U.S. industry has several advantages over the rest of the world market: modern mills, a highly skilled work force, a large domestic market, and an efficient transportation infrastructure. Major export markets for pulp are Japan, Italy, Germany, Mexico, and France. The U.S. Department of Commerce anticipates exports to grow faster than production for domestic markets through 2004. World Trade Organization (WTO) efforts to reduce tariffs include those on pulp and paper products; if these are successful, the U.S. industry expects pulp and paper export rates to increase even further.

However, pulp and paper are commodities and therefore prices are vulnerable to global competition. Countries such as Brazil, Chile, and Indonesia have built modern, advanced pulp facilities. These countries have faster-growing trees and lower labor costs. Latin American and European countries are also adding papermaking capacity. Because of this increased foreign competition, imports of paper to the U.S. market are expected to increase 3% annually through 2004.[10] In order

FIGURE 11.2 Geographic distribution of pulp, paper, and cardboard mills. (Taken from U.S. EPA, Profile of the Pulp and Paper Industry, 2nd ed., report EPA/310-R-02-002, U.S. EPA, Washington, November 2002.)

to compensate for this increasingly competitive market, pulp and paper companies have undertaken a considerable number of mergers and acquisitions between 1997 and 2002.

Historically, U.S. pulp and paper companies have invested heavily in capital improvements to their facilities. Capital investments in recent years, however, are well below historic levels due to the difficult market conditions. For the first time, industry capacity actually declined in 2001.[1] Because few new mills are being built, most capital expenditures are for plant expansions, upgrades, and environmental protection initiatives at existing facilities. Throughout the time period 1985–1999, capital improvements related to environmental protection claimed from 4% to 22% of the total investments, with significant increases in the early and late 1990s.[1]

A major movement within the pulp and paper industry has been an increased focus on the use of recovered paper. Nearly 50% of paper is now recovered and used either as recycled paper or as products such as home insulation. Furthermore, recovered paper contributes to U.S. exports; roughly ten million tons of recovered paper were exported in 2000.[1]

11.2 PROCESS DESCRIPTION

11.2.1 PROCESSES IN THE PULP AND PAPER INDUSTRY

Simply put, paper is manufactured by applying a watery suspension of cellulose fibers to a screen that allows the water to drain and leaves the fibrous particles behind in a web. Most modern paper products contain nonfibrous additives, but otherwise they fall within this general definition. Only a few paper products for specialized uses are created without the use of water, using dry forming techniques. The production of pulp is the major source of environmental impacts from the pulp and paper industry.

Processes in the manufacture of paper and paperboard can, in general terms, be divided into three steps:

1. Pulp making
2. Pulp processing
3. Paper/paperboard production

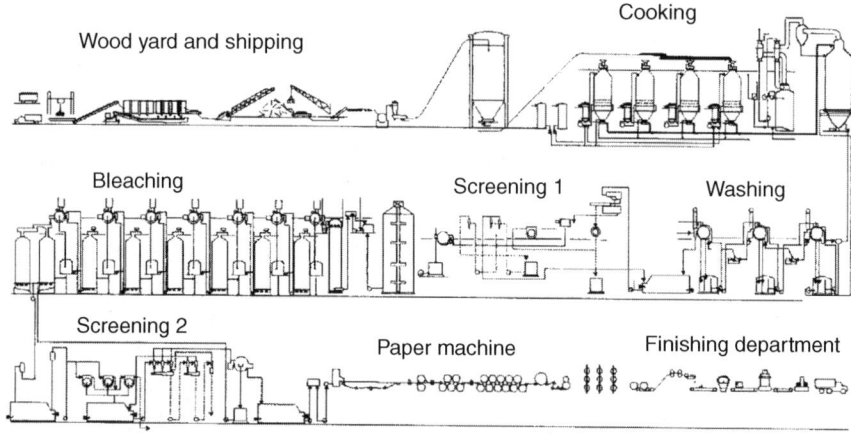

FIGURE 11.3 Simplified flow diagram of an integrated mill. (Taken from U.S. EPA, Profile of the Pulp and Paper Industry, 2nd ed., report EPA/310-R-02-002, U.S. EPA, Washington, November 2002.)

Paper and paperboard production processes are similar. After the fibers are separated and impurities have been removed, the pulp may be bleached to improve brightness and processed to a form suitable for paper-making. At the paper-making stage, the pulp can be combined with dyes, strength-building resins, or texture-adding filler materials, depending on its intended end product. Afterwards, the mixture is dewatered, leaving the fibrous constituents and pulp additives on an endless fabric belt. The fibers bond together as the web passes through a series of presses and around heated drum driers. Additional additives may be applied to the moving web. The final paper product is usually spooled on large rolls for storage (see Figure 11.3). If more information on paper making processes is desired, reference 3 is recommended.

11.2.1.1 Pulp Manufacturing

Table 11.3 presents an overview of wood pulping types by the method of fiber separation, resultant fiber quality, and percent of 1998 U.S. pulp production.[11,12] Many mills perform multiple pulping processes at the same site, most frequently nondeink secondary fiber pulping and paper-grade kraft

TABLE 11.3
General Classification of Wood Pulping Processes

Process Category	Fiber Separation Method	Fiber Quality	Examples	% of Total 1998 U.S. Wood Pulp Production
Mechanical	Mechanical energy	Short, weak, unstable, impure fibers	Stone groundwood, refiner mechanical pulp	10
Semichemical	Combination of chemical and mechanical treatments	"Intermediate" pulp properties (some unique properties)	High-yield kraft, high-yield sulfite	6
Chemical	Chemicals and heat	Long, strong, stable fibers	Kraft, sulfite, soda	84

Source: U.S. EPA, Profile of the Pulp and Paper Industry, 2nd ed., report EPA/310-R-02-002, U.S. EPA, Washington, November 2002.

pulping.[3] The following three basic types of wood-pulping processes are detailed below, followed by a discussion of secondary fiber pulping techniques:

1. Chemical pulping
2. Semichemical pulping
3. Mechanical pulping

Various technologies and chemicals are used to manufacture pulp, but most pulp manufacturing systems contain the process sequence shown in Table 11.4. Overall, most of the pollutant releases associated with pulp and paper mills occur at the pulping and bleaching stages where the majority of chemical inputs occur.

Furnish composition

According to the National Census,[13] wood is used in some form by approximately 95% of pulp and paper manufacturers. Wood can be in a variety of forms and types. Wood logs, chips, and sawdust are used to make pulp. Due to different physical and chemical properties, however, certain pulping processes are more efficient when used on specific wood types. The species of wood used has a profound influence on the characteristics of the pulp. In general, softwood fibers are longer than those from hardwood and have thinner cell walls. The longer fibers of softwood produce papers of greater strength, particularly tear strength.

Secondary fibers comprise the next most common furnish constituent. Secondary fibers consist of preconsumer fibers (e.g., mill waste fibers, which were always recycled internally) and postconsumer fiber, which is what is generally referred to as recycled paper. Postconsumer fiber sources are diverse, but the most common are newsprint and corrugated boxes. Although secondary fibers are not used in as great a proportion as wood furnish, ca. 70% of pulp and paper manufacturers use some secondary fibers in their pulp production and ca. 200 mills (40% of the total number of mills) rely exclusively on secondary fibers for their pulp furnish.[11,14] Secondary fibers must be processed to remove contaminants such as glues, coatings, or bindings, and, depending on the end product, may or may not be processed to remove ink or brighten the pulp.

Secondary fiber use is increasing in the pulp and paper industry due to the increasing prices of virgin pulp and the continuing improvement in deinking technology. Environmental concerns have led to consumer acceptance of lower brightness of products made from recycled paper, and government specifications set a minimum level of product quality. Recovered fiber accounted for 75% of the industry's increase in fiber consumption between 1990 and 2000.[15] The utilization of secondary

TABLE 11.4
Pulp Manufacturing Process Sequence

Process Sequence	Description
Fiber furnish preparation and handling	Debarking, slashing, chipping of wood logs and then screening of wood chips/ secondary fibers (some pulp mills purchase chips and skip this step)
Pulping	Chemical, semichemical, or mechanical breakdown of pulping material into fibers
Pulp processing	Removal of pulp impurities, cleaning and thickening of pulp fiber mixture
Bleaching	Addition of chemicals in a staged process of reaction and washing increases whiteness and brightness of pulp, if necessary
Pulp drying and baling (nonintegrated mills)	At nonintegrated pulp mills, pulp is dried and bundled into bales for transport to a paper mill
Stock preparation	Mixing, refining, and addition of wet additives to add strength, gloss, texture to paper product, if necessary

Source: U.S. EPA, Profile of the Pulp and Paper Industry, 2nd ed., report EPA/310-R-02-002, U.S. EPA, Washington, November 2002.

fibers, expressed as the ratio of recovered paper consumption to the total production of paper and paperboard, is ca. 39% and is climbing slowly.[1] In a resource-deficient country such as Japan, the secondary fiber utilization rate is ca. 50%, whereas the average utilization rate in Europe is ca. 40%.[16] Due to losses of fiber substance and strength during the recycling process, a 50% utilization rate is considered the present maximum overall utilization rate for fiber recycling.[12]

Until recently, secondary fiber was not used for higher quality paper products. Contaminants (e.g., inks, paper colors) are present, so production of low-purity products is often the most cost-effective use of secondary fibers. Approximately 68% of all secondary fiber in the U.S. is presently used for multi-ply paperboard or the corrugating paper used to manufacture corrugated cardboard.[15] Recently, continuing improvement of deinking processes together with the demand created by environmental concerns have resulted in an increasing use of deinked fiber for newsprint or higher-quality uses, such as office copier paper.

Other sources of fibers include cotton rags and linters, flax, hemp, bagasse, tobacco, and synthetic fibers such as polypropylene. These substances are not used widely, however, as they are typically for low-volume, specialty grades of paper.

The types of furnish used by a pulp and paper mill depend on the type of product produced and what is readily available. Urban mills use a larger proportion of secondary fibers due to the post-consumer feedstock being close at hand. More rurally located mills are usually close to timber sources and thus may use virgin fibers in a greater proportion.

Furnish preparation

Wood is prepared for pulp production by a process designed to supply a homogeneous pulping feed-stock. In the case of roundwood furnish (logs), the logs are cut to manageable size and then debarked. At pulp mills integrated with lumbering facilities, acceptable lumber wood is removed at this stage. At these facilities, any residual or waste wood from lumber processing is returned to the chipping process; in-house lumbering rejects can be a significant source of wood furnish at a facility. The bark of those logs not fit for lumber is usually either stripped mechanically or hydraulically with high pressure water jets in order to prevent contamination of pulping operations. Depending on the moisture content of the bark, it may then be burned for energy production. If not burned for energy production, bark can be used for mulch, ground cover, or to make charcoal.

Hydraulic debarking methods may require a drying step before burning. Usually, hydraulically removed bark is collected in a water flume, dewatered, and pressed before burning. Treatment of wastewater from this process is difficult and costly, however, whereas dry debarking methods can channel the removed bark directly into a furnace.[12] In part because of these challenges, hydraulic debarking has decreased in significance within the industry.[1]

Debarked logs are cut into chips of equal size by chipping machines. Chippers usually produce uniform wood pieces 20 mm long in the grain direction and 4 mm thick. The chips are then put on a set of vibrating screens to remove those that are too large or small. Large chips stay on the top screens and are sent to be recut, while the smallest chips are usually burned with the bark. Certain mechanical pulping processes, such as stone groundwood pulping, use roundwood; however, the majority of pulping operations require wood chips. Nonwood fibers are handled in ways specific to their composition. Steps are always taken to maintain fiber composition and thus pulp yield.

Chemical pulping

Chemical pulps are typically manufactured into products that have high quality standards or require special properties. Chemical pulping separates the fibers of wood by dissolving the lignin bond holding the wood together. Generally, this process involves the cooking/digesting of wood chips in aqueous chemical solutions at elevated temperatures and pressures. There are two major types of chemical pulping used in the U.S., which differ in the chemicals employed and in the waste produced:

1. Kraft/soda pulping
2. Sulfite pulping

Kraft pulping processes produced approximately 83% of all U.S. pulp tonnage during 2000 according to the American Forest and Paper Association.[1] The success of the process and its widespread adoption are due to several factors. First, because the kraft cooking chemicals are selective in their attack on wood constituents, the pulps produced are notably stronger than those from other processes (kraft is German for "strength"). The kraft process is also flexible, in so far as it can be applied to many different types of raw materials (i.e., hard or soft woods) and can tolerate contaminants frequently found in wood (e.g., resins). Lignin removal rates are high in the kraft process—up to 90%—allowing high levels of bleaching without pulp degradation. Finally, the chemicals used in kraft pulping are readily recovered within the process, making it very economical and reducing potential environmental releases.

The kraft process uses a sodium-based alkaline pulping solution (liquor) consisting of sodium sulfide (Na_2S) and sodium hydroxide (NaOH) in 10% solution. This liquor (white liquor) is mixed with the wood chips in a reaction vessel (digester). The output products are separated wood fibers (pulp) and a liquid that contains the dissolved lignin solids in a solution of reacted and unreacted pulping chemicals (black liquor). The black liquor undergoes a chemical recovery process to regenerate white liquor for the first pulping step. Overall, the kraft process converts ca. 50% of input furnish into pulp.

The kraft process evolved from the soda process. The soda process uses an alkaline liquor of only sodium hydroxide (NaOH). The kraft process has virtually replaced the soda process due to the economic benefits of chemical recovery and improved reaction rates (the soda process has a lower yield of pulp per pound of wood furnish than the kraft process).

Sulfite pulping was used for approximately 2% of U.S. pulp production in 2000.[1] Softwood is the predominant furnish used in sulfite pulping processes. However, only nonresinous species are generally pulped, particularly when a light colored pulp is required. This process is used, for example, almost exclusively for the manufacture of viscose.[17] To manufacture sulfite pulp, wood chips are boiled under pressure in large digesters with calcium sulfite, ammonium sulfite, magnesium sulfite, or sodium sulfite. The sulfite pulping process relies on acid solutions of sulfurous acid (H_2SO_3) and bisulfite ion (HSO_3^-) to degrade the lignin bonds between wood fibers. In sulfite pulping most water pollution arises from spent liquor, condensates, bleach plant effluents, and accidental discharges.

Sulfite pulps have less color than kraft pulps and can be bleached more easily; however, they are not as strong. The efficiency and effectiveness of the sulfite process is also dependent on the type of wood furnish and the absence of bark. For these reasons, the use of sulfite pulping has declined in comparison to kraft pulping over time.

Semichemical pulping

Semichemical pulping comprised 6% of U.S. pulp production in 1993.[1] Semichemical pulp is often very stiff, making this process common in corrugated container manufacture. This process primarily uses hardwood as furnish.

The major process difference between chemical pulping and semichemical pulping is that semichemical pulping uses lower temperatures, more dilute cooking liquor or shorter cooking times, and mechanical disintegration for fiber separation. At most, the digestion step in the semichemical pulping process consists of heating pulp in sodium sulfite (Na_2SO_3) and sodium carbonate (Na_2CO_3). Other semichemical processes include the Permachem process and the two-stage vapor process. The yield of semichemical pulping ranges from 55 to 90%, depending on the process used, but pulp residual lignin content is also high so bleaching is more difficult.

Mechanical pulping

Mechanical pulping accounted for 9% of U.S. pulp production in 2000.[1] Mechanically produced pulp is of low strength and quality. Such pulps are used principally for newsprint and other nonpermanent paper goods. Mechanical pulping uses physical pressures instead of chemicals to separate furnish fibers. The processes include the following:

1. Stone groundwood
2. Refiner mechanical
3. Thermo-mechanical

4. Chemi-mechanical
5. Chemi-thermo-mechanical

The stone groundwood process simply involves mechanical grinding of wood in several high-energy refining systems. The refiner mechanical process involves refining wood chips at atmospheric pressure. The thermo-mechanical process uses steam and pressure to soften the chips before mechanical refining. In the chemi-mechanical process, chemicals can be added throughout the process to aid the mechanical refining. The chemi-thermo-mechanical process involves the treatment of chips with chemicals for softening followed by mechanical pulping under heat and pressure. Mechanical pulping typically results in high pulp yields, up to 95% when compared to chemical pulping yields of 45–50%, but energy usage is also high. To offset its structural weakness, mechanical pulp is often blended with chemical pulp.

Secondary fiber pulping
Secondary fiber pulping accounted for 39% of domestic pulp production in 2000.[1] Nearly 200 mills rely exclusively on recovered paper for pulp furnish, and ca. 80% of U.S. paper mills use recovered paper in some way.[14] In addition, consumption of fiber from recovered paper is growing more than twice as fast as overall fiber consumption. Secondary fibers are usually presorted before they are sold to a pulp and paper mill. If not, secondary fibers are processed to remove contaminants before pulping occurs. Common contaminants consist of adhesives, coatings, polystyrene foam, dense plastic chips, polyethylene films, wet strength resins, and synthetic fibers. In some cases, contaminants of greater density than the desired secondary fibers are removed by centrifugal force while light contaminants are removed by flotation systems. Centri cleaners are also used to remove material less dense than fibers (wax and plastic particles).[18]

Inks, another contaminant of secondary fibers, may be removed by heating a mixture of secondary fibers with surfactants. The removed inks are then dispersed in an aqueous medium to prevent redeposition on the fibers. Continuous solvent extraction has also been used to recover fibers from paper and board coated with plastics or waxes.

Secondary fiber pulping is a relatively simple process. The most common pulper design consists of a large container filled with water, which is sometimes heated, and the recycled pulp. Pulping chemicals (e.g., sodium hydroxide, NaOH) are often added to promote dissolution of the paper or board matrix. The source fiber (corrugated containers, mill waste, and so on) is dropped into the pulper and mixed by a rotor. Debris and impurities are removed by two mechanisms: a ragger and a junker. The ragger withdraws strings, wires, and rags from the stock secondary fiber mixture. A typical ragger consists of a few "primer wires" that are rotated in the secondary fiber slurry. Debris accumulates on the primer wires, eventually forming a "debris rope," which is then removed. Heavier debris is separated from the mixture by centrifugal force and falls into a pocket on the side of the pulper. The junker consists of a grappling hook or elevator bucket. Heat, dissolution of chemical bonds, shear forces created by stirring and mixing, and grinding by mechanical equipment may serve to dissociate fibers and produce a pulp of desired uniformity.

Contaminant removal processes depend on the type and source of secondary fiber to be pulped. Mill paper waste can be easily repulped with minimal contaminant removal. Recycled postconsumer newspaper, on the other hand, may require extensive contaminant removal, including deinking, prior to reuse. Secondary fiber is typically used in lower-quality applications such as multiply paperboard or corrugating paper.

11.2.1.2 Pulp Processing

After pulp production, pulp processing removes impurities[12] such as uncooked chips, and recycles any residual cooking liquor via the washing process (Figure 11.4). Pulps are processed in a wide variety of ways, depending on the method that generated them (e.g., chemical, semichemical). Some pulp processing steps that remove pulp impurities include screening, defibering, and deknotting. Pulp may also be thickened by removing a portion of the water. At additional cost, pulp may be

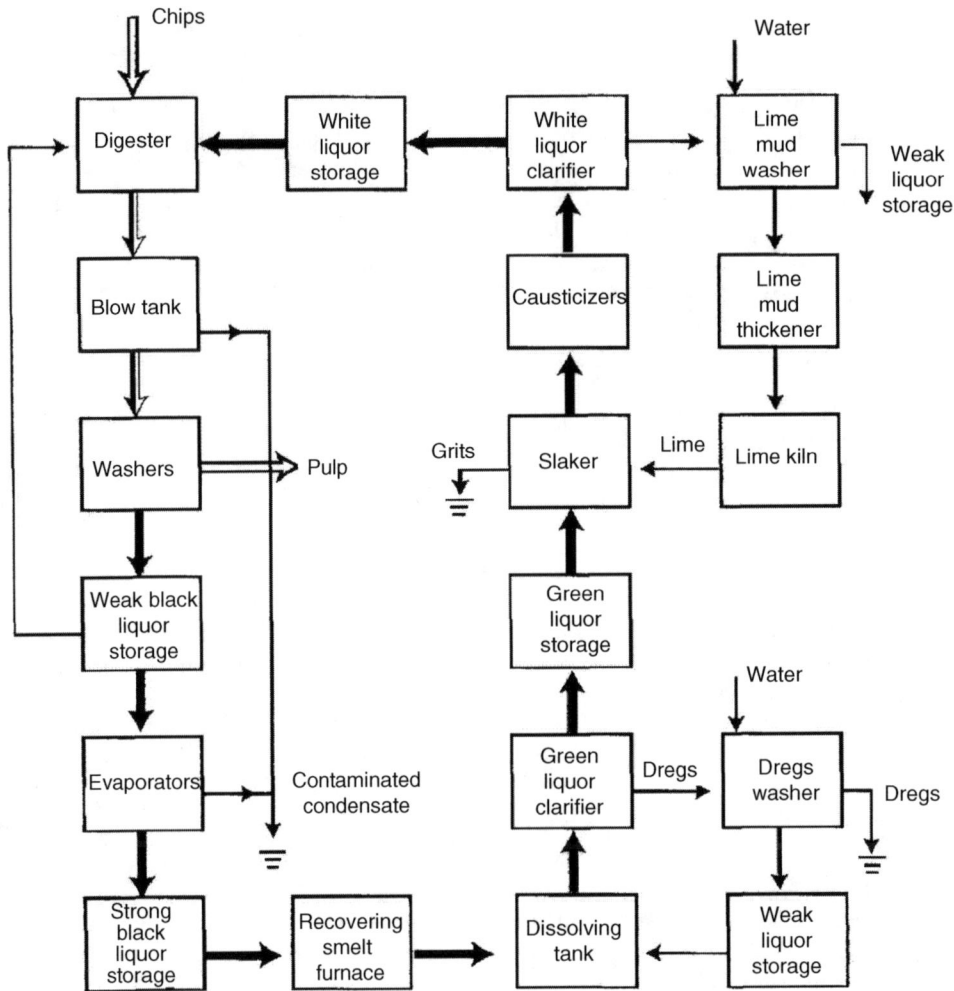

FIGURE 11.4 The kraft pulping process (with chemical recovery). (Taken from U.S. EPA, Profile of the Pulp and Paper Industry, 2nd ed., report EPA/310-R-02-002, U.S. EPA, Washington, November 2002.)

blended to ensure product uniformity. If pulp is to be stored for long periods of time, drying steps are necessary to prevent fungal or bacterial growth.[1]

Residual spent cooking liquor from chemical pulping is washed from the pulp using brown stock washers. Efficient washing is critical to maximize the return of cooking liquor to chemical recovery and to minimize the carryover of cooking liquor (known as brown stock washing loss) into the bleach plant, because excess cooking liquor increases consumption of bleaching chemicals. Specifically, the dissolved organic compounds (lignins and hemicelluloses) contained in the liquor will bind to the bleaching chemicals and thus increase bleach chemical consumption. In addition, these organic compounds are the precursors to chlorinated organic compounds (e.g., dioxins, furans). The most common washing technology is rotary vacuum washing, carried out sequentially in two or four washing units. Other washing technologies include diffusion washers, rotary pressure washers, horizontal belt filters, wash presses, and dilution/extraction washers.

Pulp screening removes the remaining oversized particles such as bark fragments, oversized chips, and uncooked chips. In open screen rooms, wastewater from the screening process receives wastewater treatment prior to discharge. In closed-loop screen rooms, wastewater from the process is reused in other pulping operations and ultimately enters the mill's chemical recovery system. Centrifugal cleaning (also known as liquid cyclone, hydrocyclone, or centricleaning) is used after screening to remove relatively dense contaminants such as sand and dirt. Rejects from the screening process are either repulped or disposed of as solid waste.

Chemical recovery systems

The chemical recovery system is a complex part of a chemical pulp and paper mill and is subject to a variety of environmental regulations. Chemical recovery is a crucial component of the chemical pulping process; it recovers process chemicals from the spent cooking liquor for reuse. The chemical recovery process has important financial and environmental benefits for pulp and paper mills. Economic benefits include savings on chemical purchase costs due to regeneration rates of process chemicals approaching 98%, and energy generation from pulp residue burned in a recovery furnace.[12] Environmental benefits include the recycle of process chemicals and lack of resultant discharges to the environment.

Kraft chemical recovery systems

Although newer technologies are always under development, the basic kraft chemical recovery process has not been fundamentally changed since the issue of its patent in 1884. The stepwise progression of chemical reactions has been refined; for example, black liquor gasification processes are now in use in an experimental phase. The precise details of the chemical processes at work in the chemical recovery process can be found in Smook's Handbook.[12] The kraft chemical recovery process consists of the following general steps:

1. *Black liquor concentration.* Residual weak black liquor from the pulping process is concentrated by evaporation to form "strong black liquor." After brown stock washing in the pulping process the concentration of solids in the weak black liquor is approximately 15%; after the evaporation process, solids concentration can range from 60 to 80%. In some older facilities, the liquor then undergoes oxidation for odor reduction. The oxidation step is necessary to reduce odor created when hydrogen sulfide is stripped from the liquor during the subsequent recovery boiler burning process. Almost all recovery furnaces installed since 1968 have noncontact evaporation processes that avoid these problems, so oxidation processes are not usually seen in newer mills. Common modern evaporator types include multiple effect evaporators as well as a variety of supplemental evaporators. Odor problems with the kraft process have been the subject of control measures.

2. *Recovery boiler.* The strong black liquor from the evaporators is burned in a recovery boiler. In this crucial step in the overall kraft chemical recovery process, organic solids are burned for energy and the process chemicals are removed from the mixture in molten form. Molten inorganic process chemicals (smelt) flow through the perforated floor of the boiler to water-cooled spouts and dissolving tanks for recovery in the recausticizing step. Energy generation from the recovery boiler is often insufficient for total plant needs, so facilities augment recovery boilers with fossil-fuel-fired and wood-waste-fired boilers to generate steam and often electricity. Industry wide, the utilization of pulp wastes, bark, and other paper-making residues supplies 58% of the energy requirements of pulp and paper companies.[11]

3. *Recausticizing.* Smelt is recausticized to remove impurities left over from the furnace and to convert sodium carbonate (Na_2CO_3) into active sodium hydroxide (NaOH) and sodium sulfide (Na_2S). The recausticization procedure begins with the mixing of smelt with "weak" liquor to form green liquor, named for its characteristic color. Contaminant solids, called dregs, are removed from the green liquor, which is mixed with lime (CaO). After the lime mixing step, the mixture, now called white liquor due to its new coloring, is processed to remove a layer of lime mud ($CaCO_3$) that has precipitated. The primary chemicals recovered are caustic (NaOH) and sodium sulfide (Na_2S). The remaining white liquor

is then used in the pulp cooking process. The lime mud is treated to regenerate lime in the calcining process.

4. *Calcining.* In the calcining process, the lime mud removed from the white liquor is burned to regenerate lime for use in the lime mixing step. The vast majority of mills use lime kilns for this process, although a few mills now use newer fluidized bed systems in which the reactants are suspended by upward-blowing air.

Sulfite chemical recovery systems

Numerous sulfite chemical pulping recovery systems are in use today. Heat and sulfur can be recovered from all liquors generated; however, the base chemical can only be recovered from magnesium and sodium base processes. See Smook's Handbook[12] for more information.

11.2.1.3 Bleaching

Bleaching is defined as any process that chemically alters pulp to increase its brightness. Bleached pulps create papers that are whiter, brighter, softer, and more absorbent than unbleached pulps. Bleached pulps are used for white or light colored paper. Unbleached pulp is typically used to produce boxboard, linerboard, and grocery bags. Of the approximately 65.5 million T (72 million tons) of pulp (including recycled pulp) used in paper production in the U.S. in 2000, about 50% is for bleached pulp.[1]

Any type of pulp may be bleached, but the type(s) of fiber furnish and pulping processes used, as well as the desired qualities and end use of the final product, greatly affect the type and degree of pulp bleaching possible. Printing and writing papers comprise ca. 60% of bleached paper production. The lignin content of a pulp is the major determinant of its bleaching potential. Pulps with high lignin content (e.g., mechanical or semichemical) are difficult to bleach fully and require heavy chemical inputs. Bleached pulps with high lignin content are subject to color reversion, loss of brightness when exposed to light. Excessive bleaching of mechanical and semichemical pulps results in loss of pulp yield due to fiber destruction. Chemical pulps can be bleached to a greater extent due to their low (10%) lignin content. For more information, the U.S. EPA reference 19 is recommended. Typical bleaching processes for each pulp type are detailed below.

Chemical pulp bleaching has undergone significant process changes since around 1990. Until that time, nearly every chemical pulp mill that had used bleaching had incorporated elemental chlorine (Cl_2) into some of its processes. Because of environmental and health concerns about dioxins, U.S. pulp mills now use elemental chlorine free (ECF) and total chlorine free (TCF) bleaching technologies. The most common types of ECF and TCF are shown in Table 11.5. The

---

TABLE 11.5

Common Chemicals Used in Elemental Chlorine Free (ECF) and Total Chlorine Free (TCF) Bleaching Processes

Bleaching Chemical	Chemical Formula	ECF/TCF
Sodium hydroxide	NaOH	ECF and TCF
Chlorine dioxide	ClO_2	ECF
Hypochlorite	HClO, NaOCl, Ca(OCl)$_2$	ECF
Oxygen	O_2	ECF and TCF
Ozone	O_3	ECF and TCF
Hydrogen peroxide	H_2O_2	ECF and TCF
Sulfur dioxide	SO_2	ECF and TCF
Sulfuric acid	H_2SO_4	ECF and TCF

Source: U.S. EPA, Profile of the Pulp and Paper Industry, 2nd ed., report EPA/310-R-02-002, U.S. EPA, Washington, November 2002.

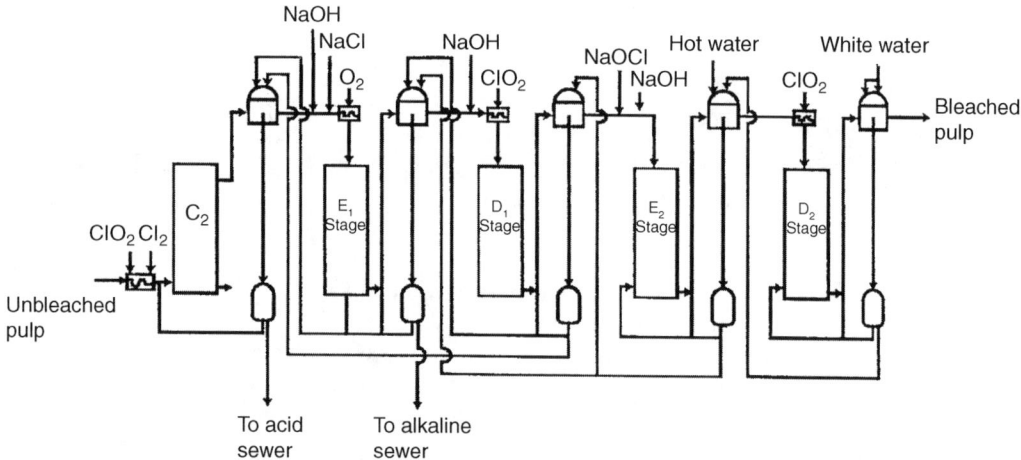

FIGURE 11.5 Typical bleach plant source. (Taken from U.S. EPA, Profile of the Pulp and Paper Industry, 2nd ed., report EPA/310-R-02-002, U.S. EPA, Washington, November 2002.)

difference between ECF and TCF is that ECF may include chlorine dioxide (ClO_2) and hypochlorite [HClO, NaOCl, and $Ca(OCl)_2$] based technologies. In 2001, ECF technologies were used for about 95% of bleached pulp production, TCF technologies were used for about 1% of bleached pulp production, and elemental chlorine was used for about 4% of production.[20]

Chemical pulp is bleached in traditional bleach plants (see Figure 11.5), where the pulp is processed through three to five stages of chemical bleaching and water washing. The desired whiteness, the brightness of the initial stock pulp, and the plant design determine the number of cycles needed.

Bleaching stages generally alternate between acid and alkaline conditions. Chemical reactions with lignin during the acid stage of the bleaching process increase the whiteness of the pulp. The alkaline extraction stages dissolve the lignin/acid reaction products. At the washing stage, both solutions and reaction products are removed. Chemicals used to perform the bleaching process must have high lignin reactivity and selectivity to be efficient. Typically, 4 to 8% of pulp is lost due to bleaching agent reactions with the wood constituents cellulose and hemicellulose, but these losses can be as high as 18%.[1,2]

Semichemical pulps are typically bleached with hydrogen peroxide (H_2O_2) in a bleach tower.

Mechanical pulps are bleached with hydrogen peroxide (H_2O_2) or sodium hydrosulfite ($NaHSO_3$). Bleaching chemicals are either applied without separate equipment during the pulp processing stage (i.e., in-line bleaching), or in bleaching towers. Full bleaching of mechanical pulps is generally not practical due to bleaching chemical cost and the negative impact on pulp yield.

Deinked secondary fibers are usually bleached in a bleach tower, but may be bleached during the repulping process. Bleach chemicals may be added directly into the pulper. The following are examples of chemicals used to bleach deinked secondary fibers: hypochlorite [HClO, NaOCl, $Ca(OCl)_2$], hydrogen peroxide (H_2O_2), and hydrosulphite ($NaHSO_3$).

11.2.1.4 Stock Preparation

At this final stage, the pulp is processed into the stock used for paper manufacture. Market pulp, which is to be shipped off-site to paper or paperboard mills, is processed little, if at all at this stage. Processing includes pulp blending specific to the desired paper product desired, dispersion in water, beating and refining to add density and strength, and addition of any necessary wet additives. Wet additives are used to create paper products with special properties or to facilitate the paper-making process. Wet additives include resins and waxes for water repellency, fillers such as clays, silicas, talc, inorganic/organic dyes for coloring, and certain inorganic chemicals (calcium sulfate, zinc sulfide, and titanium dioxide) for improved texture, print quality, opacity, and brightness.

11.2.1.5   Processes in Paper Manufacture

The paper and paperboard making process consists of the following general steps:

1. *Wet end operations*: formation of paper sheet from wet pulp
2. *Dry end operations*: drying of paper product, application of surface treatments, and spooling for storage

Wet end operations

The processed pulp is converted into a paper product via a paper production machine, the most common of which is the Fourdrinier paper machine (see Figure 11.6). In the Fourdrinier system,[3] the pulp slurry is deposited on a moving belt (made from polyester forming fabrics) that carries it through the first stages of the process. Water is removed by gravity, vacuum chambers, and vacuum rolls. This waste water is recycled to the slurry deposition step of the process due to its high fiber content. The continuous sheet is then pressed between a series of rollers to remove more water and compress the fibers.

Dry end operations

After pressing, the sheet enters a drying section, where the sheet passes around a series of steam-heated drums. It then may be calendared. In the calendar process the sheet is pressed between heavy rolls to reduce paper thickness and produce a smooth surface. Coatings can be applied to the paper at this point to improve gloss, color, printing detail, and brilliance. Lighter coatings are applied on-machine, and heavy coatings are performed off-machine. The paper product is then spooled for storage.

11.2.1.6   Energy Generation

Pulp and paper mill energy generation is provided in part from the burning of liquor waste solids in the recovery boiler, but other energy sources are needed to make up the remainder of mill energy needs. Over the last 25 years the pulp and paper industry has changed its energy generation methods from fossil fuels to a greater utilization of processes or process wastes. The increase in use of wood wastes from the wood handling and chipping processes (Table 11.6) is one example of this industry-wide movement. During the period 1972 to 1999, the proportion of total industry power generation from the combination of woodroom wastes, spent liquor solids, and other self-generation methods increased from ca. 41% to ca. 58%, while coal, fuel oil, and natural gas use decreased from ca. 54% to ca. 36%.[12,21]

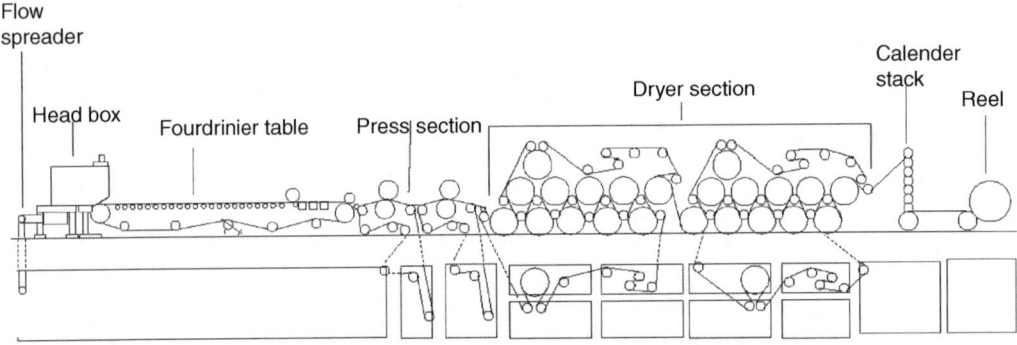

FIGURE 11.6   Fourdrinier paper machine. (Taken from U.S. EPA, Profile of the Pulp and Paper Industry, 2nd ed., report EPA/310-R-02-002, U.S. EPA, Washington, November 2002.)

TABLE 11.6
Estimated Energy Sources for the U.S. Pulp and Paper Industry

Energy Source	1972	1979	1990	1999
Purchased steam	5.4%	6.7%	7.3%	1.5%
Coal	9.8%	9.1%	13.7%	12.5%
Fuel oil	22.3%	19.1%	6.4%	6.3%
Natural gas	21.5%	17.8%	16.4%	17.6%
Other purchased energy	—	—	—	6.7%
Waste wood and wood chips (hogged fuel) and bark	6.6%	9.2%	15.4%	13.5%
Spent liquor solids	33.7%	37.3%	39.4%	40.3%
Other self-generated power	0.6%	0.8%	1.2%	1.6%

Source: U.S. EPA, Profile of the Pulp and Paper Industry, 2nd ed., report EPA/310-R-02-002, U.S. EPA, Washington, November 2002.

Power boilers at pulp and paper mills are sources of particulate emissions, SO_2, and NO_x. Pollutants emitted from chemical recovery boilers include SO_2 and total reduced sulfur compounds (TRS).

11.2.2 RAW MATERIAL INPUTS AND POLLUTION OUTPUTS IN THE PRODUCTION LINE

Pulp and paper mills use and generate materials that may be harmful to the air, water, and land:

1. Pulp and paper processes generate large volumes of wastewaters that might adversely affect freshwater or marine ecosystems.
2. Residual wastes from wastewater treatment processes may contribute to existing local and regional disposal problems.
3. Air emissions from pulping processes and power generation facilities may release odors, particulates, or other pollutants.

The major sources of pollutant releases in pulp and paper manufacture occur at the pulping and bleaching stages, respectively. As such, nonintegrated mills (i.e., those mills without pulping facilities on site) are not significant environmental concerns when compared to integrated mills or pulp mills.

11.2.2.1 Water Pollutants

The pulp and paper industry is the largest industrial process water user in the U.S.[5] In 2000, a typical pulp and paper mill used between 15,140 and 45,420 L (4000 to 12,000 gal) of water per ton of pulp produced.[4] General water pollution concerns for pulp and paper mills are effluent solids, biochemical oxygen demand (BOD), and color. Toxicity concerns historically occurred from the potential presence of chlorinated organic compounds such as dioxins, furans, and others (collectively referred to as adsorbable organic halides, or AOX) in wastewaters after the chlorination/extraction sequence. With the substitution of chlorine dioxide for chlorine, discharges of the chlorinated compounds have decreased dramatically.

Due to the large volumes of water used in pulp and paper processes, virtually all U.S. mills have primary and secondary wastewater treatment systems to remove particulates and BOD. These systems also provide significant removals (e.g., 30 to 70%) of other important parameters such as AOX and chemical oxygen demand (COD).

TABLE 11.7

Common Water Pollutants from Pulp and Paper Processes

Source	Effluent Characteristics
Water used in wood handling/debarking and chip washing	Solids, BOD, color
Chip digester and liquor evaporator condensate	Concentrated BOD; can contain reduced sulfur
"White waters" from pulp screening, thickening, and cleaning	Large volume of water with suspended solids; can have significant BOD
Bleach plant washer filtrates	BOD, color, chlorinated organic compounds
Paper machine-water flows	Solids, often precipitated for reuse
Fiber and liquor spills	Solids, BOD, color

BOD, biochemical oxygen demand.

Source: U.S. EPA, Profile of the Pulp and Paper Industry, 2nd ed., report EPA/310-R-02-002, U.S. EPA, Washington, November 2002.

The major sources of pollutants from pulp and paper mills[12] are presented in Table 11.7.

Wood processing operations in pulp mills use water for a variety of purposes. The resulting wastewaters contain BOD, suspended solids, and some color. The condensates from chip digesters and chemical recovery evaporators are sources of BOD and reduced sulfur compounds. Wastewaters containing BOD, color, and suspended solids may be generated from pulp screening operations in mills using "atmospheric" systems, although most mills have modern pressure screens that virtually eliminate such wastewaters. Kraft bleaching generates large volumes of wastewater containing BOD, suspended solids, color, and chlorinated organic compounds. From paper machines, excess white water (named for its characteristic color) contains suspended solids and BOD. Fiber and liquor spills can also be a source of mill effluent. Typically, spills are captured and pumped to holding areas to reduce chemical usage through spill reuse and to avoid loadings on facility wastewater treatment systems.

Wastewater treatment systems can be a significant source of cross-media pollutant transfer. For example, waterborne particulates and some chlorinated compounds settle or absorb onto treatment sludge and other compounds may volatilize during the wastewater treatment process.

11.2.2.2   Air Pollutants

Table 11.8 is an overview of the major types and sources of air pollutant releases from various pulp and paper processes.[12]

Water vapors are the most visible air emission from a pulp and paper mill, but are not usually regulated unless they form a significant obscurement or are a climate modifier.

Pulp and paper mill power boilers and chip digesters are generic pulp and paper mill sources of air pollutants such as particulates and nitrogen oxides. Chip digesters and chemical recovery evaporators are the most concentrated sources of volatile organic compounds. The chemical recovery furnace is a source of fine particulate emissions and sulfur oxides. In the kraft process, sulfur oxides are a minor issue in comparison to the odor problems created by four reduced sulfur gases, known collectively as total reduced sulfur (TRS): hydrogen sulfide, methyl mercaptan, dimethyl sulfide, and dimethyl disulfide. The TRS emissions are primarily released from wood chip digestion, black liquor evaporation, and chemical recovery boiler processes. TRS compounds create odor nuisance problems at lower concentrations than sulfur oxides; odor thresholds for TRS compounds are approximately 1000 times lower than that for sulfur dioxide. Humans can detect some TRS compounds in the air as a "rotten egg" odor at a level as low as 1 µg/L.

TABLE 11.8

Common Air Pollutants from Pulp and Paper Processes

Source	Type
Kraft recovery furnace	Fine particulates
Fly ash from hog fuel and coal-fired burners	Coarse particulates
Sulfite mill operations	Sulfur oxides
Kraft pulping and recovery processes	Reduced sulfur gases
Chip digesters and liquor evaporation	Volatile organic compounds
All combustion processes	Nitrogen oxides

Source: U.S. EPA, Profile of the Pulp and Paper Industry, 2nd ed., report EPA/310-R-02-002, U.S. EPA, Washington, November 2002.

Pulp and paper mills have made significant investments in pollution control technologies and processes. According to industry sources, the pulp and paper industry spent more than USD 1 billion/yr from 1991 to 1997 on environmental capital expenditures. In 1991 and 1992, this represented 20% of their total capital expenditures.[22] Chemical recovery and recycling systems in the chemical pulping process significantly reduce pollutant outputs while providing substantial economic return due to recovery of process chemicals. Chemical recovery is necessary for the basic economic viability of the kraft process. According to U.S. EPA sources, all kraft pulp mills worldwide have chemical recovery systems in place. Some sulfite mills, however, still do not have recovery systems in place. Scrubber system particulate "baghouses" or electrostatic precipitators (ESPs) are often mill air pollution control components.

11.2.2.3 Residual Wastes

The significant residual wastestreams from pulp and paper mills include bark, wastewater treatment sludges, lime mud, lime slaker grits, green liquor dregs, boiler and furnace ash, scrubber sludges, and wood processing residuals. Because of the tendency for chlorinated organic compounds (including dioxins) to partition from effluent to solids, wastewater treatment sludge has generated the most significant environmental concerns for the pulp and paper industry.

Wastewater treatment sludge is the largest volume residual wastestream generated by the pulp and paper industry. Sludge generation rates vary widely among mills. For example, bleached kraft mills surveyed as part of U.S. EPA's "104-mill study" reported sludge generation that ranged from 14 to 140 kg sludge/T pulp.[23] Total sludge generation for these 104 mills was 2.5 million dry T/yr, or an average ca. 26,000 dry T/yr/plant. Pulp making operations are responsible for the bulk of sludge wastes, although treatment of paper-making effluents also generates significant sludge volumes. For the majority of pulp and integrated mills that operate their own wastewater treatment systems, sludges are generated on site. A small number of pulp mills, and a much larger proportion of paper-making establishments, discharge effluents to publicly owned treatment works (POTWs).

Potential environmental hazards from wastewater sludges are associated with trace constituents (e.g., chlorinated organic compounds) that partition from the effluent into the sludge. It should be noted, however, that recent trends away from elemental chlorine bleaching have reduced these hazards. A continuing concern is the very high pH (>12.5) of most residual wastes. When these wastes are disposed of in an aqueous form, they may meet the RCRA definition of a corrosive hazardous waste.[24]

Landfill and surface impoundment disposal are most often used for wastewater treatment sludge, but a significant number of mills dispose of sludge through land application, conversion to sludge-derived products (e.g., compost and animal bedding), or combustion for energy recovery.[25]

11.2.2.4 Process Inputs and Pollutant Outputs

Kraft chemical pulping and traditional chlorine-based bleaching are both commonly used and may generate significant pollutant outputs. Kraft pulping processes produced ca. 83% of the total U.S. pulp tonnage during 1998 according to the American Forest and Paper Association.[11] Roughly 60% of this amount is bleached in some manner.

Pollutant outputs from mechanical, semichemical, and secondary fiber pulping are small compared to kraft chemical pulping. In the pulp and paper industry, the kraft pulping process is the most significant source of air pollutants. Table 11.9 and Figure 11.7 illustrate the process inputs and pollutant outputs for a pulp and paper mill using kraft chemical pulping and chlorine-based bleaching. Table 11.9 presents the process steps, material inputs, and major pollutant outputs (by media) of a kraft pulp mill practicing traditional chlorine bleaching. U.S. EPA resources[3,26,27] are recommended for pollutant production data (e.g., pounds of BOD per ton of pulp produced) for those pollutants presented in Table 11.9. Figure 11.7 is a process flow diagram of the kraft process, illustrating chemical pulping, power recovery, and chemical recovery process inputs and outputs.[12]

11.2.3 Management of Chemicals in Wastestreams

The Pollution Prevention Act of 1990 (PPA) requires facilities to report information about the management of TRI chemicals in wastes and efforts made to eliminate or reduce their quantities. These data on TRI have been collected annually from 1991. The data were meant to provide a basic understanding of the quantities of toxic waste handled by the industry, the methods typically used to manage them, and recent trends in these methods. TRI waste management data can be used to assess trends in source reduction within individual industries and facilities, and for specific TRI chemicals. This information could then be used as a tool in identifying opportunities for pollution prevention compliance assistance activities.

11.3 POLLUTION PREVENTION OPPORTUNITIES

The best way to reduce pollution is to prevent it in the first place. Industries have creatively implemented pollution prevention techniques that improve efficiency and increase profits while at the same time minimizing environmental impacts. This can be done in many ways, for example, by reducing material inputs, reengineering processes to reuse byproducts, improving management practices, and substituting for toxic chemicals. Some smaller facilities are able to get below regulatory thresholds just by reducing pollutant releases through aggressive pollution prevention policies.[1,2]

The chemical recovery systems used in chemical pulping processes are an example of pollution prevention technologies that have evolved alongside process technologies. An efficient chemical recovery system is a crucial component of chemical pulping mill operation. Recovery regenerates process chemicals, reducing natural resource usage and associated costs, as well as discharges to the environment, and may be used for producing energy. Many recent pollution prevention efforts have focused on reducing the releases of toxics, in particular chlorinated compounds. Pollution prevention techniques have proven to be more effective in controlling these pollutants than conventional control and treatment technologies. Most conventional, end-of-pipe treatment technologies are not effective in destroying many chlorinated compounds and often merely transfer the pollutants to another environmental medium. Efforts to prevent chlorinated releases have, therefore, focused on source reduction and material substitution techniques such as defoamers, bleaching chemical or wood chip substitution. Such source reduction efforts and material substitutions usually require substantial changes in the production process. In addition to process changes, the industry is implementing a number of techniques to reduce water use and pollutant releases (BOD, COD, and TSS); these include dry debarking, recycling of log flume water, improved spill control, bleach

TABLE 11.9

Kraft Chemical Pulped Bleached Paper Production

Process Step	Material Inputs	Process Outputs	Major Pollutant Outputs[a]	Pollutant Media
Fiber furnish preparation	Wood logs Chips Sawdust	Furnish chips	Dirt, grit, fiber, bark BOD TSS	Solid Water
Chemical pulping kraft process	Furnish chips	Black liquor (to chemical recovery system), pulp (to bleaching/ processing)	Resins, fatty acids Color BOD COD AOX VOCs [terpenes, alcohols, phenols, methanol, acetone, chloroform, methyl ethyl ketone (MEK)] VOCs (terpenes, alcohols, phenols, methanol, acetone, chloroform, MEK) reduced sulfur compounds (TRS)	Solid Water Air
	Cooking chemicals: sodium sulfide (Na_2S), NaOH, white liquor (from chemical recovery)		Organo-chlorine compounds (e.g., 3,4,5-trichloroguaiacol)	
Bleaching[b]	Chemical pulp	Bleached pulp	Dissolved lignin and carbohydrates Color COD AOX Inorganic chlorine compounds (e.g., chlorate (ClO_3^-))[c]	Water
	Hypochlorite (HClO, NaOCl, $Ca(OCl)_2$) Chlorine dioxide (ClO_2)		VOCs (acetone, methylene chloride, chloroform, MEK, chloromethane, trichloroethane)	Air/water
Papermaking	Additives, Bleached/ unbleached pulp	Paper/paperboard product	Particulate wastes Organic compounds Inorganic dyes COD Acetone	Water
Wastewater treatment facilities	Process wastewaters	Treated effluent	Sludge VOCs (terpenes, alcohols, phenols, methanol, acetone, chloroform, MEK) BOD TSS COD Color Chlorophenolics VOCs (terpenes, alcohols, phenols, methanol, acetone, chloroform, MEK)	Solid Air Water
Power boiler	Coal, wood, unused furnish	Energy	Bottom ash: incombustible fibers SO_2, NO_x, fly ash, coarse particulates	Solid Air

Continued

TABLE 11.9   (continued)

Process Step	Material Inputs	Process Outputs	Major Pollutant Outputs[a]	Pollutant Media
Chemical recovery system				
Evaporators	Black liquor	Strong black liquor	Evaporator noncondensibles (TRS, volatile organic compounds: alcohols, terpenes, phenols)	Air
			Evaporator condensates (BOD, suspended solids)	Water
Recovery furnace	Strong black liquor	Smelt energy	Fine particulates, TRS, SO_2, NO_x	Air
Recausticizing	Smelt	Regenerated white liquor	Dregs	Solids
		Lime mud	Waste mud solids	Water, solid
		Slaker grits	Solids	Solid
Calcining (lime kiln)	Lime mud	Lime	Fine and coarse particulates	Air

BOD, biochemical oxygen demand; COD, chemical oxygen demand; TSS, total suspended solids; AOX, adsorbable organic halides; VOC, volatile organic compound; MEK, methyl ethyl ketone; TRS, total reduced sulfur compounds.

[a] Pollutant outputs may differ significantly based on mill processes and material inputs (e.g., wood chip resin content).

[b] Pollutant list based on elemental chlorine-free (ECF) bleaching technologies.

[c] Chlorate only significantly produced in mills with high rates of chlorine dioxide use.

Source: U.S. EPA, Profile of the Pulp and Paper Industry, 2nd ed., report EPA/310-R-02-002, U.S. EPA, Washington, November 2002.

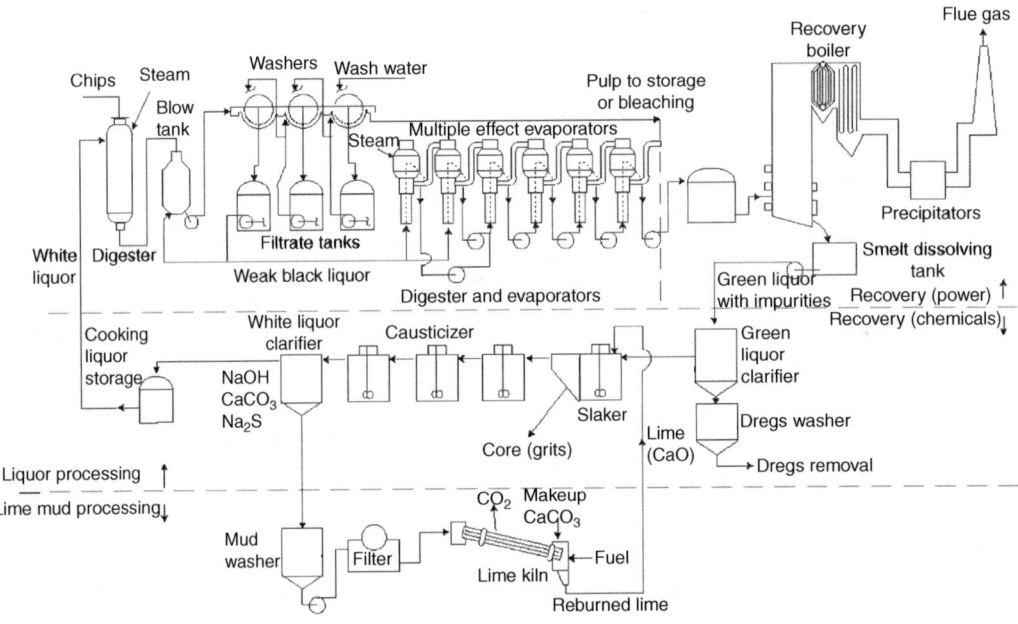

FIGURE 11.7   Kraft process flow diagram. (Taken from U.S. EPA, Profile of the Pulp and Paper Industry, 2nd ed., report EPA/310-R-02-002, U.S. EPA, Washington, November 2002.)

filtrate recycle, closed screen rooms, and improved stormwater management. The industry has also worked to increase the amount of secondary and recycled fibers used for the pulping process. According to industry sources, the pulp and paper industry set and met a 1995 goal of 40% recycling and reuse of all paper consumed in the U.S., and recovering 50% of all paper consumed in the U.S. for recycle and reuse. These figures should be compared with the utilization rate of secondary fibers (secondary fibers as a percentage of the total fibers used to make pulp), which is ca. 37% and is climbing slowly.[11] Current secondary fiber utilization rates in resource-deficient countries such as Japan are above 50%.

Because the pulp and paper industry is highly capital intensive and uses long-established technologies with long equipment lifetimes, major process-changing pollution prevention opportunities are expensive and require long time periods to implement. The pulp and paper industry is a dynamic one, however, that constantly makes process changes and material substitutions to increase productivity and cut costs. The trend towards materials substitutions is reflected in an increasing demand for alternative pulping and bleaching chemicals and in the participation of many facilities in voluntary environmental programs.

One of the factors that drove the industry towards pollution prevention much more rapidly is the integrated NESHAP (National Effluent Standards and Hazardous Air Pollutant) effluent limitation guidelines for the pulp and paper industry. These regulations were developed together in part to reduce the costs of compliance, to emphasize the multimedia nature of pollution control, and to promote pollution prevention. Many of the technology-based effluent limitation guidelines for the control of toxic releases consisted of process changes that substitute chlorine dioxide for elemental chlorine and that completely eliminate elemental chlorine in bleaching processes. The NESHAP standards also allowed hazardous air pollutant (HAP) reductions through recycling of wastewater streams to a process unit and routing pulping emissions to a boiler, lime kiln, or recovery furnace.

Brief descriptions of some pollution prevention techniques found to be effective at pulp and paper facilities are provided below. For more details on the pollution prevention options listed below and for descriptions of additional alternative pulping and bleaching processes see references 1, 2, 26, and 29–32. It should be noted that although many of the pollution prevention opportunities listed below are primarily aimed at reducing toxics releases, the process changes can often lead to reductions in conventional pollutants such as BOD_5 and TSS as well as COD and AOX, and contribute to reduced water use, a reduction in the sludge volumes and air emissions generated.

11.3.1 EXTENDED DELIGNIFICATION

Extended delignification further reduces the lignin content of the pulp before it moves to the bleach plant. Because the amount of bleaching chemicals required for achieving certain paper brightness is proportional to the amount of lignin remaining in the pulp after the pulping process, extended delignification can reduce the amounts of bleaching chemicals needed. Several different extended delignification processes have been developed. These processes include the following:

1. Increasing the cooking time
2. Adding the cooking chemicals at several points throughout the cooking process
3. Regulating the cooking temperatures
4. Carefully controlling the concentration of hydrogen sulfide ions and dissolved lignin

Most importantly, the process changes do not degrade the cellulose that would normally accompany increased cooking times. Extended delignification processes have been developed for both batch and continuous pulping processes. The lignin content of the brownstock pulp has been reduced by between 20 and 50% with no losses in pulp yield or strength using such processes. As a consequence, chlorinated compounds generated during bleaching are reduced in approximate proportion to reductions in the brownstock lignin content. In addition, the same changes have resulted in significant reductions in BOD_5, COD, and color. One study demonstrated a 29% decrease in BOD_5 resulting from an extended delignification process. Facility energy requirements have been shown to increase slightly with extended delignification. However, off-site power requirements (associated with decreased chemical use) have been estimated to more than offset the on-site increases.

11.3.2   OXYGEN DELIGNIFICATION

Oxygen delignification also reduces the lignin content in the pulp. The process involves the addition of an oxygen reactor between the kraft pulping stages and the bleach plant.[32] The brownstock pulp from the digester is first washed and then mixed with sodium hydroxide or oxidized cooking liquor. The pulp is fluffed, deposited in the oxygen reactor, steam heated, and injected with gaseous oxygen, at which point it undergoes oxidative delignification. The pulp is then washed again to remove the dissolved lignin before moving to the bleaching plant. Oxygen delignification can reduce the lignin content in the pulp by as much as 50%, resulting in a potentially similar reduction in the use of chlorinated bleaching chemicals and chlorinated compound pollutants. The process can be used in combination with other process modifications that can completely eliminate the need for chlorine-based bleaching agents. In addition, unlike bleach plant filtrate, the effluent from the oxygen reactor can be recycled through the pulp mill recovery cycle, further reducing the nonpulp solids going to the bleaching plant and the effluent load from the bleach plant. The net effect is reduced effluent flows and lower sludge generation. Facility energy requirements have been shown to increase with oxygen delignification, however, the decrease in off-site power requirements (associated with decreased chemical use) has been estimated to exceed the on-site increases, resulting in a decrease in overall energy requirements. Also, the recovered energy and reduced chemical use offset the increased cost.

11.3.3   OZONE DELIGNIFICATION

As a result of a considerable research effort, ozone delignification (ozone bleaching) is now being used in the pulp and paper industry.[32] The technology has the potential to eliminate the need for chlorine in the bleaching process. Ozone delignification is performed using processes and equipment similar to that of oxygen delignification. The ozone process, however, must take place at a very low pH (1.0 to 2.0), requiring the addition of sulfuric acid to the pulp prior to ozonation. In addition to low pH, several process conditions are critical for ozone delignification: organic materials must be almost completely washed out of the brownstock pulp; temperatures must stay at about 20 °C; and ozone-reactive metals must be removed prior to the ozonation stage. Oxygen delignification or extended delignification processes are considered a prerequisite for successful ozone bleaching. When used in combination, the two processes can result in a high-quality bright pulp that requires little or no chlorine or chlorine dioxide bleaching. Overall emissions from the combination of the oxygen and ozone processes are substantially lower than conventional processes because effluents from each stage can be recycled. Systems consisting of ozone delignification in combination with oxygen delignification and oxygen extraction have shown reductions in BOD_5 of 62%, COD of 53%, color of 88%, and organic chlorine compounds of 98%. However, ozone is unstable and will decompose to molecular oxygen, so ozone must be generated on site and fed immediately to the pulp reactor. Ozone generation systems are complex and the initial equipment is expensive. Facility energy use will increase due to the on-site production of ozone; however, this energy will be offset by the energy that would normally be used to produce chlorine and chlorine dioxide.

11.3.4   ANTHRAQUINONE CATALYSIS

The addition of anthraquinone (a chemical catalyst produced from coal tar) to the pulping liquor has been shown to speed up the kraft pulping reaction and increase yield by protecting cellulose fibers from degradation. The anthraquinone accelerates the fragmentation of lignin, allowing it to be broken down more quickly by the pulping chemicals. This lowers the amount of lignin in the prechlorination pulp, thus reducing the amount of bleaching chemicals needed. Anthraquinone catalysts are increasingly used in combination with oxygen delignification and extended delignification to overcome boiler capacity bottlenecks arising from these delignification processes.

11.3.5 BLACK LIQUOR SPILL CONTROL AND PREVENTION

The mixture of dissolved lignin and cooking liquor effluent from the pulping reactor and washed pulp is known as black liquor. Raw black liquor contains high levels of BOD, COD, and organic compounds. Spills of black liquor can result from overflows, leaks from process equipment, or from deliberate dumping by operators to avoid a more serious accident. Spills of black liquor can have impacts on receiving waters, are a source of air emissions, and can shock the microbial action of wastewater treatment systems. Black liquor losses also result in the loss of the chemical and heat value of the material. Systems needed to control black liquor spills are a combination of good design, engineering, and, most importantly, operator training. The following are a few elements of an effective spill control system:

1. Physical isolation of pieces of equipment
2. Floor drainage systems that allow spills to be collected
3. Backup black liquor storage capacity
4. Sensors that provide immediate warning of potential or actual spills
5. Enclosed washing and screening equipment

11.3.6 ENZYME TREATMENT OF PULP

Biotechnology research has resulted in the identification of a number of microorganisms that produce enzymes capable of breaking down lignin in pulp. Although the technology is new, it is believed that some mills are currently using enzyme treatment. The microorganisms capable of producing the necessary enzymes are called xylanases. Xylanases for pulp bleaching trials are available from several biotechnology and chemical companies. Because enzymes are used as a substitute for chemicals in bleaching pulp, their use will result in a decrease in the chlorinated compounds released, which is somewhat proportional to the reduction in bleaching chemicals used. Enzymes are also being used to assist in the deinking of secondary fiber. Research at the Oak Ridge National Laboratories has identified cellulase enzymes that will bind ink to smaller fiber particles, facilitating recovery of the ink sludge. Use of enzymes may also reduce the energy costs and chemical use in retrieving ink sludge from deinking effluent.

11.3.7 IMPROVED BROWNSTOCK AND BLEACHING STAGE WASHING

Liquor solids remaining in the brownstock pulp are carried over to the bleach plant and then compete with the remaining lignin in the pulp for reaction with the bleaching chemicals. Improved washing, therefore, can reduce the required amount of bleaching chemicals and lead to subsequent reductions in chlorinated compounds as well as conventional pollutants. Modern washing systems with improved solids removal and energy efficiency are beginning to replace the conventional rotary vacuum washers. State-of-the-art washing systems include the following:

1. Atmospheric or pressure diffusion washers
2. Belt washers
3. Pulp presses

Opportunities for reducing effluent flows and water use are also present in the bleaching plant. Acid filtrates from hypochlorite or chlorine dioxide stages can be used as dilution and wash water for the first bleaching stage. Similarly, second extraction stage filtrates can be used as dilution and wash water in the first extraction stage. Most new mills are designed with these countercurrent washing systems, and some mills are retrofitting their existing wash systems.

11.3.8 IMPROVED CHIPPING AND SCREENING

The size and thickness of wood chips is critical for proper circulation and penetration of the pulping chemicals. Chip uniformity is controlled by the chipper and screens that remove under- and over-sized pieces. Standard equipment does not normally sort chips by thickness, although it has been demonstrated that chip thickness is extremely important in determining the lignin content of pulp. Improper chip thicknesses can result in increased use of bleaching chemicals and the associated chlorinated compounds and conventional pollutants. Some mills have begun to incorporate equipment that will separate chips according to their thickness as well as by length and width.

11.3.9 OXYGEN-REINFORCED/PEROXIDE EXTRACTION

Oxygen-reinforced extraction (or oxidative extraction) and peroxide-reinforced extraction processes used separately or together have been shown to reduce the amount of elemental chlorine and chlorine dioxide needed in the bleaching process while increasing the pulp brightness. Gaseous elemental oxygen and aqueous hydrogen peroxide are used as a part of the first alkaline extraction stage to facilitate the solubilization and removal of chlorinated and oxidized lignin molecules. Oxygen-reinforced extraction has seen widespread adoption by the industry. It is estimated that up to 80% of mills in the U.S. are using oxygen-reinforced extraction, and that 25% of domestic mills are using peroxide extraction.[1]

11.3.10 IMPROVED CHEMICAL CONTROLS AND MIXING

The formation of chlorinated organics can be minimized by avoiding excess concentrations of chlorine-based bleaching chemicals within reactor vessels. This can be accomplished by carefully controlling the chemical application rates and by ensuring proper mixing of chemicals within the reactor. Modern chemical application control and monitoring systems and high-shear mixers have been developed that decrease the formation of chlorinated organic compounds.

11.4 APPLICABLE FEDERAL STATUTES AND REGULATIONS

The purpose of this section is to highlight and briefly describe the applicable federal requirements. For further information, readers should consult the Code of Federal Regulations and other state or local regulatory agencies.[1,2,33–36]

11.4.1 CLEAN AIR ACT (CAA)

11.4.1.1 National Ambient Air Quality Standards

At pulp and paper mills, air emissions from both process and combustion units are regulated under the National Ambient Air Quality Standards (NAAQS) and the State Implementation Plans (SIP) that enforce the standards. States may implement controls to limit emissions of particulate matter (PM), nitrogen oxides (NO_x), volatile organic compounds (VOCs), and sulfur dioxide (SO_2).

Although many limits are implemented at the state level, there are national guidelines that serve as a basis for more specific limits. Sources that are considered "major" under the CAA are subject to prevention of significant deterioration (PSD) or new source review (NSR). Both PSD and NSR are permit programs for facilities that were constructed or modified after a certain date.

Facilities in NAAQS attainment areas must follow PSD requirements by demonstrating that the construction/modification project will not cause a violation of air quality limits and will implement the best available control technology (BACT).

New or modified facilities in nonattainment areas must follow NSR requirements, which require the source to meet the lowest achievable emission rate (LAER) and to obtain emission offsets to ensure that the nonattainment problem is not made worse by the new/modified source.

In addition to the PSD/NSR preconstruction obligations, there are process-specific operational standards: New Source Performance Standards (NSPS). 40 CFR 60 lists these standards, which serve as minimum requirements in states SIPs. Individual states may impose requirements that are stricter. The following NSPSs are particularly relevant to the pulp and paper industry.

Air toxics regulations apply to several parts of the pulp and paper milling process. National Emission Standards for Hazardous Air Pollutants (NESHAP) have been developed expressly for two processes of the pulp and paper industry. These standards establish process-based maximum achievable control technologies (MACT) for "major sources," which are defined as facilities that emit or have the potential to emit 10 t per year or more of any hazardous air pollutant (HAP) or 25 t per year or more of any combination of HAPs.

11.4.1.2 Risk Management Program

Pulp and paper mills are subject to a section of the CAA that states that stationary sources using extremely hazardous substances have a "general duty" to initiate specific activities to prevent and mitigate accidental releases. The general duty requirements apply to stationary sources that produce, process, handle, or store these substances, regardless of the quantity. The general duty clause requires facilities to identify hazards that may result from accidental releases, to design and maintain a safe facility, and to minimize the consequences of releases when they occur.

Most pulp and paper mills are subject to additional, more explicit risk management requirements. Facilities that have more than a threshold quantity of any of the 140 regulated substances in a single process are required to develop a risk management program and to summarize their program in a risk management plan (RMP). All facilities meeting the RMP threshold requirements must follow Program 1 requirements:

1. An off-site consequence analysis that evaluates specific potential release scenarios, including worst-case and alternative scenarios
2. A five-year history of certain accidental releases of regulated substances from covered processes
3. An RMP, revised at least once every five years that describes and documents these activities for all covered processes

In addition, most pulp and paper facilities may be subject to the requirements of Program 2 or 3. These additional requirements include the following:

1. An integrated prevention program to manage risk. The prevention program will include identification of hazards, written operating procedures, training, maintenance, and accident investigation
2. An emergency response program
3. An overall management system to put these program elements into

11.4.1.3 Title V Permits

Title V requires that all "major sources" (and certain minor sources) obtain an operating permit. Many pulp and paper mills are required to have a Title V permit, and may be required to submit information about emissions control devices and the general process at the facility in the permit application. Permits may limit pollutant emissions and impose monitoring record keeping and reporting requirements.

11.4.1.4 Title VI Stratospheric Ozone Protection

Many pulp and paper facilities operate industrial process refrigeration units such as chillers for chlorine dioxide plants. For those units that utilize ozone-depleting chemicals, such as chlorofluorocarbons (CFCs), facilities are required under Title VI to follow leak repair requirements.

11.4.2 Resource Conservation and Recovery Act (RCRA)

The pulp and paper industry generates hazardous wastes, but most are associated with wastewater, which is rendered nonhazardous in wastewater treatment or neutralization units within the manufacturing facilities and therefore is not subject to RCRA requirements. Also, black liquor is exempt as a solid waste if it is reclaimed in a recovery furnace and reused in the pulping process.

11.4.3 Emergency Planning and Community Right-to-Know Act (EPCRA)

Three of the components of EPCRA are directly relevant to the pulp and paper industry:

1. *Emergency planning.* Businesses that produce, use, or store "hazardous substances" must (a) submit material safety data sheets or the equivalent and (b) file annual inventory report forms to the appropriate local emergency planning commission. Those handling "extremely hazardous substances" are also required to submit a one-time notice to the state emergency response commission.
2. *Emergency notification of extremely hazardous substance release.* A business that unintentionally releases a reportable quantity of an extremely hazardous substance must report that release to the state emergency planning commission and the local emergency planning commission.
3. *Release reporting.* Manufacturing businesses with ten or more employees that manufactured, processed, or otherwise used a listed toxic chemical in excess of the "established threshold" must file annually a Toxic Chemical Release form with U.S. EPA and the state. Documentation supporting release estimates must be kept for three years.

11.4.4 Clean Water Act (CWA)

There are two industry-specific components of the CWA requirements: the NPDES (National Pollutant Discharge Elimination System) permitting and pretreatment programs. Other general CWA requirements, such as those for wetlands and stormwater, may also apply to pulp and paper mills.

11.4.4.1 National Pollutant Discharge Elimination System (NPDES) Permitting

Individual NPDES requirements have been developed for several subcategories of the industry. For each of these subcategories, the regulations outline some or all of the following for facilities that discharge wastewater directly to the environment:

1. Best practicable control technology currently available (BPT) and best conventional control technology (BCT) guidelines for the control of conventional pollutants (biological oxygen demand, total suspended solids, and pH)
2. Best available technology economically achievable (BAT) guidelines for the control of nonconventional and toxic pollutants (trichlorophenol and pentachlorophenol, which are chemicals used as biocides)
3. New source performance standards (NSPS) for the control of conventional, nonconventional, and toxic pollutants from new facilities that discharge directly to the environment

11.4.4.2   Pretreatment Standards

For facilities that discharge their wastewater to a POTW, pretreatment standards may apply. In addition to general standards established by U.S. EPA that address all industries, there are Pretreatment Standards for New Sources (PSNS) and Pretreatment Standards for Existing Sources (PSES) that are specific to the pulp and paper industry. These regulate the biocides trichlorophenol and pentachlorophenol, with limits that are specified for each subcategory of the industry. In 1998, in conjunction with the development of the pulp and paper cluster rule, U.S. EPA reorganized the regulations in order to group processes that are similar.

The Cluster Rule is an integrated, multimedia regulation to control the release of pollutants to two media (air and water) from one industry. The intent of the rule is to allow individual mills in particular segments of the industry to consider all regulatory requirements at one time. This combined rule allows mills to select the best combination of pollution prevention and control technologies that provide the greatest protection to human health and the environment. Because some air requirements that reduce toxic air pollutants also reduce mill wastewater toxic pollutant loadings (and water treatment requirements can reduce air impacts), the combined rules have a synergistic effect.

Some of the features of the coordinated rule include the following:

1. Alternative emission limits
2. Varying compliance periods (3 to 8 years)
3. New and existing source controls
4. Flexibility for evolving technologies
5. Compliance dates coordinated with effluent limitations guidelines and standards

The rule sets new baseline limits for the releases of toxics and nonconventional pollutants to the air and water. There are three significant components:

1. *Air emissions standards.* New and existing pulp and paper mills must meet air standards to reduce emissions of toxic air pollutants occurring at various points throughout the mills. Specifically, U.S. EPA requires mills to capture and treat toxic air pollutant emissions that occur during the cooking, washing, and bleaching stages of the pulp manufacturing process.
2. *Water effluent limitations guidelines and standards.* New and existing standards in the bleached papergrade kraft and soda subcategory and the bleached papergrade sulfite subcategory must meet standards to reduce discharges of toxic and nonconventional pollutants. Specifically, U.S. EPA has set effluent limitations for toxic pollutants in the wastewater discharged directly from the bleaching process and in the final discharge from the mills.
3. *Analytical methods for 12 chlorinated phenolics and adsorbable organic halides (AOXs).* Samples of air emissions and water discharges from each mill must be tested using the laboratory methods included in the rule. The new methods will enable more timely and accurate measurements of releases of these pollutants to the environment and will be used to ensure compliance with air emission and water discharge permit limits.

The Cluster Rules require that mills existing as of April 15, 1998, that discharge directly to receiving streams control toxic and nonconventional pollutants at the best available technology (BAT) economically achievable level of performance. U.S. EPA established Pretreatment Standards for Existing Sources (PSES) that are based on control technologies similar to BAT for indirect dischargers.[35] As shown in Table 11.10, except for the monitoring location for AOX, the BAT

TABLE 11.10

BAT Effluent Limitations Guidelines and Pretreatment Standards for Existing Bleached Papergrade Kraft and Soda Plants

Pollutants	1-Day Maximum[a]
Bleach plant effluent	
TCDD, dioxin	< ML[c]
TCDF, furan	31.9 pg/L[d]
Chloroform[b]	1-Day maximum: 6.92 g/T[e]
	Monthly average: 4.14 g/T[e]
Trichlorosyringol	<ML
3,4,5-Trichlorocatechol	<ML
3,4,6-Trichlorocatechol	<ML
3,4,5-Trichloroguaiacol	<ML
3,4,6-Trichloroguaiacol	<ML
4,5,6-Trichloroguaiacol	<ML
2,4,5-Trichlorophenol	<ML
2,4,6-Trichlorophenol	<ML
Tetrachlorocatechol	<ML
Tetrachloroguaiacol	<ML
2,3,4,6-Tetrachlorophenol	<ML
Pentachlorophenol	<ML
Final effluent (for BAT) or bleach plant effluent (for PSES)	
AOX (adsorbable organic halides)	1-Day maximum: 0.951 kg/T[e]
	Monthly average: 0.623 kg/T[e]

BAT, best available technology economically achievable; TCDD, 2,3,7,8-tetrachlorodibenzo-p-dioxin; TCDF, 2,3,7,8-tetrachlorodibenzofuran.

[a]U.S. EPA established monthly average limitations guidelines for only chloroform and AOX.

[b]For mills that are certified to use TCF, refer to 40 CFR 430.

[c]<ML means less than the minimum level at which the analytical system gives recognizable signals and an acceptable calibration point. The MLs for each pollutant are specified in 40 CFR 430.

[d]pg = pictogram = 10^{-12} g.

[e]T = metric ton = 1000 kg.

Source: U.S. EPA, Guidance Manual for Pulp, Paper, and Paperboard and Builders' Paper and Board Mills Pretreatment Standards, U.S. EPA, Effluent Guidelines Division, WH-562, Washington, September 1984.

limitations guidelines and PSES for indirect dischargers are the same. U.S. EPA promulgated regulations for new sources (New Source Performance Standards for direct dischargers, and Pretreatment Standards for New Sources for indirect dischargers). However no new bleached kraft or papergrade sulfite mills have been constructed since 1998. Table 11.11 presents the BAT limitations guidelines and PSES for papergrade sulfite mills.

Mills in the Bleached Papergrade Kraft and Soda subcategory have additional flexibility under the Cluster Rule. Mills may comply either with the baseline regulations or with more stringent wastewater regulations under a more forgiving timetable. This latter arrangement, called the Voluntary Advanced Technology Incentives Program (VATIP), allows mills to undertake customized compliance and pollution reduction plans that further reduce environmental impacts.

Under the VATIP, each participating mill develops "Milestones Plans" for each fiber line that it enrolls in the program. Permit writers will use the Milestones Plan to incorporate enforceable

TABLE 11.11

BAT Effluent Limitations Guidelines and Pretreatment Standards for Papergrade Sulfite Existing Plants

Pollutants	Segment A Calcium, Magnesium, and Sodium Sulfite	Segment B Ammonium Sulfite[a]
Bleach plant effluent		
TCDD, dioxin	Not regulated	<ML
TCDF, furan	Not regulated	<ML
Chloroform	Not regulated	Reserved
Trichlorosyringol	Not regulated	<ML
3,4,5-Trichlorocatechol	Not regulated	<ML
3,4,6-Trichlorocatechol	Not regulated	<ML
3,4,5-Trichloroguaiacol	Not regulated	<ML
3,4,6-Trichloroguaiacol	Not regulated	<ML
4,5,6-Trichloroguaiacol	Not regulated	<ML
2,4,5-Trichlorophenol	Not regulated	<ML
2,4,6-Trichlorophenol	Not regulated	<ML
Tetrachlorocatechol	Not regulated	<ML
Tetrachloroguaiacol	Not regulated	<ML
2,3,4,6-Tetrachlorophenol	Not regulated	<ML
Pentachlorophenol	Not regulated	<ML
Final effluent (for BAT) or bleach plant effluent (for PSES)		
AOX	1-Day maximum: 2.64 kg/T[b]	Reserved
	Monthly average: 1.41 kg/T[b]	Reserved

TCDD, 2,3,7,8-tetrachlorodibenzo-p-dioxin, TCDF, 2,3,7,8-tetrachlorodibenzofuran.

[a]<ML means less than the minimum level at which the analytical system gives recognizable signals and an acceptable calibration point. The MLs for each pollutant are specified in 40 CFR 430.

[b]T = metric ton = 1000 kg.

Source: U.S. EPA, Guidance Manual for Pulp, Paper, and Paperboard and Builders' Paper and Board Mills Pretreatment Standards, U.S. EPA, Effluent Guidelines Division, WH-562, Washington, September 1984.

interim requirements into the mill's discharge permit. The three basic components of a Milestones Plan are the following:

1. A description of each technology component or process modification the mill intends to implement
2. A master schedule showing the sequence of implementing new technologies and process modifications
3. Descriptions of the anticipated improvements in effluent quality

11.4.5 STATE STATUTES

In 1986, six states (California, Kentucky, Louisianan, Maryland, North Carolina, and South Carolina) had fully U.S. EPA-approved plans to control TRS at kraft pulp mills, two states had approved TRS standards but their compliance schedules had not yet been approved (Arkansas and Georgia), and

Tennessee's and Florida's plans had been submitted for approval. Since that time, additional states have received approval of their plans. The number of states grew to 18 in 1999 (36) (Alabama, California, Florida, Georgia, Indiana, Kentucky, Maine, Michigan, Mississippi, New Hampshire, North Carolina, Ohio, Oregon, Tennessee, Texas, Virginia, Washington, and Wisconsin).

In general, PM emissions limits are established on a per ton of pulp produced basis or for specific processes (e.g., lime kilns, smelt tanks, and recovery furnaces). Certain states have also established opacity limits and performance standards for specific processes. Investigations related to the integrated rulemaking identified 17 states with regulations specific to the pulp and paper industry.

11.4.6 SUMMARY OF NATIONAL REGULATORY REQUIREMENTS

This section describes the applicable national regulatory requirements for bleached, unbleached, and dissolving kraft mills. Potential pollutants of concern for kraft pulp mills as reflected in the effluent limitations guidelines and standards promulgated by U.S. EPA and in a sampling of NPDES permits are summarized in Table 11.12.

The reader, however, should note that permit requirements will be specifically tailored for each discharging facility. Table 11.13 summarizes the discussion of regulatory requirements presented below.

Prior to the Cluster Rules, direct discharge kraft mills were regulated as shown in §11.4.4.1.

Indirect discharge kraft mills were subject to performance standards for existing sources or new sources (PSES or PSNS, as applicable) for the control of pentachlorophenol and trichlorophenol.

For kraft pulp mills, the Cluster Rules add toxic and nonconventional pollutants to the list of regulated pollutants only for bleached papergrade kraft mills. Effluent limitations guidelines and standards were added for the following BAT and PSES pollutants (and NSPS/PSNS for new sources): chloroform, 2,3,7,8-TCDD, 2,3,7,8-TCDF, 12 chlorinated phenolic compounds, and AOX. All of the

---

TABLE 11.12
Regulated Pollutant Parameters for Kraft Pulp Facilities

Effluent Guidelines/Standards	Other Potential Permit-Specific Parameters
BOD_5	Total cadmium
TSS	Total mercury
pH	Total silver
Pentachlorophenol	Total zinc
Trichlorophenol	Total copper
AOX	Lead
Chloroform	Mercury
TCDD	Temperature and thermal load
TCDF	Dissolved oxygen
Chlorinated phenols (12 pollutants)	Total phosphorous
	Ammonia
	Aluminum
	Color
	COD

AOX, adsorbable organic compounds; COD, chemical oxygen demand; TCDD, 2,3,7,8-tetrachlorodibenzo-p-dioxin; TCDF, 2,3,7,8-tetrachlorodibenzofuran.
Source: U.S. EPA, Guidance Manual for Pulp, Paper, and Paperboard and Builders' Paper and Board Mills Pretreatment Standards, U.S. EPA, Effluent Guidelines Division, WH-562, Washington, September 1984.

TABLE 11.13
Wastewater Regulations for Kraft Pulp Mills

Type of Kraft Mill	Direct or Indirect Discharger	BPT	Precluster Rules BAT	Cluster Rules BAT	Precluster Rules PSES	Cluster Rules PSES
Bleached kraft mills	Direct discharger	T	T	T		
	Indirect discharger				T	T
Unbleached kraft mills	Direct discharger	T	T			
	Indirect discharger				T	
Dissolving kraft mills	Direct discharger	T	T			
	Indirect discharger				T	

BPT, best practicable control technology; PSES, Pretreatment Standards for Existing Sources; BAT, best available technology economically achievable.

Source: U.S. EPA, Guidance Manual for Pulp, Paper, and Paperboard and Builders' Paper and Board Mills Pretreatment Standards, U.S. EPA, Effluent Guidelines Division, WH-562, Washington, September 1984.

preCluster Rules effluent limitation guidelines and standards applicable to kraft pulp mills remain in effect, although the Cluster Rules have reorganized these limits into new subcategories.

The Cluster Rules reorganized the subcategorization scheme to simplify the categories. Previously, mills were grouped by the types of products manufactured. The Cluster Rules reduced the number of subcategories by grouping mills by similar processes.

In the previous regulations, bleached kraft mills were divided into four subparts and unbleached kraft mills were divided into three subparts. As a result, the remaining preCluster Rules limits (i.e., BPT for BOD_5, TSS, and pH, and BAT and PSES for pentachlorophenol and trichlorophenol) for the four previous bleached kraft mill subparts now exist as four segments. Likewise, the remaining preCluster Rules limits for the three previous unbleached kraft subparts now exist as three segments.

11.4.7 SUMMARY OF WORLD BANK LIQUID EFFLUENTS GUIDELINES

Emissions levels for the design and operation of each pulp and paper mill project must be established through the environmental assessment (EA) process on the basis of national legislation and handbooks[1,2,36,37] as applied to local conditions. The emissions levels selected must be justified in the EA and acceptable to the World Bank Group.

The following guidelines present emissions levels normally acceptable to the World Bank Group[37] in making decisions regarding provision of World Bank Group assistance. Any deviations from these levels must be described in the World Bank Group project documentation. These emissions levels can be consistently achieved by well-designed, well-operated, and well-maintained pollution control systems. The guidelines are expressed as concentrations to facilitate monitoring. Dilution of effluents to achieve these guidelines is unacceptable. All of the maximum levels should be achieved for at least 95% of the time that the plant or unit is operating, to be calculated as a proportion of annual operating hours.

Liquid effluent requirements for direct discharge to surface waters from pulp and paper manufacturing should achieve the following maximum levels[37]:

pH: 6–9
COD: 300 mg/L and 15 kg/T for kraft pulp mills; 700 mg/L and 40 kg/T for sulfite pulp mills; 10 mg/L and 5 kg/T for mechanical and recycled fiber pulp; 250 mg/L for paper mills.
AOX: 40 mg/L and 2 kg/T (aim for 8 mg/L); 0.4 kg/L for retrofits; 4 mg/L and 0.2 kg/T for new mills; 4 mg/L for paper mills.

Total phosphorus: 0.05 kg/T
Total nitrogen: 0.4 kg/T

Molecular chlorine should not be used in the process. The effluent should not result in a temperature increase of more than 3 °C at the edge of the zone where initial mixing and dilution take place. Where the zone is not defined, 100 m from the point of discharge should be used. Solid wastes should be sent to combustion devices or disposed of in a manner that avoids odor generation and the release of toxic organics to the environment.

Solid waste treatment steps include dewatering of sludge and combustion in an incinerator, bark boiler, or fossil-fuel-fired boiler. Sludges from a clarifier are dewatered and may be incinerated; otherwise, they are landfilled.

11.5 TREATMENT OF WASTEWATER FROM PULP AND PAPER FACILITIES

According to the European Commission (EC) the best available techniques for kraft pulp mills are the following[38]:

1. Dry debarking of wood
2. Increased delignification before the bleach plant by extended or modified cooking and additional oxygen stages
3. Highly efficient brown stock washing and closed-cycle brown stock screening
4. Elemental chlorine free (ECF) bleaching with low AOX or totally chlorine free (TCF) bleaching
5. Recycling of some, mainly alkaline process water from the bleach plant
6. Effective spill monitoring, containment, and recovery system
7. Stripping and reuse of the condensates from the evaporation plant
8. Sufficient capacity of the black liquor evaporation plant and the recovery boiler to cope with the additional liquor and dry solids load
9. Collection and reuse of clean cooling waters
10. Provision of sufficiently large buffer tanks for storage of spilled cooking and recovery liquors and dirty condensates to prevent sudden peaks of loading and occasional upsets in the external effluent treatment plant
11. In addition to process-integrated measures, consider primary treatment and biological treatment as BAT for kraft pulp mills

The BAT for sulfite pulp mills, also according to the EC, are considered to be the following[38]:

1. Dry debarking of wood
2. Increased delignification before the bleach plant by extended or modified cooking
3. Highly efficient brown stock washing and closed-cycle brown stock screening
4. Effective spill monitoring containment and recovery system
5. Closure of the bleach plant when sodium-based cooking processes is being used
6. Total chlorine free (TCF) bleaching
7. Neutralizing of weak liquor before evaporation followed by reuse of most condensate in the process or anaerobic treatment
8. For prevention of unnecessary loading and occasionally upsets in the external effluent treatment due to process cooking and recovery liquors and dirty condensates, sufficiently large buffer tanks for storage are considered as necessary
9. In addition to process-integrated measures, primary and biological treatment are considered BAT for sulfite pulp mills

The BAT for recovered paper processing mills according to EC are the following[38]:

1. Separation of less contaminated water from more contaminated water and recycling of process water.
2. Optimal water management (water loop arrangement), water clarification by sedimentation, flotation or filtration techniques, and recycling of process water for different purposes.
3. Strict separation of water loops and countercurrent flow of process water.
4. Generation of clarified water for deinking plants (air flotation).
5. Installation of an equalization basin and primary treatment.
6. Biological effluent treatment — an effective option for deinked grades and, depending on the conditions, also for nondeinked grades is aerobic biological treatment and in some cases also flocculation and chemical precipitation. Mechanical treatment with subsequent anaerobic–aerobic biological treatment is the preferable option for nondeinked grades. These mills usually have to treat more concentrated wastewater because of the higher degree of water circuit closure.
7. Partial recycling of treated water after biological treatment; the possible degree of water recycling is dependent on the specific paper grades produced. For nondeinked paper grades this technique is BAT. However, the advantages and drawbacks need to be carefully investigated and will usually require additional polishing (tertiary treatment).
8. Treating internal water circuits.

Kraft pulp mills treat wastewater using primary (physical) and secondary (biological) treatment to reduce pollutant discharges to receiving waters. Kraft mills typically collect and treat the following wastewaters[36]:

1. Water used in wood handling and debarking
2. Digester, turpentine recovery, and evaporator condensates
3. Wastewater from brown stock screening
4. Bleach plant effluent
5. Paper machine white water
6. Spent pulping liquor spills from pulp processing areas

Wastewater treatment typically includes (a) neutralization, screening, sedimentation, and flotation/hydrocycloning to remove suspended solids and (b) biological/secondary treatment to reduce the organic content in the wastewater and to destroy toxic organics. Chemical precipitation is also used to remove certain cations. Fibers collected in primary treatment should be recovered and recycled. A mechanical clarifier or a settling pond may be used as primary treatment. Flocculation to assist in the removal of suspended solids is also sometimes necessary. Biological treatment systems, such as activated sludge, trickling filter, aerated lagoons, and anaerobic fermentation, can reduce BOD by over 99% and achieve a COD reduction of between 50 and 90%. Tertiary treatment may be performed to reduce toxicity, suspended solids, and color.[37]

11.5.1 PRETREATMENT

The recommended treatment option for control of toxic pollutants regulated under PSES categorical standards is chemical substitution. Although chemical substitution of sodium hydrosulfite for zinc hydrosulfite is recommended for control of zinc at groundwood mills, PSES for zinc were calculated using treatment performance data for lime precipitation.[39]

11.5.1.1 Lime Precipitation

The removal of zinc from wastewaters using zinc hydrosulfite as a bleaching agent can be achieved through both chemical coagulation and clarification or by changing to another chemical bleaching agent such as sodium hydrosulfite.

The lime application and settling process treatment consists of adding a milk of lime slurry to the wastewater to precipitate the hydroxide of the heavy metals and reduce dissolved sulfate concentrations through the formation of gypsum. Sufficient lime is needed to adjust the pH to between 10 and 11.5. Also, settling may have to be aided by adding small quantities of organic polyelectrolytes.

11.5.1.2 Chemical Substitution

It is often possible to use different process chemicals to accomplish the same goal. For example, both zinc hydrosulfite and sodium hydrosulfite can be used to bleach mechanical pulps. The substitution of the use of sodium hydrosulfite for zinc hydrosulfite was prompted, at least in part, by the establishment of effluent limitations controlling the discharge of zinc. Other opportunities exist to minimize the discharge of toxic and nonconventional pollutants through chemical substitution.

Slimicide and biocide toxic pollutants containing pentachlorophenol are used at mills in the pulp, paper, and paperboard industry. Initially, pentachlorophenol was used as a replacement for heavy metal salts, particularly mercuric types. Trichlorophenols are also used because of their availability as a byproduct from the manufacture of certain herbicides. Formulations containing organo-bromides and organo-sulfur compounds are also being used. Substitution of alternative slimicide and biocide formulations can lead to the virtual elimination of pentachlorophenol and trichlorophenol from these sources.

Ammonia is used as a cooking chemical at mills in the semichemical, dissolving sulfite pulp, and both papergrade sulfite subcategories. One method for reducing ammonia (NH_3) discharges is the substitution of a different chemical, such as sodium hydroxide, for ammonia in the cooking liquor. The equipment changes necessary to receive and feed a 50% solution of NaOH are not likely to be significant.

After conversion to the use of sodium-based chemicals, spent liquor could be incinerated, and sulfur dioxide, sodium sulfate, carbonate, or sulfide could be recovered. These compounds could be sold for use at nearby kraft mills or for other industrial uses.

Reducing smelting furnaces that produce a high-sulfidity, kraft-like green liquor are now employed at sodium-based sulfite mills. U.S. EPA anticipates that it would be necessary to replace the existing recovery boilers at ammonia-based mills if chemical substitution to a sodium base were employed. Additionally, it is likely that, because the heat value of sodium spent liquor is lower than ammonia spent liquor, evaporator modification may he required if excess capacity does not already exist.

11.5.2 Primary Treatment

Figure 11.8 shows a typical sequence of the major equipment systems in a wastewater treatment plant.[36] The function of primary treatment is to remove suspended solids from the wastewater, and then to remove organic materials by biological secondary treatment. Primary treatment processes used by kraft mills typically involve screening followed by either sedimentation or flotation.[40]

11.5.2.1 Sedimentation

Kraft mills use mechanical clarifiers[41] or, occasionally, settling ponds that provide sufficient holding time to enable suspended solids to settle. After settling occurs in the mechanical clarifier, the resulting sludge (which contains up to 6% solids) is pumped from the clarifier to sludge-handling facilities where it is dewatered prior to disposal. Mechanical clarifiers can remove as much as 80 to 90% of suspended solids.[40,42]

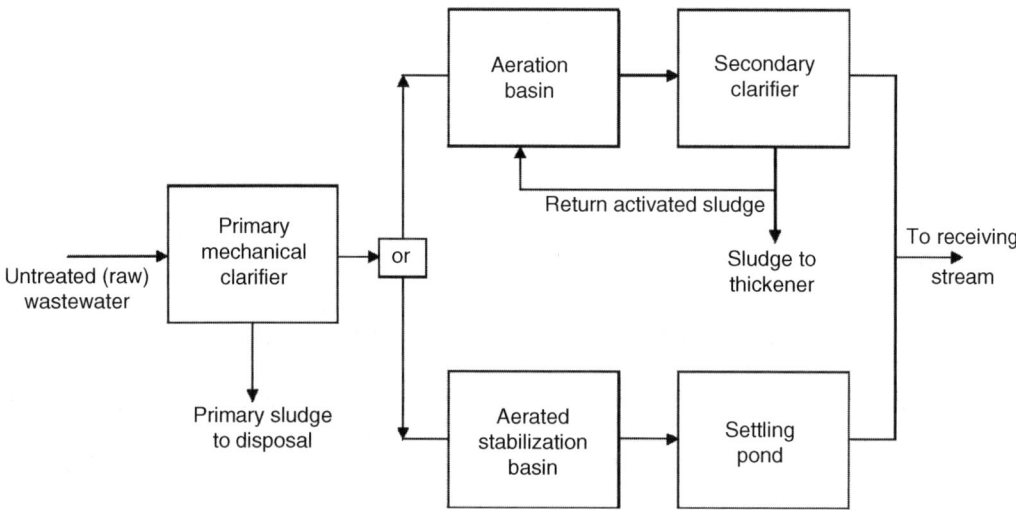

FIGURE 11.8 Typical wastewater treatment plant. (Taken from U.S. EPA, Kraft Pulp Mill Compliance Assessment Guide (CAA, CWA, RCRA and EPCRA), U.S. EPA, EPA/310-B-99-001, Washington, May 1999.)

Settling ponds, a less sophisticated alternative to mechanical clarifiers, also remove suspended solids by sedimentation. Settling ponds may be clay-lined, synthetic-lined, or unlined and earthen, and have longer retention times than clarifiers. Settling ponds produce less constant solids loadings than mechanical clarifiers, but still provide sufficient solids removal prior to secondary treatment.[40,42]

11.5.2.2 Flotation

Flotation is a solids removal process that introduces a gas, usually air, into the wastewater stream. The gas adheres to the suspended solids, reducing their density and causing them to rise to the surface of the water, where they are skimmed off. The advantage of flotation clarification over sedimentation is that lighter particles that require very long retention times to settle are removed more quickly.

A common modification of this process is dissolved air flotation (DAF), in which air under pressure is injected into the wastewater. DAF units are more efficient than conventional flotation clarifiers because more air is introduced into the wastewater, thereby removing more solids.[43–45,59]

11.5.3 SECONDARY TREATMENT

Kraft mills employ secondary treatment to reduce BOD_5 and toxicity in wastewaters. This process makes use of microorganisms (mostly bacteria and fungi) under aerobic conditions to digest the organic matter in the wastewater. The organic matter is removed as biosolids and the treated wastewater is discharged into receiving waters.[46,47] Because pulp mill wastewater is deficient in nitrogen and phosphorus relative to its high carbon load, these nutrients are usually added to the process to enhance microbial activity. Detailed information on bacterial species and biological treatment can be found in the literature.[48–52,59,87]

11.5.3.1 Aerated and Nonaerated Stabilization Basins

About 75% of U.S. kraft mills use aerated stabilization basins.[36] These basins are equipped with continuous mechanical aerators or diffusers to introduce air into the wastewater. By aerating the wastewater, an increased amount of oxygen is introduced into the wastewater stream, thereby maintaining an aerobic environment. This action significantly speeds up the biological activity compared to a nonaerated basin, so that a retention time of 5 d may achieve 90% BOD removal. The continuous aeration also provides thorough mixing, which allows mills to operate effective aeration lagoons at depths up to 7.5 m (25 ft). These basins are typically lined with clay or a combination of synthetics and clay.

Some kraft mills use basins without mechanical aerators. Known as stabilization basins, this is the simplest form of aerobic treatment. This process uses shallow basins that cover very large areas and relies on natural diffusion of air into the wastewater and algae to create aerobic conditions. At depths greater than 1.2 m (4 ft), anaerobic microorganisms will become active in lower depths; thus, stabilization basins are shallow. Typically, the basin is earthen although some are lined with compacted clay. Wastewater retention time may last up to 30 d to achieve up to 90% BOD_5 removal.

Some kraft mills use both aerated and nonaerated basins. The stabilization basin, which may precede or follow the aerated stabilization basin, serves as a "polishing" or "holding" pond to remove additional organic materials, including biological solids, or to reduce final effluent discharges to receiving waters.

11.5.3.2   Activated Sludge System Including Deep Shaft Process

This system features a microbial floc held in suspension in an aeration chamber. Soluble organic matter in the wastewater is metabolized by the microbial floc, which changes it into biosolids, thereby increasing the suspended solids load. After aeration, treated wastewater is routed to a clarifier where the biosolids are removed as sludge. A significant fraction of this sludge biosolids is recycled back to the aeration chamber to maintain the high level of microbial biomass (this is the "activated sludge").[46,83] Retention times for this system can range from 6 hr to >12 hr. The biosolids[53] that are removed may be further treated and dewatered before disposal or beneficial reuse.

The deep shaft biological treatment process,[87] which is one of the activated sludge systems, has been successfully applied to a paper mill wastewater treatment in Japan (see Case Study III, section 11.6.3).

11.5.3.3   Anaerobic and Aerobic Biological Treatment

This process, according to the manufacturer,[54] has been developed in such a way that space requirements are kept to a minimum. A BIOPAQ® IC reactor is used as the initial step in the treatment process. The name of this anaerobic reactor is derived from the "gas-lift" driven internal circulation that is generated within a tall, cylindrical vessel. These reactors have been operational in the paper industry since 1996. The second step in the purification process is a mechanically mixed and aerated tank. The aerating injectors can be cleaned in a simple way without the need to empty the aeration tank. Potential scaling materials are combined into removable fine particles. At the same time, the materials that may cause an odor nuisance are oxidized into odorless components. The process can be completed by a third and a fourth step. The third step focuses on suspended solids recovery and removal. The fourth step is an additional water-softening step with lamella separation and continuous sand filters in order to produce fresh water substitute. The benefits claimed by the manufacturer are as follows[54]:

1. Reduction in fresh water costs
2. Savings in energy costs; biogas production makes positive energy balance
3. No water discharge permit needed
4. Minimized space requirement
5. Few chemicals needed
6. Very few waste products
7. Emissions of volatile organics drastically reduced

11.5.4 TERTIARY TREATMENT

Partial recycling of treated water after biological treatment will usually require additional polishing (tertiary treatment). This may include chemical addition, flocculation, and DAF. This is BAT according to the EC.[38]

11.5.5 BIOSOLIDS MANAGEMENT PROCESSES

Some mills may perform biosolids grinding, gravity, or flotation thickening or chemical preconditioning to achieve up to 10% biosolids concentration.[53]

11.5.5.1 Gravity Thickening

Gravity thickening is a common process for dewatering and for the concentration of sludge. Gravity thickening is essentially a sedimentation process similar to what occurs in all settling tanks. The process is simple and is the least expensive of the available thickening processes.[55,56]

Gravity thickening may be classified as plain settling and mechanical thickening. Plain settling usually results in the formation of scum at the surface and stratification of sludges near the bottom. Gentle agitation is usually employed to stir the sludge, thereby opening channels for water to escape and promoting densification. A common mechanical thickener consists of a circular tank equipped with a slowly revolving sludge collector. Organic polyelectrolytes (anionic, nonionic, and cationic) have been used successfully to increase the sludge settling rates, the overflow clarity, and the allowable tank loadings.[57]

11.5.5.2 Flotation Thickening

In a DAF thickening process, air is added at pressures in excess of atmospheric pressure (2.1 to 4.9 kg/cm^2; 30 to 70 psig) either to the incoming sludge stream or to a separate liquid stream. When the pressure is reduced and turbulence is created, air in excess of that required for saturation at atmospheric pressure leaves the solution as very small bubbles of 50 to 100 μm in diameter. The bubbles adhere to the suspended particles or become enmeshed in the solids matrix. As the average density of the solids–air aggregate is less than that of water, the agglomerate floats to the surface. The floated solids build to a depth of several inches at the water surface. Skimmers continuously remove the float.[58]

Polyelectrolytes are frequently used as flotation aids, to enhance performance and create a thicker sludge blanket.[59] The advantages of a DAF thickener are as follows:

1. It provides better solids–liquid separation than a gravity thickener.
2. For many sludges, it yields higher solids concentration than gravity thickener.
3. It requires less area than a gravity thickener.
4. It has less chance of odor problems than a gravity thickener.

11.5.5.3 Belt Filter Press

Biosolids are squeezed between two porous cloth belts. The dewatered cake is scraped from the belts by blades.[60] This operation results in a typical biosolids concentration of 50% for the primary and 20% for secondary biosolids.

11.5.5.4 Vacuum Filters

Vacuum filter systems consist of a horizontal cylinder partially submerged in a tank of biosolids. A layer of porous filter media fabric or tightly wound coils covers the outer surface of the cylinder. As the cylinder surface passes through the tank, a layer of biosolids adheres to the cylinder and vacuum is applied.[61] The dewatered biosolids cake is then scraped off the fabric at up to 30% solids.

11.5.5.5 Screw Presses

Many kraft mills use screw presses that can achieve up to 55% biosolids concentration when dewatering primary biosolids. This operation does not require preconditioning to achieve high concentrations.

11.5.6 BIOSOLIDS DISPOSAL PROCESSES

Subsequent to biosolids handling processes, kraft mills dispose of biosolids by land application, landfill, or combustion.

11.5.6.1 Land Application

Biosolids from kraft mills are classified as a soil amendment because it is too low in nutrients to be of any value as a fertilizer. Owing to concerns regarding dioxin- and furan-contaminated biosolids, in 1994 U.S. EPA and AF&PA (American Forest & Paper Association) entered into an agreement governing the land disposal of biosolids.[62] In this agreement, AF&PA agreed to compile annual monitoring reports for those mills that land-apply materials with a dioxin/furan concentration equal to or greater than 0.01 µg/L. Individual mills also entered in separate agreements with U.S. EPA governing the land application of their biosolids.

11.5.6.2 Landfill

This is the most common disposal method. Kraft mills may use on-site landfills or off-site commercial landfills.[63]

11.5.6.3 Combustion

Some mills will combust the biosolids for heat recovery in a specialized biosolids incinerator, or a hogged or fossil fuel power boiler. Currently, this disposal method is less common than landfilling.[64]

11.5.7 AIR POLLUTANT EMISSIONS FROM TREATMENT PLANTS

The two main sources of air pollutants that may be emitted from basic wastewater treatment plant operations are pulping condensates and bleach plant effluent. The pulping condensates may include total reduced sulfur (TRS) compounds as well as volatile organic compounds (VOCs) such as methanol. The primary pollutants of concern for the bleach plant effluent are chloroform and methanol. Any volatile compounds that could be released as air emissions from basic wastewater treatment plant operations are relatively minor and are generally not subject to specific regulation.[65,66]

In addition, if a mill operates a sludge incinerator, there will be emissions from the incinerator. Inorganic gases (such as CO, NO_x, SO_x, and HCl) may be present, as well as particulate matter (including ash and heavy metals) and organic gases. The only compound subject to specific federal regulations for industrial wastewater sludge incinerators is mercury. In most cases, compliance with the incinerator requirements involves only an initial test to document mercury levels, with a follow-up estimate of the impact on mercury emissions if operating conditions are changed.

11.5.8 WATER POLLUTANT DISCHARGES FROM TREATMENT PLANTS

Kraft mills treat wastewater in order to minimize effluent impacts on receiving waters. Generally, treated effluent is discharged from the wastewater treatment system at a single discharge point. The following pollutants of concern exist at all kraft mills: BOD_5, total suspended solids (TSS), color, and chemical oxygen demand (COD). At kraft mills that bleach pulp with chlorine-containing compounds, additional pollutants of concern include chloroform, 2,3,7,8-TCDD (dioxin), 2,3,7,8-TCDF (furan), chlorinated phenolic compounds, and adsorbable organic halides (AOX). Each of these pollutants is discussed below[67–75]:

1. *BOD_5 and TSS.* The high concentrations of organic matter found in kraft mill wastewater result in high levels of BOD_5. Treatment of this BOD_5 results in the generation of large quantities of TSS. In general, kraft mills achieve 90% (or greater) removal of these pollutants when primary and secondary treatments are well operated.

2. *Color.* Kraft pulp mill effluents contain highly colored lignin and lignin derivatives that have been solubilized and removed from wood during pulping and subsequent bleaching operations. For kraft mill wastewaters, color is determined by spectrophotometric comparison of the sample with a 1 mg/L solution of platinum, in the form of chloroplatinate ion. The color of kraft mill wastewaters is considered to be the color of the water from which turbidity has been removed ("true" color). Further, wastewater color is highly pH dependent, so the pH of color samples is adjusted to pH 7.6. The U.S. EPA has not promulgated national regulations for color because the potential for significant aesthetic or aquatic impacts from color discharges is driven by highly site-specific conditions, such as the color of the receiving stream and the relative contribution of the mill discharge to the stream flow. However, many individual NPDES permits contain water quality-based effluent limitations on the discharge of color, developed to address local conditions.

3. *COD.* COD is a measure of the quantity of chemically oxidizable material present in wastewater. Sources of COD include the pulping area, chemical recovery area, bleaching area, and papermaking area. A portion of COD is readily biodegradable, and the rest is resistant to biodegradation (i.e., "refractory"). Although the amount and sources of refractory COD will vary from mill to mill, some portion of it is derived from black liquor; thus, COD biodegradability indicates the degree to which black liquor is recovered from brownstock pulp and kept out of the wastewater stream. Wastewater COD loads also relate to discharges of toxic organic pollutants that are not readily biodegraded. Although U.S. EPA has not established COD effluent limitations guidelines at this time, U.S. EPA is planning to do so in a future rulemaking.[36]

4. *Chloroform.* Chloroform is an extremely volatile compound that is generated during the bleaching of pulp with hypochlorite, chlorine, or chlorine dioxide. Hypochlorite bleaching results in the greatest amount of chloroform generation, and chlorine dioxide bleaching results in the least amount of chloroform generation. As chloroform is generated, it partitions to air and to bleach plant effluent (with a small fraction remaining with the pulp). Any chloroform found in bleach plant effluent that is not emitted to the air prior to reaching the wastewater treatment plant may be volatilized or degraded during secondary treatment or discharged in the effluent.

5. *Dioxin and furan.* During the late 1980s, bleaching with chlorine and hypochlorite were discovered to be a source of dioxin and furan. Although the use of chlorine dioxide (ClO_2) bleaching minimizes the formation of chlorinated pollutants, measurable quantities of 2,3,7,8-TCDF and possibly 2,3,7,8-TCDD may still be formed. Dioxin and furan are not effectively degraded during wastewater treatment; they partition to the sludge (and may be discharged with TSS into receiving waters untreated).

6. *Chlorinated phenolic compounds.* Chlorinated phenolic compounds include phenols, guaiacols, catechols, and vanillins substituted with from one to five chlorine atoms per molecule. Typically, bleaching processes that result in the formation of 2,3,7,8-TCDD and 2,3,7,8-TCDF also generate the higher substituted tri-, tetra-, and penta-chlorinated compounds. U.S. EPA has established effluent limitations guidelines and pretreatment standards for the following 12 chlorinated phenolic compounds: trichlorosyringol, 3,4,5-trichlorocatechol, 3,4,6-trichlorocatechol, 3,4,5-trichloroguaiacol, 3,4,6-trichloroguaiacol,

4,5,6-trichloroguaiacol, 2,4,5-trichlorophenol, 2,4,6-trichlorophenol, tetrachlorocatechol, tetrachloroguaiacol, 2,3,4,6-tetrachlorophenol, and pentachlorophenol.

7. *Adsorbable organic halides (AOX).* AOX is a measure of the total amount of halogens (chlorine, bromine, and iodine) bound to dissolved or suspended organic matter in a wastewater sample. In bleached kraft mill effluent, essentially all of the AOX comprises chlorinated compounds formed during bleaching with chlorine and other chlorinated bleaching agents. Inefficient application of chlorine-containing bleaching chemicals can generate increased levels of AOX. Minimizing AOX will usually have the effect of reducing the generation of chloroform, 2,3,7,8-TCDD, 2,3,7,8-TCDF, and chlorinated phenolic compounds. Some AOX is biodegraded during secondary treatment.

In addition to retaining the existing effluent limitations guidelines and standards for BOD_5, TSS, and pH, the Cluster Rules establish new effluent limitation guidelines and standards for bleached papergrade kraft mills for the other parameters described above, with the exception of color and COD. The Cluster Rules regulations require bleached kraft mills to meet limits on in-process streams and treated effluent, depending on the pollutant. See references for further sources of information on the applicable discharges and control strategies.[59,76–87]

11.5.9 BIOSOLIDS/HAZARDOUS WASTE DISCHARGES FROM TREATMENT PLANTS

Kraft pulp mills generate both primary and secondary biosolids (sludges). The collected solids may be thickened in gravity or flotation thickeners or chemically conditioned prior to dewatering. Primary solids are usually generated in greater quantities than secondary biosolids. Although the biosolids potentially can be used for alternative beneficial uses, generally dewatered biosolids are disposed of through land application, landfilling, or combustion. Because of concerns about potential contamination with dioxin, U.S. EPA was required to make a hazardous waste listing determination for solids from bleached kraft mill effluents unless the final effluent guidelines were based on the use of at least one of certain specified technologies. These technologies enable the mill to use less chlorine in bleaching pulp and thus to generate less dioxin contamination. After the promulgation of the Cluster Rules, U.S. EPA determined that the final guideline was based on the specified technologies, and thus U.S. EPA determined that it was not required to make a hazardous waste listing determination for pulp mill solids. If the solids at a particular mill exhibit a hazardous waste characteristic, the solids would be hazardous wastes even without a U.S. EPA listing determination.[36]

11.5.10 RECOVERY OF FIBERS AND TITANIUM DIOXIDE

The principal material in paper is the cellulose fiber—from wood, or less frequently from cotton—and particulate mineral filler is incorporated to enhance certain properties, especially opacity. Operations to recover the papermaking materials from wastewaters must deal with both the fiber and the filler. The most common fillers, clay and precipitated calcium carbonate, are less expensive than fiber, and so efficiency of recovery of them is of secondary importance. However, in some special types of paper, where extreme opacity is required, the much more expensive filler titanium dioxide is employed. Titanium dioxide (TiO_2) is a fine, white crystalline powder having an extremely small particle size of 0.1 to 0.4 µm that forms a negatively charged colloid in aqueous media. Because of its colloidal properties and high refractive index (ca. 2.52 for anatase and 2.76 for rutile), titanium dioxide suspensions are very stable in dilute concentrations and have an intense white to blue color. This substance is used as a filler and brightener in high-quality paper and as a white pigment for paints. As a consequence of the manufacturing process, both fibers and titanium dioxide are present in the waste effluents of paper, pulp, and other related industries. The effective and economic recovery of titanium dioxide and fibers offers both the possibility of savings in process costs and the solving of a significant pollution problem.

The case histories presented in Section 11.6.4 to 11.6.6 demonstrate that fibers and titanium dioxide can be recovered from a whitewater by DAF under full flow pressurization mode or recycle pressurization mode with or without chemical addition.[59]

11.6 CASE STUDIES

11.6.1 CASE I: INTERNATIONAL PAPER COMPANY, JAY, MAINE

This case study was a U.S. EPA initiative to evaluate the extent to which regulatory flexibility and other innovative environmental approaches could be used to achieve superior environmental performance at reduced economic and administrative burdens.

The primary goals of the project were to provide leadership in environmental stewardship and flexibility in regulation as an alternative to the command and control approach enumerated in the Cluster Rules (promulgated in 1998). The project, designated International Paper Effluent Improvements Project, was conducted at the International Paper (IP) Androscoggin Paper Mill in Jay, Maine (Figure 11.9) between the project start date of July 29, 2000, and its formal conclusion on December 29, 2004.[81]

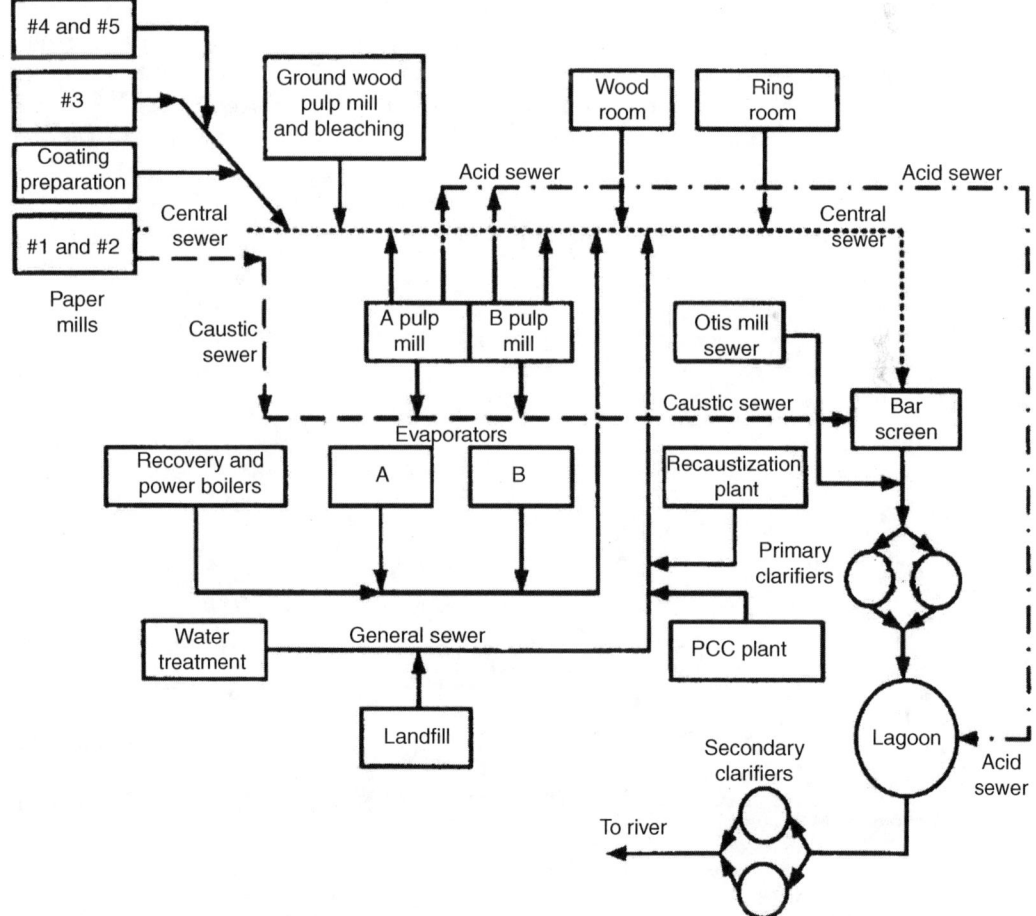

FIGURE 11.9 Schematic diagram of the IP Jay Paper Mill showing points of discharge to wastewater treatment plant. (Taken from U.S. EPA, International Paper XL-2 Effluent Improvements Project, Final Report, U.S. EPA, Maine Department of Environmental Protection, Jay, Maine, September 6, 2005.)

11.6.1.1 Description of Wastewater Collection and Treatment Plant

The wastewater treatment facility (Figure 11.10) provides primary treatment, biological treatment, and secondary clarification. The components of the facility include two coarse mechanical bar screens, two 58-m (190-ft) diameter primary clarifiers with skimmers, four influent pumps with provisions for chemical addition for pH adjustment and nutrient addition, a 265,000 m³ (70 MG) aerated lagoon, two 77.8-m (255-ft) diameter secondary clarifiers with polymer addition to enhance settling, four return sludge pumps, a 24.4-m (80-ft) diameter thickener and eight screw presses and two belt filter presses, a foam dissolving tank, an emergency spill basin with pumps (not shown), a heat exchanger, and a diffuser.

Stormwater, cooling water, water treatment backwash water, landfill leachate, and wastewater generated in the pulp and papermaking process are discharged into the mill's general sewer (caustic/neutral pH wastewater) by way of a series of collection pipes and sewers. The general sewer flows through the mechanically raked bar screens to remove large objects. The screened objects are then sent to the landfill. Process wastewater from the Otis Mill (see Figure 11.10) is combined with the general sewer after the bar screens. The combined wastewater then flows by gravity through a splitter box and into the two primary clarifiers.

Acid process wastewater is collected separately from the caustic and neutral pH range wastewater. The sanitary wastewater from the mill discharges into the process wastewater acid sewer. The sanitary waste is disinfected by reaction with the oxidants in the acid sewer coming from the bleach plant. Disinfection can also be done by using sodium hypochlorite, calcium chlorite, or other suitable oxidants when the acid sewer is unavailable for treatment. The acid wastewater, including sanitary wastewater, has few suspended solids that can be removed by screening or conventional primary clarification. Therefore, the acid sewer combines with the general wastewater effluent from the primary clarifies just downstream from the primary clarifier (Figure 11.9).

pH adjustments using lime, caustic, or sulfuric acid on the combined wastewater occur in the collection box prior to flowing to the wastewater treatment plant's influent pump station. Four

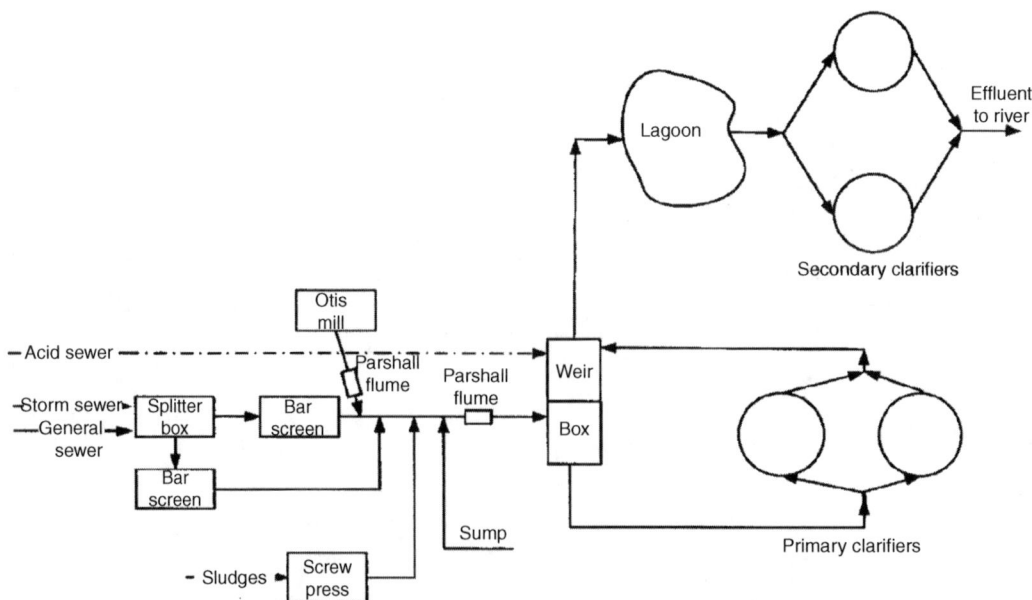

FIGURE 11.10   Schematic diagram of wastewater treatment plant, IP Company at Jay, Maine. (Taken from U.S. EPA, International Paper XL-2 Effluent Improvements Project, Final Report, U.S. EPA, Maine Department of Environmental Protection, Jay, Maine, September 6, 2005.)

centrifugal pumps lift the combined wastewater from a wet well to the aerated lagoon through a 1100-mm (42-in) force main. Before the wastewater enters the lagoon, nutrients such as phosphoric acid, urea, and other suitable nutrients are injected into the force main, as needed, to provide phosphorous and nitrogen to enhance growth of biological solids.

The lagoon at IP Jay is an irregular shaped earth-berm structure with a volume of approximately 265,000 m³ (70 MG) and an effective process volume of 90,840 m³ (24 MG). Fifty-five aerators are used to entrain air and mix the solids and liquid in the aeration lagoon to promote biological treatment of wastewater. The aerators consume about 3133 kW (4200 hp) of mechanical power.

Wastewater exits the lagoon and flows over a weir and into a splitter box where the flow is split to the two secondary clarifiers. Cationic polymer is added, as needed, before the secondary clarifiers, to enhance settling of the suspended solids. The settled solids consist of active biological matter and are returned via return sludge pumps to the lagoon through a return line that discharges from two pipes within 7.6 m (25 ft) of the influent from the lift pumps.

The waste sludge pumps also convey excess solids from the secondary clarifiers to the gravity thickener. This waste sludge is then pumped to a sludge dewatering system consisting of screw and belt filter presses. Polymer is added to the sludge prior to dewatering to increase floc size and aid in dewatering. After dewatering by the presses, the dewatered sludge is incinerated in the multifuel boiler waste fuel incinerator (WFI), or temporarily stockpiled and trucked to the mill landfill site for disposal.

Defoamer is added at the overflow from the secondary clarifiers, as necessary, as it flows to a collection box for discharge to the Androscoggin River. The effluent is monitored at the collection box for compliance with permit requirements. Before being discharged to the river, the effluent passes through a heat exchanger, which is operated during the winter months to recapture waste heat. The effluent then passes into a foam dissolving tank that allows for the physical separation of any foam from the effluent and then through a diffuser for discharge into the Androscoggin River.

11.6.1.2   Performance of the Wastewater Treatment Plant

Efficiencies for removal in the wastewater treatment plant were estimated for total and soluble BOD, total COD, soluble COD, color, total suspended and dissolved solids, and total solids. The removal efficiencies summarized in Table 11.14 are high for total BOD, soluble BOD, and suspended solids, at 96%, 96%, and 95%, respectively. The removal efficiencies for total and soluble COD were significantly lower at 76% and 66%, respectively. The removal efficiency for color was only about 38%. This value is typical for biological treatment of pulp and paper wastewater, and may be due, at least partially, to the formation of new colored groups when the bleach effluents are oxidized in the treatment system.

11.6.2   CASE II: UPGRADED TREATMENT PLANT AT A PAPER MILL IN LUFKIN, TEXAS, USING A DAF CELL

The activated sludge treatment plant at a paper mill in Lufkin, TX, treats 68,200 m³/d (18 MGD) of wastewater. The plant was designed to produce a final effluent with BOD and TSS that would not exceed 20 mg/L. However, several expansions resulted in poor effluent quality and borderline permit compliance, particularly during the periods of peak BOD loading. The first alternative to solve the plant's problems, namely increasing the aeration time by adding another aeration basin of the same size, was not a viable option. The company did not have enough land space and the capital expenditure for this conventional option is high. The alternative decision was the use of a 16.8 m (55 ft) dissolved air flotation (DAF) cell (see Figure 11.11) as a secondary clarifier that would be installed in front of the final sedimentation tanks and has a capacity to handle 30,000 m³/d (8 MGD) of flow.[45,82–84] This was accomplished at only 12% of the cost of the conventional expansion project estimate. A top view of the DAF cell is shown in Figure 11.12.

TABLE 11.14
Removal Efficiencies in Wastewater Treatment Plant

Variable	Influent T/d	Effluent T/d	Removal Efficiency %
Total BOD_5	43	1.7	96
Soluble BOD_5[a]	24	0.9	96
Total COD	153	33	78
Soluble COD[a]	75	25	66
Color	65	40	38
Total suspended solids (TSS)	97	5.2	95
Soluble solids (SS)	171	167	2.9
Estimated total solids	268	172	36
Flow			
MGD[b]	40.8	42.5	
ML/d[c]	154.4	160.9	
M^3/d	154,400	160,900	

BOD, biochemical oxygen demand; COD, chemical oxygen demand.
[a]Samples were filtered through a 0.8 μm filter.
[b]MGD = million gallons per day.
[c]ML/d = million liters per day.
Source: U.S. EPA, International Paper XL-2 Effluent Improvements Project, Final Report, U.S. EPA, Maine
Department of Environmental Protection, Jay, Maine, September 6, 2005.

The sludge return to the aeration basin from the flotation cell at 2% concentration is five times thicker than the 0.4% sludge return from the final settling tanks. The resulting reduction in the volume of recycle to the aeration basin by 9500 m^3/d (2.5 MGD) provides an extra 10% hydraulic capacity for aeration. The solids removed from the 30,000 m^3/d flow processed by the flotation cell reduced the solids flowing to the final clarifiers by at least 30% so that no violations of the discharge limits have occurred since installation. The net results were reduced solids loading to the final clarifiers, increased hydraulic capacity and retention time of the aeration basin, threefold increase in overall concentration of biosolids, more active recycled sludge, better effluent quality, and no biosolids bulking problems. More case histories of waste treatment in the pulp and paper industry using flotation technology can be found from the literature.[85,86]

11.6.3 CASE III: DEEP SHAFT PLANT AT OHTSU PAPER COMPANY IN OHTSU, JAPAN

The Deep Shaft plant at Ohtsu Paper Co. came on line in 1980. It treats the wastewater generated by a cardboard recycling facility located within 18.3 m (60 ft) of a residential area inside the city of Ohtsu. The plant discharges treated wastewater to a beautiful recreational body of water named Lake Biwa.

11.6.3.1 Plant Description

Flow to the deep shaft biological wastewater treatment plant is screened and goes through a DAF unit for fiber removal prior to entry into the deep shaft. The Ohtsu plant consists of one shaft 2.79 m (110 in.) in diameter by 100 m (330 ft) deep. The shaft design incorporates one downcomer and one riser where the downcomer is located concentrically within the shaft with the resultant annular volume serving as the riser. Mixed liquor in the shaft is maintained at approximately 5000 mg/L and the hydraulic detention time in the shaft is 1 h. Mixed liquor enters the head tank at the top of the shaft where gas disengagement occurs. The head tank is 6 m × 12 m × 3 m

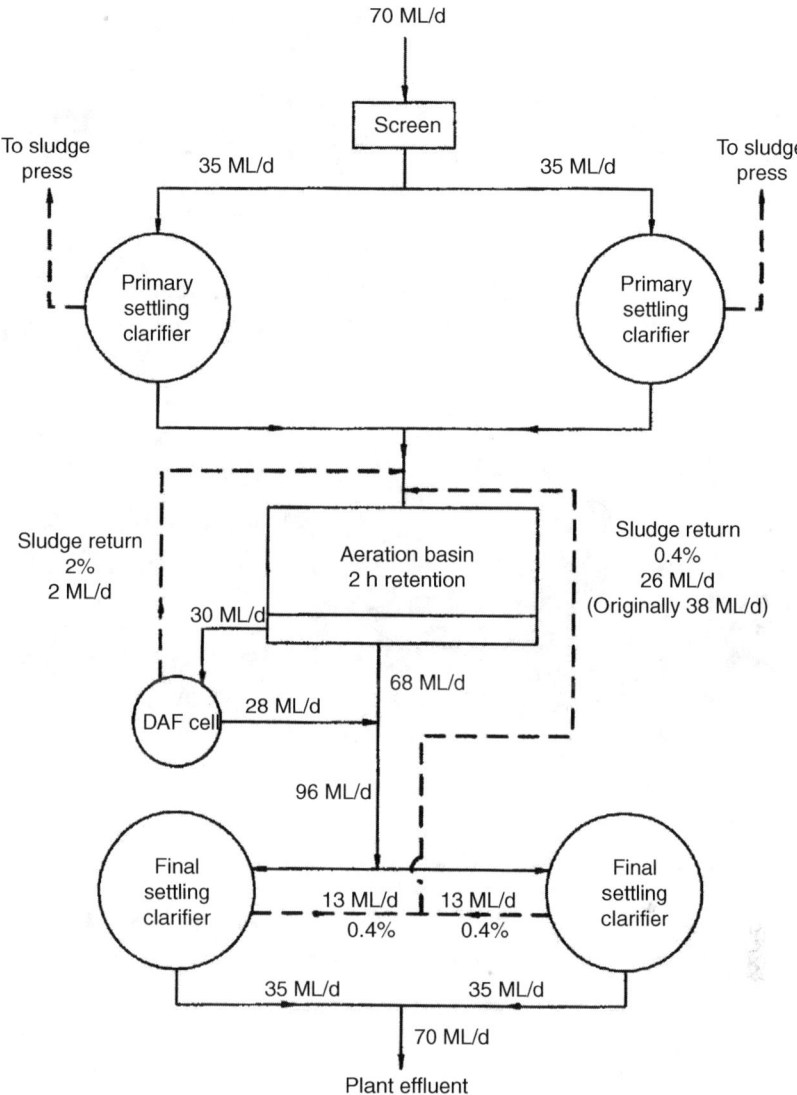

FIGURE 11.11 Process flow diagram of upgrade activated sludge plant at a paper mill using a DAF cell.

(20 ft × 40 ft × 10 ft). A portion of the mixed liquor overflows the head tank to a vacuum degasser, which is 3.8 m (12.5 ft) diameter by 10.1 m (33 ft) high equipped with one operating vacuum pump of 14.9 kW (20 hp) capacity. The degasser overflows by gravity into a holding tank that splits the flow equally to two sedimentation clarifiers, both of which are 25 m (82 ft) in diameter and have 2.7 m (9 ft) sidewater depth. Sludge is wasted at 1% concentration to a sludge holding tank. Waste sludge is subsequently pumped to a belt filter press and dewatered to a 40% by weight solids content prior to disposal.[87]

The aeration requirements of the deep shaft (vertical shaft bioreactor using flotation technology) are provided by two, 100 hp rotary screw compressors rated at a pressure of 7 kg/m² (100 psig). Dissolved oxygen (DO) levels of 4 mg/L are maintained in the head tank, and during the startup phase of the plant a DO meter measured a dissolved oxygen concentration of 25 mg/L at the shaft bottom.

FIGURE 11.12 Top view of DAF cell.

Operating staff at the plant consist of two operators per shift plus one maintenance person on dayshift. The wastewater is nutrient deficient and a ratio of 100:8:1 of BOD:N:P is maintained by the addition of diammonium phosphate and phosphoric acid for nitrogen and phosphorous requirements, respectively.

11.6.3.2 Performance

The plant has performed very well over its operating lifetime since 1980. Problems experienced have been a transmittance problem on one occasion, which was resolved by the addition of bentonite clay to the mixed liquor suspended solids (MLSS). This provided an inexpensive solution and the problem has never reoccured. Bentonite addition is still being practiced as an added insurance. Septage collecting in the degasser caused an odor problem during low flow periods and it was corrected by the installation of an air line in the base of the degasser.

Table 11.15 illustrates the mean operating performance of the plant. The regulatory standard for the plant is based on effluent COD, which is monitored three times per day. A BOD_5 test is done once per month. The BOD/COD correlation for the plant effluent is $BOD = COD - 25$.[87]

11.6.4 CASE IV: COTTON FIBER RECOVERY FLOTATION CELL OF KROFTA ENGINEERING CORPORATION, LENOX, MASSACHUSETTS

Cotton paper (100%) obtained from Mead Corporation in South Lee, Massachusetts, was pulped at Krofta Engineering Corporation (KEC), Lenox, MA.[59] A known amount of pulp was suspended in tap water to determine the percent recovery by a circular DAF cell (Model Supracell Type 3; diameter = 0.91 m [3 ft]; depth = 55.88 cm [22 in.]; flow = 45 L/min [12 gal/min]). The initial total

TABLE 11.15

Performance of Ohtsu Paper Company's Deep Shaft Biological Treatment Plant in Japan

Parameter	Influent[a] (mg/L)		Effluent (mg/L)	
	Design	Operation range	Design	Operation range
BOD_5	200	100–215	<10	<10
COD	30	120–260	34	21–30
TSS	22	10–60	20	10–18

BOD, biochemical oxygen demand; COD, chemical oxygen demand; TSS, total suspended solids.

[a]Flow range = 18,000–23,000 m^3/d; temperature range = 10–29°C.

Source: Adapted from Daly, P.G. and Shen, C.C., The deep shaft biological treatment process, in *Proceedings of the 43rd Purdue Industrial Waste Conference*, May 1988. With permission.

suspended solids concentration (TSS) of raw pulp was 1260 mg/L. In two separate continuous full flow pressurization operations, 95.5% fiber was recovered without any chemical addition, and 98% fiber was recovered with the addition of 1 mg/L polymer Betz 1260. It was concluded that recovery of cotton fiber by a DAF clarifier can be successfully achieved even without chemical addition. It is also important to note that rectangular DAF cell should be equally effective for fiber recovery although the exact fiber recovery efficiency must be demonstrated by a pilot plant testing.

11.6.5 CASE V: FIBER AND TITANIUM DIOXIDE RECOVERY FACILITY AT MEAD CORPORATION, SOUTH LEE, MASSACHUSETTS

Almost all fiber and partial titanium dioxide can be recovered from white water by DAF under full flow pressurization mode[43] with chemical addition. On June 10, 1982, at Mead Corporation, pulp was prepared with 40% cotton fiber and 60% wood fiber. The loading of titanium dioxide was about 50% (i.e., 273 kg TiO_2 per 600 kg total pulp). The white water from No. 2 machine was fed to a DAF cell (diameter = 3 m) at 15.8 L/sec (250 gal/min) under full flow pressurization mode. Turkey red oil (TRO) was dosed as a flotation aid at 80 mL/min. The influent white water (before TRO addition), DAF effluent, and floated scum were sampled for analysis. The DAF influent had 98 mg/L of TSS, and 650 NTU of turbidity at pH 9.27. The DAF effluent had 15 mg/L TSS and 550 NTU of turbidity at pH 9.25. Although TSS (fiber and titanium dioxide) recovery rate was 85%, the ash content (titanium dioxide) of the recovered TSS was very low. Therefore, using a DAF clarifier under full flow pressurization mode and TREO, the majority of fibers in white water but only about half of titanium dioxide can be recovered.

11.6.6 CASE VI: RESOURCE RECOVERY FACILITY OF LENOX INSTITUTE OF WATER TECHNOLOGY (LIWT), LENOX, MASSACHUSETTS

Both fibers and titanium dioxide can be almost totally recovered by DAF under recycle flow pressurization mode[59] when using adequate coagulant. Various operational modes of DAF can be found in the literature.[43] White water containing 500 mg/L titanium dioxide and 1000 mg/L cotton fiber was continuously fed to a LIWT research facility (circular high rate DAF cell; diameter = 0.9 m) at 45 L/min (12 gal/min) under 33.3% recycle flow pressurization mode. After one hour of continuous operation and at steady state, the influent, effluent and floated scum were sampled for analysis. It was found that over 99% of titanium dioxide and fibers was recovered when 100 mg/L of magnesium carbonate, 120 mg/L of calcium hydroxide, and 0.3 mg/L of polymer Magnifloc 1563 C were dosed at pH 11. Initial DAF influent feed was milky. The DAF effluent became crystal clear having a turbidity of 2 NTU. The floated scum was 3.9% in consistency. Titanium dioxide concentrations were measured by both atomic absorption spectrometry and ash content.

A separate continuous DAF operation conducted by Krofta and Wang[59] under 33.3% recycle flow pressurization mode demonstrated that aluminum sulfate, sodium aluminate, and polyelectrolyte combination at pH 6.2 also effectively recovered both fibers and titanium dioxide from the same white water containing 500 mg/L of titanium dioxide and 1000 mg/L of cotton fibers.

In practical applications, adequate coagulants should be chosen based on the quality of the floated scum (i.e., recovered titanium dioxide and fiber mixture), which is intended to be reused in the paper manufacturing process. The reused titanium dioxide and fibers should not adversely affect the quality of the paper.

Additional research conducted by LIWT[88–91] has shown that the wastestreams, such as those shown in Figure 11.11, can be effectively treated by the two-stage biological-physicochemical process system or two-stage DAF-DAFF (dissolved air flotation-filtration) process system. The readers are referred to the literature[88–91] for details.

NOMENCLATURE

AF&PA	American Forest & Paper Association
AOX	Adsorbable organic halides
BAT	Best available technology economically achievable
BCT	Best conventional pollutant control technology
BOD	Biochemical oxygen demand
BPT	Best practicable control technology
CAA	Clean Air Act
CAAA	Clean Air Act Amendments of 1990
CERCLA	Comprehensive Environmental Response, Compensation and Liability Act
CFCs	Chlorofluorocarbons
CFR	Code of Federal Regulations
COD	Chemical oxygen demand
CWA	Clean Water Act
ELGs	Effluent Limitations Guidelines and Standards
EPCRA	Emergency Planning and Community Right-to-Know Act
F/M	Food/microorganisms ratio
HAPs	Hazardous Air Pollutants (CAA)
LDR	Land Disposal Restrictions (RCRA)
LEPCs	Local Emergency Planning Committees
MACT	Maximum achievable control technology (CAA)
MCLGs	Maximum contaminant level goals
MCLs	Maximum contaminant levels
ML	Minimum level
NAAQS	National Ambient Air Quality Standards (CAA)
NCP	National Oil and Hazardous Substances Pollution Contingency Plan
NESHAP	National Emission Standards for Hazardous Air Pollutants
NOX	Nitrogen oxides
NPDES	National Pollutant Discharge Elimination System (CWA)
NPL	National Priorities List
NSPS	New Source Performance Standards (CAA)
OPA	Oil Pollution Act
OSHA	Occupational Safety and Health Administration
PAC	Polycyclic aromatic compounds
POTW	Publicly owned treatment works

PSES   Pretreatment Standards for Existing Sources
PSNS   Pretreatment Standards for New Sources
RCRA   Resource Conservation and Recovery Act
SARA   Superfund Amendments and Reauthorization Act
SDWA   Safe Drinking Water Act
SIC   Standard Industrial Classification
SOX   Sulfur oxides
T   Metric ton = 1000 kg
t   English ton = 2000 lb
TRI   Toxic release inventory
TSCA   Toxic Substances Control Act
TSS   Total suspended solids
U.S. EPA   U.S. Environmental Protection Agency
UST   Underground storage tanks (RCRA)

APPENDIX

U.S. Army Corps of Engineers Civil Works Construction Yearly Average Cost Index for Utilities (for conversion of USD costs in terms of 2008 USD).

TABLE A11.1

U.S. Army Corps of Engineers Civil Works Construction Yearly Average Cost Index for Utilities

Year	Index	Year	Index
1967	100	1988	369.45
1968	104.83	1989	383.14
1969	112.17	1990	386.75
1970	119.75	1991	392.35
1971	131.73	1992	399.07
1972	141.94	1993	410.63
1973	149.36	1994	424.91
1974	170.45	1995	439.72
1975	190.49	1996	445.58
1976	202.61	1997	454.99
1977	215.84	1998	
	459.40		
1978	235.78	1999	460.16
1979	257.20	2000	468.05
1980	277.60	2001	472.18
1981	302.25	2002	484.41
1982	320.13	2003	495.72
1983	330.82	2004	506.13
1984	341.06	2005	516.75
1985	346.12	2006	528.12
1986	347.33	2008	552.16
1987	353.35		

Source: U.S. ACE. Yearly average Cost Index for Utilities, in *Civil Works Construction Cost Index System Manual*, 110-2-1304, U.S. Army Corps of Engineers, Washington, p. 44, 2008. Available at http://www.nww.usace.army.mil/cost.

REFERENCES

1. U.S. EPA, Profile of the Pulp and Paper Industry, 2nd ed., report EPA/310-R-02-002, U.S. EPA, Washington, November 2002.
2. U.S. EPA, Profile of the Pulp and Paper Industry, report EPA/310-R-95-015, U.S. EPA, Washington, September 1995.
3. U.S. EPA, Development Document for Proposed Effluent Limitations Guidelines and Standards for the Pulp, Paper, and Paperboard Point Source Category, EPA-821-R-93-019, U.S. EPA, Washington, October 1993.
4. Pulp & Paper Magazine, *Paper Help Online Encyclopedia*, 2006. Available at http://www.paperloop.com/pp_mag/paperhelp/homepage.shtml.
5. U.S. Department of Commerce/International Trade Administration, *U.S. Industry & Trade Outlook 2000*, U.S. Department of Commerce, McGraw-Hill, New York, 2000.
6. U.S. Department of Energy, *Forest Products Project Fact Sheet: Closed-Cycle Bleach Kraft Pulp Production*, Office of Industrial Technologies, Washington, October 2000.
7. U.S. Census Bureau, *1997 County Business Patterns for the U.S.*, U.S. Census Bureau, 1998.
8. U.S. Census Bureau, *1997 Economic Census: Comparative Statistics for the U.S.*, U.S. Census Bureau, 2000.
9. U.S. EPA, *Toxics Release Inventory Database*, U.S. EPA, Washington, 1999.
10. U.S. Department of Commerce, *U.S. Industry & Trade Outlook 2000*, McGraw-Hill and U.S. Department of Commerce/International Trade Administration, Washington, 2000.
11. AF&PA, *1999 Statistics, Data through 1998*, American Forest and Paper Association, Washington, 1999.
12. Smook, G.A., *Handbook for Pulp & Paper Technologists*, 2nd ed., Angus Wilde Publications, Vancouver, Canada, 1992.
13. U.S. EPA, *1990 National Census of Pulp, Paper, and Paperboard Manufacturing Facilities*, U.S. EPA, Washington, 1990.
14. AF&PA, *Paper Recycling Facts*, American Forest and Paper Association, Washington, 2006. Available at http://www.afandpa.org/recycling/Rec_paperrecfacts_open.html.
15. AF&PA, *Paper Recovery Progress Report*, American Forest and Paper Association, Washington, May 2000.
16. VDP, *Papier '97—Ein Leistungsbericht*, Verband Deutscher Papierfabriken, Bonn, Germany, 1997.
17. Kenny, R.G. and Chitchumroonchokchai, P., *Wastewater Treatment in Pulp and Paper Mill*, 2006. Available at http://library.kmitnb.ac.th/article/atc38/atc00027.html.
18. AF&PA, Personal Communication, American Forest and Paper Association, Washington, 1995.
19. U.S. EPA, *Summary of Technologies for the Control and Reduction of Chlorinated Organics from the Bleached Chemical Pulping Subcategories of the Pulp and Paper Industry*, U.S. EPA, Washington, 1990.
20. Alliance for Environmental Technology, *Trends in World Bleached Chemical Pulp Production: 1990–2000*, January 2001. Available at http://www.aet.org/reports/market/aet_trends_2000.html.
21. AF&PA, *2000 Recovered Paper Statistical Highlights*, American Forest and Paper Association, Washington, 2000. Available at http://www.afandpa.org/recycling/Rec_introduction.html.
22. AF&PA, *1994 Statistics, Data through 1993*, American Forest and Paper Association, Washington, 1994.
23. U.S. EPA, *104-Mill Study*, U.S. EPA, Washington, 1988.
24. U.S. EPA, Personal communication from Jacquelyn Vega, National Enforcement Investigations, U.S. EPA, Washington, 2002.
25. AF&PA, Written comments from Richard Wasserstrom to Seth Heminway, EPA Office of Compliance, American Forest and Paper Association, Washington, March 7, 2002.
26. U.S. EPA, *Pollution Prevention Technologies for the Bleached Kraft Segment of the U.S. Pulp and Paper Industry*, U.S. EPA, Washington, 1993.
27. U.S. EPA, *Pulp, Paper and Paperboard Industry—Background Information for Proposed Air Emission Standards: Manufacturing Processes at Kraft, Sulfite, Soda, and SemiChemical Mills* (NESHAP), U.S. EPA, Washington, 1993.
28. Enviro Science, *An Industry Overview of Pulp and Paper Mills*, 2006. Available at http://es.epa.gov/techinfo/facts/pulppapr.html.
29. Allen, D.G., Reeve, D.W., and Liss, S.N., Year Four Report, Pulp & Paper Centre, *Research Consortium on Minimizing the Impact of Pulp and Paper Mill Discharges*, University of Toronto, April 2004.

30. Jortama, P., Implementation of a Novel Pigment Recovery Process for a Paper Mill, Dissertation, University of Oulu, Oulu University Press, Oulu, Finland, 2003.

31. Chisholm, D.G., Project Profile, Green Industrial Analysis, Ministry of Environment and Energy, Provincial Papers Mill, Thunder Bay, Ontario, Canada, January 1997.

32. Air Products & Chemicals Inc., *Innovative Solutions for the Pulp and Paper Industry*, 2006. Available at www.airproducts.com/Markets/PulpandPaper/SolutionsForthePulpandPaperIndustry.htm.

33. U.S. Federal Register, Proposed Rules, Part II, December 17, 1993, pp. 66078–66216.

34. EKONO, Environmental Performance, Regulations and Technologies in the Pulp and Paper Industry, EKONO Inc. Strategic Study, Bellevue, WA, August 2005.

35. U.S. EPA, Pulp, Paper, and Paperboard Detailed Study, preliminary report EPA-821-B-05-007, U.S. EPA, Washington, August 2005.

36. U.S. EPA, Kraft Pulp Mill Compliance Assessment Guide (CAA, CWA, RCRA and EPCRA), U.S. EPA, EPA/310-B-99-001, Washington, May 1999.

37. World Bank, *Pulp and Paper Mills, Pollution Prevention and Abatement Handbook*, World Bank Group, July 1998.

38. EC, Integrated Pollution Prevention and Control (IPPC), *Reference Document on Best Available Techniques in the Pulp and Paper Industry*, European Commission, 2001.

39. U.S. EPA, Guidance Manual for Pulp, Paper, and Paperboard and Builders' Paper and Board Mills Pretreatment Standards, U.S. EPA, Effluent Guidelines Division, WH-562, Washington, September 1984.

40. Water and Agro Industry, Efficient Water use in Agro-Based Industries: Pulp and Paper, February 2006. Available at http://www.waterandagroindustry.org/pulp_paper.htm.

41. Shammas, N.K., Kumar, I.J., and Chang, S.Y., Sedimentation, in *Physicochemical Treatment Processes*, Wang, L.K., Hung, Y.T., and Shammas, N.K., Eds., Humana Press, Totowa, NJ, 2005, pp. 379–430.

42. P²AD, An Analysis of Pollution Prevention Opportunities and Impediments in the Pulp and Paper Manufacturing Sector in Georgia, Pollution Prevention Assistance Division, Georgia Department of Natural Resources, April 1, 1996.

43. Wang, L.K., Fahey, E.M., and Wu, Z., Dissolved air flotation, in *Physicochemical Treatment Processes*, Wang, L.K., Hung, Y.T., and Shammas, N.K., Eds., Humana Press, Totowa, NJ, 2005, pp. 431–500.

44. Shammas, N.K., Wang, L.K., and Hahn, H.H., Fundamentals of wastewater flotation, in *Flotation Technology*, Wang, L.K., Shammas, N.K., Selke, W.A., and Aulenbach, D.B., Eds., Humana Press, Totowa, NJ, 2008.

45. Krofta, M., and Wang, L.K., *Flotation Engineering*, 1st ed., Lenox Institute of Water Technology, Lenox, MA, Technical Manual Lenox/1-06-2000/368, 2000.

46. Wang, L.K., Pereira, N.C., and Hung, Y.T., Eds., and Shammas, N.K., Consulting Ed., *Biological Treatment Processes*, Humana Press, Totowa, NJ, 2008.

47. Wang, L.K., Shammas, N.K., and Hung, Y.T., Eds., *Advanced Biological Treatment Processes*, Humana Press, Totowa, NJ, 2008.

48. Tezel, U., Guven, E., Erguder, T.H., and Demirer, G.N., Sequential (anaerobic/aerobic) biological treatment of Dalaman SEKA pulp and paper industry effluent, *Waste Manage.*, 21, 717–724, 2001.

49. Reddy, P., Pillay, V.L., Kunamneni, A., and Singh, S., Degradation of pulp and paper-mill effluent by thermophilic micro-organisms using batch systems, *Water SA*, 31, 2005. Available at http://www.wrc.org.za.

50. Kirkwood, E., Nalewajko, C., and Fulthorpe, R.R., The occurrence of cyanobacteria in pulp and paper waste treatment systems, *Can. J. Microbiol.*, 47, 761–766, 2001.

51. Kirkwood, E., Nalewajko, C., and Fulthorpe, R.R., Physiological characteristics of *Cyanobacteria* in pulp and paper waste treatment systems, *J. Appl. Phycol.* 15, 324–335, 2003.

52. Baker, C.J.O., Fulthorpe, R.R., and Gilbride, K.A., An assessment of variability of pulp mill wastewater treatment system bacterial communities using molecular methods, *Wat. Qual. Res. J.* (Canadian), 38, 227–242, 2003.

53. Wang, L.K., Shammas, N.K., and Hung, Y.T., Eds., *Biosolids Treatment Processes*, Humana Press, Totowa, NJ 2007.

54. Pulp & Paper Technology, PAQUES—*Water Treatment for the Pulp and Paper Industry*, 2006. Available at http://www.pulpandpaper-technology.com/contractors/environmental/paques.

55. McMillon, R.T., Rockers, G.F., and Lewis, W.R., Biosolids treatment and disposal practices survey, *14th Annual Residuals and Biosolids Management Conference*, February/March 2000.

56. Wayne, P. and Shonali, L., Biosolids and sludge management, *Water Environ. Res. Literature Rev.*, September/October 2003.
57. Shammas, N.K. and Wang, L.K., Gravity thickening, in *Biosolids Treatment Processes*, Wang, L K., Shammas, N.K., and Hung, Y.T., Eds., The Humana Press, Totowa, NJ, 2007.
58. Shammas, N.K. and Wang, L.K., Flotation thickening, in *Biosolids Treatment Processes*, Wang, L.K., Shammas, N.K., and Hung, Y.T., Eds., The Humana Press, Totowa, NJ, 2007.
59. Krofta, M. and Wang, L.K., Pollution Abatement Using Advanced flotation Technology in the Paper and Pulp Industry, Powder and Bulk Solids Conference, Rosemont, Chicago, IL, May 1985.
60. Shammas, N.K. and Wang, L.K., Belt filter presses, in *Biosolids Treatment Processes*, Wang, L.K., Shammas, N.K., and Hung, Y.T., Eds., Humana Press, Totowa, NJ, 2007.
61. Shammas, N.K. and Wang, L.K., Vacuum filtration, in *Biosolids Engineering and Management*, Wang, L.K., Shammas, N.K., and Hung, Y.T., Eds., Humana Press, Totowa, NJ, 2008.
62. U.S. EPA, Memorandum of Understanding between the American Forest & Paper Association and the U.S. Environmental Protection Agency, Regarding the Implementation of Land Application Agreements Among AF&PA Member Pulp and Paper Mills and the U.S. Environmental Protection Agency, executed April 14, 1994.
63. Shammas, N.K. and Wang, L.K., Land application of biosolids, in *Biosolids Treatment Processes*, Wang, L.K., Shammas, N.K., and Hung, Y.T., Eds., Humana Press, Totowa, NJ, 2007.
64. Wang, L.K., Shammas, N.K., and Hung, Y.T., Eds., *Biosolids Engineering and Management*, Humana Press, Totowa, NJ, 2008.
65. U.S. EPA, AIRS Database, Office of Air and Radiation, U.S. EPA, Washington, November 2001.
66. U.S. EPA, *Toxics Release Inventory Database*, U.S. EPA, Washington, 2001.
67. U.S. EPA, Pulp and Paper NESHAP: A Plain English Description, U.S. EPA, Washington, November 1998.
68. U.S. EPA, Kraft Pulp Mill Compliance Assessment Guide, U.S. EPA, Washington, May 1999.
69. U.S. EPA, Questions and Answers for the Pulp and Paper NESHAP, U.S. EPA, Washington, September 1999.
70. U.S. EPA, National Emission Standards for Hazardous Air Pollutants for Source Category: Pulp and Paper Production, 40 CFR Part 430, U.S. EPA, Washington, 1993.
71. Domtar Inc., Chlorine-Free Bleaching of Kraft Pulp: Feasibility Study, Ontario Ministry of the Environment, and Environment Canada, June 1993.
72. Deal, H., Environmental Pressure Causes Changes in Bleaching Technologies, Chemicals, *Pulp & Paper*, November 1991.
73. Fleming, B., Alternative and Emerging Non-Kraft Pulping Technologies, EPA-744R-93-002, U.S. EPA, Washington, 1993.
74. NCASI Technical Workshop, Effects of Alternative Pulping and Bleaching Processes on Production and Biotreatability of Chlorinated Organics, NCASI Special Report 94-01, February 1994.
75. U.S. EPA, Pollution Prevention Technologies for the Bleached Kraft Segment of the U.S. Pulp and Paper Industry, EPA/600/R-93/110, U.S. EPA, Washington, 1993.
76. U.S. EPA, Fact Sheet: EPA's Final Pulp, Paper, and Paperboard "Cluster Rule–Overview", U.S. EPA, Washington, November 1997.
77. U.S. EPA, Proposed Technical Development Document for the Pulp, Paper and Paper board Category and Effluent Limitations Guidelines, Pretreatment Standards, and New Source Performance Standard, U.S. EPA, Washington, October 1993.
78. U.S. EPA, Supplemental Technical Development Document for Effluent Limitations Guidelines and Standards for the Pulp, Paper, and Paperboard Category Subpart B (Bleached Papergrade Kraft and Soda) and Subpart E (Papergrade Sulfite), U.S. EPA, Document Control 14487, Washington, October 15, 1997.
79. U.S. EPA, Data Available for Limitations Development for Toxics and Nonconventional Pollutants, U.S. EPA, Washington, November 12, 1997.
80. Eastern Research Group, Final Analysis of Data Available for Development of COD Limitations, prepared for U.S. EPA, August 25, 1997.
81. U.S. EPA, International Paper XL-2 Effluent Improvements Project, Final Report, U.S. EPA, Maine Department of Environmental Protection, Jay, Maine, September 6, 2005.
82. Shammas, N.K., Wastewater renovation by flotation, in *Flotation Technology*, Wang, L.K., Shammas, N.K., Selke, W.A., and Aulenbach, D.B., Eds., Humana Press, Totowa, NJ, 2008.
83. Krofta, M. and Wang, L.K., Flotation technology and secondary clarification, *TAPPI J.*, 70, 92–96, April 1987.

84. Krofta, M. and Wang, L.K., Development of a total closed water system for a deinking plant, *Proc. Am. Water Works Assoc., Water Reuse Symposium III*, 2, 881–898, 1984.

85. Krofta, M. and Wang, L.K., Pollution abatement using dissolved air flotation technology in the paper and pulp industry, in *Proc. 1985 Powder and Bulk Solids Conference*, Chicago, IL, May 1985, p. 28.

86. Krofta, M. and Wang, L.K., Total closing of paper mills with reclamation and deinking installations, in *Proc. 43rd Industrial Waste Conference*, Purdue University, West Lafayette, IN, 1989, p. 673.

87. Daly, P.G. and Shen, C.C., The deep shaft biological treatment process, in *Proceedings of the 43rd Purdue Industrial Waste Conference*, May 1988.

88. Wang, L.K., Yoo, S.H., and Hong, Y.N., Development of Two-Stage Physical-Chemical Process System for Treatment of Pulp Mill Wastewater, Lenox Institute of Water Technology, Lenox, MA, Technical Report LIR/05-84/2, 1984, 64 p.

89. Krofta, M. and Wang, L.K., Total Waste Recycle System for a Wastewater Treatment System Using Pulp Mill Chemicals Containing Aluminum and Calcium, Lenox Institute of Water Technology, Lenox, MA, Technical Report LIR/10-84/, 1984, 524 p.

90. Krofta, M. and Wang, L.K., Design of Dissolved Air Flotation systems for Industrial Pretreatment and Municipal Water and Wastewater Treatment, *National Meeting of American Institute of Chemical Engineers*, Houston, Texas, March 1983.

91. Krofta, M. and Wang, L.K., Wastewater treatment by biological-physicochemical two-stage process system, *Proc. 41st Purdue Industrial Waste Conference*, Lewis Publishers, Chelsea, MI, May 1986, p. 67–72.

12 Treatment of Nickel-Chromium Plating Waste

Nazih K. Shammas, Lawrence K. Wang,
Donald B. Aulenbach, and William A. Selke

CONTENTS

12.1 INTRODUCTION

Applicable local, state, and federal environmental laws require that the waste generated by the nickel-chromium plating process be pretreated to provide a discharge acceptable to the public wastewater treatment system.

  The specific purpose of this chapter is to describe the chemical and physical pretreatment methods required for nickel-chromium plating wastewater, to describe the upgrades needed by a municipal wastewater treatment system to manage this waste, and to relate the methods and upgrades to the operation of the total treatment system. Special emphasis is placed on presentation of the following:

1. The chemistry of nickel-chromium plating and waste generation
2. The type of pollutants and their sources
3. Waste minimization
4. Recovery and recycling
5. Conventional reduction–precipitation treatment systems
6. Modified reduction–flotation treatment systems
7. Innovative flotation–filtration treatment systems

12.2 THE NICKEL-CHROMIUM PLATING PROCESS

The nickel-chromium plating process includes the steps in which a ferrous base material is electroplated with nickel and chromium. The electroplating operations for plating the two metals are basically oxidation–reduction reactions. Typically, the part to be plated is the cathode, and the plating metal is the anode.

12.2.1 NICKEL PLATING

To plate nickel on iron parts, the iron parts form the cathodes, and the anode is a nickel bar. On the application of an electric current, the nickel bar anode oxidizes, dissolving in the electrolyte:

$$Ni \longrightarrow Ni^{2+} + 2e^- \tag{12.1}$$

The resulting nickel ions are reduced at the cathode (the iron part) to form a nickel plate:

$$Ni^{2+} + 2e^- \longrightarrow Ni \tag{12.2}$$

  Nickel plating can also be accomplished by an electroless plating technique involving deposition of a metallic coating by a controlled chemical reduction that is catalyzed by the metal or alloy being deposited. A special feature of electroless plating is that no external electrical energy is required. The following are the basic ingredients in electroless plating solutions:

1. A source of metal, usually a salt
2. A reducer to reduce the metal to its base state
3. A chelating agent to hold the metal in solution so the metal will not plate out indiscriminately
4. Various buffers and other chemicals designed to maintain stability and increase bath life

  Nickel electroless plating on a less noble metal is common.[1–7] For example, the source of nickel can be nickel sulfate. The reducer can be an organic substance, such as formaldehyde. A chelating agent (tartrate or equivalent) is generally required. The nickel salt is ionized in water:

$$NiSO_4 \longrightarrow Ni^{2+} + SO_4^{2-} \tag{12.3}$$

There is then a redox reaction with the nickel and the formaldehyde:

$$Ni^{2+} + 2H_2CO + 4OH^- \longrightarrow Ni + 2HCO_2^- + 2H_2O + H_2 \tag{12.4}$$

The base metal nickel now begins to plate out on an appropriate surface, such as a less noble metal.

12.2.2 CHROMIUM PLATING

In chromium plating, the chromium is supplied to the plating baths as chromic acid. For example, plating baths can be prepared by adding hexavalent chromium in the form of either sodium dichromate ($Na_2Cr_2O_7 \bullet H_2O$) or chromium trioxide (CrO_3). When sodium dichromate is used it dissociates to produce the divalent dichromate ion ($Cr_2O_7^{2-}$). When chromium trioxide is used, it immediately dissolves in water to form chromic acid according to the following reaction[8–15]:

$$CrO_3 + H_2O \longrightarrow H_2CrO_4 \tag{12.5}$$

Chromic acid is considered a strong acid, although it never completely ionizes. Its ionization has been described as follows:

$$H_2CrO_4 \longrightarrow H^+ + HCrO_4^- \text{ (acid chromate ion)} \tag{12.6}$$
$$K_a = 0.83 \text{ at } 25°C$$

$$HCrO_4^- \longrightarrow H + CrO_4^{2-} \text{ (chromate ion)} \tag{12.7}$$
$$K_a = 3.2 \times 10^{-7} \text{ at } 25°C$$

Moreover, the dichromate ion ($Cr_2O_7^{2-}$) will exist in equilibrium with the acid chromate ion as follows:

$$Cr_2O_7^{2-} + H_2O \longrightarrow 2HCrO_4^- \tag{12.8}$$
$$K_a = 0.0302 \text{ at } 25°C$$

Theoretically, $HCrO_4^-$ is the predominant species between pH 1.5 and 4.0, $HCrO_4^-$ and CrO_4^{2-} exist in equal amounts at pH 6.5, and CrO_4^{2-} predominates at higher pH values. Chromium plating wastewater is generally somewhat acid, and the acid chromate ion $HCrO^-$ is predominant in this wastewater.

Chromating is one of the chemical conversion coating technologies. Chrome coatings are applied to previously deposited nickel for increased corrosion protection and to improve surface appearance. Chromate conversion coatings are formed by immersing the metal in an aqueous acidified chromate solution consisting substantially of chromic acid or water-soluble salts of chromic acid, together with various catalysts or activators.

12.3 SOURCES OF POLLUTION

A conceptual arrangement of the nickel-chromium plating process can be broken down into three general steps:

1. Surface preparation involving the conditioning of the base material for plating
2. Actual application of the plate by electroplating
3. The posttreatment steps

The major waste sources during normal nickel-chromium plating operations are alkaline cleaners, acid cleaners, plating baths, posttreatment baths, and auxiliary operation units.

The wastestreams generated by the plating process can be subdivided and classified into eight categories[1,5,6,15]:

1. Concentrated acid wastes
2. Concentrated phosphate cleaner wastes
3. Acid rinsewater
4. Alkaline rinsewater
5. Chromium rinsewater
6. Nickel rinsewater
7. Concentrated nickel wastes
8. Concentrated chromium wastes

In the above categories, there are seven major types of aqueous pollutants that must be pretreated and removed[5,15]:

1. Acidity
2. Alkalinity
3. Nickel
4. Chromium
5. Iron
6. Organics (COD, BOD)
7. Suspended solids

The environmental impact of the two most toxic pollutants, nickel and chromium, is briefly presented in the following[1,16,17]. Significant concentrations of these elements pass through conventional treatment plants.

12.3.1   Environmental Impact of Nickel

Nickel is toxic to aquatic organisms at levels typically observed in POTW (publicly owned treatment works) effluents:

1. 50% reproductive impairment of *Daphnia magna* at 0.095 mg/L
2. Morphological abnormalities in developing eggs of *Limnaea palustris* at 0.230 mg/L
3. 50% growth inhibition of aquatic bacteria at 0.020 mg/L

Because surface water is often used as a drinking water source, nickel passed through a POTW becomes a possible drinking water contaminant.

A U.S. Environmental Protection Agency (U.S. EPA) study of 165 sludges showed nickel concentrations ranging from 2 to 3520 mg/kg (dry basis).[18] Nickel toxicity may develop in plants from application of municipal wastewater biosolids on acid soils. Nickel reduces yields for a variety of crops including oats, mustard, turnips, and cabbage.

12.3.2   Environmental Impact of Chromium

Chromium can exist as either trivalent or hexavalent compounds in raw wastewater streams. The chromium that passes through the POTW is discharged to ambient surface water. Chromium is toxic to aquatic organisms at levels observed in POTW effluents[15]:

1. Trivalent chromium significantly impaired the reproduction of *Daphnia magna* at levels of 0.3 to 0.5 mg/L
2. Hexavalent chromium retards growth of chinook salmon at 0.0002 mg/L. Hexavalent chromium is also corrosive and a potent human skin sensitizer.

   Besides providing an environment for aquatic organisms, surface water is often used as a source
of drinking water. The National Primary Drinking Water Standards are based on total chromium,
the limit being 0.1 mg/L.[19]

   A U.S. EPA study of 180 municipal wastewater sludges showed that municipal wastewater
sludge contains 10 to 99,000 mg/kg (dry basis) of chromium. Most crops absorb relatively little
chromium even when it is present in high levels in soils, but chromium in sludge has been shown to
reduce crop yields in concentrations as low as 200 mg/kg.[18]

12.4   WASTE MINIMIZATION

All metal finishing facilities have one thing in common—the generation of metal-containing hazardous waste from the production processes. Reducing the volume of waste generated can save
money and at the same time decreases future liabilities. Typical wastes generated are as follows:

1. Industrial wastewater and treatment residues
2. Spent plating baths
3. Spent process baths
4. Spent cleaners
5. Waste solvents and oil

   This section identifies areas for reducing waste generation. It also suggests techniques available
to metal finishers for waste reduction and is intended to help metal finishing shop owners decide
whether waste reduction is a possibility.

   Both state (Health and Safety Code) and federal (40 CFR, Part 262, Subpart D) regulations
require that generators of hazardous waste file a biennial generator's report. Among other things,
this report must include a description of the efforts undertaken and achievements accomplished
during the reporting period to reduce the volume and toxicity of waste generated. The Uniform
Hazardous Waste Manifest requires that large generators certify that they have a program in place
to reduce the volume and toxicity of waste generated that is determined to be economically practicable. Small-quantity generators must certify that they have made a good faith effort to minimize
waste generation and have selected the best affordable waste management method available.

   As waste reduction methods reduce the amount of waste generated, and also the amount subject
to regulation, these practices can help a shop comply with the requirements while also saving money.
The shop's owner or manager must be committed to waste reduction and pass that commitment on
to the employees, establish training for employees in waste reduction, hazardous material handling
and emergency response, and establish incentive programs to encourage employees to design and
use new waste reduction ideas. The following is a list of some common waste reduction methods
for metal finishing electroplating shops.[20,21]

12.4.1   ASSESSMENT OF HAZARDOUS WASTE

Waste assessments are used to list the sources, types, and amounts of hazardous waste generated to
make it easier to pinpoint where wastes can be reduced.

   Source reduction is usually the least expensive approach to minimizing waste. Many of these
techniques involve housekeeping changes or minor inplant process modifications.

12.4.2   IMPROVED PROCEDURES AND SEGREGATION OF WASTES

These may be summarized as follows:

1. Good housekeeping is the easiest and often the cheapest way to reduce waste. Keep work
   areas clean.

2. Improve inventory procedures to reduce the amount of off-specification materials generated.
3. Reduce quantities of raw materials to levels where materials will be used up just as new materials are arriving.
4. Designate protected raw material and hazardous waste storage areas with spill containment. Keep the areas clean and organized and give one person the responsibility for maintaining the areas.
5. Label containers as required and cover them to prevent contact with rainfall and avoid spills.
6. Use a "first-in, first-out" policy for raw materials to keep them from becoming too old to be used. Give one person responsibility for maintaining and distributing raw materials.
7. Use bench-scale testing for samples rather than process baths.
8. Designate one person to accept chemical samples and return unused samples to suppliers.
9. Limit bath mixing to trained personnel.
10. Segregate wastestreams for recycling and treatment, and keep nonhazardous material from becoming contaminated.
11. Prevent and contain spills and leaks by installing drip trays and splash guards around processing equipment.
12. Conduct periodic inspections of tanks, tank liners, and other equipment to avoid failures. Repair malfunctions when they are discovered. Use inspection logs to follow up on repairs.
13. Inspect plating racks for loose insulation that would cause increased dragout.
14. Use dry cleanup where possible to reduce the volume of wastewater.

12.4.3 MATERIAL SUBSTITUTION

In summary:

1. Use process chemistries that are treatable or recyclable on site.
2. Use deionized water instead of tap water in process baths or rinsing operations to reduce chemical reactions with impurities in the tap water, which would increase sludge production.
3. Use nonchelated process chemistries rather than chelated chemistries to reduce sludge volume.
4. Replace cyanide process baths with noncyanide process baths to simplify the treatment required.
5. Use alkaline cleaners instead of solvents for degreasing operations; they can be treated on site and usually discharged to the sewer with permit authorization.

12.4.4 EXTENDING PROCESS BATH LIFE

This may be achieved with the following procedures:

1. Treatment of process baths can extend their useful life.
2. Bath replenishment extends the useful life of the bath.
3. Monitoring (using pH meters or conductivity meters) the process baths can determine the need for bath replenishment.

12.4.5 DRAGOUT REDUCTION

Dragout reduction is achieved using the following steps:

1. Minimize bath concentrations to the lower end of their operating range.
2. Maximize bath operating temperatures to lower the solution's viscosity.
3. Use wetting agents (which reduce the surface tension of the solution) in process baths to decrease the amount of dragout.
4. Withdraw workpieces from tanks slowly to allow maximum drainage back into process tank.

5. Use air knives or spray rinses above process tanks to rinse excess solution off a workpiece and into the process bath.
6. Install drainage boards between process tanks and rinse tanks to direct dragout back into process tank.
7. Use dedicated dragout tanks after process baths to capture dragout.
8. Install rails above process tanks to hang workpiece racks for drainage prior to rinsing.
9. Use spray rinses as the initial rinse after the process tank and before the dip tank.
10. Use air agitation or workpiece agitation to improve rinse efficiency.
11. Install multiple rinse tanks (including counterflow rinse tanks) after process baths to improve rinse efficiency and reduce water consumption.

12.4.6 REACTIVE RINSES

The following steps should be applied:

1. Reuse the acid rinse effluent as influent for the alkaline rinse tank, thus allowing the fresh water feed to the alkaline rinse tank to be turned off (reactive rinsing). This can also be applied to process tank rinses.
2. Treat rinsewater effluent to recover process bath chemicals. This allows the reuse of the effluent for rinsing or neutralization prior to discharge.
3. Reuse the spent reagents from the process baths in the wastewater treatment process.
4. Recycle spent solvents on site or off site.
5. Use treatment technologies to recycle rinsewaters in a closed loop or open loop system.
6. Some recycling and most treatment processes require a permit. Be sure to contact the local Department of Health Services regional office to determine if there is a need for a permit to treat or recycle the wastes.
7. Pretreat process water to reduce the natural contaminants that contribute to the sludge volume.
8. Use treatment chemicals that reduce sludge generation (e.g., caustic soda instead of lime).
9. Use sludge dewatering equipment to reduce sludge volume.
10. Use treatment technologies (such as ion exchange, evaporation, and electrolytic metal recovery) that do not use standard precipitation/clarification methods that generate heavy metal sludges.

12.5 MATERIAL RECOVERY AND RECYCLING

Unlike the 1970s and 1980s when waste management costs were relatively inexpensive, today's metal finishers are facing increasingly higher disposal costs. This change is due in part to a decrease in the volume of available landfill space, which has resulted in escalating landfill fees and more stringent federal and state environmental regulations that mandate treatment prior to landfilling.

Metal finishers are seeing their profits shrink as waste management costs increase. To control waste disposal costs, metal finishers must focus on developing and implementing a facility-wide waste reduction program. In other words, as discussed in Section 12.4, metal finishers must consciously seek out ways to decrease the volume of waste that they generate.

One approach to waste reduction is to recover process materials for reuse. Materials used in metal finishing processes can be effectively recovered using available technologies such as dragout, evaporation, reverse osmosis, ion exchange, electrodialysis, and electrolytic recovery.[22–26]

12.5.1 DRAGOUT RECOVERY

Dragout recovery is a simple technology used by metal finishers to recover plating chemicals. It involves using drain boards, drip tanks, fog-spray tanks, or dragout tanks separately, or in combination, to capture plating chemicals dragged out of plating tanks from parts being plated.

Drain boards are widely used throughout the metal industry to capture plating solutions. Boards are suspended between process tanks and are constructed of plastic, plain or teflon-coated steel. Solutions drip on the boards and drain back into their respective processing tanks.[22,27]

In contrast, a drip tank recovers process chemicals by collecting dragout into a separate tank, from which it can be returned to the process as needed.

In a fog-spray tank, plating chemicals clinging to parts are recovered by washing them with a fine water-mist. The solution that collects in the fog-spray tank is returned to the process tank as needed. The added water helps to offset evaporative losses from the process tanks.

Dragout tanks are essentially rinse tanks. Dragout chemicals are captured in a water solution, which is returned to the process tank as needed.

The presence of airborne particles and other contaminants in recovered plating chemicals may necessitate treatment of the collected solution to remove the contaminants prior to solution reuse.

There are advantages and disadvantages to dragout recovery. Depending upon the solution, up to 60% of the materials carried out of a plating tank can be recovered for reuse; thus dragout can affect metal deposition and surface finish quality. Impurities can concentrate in the solutions causing a deteriorating effect on the plating process when returned to the plating bath.

12.5.2   Evaporative Recovery

A widely used metal salt recovery technique is evaporation. With evaporation, plating chemicals are concentrated by evaporating water from the solution. Evaporators may use heat or natural evaporation to remove water.[22,28] Additionally, evaporators may operate at atmospheric pressure or under vacuum.

Atmospheric evaporators are more commonly used. They are open systems that use process heat and warm air to evaporate water. These evaporators are relatively inexpensive, require low maintenance and are self-operating. Under the right conditions, they can evaporate water from virtually any plating bath or rinse. A packed-bed evaporator is an example of an atmospheric evaporator.

Vacuum evaporators are also used to recover plating chemicals. They are closed systems that use steam heat to evaporate water under a vacuum. This results in lower boiling temperature, with a reduction in thermal degradation of the solution. Like atmospheric evaporators, they require low maintenance and are self-operating. A climbing file evaporator is an example of a vacuum evaporator.

A typical evaporative recovery system consists of an evaporator, a feed pump, and a heat exchanger. Plating solution or rinsewater containing dilute plating chemicals is circulated through the evaporator. The water evaporates and concentrates the plating chemicals for reuse. In open evaporator systems, the water evaporates and mixes with air and is released to the atmosphere. It may be necessary to vent the contaminated airstream to a ventilation/scrubber treatment system prior to release. In enclosed evaporators the water is condensed from the air and can be reused in rinses, which further increases savings. Water reuse is preferred whenever possible.

As with all process equipment, the design size of an evaporator system is dependent upon volumetric flow, specifically the rinsewater flow rate required and the volume of process solution dragout. When operated properly, a commercial evaporator can attain a 99% material recovery rate.

There are drawbacks to using an evaporator to recover plating chemicals. For instance, impurities are concentrated along with recovered plating chemicals. These impurities can alter desired deposited metal characteristics, including surface finish quality. Vacuum evaporation can be used to avoid degradation of plating solutions containing additives that are sensitive to heat.

The evaporative recovery is a very energy-intensive process. Approximately 538 chu (970 Btu) are required to evaporate 1 lb of water at standard atmospheric pressure. Additional energy is required to raise the temperature of the solution to its boiling point.

12.5.3   Reverse Osmosis

Reverse osmosis (RO) recovers plating chemicals from plating rinsewater by removing water molecules with a semipermeable membrane. The membrane allows water molecules to pass through, but blocks metallic salts and additives.[29]

Like evaporators, RO works on most plating baths and rinse tanks. Most RO systems consist of a housing that contains a membrane and feed pump. There are four basic membrane designs: plate-and-frame, spiral-wound, tubular, and hollow-fiber. The most common types of membrane materials are cellulose acetate, polyether/amide, and polysulfones.[29]

Diluted or concentrated rinsewaters are circulated through the membrane at pressures greater than aqueous osmotic pressure. This action results in the separation of water from the plating chemicals. The recovered chemicals can be returned to the plating bath for reuse, and the permeate, which is similar to the condensate from an evaporator, can be used as make-up water. RO units work best on dilute solutions.[30]

The design and capacity of an RO unit is dependent upon the type of chemicals in the plating solution and the dragout solution rate. Certain chemicals require specific membranes. For instance, polyamide membranes work best on zinc chloride and nickel baths, and polyether/amide membranes are suggested for chromic acid and acid copper solutions. The flow rate across the membrane is very important. It should be set at a rate to obtain maximum product recovery. RO systems have a 95% recovery rate with some materials and with optimum membrane selection.[22]

There are advantages to using RO. Energy usage is much lower than for other recovery systems and plating chemicals can be recovered from temperature-sensitive solutions. However, RO also has limitations. The membrane is susceptible to fouling, which is often caused by the precipitation of suspended and dissolved solids that plug the membrane's pores. Also, as with evaporators, RO can concentrate impurities along with plating chemicals, which degrade plating quality.

12.5.4 Ion Exchange

Ion exchange is a molecular exchange process where metal ions in solution are removed by a chemical substitution reaction with an ion-exchange resin.[31] Ion exchange can be used with most plating baths. Metal cations exchange sites with sodium or hydrogen ions and anions (such as chromate) with hydroxyl ions. The exchange resin can generally be regenerated with an acid or alkaline solution and reused. When a cation exchange resin is regenerated, it produces a metal salt. For example, copper is removed from an ion exchange resin by passing sulfuric acid over the resin, producing copper sulfate. This salt can be added directly into the plating bath.[23,32]

The required size of an ion-exchange unit is dependent upon the composition and volume of plating dragout. Each ion-exchange resin has a maximum capacity for recovery of specific ions. The ion-exchange unit's size (volume of resin) is determined by the amount of metal to be removed from the recovered solutions.

Ion exchange has its drawbacks. Most commercially available resins are nonselective and, therefore, similarly charged ions can be exchanged by a given resin whether desired in the process or not. This means that certain contaminants cannot be removed by ion exchange and are returned to the plating tank with the metal salt.[22] The metal salt solution produced after regeneration is often a dilute solution that can only be put back into the process bath if evaporation is used to make room in the process tank. In addition, ion exchange is not a continuous process and system sizing must take into account resin regeneration time.

12.5.5 Electrodialysis

Electrodialysis units recover plating chemicals differently from the recovery units discussed thus far. In electrodialysis, electromotive forces selectively drive metal ions through an ion-selective membrane (in RO, pressure is the driving force; in ion exchange, the driving force is chemical attraction). The membranes are thin sheets of plastic material with either anionic or cationic characteristics.[33]

Electrodialysis units are constructed using a plate-and-frame technique similar to filter presses. Alternating sheets of anionic and cationic membranes are placed between two electrodes. The plating or rinse solution to be recovered (electrolyte) circulates past the system's electrodes. Hydrogen and oxygen evolve. Positive ions travel to the negative terminal and negative ions travel to the positive terminal. The electrolyte also provides overall electrical conductivity to the cell. In some units, the current is periodically reversed to reduce membrane fouling.

Electrodialysis is compatible with most plating baths, and the design size of a unit is dependent upon the rinsewater flow rate and concentration.[22]

Electrodialysis has advantages and disadvantages. For instance, the process requires very little energy and can recover highly concentrated solutions. On the other hand, similarly to other membrane processes, electrodialysis membranes are susceptible to fouling and must be regularly replaced.

12.5.6 ELECTROLYTIC RECOVERY

Electrolytic recovery (ER) is the oldest metal recovery technique. Metal ions are plated-out of solution electrochemically by reduction at the cathode.[34] There are essentially two types of cathodes used for this purpose: a conventional metal cathode and a high surface area cathode (HSAC). Both cathodes can effectively plate-out metals, such as gold, zinc, cadmium, copper, and nickel.[22]

Electrolytic recovery systems work best on concentrated solutions. For optimal plating efficiency, recovery tanks should be agitated ensuring that good mass transfer occurs at the electrodes. Another important factor to consider is the anode/cathode ratio. The cathode area (plating surface area) and mass transfer rate to the cathode greatly influence the efficiency of metal deposition.

Electrolytic recovery can be used with most plating baths. The amount of metal to be plated per square meter of cathode determines the electrolytic recovery unit's design capacity. Therefore, the volume and concentration of plating dragout greatly influences system design and size.[22,35]

There are advantages to the electrolytic recovery process. For instance, ER units can operate continuously, and the product is in a metallic form that is very suitable for reuse or resale. Electrolytic units are also mechanically reliable and self-operating. Very importantly, contaminants are not recovered and returned to the plating bath. Thus, electrolytically recovered metals are as pure as "virgin" plating raw material.

The major disadvantage to electrolytic recovery is high energy cost. Energy costs will vary, of course, with cathode efficiencies and local utility rates.[22]

12.5.7 DEIONIZED WATER

Using deionized water to prepare plating bath solutions is an effective way of preventing waste generation. Some groundwater and surface waters contain high concentrations of calcium, magnesium, chloride, and other soluble contaminants that may build up in process baths.[22] By using deionized water, buildup of these contaminants can be more easily controlled. Technologies such as RO and ion exchange can also be used to effectively remove soluble contaminants from incoming water.[36]

12.6 CHEMICAL TREATMENT

Treatment for the removal of chromium and nickel from electroplating wastewater involves neutralization, hexavalent chromium reduction, pH adjustment, hydroxide precipitation, and final solid–liquid separation.[15,37–48]

12.6.1 NEUTRALIZATION

Excess acidity and alkalinity may be eliminated by simple neutralization by either a base or an acid. This is a simple stoichiometric chemical reaction of the following type[5,15,49]:

$$\text{Strong base} + \text{strong acid} \longrightarrow \text{salt} + \text{water} \tag{12.9}$$

Examples of this include the following:

1. Alkali

$$NaOH + H_2SO_4 \longrightarrow Na_2SO_4 + 2\ HOH \qquad (12.10)$$
$$\text{Base} \qquad \text{Acid} \qquad \text{Salt} \qquad \text{Water}$$

2. Acid

$$HNO_3 + NaOH \longrightarrow NaNO_3 + HOH \qquad (12.11)$$
$$\text{Acid} \qquad \text{Base} \qquad \text{Salt} \qquad \text{Water}$$

A slight excess of base may be titrated in the previous reactions to shift the pH to a slight basic condition. This is important for the precipitation of certain metal salts (such as nickel, iron, and trivalent chromium) as hydroxides.

12.6.2 HEXAVALENT CHROMIUM REDUCTION

Chemical treatment of chromium wastewater is usually conducted in two steps. In the first step hexavalent chromium is reduced to trivalent chromium by the use of a chemical reducing agent. The trivalent chromium is precipitated during the second stage of treatment.[15]

Sulfur dioxide (SO_2), sodium bisulfite ($NaHSO_3$), and sodium metabisulfite ($Na_2S_2O_5$) are commonly used as reducing agents.[15,50] All these compounds react to produce sulfurous acid when added to water, according to the following reactions:

$$SO_2 + H_2O \longrightarrow H_2SO_3 \qquad (12.12)$$
$$Na_2S_2O_5 + H_2O \longrightarrow 2\ NaHSO_3 \qquad (12.13)$$
$$NaHSO_3 + H_2O \longrightarrow H_2SO_3 + NaOH \qquad (12.14)$$

It is the sulfurous acid produced from these reactions that is responsible for the reduction of hexavalent chromium. The reaction is shown in the following equation:

$$2\ H_2CrO_4 + 3\ H_2SO_3 \longrightarrow Cr_2(SO_4)_3 + 5\ H_2O \qquad (12.15)$$

The typical amber color of the hexavalent chromium solution will turn to a pale green once the chromium has been reduced to the trivalent state. Although this color change is a good indicator, redox control is usually employed.

The theoretical amount of sulfurous acid required to reduce a given amount of chromium can be calculated from the above equation. The actual amount of sulfurous acid required to treat a wastewater will be greater than this because other compounds and ions present in the wastewater may consume some of the acid. Primary among these is dissolved oxygen, which oxidizes sulfurous acid to sulfuric acid according to the following reaction:

$$H_2SO_3 + 0.5\ O_2 \longrightarrow H_2SO_4 \qquad (12.16)$$

Each part of dissolved oxygen initially present in the wastewater produces 6.1 parts of sulfuric acid.

Undissociated sulfurous acid is responsible for the reduction of hexavalent chromium. Consequently, the reduction reaction is strongly pH-dependent because of the effect of pH on acid dissociation:

$$H_2SO_3 \longrightarrow H^+ + HSO_3^- \qquad (12.17)$$
$$K_a = 1.72 \times 10^{-2}\ \text{at } 25°C$$

$$HSO_3^- \longrightarrow H^+ + SO_3^{2-} \qquad (12.18)$$
$$K_a = 1.0 \times 10^{-7} \text{ at } 25°C$$

The dissociation as a function of pH and the effect of pH on reaction rate is shown in Figure 12.1 and Figure 12.2, respectively.[15] Obviously, the reaction proceeds much faster at low pH values, where the concentration of undissociated sulfurous acid is highest. As a result, chromium reduction processes are generally conducted at pH values of 2 to 3 to maximize reaction rates and minimize the volume of reaction vessels. Sulfuric acid is generally added to reduce the pH of the wastewater to the desired level and to maintain it at that level throughout treatment. If the pH is not maintained at the desired level but is allowed to increase during treatment, the reaction may not go to completion in the retention time available, and unreduced hexavalent chromium may exist in the effluent. The amount of acid required to depress the pH to the level selected for chrome reduction will depend on the alkalinity of the wastewater being treated. This acid requirement can be determined by titrating a sample of wastewater with sulfuric acid to the desired pH in the absence of a reducing agent.

In addition to the sulfuric acid required for pH adjustment, some amount of acid is consumed by the reduction reaction (Equation 8.15). If sulfur dioxide is used as the reducing agent, it will provide all the acid consumed by this reaction, and additional acid will not be required. However, if sodium bisulfite or sodium metabisulfite is used, additional acid must be supplied to satisfy the acid demand. This acid requirement is stoichiometric and can be calculated from Equations 12.19 to 12.22.

At pH 3.0 to 4.0:

$$3\,NaHSO_3 + 1.5\,H_2SO_4 + 2\,H_2CrO_4 \longrightarrow Cr_2(SO_4)_3 + 1.5\,NaSO_4 + 5\,H_2O \qquad (12.19)$$
$$1.5\,Na_2S_2O_5 + 1.5\,H_2SO_4 + 2\,H_2CrO_4 \longrightarrow Cr_2(SO_4)_3 + 1.5\,Na_2SO_4 + 3.5\,H_2O \qquad (12.20)$$

At pH 2.0:

$$3\,NaHSO_3 + 2\,H_2SO_4 + 2\,H_2CrO_4 \longrightarrow Cr_2(SO_4)_3 + Na_2SO_4 + NaHSO_4 + 51\,H_2O \qquad (12.21)$$
$$1.5\,Na_2S_2O_5 + 2\,H_2SO_4 + 2\,H_2CrO_4 \longrightarrow Cr_2(SO_4)_3 + Na_2SO_4 + NaHSO_4 + 3.5\,H_2O \qquad (12.22)$$

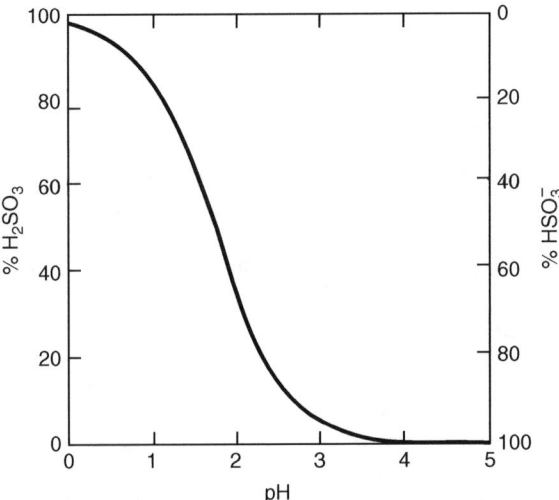

FIGURE 12.1 Relationship between H_2SO_3 and HSO_3^- at various pH values. (Taken from Krofta, M. and Wang, L.K., *Design of Innovative Flotation–Filtration Wastewater Treatment Systems for a Nickel-Chromium Plating Plant*, U.S. Department of Commerce, National Technical Information Service, Springfield, VA, Technical Report PB-88-200522/AS, January 1984. *Source*: U.S. EPA.)

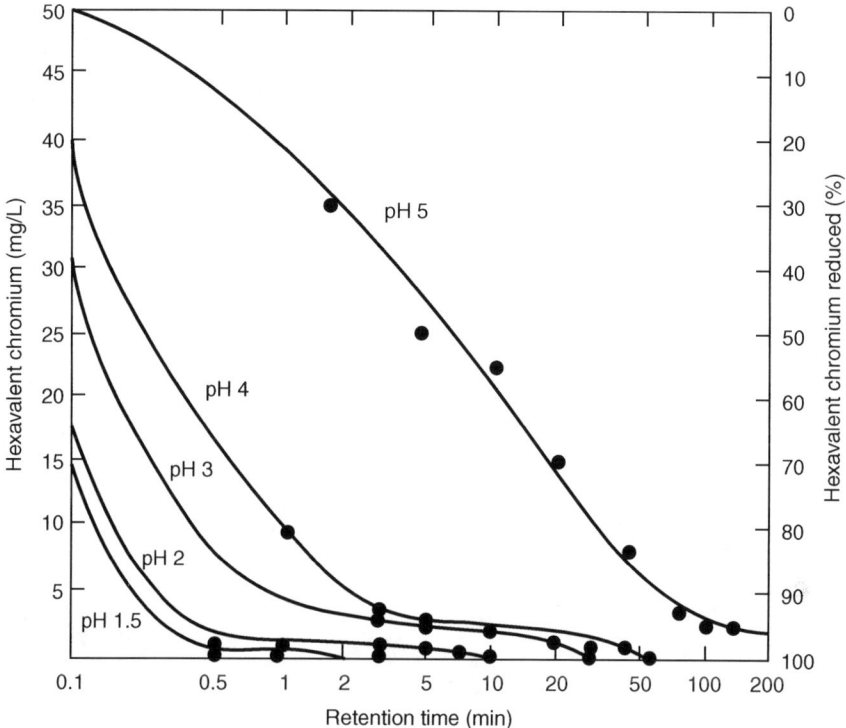

FIGURE 12.2   Rate of reduction of hexavalent chromium in the presence of excess SO_2 at various pH levels. (Taken from Krofta, M. and Wang, L.K., *Design of Innovative Flotation–Filtration Wastewater Treatment Systems for a Nickel-Chromium Plating Plant*, U.S. Department of Commerce, National Technical Information Service, Springfield, VA, Technical Report PB-88-200522/AS, January 1984. *Source*: U.S. EPA.)

Similar equations can be developed for pH values between 2 and 3 as a function of the SO_4^{2-} and HSO_4^- distribution.

12.6.3   pH Adjustment and Hydroxide Precipitation

Wastewater pH is adjusted by addition of an acid or an alkali, depending on the purpose of the adjustment. The most common purposes of wastewater pH adjustment are the following:

1. Chemical precipitation of dissolved heavy metals, as illustrated by Figure 12.3
2. Pretreatment of metal-bearing wastewater before sulfide precipitation so that the formation of hazardous gaseous hydrogen sulfide does not occur
3. Neutralization of wastewater before discharge to either a stream or a sanitary sewer[37-48]

To accomplish hydroxide precipitation, an alkaline substance such as lime or sodium hydroxide is added to the wastewater to increase the pH to the optimum range of minimum solubility at which the metal precipitates as a hydroxide[51]:

$$M(II)^{2+} + Ca(OH)_2 \longrightarrow M(II)(OH)_2 + Ca^{2+} \tag{12.23}$$
$$2\,M(III)^{3+} + 3\,Ca(OH)_2 \longrightarrow 2\,M(III)(OH)_3 + 3\,Ca^{2+} \tag{12.24}$$

where M(II) = divalent metal and M(III) = trivalent metal.

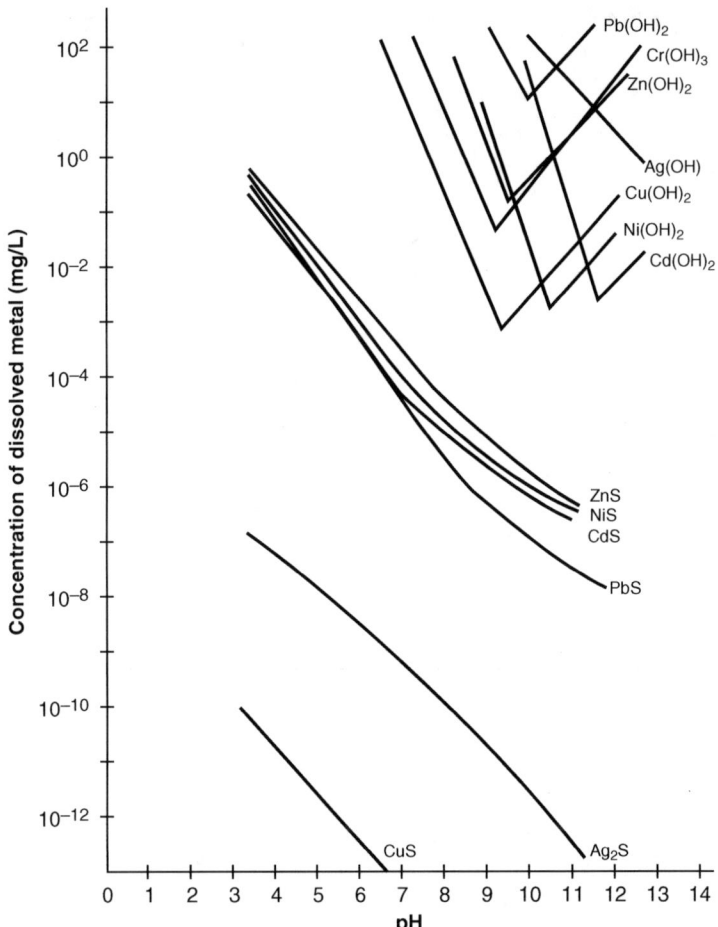

FIGURE 12.3 Solubility of metal hydroxides and sulfides. (Taken from Krofta, M. and Wang, L.K., *Design of Innovative Flotation–Filtration Wastewater Treatment Systems for a Nickel-Chromium Plating Plant*, U.S. Department of Commerce, National Technical Information Service, Springfield, VA, Technical Report PB-88-200522/AS, January 1984. *Source*: U.S. EPA.)

The precipitated metal hydroxide can then be removed from the wastewater by clarification or other solid–water separation techniques.[52]

As a practical example, following the reduction of hexavalent chromium, sodium hydroxide, lime, or sodium hydroxide can be added to the wastewater to neutralize the pH and precipitate the trivalent chromium, nickel, iron, divalent, and other heavy metals. If lime is used, lime will react with heavy metals and with any residual sodium sulfate, sulfurous acid, or sodium bisulfite. The following reactions apply:

$$NiCl_2 + Ca(OH)_2 \longrightarrow Ni(OH)_2 + CaCl_2 \tag{12.25}$$

$$NiSO_4 + Ca(OH)_2 \longrightarrow Ni(OH)_2 + CaSO_4 \tag{12.26}$$

$$2\ Fe_2(SO_4)_3 + 6\ Ca(OH)_2 \longrightarrow 4\ Fe(OH)_3 + 6\ CaSO_4 \tag{12.27}$$

$$Cr_2(SO_4)_3 + 3\ Ca(OH)_2 \longrightarrow 2\ Cr(OH)_3 + 3\ CaSO_4 \tag{12.28}$$

$$H_2SO_4 + Ca(OH)_2 \longrightarrow CaSO_4 + 2\ H_2O \tag{12.29}$$

$$2\ NaHSO_4 + Ca(OH)_2 \longrightarrow CaSO_4 + Na_2SO_4 + 2\ H_2O \tag{12.30}$$

$$H_2SO_3 + Ca(OH)_2 \longrightarrow CaSO_3 + 2\ H_2O \tag{12.31}$$

$$2\ NaHSO_3 + Ca(OH)_2 \longrightarrow CaSO_3 + Na_2SO_3 + 2\ H_2O \tag{12.32}$$

Chromium hydroxide is an amphoteric compound and exhibits minimum solubility in the pH range of 7.5 to 10.0. Effluents from chromium reduction processes should be neutralized to the range of zero solubility (pH 8.5 to 9.0) to minimize the amount of soluble chromium remaining in solution.

It should be noted that if sodium hydroxide is used instead of lime, the chemical cost will be higher, less sludge will be produced, and effluent sulfate concentration will be higher.[15]

12.6.4 REDUCTION AND FLOTATION COMBINATION

Alternatively, hexavalent chromium can be reduced, precipitated, and floated by ferrous sulfide. By applying ferrous sulfide as a flotation aid to a plating waste with an initial hexavalent chromium concentration of 130 mg/L and total chromium concentration of 155 mg/L, an effluent quality of less than 0.05 mg/L of either chromium species can be achieved if a flotation–filtration wastewater treatment system is used.[15]

Ferrous sulfide acts as a reducing agent at pH 8 to 9 for reduction of hexavalent chromium and then precipitates the trivalent chromium as a hydroxide in one step without pH adjustment.[51,62] So, the hexavalent chromium in the nickel-chromium plating wastewater does not have to be isolated and pretreated by reduction to the trivalent form. The new process is applicable for removal of all heavy metals. All heavy metals other than chromium are removed as insoluble metal sulfides, M(II)S.

$$FeS + M(II)^{2+} \longrightarrow M(II)S + Fe^{2+} \qquad (12.33)$$
$$6\ Fe^{2+} + Cr_2O_7^{2-} + 14\ H^+ \longrightarrow 2\ Cr^{3+} + 6\ Fe^{3+} + 7\ H_2O \qquad (12.34)$$
$$Cr^{3+} + 3\ OH^- \longrightarrow Cr(OH)_3 \qquad (12.35)$$
$$Fe^{3+} + 3\ OH^- \longrightarrow Fe(OH)_3 \qquad (12.36)$$

M(II)S, $Cr(OH)_3$, and $Fe(OH)_3$ are all insoluble precipitates, which can be floated by dissolved air flotation (DAF).

This new method can eliminate the potential hazard of excess sulfide in the effluent and the formation of gaseous hydrogen sulfide. In operation, the FeS is added to wastewater to supply sufficient sulfide ions to precipitate metal sulfides that have lower solubilities than FeS. Typical reactions include the following[51,62]:

$$FeS + Ni^{2+} \longrightarrow NiS + Fe^{2+} \qquad (12.37)$$
$$FeS + Zn^{2+} \longrightarrow ZnS + Fe^{2+} \qquad (12.38)$$
$$FeS + Pb^{2+} \longrightarrow PbS + Fe^{2+} \qquad (12.39)$$
$$FeS + Cd^{2+} \longrightarrow CdS + Fe^{2+} \qquad (12.40)$$
$$FeS + Cu^{2+} \longrightarrow CuS + Fe^{2+} \qquad (12.41)$$
$$FeS + 2Ag^+ \longrightarrow Ag_2S + Fe^{2+} \qquad (12.42)$$

Ferrous sulfide can also react with metal hydroxide to form insoluble metal sulfide:

$$FeS + M(II)(OH)_2 \longrightarrow Fe(OH)_2 + M(II)S \qquad (12.43)$$

Ferrous sulfide itself is also a relatively insoluble compound. Thus, the sulfide ion concentration is limited by its solubility, which amounts to only about 0.02 g/L, and the inherent problems associated with conventional sulfide precipitation are significantly minimized.

The newly developed flotation–filtration process involving the use of ferrous sulfide as a flotation aid offers a distinct advantage in the treatment of nickel-chromium plating wastewater that contains hexavalent chromium, nickel, iron, and other metals.

12.7 CONVENTIONAL REDUCTION–PRECIPITATION SYSTEM

A conventional system for treatment of nickel-chromium plating wastewater involves the use of the following unit processes[37–48]:

1. Neutralization
2. Chromium reduction
3. pH adjustment and hydroxide precipitation
4. Clarification (either sedimentation or DAF)
5. Sludge treatment (filter press and final disposal)

Figure 12.4 shows an example of an existing plating facility and its conventional reduction–precipitation wastewater treatment system in New Britain, TN.[15]

Initially the nickel-chromium plating process is designed to minimize the liquid loading to the waste treatment system. Counterflow rinsing, spray rinsing, and stagnant rinse recovery methods are employed in order to minimize the amount of wastes to be treated and allow as much treatment or retention time in the waste treatment system as is possible.

In the application of the previous chemical methods, a certain amount of steady-state continuity has been built into the system. To accomplish this, initial concentrated alkaline and acid rinse wastewaters are retained after dumping in the waste holding tank [T-91] (Figure 12.4) and acid chromium plating wastewater is stored in the waste holding tank [T-51]. Extremely concentrated chromium plating wastewater from rinse step No. 1 is sent to an evaporation tank [T-40] for

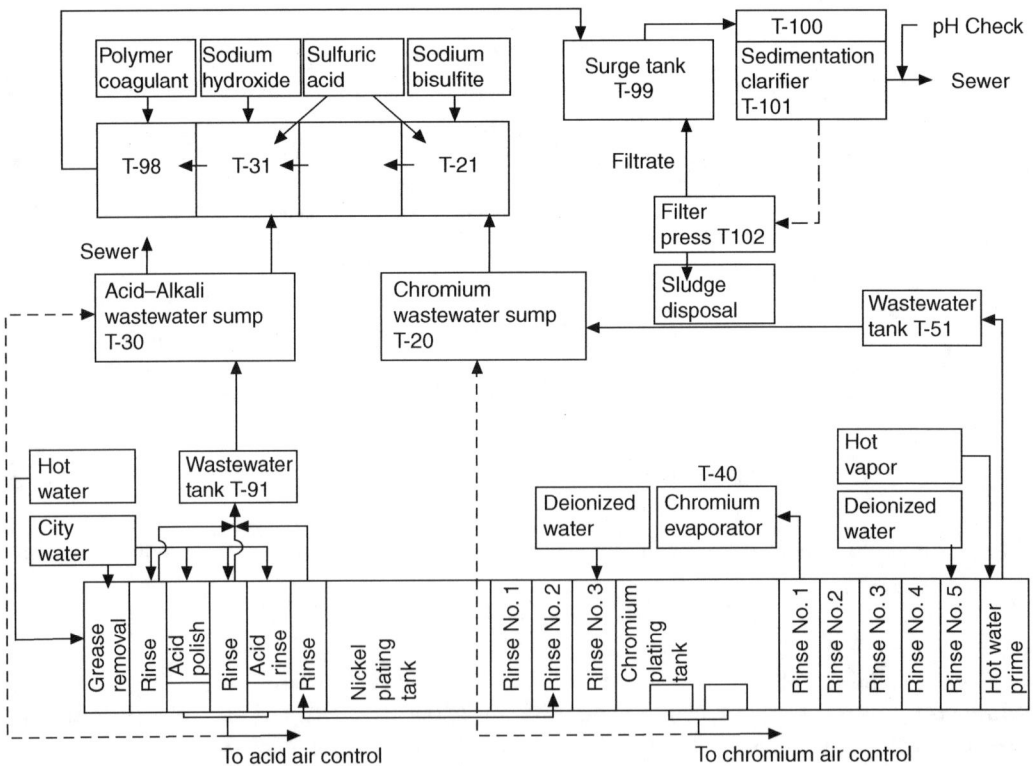

FIGURE 12.4 Conventional reduction–precipitation wastewater treatment system.

chromium recovery. In the case of the wastewater tank [T-51], the waste is slowly bled into the chromium wastewater sump [T-20] to minimize overloading of the total system. The alkaline and acid wastes in [T-.91] are neutralized and slowly bled directly to an acid–alkali wastewater sump [T-30]. It should be noted that the concentrated alkaline wastes are the result of alkaline cleaner replenishment and do not contain heavy metals.

Hexavalent chromium wastes resulting from rinsewater and the concentrated acid bleed accumulate in the chromium waste sump [T-20]. The chromium wastes are then pumped into the chromium treatment module [T-21] for reduction to the trivalent form. This pump is activated only if the oxidation–reduction potential (ORP) and pH are at the proper levels and if the level in the chromium wastewater sump [T-20] is sufficiently high.

Liquid flowing into the chromium treatment module [T-21] is monitored by a pH instrument that controls a feed pump to add the required amount of sulfuric acid from a storage tank. The sulfuric acid is needed to lower the pH to 2.0 to 2.5 for the desired reduction reaction to occur. An ORP instrument controls the injection rate of sodium metabisulfite solution from a metering pump to reduce hexavalent chromium (Cr^{6+}) to the trivalent state (Cr^{3+}).

The acid and alkali wastes are pumped from the acid–alkali wastewater sump [T-30] into the acid–alkali treatment module [T-31]. Metering pumps controlled by pH instruments feed either acid or caustic to the module as required to maintain an acceptable alkalinity for the formation of metal hydroxides prior to discharge to the precipitator consisting of a mixing tank [T-98], a surge tank [T-99], and a sedimentation clarifier [T-101]. The pH is adjusted to a value of 8.5 for optimum metal hydroxide formation and removal.

An ultrasonic transducer is installed on the pH probe mount in the acid–alkali treatment module [T-3l]. This prevents fouling of the electrodes and provides a more closely controlled pH in the effluent discharged to the precipitator.

The first step in the precipitator is the addition of polyelectrolyte solution in the flash mix tank [T-98], surge tank [T-99], and then into the slow mix unit [T-100] containing a variable speed mixing paddle. The purpose of this unit is to coagulate and flocculate[53] the metal hydroxide precipitates.

From the slow mix unit [T-100], the waste flows into the lamellar portion of the sedimentation clarifier [T-101].[54,55] The lamella in the clarifier concentrates the metal hydroxide precipitates. Clarified effluent can be discharged to the sewer.

Concentrated metal hydroxide sludge is pumped from the clarifier to a polypropylene plate filter press [T-102]. The plate filter press[56] is of sufficient capacity without any buildup in the lamellar portion of the unit. This also prevents any overflow of precipitate to the sewer system. The metal hydroxides form a dense sludge cake suitable for disposal in an approved landfill. The liquid effluent from the plate filter is returned to the surge tank [T-99].

A sampling station is provided on the rear exterior wall of the facility for flow measurement and monitoring of the effluent stream.

12.8 MODIFIED REDUCTION–FLOTATION WASTEWATER TREATMENT SYSTEM

A modified reduction–flotation system (Figure 12.5) is very similar to the existing conventional reduction–precipitation system (Figure 12.4), except that a DAF clarifier [T-101F] is used for clarification[15,57] instead of using a conventional sedimentation clarifier (Tank T-101, Figure 12.4).

The flotation system consists of four major components: air supply, pressurizing pump, air dissolving tube, and flotation chamber.[57,58] According to Henry's Law, the solubility of gas (such as air) in aqueous solution increases with increasing pressure. The influent feedstream can be saturated at several times atmospheric pressure, 1.8 to 6 kg/cm^2 (25 to 85 psig), by a pressurizing pump. The pressurized feedstream is held at this high pressure for about 0.5 min in an air dissolving tube (i.e., a pressure vessel) designed to provide sufficient time for dissolution of air into the stream to be

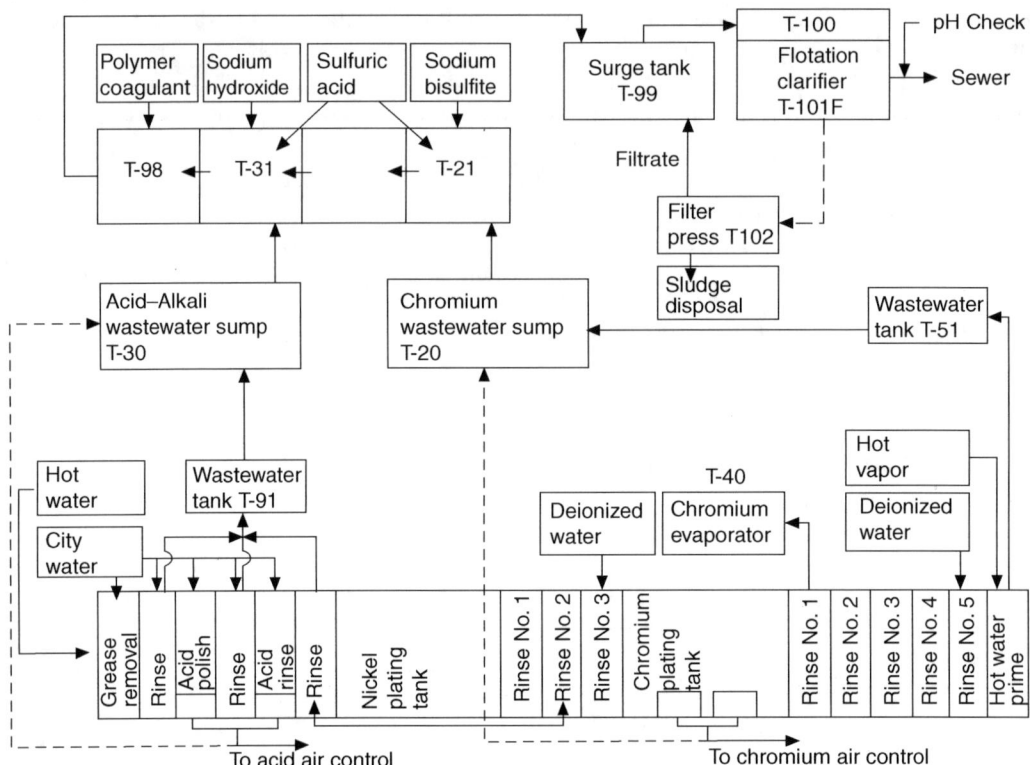

FIGURE 12.5 Modified reduction–precipitation wastewater treatment system.

treated. From the air dissolving tube, the stream is released back to atmospheric pressure in the flotation chamber.[15,57]

Most of the pressure drop occurs after a pressure reducing valve and in the transfer line between the air dissolving tube and the flotation chamber so that the turbulent effects of the depressurization can be minimized. The sudden reduction in pressure in the flotation chamber results in the release of microscopic air bubbles (average diameter 50 μm or smaller) which nucleate on suspended or colloidal particles in the process water in the flotation chamber. This results in agglomeration that, due to the entrained air, gives a net combined specific gravity less than that of water, causing the flotation phenomenon. The vertical rising rate of air bubbles ranges between 15 and 60 cm/min (0.5 to 2.0 ft/min). The floated material rises to the surface of the flotation chamber to form a floated layer. Specially designed flight scrapers or other skimming devices continuously remove the floated material. The surface sludge layer can in certain cases reach a thickness of many inches and can be relatively stable for a short period. The layer thickens with time, but undue delays in removal will cause a release of particulates back to the liquid. Clarified subnatant water (effluent) is drawn off from the flotation chamber and either recovered for reuse or discharged.

The retention time in the flotation chamber is usually about 3 to 5 min, depending on the characteristics of the process water and the performance of the flotation unit. The process effectiveness depends upon the attachment of air bubbles to the particles to be removed from the process water.[57] The attraction between the air bubbles and particles is primarily a result of the particle surface charges and bubble size distribution. The more uniform the distribution of water and microbubbles, the shallower the flotation unit can be.

A high-rate DAF unit with only 3 min of retention time can treat water and wastewater at an overflow rate of 2.4 L/sec/m² (3.5 gal/min/ft²) for a single unit and up to 7.2 L/sec/m² (10.5 gal/min/ft²) for triple stacked units. The comparison between a flotation clarifier and a settler shows the following[59,60]:

1. The DAF unit floor space requirement is only 15% of the settler.
2. The DAF unit volume requirement is only 5% of the settler.
3. In DAF, higher biosolids densities are obtained than in sedimentation. Even in shallow flotation clarifiers a satisfactory biosolids density is attainable.
4. The degrees of clarification of both clarifiers are the same with the same flocculating chemical addition.
5. The operational cost of the DAF clarifier is slightly higher than that for the settler, but this is offset by the considerably lower cost of the installation's financing.
6. DAF clarifiers are mainly prefabricated in stainless steel for erection cost reduction, corrosion control, better construction flexibility, and possible future upgrades, contrary to *in situ* constructed heavy concrete sedimentation tanks.

It should be noted that the chemical reactions of the conventional reduction–precipitation system (Figure 12.4) and the modified reduction–flotation system are identical.

Comparatively, the modified reduction–flotation system will have lower annual total cost (amortized capital cost plus O&M cost) and will require less space, because the flotation unit is very shallow in depth and thus can be elevated. It is expected, however, that the treatment efficiency of the modified system will be higher due to the fact that the DAF clarifier can separate not only the suspended solids but also organics such as oil and grease, detergent, and so on.[57,58,61] Conventional sedimentation clarifiers can separate only insoluble suspended solids.

12.9 INNOVATIVE FLOTATION–FILTRATION WASTEWATER TREATMENT SYSTEMS

12.9.1 FLOTATION–FILTRATION SYSTEM USING CONVENTIONAL CHEMICALS

There are two innovative flotation–filtration wastewater treatment systems that are technically feasible for the treatment of the nickel-chromium plating wastewater.

The first system, shown in Figure 12.6, is identical to the conventional reduction–precipitation in chemistry (i.e., neutralization, chromium reduction, pH adjustment, metal hydroxide precipitation, and so on). However, a flotation–filtration clarifier (Tank T101SF, as shown in Figure 12.6) is used. The unit consists of rapid mixing, flocculation, high-rate DAF, and sand filtration.[15,57]

The treatment efficiency of this system (Figure 12.6) is much higher than that of the conventional reduction–precipitation wastewater treatment system (Figure 12.4).[15]

12.9.2 FLOTATION–FILTRATION SYSTEM USING INNOVATIVE CHEMICALS

Another innovative flotation–filtration wastewater treatment system adopts the innovative use of the chemical ferrous sulfide (FeS), which reduces the hexavalent chromium and allows separation of chromium hydroxide, nickel hydroxide, and ferric hydroxide in one single step at pH 8.5. Figure 12.7 illustrates the entire system. Again, a DAF–filtration clarifier plays the most important role in this wastewater treatment system.

It is seen from Figure 12.7 that this system is much simpler, more cost-effective, and easier to operate in comparison with all other process systems discussed earlier. The treatment efficiency of the new flotation–filtration system is expected to be higher than that of the conventional reduction–precipitation system. The new flotation–filtration system also requires much less land space.[15]

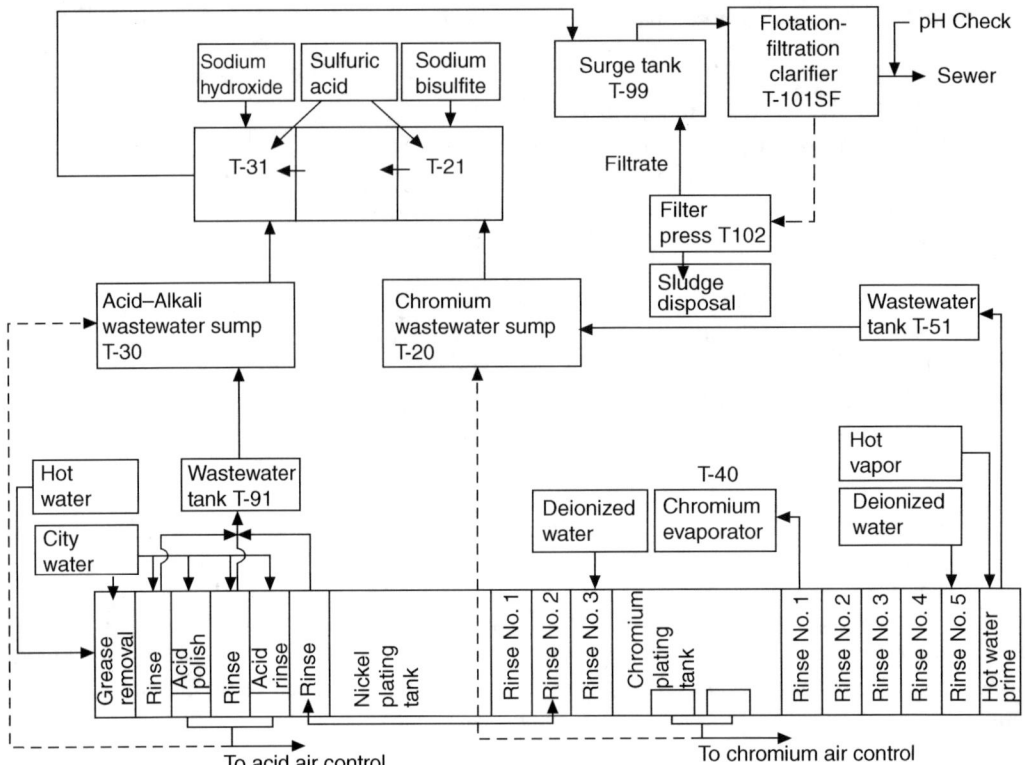

FIGURE 12.6   Innovative flotation–filtration wastewater treatment system using conventional chemicals.

12.9.3   Flotation–Filtration Systems

12.9.3.1   Combined Flotation–Filtration Unit

A combined flotation–filtration unit, shown in Figure 12.8, is an advanced water clarification system, using a combination of chemical flocculation, DAF, and rapid sand filtration in one unit. The average processing time from start to finish is less than 15 min.[15,57,58]

Its unique compact and efficient design is made possible by the use of the principle of zero velocity eliminating internal water turbulence (see below). The flocculated water thus stands still in the flotation tank for optimum clarification. The unit is complete with automatic backwash filter in which dirty backwash water is recycled back to the unit inlet for reprocessing. The average waste flow from the process is less than 1.0% of the incoming raw water.

The flotation unit maximum loading is 2.1 L/sec/m² (3.1 gal/min/ft²). The maximum filtration rate is 1.7 L/sec/m² (2.5 gal/min/ft²). Each filter compartment is backwashed at or more than 10.2 L/sec/m² (15 gal/min/ft²) during the backwash operation. The single-medium backwash filter consists of 28 mm (11 in.) high-grade silica sand. The effective size and uniformity coefficient for the sand are 0.35 mm and 1.55, respectively.

The following paragraphs briefly describe how the flotation–filtration unit shown in Figure 12.8, works.[15,57,58]

The influent raw water or wastewater enters the unit at the center near the bottom [1] and flows through a hydraulic rotary joint [2] and an inlet distributor [3] into the rapid mixing section of the slowly moving carriage. The entire moving carriage consists of a rapid mixer [3], flocculator [4], backwash pumps [5 & 6] and sludge discharge scoop [7]. To flocculate colloids and suspended solids, chemicals [8] are added at the inlet [1].

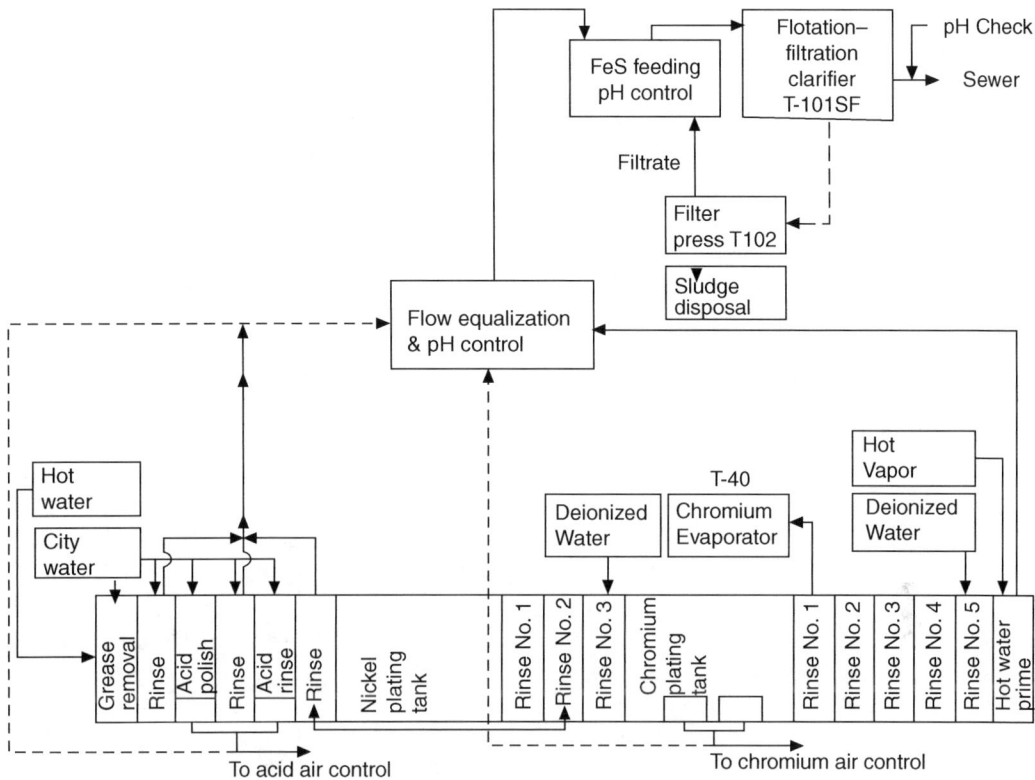

FIGURE 12.7   Innovative flotation–precipitation wastewater treatment system using an innovative chemical.

From the rapid mixing section [3] water enters the hydraulic flocculator [4], gradually building up the flocs by gentle mixing. The flocculated water moves from the flocculator into the flotation tank [9], clockwise, with the same velocity as the entire carriage including flocculator simultaneously moves counterclockwise. The outgoing flocculator effluent velocity is compensated by the opposite velocity of the moving carriage, resulting in a zero horizontal velocity of the flotation tank influent. The flocculated water thus stands still in the flotation tank for optimum clarification.

At the outlet of the flocculator [10], a small percentage of chemically pretreated raw water with microscopic air bubbles is added to the flotation tank [9] in order to float the insoluble flocs and suspended matter to the water surface. The floating sludge accumulated on the water surface is scooped off by a sludge discharge scoop [7] and discharged into the center sludge collector [14] where there is a sludge outlet [15] to an appropriate sludge treatment facility.

The small microscopic air bubbles are the product of raw water pressurized to 4 to 6 kg/cm^2 (55 to 85 psi) in the air dissolving tube (ADT) [32]. Water enters the ADT tangentially [33] at one end and is discharged at the opposite end. During its short passage, the water cycles inside the tube and passes repeatedly by an insert fed by compressed air. Very thorough mixing under pressure dissolves the air into the water.

The bottom of the unit is composed of multiple sections of sand filter [11] and clearwell [12]. The clarified flotation effluent passes through the sand filter downward and enters the clearwell through the circular hole underneath each sand filter section; the filter effluent then enters the center portion of the clearwell where there is an outlet for the effluent.[13]

For backwashing of the sandbeds, two pumps [5 & 6] are placed on the carriage. One pump [5] is at the center of the clearwell for pumping washing water during the backwash cycle through the individual clearwell compartments. The turbid backwash water is collected in a traveling hood [16],

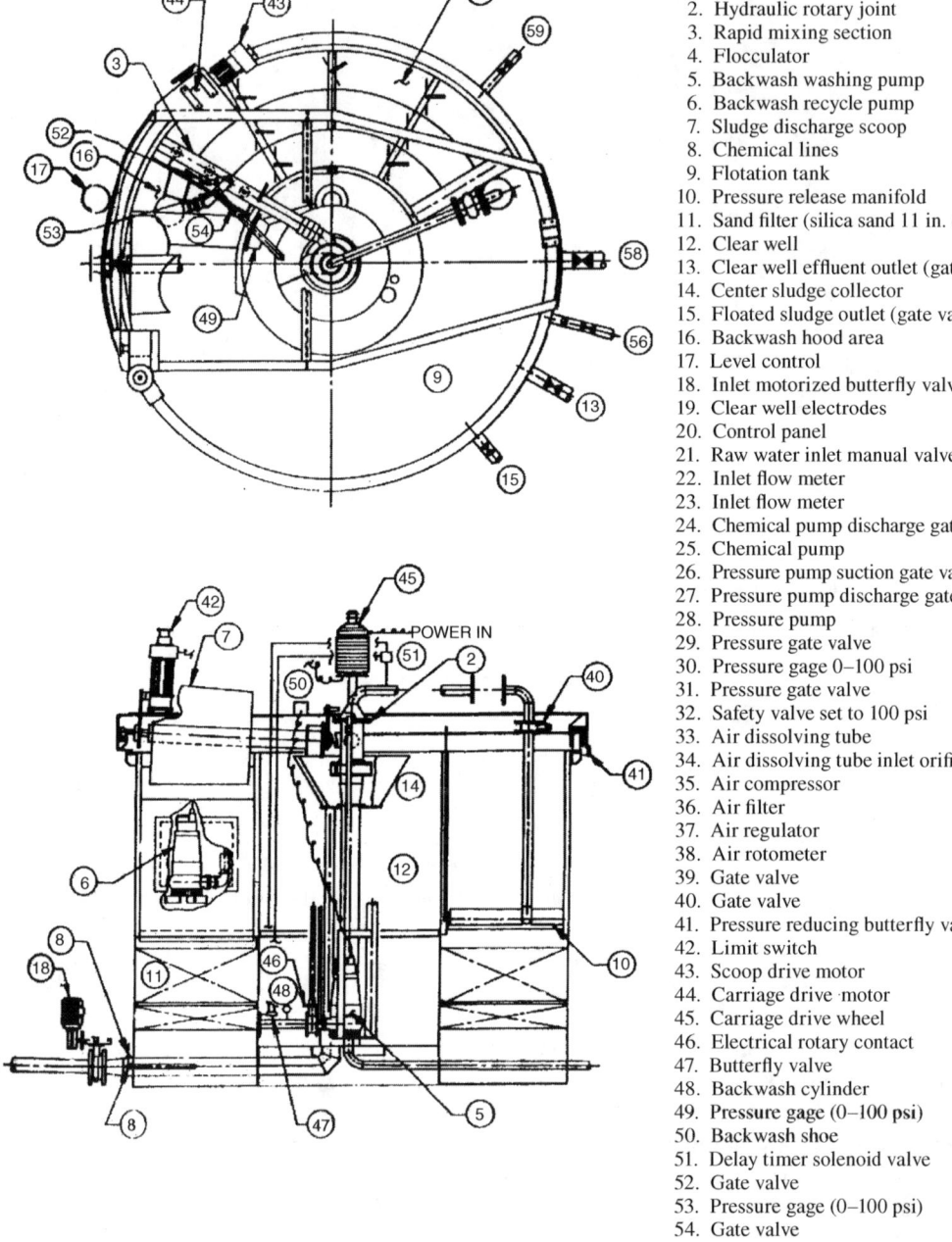

1. Raw water inlet
2. Hydraulic rotary joint
3. Rapid mixing section
4. Flocculator
5. Backwash washing pump
6. Backwash recycle pump
7. Sludge discharge scoop
8. Chemical lines
9. Flotation tank
10. Pressure release manifold
11. Sand filter (silica sand 11 in. depth)
12. Clear well
13. Clear well effluent outlet (gate valve)
14. Center sludge collector
15. Floated sludge outlet (gate valve)
16. Backwash hood area
17. Level control
18. Inlet motorized butterfly valve
19. Clear well electrodes
20. Control panel
21. Raw water inlet manual valve
22. Inlet flow meter
23. Inlet flow meter
24. Chemical pump discharge gate valve
25. Chemical pump
26. Pressure pump suction gate valve
27. Pressure pump discharge gate valve
28. Pressure pump
29. Pressure gate valve
30. Pressure gage 0–100 psi
31. Pressure gate valve
32. Safety valve set to 100 psi
33. Air dissolving tube
34. Air dissolving tube inlet orifice
35. Air compressor
36. Air filter
37. Air regulator
38. Air rotometer
39. Gate valve
40. Gate valve
41. Pressure reducing butterfly valve
42. Limit switch
43. Scoop drive motor
44. Carriage drive motor
45. Carriage drive wheel
46. Electrical rotary contact
47. Butterfly valve
48. Backwash cylinder
49. Pressure gage (0–100 psi)
50. Backwash shoe
51. Delay timer solenoid valve
52. Gate valve
53. Pressure gage (0–100 psi)
54. Gate valve
55. Clear well site tube
56. Clear well drain (gate valve)
57. Bleed off vent
58. Pressurized inlet (gate valve)
59. Pressurized suction pipe (gate valve)

FIGURE 12.8 Top and side views of the flotation–filtration unit. (Taken from Krofta, M. and Wang, L.K., Flotation Engineering, Technical Manual Lenox/1-06-2000/368, Lenox Institute of Water Technology, Lenox, MA, 2000. With permission.)

where the second backwash pump [6] collects the water and discharges it into the rapid mix inlet section [3] for reprocessing.

The flotation–filtration unit can be either manually operated or completely automated with a level control [17] that operates the inlet flow valve [18]. Filter backwashing can also be automated by a timer or head loss control [19].

12.9.3.2   Separate Flotation and Filtration Units

It is important to note that all flotation clarifiers[63,64] may be used for treatment of nickel-chromium plating wastes regardless of their shapes (rectangular or circular) or manufacturers. A filtration unit is an optional step for final polishing. The treatment efficiency of separate flotation and filtration units[65] will be similar to that of a combined flotation–filtration unit (Figure 12.8).

The authors of this chapter are introducing a modern technology involving the use of flotation and filtration for treating nickel-chromium plating wastes. The authors are not endorsing any manufacturer's products.

12.10   SUMMARY

Waste reduction methods reduce the amount of waste generated and also the amount subject to environmental regulations. Hence, these practices can help an electroplating shop comply with requirements and save money. The shop's owner or manager must be committed to waste reduction and pass that commitment on to the employees.

Technologies exist for capturing and reclaiming metal finishing waste, including rinsewaters. It is important to treat each recovery system with as much care as the plating baths. Regular maintenance and the use of trained operators will help ensure that the recovery system performs at its optimum design capacity. The recovery and reuse of the chemical solutions will add dollars into the metal finisher's pockets.

All four wastewater treatment systems introduced in this chapter are technically feasible for treating nickel-chromium plating wastewater in order to meet the maximum permissible concentrations shown in Table 12.1 for industrial wastewater discharge into a municipal sewerage system[15] or Table 12.2 for discharge to surface waters.[21]

TABLE 12.1
Maximum Permissible Concentrations of Industrial Wastewater Discharge into Municipal Systems

Parameter	Maximum Permissible Concentration (mg/L)
Arsenic	0.1
Boron	10.0
Barium	0.5
Cadmium	0.2
Copper	0.1
Cyanide	0.5
Lead	1.0
Mercury	0.5
Nickel	0.5
Silver	0.03
Chromium (total)	0.5

Continued

TABLE 12.1   (continued)

Parameter	Maximum Permissible Concentration (mg/L)
Vanadium	0.5
Zinc	0.5
Chloroform	1.0
BOD	1000
TSS	1000
COD	1500
Oil and grease (nonmineral)	300
Oil and grease (mineral)	100
Chlorinated hydrocarbons	0.02
Phenolic compounds	1.0
pH	5.5–9.5
Temperature	55.5°C

BOD, biochemical oxygen demand; COD, chemical oxygen demand; TSS, total suspended solids.

Source: Krofta, M. and Wang, L.K., *Design of Innovative Flotation–Filtration Wastewater Treatment Systems for a Nickel-Chromium Plating Plant*, U.S. Department of Commerce, National Technical Information Service, Springfield, VA, Technical Report PB-88-200522/AS, January 1984. *Source*: U.S. EPA.

TABLE 12.2
Maximum Permissible Concentrations of Electroplating Wastewater Discharge to Surface Waters

Parameter	Maximum Permissible Value (mg/L)
pH	7–10
TSS	25
Oil and grease	10
Arsenic	0.1
Cadmium	0.1
Chromium (hexavalent)	0.1
Chromium (total)	0.5
Copper	0.5
Lead	0.2
Mercury	0.01
Nickel	0.5
Silver	0.5
Zinc	2
Total metals	10
Cyanides (free)	0.2
Fluorides	20
Trichloroethane	0.05
Trichloroethylene	0.05
Phosphorus	5

TSS, total suspended solids.
Source: See Table 12.1.

TABLE 12.3

Characteristics of Typical Wastewater Discharge of Conventional Reduction–Precipitation System[a]

Characteristic	Concentration (mg/L)	Amount of Pollutant Discharged (kg/d)
Chromium, C_r^{6+}	<0.006	<0.0003
Chromium, total	0.79	0.04
Iron	0.15	0.008
Nickel	0.21	0.011
TSS	<0.10	<0.51
Settleable solids	<0.1 mL/L	
pH	7.6–8.2	

TSS, total suspended solids.
[a] See Figure 12.4.

Table 12.3 shows the characteristics of a typical effluent discharge from a conventional reduction–precipitation system. The effluent quality meets industrial pretreatment requirements.

The modified reduction–flotation wastewater treatment system (Figure 12.5) will be very attractive if all or most of an existing wastewater treatment facilities are to be reused. The high-rate DAF clarifier is a very low-cost clarification unit.

The treatment efficiencies of the two innovative flotation–filtration wastewater treatment systems (Figures 12.6 and 12.7) are expected to be higher than those of the conventional reduction–precipitation system.

The innovative flotation–filtration wastewater treatment system (Figure 12.6) using conventional chemicals has the highest flexibility and best performance. When desirable, the innovative chemical FeS or equivalent can also be used.

Another innovative flotation–filtration wastewater treatment system using FeS (Figures 12.7 and 12.8) is highly recommended if a totally new system is to be designed and installed for treatment of nickel-chromium plating wastewater. This system is extremely compact, easy to operate, and cost-effective. Treatment efficiency is also excellent.

All flotation clarifiers[63] may be used for the treatment of nickel-chromium plating wastes regardless of their shapes (rectangular or circular) or manufacturers. A filtration unit is an optional step for final polishing. The treatment efficiency of separate flotation and filtration units[64] will be similar to that of a combined flotation–filtration unit (Figure 12.8).

REFERENCES

1. Elsevier Inc., *Metal Finishing Guidebook & Directory*, Elsevier, New York, 2005.
2. Dini, J.W., *Electrodeposition — The Materials Science of Coating and Substrates*, Noyes Publishing, Park Ridge, NJ, 1993.
3. American Society for Metals, *Surface Engineering*, ASM Metals Handbook, American Society for Metals, Vol. 5, Materials Park, OH, 1994.
4. Graves, B.A., *Hardware for Plating Hand-Tools*, Danaher Tool Group, Products Finishing Online, 2005. Available at http://www.pfonline.com/articles/049903.html.
5. Parthasaradhy, N.V., *Practical Electroplating Handbook*, Prentice-Hall, New York, 1989.
6. Durney, L.J., Ed., *The Electroplating Engineering Handbook*, American Society of Electroplated Plastics, 4th ed., Washington, DC, 1984.
7. Brenner, A., *Electrodeposition of Alloys*, Academic Press, New York, 1991.
8. Mandich, N.V., Important practical considerations in chromium plating—Part IV, *Met. Finish.*, 97, 9, 1999.

9. Mandich, N.V., Important practical considerations in chromium plating—Part V, *Met. Finish.*, 97, 10, 1999.

10. Mandich, N.V., Important practical considerations in chromium plating—Part VI, *Met. Finish.*, 98, 3, 2000.

11. Mandich, N.V. and Vyazovikina, N.V., Kinetics and mechanisms of the chromium anodic dissolution in the chromium plating solution in the transpassive range, *47th Meeting of International Society for Electrochemistry*, Veszprem, Hungary, September 1996.

12. Mandich, N.V., Chemistry and theory of chromium deposition—Part I: Chemistry, *Plating Surf. Finish.*, 84, 5, 1997.

13. Mandich, N.V., Chemistry and theory of chromium deposition—Part II: Theory of deposition, *Plating Surf. Finish.*, 84, 6, 1997.

14. Finishing.com, *Equipment Needed to Get Started in Chroming Business*, The home page of the finishing industry, 2005. Available at http://www.finishing.com/197/26.shtml.

15. Krofta, M. and Wang, L.K., *Design of Innovative Flotation–Filtration Wastewater Treatment Systems for a Nickel-Chromium Plating Plant*, U.S. Department of Commerce, National Technical Information Service, Springfield, VA, Technical Report PB-88-200522/AS, January 1984.

16. UNEP, *Environmental Aspects of the Metal Finishing Industry: A Technical Guide*, United Nations Environment Program, Paris, 1992.

17. U.S. EPA, Monitoring Industrial Wastewater, U.S. EPA, National Technical Information Service, Springfield, VA, 1973.

18. U.S. EPA, Process Design Manual for Sludge Treatment and Disposal, Report EPA 6251-174-006, U.S. EPA, National Technical Information Service, Springfield, VA, 1974.

19. SFDA, National Primary Drinking Water Regulations, Safe Drinking Water Act, Update March 2002.

20. CDHS, Hazardous Waste Generated by Metal Refinishing Facilities, Fact Sheet, California Department of Health Services, Toxic Substances Control Program, Alternative Technology Division, April 1990.

21. World Bank, Electroplating, in *Pollution Prevention Abatement Handbook*, World Bank Group, July 1998.

22. DCNR, Metal Recovery Technologies for the Metal Finishing Industry, Fact Sheet, Office of Waste Reduction Services, Environmental Services Division, State of Michigan, Departments of Commerce and Natural Resources (DCNR), Lansing, MI, November 1993.

23. Gouthro, R.P. and Vaz, L., Recovery and purification of nickel salts and chromic acid using the RECOFLO system, Eco-Tec, Technical Paper 145, presented at the Metal Finisher's Association of India, Mumbai, India, September 1999. Available at http://www.eco-tec.com/main/electroplate.htm.

24. Lowenheim, F.A., *Electroplating*, McGraw-Hill, New York, 1978.

25. Montgomery, D., *Light Metals Finishing Process Manual*, American Electroplaters & Surface Finishers Society, Orlando, FL, 1990.

26. Nordic Council of Ministers, *Possible Ways of Reducing Environmental Pollution from the Surface-Treatment Industry*, Oslo (1993).

27. Kushner, J.B. and Kushner, A.S., *Water and Waste Control for the Plating Shop*, Hanser Gardner, 3rd ed., (1994).

28. Wang, L.K., Shammas, N.K., Williford, C., Chen, W.-Y., and Sakellaropoulos, G.P., Evaporation processes, in *Advanced Physicochemical Treatment Processes*, Wang, L.K., Hung, Y.T., and Shammas, N.K., Eds., The Humana Press, Totowa, NJ, 2006, pp. 549–580.

29. Chian, E.S.K., Chen, J.P., Sheng, P.X., Ting, Y.P., and Wang, L.K., Reverse osmosis technology for desalination, in *Advanced Physicochemical Treatment Technologies*, Wang, L.K., Hung, Y.T., and Shammas, N.K., Eds., The Humana Press, Totowa, NJ, 2007, pp. 327–366.

30. Golomb, A., *Application of Reverse Osmosis to Electroplating Waste Treatment*, Ontario Research Foundation, Ontario, Canada, 1995.

31. Chen, J.P., Yang, L., Ng, W.J., Wang, L.K., and Thong, S.L., Ion exchange, in *Advanced Physicochemical Treatment Processes*, Wang, L.K., Hung, Y.T., and Shammas, N.K., Eds., The Humana Press, Totowa, NJ, 2006, pp. 261–292.

32. CAT, *Treatment of Electroplating Waste*, Center for Advanced Technology, Indore, India, 2005. Available at http://www.cat.ernet.in/technology/accel/indus/wsa/ctl/chem_effulent.html.

33. Marquínez, R., Pourcelly, G., Bauer, B., Ochoa, J.R., López, R., Viala, S., Mahiout, A., and Leinonen, H., Chromic acid recycling from rinsewater in galvanic plants by electro-electrodialysis, in *Waste Treatment, and Clean Technology*, The Global Symposium on Recycling, REWAS '04, Vol. II, Gaballah, I., Mishra, B., Solosabal, R., and Tanaka, M., Eds., Madrid, Spain, September 26–29, 2004.

34. Chen, J.P., Chang, S.Y., and Hung, Y.T., Electrolysis, in *Physicochemical Treatment Processes*, Wang, L.K., Hung, Y.T., and Shammas, N.K., Eds., The Humana Press, Totowa, NJ, 2005, pp. 359–378.

35. Van De Putte, B. Verhaege, M., Brughmans, S., and Vanrobaeys, D., Electrochemical recovery of nickel from electroplating effluents by an electro-active mixed bed resin cathode, in *Waste Treatment, and Clean Technology*, The Global Symposium on Recycling, REWAS '04, Vol. II, Gaballah, I., Mishra, B., Solosabal, R., and Tanaka, M., Eds., Madrid, Spain, September 26–29, 2004.

36. Wang, L.K., Hung, Y.T., and Shammas, N.K., Eds., *Advanced Physicochemical Treatment Processes*, The Humana Press, Totowa, NJ, 2006, 690 p.

37. Ayers, D.M., Davis, A.P., and Gietka, P.M., Removing Heavy Metals from Wastewater, Report, University of Maryland, Engineering Research Center, 1994.

38. NETCSC, Industrial Pretreatment and Hazardous Material Recognition for Small Communities, The National Environmental Training Center for Small Communities, West Virginia University, Morgan Town, WV, 1996.

39. U.S. EPA, Industrial Waste Treatment, U.S. EPA, Office of Water Programs, V1 and V2, 1995, 1996.

40. Wang, L.K., Hung, Y., Lo, H.H., and Yapijakis, C., Eds., *Handbook of Industrial and Hazardous Wastes Treatment*, 2nd ed., Marcel-Dekker, New York, 2004, 1345 p.

41. Guyer, H.H., *Industrial Processes and Waste Stream Management*, Wiley, New York, 1998.

42. Cheremisinoff, P.N., *Handbook of Water and Wastewater Treatment Technology*, Marcel-Dekker, 1994.

43. Cushnie, G.C., Jr, *Electroplating Wastewater Pollution Control Technology*, Noyes Publishing, Park Ridge, NJ, 1985.

44. Patterson, J.W., *Industrial Wastewater Treatment Technology*, 2nd ed., Butterworth, Boston, MA, 1985.

45. Johannes, R.D. et al., Electroplating/metal finishing wastewater treatment: Practical design guidelines, *Proc. 43rd Purdue Industrial Wastes Conference*, West Lafayette, IN, 1988.

46. Pattanayak, J., Mandich, N.V., Mondal, K., Wiltowski, T., and Lalvani, S.B., Removal of iron and nickel from solutions by applications of electric field, *Environ. Technol.*, 20, 317, 1999.

47. Pattanayak, J., Mandich, N.V., Mondal, K., Wiltowski, T., and Lalvani, S.B., Recovery of metallic impurities from plating solutions by electromigration, *Met. Finish.*, 98, 3, 2000.

48. Medina, B.Y., Torem, M.L., and De Mesquita, L.M.S., Removal of chromium III from liquid effluent streams by precipitate flotation, in *Waste Treatment, and Clean Technology*, The Global Symposium on Recycling, REWAS '04, Vol. II, Gaballah, I., Mishra, B., Solosabal, R., and Tanaka, M., Eds., Madrid, Spain, September 26–29, 2004.

49. Goel, R.K., Flora, J.R.V., and Chen, J.P., Flow equalization and neutralization, in *Physicochemical Treatment Processes*, Wang, L.K., Hung, Y.T., and Shammas, N.K., Eds., The Humana Press, Totowa, NJ, 2005.

50. Chamberlain, N.S. and Day, R.V., Technology of chrome reduction with sulfur dioxide, *Proc. Eleventh Purdue Industrial Waste Conference*, West Lafayette, IN, 129, 1956.

51. Wang, L.K., Vaccari, D.A., Li, Y., and Shammas, N.K., Chemical precipitation, in *Physicochemical Treatment Processes*, Wang, L.K., Hung, Y.T., and Shammas, N.K., Eds., The Humana Press, Totowa, NJ, 2005, pp. 141–198.

52. Wang, L.K., Hung, Y.T., and Shammas, N.K., Eds., *Physicochemical Treatment Processes*, The Humana Press, Totowa, NJ, 2005, 723 p.

53. Shammas, N.K., Coagulation and flocculation, in *Physicochemical Treatment Processes*, Wang, L.K., Hung, Y.T., and Shammas, N.K., Eds., The Humana Press, Totowa, NJ, 2005, pp. 103–140.

54. Shammas, N.K., Kumar, I.J., and Chang, S.Y., Sedimentation, in *Physicochemical Treatment Processes*, Wang, L.K., Hung, Y.T., and Shammas, N.K., Eds., The Humana Press, Totowa, NJ, 2005, pp. 379–430.

55. U.S. EPA, Process Design Manual for Suspended Solids Removal, Report EPA 625/1-75-003a, U.S. EPA, National Technical Information Service, Springfield, VA, 1975.

56. Wang, L.K., Shammas, N.K. and Hung, Y.T., Eds., *Biosolids Engineering and Management*, The Humana Press, Totowa, NJ, 2008, 788 p.

57. Wang, L.K., Fahey, E.M., and Wu, Z., Dissolved air flotation, in *Physicochemical Treatment Processes*, Wang, L.K., Hung, Y.T., and Shammas, N.K., Eds., The Humana Press, Totowa, NJ, 2005, pp. 431–500.

58. Krofta, M. and Wang, L.K., Flotation Engineering, Technical Manual Lenox/1-06-2000/368, Lenox Institute of Water Technology, Lenox, MA, 2000.

59. Shammas, N. and DeWitt, N., Flotation: a viable alternative to sedimentation in wastewater treatment plants, *Water Environment Federation 65th Annual Conf., Proc. Liquid Treatment Process Symposium*, New Orleans, LA, September 20–24, 1992, pp. 223–232.

60. Krofta, M. and Wang, L.K., Flotation Replaces Sedimentation in Water and Effluent Clarification, Report, Krofta Engineering Corp., Lenox, MA, 1990.
61. Uribe-Salas, A., Tinoco-Elvir, J.F., Pérez-Garibay, R., and Nava-Alonso, F., Experimental study on the flotation of Pb^{2+} and Ni^{2+} with sodium dodecylsulfate, in *Waste Treatment, and Clean Technology*, The Global Symposium on Recycling, REWAS '04, Vol. II, Gaballah, I., Mishra, B., Solosabal, R., and Tanaka, M., Eds., Madrid, Spain, September 26–29, 2004.
62. Wang, L.K., Kurylko, L., and Wang, M.H.S. Water and Wastewater Treatment, U.S. Patent 5,240,600, August 31, 1993.
63. Dissolved Air Flotation Equipment—Annual 2005–2006 Buyer's Guide, *Water & Waste Digest*, 45, 73–76, 2005.
64. Flotation Equipment—2005 Buyer's Guide, *Environ. Protect.*, 16, 79, 2005.
65. Olson, S., Dissolved air flotation—new applications, *J. NEWWA*, 133–151, 1994.

Index

A

Abandoned solid waste, 3
ABF. *See* Automatic backwash filtration (ABF)
Abiotic hydrolysis, 438–439
Aboveground gasoline recovery, 316
Abrasive jet machining, 141
Absolute viscosity, 355
    air, 354
    water, 355
Acid base reactions, 398–399
Acid hydrolysis
    cyanates, 167
Acidic wastes
    neutralization, 401
Activated carbon
    adsorption, 246–247, 329–331
    air stripping, 331–332
Activated sludge systems, 498
Activation, 406, 407
Active gas extraction, 214
Active interior gas collection/recovery system, 214
Active perimeter gas control systems, 212
Acutely hazardous wastes, 17
Adhesive factory wastewater, 502
    industrial plant, 379
Adsorbable organic halides (AOX), 501
    analytical methods, 489
Adsorption, 399, 431–437
    flotation, 336
    gasoline, 307–308
    gasoline compounds, coefficients, 308
    isotherms, 431–432
    prediction deep-well environment, 432
    underground storage tank releases, contamination
        remediation, 307–308
Adsorptive force
    subsurface liquid transport, 301
    underground storage tank releases, contamination
        remediation, 301
AEA. *See* Atomic Energy Act (AEA)
Aerated stabilization basins
    kraft mills, 497
Aeration, 246
    method, 323
    schematic diagram, 325
    VOC, 246
Aerobic biodegradation, 184
Aerobic biofilm reactor
    bioremediation leachate treatment, 187–188
Aerobic biological treatment, 498
    characteristics, 318
Aerobic completely stirred tank reactor
    bioremediation leachate treatment, 184–186

Aerobic degradation
    formaldehyde, 370
Aerobic treatment
    bioremediation leachate treatment, 184–186
AFPA. *See* American Forest & Paper Association
    (AF&PA)
Agency for Toxic Substances and Disease
    Registry (ATSDR), 206
Agricultural waste, 11
Air
    absolute viscosity, 354
    diffusion method, 323
    emissions standards, 489
    kinetic viscosity, 354
    knives, 157–158
Airflow equation, 43–44, 46
Airflow rates, 45
Air injection
    depth, 60
    pressure and flow rate, 60–61
    system, 240
Air pollution
    pulp and paper mill, 478
    pulp and paper processes, 479
    VOC, 333
Air sparging, 39, 237
    aquifers, 224
    systems, 56–61
    technology, 39
Air stripping, 230, 323
    activated carbon, 331–332
    applicability, 328–329
    limiting factors, 329
    mass transport, 326
    tower, 327
Alabama
    north–south geologic section, 443
Algorithm
    defined, 428
Alkali-earth metals, 423
Alkaline chlorination, 165
    cyanide reduction, 166
Allied Signal Immobilized Cell Bioreactor, 332, 333
Alternative component case example, 251
Alternative detailed analysis
    criteria, 209
    procedure, 210
Alternative water supply, 247
American Cyanamid Company
    chemical processes, 446
    hazardous waste deep-well injection, 444–445
    injection/confining-zone lithology and
        chemistry, 445–446
    injection facility, 444–445